CONVERGENCE OF MOBILE AND STATIONARY NEXT-GENERATION NETWORKS

CONVERGENCE OF MOBILE AND STATIONARY NEXT-GENERATION NETWORKS

Edited by
Krzysztof Iniewski

WILEY
A JOHN WILEY & SONS, INC., PUBLICATION

Published by John Wiley & Sons, Inc., Hoboken, New Jersey
Published simultaneously in Canada

Library of Congress Cataloging-in-Publication Data:

Iniewski, Krzysztof.
Convergence of mobile and stationary next-generation networks / edited by Krzysztof Iniewski.
p. cm.
Summary: "Filled with illustrations and practical examples from industry, this book provides a brief but comprehensive introduction to the next-generation wireless networks that will soon replace more traditional wired technologies. Written by a mixture of top industrial experts and key academic professors, it is the only book available that covers both wireless networks (such as wireless local area and personal area networks) and optical networks (such as long-haul and metropolitan networks) in one volume. It gives engineers and engineering students the necessary knowledge to meet challenges of next-gen network development and deployment"—Provided by publisher.
Summary: "This book covers wireless networks such as wireless local area networks (WLAN), wireless personal area networks (WPAN), wireless access, 3G/4G cellular, and RF transmission, as well as optical networks like long-haul and metropolitan networks, optical fiber, photonic devices, VLSI chips"—Provided by publisher.
Includes bibliographical references and index.
ISBN 978-0-470-54356-6
1. Wireless LANs. 2. Optical fiber communication. 3. Internetworking (Telecommunication) I. Title.
TK5105.78.I535 2010
004.6'8—dc22

2010016920

Printed in Singapore.

10 9 8 7 6 5 4 3 2 1

CONVERGENCE OF MOBILE AND STATIONARY NEXT-GENERATION NETWORKS

Edited by
Krzysztof Iniewski

WILEY

A JOHN WILEY & SONS, INC., PUBLICATION

Published by John Wiley & Sons, Inc., Hoboken, New Jersey
Published simultaneously in Canada

Library of Congress Cataloging-in-Publication Data:

Iniewski, Krzysztof.
Convergence of mobile and stationary next-generation networks / edited by Krzysztof Iniewski.
p. cm.
Summary: "Filled with illustrations and practical examples from industry, this book provides a brief but comprehensive introduction to the next-generation wireless networks that will soon replace more traditional wired technologies. Written by a mixture of top industrial experts and key academic professors, it is the only book available that covers both wireless networks (such as wireless local area and personal area networks) and optical networks (such as long-haul and metropolitan networks) in one volume. It gives engineers and engineering students the necessary knowledge to meet challenges of next-gen network development and deployment"—Provided by publisher.
Summary: "This book covers wireless networks such as wireless local area networks (WLAN), wireless personal area networks (WPAN), wireless access, 3G/4G cellular, and RF transmission, as well as optical networks like long-haul and metropolitan networks, optical fiber, photonic devices, VLSI chips"—Provided by publisher.
Includes bibliographical references and index.
ISBN 978-0-470-54356-6
1. Wireless LANs. 2. Optical fiber communication. 3. Internetworking (Telecommunication) I. Title.
TK5105.78.I535 2010
004.6'8—dc22

2010016920

Printed in Singapore.

10 9 8 7 6 5 4 3 2 1

ERRATA

Convergence of Mobile and Stationary Next-Generation Networks
Edited by Krzysztof Iniewski
 ISBN 978-0-470-54356-6

The following corrections and additions are made to the followings pages:

On pages vi and xi, delete the author, *Adam Bojanczyk.*

In Chapter 11, entitled "Power-Line Communications and Smart Grid":

Page 317, remove the author Adam Bojanczyk.

Page 319, insert the following paragraph after the Subsection 11.2.1.1:

Real-Time Energy Management Systems As an example of Smart Grid functionalities illustrating the requirements for the communication network, a PLC network is proposed to be a very attractive solution. Assuming that every household is equipped with such a real-time pricing meter, the PLC network between meters and the common transformer station can accommodate thousands of nodes or more. It is therefore essential to enable fast access to individual meters for better quality of service on the customer side [48].

Another application of PLC system in interaction with Smart Grid functionalities is the management of energy distribution using a supervisory control and data acquisition (SCADA) system. SCADA sensors monitor the grid load and forward to its fusion center, where two-way commands adapt the distribution structure to the dynamics of energy generation and consumption. Thus, the SCADA system imposes strong reliability and real-time equipments on PLC.

On page 346, add the following acknowledgment before the Reference section:

Acknowledgment

I thank Dr. Adam Bojanczyk, Professor of Electrical and Computer Engineering at Cornell University and Dr. Ki-Il Kim, Professor of Department of Informatics at Gyeongsang National University, Korea, for their valuable and helpful comments during reviewing and proofreading of this chapter.

On page 348, add the reference:

[48] G. Bumiller, L. Lampe, and H. Hrasnica, Power Line Communication Networks for Large-Scale Control and Automation Systems, *IEEE Communication Magazine*, Vol. 48, No. 4, pp. 106–113, April 2010.

CONTENTS

Part V Photonic and Electronic Component Technology

PREFACE

The optical networking technology that suffered in the post-dot-com crash several years ago has since recovered and is once again poised for rapid growth due to the exhaustion of available bandwidth. Today, photonics networks transport Internet data over large distances in long-haul and metropolitan networks. Improvements in photonics components and silicon chips have enabled several new technologies that are changing how these networks are built and operated. While the network core has always been optical Internet access traditionally secured through wireline access networks, various DSL (ADSL, VDSL, VDSL2), cable (DOCSIS 2.0, DOCSIS 3.0), and passive optical networks (BPON, GPON, EPON) have been used. The challenge in the YouTube and Facebook era is to manage the amount of traffic and service growth while securing or preferably growing revenue. In particular, dynamic bandwidth allocation (DBA) in passive optical networks (PON) presents a key issue for providing efficient and fair utilization of the PON upstream bandwidth while supporting the quality of service (QoS) requirements for different traffic classes.

Wireless networks have been booming largely independently of changes in photonic and wireline networks. WLAN (IEEE 802.11), Zigbee (IEEE 802.15.4), WiMax (IEEE 802.16), and 3G/4G cellular telephony are growing quickly, while 60 GHz, wireless sensor networks and cognitive radios are starting to be considered for volume deployment. In the next 10 years, Internet access will likely become dominated by mobile wireless terminals. The fourth-generation wireless system (4G) is seen as an evolution and an integration of existing wireless network architectures such as 2G and 3G with new ones such as mobile ad hoc networks (MANETs). There are several challenges ahead to make this integration happen. Many issues in 4G related to provisioning of ubiquitous and seamless service access with different underlying wireless technologies remain to be solved.

The main objectives of next-generation networks are to efficiently provide adequate network quality to multimedia applications with high bandwidth and strict QoS requirements and to seamlessly integrate mobile and fixed architectures. These objectives are becoming increasingly relevant due to the huge increment of multimedia applications that require better quality than plain best effort. Wireless and wireline next-generation networks that access the photonic core will be as ubiquitous as traditional telephone networks, and today's engineering students must be prepared to meet the challenges of their development and deployment.

Filled with illustrations and practical examples from industry, this book provides a brief but comprehensive introduction to these technologies. A unique feature of this book is coverage of wireless, wireline, and optical networks in one volume. It describes access and transport network layer technologies while also discussing the network and services aspects. This text attempts to explain how the network will accommodate the foreseen tenfold increase in traffic over the next few years. I hope it will become an invaluable reference to engineers and researchers in industry and academia.

Krzysztof (Kris) Iniewski

Vancouver, British Columbia, Canada
May 2010

CONTRIBUTORS

JAWWAD AHMED, School of ICT, Royal Institute of Technology (KTH), Stockholm, Sweden

SANTHANAKRISHNAN ANAND, Stevens Institute of Technology, Hoboken, New Jersey

NEO ANTONIADES, Department of Engineering Science, The College of Staten Island/City University of New York, Staten Island, New York

ENRIQUE AREIZAGA, Broadband Networks at Tecnalia Telecom, Zamudio, Spain

KRISHNA BALA, Oclaro Inc., Morris Plains, New Jersey

MARIO BALDI, Department of Control and Computer Engineering, Politecnico di Torino (Technical University of Turin), Turin, Italy

MIGUEL BERG, Ericsson Research, Ericsson AB, Stockholm, Sweden

LARS TORSTEN BERGER, Design of Systems on Silicon (DS2), Paterna, Valencia, Spain

ADAM BOJANCZYK, Cornell University, Ithaca, New York

PER OLA BÖRJESSON, Department of Electrical and Information Technology, Lund University, Lund, Sweden

CHRISTIAN van den BRANDEN, EMC Corporation, Alexandria, Virginia

DIVANILSON R. CAMPELO, Department of Electrical Engineering, University of Brasilia (UnB), Brasilia, Brazil

ANTHONY CHAN CARUSONE, University of Toronto, Toronto, Ontario, Canada

RAJARATHNAM CHANDRAMOULI, Stevens Institute of Technology, Hoboken, New Jersey

BIAO CHEN, School of ICT, Royal Institute of Technology (KTH), Stockholm, Sweden and Department of Optical Engineering, Zhejiang University, Hangzhou, China

JIAJIA CHEN, School of ICT, Royal Institute of Technology (KTH), Stockholm, Sweden

JOHN M. CIOFFI, CEO and Chairman, Board of Directors, ASSIA Inc., Redwood City, California

JUSTIN P. COON, Toshiba Research Europe, Bristol, United Kingdom

HOOMAN DARABI, Broadcom Corporation, Irvine, California

GEORGIOS ELLINAS, Department of Electrical and Computer Engineering, University of Cyprus, Nicosia, Cyprus

CHUNYAN FU, Ericsson, Montreal, Quebec, Canada

JOSÉ ENRÍQUEZ GABEIRAS, Telefónica S.A., Madrid, Spain

NAVID GHAZISAIDI, Optical Zeitgeist Laboratory, Institut National de la Recherche Scientifique, Montreal, Quebec, Canada

ROCH GLITHO, Concordia University, Montreal, Quebec, Canada

BERNHARD GOLL, Vienna University of Technology, Vienna, Austria

MAREK HAJDUCZENIA, ZTE Corporation, Lisbon, Portugal

JONATHAN HOMA, Oclaro Inc., Morris Plains, New Jersey

STEFAN HÖST, Department of Electrical and Information Technology, Lund University, Lund, Sweden

EDUARDO JACOB, Department of Electronics and Telecommunications, University of the Basque Country, Spain

EDWARD W. JANG, McKinsey & Company, Seoul, Korea

HENRIK JENSEN, Broadcom Corporation, Irvine, California

YUEFENG JI, Beijing University of Posts and Telecommunications, Beijing, China

YUICHI KADO, NTT Microsystem Integration Laboratories, Tokyo, Japan

OSAMU KAGAMI, NTT Network Innovation Laboratories, Tokyo, Japan

LEONID G. KAZOVSKY, Department of Electrical Engineering, Stanford University, Palo Alto, California

FERHAT KHENDEK, Concordia University, Montreal, Quebec, Canada

SHIGERU KUWANO, NTT Network Innovation Laboratories, Tokyo, Japan

JUNGWON LEE, Georgia Institute of Technology, Atlanta, Georgia

XIN LIU, Beihang University, Beijing, China

HAO LONG, HUAWEI, Shenzhen, China

HUI-LING LOU, Marvell Semiconductor, Inc., Santa Clara, California

MARTIN MAIER, Optical Zeitgeist Laboratory, Institut National de la Recherche Scientifique, Quebec, Canada

THOMAS MAGESACHER, Department of Electrical and Information Technology, Lund University, Lund, Sweden

MASATO MATSUO, NTT Network Innovation Laboratories, Tokyo, Japan

FAISAL A. MUSA, University of Toronto, Toronto, Ontario, Canada

RON NADIV, Ceragon Networks, Ltd., Tel Aviv, Israel

GEORGE NG, University of Toronto, Toronto, Ontario, Canada

PER ÖDLING, Department of Electrical and Information Technology, Lund University, Lund, Sweden

EUGENE ORTENBERG, EMC Corporation, Alexandria, Virginia

SCOTT REYNOLDS, Zeugma Systems, Richmond, British Columbia, Canada

MARÍA ÁNGELES CALLEJO RODRÍGUEZ, Telefónica S.A., Madrid, Spain

HIROSHI SAITO, NTT Service Integration Laboratories, Tokyo, Japan

MAGNUS SANDELL, Toshiba Research Europe, Bristol, United Kingdom

ARYAN SAÈD, PMC-Sierra, Santa Clara, California

SHAMIK SENGUPTA, John Jay College of Criminal Justice, City University of New York, New York, New York

JONATHAN SEWTER, University of Toronto, Toronto, Ontario, Canada

HENRIQUE J. A. DA SILVA, Universidade de Coimbra, Coimbra, Portugal

BJÖRN SKUBIC, Ericsson Research, Ericsson AB, Stockholm, Sweden

TAE EUNG SUNG, Cornell University, Ithaca, New York

ROBERT SWOBODA, Vienna University of Technology, Vienna, Austria

DIMITRIS TOUMPAKARIS, University of Patras, Patras, Greece

DAISEI UCHIDA, NTT Network Innovation Laboratories, Tokyo, Japan

CHERAN VITHANAGE, Toshiba Research Europe, Bristol, United Kingdom

SHING-WA WONG, Department of Electrical Engineering, Stanford University, Palo Alto, California

YUE WANG, Toshiba Research Europe, Bristol, United Kingdom

LENA WOSINSKA, School of ICT, Royal Institute of Technology (KTH), Stockholm, Sweden

HORST ZIMMERMANN, Vienna University of Technology, Vienna, Austria

ALIREZA ZOLFAGHARI, Broadcom Corporation, Irvine, California

PART I

ACCESS AND BACKHAUL NETWORKS

1

ROADMAP FOR NEXT-GENERATION COMMUNICATIONS NETWORKS

María Ángeles Callejo Rodríguez and José Enríquez Gabeiras

In recent years, multiple initiatives to progress on the building of the Future Internet have been launched worldwide (such as GENI [1] and FIND [2] in the United States, the "Future Internet Assembly" [3] just built in Europe, and the Akari Project [4] in Japan). In all these innovation programs, the research community is proposing new evolution strategies for the present Internet that, from the point of view of maintaining the present status quo, are either revolutionary or evolutionary. In the case of revolutionary approaches, the so-called "clean slate" approach is proposed in order to consider new requirements since the initial phases of the design of new networks (such as security or new network virtualization techniques), disregarding any demand for compatibility with the present infrastructure. On the other hand, the evolutionary path has as a starting point the current Internet infrastructure, transforming the architecture of present Next-Generation Communications Networks to meet the requirements of the services of the future.

This chapter aims to provide an (a) overview of how this evolution of Next-Generation Communication Networks is being done and (b) a summary of its role to build up the Internet of the Future. It aims to identify the main requirements for the network of the future, and it discusses how the improvement of the

Convergence of Mobile and Stationary Next-Generation Networks, edited by Krzysztof Iniewski

current Next-Generation Network (NGN) capabilities should be the basis to provide them. In order to fulfill this view, the chapter first presents the set of goals to be met by future networks considering the users' behavior and the evolution of Internet traffic, then analyzes the main problems and solutions associated with QoS (Quality of Service) provision, and finally proposes a roadmap to steer the evolution of present networks to the multiaccess, multitransport Future Internet.

1.1 REQUIREMENTS FOR NGN AND THE ROLE OF QoS FOR THE FUTURE INTERNET

As a first step to build the Internet of the future, it is mandatory to analyze the expected evolution of the Internet users' behavior. One of the main characteristics of present-day network planning is the high uncertainty of the users' demand evolution. Nowadays, the main characteristic of the end users is their diversity: There are multiple applications with heterogeneous requirements (Peer-to-Peer, streaming, voice over IP, blogs, social networks, chats, gaming, etc.), which can be accessed from multiple devices (mobile devices, PCs, game consoles, etc.) using different types of connectivity (mobile of different types, fixed by different media).

There have been multiple attempts to evaluate how the different Internet users behave, but, probably, in order to define the requirements for the Internet of the Future, a classification of the end users' behavior according to their age could be the best approach to foresee what is expected in the future from the end users' perspective. In this sense, the popular generation tagging has identified the following generations:

- Generation X (born in 1960s and 1970s) uses the Internet for web navigation and access to mail.
- Generation Y (born in 1980s) is all day connected, so this generation considers ubiquity and reliability in their Internet connections.
- Finally, Generation Z or digital natives (born in 1990s and 21st century) is used to the technical changes and has much more knowledge about the technology [5]. This generation is highly connected and makes a lifelong use of the communications and media technologies.

It is clear that in a short time span, different patterns of usage of the network have arisen, due to the ever richer offer of services being offered and to the lower entry age to the Internet. Therefore, regarding connectivity, many new requirements must be considered in the evolution of NGN: future networks must support new traffic demands, reliability to allow the end users to trust the availability of their connections, ubiquity of the access, security in the service usage, flexibility to adapt to the different requirements, neutrality and openness to allow the development of new services, and ability to provide advanced services which combine all these characteristics.

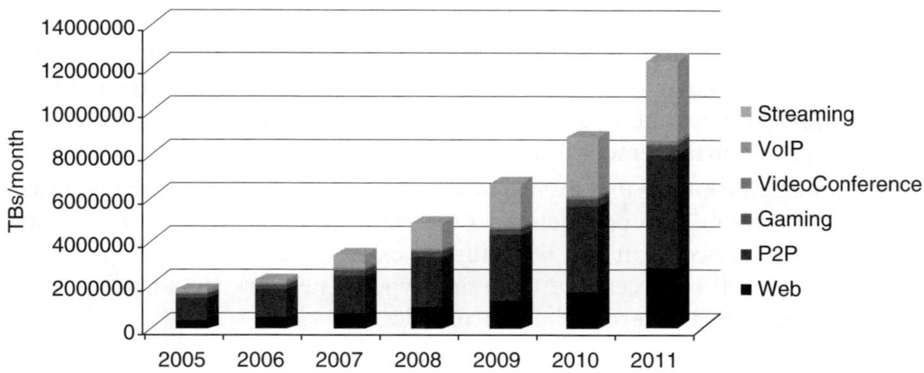

Figure 1.1. Evolution of Internet traffic per application type [6].

An important indicator to evaluate the capacities to be provided by future networks is to evaluate how the traffic will evolve in the coming years. As stated by Cisco [6] and represented in Figure 1.1 [6], it is clear that we are witnessing a huge increment of the demand of new multimedia applications that require better network performances or guaranteed QoS, such as online gaming, video streaming, or videoconference.

Moreover, this will be especially demanding with the above-mentioned new generation of young people that is "always on" in Internet, is also able to create their own services, and values the connectivity as an important service, for which they are keen to pay to have the possibility to access the wide set of Internet services. The provisioning of advanced QoS connectivity services will become a key driver for the operators' business role in the Future Internet.

In this process to build the NGN, due to the strategic role played by the standards in innovation, competition, and regulation, it is important to identify any standardization gap in order to guarantee the fair play of the different players while building the technology roadmap. In this context, the specification of the NGN architectures will play a key role. But, what are the main objectives of the NGN? It is generally agreed that the main focus of the NGN could be summarized in the following key topics:

- *To provide better access:* In the context of NGN, we include the evolution of access and core technologies which are able to provide higher bandwidths in both fixed and mobile technologies. The evolution of all these technologies is one of the main topics addressed by this book in next chapters.
- *To be able to efficiently carry different services:* One important topic being fostered by the NGN architecture is the integration of multiple services into IP networks. All these services must be integrated in such a way that carrier-class capabilities should be also provided. This would allow the operators to provide their services (both corporate and residential) over the same network; moreover, this could represent an opportunity to offer

these capabilities to other service providers that do not own network infrastructure. In order to effectively make this scenario possible, all the features related to the provisioning of different services in the NGN networks should be implemented in such a way that simplicity for end users' and operators' management is a given.

- *To integrate mobile and fixed architectures and services:* Since the end users are accustomed to connecting to the Internet from multiple devices and types of networks, it can be naturally expected that in the future all the services will be accessible from any type of network by adapting to the transmission and terminal characteristics.

This chapter will focus on how the NGN control capabilities should evolve in order to make possible the provisioning of QoS and the convergence of network services in an efficient way, independently on the usage of fixed or mobile access, which means to avoid introducing unnecessary complexity that could make the solutions practically unfeasible. This chapter does not focus on the evolution of the network technologies that could make possible the provisioning of more advanced communication features (this, in fact, is the main purpose of the rest of the book), but is more focused on how the different mechanisms to control all these features should be implemented and standardized in order to really make possible the NGN objectives.

In order to evaluate how these control mechanisms for NGN should be implemented, first and foremost, it would be important to take a look at how the NGN is structured. In ITU-T [8], a draft architecture of the NGN is depicted and described. In this architecture the following strata can be distinguished:

- The Transport Stratum, including:
  - The transport functions (in the access, metro, and core network). The evolution of all these functionalities are related to the evolution of the network technologies itself (new optical solutions for the core networks, FTTH, new wireless mechanisms, etc.).
  - The transport control functions (resource and admission control functions and network attachment control functions). Currently there are several standardization bodies in charge of leading the evolution of this control plane: ETSI/TISPAN [9], 3GPP [10], and ITU-T [11].
- The Service Stratum, which includes the service control functions (including service user profile functions) and the application and service support functions. In principle, any service stratum can use the transport stratum capabilities, but the most clear standardization (and also commercial) initiative proposed so far is the IMS (IP Multimedia Subsystem), fostered by the 3GPP, which specifies an environment where the network operator is in charge of providing the services, which also takes advantage of the control functionalities.

Therefore, taking into account the foreseen users' requirement for improved QoE, and considering the various technology-centered solutions proposed in the realm of NGN to provide QoS, one of the major challenges is to guarantee the users' QoS requirements between the end points involved in a communication that spans over several network segments. As explained in ITU-T [7] and [8], a new architecture must be designed in order to address this goal, whose main feature is to integrate and synchronize the tasks performed in the different planes of the networks along the e2e (end-to-end) path. Moreover, the solution must also consider the interaction with the Home Gateway, since this entity is in charge of managing the Home Network, which represents the first and/or last part of the entire network chain and will play a key role in the provision of quality communications to the end users.

In this sense, the role of the transport control functionalities is the key for the success of this e2e QoS provisioning. Even though a multiple standardization effort has been invested in order to design and implement this control function in ETSI/TISPAN [9], 3GPP [10], and ITU-T [11], there is still not a common solution, since each standardization body is focused on different technologies, and the proposed solutions are offering different interfaces for resource reservation and enforcement. Therefore, up to now only partial solutions to provide QoS in specific underlying technologies are implemented, and these do not have e2e significance. Besides, these solutions usually require a manual administrator configuration (which in fact leads to high operational effort) and are therefore hard to reconfigure according to the online users requests.

In the next subsection, an overview of an experimental system to address this problem is provided (the EuQoS system). This system was designed to build a framework able to provide e2e QoS over heterogeneous networks. The design and implementation of this system has allowed the authors to identify the problems that exist today in current specification of NGN transport control functionalities and to provide a set of recommendations to support the development of an architecture able to make possible the integration of fixed and mobile access to provide any carrier-class IP service.

1.1.1 The EuQoS System as a Solution to Provide End-to-End QoS

The EuQoS Project [12] main achievements have been the design, development, integration, testing and validation of QoS mechanisms over heterogeneous networks while preserving the Internet openness principle. In a nutshell, the EuQoS system has provided a new approach that allows the Network Operator to take advantage of the requirements of new Internet Services as the driver for a new commercial offer based on advanced connectivity requested by the end user.

The EuQoS system architecture that is presented in Callejo et al. [13] allows the network operators to provide e2e QoS connections over heterogeneous network technologies. This approach considers that Net Neutrality will be both a users' and regulators' requirement in the Future Internet, and therefore the system must enable the end user to request a specific type of connectivity service to the network (Real Time, Non-Real Time, etc.), regardless of the Service Provider particular end-services, which means that no linkage between applications and service levels can be set beforehand. The EuQoS system is able to effectively guarantee e2e QoS according to users' demands by means of coordinating the QoS mechanisms available in the different network technologies along the communications path. This is done by specifying a set of well-known e2e Classes of Services (known by the end users) that are mapped to the different underlying network mechanisms in each communication segment. In order to ensure the scalability of the solution, two timescales are specified: the long timescale process of the provisioning (where resources are reserved per aggregate according to the dimensioning of the network and expected users' demand) and the short timescale of session setup by the end users (where the processes available in the access technologies should be triggered in order to use the provisioned paths in the longer timescale).

In order to build, use, and monitor the QoS guaranteed paths, an architecture based on different planes is proposed in Callejo et al. [13] and is depicted in Figure 1.2, where the main planes and interfaces are shown. Next, the implemented functionalities are briefly described:

- The Service Plane provides the *QoS on demand* interface that allows the end users to request QoS guarantees for their applications. Moreover, this Service Plane also implements the AAA (Authentication, Authorization and Accounting) and charging functionalities. This interface provides QoS as a service that can be used by any application/service without the need of integrating the full stack of the application signaling in this Service Plane.
- The Control Plane is in charge of the control procedures to ensure the provisioning of the QoS in both provisioning and invocation phases. This plane is split in two different levels:

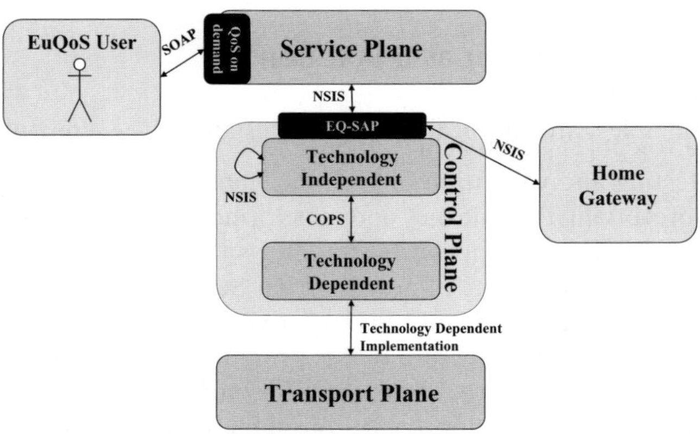

Figure 1.2. EuQoS architecture (reference points and protocols).

- The Network Technology Independent level provides a reference point that is used by the Service Plane and by the Home Gateway to request the reservation of resources; this level manages an abstraction of the domain topology, maintains a set of operator's policies, the users' localization, and contacts other domains involved in the e2e QoS guaranteed path using the interface provided by other Network Technology-independent levels.
- The Network-dependent level provides a well-known interface to the independent level and maps the e2e Classes of Services to the specific underlying network mechanisms, applies specific admission control algorithms (i.e., in mobile networks it can consider physical parameters, while in fixed networks this could be optional), and interacts with the network equipment in order to configure the QoS policies.

According to the design and evaluation of the EuQoS system, the following design principles must be maintained during the specification of the control capabilities of the networks of the future:

1. A set of well-known technology-independent classes of services must be provided. These e2e classes of services would allow carrying out the same service over different technologies without the need to deploy specific solutions per network type. These classes of services are presented in Reference 14.
2. There must be a clear distinction between the different planes and clear specification of the reference points at each plane and layer. This permits the configuration of multiple scenarios over the same infrastructure (a Home Gateway or a service provider accessing the control plane capabilities directly or the end users using the QoS on demand interface).
3. There must be a distinction between network technology-dependent and -independent mechanisms. This makes the deployment of the solution easier: The ISP or the vendors just need to provide a common interface to make use of the QoS capabilities in their control systems. Moreover, if, in the future this interface is provided by some network equipment, the development of ad hoc control systems could skipped.

All these principles are applicable to the design of the control capabilities to provide QoS guarantees, but some of them could be also applied to other features, such as to ensure the design of management systems.

1.2 PROBLEMS AND RECOMMENDATIONS FOR NGN EVOLUTION

Based on the experience of the authors in the design, implementation, and validation of a system able to provide QoS over heterogeneous networks, this section identifies a set of problems to develop the NGN capabilities and presents a set of recommendations that could feed the standardization process of some ITU-T initiatives for NGN evolution to support e2e management of the services, such as the AMS (Advanced Multimedia System) or any other standardization process.

1.2.1 Problems to Provide QoS in an Efficient Manner

Multiple solutions and architectures were studied during the referred research; as a result, possible weaknesses in present standards and commercial solutions were detected.

Problem 1: Application Signaling Integration with NGN. Most of the current NGN specifications propose the integration of the application signaling; this means that, in order to provide QoS for specific services, the NGN must not only be aware of the application signaling but also must be an essential part of the service negotiation (e.g., for codecs selections, user discovery, etc.).

A widely known example is IMS, which specifies the usage of SIP (Session Initiation Protocol) as the only way to interact with the P-CSCF (Proxy Call Session Control Function, which is the first point of contact of the end user with the IMS control entities). If SIP is discovered as a protocol not suitable to deal with different types of applications/services, the core of the IMS control will become useless as the Service Plane (or Application Level) of the NGN in the Future Internet networks for any kind of application. In addition, there are some complaints from application developers due to the lack of specification about the actions to be taken when an application signaling event is detected.

Moreover, taking into account the wide variety of application protocols that are currently being used in the Internet (e.g., MSN, Skype, P2P Streaming applications, etc.), if this design principle is maintained, this could lead to complex systems where several gateways should be integrated to interwork with the users' favorite applications, not necessarily SIP-based. Finally, if we take into account that in the Future Internet the users will not be only service consumer but also service providers/creators, a wide variety of non-IMS applications can be expected to coexist in the future.

To sum up, the requirement of the integration of the application signaling in the NGN structure would lead to two main scenarios:

- Complex systems in charge of managing multiple signaling protocols or with several gateways where the provisioning of advanced connectivity services for new (and probably users favorite) applications will be delayed. This scenario could lead to a solution difficult to manage due to its complexity and, maybe, lack of scalability.
- Walled gardens where only specific services will be provided with QoE, losing the openness as a main principle of the Internet. Probably, this option will not even be attractive for the operators, since they will not be able to provide advanced connectivity services to their end users (but the users appreciate the good connectivity service provided by their ISP as

the main quality criteria) nor will they be able to provide their network capacities to third-party applications.

Problem 2: Weak Specification of the Interfaces. In current recommendations and/or specifications, there is a clear weakness in the specification of the interfaces and reference points. This problem is reflected in the following points:

- There are three standardization bodies specifying the transport control functionalities. Each body proposes different interfaces that, in fact, could lead to interoperability problems. This is something well-identified, and in fact the standardization bodies are trying to converge in a common solution (e.g., there is a clear integration attempt between 3GPP and ETSI/TISPAN).
- Some interfaces are not specified and are left for further study (such as, e.g., the interface between trusted CPE and the RACF in Y.2111, which will be essential in the full integration of the Home Gateway as an extension of the Operators Control plane).
- Other interfaces are only specified in terms of methods, parameters, and requirements for the transactions. For these interfaces, there is no clear choice of the protocols.

This lack of clear specification of the reference points could lead the different vendors to provide their own solutions for the whole system. This will lead to interworking problems between the different vendor equipments that, indeed, result in interworking problems for multivendor solutions (i.e., control capabilities of one vendor able to interwork with the transport capacities provided by two different equipment providers) or interoperability problems across different network domains.

As an example, it can be mentioned that in Recommendation ITU-T Y.2111, the details of the interaction between different RACFs (reference point Ri) is left for further study. This interaction is mandatory if QoS information is shared by different domains involved in the end-to-end path; it is being addressed in other standardization bodies and implemented by some vendors, in particular the following:

- ETSI/TISPAN has provided a draft version of the RCIP (Resource Connection Initiation Protocol) which is specified to allow the interaction between the ETSI/TISPAN RACSs (similar to ITU-T NGN RACF) during the reservation of the resources to ensure a specific QoS level.
- At the time of this writing, a telecommunications equipment manufacturer[1] has released a commercial implementation of a RACS (ETSI/TISPAN module similar to the ITU-T NGN RACF), which provides an RCIP-based interface to allow the communication between different resource managers.

[1]Huawei, through the RM9000-Resource Manager.

In this scenario, it is very likely that the final specification of the RCIP-based interactions will not meet the real implementation of the commercial equipment produced today, becoming the source of future interworking problems.

Problem 3: Weak Specification of the Modules Functionalities. The current specification of some modules does not really specify a general state machine of the module and functions. This scenario could lead to vendor-specific solutions, resulting in competitive problems that are not only due to the interworking problems between different vendor equipments but also due to the overlapping of functionalities and/or lack of them.

Therefore, it is felt that the implementation guidelines to support the specification of the interfaces and the development process are missing. This would not mean the specification and standardization of the algorithms that implement the functionalities and that, in fact, could constitute the competitive difference between different providers, but at least to clearly identify input and output parameters as well as the raw description of the processes that should be triggered in each module.

Problem 4: Non-open Interfaces to Configure the Network Equipments. One of the key points in the provisioning of QoS is the coordination of the different QoS mechanisms available in different network technologies. In order to do that, it is important to have access to the Network Equipments involved in the end-to-end path as well as to have the availability of mechanisms to provide configuration commands to the different equipments.

For instance, during the integration of the EuQoS system with several network technologies, the project had to address several integration problems due to the lack of common reference points in the different network elements. For example, in order to integrate UMTS technology, there were different reference points to interact with the GGSN depending on the vendor provider. This issue resulted in the need for the EuQoS system to relay on the UMTS user interface to setup PDP context as the only possible standard compliant solution to integrate QoS UMTS built-in mechanisms.

Similar problems were faced to integrate the Ethernet technology, where different strategies to interact with the switches had to be followed depending on the equipment vendor.

If non-open interfaces are available in the network equipments, the provisioning of QoS guarantees will be hard to deploy due to the high dependence on specific vendor solutions that will probably try to provide their network control platforms for the new equipments. If this issue is added to the lack of a clear specification of the interface between different control layers (in particular, between the RACS deployed in different domains), this could lead to a general interoperability problem (both between technologies and between domains), rendering the end-to-end QoS provision almost impossible.

Problem 5: The Regulatory Environment Is Not Clear. Nowadays, multiple regulatory scenarios are being defined across the world with a clear trend toward enforcing some way of functional separation between services and networks operations. In this scenario, all those systems based on the vertical integration of services and networks will probably not be viable. Therefore, a clear specification of the interfaces between service layer and network layer will be necessary in order to ensure the validity of the NGN proposals in different scenarios. This means that NGN specification must meet the requirements imposed by the different roles that could arise in the different business models that can be foreseen for the near future.

In this context, the clear specification of reference points will be mandatory.

1.2.2 Recommendations and Proposals to Provide QoS in NGN

According to the problems exposed in Section 1.2.1 and according to our experience, a set of recommendations are provided in this section in order to allow the integration of end-to-end QoS capabilities in ITU-T NGN.

All the recommendations are made taking into account that Net Neutrality will be a requirement from the end users (who are willing to improve their QoE in Internet) and from the regulators (whose position is clearly against "walled gardens").

Recommendation 1: Clear Analysis of the Users' Requirements and Knowledge. The end users' behavior has become a moving target in the Internet era. The operators have seen an evolution of a landscape of a single service with a very predictable demand to a situation in which there are a myriad of services, most of them generated by the users, with a demand that is highly unpredictable. Therefore we need to assess the user demand and evolution at this point in time. It is important to carry out market studies in order to know the end-users' expectations and how they can use the new QoS capabilities with the new services. In particular, at least the following questions must be addressed:

- What is the end-users' knowledge about QoS? What do they really know about this concept?
- Which end-to-end attributes (such as security, reliability, availability, fail recovery, etc.) would be required by the end users?
- What are the most used end-users devices (IPhone, PDAs, laptops, etc.)? How many attributes should be provided by a common user?
- What Internet Services are more in demand by the end users? What is the added value of operators' services perceived by the end users?

With this study, we could characterize the current Internet usage and infer the Future Internet access requirements. Taking into account this characterization, it is important to provide the following conclusions:

- Identification of the current Internet access limitations in terms of QoS as it is perceived by the end user.
- Specification of the new requirements for Network Performance, especially taking into account new services such as High-Definition TV (with its associated QoS requirements), 3D Internet applications, or P2P applications. This could be the starting point of the specification of the Future Internet Classes of Services in terms of e2e QoS guarantees (IPTD, IPDV, and IPLR) and other performance metrics (availability, security, fail recovery time, etc.)
- First draft of the end-user Interfaces that could allow the invocation of advanced network capabilities services.

This analysis will be mandatory in the specification of the AMS end-users reference points; this must be comprehensible by the end user and must be able to support requests for different Internet applications with different QoS requirements.

Recommendation 2: QoS Must Not Be Against Net Neutrality. The evolution of NGN transport technologies offered as network services create an excellent opportunity for the innovation, but not just for the operators (to provide their own services) but also for end users and service providers. If these capabilities are offered in a fair way, the QoS will be a clear driver for the development of guaranteed services to any party.

Effectively, a framework to provide guaranteed QoS is not necessary for just discrimination or filtering for other non-QoS applications. In order to meet this requirement, it is important that this framework provides to the end users and service providers a (set of) clear interface that could allow the user to decide which QoS level is required for each of his/her flows. In this way, QoS will be provided not only to operators' services but also to other Internet services according to the end-users' demands.

If the specified framework meets this requirement, it could allow the network operators to provide their own services and also to take advance on third-party applications (Internet services) as a driver to offer advanced QoS connectivity services. In a scenario where the QoS is offered as a service, it is necessary to provide mechanisms to ensure that the capabilities are effectively provided, and therefore these monitoring capabilities should be linked to the development of the system itself.

Recommendation 3: New Business Models Must Be Drafted. Clark et al. [15], state that:

> One can learn from the past. To some of us in the research community, a real frustration of the last few years is the failure of explicit QoS to emerge as an open end-to-end service. This follows on the failure of multicast to emerge as an open end-to-end service. It is instructive to do a post mortem on these failures.

> Here is one hypothesis. For the ISPs to deploy QoS, they would have to spend money to upgrade routers and for management and operations. So there is a real cost. There is no guarantee of increased revenues. Why risk investment in this case?
>
> If the consumer could exercise effective competitive pressure in ISP selection, fear and greed might have driven ISPs to invest, but the competitive pressures were not sufficient. On the other hand, if ISPs use the new QoS mechanisms in a closed way, rather than an open way, they greatly enhance revenue opportunities. Thus, for example, if they deploy QoS mechanisms but only turn them on for applications that they sell, they reduce the open nature of the Internet and create opportunities for vertical integration. If Internet Telephony requires QoS to work, and they only turn on QoS for their version of Internet Telephony, then they can price it at monopoly prices.

This statement lets us think that a possible risk is that the current Internet model could evolve according to the next two threads: a classical best effort Internet (where carriers will limit their investments) and a premium Internet (where ISPs will invest in NGN equipment to ensure high QoS capacities but at high prices).

From the social point of view, this would result in the lack of the universality of the Internet. In this context, the specification of business models to support a better resource usage and economical revenues is a must.

In this context, the study of the evolution of the society will have to be taken into account. In this kind of study, a new generation has been identified as being "always connected" to use a wide variety of services and applications in the Internet (social networks, P2P file sharing, video streaming, videoconference, gaming, etc.); thus, these people would highly value all those connectivity services that are neutral, reliable, ubiquitous, and able to support multiple traffic profile demands. This could be the starting point for the specification of those business models that could guarantee the deployment of end-to-end QoS.

On the other hand, some service providers not integrated with the network providers could have the need to collaborate with the network providers (that manage the last mile of the network) in order to provide carrier-class services. This could be interesting for streaming-based services[2] or gaming applications.[3]

It should be noted that the specification of these business models should also consider the evolution of interconnection agreements.

Recommendation 4: To Promote the Standardization and Implementation of Reference Points in Commercial Equipments. As stated before, one of the key points in the provisioning of QoS is the coordination of the different QoS mechanisms available in different technologies. In order to do that, it is important to have access to the network equipments involved in the end-to-end path as well

[2]In http://www.layer3media.com/joost/joost-network.pdf, Joost presents its architecture to provide a VoD service based on P2P systems, but at the end it is clearly stated that a possible collaboration with network providers could be beneficial in order to control the capabilities available in the last mile (that is, not under the control of the service providers).

[3]In the Xbox, LIVE service could take advantage of the QoS capabilities in order to improve the end-user QoE.

as to have the opportunity to introduce some minor modifications in the different equipments.

In order to achieve this goal, the specification of reference points in the different network equipments is mandatory in order to reuse the built-in mechanisms to provide end-to-end QoS. This would avoid some of the same integration problems as those faced in EuQoS during the integration of the UMTS or Ethernet technologies that were described in the previous section.

As a first step in this direction, some companies[4] have recently opened their interfaces and operating system in order to allow third parties to implement different applications, such as bandwidth management strategies.

Recommendation 5: To Design a Common Framework to Provide End-To-End QoS: IP Interconnection Models. In order to really meet the QoS requirement, it is essential to ensure the QoS in the end-to-end path. Therefore the coordination between the different domains and technologies will be required, at least in the different access technologies. As a consequence, it is necessary to promote the specification of a common framework for IP interconnection to be used as the basis for the synchronization of the QoS mechanisms available in different domains. The impact on the routing protocols must be studied. In reference 8, the EuQoS system presents its interdomain routing strategy based on EQBGP that allows the different domains to announce their QoS capabilities.

Therefore, in order to guarantee the provisioning of end-to-end QoS in any network, it is important to define a set of end-to-end Classes of Services (CoS) well known by all the domains (that follow their own strategy to implement each CoS) in order to define a converged policy control infrastructure.

Recommendation 6: Implementation of Preliminary Version of Some Interfaces. As was stated in the previous section, one of the major problems in the development of NGN architectures is the interfaces specification and implementation. In order to really success in the implementation of the interfaces, it is mandatory to be able to perform interaction tests. For this reason, it would be recommendable to build basic modules in order to launch compatibility tests.

Recommendation 7: To Build a Common NGN Roadmap. In order to be a reference in the standardization process for NGN, ITU-T, ETSI/TISPAN and 3GPPP should provide a clear roadmap of the technologies, business models, users' requirements, etc., that are being covered or are expected to be covered in the near future. This would allow the alignment of the research efforts in the standard fora and, indeed, in the different implementation efforts done by the different main vendors.

[4]Juniper and Cisco have recently released SDK for its operating systems.

To sum up, this roadmap should let the different players (operators, vendors, regulation bodies, etc.) know when the different standards are expected.

1.3 NGN ROADMAP

According to the recommendations presented in the previous section, the specification of a roadmap to define the evolution of the NGN technologies is an important requirement to make possible the synchronized and effective evolution of the NGN as a key construct of the Future Internet.

In the scope of the EuQoS project, such a roadmap was developed considering the status of the technology in such a moment. This roadmap was presented in Callejo and Enriquez [16], and it is updated (see Figure 1.3) considering the recent developments.

The technology roadmap has been built considering both business and technological perspectives. In particular, the following evolution threads are considered: business models, user requirements, service plane, control plane, underlying network technologies, and operation capabilities.

Evolution Thread 1: Analysis of the Different Business Models. The analysis of suitable business models is a must in order to define the evolution of NGN networks. It is important to clearly identify how the different stakeholders could get incentives from the different features to be developed.

In order to build the Internet of the Future with additional capabilities, it is important to understand which party will take advantage of it and how much the party is willing to pay both directly (i.e., if the customer pays directly) or indirectly (i.e., incomes coming from publicity). This would result in the specification of open interfaces requirements (from the economic point of view) and in an evolution of the current interconnection models.

Evolution Thread 2: Analysis of the End-Users' Requirements. As stated in the recommendations, the analysis of the end-user behavior and an estimation of their future preferences is a must for the success of the NGN. This requirement could cover multiple issues, such as security, usability of the interfaces, and so on.

Evolution Thread 3: Evolution of the Service Plane. This evolution thread aims to analyze the features to be covered in the Service Plane of NGN. These technical features are developed according to the business models and the specification of the users' requirements.

Important aspects to be covered here are all the issues related to the management of the user profile and the mechanisms for AAA.

Evolution Thread 4: Evolution of the Control Plane. The mechanisms and features to be integrated and or developed for the deployment of the QoS guarantees should be analyzed. The evolution of this plane must consider

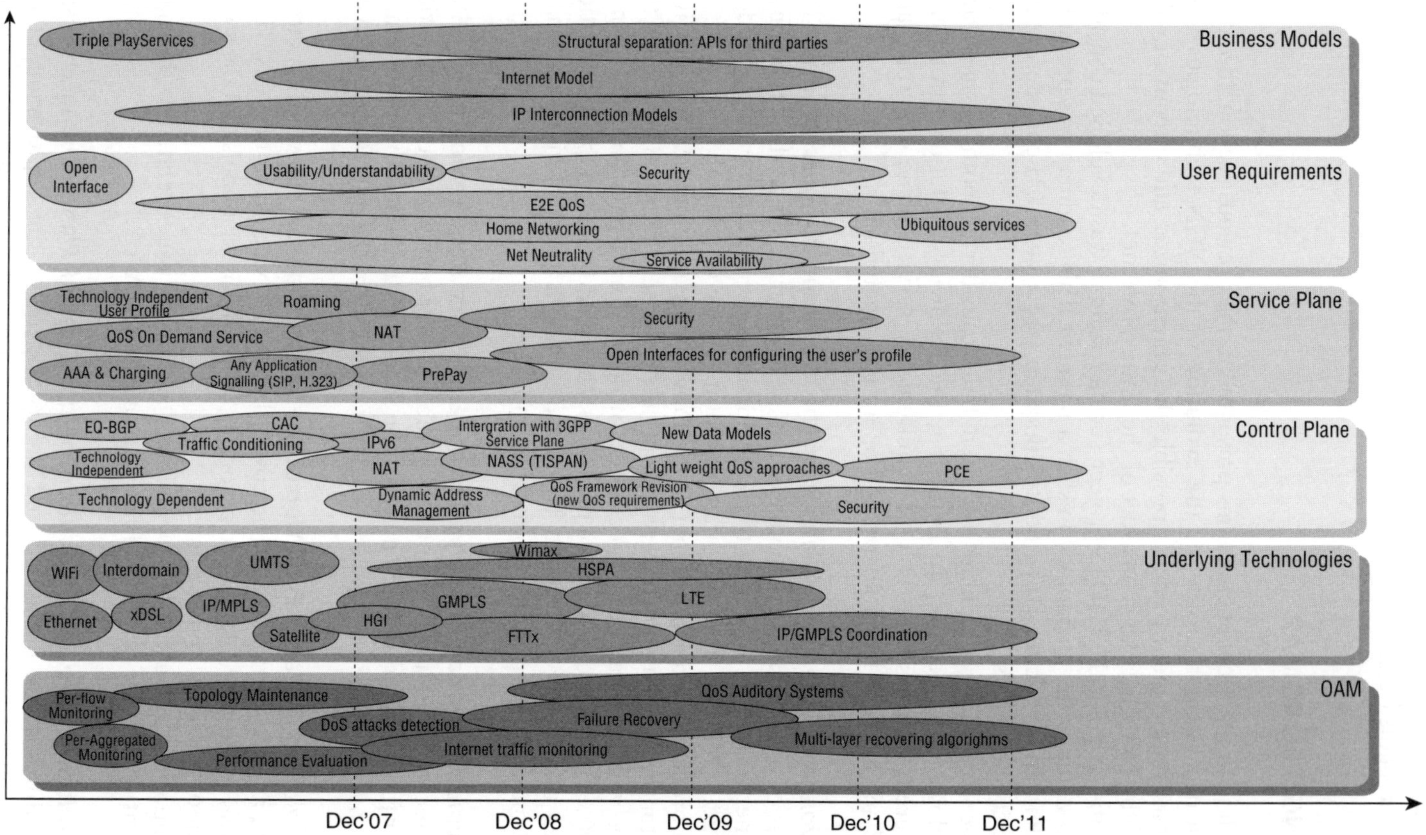

Figure 1.3. NGN roadmap.

operations at different timescales (i.e., for the reservation per aggregate flows and the actions to be done according to end-users' requests).

Evolution Thread 5: Evolution of the Underlying Network Capabilities (Transport Capacities). As far as new network technologies appear, it is important to identify when there will be solutions able to interoperate with the new network equipment and to reuse their built-in QoS mechanisms.

Moreover, in this thread, it is very important to ensure the interoperability and cooperation between different network technologies. A clear example would be the coordination between optical capabilities and the IP core routers.

Evolution Thread 6: Evolution of the OAM (Operation, Administration, and Maintenance). OAM (Operation, Administration, and Maintenance) includes all those features that should be required in a commercial system, especially focusing on security and auditory capabilities. This is a key requirement in order to ensure the traceability of the delivery of QoS-based services.

REFERENCES

1. http://www.geni.net/.
2. http://www.nets-find.net/.
3. http://www.future-internet.eu/.
4. http://akari-project.nict.go.jp/eng/index2.htm.
5. The generation Z connection: Teaching information literacy to the newest net generation, *Teacher Librarian*. Available online. February 2006.
6. Cisco, Global IP traffic forecast and methodology 2006–2011, January 2008.
7. ITU-T Y.2001, General overview of NGN, December 2004.
8. ITU-T Y.2012, Functional requirements and architecture of the NGN release 1, June 2006.
9. http://www.etsi.org/tispan/.
10. http://www.3gpp.org/.
11. http://www.itu.int/en/pages/default.aspx.
12. http://www.euqos.eu/.
13. M. A. Callejo, J. Enríquez, et al., EuQoS: End-to-end QoS over heterogeneous networks, *ITU-T Innovations in NGN—Future Network and Services*, Geneva, pp. 177–184. May 2008.
14. X. Masip, J. Enríquez, M. A. Callejo, et al., The EuQoS system: A solution for QoS routing in heterogeneous networks. *IEEE Communi. Maga.*, Vol. 45, pp. 96–103. February 2007.
15. D. d. Clark, J. Wroclawski, K. Sollins, and R. Braden, Tussle in cyberspace: Defining tomorrow's Internet, SIGCOMM 2002.
16. M. A. Callejo and J. Enríquez, Bridging the standardization Gap to provide QoS in current NGN architectures, *IEEE Communi. Maga.*, Vol. 46, pp. 132–137, October 2008.

2

WIDE-AREA UBIQUITOUS NETWORK: INFRASTRUCTURE FOR SENSOR AND ACTUATOR NETWORKING

Hiroshi Saito, Masato Matsuo, Osamu Kagami, Shigeru Kuwano, Daisei Uchida, and Yuichi Kado

2.1 INTRODUCTION

The growth of Internet traffic has been at a sustainable rate because of the increase of the rate of wired/wireless access lines, which is a product of the high-speed competition among network providers—for example, from ADSL to FTTH or from a third-generation cellular network to long-term evolution. As a result of this intense competition, the number of broadband subscribers has increased rapidly, and the price of communications services has decreased, although the speed of the services has increased (Figure 2.1). In addition, many network providers have introduced fixed-rate charging. Fixed-rate charging was used for wired network services but cellular network providers have also introduced it, mainly for data services (Figure 2.2). Under this charging system, the network operator's revenue does not increase even if the traffic drastically increases. However, it is difficult for the network providers that have already introduced fixed-rate charging to discontinue it because it is attractive to heavy users and is an essential tool for keeping these customers.

Fixed-rate charging and lowering prices imply the possibility of decreased revenue for network providers when the increase of broadband access subscribers

Convergence of Mobile and Stationary Next-Generation Networks, edited by Krzysztof Iniewski

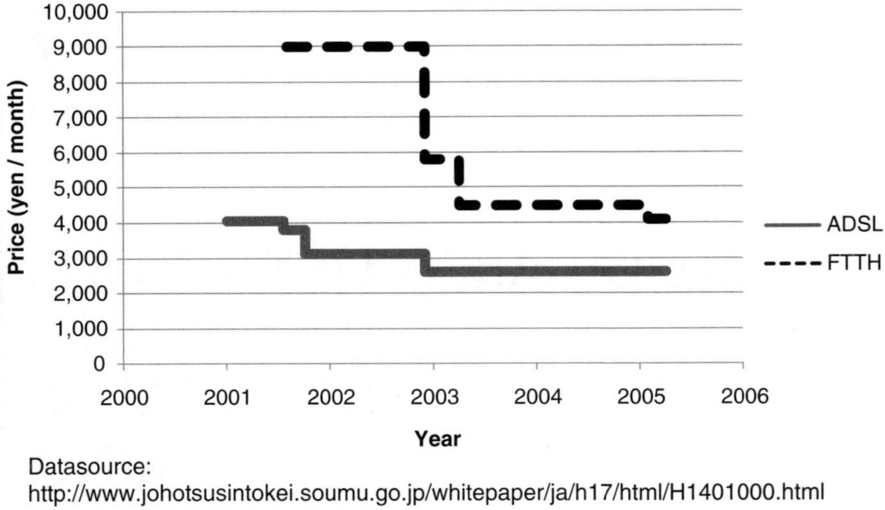

Figure 2.1. Price of broadband access in Japan. (This price is that of the typical ADSL/FTTH services provided by NTT-East.) Data source: http://www.johotsusintokei.soumu.go.jp/whitepaper/html/H1401000.html.

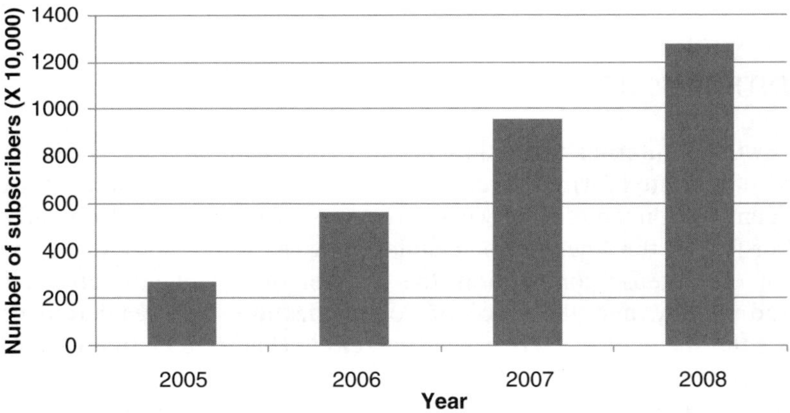

Figure 2.2. Number of fixed-charge subscribers. (Number of subscribers of NTT docomo's "pake-houdai" service, in which charging is fixed and independent of the number of data communication packets.) Data source: http://www.ntt.co.jp/ir/library/annual/pdf/08/p9.pdf.

stops. Therefore, network providers need to find a new source of revenue. The new revenue source will require connectivity to the network through a new mechanism, and this connectivity should not be through the existing broadband access. This is because it cannot be a new revenue source if it is already possible.

The "ubiquitous computing" concept [1] has been studied as an aspect of computer technology. The concept requires computers to be out in the real world with people. In a society where this concept is actualized, these computers are embedded in many products/objects with sensors or actuators, but people are not aware of them. The computers calmly work and communicate with each other to offer a variety of services. The concept introduced in the 1980s is no longer only a dream. Recent developments in computer technologies as well as electronics and micromechanics enable us to make small, low-cost, low-power internal battery-powered sensor (actuator) nodes with computing functions. These nodes can be attached to anything and placed anywhere. We will build networks with these sensor/actuator nodes, capable of providing communication anytime through a wireless link, to sense and detect events of interest and to operate machines and products [2–4]. Instigating this ubiquitous networked society is a technical challenge for network engineers as well as computer engineers, and the necessary technology for achieving this society has great potential as a new revenue source for network operators.

This chapter describes a network called the wide area ubiquitous network (WAUN) [5]. The objective of this network is to globally provide a networked infrastructure of sensors and actuators to implement the ubiquitous networked society.

2.2 TARGET APPLICATIONS AND MARKET

Machine-to-machine applications where WAUN may be helpful for implementing have been emerging. The markets include the following:

1. *Security:* A house equipped with sensors detecting intrusion via the garden and the breaking of a window.
2. *Health Management:* A house with sensors detecting motion and counting the heartbeats of occupants to enable a rapid response to sudden sickness, such as a heart attack in the bath.
3. *Nursing:* A nursing home with (a) sensors detecting the location of each resident and sounding an alarm if the resident enters a dangerous place or leaves the home and (b) sensors detecting residents falling or experiencing incontinence.
4. *Inventory Management:* Tanks (such as gasoline tanks and beer tanks) with sensors measuring the quantity remaining and sending data to an inventory management system, which determines the schedule for refilling the tanks.
5. *Environment Protection:* Sensors for detecting temperature, moisture, and chemical substances in a forest to detect a forest fire and environmental destruction and to calculate the carbon dioxide processing capacity of the trees.

6. *Disaster Management:* Acceleration sensors and strain sensors detecting landslides and earthquakes to shut down gas supply pipelines, stop high-speed trains, or set traffic lights to red.
7. *Infrastructure Management:* Sensors attached to buildings, bridges, tunnels, and highways to detect structural deterioration due to age, enabling them to be repaired.
8. *Logistics Management:* Sensors detecting locations of packages and monitoring conditions such as temperature for frozen packages.
9. *Car Maintenance:* Sensors to monitor the conditions and to recommend the renewing of car parts.
10. *Child Care:* Sensors detecting the location of children and delivering information such as "arriving at school."
11. *Parking Lot Management:* Sensors monitoring the use of parking lots to lead a car to an empty lot.

Most of these applications can be implemented via existing networks, but some of them are not widely deployed because of high networking cost, short battery life, and limited network coverage. If WAUN could overcome such problems, these above would be good initial markets.

In addition to these recently conceived applications, many new services would be possible if products, including daily commodities, were networked [6].

1. Medicine dosage could be managed by detecting the removal of tablets from a package and sending that information to a patient's doctor. The doctor could check that the patient takes the medicine correctly, and consequently he/she could check the effectiveness of the medicine. Dosage management is appropriate particularly for elderly people who may forget to take their medicine. In Japan, the market size for pharmaceutical drugs is about 6.7 trillion yen a year [7]. If medicine dosage management were applied to 10% of this market, this would equal 670 billion yen. This dosage management would help to reduce medical costs.
2. Business cards that have already been distributed could be updated. Business cards are distributed at business meetings, but their information may become obsolete due to, for example, restructuring in a company. Business cards with a simple e-paper display having a wireless transceiver and a thin-film battery would enable updating of the information.
3. Name stamps could be managed by detecting their use. In many business as well as private situations, Japanese people use name stamps in the manner that signatures are used in Western countries. In particular, if we could detect and manage the use of name stamps in business situations, great progress could be made in the computerization of office work. In addition to name stamp management, location management of important business tools including valuable documents is also a promising application.

4. Usage evidence could be obtained by monitoring the opening of a package. This would enable us to easily take stock of inventory that has or has not been used. It could be used as evidence for a tax agent and could help in inventory management. (About 3 billion packages are delivered by parcel delivery services in Japan per year [8], and more than 100 million pieces of registered mail are delivered by the Japan Postal Services Corporation per year [9].)

Here, we should note that these new applications could be implemented using very small simple sensors or actuators that are networked. Sensors and actuators are very low-end telecommunication terminals. This would result in a different future network vision from those based on the assumption that most terminals will have high performance and rich functions, such as personal computers. However, these applications could require an extremely large number of terminals and open up new markets for network providers because they are not implemented through the existing wired/wireless broadband networks.

2.3 REQUIREMENTS, CONCEPT, AND TOTAL SYSTEM

To support the applications mentioned in the previous section and to obtain a new revenue source, we propose the wide-area ubiquitous network (WAUN). WAUN needs to satisfy the following requirements before being used for the mentioned applications: low cost, low power consumption (long battery life), mobility support, support of low-end terminals, security, scalability regarding the number of terminals, and wide (ubiquitous) coverage.

To satisfy the requirements, our proposed architecture uses a long-range wireless link dedicated for use by the first hop of the wireless terminal (WT), although most research efforts focus on setting up sensor networks by using multiple wireless hops in an ad hoc manner. We chose this architecture mainly because the ad hoc approach still has problems related to the mobile ad hoc network, such as the amount of power consumed when a terminal is used as a mobile switching node, the security threat to mobile switching nodes, and the unstable service area or routing due to too many mobile nodes having excessive freedom of movement. In addition, in the low-end terminals, complicated routing protocols do not work.

Because we choose architecture that does not use the multiple hops of the wireless link, a long-range wireless link (large cell) is essential for economical wide area coverage. The actual target range is several kilometers. There are two main reasons for this. First, as the cell radius, r, increases, the number of access points (APs) can be reduced to $1/r^2$ and the capital expense of the wireless system is reduced to nearly $1/r^2$ because the cost of the AP dominates this expense. Second, most network providers have their own buildings containing data cables and electric power supply cables every few kilometers. Thus, if the radius of a cell of WAUN is more than a few kilometers, an existing network provider can fully use it to cable the APs, resulting in minimized capital expenses. However, longer-range radio transmission needs terminals with more power. Thus, we need to satisfy the contradictory requirements by using the technologies shown in this chapter. Our specific R&D target is a "5-km cell radius using 10-mW transmission power terminals with 10-year battery life."

Scalability is also a major issue. We introduce the following rules to maintain the scalability of WAUN: The end-to-end communications are between a pair of a WT and a wired terminal; for each WT, the corresponding wired terminal is unique and fixed. (In actuality, one virtual wired terminal can consist of multiple distributed wired terminals.) That is, WAUN does not support public switching services between arbitrary terminals. This is because many applications do not require communication with arbitrary terminals but with terminals in a private company or in a community. In addition, the security improvement resulting from implementation of these rules is large. (In practice, a higher layer service supports communications with other terminals.) As a result, WAUN supports many private sensor networks. The total WAUN system is depicted in Figure 2.3.

WAUN has the following features to satisfy the requirements.

1. WAUN works as a middle box between a WT and a wired terminal. WAUN does not provide a transparent session between them.
2. The wireless link has a large range (about 5 km). The range will be kept large by using reception diversity on the basis of combined maximal ratios. This will enable a network provider to cover a wide area with a small number of APs and thus offer services at a reasonable cost.
3. WTs are not Internet protocol (IP) terminals and do not use transmission control protocol (TCP)/IP because TCP/IP has too much overhead and WTs have low-performance central processing units (CPUs) and little memory. The APs offered by the network provider convert the wireless link protocol dedicated to WAUN WTs into protocols developed in the IP community. APs also convert identification numbers (IDs) between

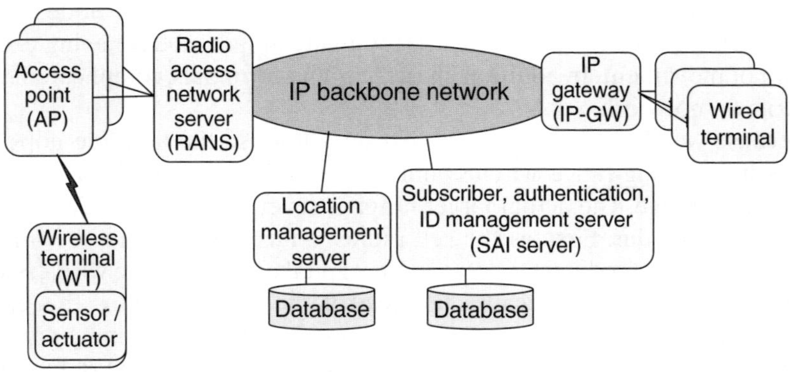

Figure 2.3. Wide-area ubiquitous network (WAUN).

those used at the wireless link and those used in the core (fixed) network to identify a WT. The radio access network gateway servers (RANSs) distinguish the signal (e.g., user data and authentication request) sent from each WT and switch it to the appropriate destination on the basis of the distinguished result.

4. WTs can move. The mobility management function maintains the area in which a WT exists. This function is implemented by location registration technologies similar to those used in cellular networks.
5. Wired terminals are IP terminals that communicate with WTs in WAUN and are accommodated through the IP gateway (IP-GW). Their interface with WAUN is a widely used common interface such as TCP/IP, but wired IP terminals are prevented from directly accessing WAUN network entities such as RANSs for security reasons. WAUN does not support mutual communication between wired IP terminals.
6. In WAUN, several IDs are used to make the service convenient, secure, and efficient. WAUN offers security functions and ID conversion/resolution. In particular, to prevent tracking by a stranger, the ID of the WT is assigned temporarily and often updated. This ID management with mutual authentication between the terminal and the network enables us to achieve secure communication. The subscriber authentication and ID management (SAI) server stores the subscriber profile information including the wireless/wired terminal information for authentication and ID management.

2.4 CORE NETWORK

The WAUN core network controls communication between pairs of a wired terminal and a mobile WT and offers functions of ID management, location management, security, and access control. It needs to have enough scalability to process these functions for huge numbers of WTs. There are two important points in designing the WAUN core network. First, the processing necessary for communication control in a WT should be reduced as much as possible because the WT has low power consumption and low capability. Second, a limited radio-frequency band should be shared efficiently among many WTs. Thus, we need to decrease the communication overhead to increase the number of WTs that WAUN can accommodate.

2.4.1 Communication Protocol

The WAUN communication protocol for data transmission between a WT and a wired terminal is shown in Figure 2.4. An AP converts the wireless link protocol for a WT to TCP/IP, used in the WAUN core network, and vice versa. A light communication protocol appropriate for low-power-consumption low-capability WTs must be used between the WT and AP, but a commonly used protocol such as TCP/IP is appropriate for the economical implementation of the WAUN core network.

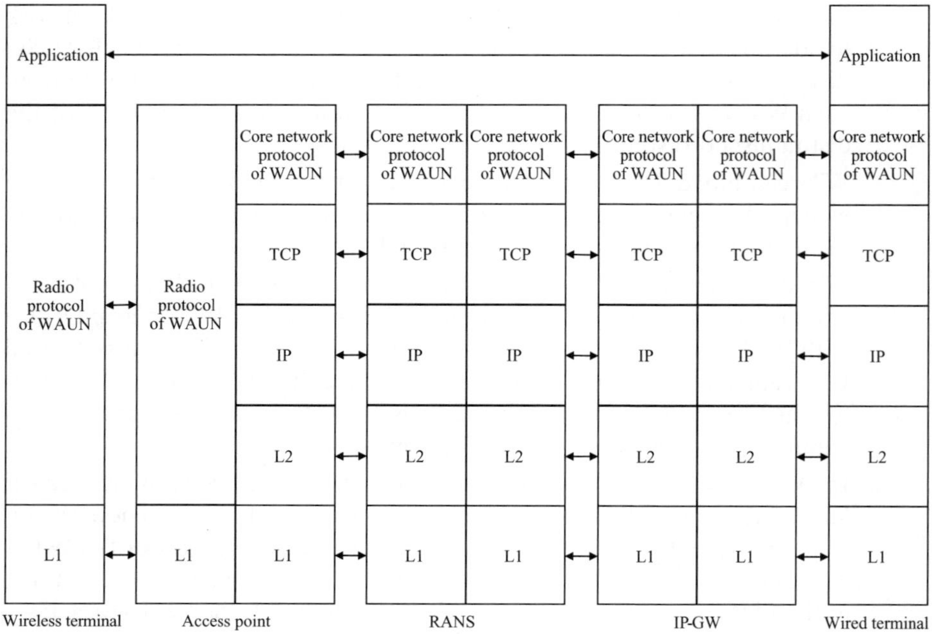

Figure 2.4. WAUN communication protocol.

The WAUN core network decides the destination of a packet from a WT on the basis of the WT ID because the corresponding wired terminal is fixed and registered in the subscriber authentication and ID management (SAI) server. Therefore, it is not necessary to convey the destination address (when a WT sends a packet to a wired terminal) or the source address (when a wired terminal sends a packet to a WT) in the packet. This brings three advantages. First, the load related to communication by the low-power-consumption low-capability WT is reduced and intentional/unintentional transmissions to an incorrectly accessed wired terminal can be prevented. Second, the radio channel can be used effectively. For example, the address of IPv6 has 128 bits. In WAUN, where many terminals share a limited band, this 128-bit address imposes a burden on the limited bandwidth of the wireless link. Third, the corresponding wired terminal can be replaced without changing the program in the WT. Otherwise, we might need to change the program in a large number of WTs scattered in various places when the wired terminals must be replaced for some reason.

2.4.2 ID Management

WAUN transmits data from/to a WT on the basis of its ID. WAUN uses the following three kinds of IDs for a WT according to the purpose. (1) A permanent ID is allocated for the entire service period to uniquely identify a WT. It is used only inside the WAUN core network for security. Instead of using its permanent

ID, a WT can use a temporary ID to identify itself, and a wired terminal uses a service ID to identify the WT. Therefore, an AP performs the conversion between the temporary ID and the permanent ID, and an IP-GW performs the conversion between the service ID and the permanent ID. The SAI server manages the relations of the three IDs. (2) A temporary ID is used to authenticate a WT before a wireless communication link is established. Thus, both the WT and the SAI server manage this ID. The temporary ID is often updated synchronously in the WT and the SAI server to avoid tracking by unauthorized people. Because this ID can be reused if it is unique at the update, its length can be shortened more than the permanent ID for efficient use of a limited wireless band. (3) A service ID is used for a wired terminal to identify a WT. Using this ID can conceal the replacement of the WT due to, for example, a breakdown.

The huge number of these IDs must be managed in SAI servers, and the conversion among IDs is performed in APs and IP-GWs for each communication. Therefore, the data management and conversions should be performed in a decentralized manner.

2.4.3 Security

The existing security technology with high reliability in an IP community, such as a virtual private network (VPN), can be applied to the wired portion of WAUN. However, special efforts are needed for the wireless link because its bandwidth is limited and a WT has low capability and low power consumption. Similar to the cellular phone network, WAUN offers security functions that prevent tapping, tracking, identity theft, and falsification of identity. However, WAUN achieves these functions in less than half of the total length of messages used in the cellular phone network. For example, to update a temporary ID, the cellular phone network method transmits the new encrypted temporary ID to the WT, while WAUN transmits not the temporary ID itself but reduced amounts of information for mutual updating.

A WT is authenticated by a RANS caching its temporary ID managed in a SAI server whenever the wireless communication link is established to transmit data between the WT and an AP. The RANS authenticates the WT when it is switched on and when it moves beyond the paging area boundary. Similarly, a wired terminal is authenticated by an IP-GW, to which the wired terminal is a priori assigned.

2.4.4 Access Control (Authorization)

WAUN charges can be based on the number or frequency of communications and may limit communication frequency, such as to once an hour or once a day, for a WT. The allowed communication frequency for each WT is originally determined by a subscriber contract and registered in the SAI server. A RANS and an IP-GW temporarily maintain this information on the limit, which they obtained at the authentication, and check the communication frequency. If the frequency of the WT sending and receiving goes beyond the limit, the RANS and IP-GW reject the communication request.

2.4.5 Upload and Download Protocols

The protocols for uploading data from a WT to a wired terminal and for downloading data from a wired terminal to a WT are shown in Figures 2.5 and 2.6 [10],

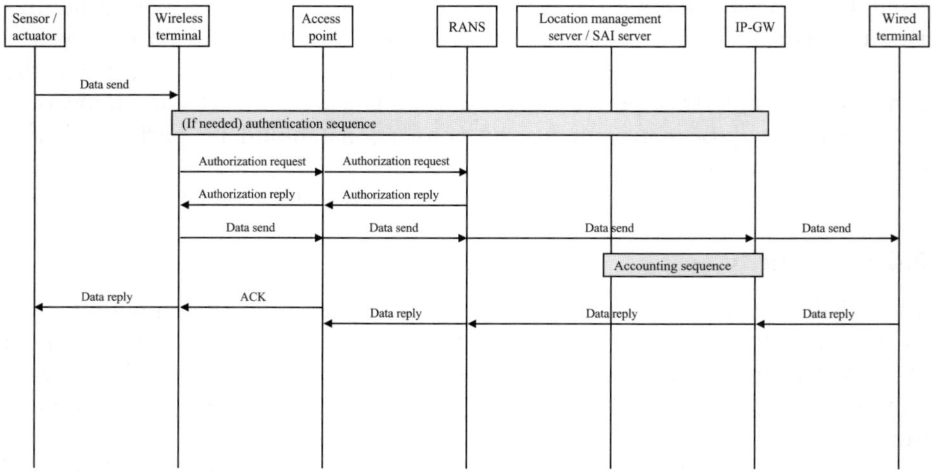

Figure 2.5. WAUN upload protocol.

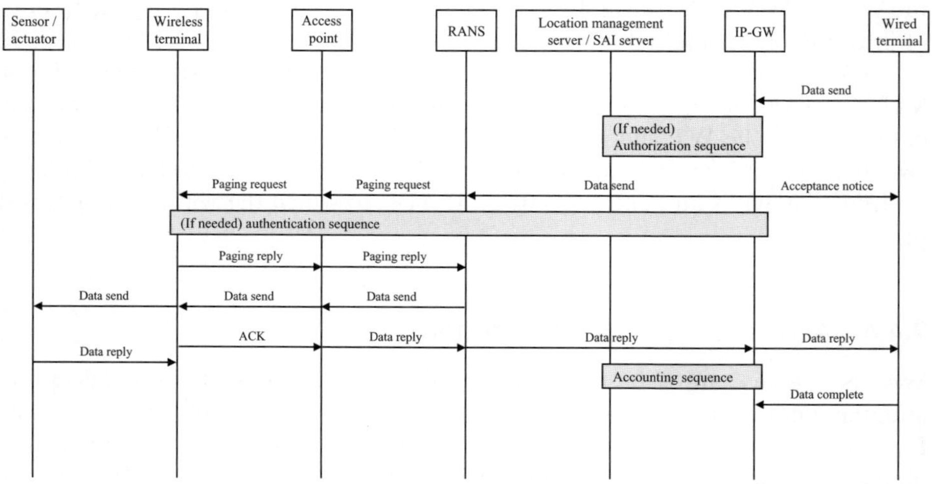

Figure 2.6. WAUN download protocol.

respectively. In the upload protocol, the authentication and authorization procedure are performed by a RANS before every data transmission.

The authentication sequence in Figures 2.5 and 2.6 includes a procedure to update the WT information. After the authentication, the RANS updates the WT information (including the paging area in which it resides) in the location management server and caches the information locally. The RANS can maintain the information of the WT in the coverage area of an AP under the RANS. Then, the RANS informs the IP-GW that accommodates the wired terminal corresponding to the WT of the updated paging area information including this RANS address. (The IP-GW caches this information, which enables selection of the RANS when the download protocol is executed without retrieving the residing paging area information. Thus, the use of the cached information reduces the load on the location management server and the SAI server and shortens the time needed for communication.)

In the download protocol, WAUN uses a four-way sequence between a wired terminal and an IP-GW. An "acceptance notice" message indicates that the IP-GW has authorized the transmission and started the procedure for transmitting data. A "data reply" message indicates that the WT has been authenticated and has received the data.

2.5 WIRELESS ACCESS NETWORK [35]

2.5.1 Background

Because WAUN is a new type of network, the requirements [6] for wireless access systems, such as scalability with respect to the number of terminals, terminal mobility, and support for low-performance terminals, cannot be adequately met by current wireless access technologies. New techniques should be developed to establish a wireless access system for WAUNs. For the network infrastructure, a cellular configuration is more suitable than a multihop one, considering stable operation and power consumption of WTs [6]. A fundamental link analysis [6] predicted that a 5-km cell radius could be used for WTs having 10-mW transmission power. In this case, the reception power is extremely low, so making a wireless access system by using current wireless technologies is difficult. Therefore, a sophisticated technology that combines wireless techniques, such as modulation/demodulation, error correction, and diversity, should be developed. In particular, because the WTs themselves should have a simple configuration and simple and low-power operation, the wireless AP should support complex operations to compensate for the simplicity of the WTs. Medium access control technology that can handle tens of thousands of terminals with various service levels while suppressing the power consumption of WTs by supporting intermittent operation is also important. Moreover, a network control function should be developed for stable and efficient operation of the wireless access system and to ensure terminal mobility and connectivity to the backbone network.

The following subsections describe the system architecture and key technologies for the WAUN wireless access network.

2.5.2 System Architecture

A schematic of the wireless access network is shown in Figure 2.7. The wireless access network consists of two network elements: an AP and a WT. They are connected by wireless links. The AP and WT both have three functions: physical layer interface (PHY), medium access control (MAC), and network control (CNT) functions. PHY provides wireless modulation/demodulation and transmission/reception functions. The MAC function is implemented to provide multiple access control for the wireless system, and it should support various levels of quality of service (QoS) for the ubiquitous network infrastructure. The AP is connected to the WAUN Internet protocol (IP) backbone network via a radio access network server (RANS) using CNT, and the WT connection to a sensor or actuator is established using CNT. The CNT function also manages the PHY and MAC functions of the AP and WT. The key issues concerning these functions are described in the next subsection.

An example of the cell architecture is shown in Figure 2.8. Each cell is a hexagon with a radius of a few kilometers, and the cells form a honeycomb structure. Three-sector antennas are deployed at three of the apexes of each cell, and radio-frequency (RF) signals are emitted inwards toward the center of the cell. To avoid interference from neighboring cells, different frequency channels are assigned to adjacent cells. Space and site diversity techniques are used simultaneously to achieve a high coverage probability in the cell. The PHY function in the cell is physically divided into two types of modules: the modulation/demodulation module and the RF transmission/reception module. The first module is centralized as the master equipment of the AP (AP-M), and the second is deployed with

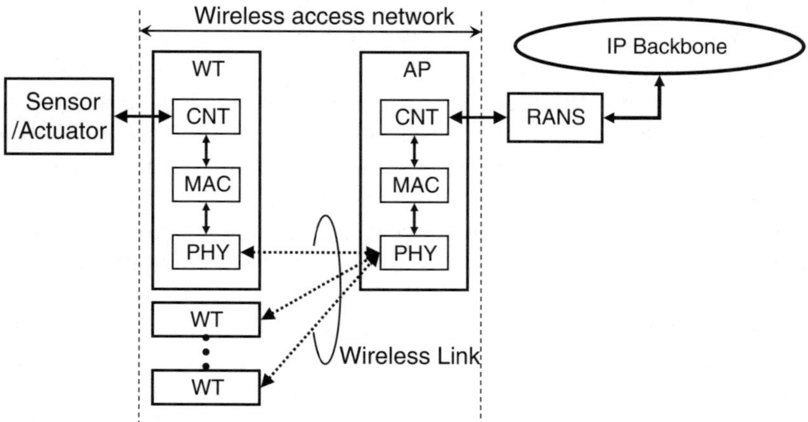

Figure 2.7. Schematic of wireless access network.

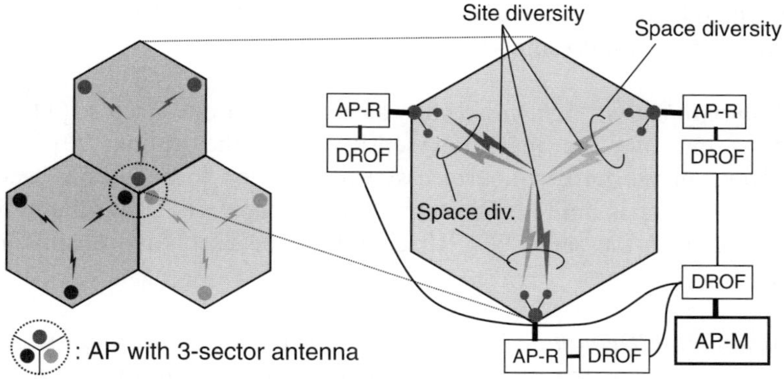

Figure 2.8. Wireless cell configuration.

an antenna as remote equipment of the AP (AP-R). Diversity signal processing is implemented at the AP-M. The same downlink RF signals are delivered to the AP-Rs, and the uplink RF signals at the AP-Rs are collected at the AP-M. To transmit RF signals between the AP-M and AP-Rs, we use a digitized radio over fiber (DROF) subsystem [11]. Digitized RF signals are transmitted between AP-M and AP-Rs over the Ethernet.

2.5.3 Key Issues

2.5.3.1 Physical Layer Interface (PHY)

2.5.3.1.1 PHY Requirements. The PHY is an essential part of the wireless access network. Its requirements are summarized below.

1. *Long-Range Transmission.* The system target is a 5-km cell radius with 10-mW WT transmission power.
2. *Use of a Modulation Scheme with High Frequency Utilization.* A huge number of WTs should be supported with limited frequency resources.
3. *Short Overhead.* Because WAUN data traffic may be dominated by short packets, the PHY overhead of each packet should be as short as possible to avoid degrading the transmission efficiency.
4. *Compact, Low-Power-Consumption WTs.* WTs should be compact and have the longest possible battery life and lowest possible cost.

2.5.3.1.2 PHY Functions. To achieve the above requirements, PHY functions are designed as follows.

1. RF Specifications. The VHF or UHF band is selected for its low propagation and shielding loss characteristics. The WT transmission power is limited to

10 dBm, which is the limit for unlicensed systems in Japan. The AP is a licensed station and can transmit signals with higher power.

2. Communication Scheme. Time division multiple access (TDMA) is selected to improve the reception sensitivity of the uplink (UL) because the receiver can use narrow-aperture synchronization in conditions where the transmission power is limited. The time division duplex (TDD) scheme is used because it enables the same WT RF circuits to be used for transmission and reception.

3. Modulation/Demodulation Scheme. π/4-shift quadrature phase shift keying (QPSK) is selected as the modulation scheme for better frequency utilization. As the demodulation scheme, differential detection is used in the WTs because it can be implemented by a simple circuit, and coherent detection is used in the AP for its high sensitivity. For forward error correction, convolutional coding and soft decision Viterbi decoding are selected for their high coding gain and relatively low hardware implementation complexity.

4. PHY Burst Structure. The PHY burst structure is shown in Figure 2.9. The downlink (DL) burst consists of a preamble and data symbols. The preamble is located at the front of the DL burst and used for burst synchronization. The UL burst consists of multiple pilot symbols and data symbols. The pilot symbols are uniformly inserted into the UL burst and used for burst synchronization and channel tracking for coherent detection.

5. Downlink Synchronization. The WT's synchronization with the received signal is established by differential detection and cross-correlation of the preamble. More specifically, frequency synchronization is established by estimating the carrier frequency offset from the phase component of the cross-correlation value, and the clock and frame synchronization are established by estimating their time position from the peak of the cross-correlation value. Differential detection enables calculation of the carrier frequency offset even when the phase shift due to the carrier frequency offset within the preamble exceeds π; this reduces the cost of the WT oscillator. The cross-correlation scheme can detect the time positions in the case of a low carrier-to-noise ratio. Therefore, this

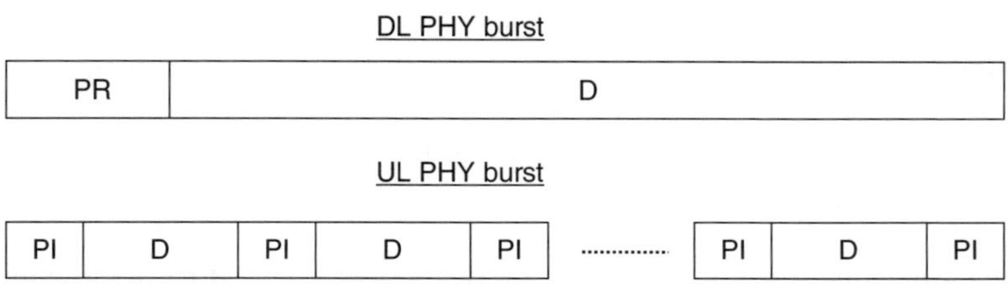

Figure 2.9. PHY burst structure. PR, preamble; PI, pilot symbol; D, data symbol.

method establishes synchronization with high accuracy within one burst. This shortens the WT receiving span and reduces the DL overhead.

6. Uplink Synchronization. A transmission automatic frequency control technique is used at the WTs to eliminate the UL overhead for the frequency synchronization [12]. The AP stores the entire burst within the narrow aperture, and clock and frame synchronization are established by differential detection and cross-correlation of the training symbols for channel tracking needed for coherent detection. This can also eliminate the UL overhead for the exclusive use of the clock and frame synchronization, and a highly efficient transmission can be achieved.

7. Diversity Scheme. The AP has transmission and reception diversity functions to enhance the reception sensitivity of the DL and UL in the fading and shadowing environment. In transmission and reception diversity, space and site diversity techniques are used to offset fading and shadowing, respectively. The frequency offset transmission diversity technique, which is used in the Japanese paging system, is used for the DL. With this technique, the same signals with slightly different carrier frequencies are transmitted from the antennas, so the reception level varies with time everywhere in the cell due to interference between the RF signals from different antennas and the cell does not have any dead spots. This technique offers a transmission diversity gain that corresponds to the number of AP antennas. In the UL, the maximal ratio combining (MRC) diversity technique is used. This technique enhances the receiving diversity gain simply by increasing the number of AP antennas.

2.5.3.2 MAC Functions. The MAC protocol manages tens of thousands of WTs, maintains the communication quality using automatic repeat request (ARQ) regardless of radio link bit errors, and achieves QoS control using random access backoff window size control and dynamic slot assignment [13]. Furthermore, the MAC protocol efficiently accommodates WTs that send small amounts of data at very long intervals. It has a function for supporting intermittent reception by WTs in accordance with the traffic demand. This reception scheme suppresses the power consumption of WTs and leads to a long battery life.

2.5.3.2.1 LOGICAL CHANNEL AND FRAME STRUCTURE. The MAC frame structure of the wireless system is depicted in Figure 2.10. Logical channels used in the MAC layer are listed in Table 2.1. To achieve good frame synchronization performance even though WTs experience poor radio link conditions, the wireless system uses TDMA/TDD in the MAC layer. The length of the MAC frame was chosen to be 9600 symbols, considering throughput efficiency and transmission delay performance. The first part of the frame is for the downlink, and the second part is for the uplink. The boundary between them changes dynamically in accordance with the results of burst assignment scheduling, which is achieved frame by frame.

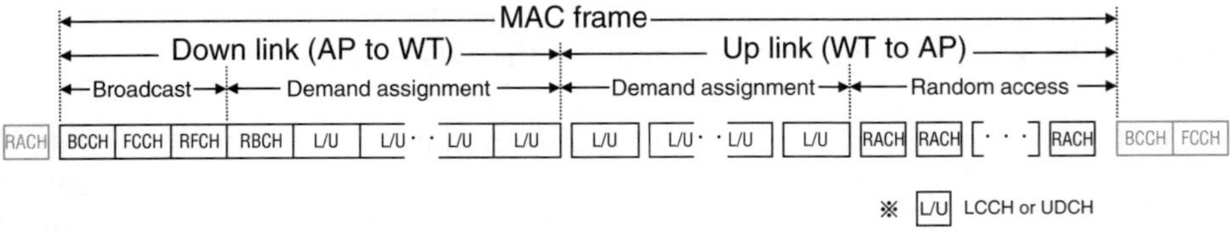

Figure 2.10. MAC frame structure.

TABLE 2.1. Logical Channels

Name	Usage
BCCH	Broadcast control channel is used in the downlink direction and conveys broadcast control channel information related to the entire cell. It contains a fixed amount of data.
FCCH	FCCH is sent in the downlink direction and conveys information that describes the structure of the MAC frame at the air interface.
RFCH	The purpose of the random access feedback channel is to inform the terminals regarding the random access parameters.
RBCH	RLC broadcast channel is used in downlink and conveys broadcast control information related to the entire cell.
UDCH	User data channel is used to transmit user data between the AP and a WT.
LCCH	Link control channel is bidirectional and is used to transmit ARQ feedback.
RACH	Random access channel is defined for the purpose of giving a WT the opportunity to send control information to the AP using random access.

The MAC frame consists of the broadcast channel (BCCH), frame control channel (FCCH), random access feedback channel (RFCH), radio link control broadcast channel (RBCH), user data channel (UDCH), logical control channel (LCCH), and random access channel (RACH). BCCHs are used to report the attributes of the AP and are sent at the beginning of each MAC frame. FCCHs indicate the MAC frame structure—that is, the position and length of other channels following the FCCH. RFCHs are used to send information associated with the random access—that is, the results of random access in the previous MAC frame, backoff window size of each QoS class, position of random access slots, and number of random access slots in the current MAC frame. RBCHs are used to broadcast radio link control messages. UDCHs are used to send user data and LCCHs are used to send MAC control information—for example, ARQ-ACK (ACK: acknowledgment).

2.5.3.2.2 Access Sequences. The user data transmission sequence using random access and demand assignment when the user data is sent from a WT to the AP is shown in Figure 2.11. In this wireless system, the AP sends a BCCH, FCCH, and RFCH in order at the beginning of a MAC frame. When a WT has

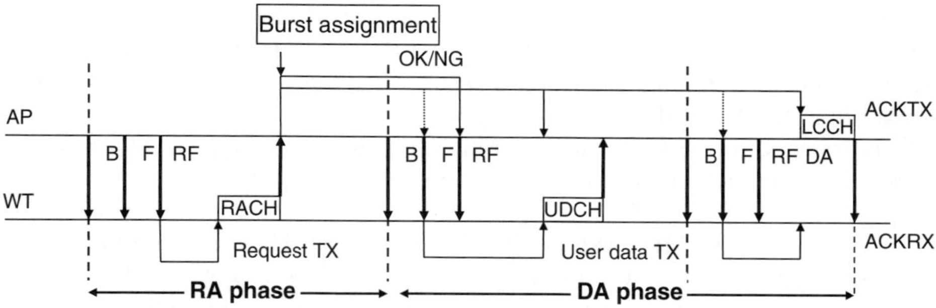

Figure 2.11. User data transmission sequence. B, BCCH; F, FCCH; RF, RFCH.

user data to transmit, it receives the RFCH to determine the start position and the number of random access slots. To avoid collision with the RACHs of other WTs, the WT executes a backoff. The number of backoff slots in the random access is randomly selected by the WT within the range of zero to the window size. At the first random access attempt, the window size is set to the value provided by the AP as the initial backoff window size through the RFCH. After performing the backoff, the WT sends the RACH in order to request the AP to assign a UDCH for transmitting its user data. The RACH contains a MAC-ID. A unique MAC-ID is assigned to each WT, and it enables the AP to identify the WTs. The length of the MAC-ID is 16 bits in the prototype. Therefore, the AP can manage more than 60,000 WTs.

After transmitting the RACH, the WT receives the RFCH again in the next MAC frame. If the random access was successful in the previous MAC frame, the AP sends back an ACK to the WT through the RFCH. Otherwise, the AP sends nothing. When the WT fails to receive an ACK, it doubles the window size and calculates the backoff slot size according to the window size. After the backoff, the WT tries to send the RACH again. The RACH is retransmitted until it is successfully received or until the sending count reaches the predefined limit.

After receiving an ACK from the AP, the WT waits for the UDCH assignment notified by FCCH. When the UDCH has been assigned to the WT, it transmits the UDCH containing the user data. When the AP receives the UDCH, it sends the LCCH back with an ARQ-ACK if it receives the UDCH without any error. Otherwise, it sends the LCCH back with an ARQ-NAK (negative acknowledgment), which informs the WT of a UDCH transmission failure.

2.5.3.2.3 QoS Control. Three QoS classes are defined in this system. The AP offers services based on these three QoS classes, and WTs can receive these services at the same time. The services can be distinguished by the data link control (DLC) connection, and each WT establishes three or fewer DLC connections. Moreover, in each DLC connection, ARQ is executed to correct bit errors in the radio link and to maintain the communication quality. QoS control in the uplink is achieved through AP control of the backoff window size for random access. That is, the AP decides the backoff window size appropriate for each QoS class and notifies WTs through the RFCH. WTs then execute random access using the notified backoff window size when they try to transmit their user data. When the backoff window sizes are decided in the AP, a smaller window size is assigned to the QoS class with high priority and a bigger window size is assigned to the QoS class with low priority.

2.5.3.2.4 Sleep Mechanism. WTs have two BCCH reception modes: successive reception and intermittent reception. The WT changes to the BCCH intermittent reception mode when there is no transmitted data. In this mode, the WT receives a BCCH only once every N times (N = 2–128). Therefore, the WT decreases its power consumption in order to achieve a longer battery life. An example of WT transition to the BCCH intermittent reception mode is shown in Figure 2.12. In the case shown here, the WT is set to change to BCCH intermittent reception mode if there is no burst assignment, such as UDCH and LCCH, to the WT over three successive MAC frames. The WT is set to receive the BCCH once every four times in the BCCH intermittent reception mode.

2.5.3.3 Network Control Function (CNT)

2.5.3.3.1 CNT Configuration. The network control function of the AP (AP-CNT) manages the AP wireless link and a link to the WAUN backbone network via the RANS. WTs also have CNT (WT-CNT), and this manages WT wireless links and interworks with RF and external equipment such as sensors and/or actuators.

The AP-CNT has a TCP/IP connection to the RANS. The AP-CNT transfers user messages and control messages through this TCP/IP link. In the case of WAUN, the WT hardware must be compact and have low power consumption,

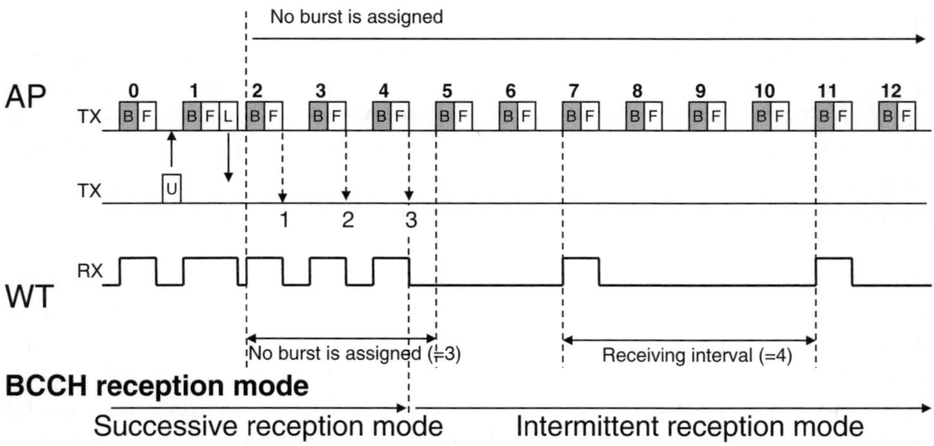

Figure 2.12. Transition of BCCH reception mode. B, BCCH; F, FCCH; U, UDCH; L, LCCH.

so the TCP/IP protocol stack cannot be implemented on the WT and the TCP/IP overhead is too large for the narrow wireless link of WAUN. Therefore, the AP should function as a translator between the TCP/IP-defined backbone network and the wireless access network.

2.5.3.3.2 CNT Function

1. Hardware Control. AP-CNT controls the AP hardware. When the AP is powered on, the AP-CNT attempts to connect to the RANS. After connecting to the RANS, the AP-CNT starts broadcasting the AP information on the control carrier. At the end of the broadcast, WTs can search for the AP to which they should establish a link.

2. Carrier Management. Because the AP can transmit multiple carriers, AP-CNT must have a management function for handling multiple carriers. An AP has one control carrier and multiple communication carriers. The control carrier transmits the operational information such as the ID of the AP. When a WT finds the AP's control carrier, it obtains the ID of the AP, traffic loads of communication carriers, and other information from the BCCH. The communication carrier is used to transfer messages between terminals and the backbone network. When a WT that has selected a communication carrier requests a UDCH so that it can communicate with the backbone network, the RACH of the communication carrier is used to send the request. If the request is acceptable, the UDCH is assigned to the WT on the same carrier. The AP-CNT also manages traffic information for each carrier, such as the frequency, traffic load, error information, and interference level. The WT searches for the control carrier first and synchronizes with it. Then, it selects the AP to which it should connect.

3. Terminal Management. AP-CNT manages WT information using a terminal management table stored in its memory. First, the WT information is set in the table when it is authenticated. Then, the elements of the table needed to establish a link are set, such as the terminal ID, link ID, information pertaining to security, and terminal status. The AP obtains these elements from the backbone network when the WT is authenticated and removes them when the link assigned to the WT is purged. Several elements are updated with appropriate timing because of security issues.

4. Connection Management. The AP-CNT and WT-CNT make a connection between the backbone network and terminal when data arrive from the backbone network or terminals.

Initially, the AP-CNT evaluates the wireless network capacity usage. The results of the evaluation are advertised on the broadcast channel to the WTs. The WT-CNT, which attempts to connect to the backbone network, can select the AP on the basis of its traffic class and advertised traffic capacity. If the used capacity exceeds the established threshold set for generating a new link for a WT, the

WT-CNT declines to generate a new link to the AP and chooses one from among other available links.

Next, in the case of an upload, the WT-CNT that is trying to send an upload packet attempts to establish a link. If it has already obtained a link, it requests a communication channel. Otherwise, the WT needs authentication to establish a link. Even if it has a link, subscriber authentication and an ID management server in the backbone network are needed to allocate a communication channel. The subscriber authentication and ID management server evaluate whether the AP should allocate the communication channel requested by the WT. On the basis of the evaluation results, the AP-CNT permits or denies the WT access to the network through the communication channel. Only terminals that have been authenticated and that have a radio link and a communication channel can send their own data packets to the AP.

In the download case, the AP-CNT waits for a data or paging message from the backbone network. When a message arrives, the AP-CNT checks the WT management table to find the link through which the AP-CNT should transfer the message. If the WT is found in the table, the AP-CNT allocates a communication channel for sending the message to the WT. However, if it is not found, the AP-CNT attempts to page the WT through the broadcast channel. A WT-CNT that monitors the broadcast channel but does not have a radio link detects this paging message. Then, the WT-CNT tries to establish a link.

2.6 WIRELESS TERMINAL (WT)

This section describes methods for reducing power consumption in WTs for WAUN. The importance of low-power operation and the impact of intermittent operation on WT power consumption are discussed. A multithreshold complementary metal oxide semiconductor (CMOS) circuit scheme that greatly reduces power consumption is presented. It can extend battery life significantly.

2.6.1 Requirements for Wireless Terminal

A simplified block diagram of a WT for the WAUN [14] is shown in Figure 2.13. The WT consists of radio-frequency [transmitter (Tx) and receiver (Rx)] circuits, a phase-locked loop (PLL), a clock, baseband digital circuits, a battery, an interface for a sensor or actuator, and an antenna. To make a low-cost WT, we must integrate almost all of the active circuits into a single-chip large-scale integration (LSI) circuit. Integrated wireless transceiver LSI technology enables the elimination of many external components, so it is promising for making small, low-cost WTs. Components that cannot be integrated are the battery, sensor or actuator, and antenna. The antenna can be built on a board together with the LSI because the designed antenna gain is about −15 dBi, which can be attained by using a printed pattern on the board. The interface to the sensor or actuator is typically a serial one having a low bit rate, such as 9600 bit/s. Provided that it is small,

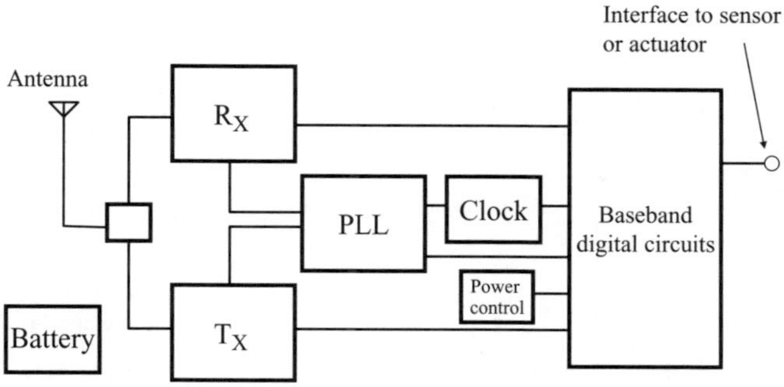

Figure 2.13. Simplified block diagram of WAUN terminal.

a sensor or an actuator (e.g., a single-chip temperature sensor) can be built on the WT's board through customization of the interface. Due to its small size, a WAUN WT containing a sensor or actuator could be installed in various places. However, if the sensor or actuator has complex functions and is not small, the WAUN WT should be built into the sensor or actuator and could be connected using an adapter that converts the protocols used by the sensor/actuator and WAUN WT. In this case, the WAUN WT itself must be small enough to build into another device. Minimizing WAUN WT size is therefore very important regardless of whether the sensor or actuator is simple or large and complex. Our target size for WAUN WTs is $10\,cm^3$ or less. The main obstacle to such miniaturization is the size of the battery. In WAUN, an extremely large number of WTs will be widely distributed in various environments, including outdoors. Thus, the WTs must have small-sized power supplies, such as coin batteries or thin-film batteries with a life of several years, because the batteries cannot be recharged or replaced. Thus, lowering power consumption (and hence decreasing battery size) is a crucial issue in WAUN WT development.

2.6.2 Intermittent Operation

Clearly, extremely low-power operation of WAUN WTs cannot be attained if operation is continuous. WAUN WTs do not need to communicate frequently, like terminals in cellular phone systems do, so we can reduce power consumption by decreasing the activity ratio. We set the activity ratio a few orders of magnitude smaller than that of conventional cellular phone systems, which is usually around 10^{-2}. Our target activity ratio is less than 10^{-4}, which corresponds to a few seconds of communication per day. Typical intermittent operation is shown in Figure 2.14, where P_{act} and P_{stb} represent the power consumption in the active and standby periods, respectively, and T_{act} and T_{stb} are the lengths of the active and standby periods, respectively. Average power consumption P_{av} is calculated as

$$P_{av} = R_{act} \times P_{act} + P_{stb} \qquad (2.1)$$

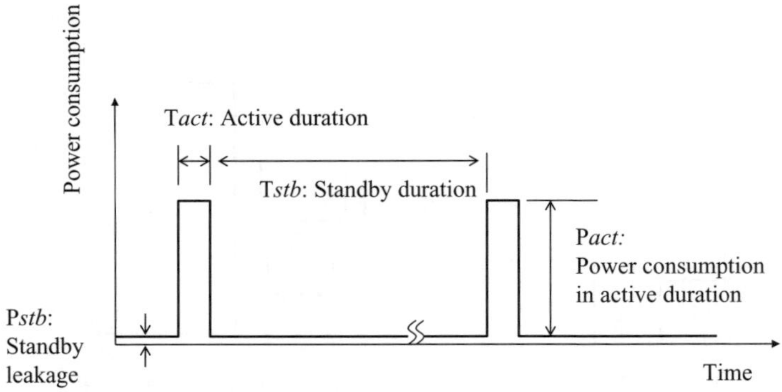

Figure 2.14. Intermittent operation. (If standby leakage can be neglected, averaged power consumption is proportional to activity ratio.)

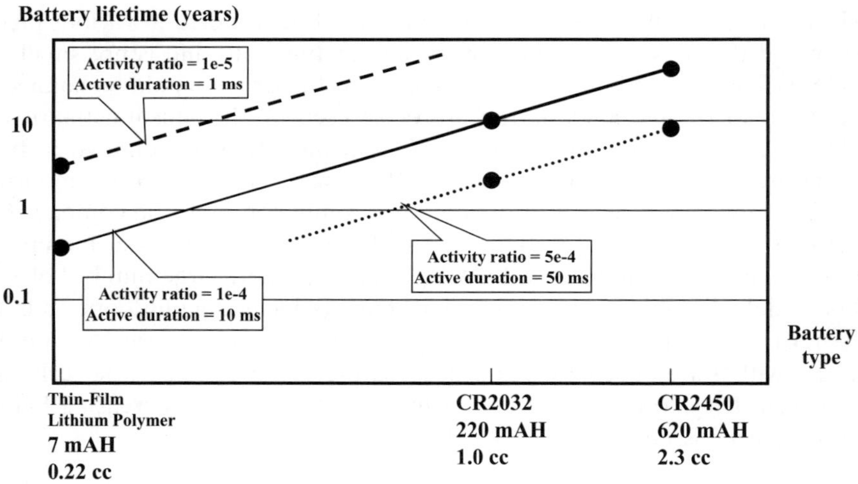

Figure 2.15. Battery lifetime for various battery types with activity ratio as parameter. (Power consumption of 50 mW in active mode is assumed.)

Here, the activity ratio R_{act} is defined as $T_{act}/(T_{act} + T_{stb})$, which is an important parameter in our work. For example, if we assume that T_{act} is 10 ms and T_{stb} is 100 s, then P_{av} and R_{act} become 5 µW and 10^{-4}, respectively. Here, we assume that P_{act} is 50 mW, which is a typical value for low-power WTs. Note that the calculation is valid only if P_{stb} is small enough; it must be less than a few microwatts. On the basis of this calculation, we plot in Figure 2.15 the battery lifetimes of various batteries acceptable for WAUN WTs, where an acceptable battery is considerably smaller than the target WT size of 10 cm^3. It is clear that the activity ratio should be set sufficiently low to obtain a battery life of several years. In Figure 2.15, we assume that ($T_{act} + T_{stb}$) is a constant with a value of 100 s, which corresponds to the interval of intermittent operation. Thus, if we want to decrease R_{act}, we must decrease T_{act}. To enable a thin-film battery to be used for a WAUN WT, T_{act} should be as short as 1 ms. One technique that enables such short T_{act} is a fast-locking PLL circuit scheme [15].

The above scenario, in which P_{stb} is negligible, is an ideal one and is currently beyond the capability of conventional technology. In consideration of this, the next section focuses on a key technique—a multithreshold complementary metal oxide semiconductor (MTCMOS) [16] circuit scheme—for reducing P_{stb}.

2.6.3 MTCMOS

Power consumption is generally expected to be proportional to the activity ratio R_{act}. However, we cannot rely on this expectation for our target R_{act} because static leakage current could become the dominant component of the power consumption. It will be very difficult to achieve the target R_{act} by using conventional CMOS circuit technologies. The target R_{act} and the target leakage level are shown in Figure 2.16, where we again assume that power consumption during the active period is 50 mW. In some cases, a WT should have a kind of slow clock that continues to operate during standby, which could affect the standby power. However, slow-clock circuits usually consist of a small number of gates, which can have a fairly small current consumption of 0.1 μA [17]. In contrast, the main circuits in WTs have a large number of gates because they must provide complicated functions. Consequently, their power leakage is large. In addition, the static leakage always increases by one or two orders of magnitude when the ambient temperature rises from room temperature to about 85°C [18]. Because outdoor

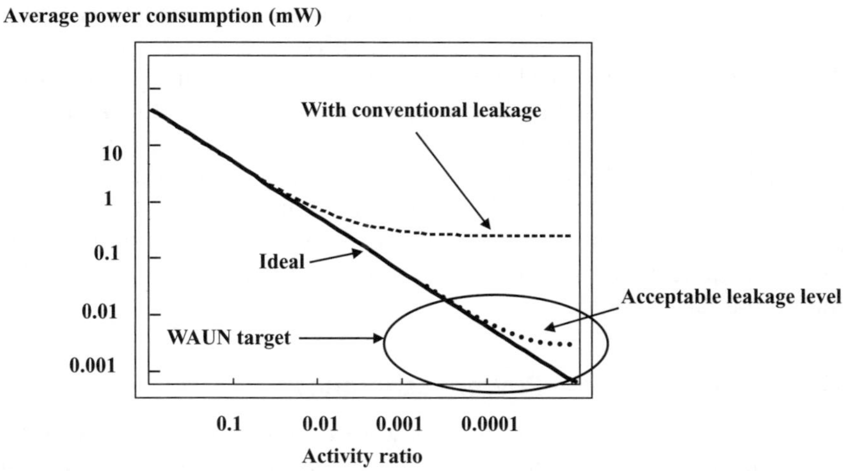

Figure 2.16. Power consumption of WAUN terminal and its lower limit determined by leakage current. (Power consumption of 10 mW is assumed in active mode.)

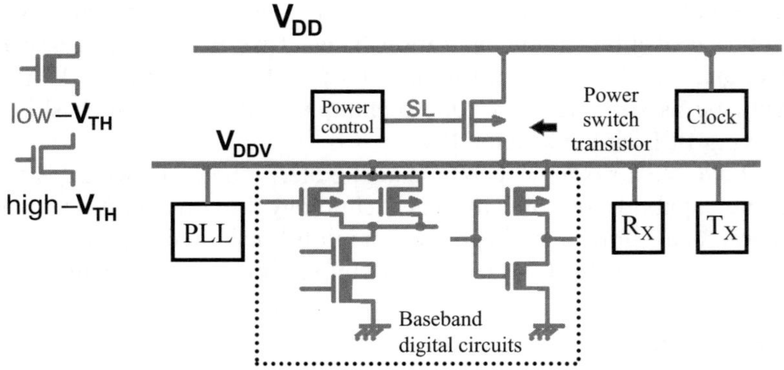

Figure 2.17. MTCMOS circuit scheme. (Slow clock operating continuously is connected to VDD directly.)

WTs may be set at various points, robustness against environmental temperature is a major issue. This means that we must keep the leakage low with a margin of at least two orders of magnitude.

One of the most promising solutions is MTCMOS on silicon-on-insulator (SOI) technology [16, 19]. In this technology, low- and high-threshold-voltage metal oxide semiconductor field effect transistors (MOSFETs) are integrated in a single LSI. The low-threshold-voltage ones enhance speed performance, especially in RF circuits, while the high-threshold-voltage ones suppress standby leakage current during the sleep period. A power-switch transistor supplies the operating current to circuits in the active mode and cuts the leakage current in the sleep mode. The basic MTCMOS circuit scheme for WTs is shown in Figure 2.17. The main circuits are composed of MOSFETs with a low threshold voltage. The circuits are not connected directly to the power supply lines (V_{DD}), but rather to virtual power supply lines (V_{DDV}). The real and virtual power lines are linked by a power-switch MOSFET, whose threshold voltage is high enough to make the standby leakage current extremely low when the switch is off. We experimentally examined the leak-cut performance using a power switch fabricated in a CMOS/SOI LSI, which exhibited leakage of less than 1 nA at room temperature with sufficiently high current drivability of more than 30 mA. Its measured leakage current is shown in Figure 2.18 as a function of temperature. These results indicate that this switch has a sufficiently large margin of 1 μA with respect to the target value.

2.6.4 Summary of Low-Power Techniques

A key circuit technique for WAUN WTs was presented. Because a WAUN WT operates intermittently, the total power consumption of the terminal strongly depends on the standby leakage current, so minimizing it is essential. We showed that a multithreshold CMOS circuit scheme can attain sufficiently low standby

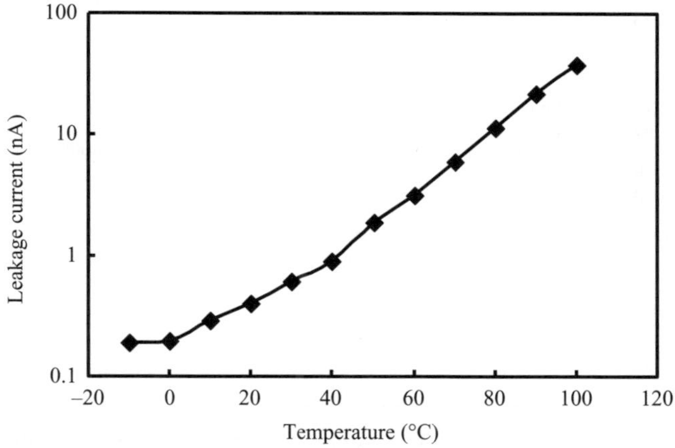

Figure 2.18. Measured leakage current of fabricated power switch as a function of temperature.

leakage power of less than 1 μW. The scheme is highly effective with intermittent operation, and use of it can extend battery life to over 10 years, even if a coin or thin-film battery is used.

2.7 APPLICATION PROGRAMMING INTERFACE (API) AND MIDDLEWARE

WAUN connects many various kinds of sensors and actuators distributed in a large area. This opens up the possibility of a world in which ubiquitous application programs input and output the physical state of the world by uploading data from the sensors and downloading data to the actuators. The WAUN application programs are allocated both to a wired terminal (a server program) and a WT equipped with a sensor/actuator (a device program) and operate by mutually exchanging data. The server program on a wired terminal collects, processes, and analyzes the sensory data acquired by the device programs, or it controls the device programs. To facilitate the development of these programs, a software library and middleware are provided.

2.7.1 Communication Library in Wired/Wireless Terminals

Application programmers do not need to become familiar with the protocol of WAUN because the WAUN communication library provides a simple and convenient communication interface for the server and device programs to send/receive data. One usage example is shown in Figure 2.19. The "ubi_send_blk" function invokes the WAUN download sequences. This function sends the data in "msg.payload" to the device program indicated by the service ID in "msg.

```
char* buffer;
struct send_message msg;
 int ret;

 buffer = (char*)malloc( BUFFER_SIZE + 1 );
 strcpy( msg.service_id,
         "0x010000000000000000000000000000000000000000001" );
 msg.len = BUFFER_SIZE;
 msg.payload = buffer;

 ret = ubi_send_blk( UBI_DL_IMM, &msg );
 if ( ret < 0 )
   {
    fprintf( stderr, "ubi_send_blk NG: %d\n", ret );
    ubi_finalize();
    free( buffer );
    exit( 1 );
   }
```

Figure 2.19. Sample code of server program using WAUN communication library.

service_id" and returns after it succeeds in receiving the "Data replay" message from an IP-GW, as shown in Figure 2.6. The "ubi_send_nbk" function returns without waiting for the reply messages from the IP-GW. The server program receives the reply messages by using the "ubi_recv_nbk" function.

The library also offers the following optional functions to improve convenience and the quality of the programs: (1) The pseudo-multisession function queues data transmission requests from the program and serially processes the requests to enable the program to not need to manage the session. (2) If the data transmission fails, the retransmission and duplication deletion function automatically resends it (and deletes the duplicate received data) to improve the reliability of the communication. (3) Fragmentation and reassembling is a function for transmitting data longer than the maximum transmission unit (MTU). (4) Packing and unpacking is a function to reduce the frequency of transmitting small amounts of data. This is effective for WTs in which communication frequency is limited by WAUN access control. We have implemented the application programming interface (API) of the C language version for both terminals and the API of the nesC [20] language version for the TinyOS [21] WT.

2.7.2 Middleware: uTupleSpace

Development of the server program may be complicated for various reasons. First, the data formats used to communicate with each device program might be different because the format often depends on the sensor and actuator devices. Second, one device program might be used at the same time by two or more server programs. Third, device programs cannot always communicate because the WTs work intermittently and might move outside of the service area. In addition, not all addresses of available device programs may always be known because the device programs are added or removed dynamically. Therefore, we need the middleware that enables server programs to uniformly and equally communicate

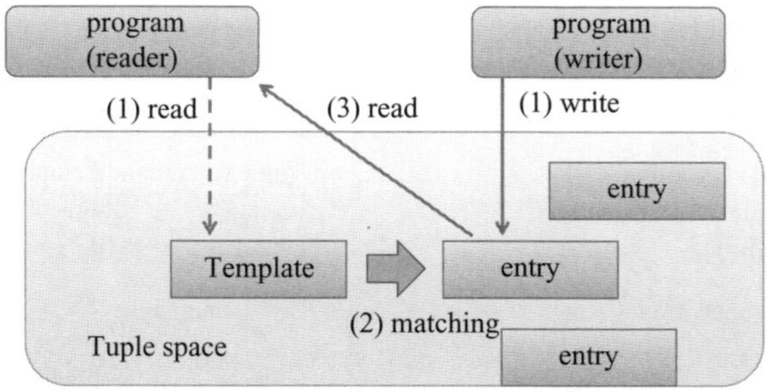

Figure 2.20. Tuple space model.

with the various types of device programs and transparently communicate with the device programs without knowing their existence and addresses even if they dynamically enter and leave the service area. We have proposed uTupleSpace [22] as a middleware that satisfies these requirements.

Our uTupleSpace is based on the tuple space model [23], a concept used in parallel/distributed computing. The "u" indicates ubiquitous. The tuple space model is a shared memory model, in which programs communicate with one another by reading/writing the data called "tuple" from/to the logically shared memory called "tuple space," as shown in Figure 2.20. The writer program to the tuple space registers the tuple called "entry" that contains values while the reader program registers the tuple called "template," which contains patterns matched against values and reads the matched entries from the tuple space. This data matching achieves uniform, equal, and indirect communication without knowing the recipient's existence and address.

Many applications of the tuple space to a ubiquitous environment have been researched [24–27]. Though they support a selective read, a selective write is not supported in the global environment. The selective write is essential for operating an actuator because the selective write can specify the conditions of the desired readers to prevent unauthorized actuator programs from reading the entry. In addition, the template in the original proposal cannot specify the spatial–temporal range of device programs (WTs). The position and time are, however, important for ubiquitous application programs. The uTupleSpace model, therefore, extends the tuple space model to support the selective write and the efficient range search for multidimensional keys.

The uTupleSpace enhances the tuple to a uTuple that has a metadata part in addition to a data part corresponding to the original tuple. The metadata part contains information about the device type, address (service ID of the WT), position and time at which the reader/writer (the WT) resides, and the data type of the data part. In addition, the uTupleSpace supports two types of communication: Event communication achieves selective read by matching the reader's template with the writer's entries, while command communication achieves selective write by matching the writer's template and the reader's entries. These communication models are shown in Figure 2.21. In event communication, the writer (sensor

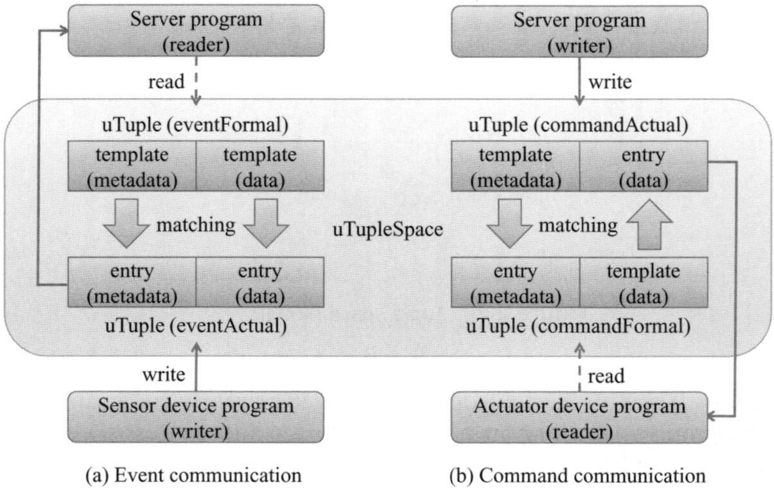

Figure 2.21. Communication model of uTupleSpace.

device program) registers the uTuple (eventActual) that consists of the writer's own entries in each part. Then the reader (server program) registers the uTuple (eventFormal) that consists of templates in each part to match the desired data and writers. In command communication, the reader (device program) registers the uTuple (commandFormal) that consists of the reader's own entry in a metadata part and a template in a data part to match the desired command data. Then the writer (server program) registers the uTuple (commandActual) that consists of an entry in a data part and a template in a metadata part to match the desired readers.

To apply the uTupleSpace to WAUN, it is implemented as uTupleServers in wired terminals, as shown in Figure 2.22. This is because the wired terminals have enough resources and all sensory data are uploaded to those terminals. The programs that write/read uTuples to/from uTupleServers are implemented with a uTupleClient that offers an API of the uTupleSpace.

Device programs in WTs are connected by WAUN upload/download data to/from a proxy called uRelayAgent in the upper tier to write/read uTuples. The uRelayAgent uses the communication library to communicate with device programs and manages their device information to make the metadata part of an eventActual and a commandFormal for each device program. This information is registered at uRelayAgent beforehand with the corresponding service ID and updated by using information obtained from WAUN. For uploaded data from the device program, the uRelayAgent generates an eventActual and writes it in the uTupleServer. To download data to the device program, uRelayAgent works as follows. The uRelayAgent generates the commandFormal, reads the commandActual matched with this commandFormal, and maintains this set of commandFormal and commandActual, along with their corresponding service ID. This

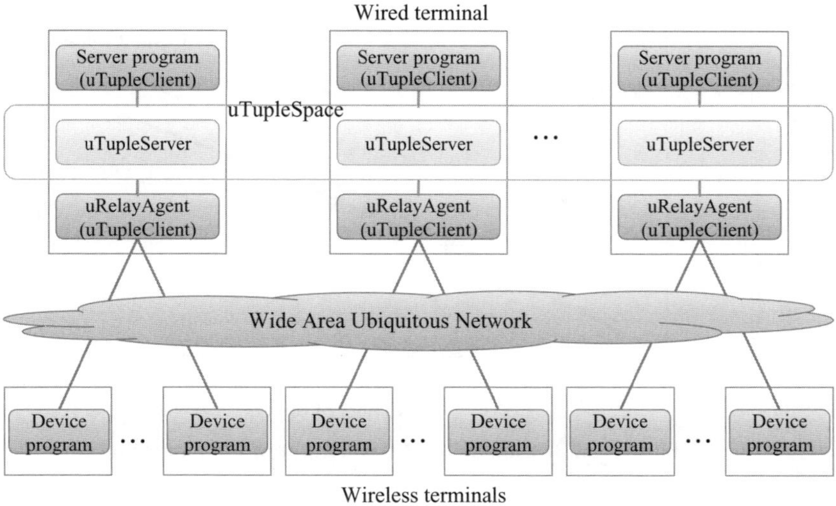

Figure 2.22. Architecture of uTupleSpace middleware.

service ID is the ID of the destination WT of the download. Using this service ID, the uRelayAgent generates and sends the download data. This mechanism enables server programs to communicate with the device programs in the desired WTs via WAUN only by writing and reading uTuples in the uTupleServer.

Although the tuple space model offers flexible communication in a ubiquitous environment, it must have scalability to accommodate many programs and a huge amount of data because data matching occurs at each communication. To achieve scalability, the tuple space is constructed by using many servers with the following functions for distributing the data storage and the load of match processing. That is, the uTupleSpace distributes uTuples to uTupleServers by using a distributed hash table (DHT) for fast matching of a large amount of diverse data. The match processing is distributed to the uTupleServers in accordance with the distribution key, which consists of the device-type and the data-type information in a metadata part of a uTuple. Therefore, multidimensional range search using the position and time, address, and values in the data part for uTuples that have the same device type and data type can be performed in a single uTupleServer [28]. The scale-out technology is also applied to uTupleServers for improving throughput of the data matching even if the number of writing and reading uTuples increases [29]. When the amount of written uTuples in a certain uTupleServer increases, another new uTupleServer automatically shares those uTuples. When data matching for reading uTuples in a certain uTupleServer overloads, another new uTupleServer copies the uTuples from the original uTupleServer and performs data matching in parallel.

One of the reasons that the concept "tuple space" has not been widely used is its scalability. We measured the processing performance of the prototype

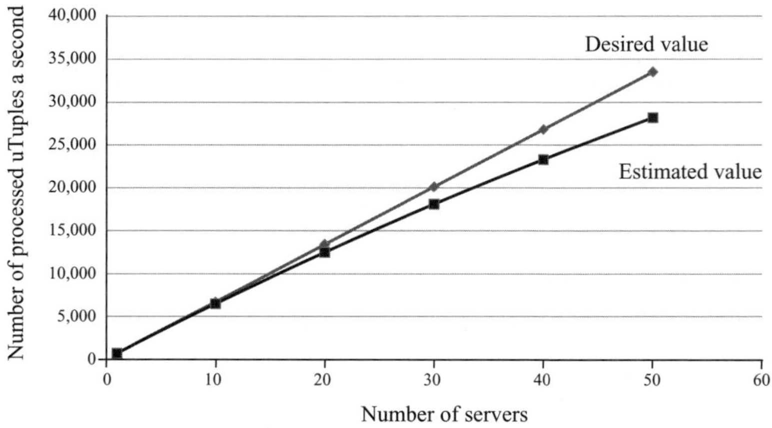

Figure 2.23. Estimation of throughput of data matching.

system constructed by using Mac OS X Server (2.8 GHz Quad-Core Xeon, 2-GB memory) and then estimated the scalability based on the measurement results. The estimation result is shown in Figure 2.23. The number of processed uTuples a second from the registration of uTuple to the retrieval of the results of the matching is shown. The assumed application supports only the event communication and has 10 million WTs (device programs) that write 500 million uTuples (eventActuals) a day. The uTupleServers have already stored 3.5 billion eventActuals and 1000 eventFormals and perform the matching against the 6000 eventActuals that are newly registered every second and the 1 eventFormal that is newly registered every 10 secs. Each of the new eventActuals matches one eventFormal already registered and each of the new eventFormals matches 100 eventActuals already registered. Figure 2.23 shows that the uTupleSpace can achieve scalability against an increase in the number of data and matching processes.

2.8 EXPERIMENTS [36]

2.8.1 Background of Experiments

The uplink of the WAUN is the key technical barrier because the transmission power of a WAUN WT is greatly limited to reduce wireless access cost. Thus, our studies on long-range wireless transmission techniques have focused on uplink (UL) rather than downlink (DL) performance [12]. Our immediate goal is to achieve a 5-km cell radius with 10-mW WT transmission power, which is the license-free power limit in Japan, in the outdoor environment [30].

To achieve this goal, we believe that space diversity reception must be supported by site diversity reception at the AP to compensate for the degradation of the performance in a fading and shadowing environment. The conventional approach to site diversity reception is to use the selection combining scheme [31, 32]. However, WAUN must have a lower outage probability than the conventional cellular system because it must support mainly immobile objects. Therefore, we believe that the maximum ratio combing (MRC) scheme needs to be used as the site diversity reception scheme. To achieve this, our approach is to use the digital radio over fiber (DROF) subsystem to achieve site diversity [11]. It can link radio-frequency (RF) components and the baseband (BB) component of the AP through optical fiber. These functions allow the multiple RF components to be located far from each other.

To confirm the validity of the DROF subsystem and the feasibility of site diversity, we conducted field tests in Tokyo, Japan. This section describes the results of those field tests. First, before conducting the field tests, we theoretically evaluated the feasibility of the 5-km outdoor cell with 10-mW WT transmission power. Then, we evaluated the reception level of our prototypes developed on the basis of wired experiments. We also evaluated the reception levels and transmission performances of the site diversity in field tests by using the prototype. We found that the MRC scheme was completely feasible as a means of providing site diversity. From our investigations, we conclude that the AP can implement both site diversity reception based on MRC and space diversity reception based on MRC. Thus, a cell radius of 5 km and a WT transmission power of 10 mW in an outdoor environment are feasible.

2.8.2 Wireless Link Design in UL for WAUN

The wireless link design in UL for WAUN is listed in Table 2.2. The very-high-frequency (VHF) or ultra-high-frequency (UHF) band is selected as the carrier frequency band because the propagation loss is relatively small and WAUN does not need a large bandwidth. We envisage that the AP antenna will be located on

TABLE 2.2. Specifications of Wireless Link Design in UL for WAUN

Carrier frequency band:	VHF / UHF
Transmission power of WT:	10 mW
Antenna gain:	AP: 10 dBi WT: −10 dBi (internal antenna), 0 dBi (external antenna)
Noise figure:	4 dB (at the AP)
Modulation/demodulation:	π/4-shift QPSK/Coherent detection
Forward error correction:	Convolutional coding (Constraint length of 7, Coding rate of 1/2) Soft-decision Viterbi decoding
Interleaver depth:	16 bit
Transmission rate:	9600 bit/s
Required reception level:	−126 dBm (to achieve the PER of 0.01 under AWGN)
Diversity scheme:	Site diversity reception and space diversity reception

the rooftop of NTT buildings because larger antenna gain requires a larger antenna size. Thus, the AP antenna gain is set to 10 dBi, considering a practical antenna size for use for VHF or UHF bands. The WT antenna gain is set to be –10 dBi for the internal antenna and 0 dBi for the external antenna. In addition, the noise figure (NF) is set to be 4 dB, a value that can be achieved by locating the low-noise amplifier (LNA) next to the antenna on the rooftop to compensate for any propagation loss by a feeder cable. The modulation scheme and FEC scheme is π/4-shift QPSK and convolution coding/soft-decision Viterbi decoding, which is introduced in Section 2.5. The transmission rate is 9600 bit/s. We believe that this speed is sufficient for WAUN services. This bandwidth is narrow and thus causes a flat fading environment. Therefore, an interleaver–de-interleaver scheme is used to maximize the effect of FEC gain in the flat fading environment. The required reception level is –126 dBm, which is the reception level for achieving a packet error rate (PER) of 0.01 under additive white Gaussian noise (AWGN). Furthermore, to compensate for the degradation of the performance in a shadowing and fading environment, we use site and space diversity receptions at AP.

2.8.3 Theoretical Evaluation for Reality of 5-km Outdoor Cell

2.8.3.1 Cell Configuration. To implement both site and space diversity reception at the AP, we use the cell configuration shown in Figure 2.8 and described in Section 2.5. The RF components of AP (denoted AP-R for AP-remote) are placed at three roughly equidistant sites on the edge of a cell with angular intervals of 120°. The BB component of AP (denoted AP-M for AP-master) can be located anywhere because all AP-Rs are linked to the AP-M by DROF. Each site has three AP-Rs, and each AP-R has a sector antenna with a half-power beamwidth of 120°; each of the three AP-Rs in a site uses a different frequency and covers a 120° area of a different cell. This configuration can achieve three-site diversity reception without any increase in the number of buildings with AP-Rs installed in them. Moreover, space diversity reception can be used at the AP if each AP-R has multiple sector-type antennas with the same antenna direction.

2.8.3.2 Theoretical Evaluation. We theoretically evaluated the feasibility of 5-km wireless transmission with 10-mW WT transmission power by calculating the cumulative distribution function (CDF) of the reception level. We assume that signals received at each AP-R of a site are combined using MRC when the APu-R has multiple sector-type antennas, which means that AP can implement space diversity reception. Under this assumption, we calculate the CDF of the reception levels for two diversity reception schemes: (i) The combined signals of all sites in a cell are additionally combined using MRC (three-site MRC), and (ii) the combined signal having the highest receiving level among all sites in a cell is selected (three-site selection). (iii) Furthermore, for reference, we also calculate the CDF of the reception level when all antennas of the AP are located at the center of the cell; the AP antennas are omnidirectional, and the received signals are combined using MRC (one site) when the AP has multiple antennas. We assume that in (i) and (ii) each AP-R has either one or two antennas (with one or two branches per site using three-site MRC (1B3S-MRC or 2B3S-MRC) or three-site selection (1B3S-sel or 2B3S-sel)) and that in (iii) the AP has three or six antennas (three or six branches per site (3B1S or 6B1S)). We also assume that the carrier frequency is 280 MHz. The AP and WT antenna heights are assumed to be 30 and 1 m, respectively. The WT antenna gain is assumed to be −10 dBi (internal antenna). As the propagation model, we examined the Okumura–Hata [33], log-normal, and Rayleigh models as long-distance path, shadowing, and fading models, respectively. Note that we also conducted field research on wireless propagation in the same area of Tokyo. The propagation values identified in the field research were used for the following parameters: standard variation of the log-normal, shadowing correlation, and fading correlation. The evaluated CDFs are shown Figure 2.24. We assumed that a required outage probability was 1%, which is lower than the value of about 10% of the conventional cellular network [34]. As shown in Figure 2.25, the 2B3S-MRC offers 6-dB improvement over 6B1S when both configurations have the same number of antennas. Moreover, 2B3S-MRC achieves improvements of 4.5 and 2.5 dB compared with 1B3S-MRC and 2B3S-sel, respectively. Therefore, using both site diversity based on MRC and space diversity based on MRC can remarkably enhance the cell coverage.

Next, we evaluate the AP antenna height required to meet the required outage probability under the cell configuration (i), three-site MRC. Specifically, we calculated the outage probability at the required reception level as a function of AP antenna height. Moreover, in this evaluation, we consider each AP-R as

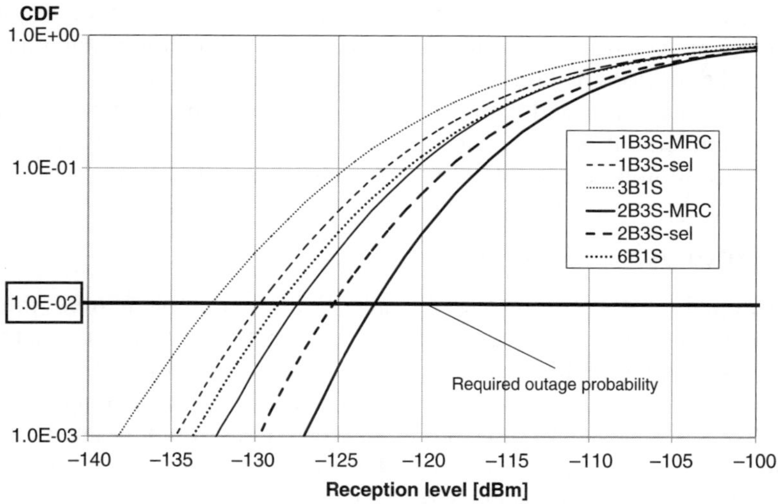

Figure 2.24. CDF of reception level.

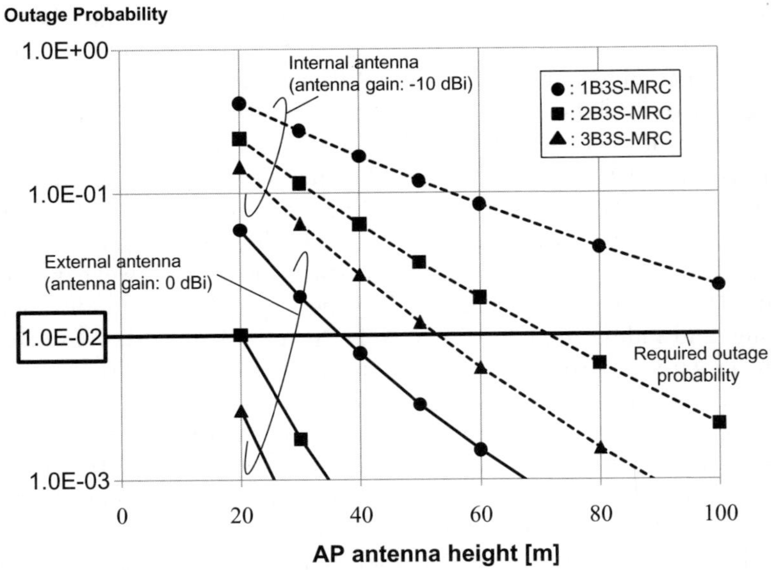

Figure 2.25. Outage probability performances as function of AP antenna height.

having one, two, or three antennas (1B3S-MRC, 2B3S-MRC, or 3B3S-MRC, respectively) and WT antenna gains of 0 and −10 dBi for the external and internal antennas, respectively. The outage probability is shown as a function of AP antenna height in Figure 2.25. With the external WT antenna, 2B3S-MRC can achieve outage probability of 1% at an AP antenna height of 20 m, which is typical for the height of an NTT building. Furthermore, with an internal WT antenna, 2B3S-MRC can achieve the outage probability of 1% at an AP antenna height of 70 m, which is typical for the height of NTT buildings including steel towers on the rooftops. Therefore, a 5-km outdoor cell with 10-mW WT transmission power can be achieved with the 2B3S-MRC cell configuration. This means that if we implement site diversity reception based on MRC as well as space diversity reception based on MRC, we can achieve a cell radius of 5 km.

2.8.4 Prototype Evaluation

We developed prototypes of the AP and WTs that implemented the PHY functions as described in Section 2.5. Their specifications are summarized in Table 2.3. We evaluated the reception sensitivity of the prototypes based on wired experiments through the transmission of 16-byte data packets. Single- and two-branch diversity schemes were tested in an AWGN environment and in a fading environment with independent and identically distributed (i.i.d.) Rayleigh channels, respectively. The packet error ratio (PER) characteristics are shown in Figure 2.26. For reference, the results of a computer simulation are indicated by

TABLE 2.3. Specifications of Prototype

Specifications	AP	WT
Carrier frequency	286.4625 MHz	
Transmission power	100 mW	10 mW
Symbol rate	9600 baud	
Data rate	9600 bit/s	
Communications scheme	TDMA-TDD	
Modulation scheme	π/4-shift QPSK	
Demodulation scheme	Coherent detection	Differential detection
Synchronization overhead	Only pilot symbols	Only one preamble
Forward error correction	Convolutional coding + Soft-decision Viterbi decoding (constraint length of 7, Coding rate of 1/2)	
Diversity scheme	UL: MRC DL: frequency offset	

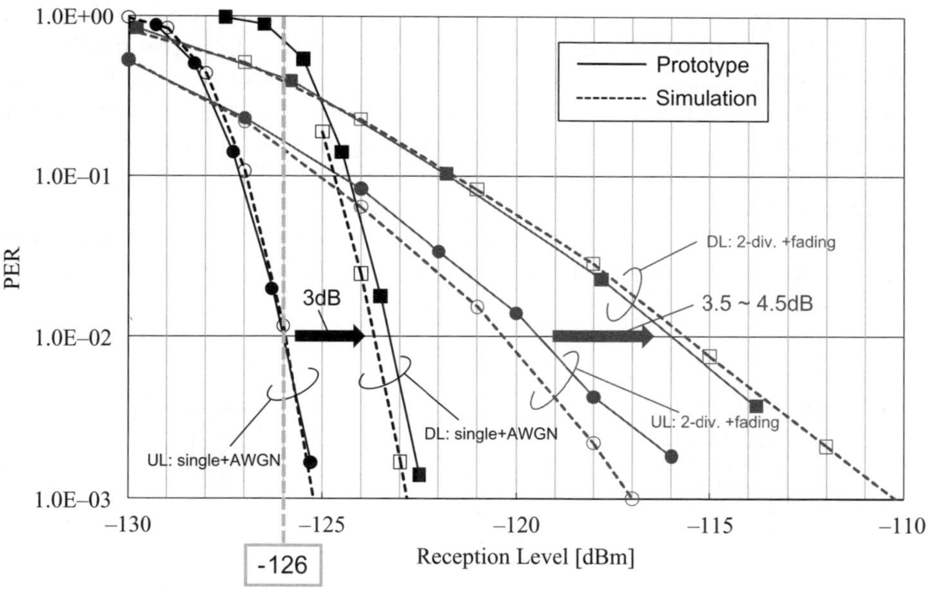

Figure 2.26. Packet error ratio performance of prototype.

dashed lines, where the noise figure was set to 4 dB, which is the value we used in designing the wireless link. The required reception level is defined as the level needed to achieve a PER of 0.01 in an AWGN environment. As shown in Figure 2.26, the AP reception sensitivity of the prototype matched the simulation results, and the reception level of UL to achieve a PER of 0.01 is the required reception level of –126 dBm in wireless link designs as shown in Table 2.2.

Therefore, the transmission performance of the prototype can meet the requirements. Moreover, Figure 2.26 shows that the DL performance was several decibels worse than the UL performance in both the single- and two-branch diversity reception cases. These results are due to the inherent difference in sensitivity between the differential and coherent detection schemes used in the DL and UL, respectively. Therefore, the AP transmission power must be greater than the WT transmission power.

2.8.5 Field Test

2.8.5.1 Test Outline. We used our prototype in a field test. The DROF subsystem was used as a feeder network linking the AP-M to AP-Rs. A schematic of the DROF subsystem used in the field test is shown in Figure 2.27. One AP-R was located at each of three NTT buildings (in Shirahige, Joto, and Koiwa) in the Tokyo metropolitan area; the AP-M was located at the Joto building. These buildings were connected by the DROF subsystem on a commercial wide-area Ethernet service. In the DROF subsystem, the digitized radio BB signals from the radio interfaces were loaded into Ethernet packets and transmitted between master DROF (DROF-M) and remote DROF (DROF-R) bidirectionally. The DROF-M's reference clock was fed to the AP-Rs to establish time and frequency synchronizations between DROF-M and DROF-Rs. The time and frequency synchronization of the AP-Rs allowed precise alignment of the reception timing of the RF signals captured by the AP-Rs and the AP-M. Therefore, MRC-based site diversity reception could be achieved by using the DROF subsystem.

The three AP-Rs and the AP-M were used to evaluate the reception level of three-site diversity without space diversity (1B3S-MRC). Moreover, two of the

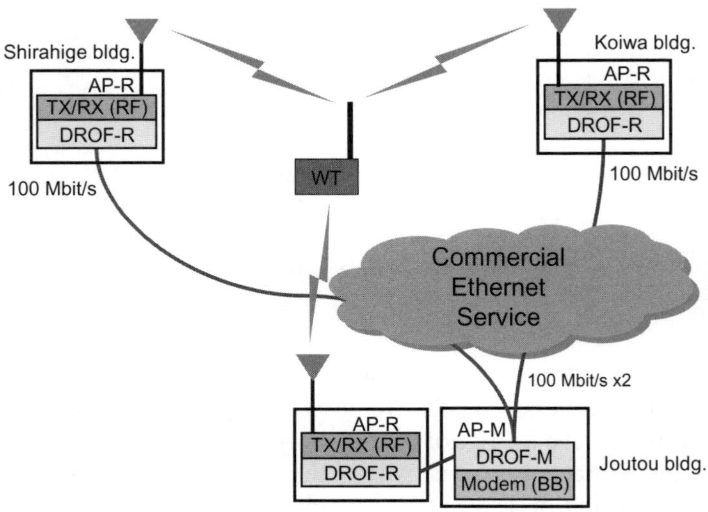

Figure 2.27. Schematic of DROF subsystem and site diversity system used in field test.

AP-Rs (at the Shirahige and Koiwa buildings) and the AP-M were used to evaluate the PER performances of two-site diversity without space diversity (1B2S-MRC). All the AP antennas were placed on the rooftops of the NTT buildings. Each antenna was a corner reflector antenna with antenna gain of 7 dBi; the antenna heights were 35, 29, and 25 m on the Shirahige, Joto, and Koiwa buildings, respectively. The WT was set in a measurement vehicle, and its dipole antenna with antenna gain of 2 dBi was set on the roof of the vehicle. We varied the loss between the WT antenna and WT equipment by using a step attenuator to make the total system gain of this field test equal to that of the wireless link design. The WT antenna was an internal type with antenna gain of −10 dBi. The measurement vehicle was driven around Tokyo City while transferring a fixed quantity of data (16 bytes) with and without the site diversity reception based on MRC. A map of the field test area is shown in Figure 2.28. The circle indicates the measurement area for the reception level. The actual test course is indicated by the dotted line. The test course was selected so as to roughly equalize the reception levels at the Shirahige and Koiwa buildings.

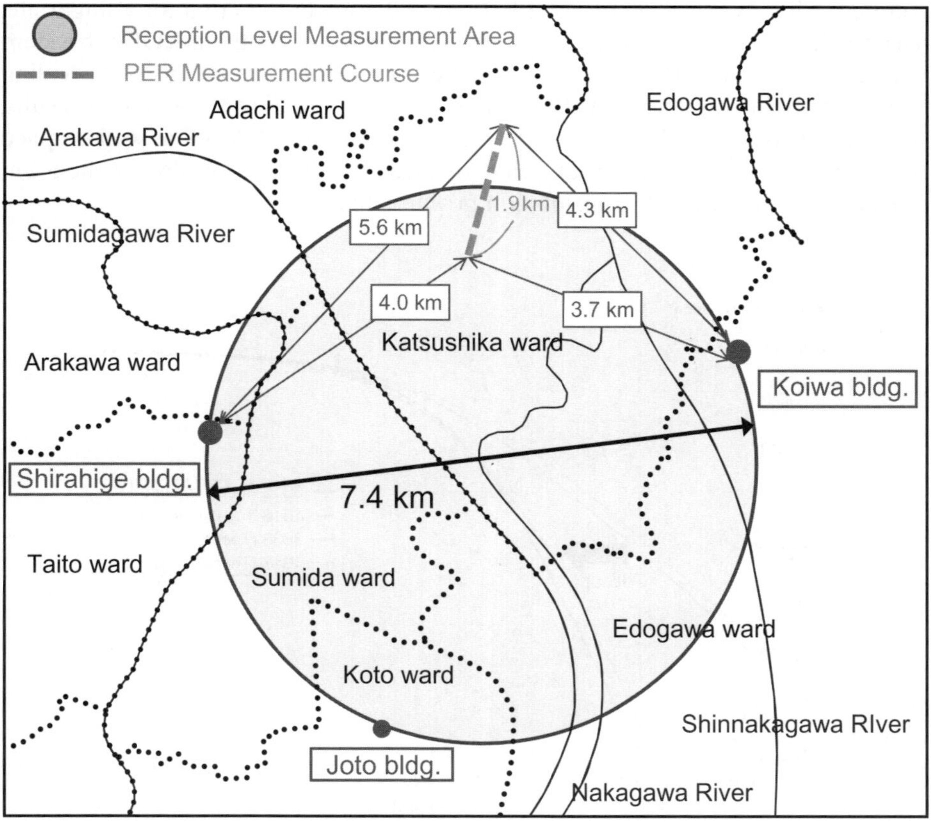

Figure 2.28. Map of reception level and PER measurement area in field test.

2.8.5.2 Field Test Results. The CDFs of the reception levels are shown in Figure 2.29. The thick and thin solid lines show the measured CDFs of 3B1S-MRC and 1B1S-MRC, respectively, where AP-R in 1B1S-MRC was located at the Shirahige building. For reference, both the dashed lines show the theoretical CDFs evaluated by the same computer simulations used for Figure 2.24. The measured CDFs matched the theoretical ones. This confirms that the theoretical CDFs of reception levels shown in Figure 2.24 and the theoretical outage probability shown in Figure 2.25 are valid.

Next, the PER performances as a function of reception level are shown in Figure 2.30. Here, the reception level of site diversity is defined to be the mean value for both buildings. For reference, the PER performances of MRC-based site diversity and of a single site (evaluated by computer simulation) are shown by the solid and dashed lines, respectively. The noise figure was assumed to be 6 dB, and an i.i.d. Rayleigh channel with the same power on all paths was used as the fading model. As can be seen in Figure 2.30, the measured PER performances basically match those obtained by the computer simulation for both the site diversity and single site cases. This confirms that DROF based on site diversity reception can offer theoretical MRC diversity gain in an actual propagation environment. Site diversity can offer a PER one order lower than that for a single site. The results confirm the feasibility of site diversity based on the DROF subsystem and show that the MRC scheme is completely feasible as a means of providing site diversity in conjunction with the DROF subsystem. From the above results, we conclude that an AP can run site diversity reception with MRC as well as space diversity reception with MRC and that, by means of these technologies, the 5-km outdoor cell with 10-mW transmission power is feasible.

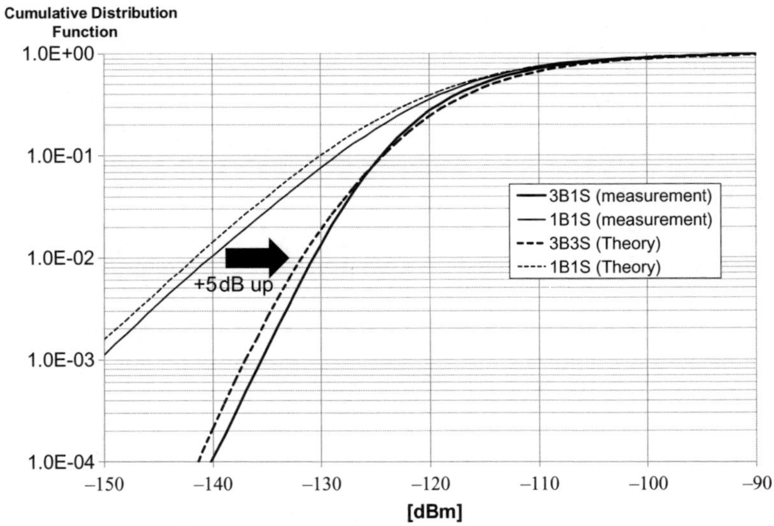

Figure 2.29. CDFs of reception levels in field test.

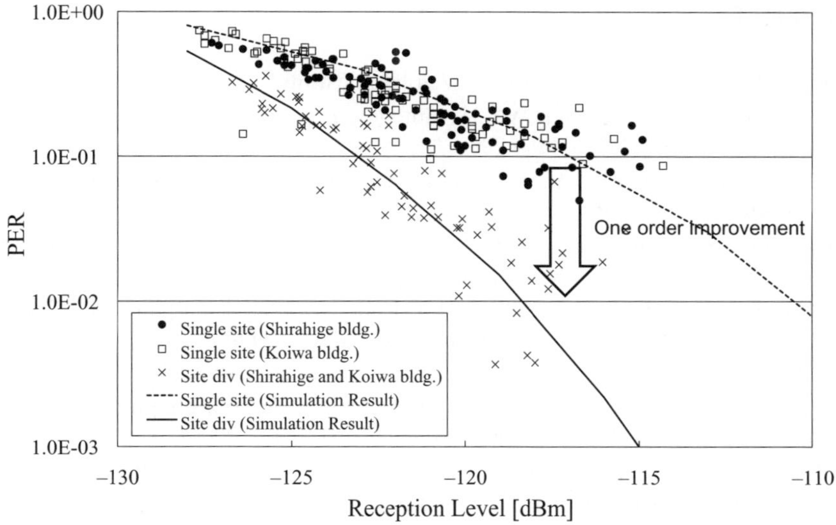

Figure 2.30. PER performances versus reception level in field test.

2.9 CONCLUSION

This chapter described our proposed wide-area ubiquitous network (WAUN), which is a network infrastructure dedicated to sensors and actuators. We have developed a prototype system and confirmed that we can achieve the R&D target of a 5-km cell radius using 10-mW wireless transmission power terminals with 10-year battery life. We believe that there is a large promising market for sensor/actuator networks or machine-to-machine communications and that this market is essential for the growth of network operators. To cultivate the market, we are conducting application service tests with our business partners by using our prototype system.

Furthermore, we have proposed our network to standardization bodies to increase our partners. For example, ITU-R/WP-5A held on October 23–November 6, 2008 started discussion on the preliminary draft new question "Mobile wireless access systems providing communications to a large number of ubiquitous sensors and/or actuators scattered over wide areas in the land mobile service," which covers WAUN. Through international standardization and discussion for collaboration with business partners, we will commercialize our proposed system of WAUN for global deployment.

We also believe that wide use of WAUN will initiate a secondary user market where users can buy sensory data collected by the primary users of WAUN. These secondary users could create products or services with added value by using the data. In essence, WAUN is a step toward opening the gateway to a new information distribution market.

REFERENCES

1. http://www.ubiq.com/hypertext/weiser/UbiHome.html.
2. G. J. Pottie and W. J. Kaiser, Wireless integrated network sensors, *Commun. ACM*, Vol. 43, No. 5, pp. 551–558, May 2000.
3. I. F. Akyildiz, W. Su, Y. Sankarasubramaniam, and E. Cayirci, A survey on sensor networks, *IEEE Commun. Mag.*, Vol. 40, No. 8, pp. 102–114, 2002.
4. B. W. Cook, S. Lanzisera, and K. S. J. Pister, SoC issues for RF smart dust, *Proce. IEEE*, Vol. 94, No. 6, pp. 1177–1196, June 2006.
5. H. Saito, O. Kagami, M. Umehira, and Y. Kado, Wide area ubiquitous network: The network operator's view of a sensor network, *IEEE Commun. Mag.*, Vol. 46, No. 12, pp. 112–120, 2009.
6. H. Saito, M. Umehira, and T. Ito, Proposal of the wide area ubiquitous network, in *Telecommunications World Congress*, Budapest, Hungary, 2006.
7. http://www.kantei.go.jp/jp/singi/bt/dai2/2siryou10-3-3-4.pdf (in Japanese).
8. http://www.mlit.go.jp/kisha/kisha07/09/090704_.html (in Japanese).
9. http://www.zaimu.japanpost.jp/tokei/2004/excel/yuubin/ya040002.xls (in Japanese).
10. H. Toshinaga, K. Mitani, H. Shibata, K. Takasugi, M. Ishizuka, S. Kotabe, S. Ishihara, H. Tohjo, and H. Saito, Wide area ubiquitous network service system, *NTT Technical Rev.*, Vol. 6, No. 3, 2008. https://www.ntt-review.jp/
11. S. Kuwano, Y. Suzuki, Y. Yamada, and K. Watanabe, Digitized radio-over-fiber (DROF) system for wide-area ubiquitous wireless network, in *Proceedings, Topical Meeting on Microwave Photonics*, Grenoble, France, October 2006.
12. T. Fujita, D. Uchida, Y. Fujino, O. Kagami, and K. Watanabe, A short-burst synchronization method for narrowband wireless communications systems, *IEEE ISWPC 2007*, Puerto Rico, February 2007.
13. F. Nuno, Y. Shimizu, and K. Watanabe, A new QoS control scheme using dynamic window size control for wide area wireless networks, *ICSNC 2007*, French Riviera, France, August 2007.
14. H. Saito, M. Umehira, and M. Morikura, Considerations of global ubiquitous network infrastructure, *IEICE Trans. Commun.*, Vol. J88-B, No. 11, pp. 2128–2136, 2005 (in Japanese).
15. M. Nakamura, A. Yamagishi, M. Harada, M. Nakamura, and K. Kishine, Fast-acquisition PLL using fully digital natural-frequency-switching technique, *Electron. Lett.*, Vol. 44, No. 4, pp. 267–269, 2008.
16. S. Mutoh, T. Douseki, Y. Matsuya, T. Aoki, S. Shigematsu, and J. Yamada, 1-V power supply high-speed digital circuit technology with multithreshold-voltage CMOS, *IEEE J. Solid-State Circuits*, Vol. 30, No. 8, pp. 847–855, 1995.
17. http://www.okisemi.com/eu/Products/RTC/ml907x.html
18. T. Douseki, M. Harada, and T. Tsuchiya, Ultra-low-voltage MTCMOS/SIMOX technology hardened to temperature variation, ScienceDirect, *Solid-State Electron.*, Vol. 41, No. 4, pp. 519–525, 1997.
19. T. Ohno, Y. Kado, M. Harada, and T. Tsuchiya, Experimental 0.25-μm-gate fully depleted CMOS/SIMOX process using a new two-step LOCOS isolation technique, *IEEE Trans. Electron Devices*, Vol. 42, No. 8, pp. 1481–1486, 1995.

20. http://nescc.sourceforge.net/.
21. http://www.tinyos.net/.
22. T. Nakamura, M. Nakamura, A. Yamamoto, K. Kashiwagi, Y. Arakawa, M. Matsuo, H. Minami, uTupleSpace: A bi-directional shared data space for wide-area sensor network, in *Proceedings of the 2009 International Conference on Parallel and Distributed Computing, Applications and Technologies (PDCAT2009)*, pp. 396–401, December 2009.
23. D. Gelernter, Generative communication in Linda, *ACM Trans. Prog. Lang. Syst.*, Vol. 7, No. 1, pp. 80–112, January 1985.
24. C. Curino, M. Giani, M. Giorgetta, A. Giusti, A. Murphy, and G. Picco, TinyLIME: Bridging mobile and sensor networks through middleware, in *Proceedings of the Third IEEE International Conference on Pervasive Computing and Communications (PerCom2005)*, pp. 61–72, March 2005.
25. P. Costa, L. Mottola, A. L. Murphy, and G. P. Picco, TeenyLIME: Transiently shared tuple space middleware for wireless sensor networks, in *Proceedings of the International Workshop on Middleware for Sensor Networks (MidSens '06)*, ACM, pp. 43–48, 2006.
26. G. Castelli, A. Rosi, M. Mamei, and F. Zambonelli, A simple model and infrastructure for context-aware browsing of the world, in *Proceedings of the Fifth Annual IEEE International Conference on Pervasive Computing and Communications (PerCom2007)*, pp. 229–238, March 2007.
27. G. Hackmann, C.-L. Fok, G.-C. Roman, and C. Lu, Agimone: Middleware support for seamless integration of sensor and IP networks, in *Lecture Notes in Computer Science*, Vol. 4026, Springer, Berlin, pp. 101–118, 2006.
28. Y. Arakawa, A. Yamamoto, H. Minami, M. Matsuo, H. Tohjo, and H. Saito, Implementation of wide-area ubiquitous platform considering scalability, in *Proceedings of IEICE General Conference '08*, B-7-149, 2008 (in Japanese).
29. Y. Arakawa, K. Kashiwagi, T. Nakamura, M. Nakamura, and M. Matsuo, Evaluation of dynamic scaling method for real-world data sharing mechanism, IEICE Technical Report IN2009-21 (in Japanese).
30. M. Umehira, H. Saito, O. Kagami, T. Fujita, and Y. Fujino, Concept and feasibility study of wide area ubiquitous network for sensors and actuators, *IEEE VTC-Spring*, pp. 165–169, 2007.
31. Y. Yeh, J. C. Wilson, and S. C. Schwartz, Outage probability in mobile telephony with directive antennas and macrodiversity, *IEEE Trans. Vehicular Technol.*, Vol. vt-33, No. 3, pp. 123–127, 1984.
32. S. Fukumoto, K. Higuchi, A. Morimoto, M. Sawahashi, and F. Adachi, Combined effect of site diversity and fast transmit power control in W-CDMA mobile radio, *IEEE VTC-Spring*, pp. 1527–1534, Tokyo, 2000.
33. M. Hata, Empirical formula for propagation loss in land mobile radio services, *IEEE Trans. Vehicular Technol.*, Vol. 29, pp. 317–325, 1980.
34. M. Sakamoto, Location probability estimation of services availability in portable radio telephone systems, *IEEE VTC*, pp. 575–581, 1988.
35. S. Kuwano, D. Uchida, F. Nuno, and M. Takahashi, Wireless access system for wide area ubiquitous networks, *NTT Technical Rev.*, Vol. 6, No. 3, 2008. https://www.ntt-review.jp/archive/ntttechnical.php?contents=ntr200803sp3.html
36. D. Uchida, S. Kuwano, T. Fujita, and Y. Fujino, Field test results for wide area ubiquitous networks, *NTT Technical Rev.*, Vol. 6, No. 3, 2008. https://www.ntt-review.jp/archive/ntttechnical.php?contents=ntr200803sp3.html.

3

WIRELINE ACCESS NETWORKS

Scott Reynolds

3.1 INTRODUCTION

Wireline access networks refer to the collection of "last-mile" data transmission technologies that connect businesses and residences to a public communications network. Historically, access networks were service-specific; everyone is familiar with the copper twisted-pair loop used to carry analog telephony, and many people continue to receive video entertainment programming through an RF-based coaxial distribution network. Broadband or high-speed Internet (HSI) access has motivated access network providers (ANPs) to upgrade and evolve their last-mile networks. In the later part of the last century, driven by the popularity of the Web, access network operators recognized that their networks needed to support more than the single service their networks were offering. The wireline access network was now a conduit into homes and businesses in which a portfolio of services could be delivered and charged for. This portfolio is typically referred to as "triple play," consisting of voice, video, and HSI access. ANPs have embarked upon a large-scale upgrade to their deployed networks focusing on delivering increased bandwidth to the subscriber based on the assumption that bandwidth is a proxy for revenue; the higher the bandwidth, the higher the average revenue per user (ARPU). Telephony networks were upgraded to support HSI using

Convergence of Mobile and Stationary Next-Generation Networks, edited by Krzysztof Iniewski

digital subscriber line (DSL) technology, and the hybrid-fiber coaxial (HFC) network was re-architected to support full duplex communications enabling HSI access. To some network operators, incremental enhancements to existing last-mile networks are insufficient, and the only viable path forward is to "re-wire" the entire last mile with fiber optics, promising near-unlimited bandwidth, directly to homes and businesses.

This chapter will examine three common wireline access technology: digital subscriber line (DSL), hybrid-fiber coaxial (HFC), and the emergent passive optical network (PON) found in the latest fiber-to-the-home (FTTH) networks. We will describe the foundational technology, detail the current "state of the art," and examine the future trends for each.

We are in the midst of an exciting time in information technology. The access network and the Internet have formed a symbiotic relationship, where the availability of plentiful and affordable bandwidth creates opportunities for new services and applications to exploit the upgraded infrastructure. Music sharing and commercial services like iTunes™ would not have been possible without the large-scale deployment of broadband Internet connectivity. The wireline networks of today are being "re-factored" to accommodate streaming video through YouTube™, Netflix, and other similar services. High-definition programming is pushing bandwidth demands higher still. But one thing is for certain: This trend shows no sign of abating. Bandwidth is an enabling technology: The services and applications we have access to tomorrow will only be possible because of the bandwidth that access networks provide today.

3.2 COPPER-BASED ACCESS NETWORKS

Few would have guessed that Alexander Graham Bell's invention and subsequent patenting in 1881 of the copper twisted-pair phone loop [1] would have led to a near-ubiquitous access network for telephony and data communications that is still being exploited and enhanced more than a century later. Over 650 million copper loops exist worldwide [2]. With an enormous replacement cost, access network providers (ANPs) are motivated to migrate from low-bit-rate highly reliable circuit-based voice and unlock the bandwidth hidden in the copper loop to enable new services and generate new revenues. Mining the "last mile" for additional bandwidth comes through exploiting the unused spectrum using advanced signal processing techniques to encode multiple bits of digital information into the analog waveform, a process made all the more difficult in a noise limited medium such as twisted pair. However many obstacles have been overcome, increasing the bandwidth available from a few tens of kilobits per second, to tens of megabits in advanced networks of today with the promise of hundreds of megabits per second in the not-too-distant future [3]. The century-old twisted-pair wiring plant is now capable of delivering multiple high-definition television (HDTV) broadcasts, Video on Demand (VOD), and other multimedia-rich services directly into the home and businesses the world over. Such extensibility will ensure that the humble copper twisted pair will remain a dominant networking technology for the foreseeable future.

3.2.1 Copper Plant Overview

A copper twisted pair is a form of cabling in which two insulated copper conductors, typically 24 or 26 AWG (0.51 or 0.40 mm) in diameter, are twisted to reduce electromagnetic interference. Initially the conductors were wrapped in paper for insulation, but the lack of waterproofing made this material unsuitable for outside applications; eventually polyethylene became the dominant dielectric material and continues to be used [4]. Multiple insulated twisted pairs are grouped into a cable called a "binder group," which emanates from Access Network Providers' central office (CO) to residential areas and businesses. The lengths of the loops vary based on the population density of the served areas. Urban loop lengths tend to be short (less than 3000 ft), and suburban areas range in length from 3000 to 9000 ft and increase up to 18,000 ft typical in rural settings [4]. A single loop between the CO and customer is usually made up of pairs from sections of several binder cables, leading to different gauges and bridge taps. Bridge taps are open-circuit pairs, either intentionally placed along the main cable route in anticipation of new service connections or resulting from past disconnections and rearrangements (Figure 3.1). The changes in wire gauge and the presence of bridge taps impact the frequency response of the loop, potentially affecting the maximum bandwidth that an individual loop can support. Although ANPs have progressively remediated their copper loop infrastructure, a large variance in the length and quality still exists, resulting in a challenging environment to provide universal access to the highest bandwidths.

The twisted pair is often referred to a noise-limited transmission media due to the number and variety of impairments that are inflicted upon it (Figure 3.2), including:

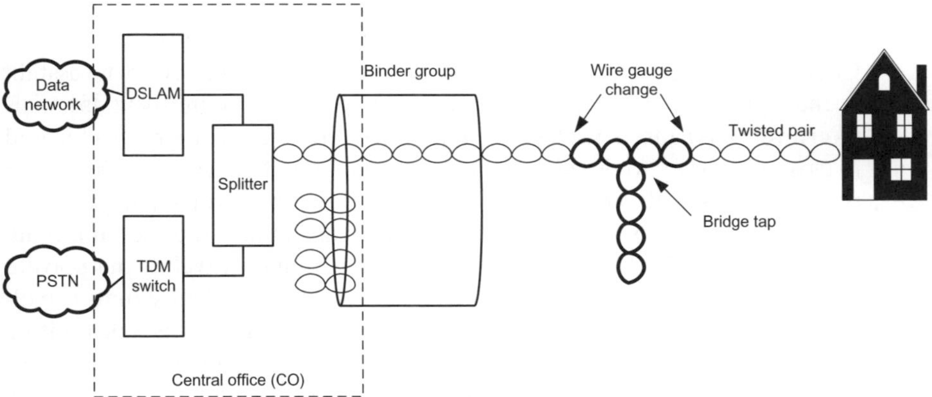

Figure 3.1. Copper twisted-pair wiring.

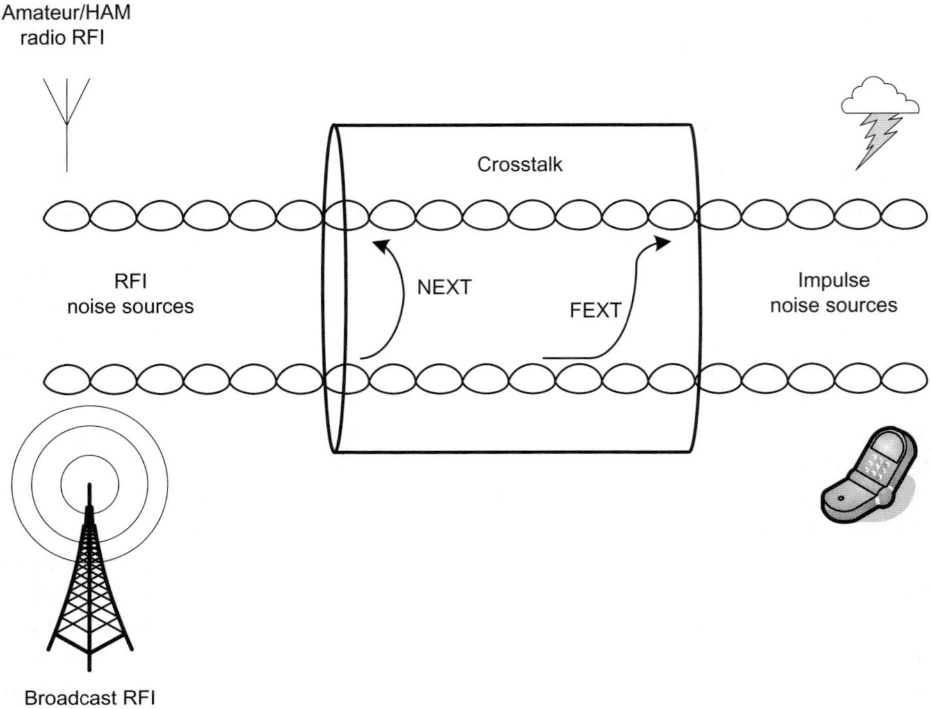

Figure 3.2. Impairment environment for a twisted pair.

- Attenuation, which is dependent on the type of dielectric material used, the wire gauge, type of twisting, and overall loop length. Attenuation increases with both signal frequency and loop length.
- Nonlinear phase response causing intersymbol interference.
- Echo resulting from full duplex communications across a single pair of conductors.
- Bridge taps that produce notches in the line transfer function.
- Crosstalk interference between pairs in the same binder group. Crosstalk is the dominant impairment experienced in DSL transmission [5]. This interference is caused by electromagnetic radiation from other pairs located in close proximity to the victim pair. The coupling of energy into the victim pair increases with frequency, making this more harmful for higher data rates. Two modes of crosstalk interference can impact the victim circuit: near-end crosstalk (NEXT), which is caused by signals traveling in opposite directions, and far-end crosstalk (FEXT), which is caused by signals traveling in the same direction. For longer loop lengths, FEXT is self-limiting due to the line attenuation; however, to support the new generation of high-bandwidth services, loop lengths are being systematically reduced, thereby increasing the negative effects of FEXT.

- Radio-frequency interference (RFI) is noise generated by radio transmitters such as amateur/HAM radios. These transmission frequencies are regulated and well known, allowing accommodation to be made in spectrum planning for high-speed data transmission over copper loops.
- Background noise is additive white Gaussian noise and represents the "noise floor."
- Impulse noise is randomly occurring discrete noise "spikes," narrowband or wideband in nature.

International standards bodies such as the American National Standards Institute (ANSI), the European Telecommunications Standards Institute (ETSI), and the International Telecommunications Union (ITU) characterize impairments and provide mathematical models to aid in the evaluation of the noise performance of various DSL transmission schemes [6].

3.2.2 Digital Subscriber Line Overview

A digital subscriber line (DSL) is a modem technology that uses a copper twisted pair to transport high-bandwidth data. The term "xDSL" covers a number of technology variants that offer a competing blend of data rate, reach, and spectral compatibility. Examples are asymmetric DSL (ADSL), symmetric DSL (SDSL), high-bit-rate DSL (HDSL), and very-high-bit-rate DSL (VDSL). xDSL variants, speeds, distances, and standardization are shown in Table 3.1.

TABLE 3.1. DSL Comparison

Technology	ITU-T Standard/ Ratified	Maximum Upstream	Maximum Downstream (Distance)	Maximum Distance[a]	Frequency Range
ADSL	G.992.1 (1999)	1.3 Mbit/s	12 Mbit/s (1,000 ft)	~18,000 ft	1.1 MHz
ADSL2	G.992.3 (2002)	1 Mbit/s	12 Mbit/s (1,000 ft)	8,000 ft	1.1 MHz
ADSL2+	G.992.5 (2005)	1 Mbit/s	26 Mbit/s (1,000 ft)	12,000 ft	2.2 MHz
ADSL-RE	G.992.3 (2003)	0.8 Mbit/s	5 Mbit/s	23,000 ft	~1.8 MHz
SHDSL	G.991.2 (2002)	5.6 Mbit/s	5.6 Mbit/s	9,000 ft	1.1 MHz
VDSL	G.993.1 (2004)	16 Mbit/s	52 Mbit/s (1,000 ft)	4,000 ft	12 MHz
VDSL2	G.993.2 (2005)	100 Mbit/s	100 Mbit/s	15,000 ft	30 MHz

[a]Maximum distances obtainable, not maximum distance at which maximum speed is obtained. In practice, maximum distance at maximum speed is far less than maximum distance. For example, VDSL2 supports 100-Mbit/s operation at ~1,500 ft, 50 Mbit/s at 3,000 ft and around 1 Mbit/s at 15,000 ft.

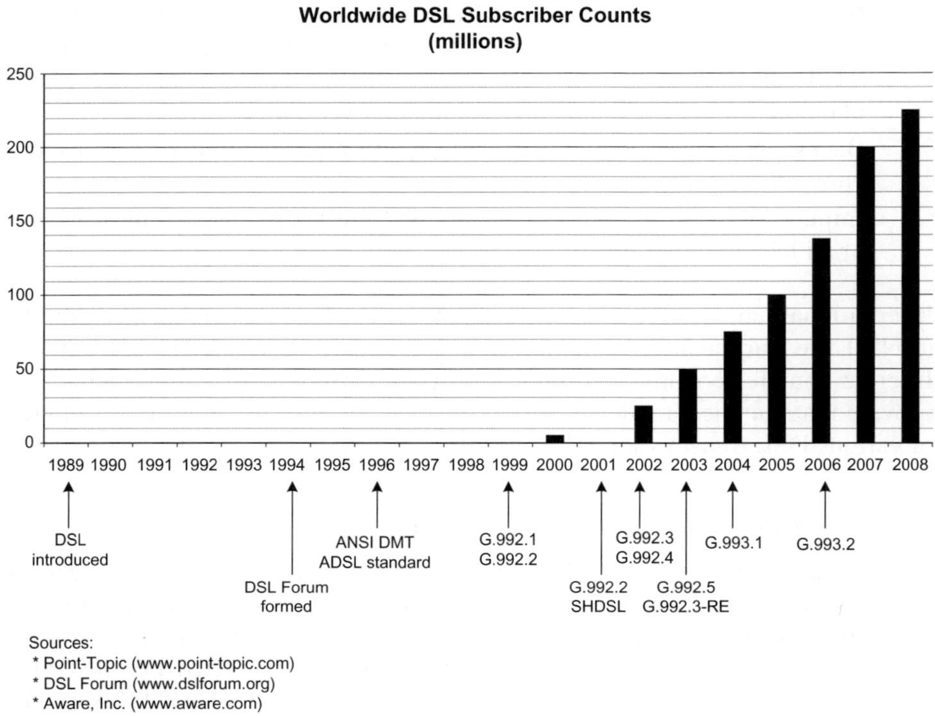

Figure 3.3. DSL standardization timeline and worldwide deployment.

ADSL was the first internationally standardized and widely deployed DSL technology. The timeline illustrating the standards activities and corresponding deployment is shown in Figure 3.3. Motivated by video delivery, ADSL provided more downstream bandwidth than upstream, a characteristic present in many subsequent DSL variants. The asymmetric bandwidth distribution reduces NEXT, increasing practical loop lengths, which in turn made the service more economically viable and accelerated its deployment. Another important factor with ADSL deployment was the coexistence with the Plain Old Telephone Service (POTS). ADSL coexists with POTS (and narrowband ISDN) through frequency domain multiplexing, where the POTS service occupies frequencies between DC and 4 kHz, while the ADSL service occupies the upper frequencies from 25 kHz to 1.1 MHz.[1] Through the use of frequency splitters at both the residence and central office (CO), the low-frequency POTS service is routed to the PSTN and the high-frequency DSL service is directed to the digital subscriber line access multiplexer (DSLAM). Frequency division duplexing (FDD) allocates the available ADSL bandwidth into upstream and downstream components. The upstream

[1]Exact frequency usage is dependent on the type of installed hardware and local deployment conditions.

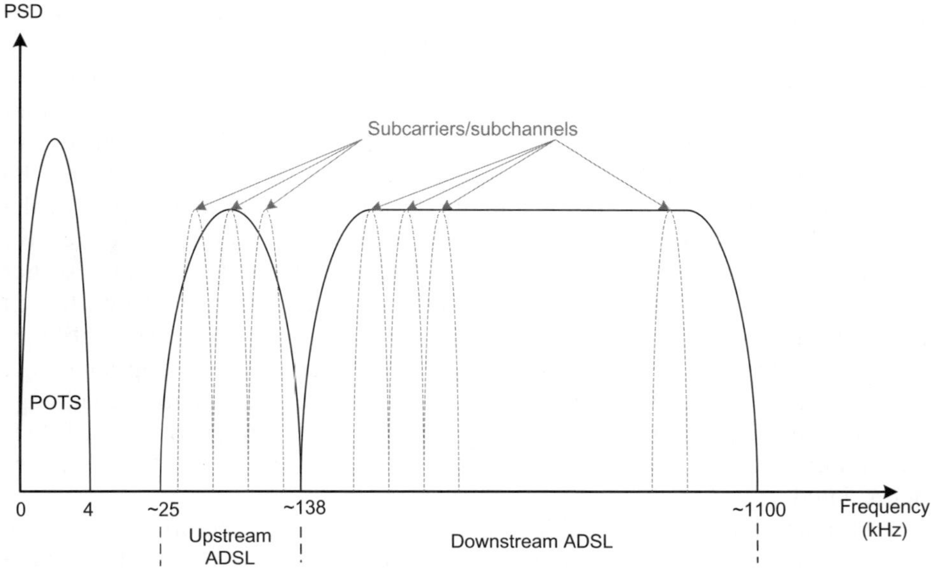

Figure 3.4. ADSL spectrum.

transmissions occupy the lower-frequency bands, from 25 kHz to 138 kHz, with the downstream bands beginning at 138 kHz and progressing up to 1.1 MHz, as shown in Figure 3.4. Given that attenuation increases with loop length and frequency, the downstream frequencies are more severely impacted as the loop length increases, reducing the downstream bandwidth.

Early ADSL implementations contained different and incompatible modulation techniques. Two examples were carrierless amplitude phase (CAP) modulation and discrete multitone (DMT) modulation. CAP is a nonstandard variation of quadrature amplitude modulation (QAM). Whereas QAM modulates the amplitude of two quadrature carrier waves,[2] CAP combines two pulse amplitude modulation signals that are filtered to produce a QAM signal. CAP uses a wide pass band, dividing the available spectrum into three regions: POTS (DC to 4 kHz), upstream (25 kHz to 160 kHz), and downstream (240 kHz to 1.1 MHz). Conversely, DMT provides multiple narrowband channels (224 downstream and 25 upstream in the case of ADSL), each with their own carrier. Each channel is individually modulated and combined using a QAM-like modulation, allowing multiple bits to be represented with a single QAM symbol. In QAM, each combination of amplitude and phase represents a unique combination of bits. A 64-symbol constellation (64-QAM) contains 64 unique combinations of amplitude and phase capable of representing any binary value consisting of 6 bits. Each channel has a 4-kHz frequency range and is capable of transporting up to 32 kbit/s.

[2]Quadrature carriers are sinusoid waves that are out of phase by 90°.

DMT provides advantages over competing modulation techniques such as simpler channel equalizers, bandwidth efficiency through the redistribution of bits to other subchannels, filterless frequency division multiplexing, good immunity properties to narrowband interference and performance in the presence of other line impairments such as wire gauge variations and bridged taps. The use of adaptive filtering leads to the principal disadvantage of DMT, that being the need to train the filter to ensure stability. Training is performed as part of the line initialization procedure prior to the line being ready to transmit or receive user data and may reoccur periodically if line conditions deteriorate. Further advanced signal processing techniques such as Trellis coding and interleaved forward error correcting (FEC) codes are used to improve noise susceptibility, leading to improved signal-to-noise ratios and longer practical loop lengths at nominal bandwidths. Although Bell Labs first developed digital technology using copper loops in the mid-1980s, it was not until the ratification of the DMT-based international standard G.992.1 [7] where CAP-based ADSL systems depreciated.

Although motivated by video delivery, ADSL did not offer bandwidth sufficient to support the prevalent video codecs of the day. To address these bandwidth limitations, G.992.3 [8] ADSL2 improves the data rates and reach, achieving as high as 12 Mbit/s at 600 ft while also being capable of supporting lower-bit-rate services out to 15,000 ft. ADSL2 includes features to improve line efficiency through reduced framing overhead [providing an all-digital (no POTS) mode of operation], supports mechanisms to transport both asynchronous transfer mode (ATM) and synchronous traffic types, and offers further increases in bandwidth through bonding of lines using inverse multiplexing for ATM (IMA). However, ADSL2 remained spectrally incompatible with ADSL, preventing the combination of ADSL and ADSL2 pairs in the same binder group. Spectral compatibility was later addressed in ADSL2+ [9], which also increased the downstream frequency band to 2.2 MHz with 512 subcarriers, resulting in 24-Mbps downstream bandwidth out to 3000 ft.

VDSL [10] and VDSL2 [11] represent the latest installment in the continuing evolution of DSL. Both standards provide a dramatic increase in potential data rates, with VDSL increasing the usable frequency spectrum to 12 MHz and VDSL2 pushing the upper end to 30 MHz. Such an increase in frequency affords downstream bit rates in excess of 50 Mbps, sufficient for multiple HD and SD video streams. As illustrated in Figure 3.5, VDSL introduced the use of segmented spectrum, where the upstream and downstream frequency ranges are no longer contiguous, in order to make VDSL more spectrally compatible with existing xDSL deployments.

The move to native Ethernet encapsulation in VDSL, leveraging the work undertaken by IEEE 802.3ah Ethernet in the First Mile (EFM) taskforce, recognizes Ethernet's growing importance outside of the LAN, reducing the interworking burden between the home network and Ethernet-based aggregation and backhaul networks. Asynchronous transfer mode (ATM) and synchronous transport remain legacy options. Although VDSL and VDSL2 data rates are

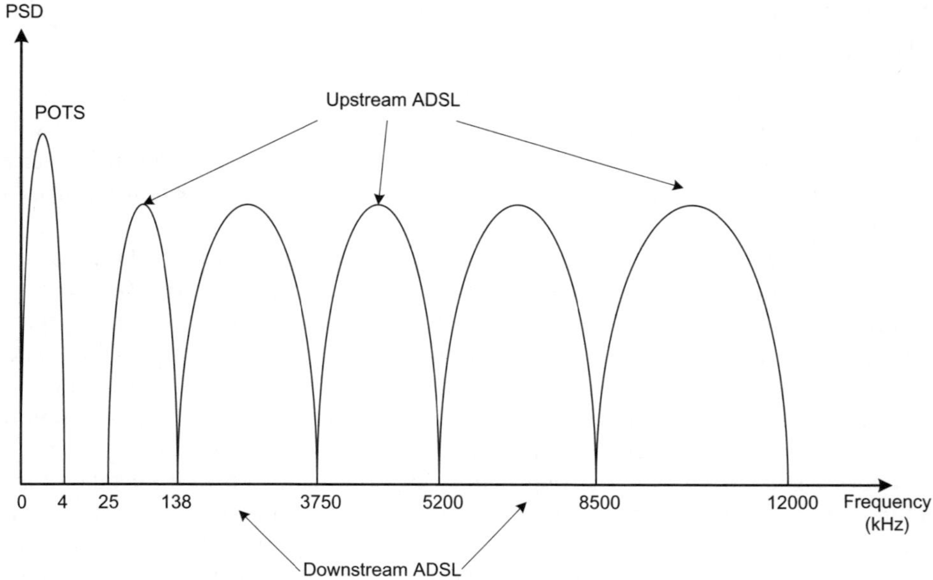

Figure 3.5. VDSL spectrum.

attractive, they require short loop lengths to achieve their impressive performance. In both schemes, data rates drop considerably after 4000 ft [12]; and given that 18,000 ft is not uncommon in rural POTS lines, loop length reduction, through the deployment of active electronics deeper into the distribution network, is a necessary network evolution in order to deliver rates close to these stated maximums.

VDSL2 was approved by the International Telecommunications Union (ITU) in May 2006 as G.993.2 and is designed to provide both increases in data rates and reach compared to what is achievable with prior technologies. VDSL2 offers impressive raw data bandwidths, providing in excess of 25 Mbit/s over longer loops (4000–6000 ft) and symmetric data rates of 100 Mbit/s over short loops (less than 1000 ft) [12], an important data rate to address high-speed business services. Many of the improvements first contained in ADSL2+, including advanced diagnostics, the ability to dynamically alter bandwidth per channel, and superior impulse noise immunity, are also included in VDSL2. Impulse noise immunity is a particularly important property when transporting packetized video. Importantly, VDSL2 offers interoperability with a wide range of prior technology variants, including the original ADSL, ADSL2, and ADSL2+. VDSL2 utilizes the same modulation scheme as ADSL, discrete multitone (DMT), providing an increase to 4096 channels. A critical aspect of VDSL2 standardization has been the development of band plans that reflect the regional differences in frequency allocations while supporting both symmetric and asymmetric bandwidth services in same binder group.

3.2.3 Digital Subscriber Line (DSL) Reference Network Architecture

Although the copper loop is a crucial element in today's wireline access network, it is not the only component required to deliver DSL service. This section will introduce a more network-centric view of emerging DSL networks, describing the equipment and companion networks necessary to deliver residential broadband services.

DSL network architectural decisions fall under the umbrella of the DSL Forum [13], now known as the Broadband Forum, an industry body responsible for promoting DSL technology worldwide through the development of technical specifications. With a membership consisting largely of network service providers and equipment vendors, the Broadband Forum develops specifications covering the requirements for equipment used in broadband DSL networks and defines common network reference architectures that guide service providers in the design of such networks. The forum has a strong history of producing relevant specifications spanning the entire evolution of broadband DSL technology. This section will examine in more detail one specification, Technical Report (TR) 101, "Migration to Ethernet-Based DSL Aggregation" [14], because it provides an overview of network architectural trends in residential wireline broadband deployments.

Conventional residential broadband DSL deployments have utilized asynchronous transfer mode (ATM) as an aggregation technology. Each subscriber would communicate with the Internet through Internet protocol (IP) connections encapsulated in one or more ATM-based permanent virtual circuits. Traffic from many thousands of subscribers would be aggregated by a network of ATM switches that would backhaul the traffic to be processed by a broadband remote access concentrator (BRAS). A BRAS is a centralized subscriber policy decision point in the network, responsible for subscriber authentication, authorization, and accounting (AAA) functions. The BRAS represented the first IP layer element in the network, the default gateway. Figure 3.6 illustrates an exemplary residential broadband deployment with ATM aggregation.

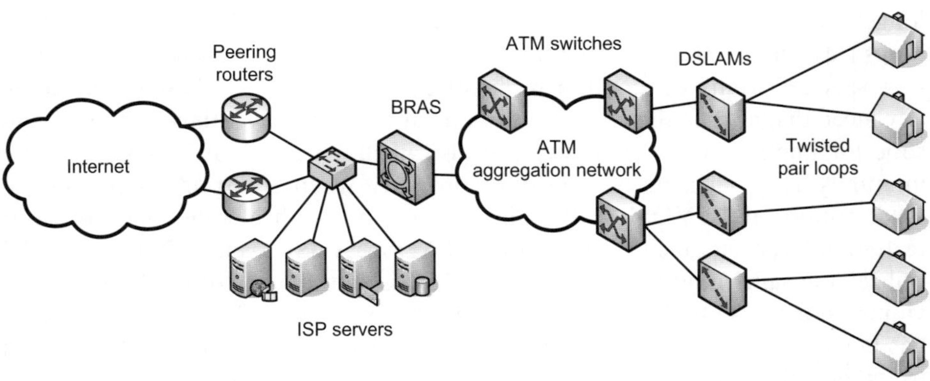

Figure 3.6. ATM-based residential DSL network architecture.

The push to support video broadcast services has driven the need to increase the bandwidth in the copper loop, which has consequential implications for the aggregation network in particular. To efficiently transport video, both the access and aggregation portions of the network require large quantities of bandwidth, quality of service guarantees, and efficient bandwidth utilization for wide-scale packetized video distribution. Ethernet has evolved to provide high connection speeds, packet-based quality of service, simple and efficient provisioning, native multicasting capabilities, and network redundancy suitable for deployments in carrier-grade networks. The technical capabilities of modern Ethernet transport and its associated compelling cost benefits has resulted in a wholesale migration away from ATM to Ethernet transport in new and upgraded residential broadband networks. The Broadband Forum specification TR-101 defines a reference network architecture (shown in Figure 3.7) and requirements for the network elements referenced in the architecture.

The reference network consists of a number of network elements (broadband network gateway, access node, network interface device, etc.), networks (regional, aggregation, access, and customer premise) and reference points (T, U, V, and A10). The reference points represent clear boundaries: either demarcation points, administrative boundaries, or network-specific traffic congestion points.

The broadband network gateway (BNG) is an Ethernet-centric IP router with subscriber AAA, quality of service, multicast and security functions. Additionally, the BNG is required to implement many of the legacy features commonly found in BRAS devices today, including point-to-point protocol (PPP)-based subscriber sessions over Ethernet via PPPoE [15] and bandwidth wholesaling using Layer 2 Tunneling Protocol (L2TP)-based tunneling [16]. The BNG occupies a critical location in the network, processing all the upstream and downstream subscriber traffic, and is evolving into a centralized policy decision-making and enforcement point. Such a control point performs service session identification and traffic management functions ensuring that no point in the

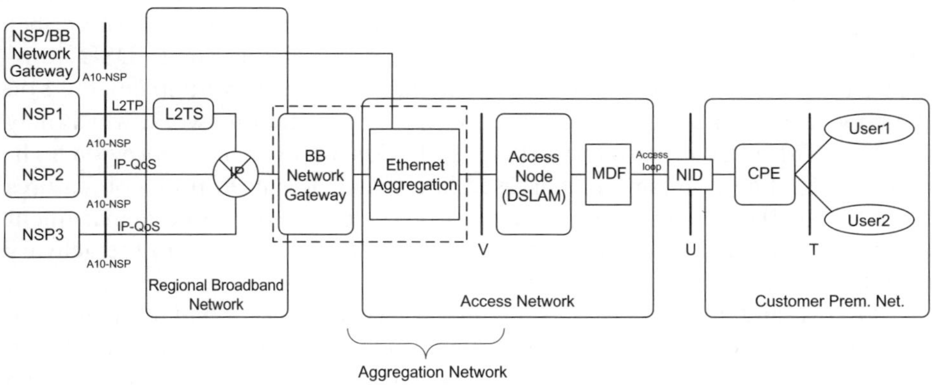

Figure 3.7. TR-101 Ethernet aggregation network architecture.

aggregation and access network, including the copper loop, is overcommitted, thereby ensuring that the subscribers' applications receive the required service and acceptable quality of experience.

The principal purpose of the aggregation network is unchanged in a native Ethernet deployment, that being to combine traffic from potentially thousands of subscribers over a residential neighborhood and backhaul the aggregated traffic to the BNG. The aggregation network consists of a series of interconnected Ethernet switches, connecting the BNG and digital subscriber line access multiplexer (DSLAM) devices. The Ethernet switches offer vastly superior switching capacities compared to the ATM equivalents and at a reduced relative cost. The Ethernet switches commonly used in such deployments provide IGMP snooping functionality [14], where the switch selectively reconfigures their multicast replication tables based on IGMP join/leaves sent upstream from the subscriber. The Internet Group Management Protocol (IGMP) is used in IP video broadcast applications to signal "channel change" events from a subscriber's set-top box device. By providing snooping functionality in the aggregation network, bandwidth efficiencies may be realized by selectively replicating channels (IGMP groups) that are being watched by one or more subscribers. The aggregation network may serve as a point where "walled garden" content is injected into the network—for example, carrier branded video traffic. This is typically accomplished through the direct connection of video servers into the aggregation network. Traffic injected downstream of the BNG potentially invalidates policy decisions made at the BNG since that traffic is not visible to that network element. In such architectures it is necessary to separate the "walled garden" traffic from the high-speed Internet traffic through the use of multiple VLANs and traffic engineering to ensure that high-speed Internet traffic cannot interfere with the premium video content during periods of peak usage and congestion. The converse architecture is to carry all traffic, including the "walled garden" traffic, through the BNG, where it can be properly accounted and the correct policy decisions applied ensuring no downstream congestion.

The digital subscriber line access multiplexer (DSLAM) is the network element that bridges the service providers' internal network and the subscriber through the copper loop. The DSLAM provides the physical layer DSL transceiver technology and adapts between the link layer protocol implemented on the copper loop and the protocol used in the aggregation network, which may involve ATM to Ethernet interworking. DSLAMs are evolving to provide high-density DSL line termination, increased line speed, advanced protocol features (such as IGMP snooping and proxy reporting), advanced remote management and diagnostic capabilities, and quality of service features to support the delivery of multiple services across the copper loop.

TR-101 and its predecessors define the U-reference point as being the customer premise located end of the copper loop. The network interface device (NID) implements the DSL physical layer and link layer encapsulation, such

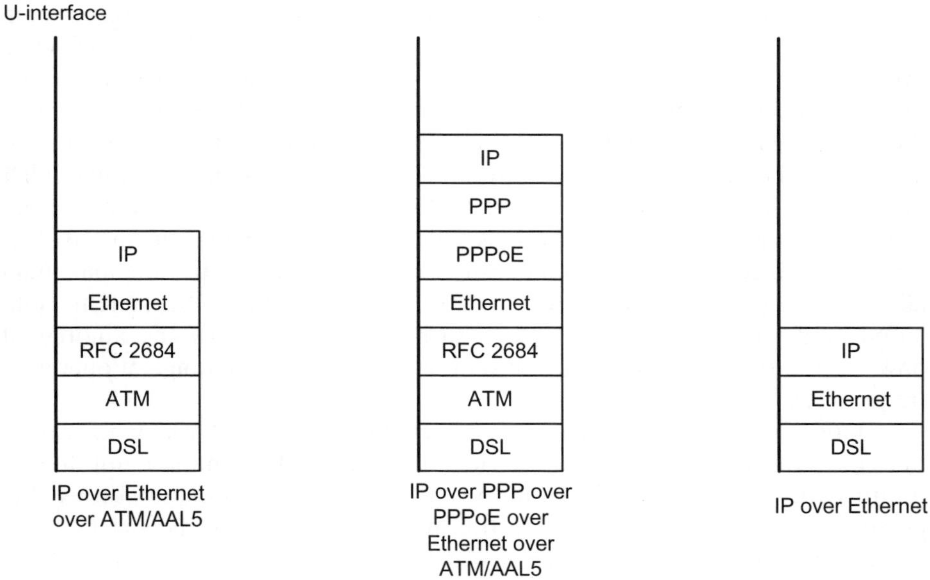

Figure 3.8. Exemplary packet protocol stacks.

as IP over PPP over ATM or IP over PPP over Ethernet over ATM or simply IP over Ethernet. Figure 3.8, taken from TR-101 illustrates several possibilities.

The U-reference point represents the demarcation point between the service providers network and the customer premise network. The CPE device interfaces with elements of the home network through Ethernet or some other preexisting wiring plant (twisted pair, coax, power, or wireless). The CPE may connect to several appliances in the home, including a video set-top box for video services, a phone (either PSTN via a gateway device or directly with VoIP phone), and the home computer or router-gateway device.

Often DSL networks are seen as simply the copper loop, but in reality they are sophisticated networks of networks, combining many discrete and heterogeneous networking technologies and elements in order to deliver services over a converged IP transport.

3.2.4 Futures

The last decade has seen remarkable increases in bandwidth delivered through century-old copper loop facilities, but surely we're at the end of what can be reasonably extracted from this aging technology? Fortunately, some disagree and are actively investigating techniques to continue to mine copper loops' hidden spectrum, even promising data rates greater than current and planned passive optical networks [3, 5].

The key strategy to increasing the available bandwidth in the copper loop is to minimize the impact of noise, particularly crosstalk. This is achieved through managing the spectrum of all transmitters in a binder group. There is a range of spectrum management techniques that deliver increasing benefits at the cost of increasing complexity. The simplest technique is static spectrum management, which defines rules that are applied to ensure spectral compatibility between all services and technologies that use pairs in the same binder group using worst-case spectra assumptions. Static spectrum management involves using predefined power spectrum density (PSD) masks to control the maximum allowed transmit power levels for any given frequency in a DSL system. Both VDSL and VDSL2 standards specify such masks, controlling the amount of power generated by a compliant transmitter across the entire range of operating frequencies.

Static spectrum management is a technique employed today to ensure that the current generation of services are deployable and do not hinder future DSL technology variants; however, it does not attempt to address the possibility of dynamically managing transmit power based on the specific local deployment conditions. Dynamic spectrum management [17] is a form of adaptive spectrum control that globally optimizes the spectra of different DSL systems deployed in one binder by tuning each transmitter's power output to ensure that the bandwidth requirements are met with acceptable margin. The objective is to reduce transmission power on loops that can acceptably function with lower output power, coupling less energy into other victim loops, reducing crosstalk.

Dynamic spectrum management requires the ability to characterize the transmission properties of the loop such as line margin, transmit power, bits/tone tables, and insertion loss per tone. These parameters, combined with other line knowledge, such as loop length, bridge taps, and binder group identification, are input into a decision-making "authority" that performs the dynamic spectrum management function, predicting expected data rates and recommended margins and generating individual line control parameters such as PSD mask values and forward error correction (FEC) parameters. This involves calculating the desired transmission power necessary to cover the noise-to-signal curve as a function of frequency (NSF(f)). Increasing transmission power past this "optimal value" provides for more margin but offers no additional bandwidth gain; rather it translates to more energy coupled into the victim loop in the form of crosstalk interference. This approach results in the largest benefits on shorter loops, which require less power to achieve the desired bit rates. The final step in dynamic spectrum management is to configure each transmitter with the calculated control parameters. This three-step "characterize–calculate–update" procedure may be performed during line initialization and then executed periodically, modifying parameters as line conditions change. Since dynamic spectrum management automates much of the provisioning, maintenance, and operations of the DSL line, it has the secondary benefit of reducing the operating costs associated with the line [17].

Finally, an emerging DSL transmission technique called vectored transmission [3, 5, 17] aims to eliminate the effect of crosstalk all together, effectively creating a noise-free copper transmission medium potentially offering enormous data rates, limited only by analog and analog–digital convertor (ADC) circuit technology. The goal of vectored transmission is to eliminate far-end crosstalk (FEXT) introduced by other DSL transmissions through employing joint signal processing. FEXT introduced by other DSL systems is the dominant impairment constraining performance—especially as loop lengths decrease, resulting in less line attenuation. Vectored transmission coordinates both downstream and upstream transmissions, creating a vector of downstream transmitted signals and receiving a vector of upstream signals, conceptually treating the binder group pairs as a multiple-input–multiple-output (MIMO) transmission system. This coordinated transmission has been shown to provide in excess of 400 Mbps on loop lengths approaching 1200 ft [3].

Vectored transmissions rely on the ability to coordinate (in the time domain) all the downstream transmissions. All modems located on the DSLAMs DSL line card are synchronized to a common discrete multitone (DMT) symbol clock. In the upstream direction, through exploitation of the network timing reference transport in DSL, all customer premise modems synchronize their transmissions to the common symbol clock. The receivers on the DSLAM DSL line card can cancel both the upstream far-end and near-end crosstalk introduced by other DSL sources by performing a MIMO decision feedback within each tone. In the upstream direction, one of the users will have no crosstalk whatsoever on each tone. This user is decoded first. This users influence on the next user is constructed and removed from all subsequent users for each tone in the DMT transmission. Crosstalk is eliminated in the downstream direction by calculating the noise contribution of the other (coordinated) transmitters and pre-distorting the transmitted signal. Vectored transmission requires detailed knowledge of the exact crosstalk and noise autocorrelation matrix for the binder group in question, which requires extensions to the existing training sequence in order to generate these data in addition to the insertion loss and gain/phase information resulting from the line initialization procedure.

Vectored transmission is not the same as bonding mechanisms standardized in ADSL2. Link bonding is a data link layer multiplexing and demultiplexing technique that aggregates the bandwidth of multiple physical links to create a single virtual link with bandwidth aggregate of the constituent links. Vectoring is cogeneration and co-processing of the physical layer signals and is therefore independent of any bonding scheme employed.

3.2.5 Hybrid Fiber/Copper Networks

In order to deliver the high bandwidths required for video-rich triple-play services, access network providers (ANPs) must reduce the lengths of the copper loop, necessitating the deployment of active elements in the distribution network. This results in pushing fiber from the central office (CO) to DSLAMs located closer to the customer premise. These networks are commonly called fiber-to-the-node (FTTN), fiber-to-the-curb (FTTC), or fiber-to-the-building (FTTB), identifying the location of the DSLAM as being a curbside cabinet or in the basement of a multitenant building. Short copper loops then connect the DSLAM to the residence. Figure 3.9 illustrates the various FTTx models.

The decision to deploy a FTTN network versus a complete fiber-to-the-home (FTTH) network is extremely dependent on the business case, which takes into account the installed cost, operational cost, and expected service revenues. However, in some cases, it may not be possible to deploy a new building cabling system due to aesthetic or historical considerations. As a general observation, FTTN with VDSL2 loops, allowing the reuse of copper pairs, is more cost effective with regard to existing network infrastructure upgrades and allows faster time to market. FTTH can be more cost effective in completely new network deployments and promises lower operating costs [18].

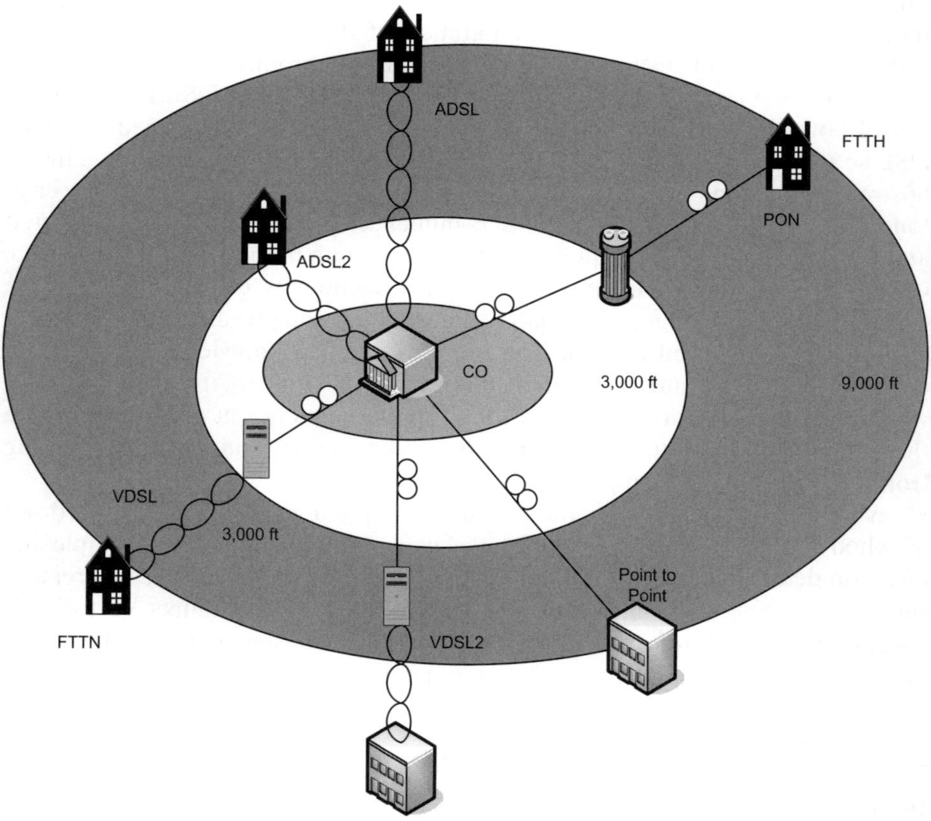

Figure 3.9. FTTx models.

One of the principal challenges in a mixed xDSL/FTTN network architecture is spectrum management—in particular, preventing the mixture of existing (longer-loop) DSL service and new, shorter-loop DSL service from a fiber-fed DSLAM in the same binder group [17]. Unless some form of spectrum management practices are adopted, the shorter high-speed loops create large amounts of crosstalk, which interfere with the longer, lower-rate CO feed loops. Without dynamic spectrum management, the transmitter on the shorter loop may be using orders of magnitude more power needed to achieve the required data rate.

Other considerations in FTTN deployments are less theoretical and more mundane. How does one power the remote DSLAMs? What environmental controls (if any) are required? How is lifeline POTS provided in the event of a power outage? Central office (CO) located equipment is deployed within a secure facility, with advanced heating and air conditioning, backup power supplies, and other building management systems, none of which are available in a sheet metal cabinet bolted to a concrete footing, located on a street corner. The challenge with deploying such complex, active electronics outside the controlled environment of the CO are numerous and require equipment specifically designed to withstand the environmental conditions found in non-air-conditioned cabinets or when mounted on poles or pedestals.

Finally, one often overlooked consideration is the additional burden placed on the operational divisions of the network operator. Deploying many smaller DSLAMs in the field may lead to higher operational costs associated with management and provisioning and in maintenance. These operational costs tend to be proportional to the number of network elements deployed. Operational costs are already a significant expense, with ANPs looking at reducing their effect on the business.

Even with these challenges, FTTN architectures represent a common network evolution path to enable to the bandwidth necessary for future services delivered over the existing copper loop.

The twisted-pair copper loop has formed the backbone of wireline access networks for more than a century. This enduring medium, coupled with enormous advances in signal processing technology, has evolved from providing basic telephony to being capable of delivering high-definition television broadcasts, video on demand, and other multimedia-content-rich services ensuring that the longevity of this particular type of wireline access network continues.

3.3 PASSIVE OPTICAL NETWORKS

This section will introduce a new variety of wireline access networks—the passive optical network (PON). Passive optical networks are topological tree point-to-multipoint (P2MP) optical distribution networks that have no active elements in their transmission paths. Their plentiful bandwidth and intrinsic P2MP nature provides for efficient video broadcasting, making PONs an ideal technology for next-generation wireline access networks. This section will introduce and describe a number of PON variants, with specific emphasis on Ethernet PON (EPON) technology. PON technology has been standardized in both the ITU-T [19,20–23] and IEEE [24] organizations providing alternatives optimized for the carriage of legacy traffic or Ethernet frames. The IEEE's EPON and the emerging 10 Gbit/s EPON [25] offer copious bandwidth while preserving the familiarity and cost benefits associated with Ethernet—arguably the world's most successful networking technology.

3.3.1 Passive Optical Network Fundamentals

PON networks consist of fiber-optic cables for transmission, optical couplers for distribution, and active elements responsible for converting between the optical and electrical domains. Optic fiber is a very thin filament of glass, which acts as a waveguide allowing light to propagate with very little loss due to the principal of total internal reflection. The optical signal is distributed into branches using *splitters*, which replicate the optical signals received on an input across multiple outputs. Since splitters are passive elements, the optical power of any output signal is only a fraction of the optical power received on the input port. This reduction in optical power limits the fan-out of the network. *Combiners* perform the inverse function by combining optical signals from multiple inputs and transmitting the accumulated signal upstream. Combiners are highly directional, leaking very little optical power across input ports, a property referred to as *directivity*. The splitting and combining functions are integrated into a single element called an *optical coupler*. The final elements in a basic PON architecture are the *optical line terminal* (OLT) and *optical network unit* (ONU).[1] These active elements are deployed within the central office and close to the residential or business premise respectively. The OLT and ONU "book-end" the optical distribution network, interfacing between the electrical and optical domains and performing functions typically required of the data link layer of the OSI reference model. The basic PON network architecture is shown in Figure 3.10.

Most PON technologies share a remarkable technical similarity with the selection of bearer protocol being their principal distinction. The ITU-T has standardized two PON architectures: broadband PON (BPON) based on asynchronous transfer mode (ATM) and more recently gigabit PON (GPON) that employs the Generic Encapsulation Method (GEM) for the native transport of time division multiplex (TDM), ATM, and the lightweight encapsulation of Ethernet frames. Ethernet PON, as standardized in the IEEE organization, provides a completely 802.3 compliant MAC interface, for the transmission and reception of Ethernet frames.

[1] Also known as a optical network terminal (ONT) in ITU terminology.

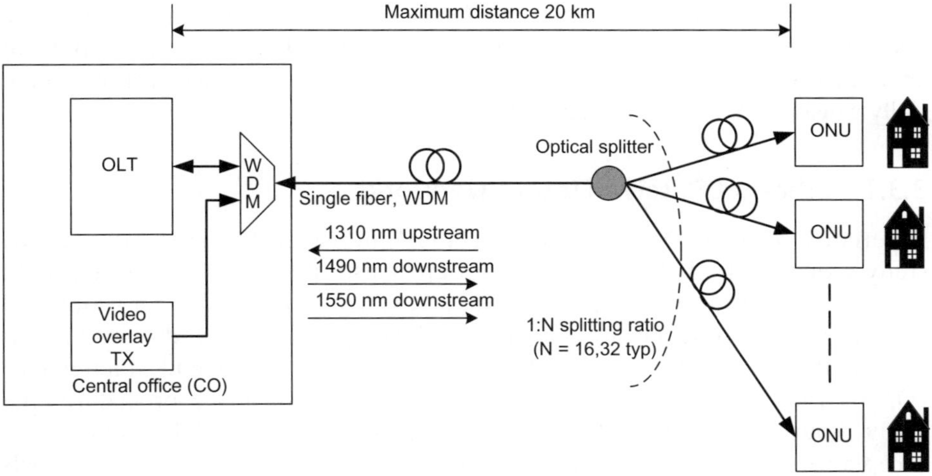

Figure 3.10. Basic PON distribution network architecture.

Due to the optical replication properties and directivity of the couplers, PONs are asymmetric in nature. In the downstream direction, the OLT frame transmission is broadcasted to all ONUs. Only frames explicitly addressed to an ONU are extracted and processed further. Using Ethernet as an analogy, in the downstream direction, a PON appears similar to a shared medium LAN. In the upstream direction, bandwidth is also shared between the ONUs, due to the optical couplers combining the ONU optical transmissions. This requires an arbitration scheme to coordinate access to the upstream transmission bandwidth to prevent collisions. Due to the directivity of the optical couplers, an ONU is unable to detect a collision, so although the upstream of a PON behaves as a shared medium, conventional contention-based mechanisms for resource arbitration such as carrier sense multiple access with collision detection (CSMA/CD) and carrier sense multiple access with collision avoidance (CSMA/CA) are difficult to implement. Rather than develop a cumbersome contention-based algorithm, both the ITU-T and IEEE have specified a time division multiple access (TDMA) scheme to avoid collisions at the optical level. Although differences exist in the details, BPON, GPON, and EPON share a common approach in dividing the upstream bandwidth into timeslots. Each timeslot represents a transmission window specified by a start time and either duration or stop time. The transmission timeslot is then granted by the OLT to the ONU using a specific bandwidth allocation algorithm. The OLT is free to allocate bandwidth to ONUs in any practical fashion, ranging from a static allocation to a dynamic one, based on the quantity of data that an ONU has to transmit. The relative merits of these algorithms are considered later in this chapter.

Once an ONU has been granted a transmission window, the ONU transmits up to the window size in a burst at the full physical layer data rate (1 Gbps in the case of EPON). When complete, the ONU ceases transmission including the disabling of the laser. The laser remains off during silence periods to prevent spontaneous emission noise from being injected in the upstream direction, potentially causing bit errors to other valid transmissions.

3.3.2 Ethernet Passive Optical Networks

Ethernet PON (EPON) emerged from the Ethernet in the First Mile (EFM) study group of the IEEE. The study group, constituted in 2001, was chartered with extending Ethernet technology into access networks suitable for residential and business subscribers, and it covered point-to-point (P2P) and point-to-multipoint (P2MP) architectures and Operations, Administration, and Maintenance (OAM) aspects of Ethernet networks. In accordance with IEEE procedures, the study group became the 802.3ah task force in September 2001 following the acceptance of the Project Authorization Request (PAR) [26]. One of the goals of the taskforce was to provide a 1-Gbps Ethernet service using P2MP passive optical networking supporting a minimum splitting ratio of 1:16, while preserving the existing frame format, media access control (MAC) layer and media independent interface (MII) of standard 802.3 Ethernet for distances of 10 km or more (Table 3.2).

Adhering to these goals and minimizing the modifications to the physical and physical medium-dependent (PMD) sublayers would promote rapid adoption through the use of high-volume, low-cost 1-Gbps optical and semiconductor components [26].

The 802.3ah taskforce successfully completed their standards activity with the production of the 802.3ah-2004 document. This work has since been subsumed into Section 5 of the 2005 Edition of IEEE 802.3 standard [24] (Figure 3.11). The 802.3ah EPON standardizes a 1-Gbit/s symmetric (1 Gbps downstream and 1 Gbit/s upstream) PON network that employs two wavelengths over a single fiber (1490 nm downstream and 1310 nm upstream) for data transmission, with the option of a further downstream wavelength (1550 nm) reserved for additional services such as analog video broadcast. With an optical power

TABLE 3.2. EPON PAR Goals

Technical	Objective
Frame format	802.3 standard format and encoding
Line rates	Standard Ethernet rates, 1000 Mbit/s
Fiber	Single fiber, single mode (SMF)
Distances	10 km minimum
PMD	Investigate 1310/1310, 15xx/1310, and ITU 983.3
Splitting ratio	1:16 min
Connector	SC, investigate high-density connectors like LC

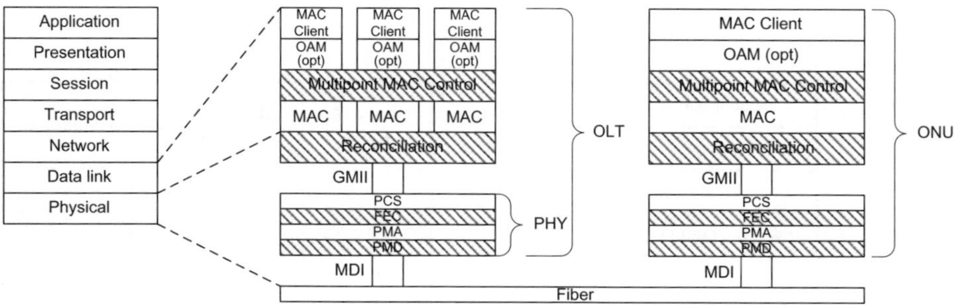

Description	1000BASE-PX10-U	1000BASE-PX10-D	1000BASE-PX20-U	1000BASE-PX20-D	Unit
Fiber type	B1.1, B1.3 SMF				
Number of fibers	1				
Nominal transmit wavelength	1310	1490	1310	1490	nm
Transmit direction	Upstream	Downstream	Upstream	Downstream	
Minimum range	0.5 m to 10 km		0.5 m to 20 km		
Maximum channel insertion loss	20	19.5	24	23.5	dB
Minimum channel insertion lost	5		10		dB

Figure 3.11. IEEE EPON protocol layers and network capabilities.

budget of 24 dB, distances of 20 km can be obtained, with a splitting ratio of 1:16. Frames are encapsulated in the standard 802.3 format and are encoded using 8b/10b line codes.

The standard 802.3 Ethernet MAC supports two operating modes: a shared-medium, single-collision domain and a full-duplex point-to-point topology. Upstream characteristics of a PON exhibit aspects of both modes. The IEEE has specified a TDMA scheme, coupled with a P2P logical topology emulation to simulate a P2P connection based upon the shared upstream bandwidth of the PON. The key extension proposed by the taskforce and adopted by the standard is the creation of an additional MAC control protocol: Multipoint MAC Control Protocol (MPCP). MPCP is a real-time control protocol that is responsible for manipulation of the MAC sublayer operation. MPCP extends the existing pause-based flow control of the MAC Control sublayer to include facilities for network clock synchronization, ONU resource management, and the discovery, registration, and initialization of ONUs. MAC Control Protocol frames are distinguished from other MAC frames by a specific value (88-08 hexadecimal) in the length/type field. MPCP frames are further distinguished by a specific 16-bit opcode value that follows the length/type field. This is shown in Figure 3.12.

MPCP defines five additional MAC control frames:

- GATE: A grant of a transmission window made to the recipient.
- REPORT: Notification of a pending transmission from the sender.

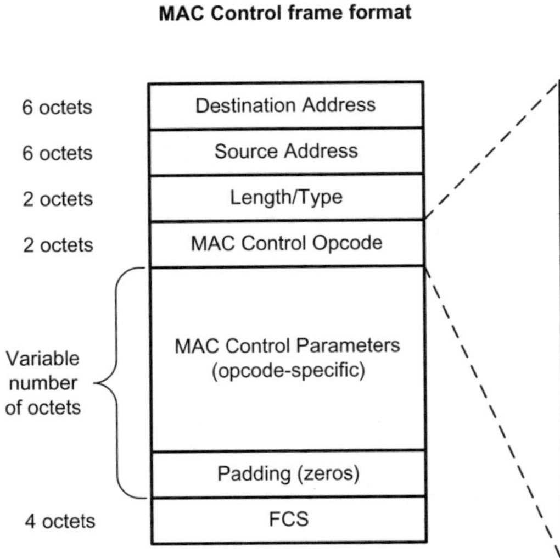

Opcode (hex)	MAC Control function	Timestamp
00-00	Reserved	
00-01	PAUSE	No
00-02	GATE	Yes
00-03	REPORT	Yes
00-04	REGISTER_REQ	Yes
00-05	REGISTER	Yes
00-06	REGISTER_ACK	Yes
00-07 thru FF-FF	Reserved	

Figure 3.12. MAC control frame format.

- REGISTER_REQ: A request for the sender to be recognized as participating in the gated transmission protocol.
- REGISTER: A notification that the recipient is recognized to participate in the gated transmission protocol.
- REGISTER_ACK: A notification that the sending station acknowledges its participation in the gated transmission protocol.

TDMA bandwidth allocation mechanisms require all transmitting nodes to be synchronized to a common time base in order to avoid collisions. Given that each ONU stores frames received from a subscribers' network, waiting for a granted transmission opportunity, each ONU must have a common and consistent view of "time." The ONUs achieve clock synchronization through a timestamp that is inserted in the MPCP GATE message that is transmitted by the OLT. The timestamp is a monotonically increasing 32-bit counter, incremented every 16 ns (referred to as the *time_quanta*). The timestamp field is populated by the OLT and represents the transmission time of the first byte of the destination address of the GATE message. When the GATE message is received, the ONU sets the MPCP local clock to the value contained in the timestamp field. The MPCP local clock is incremented by a clock recovered from the incoming data stream. Since the line contains IDLE characters during periods of silence, the clock can be continuously recovered, minimizing wander. Although this technique requires strict control of the delay variation within the OLT, which can be no more than 16-bit times (1 *time_quanta*) through the RS, PCS, and PMA sublayers, it does allow the OLT to continue to use a conventional +/–100-ppm clock source.

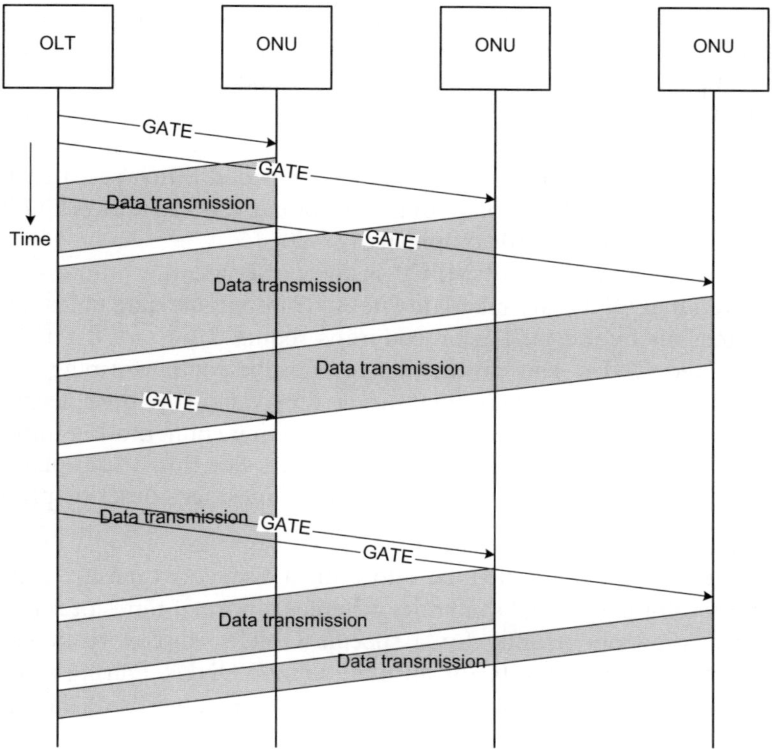

Figure 3.13. Pipelined GATE/data transmission.

Resource management consists of the assignment of upstream bandwidth to an ONU, upstream scheduling to decide ONU transmission order, and upstream bandwidth allocation to determine timeslot length. MPCP uses the GATE and REPORT MPCP control messages to perform resource management and coordinate access to upstream bandwidth. The GATE message is sent downstream from the OLT to a specific ONU granting a defined transmission opportunity to the ONU. The timeslot is defined by start time and transmission duration. On reception of the GATE, the ONU synchronizes its MPCP local clock and selects a frame (or sequence of frames) for transmission. A single GATE message can specify up to four independent granted timeslots. Timeslot allocation is carefully pipelined in order to minimize the time to walk all ONUs and maximize network utilization. The pipelining effectively eliminates the overhead of the GATE message transmission and processing; however, it does require knowledge of the round-trip time (RTT) to each ONU. The ONU transmits the REPORT message to the OLT to request subsequent transmissions opportunities. The pipelined transmission timing of the GATE and REPORT messages is illustrated in Figure 3.13.

The REPORT message contains queue occupancy (in units of *time_quanta*) information indicating the amount of transmission bandwidth that the ONU is requesting in the next scheduling epoch. The OLT may use this information in the execution of the bandwidth allocation algorithm. A report message can specify occupancy of up to 8 queues within a grouping, called a "queue set." A single REPORT message can contain up to 13 separate "queue sets." The actual bandwidth allocation algorithm is outside the scope of the IEEE standard and remains implementation-dependent. This has provided a rich research topic as studies have tried to balance competing goals of maximizing network utilization, service quality assurances and complexity [27–30].

The third responsibility of MPCP is to coordinate the auto-discovery of ONUs. Given that ONUs are mandated not to transmit any data unless explicitly granted timeslots by the OLT, a procedure is required for an OLT to discover newly activated ONUs. This protocol is responsible for performing the initial discovery, activation handshake, measuring the round-trip time, learning the ONUs individual 48-bit MAC address, and providing configuration information to enable successful bidirectional communications. For this, a four-phase handshake procedure is built using the MPCP control messages describe prior.

- The first step involves the OLT allocating a discovery timeslot. The discovery timeslot has special semantics, whereas a normal timeslot represents a transmission opportunity for a specific ONU; a discovery timeslot is a transmission opportunity for all uninitialized ONUs. During a discovery timeslot, the OLT broadcasts a discovery GATE, advertising the start time and duration of the discovery timeslot. The discovery GATE also contains the OLT timestamp.
- All known and initialized ONUs will discard the discovery GATE message. An uninitialized ONU will accept the broadcasted discovery GATE, synchronizing its MPCP local clock and wait until the designated transmission time. To avoid the expected collisions from more than one uninitialized ONU transmitting at the discovery timeslot start time, the protocol requires each ONU to wait an additional random amount of time before commencing transmission. The ONU transmits a REGISTER_REQ message that contains the source address and the ONU timestamp of the first byte of destination MAC address. The REGISTER_REQ message is received by the OLT, which allows it to calculate the round-trip time (RTT) for the ONU. The RTT is the numerical difference between the time of reception of the REGISTER_REQ and the timestamp included in the message. This is shown in Figure 3.14.
- The OLT replies to the REGISTER_REG with a REGISTER message assigning a logical link identifier (LLID). The LLID allows the ONU to discriminate unicast frames addressed to it. The OLT immediately sends a unicast GATE message to the newly activated ONU.
- Once the ONU receives the REGISTER and the GATE message, it replies with a REGISTER_ACK (in the timeslot allocated by the GATE message) to complete the process.

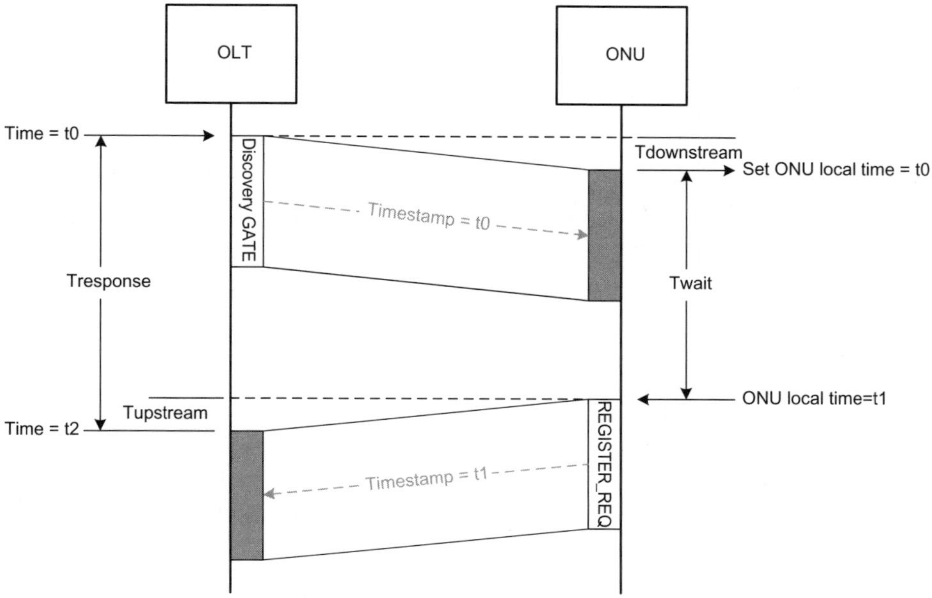

Tdownstream = downstream propagation delay
Tupstream = upstream propagation delay
Twait = wait time at ONU = t1 –t0
Tresponse = response time at OLT = t2-t0

Round-Trip Time (RTT) = Tdownstream + Tupstream = Tresponse -Twait = (t2 –t0) –(t1 –t0) = t2 –t1

Figure 3.14. Round-trip calculation.

Figure 3.15 graphically represents the autodiscovery procedure.

The discovery process is an overhead that consumes bandwidth from the overall network transmissions, thereby reducing utilization. The size and periodicity of the discovery timeslots must balance the time to discover and initialize a new ONU versus the overall decrease in network utilization due to the lost transmission opportunities within the discovery timeslot periods.

The IEEE standard [24] requires a logical topology emulation (LTE) function in order to fully comply with the requirements outlined in 802.1D concerning Ethernet bridging devices. In addition to the shared-medium/single-collision domain and full-duplex point-to-point operating modes of the 802.3 MAC, 802.1D compliant bridges do not forward frames back out the source port. If an OLT is to be considered an 802.1D-compliant bridge, the PON must be considered either a full duplex point-to-point topology or a completely shared-medium/single-collision domain topology. The 802.3 standard [24] defines a logical topology emulation function that resides below the MAC sublayer (thereby being transparent to the MAC layer), which allows a PON to mimic either topology. The 802.3 standard [24] only specifies the behavior of the full-duplex point-to-point model but is unwilling to lose the enormous benefit of single-copy broadcast in the downstream; it adds a single-copy broadcast (SCB) MAC

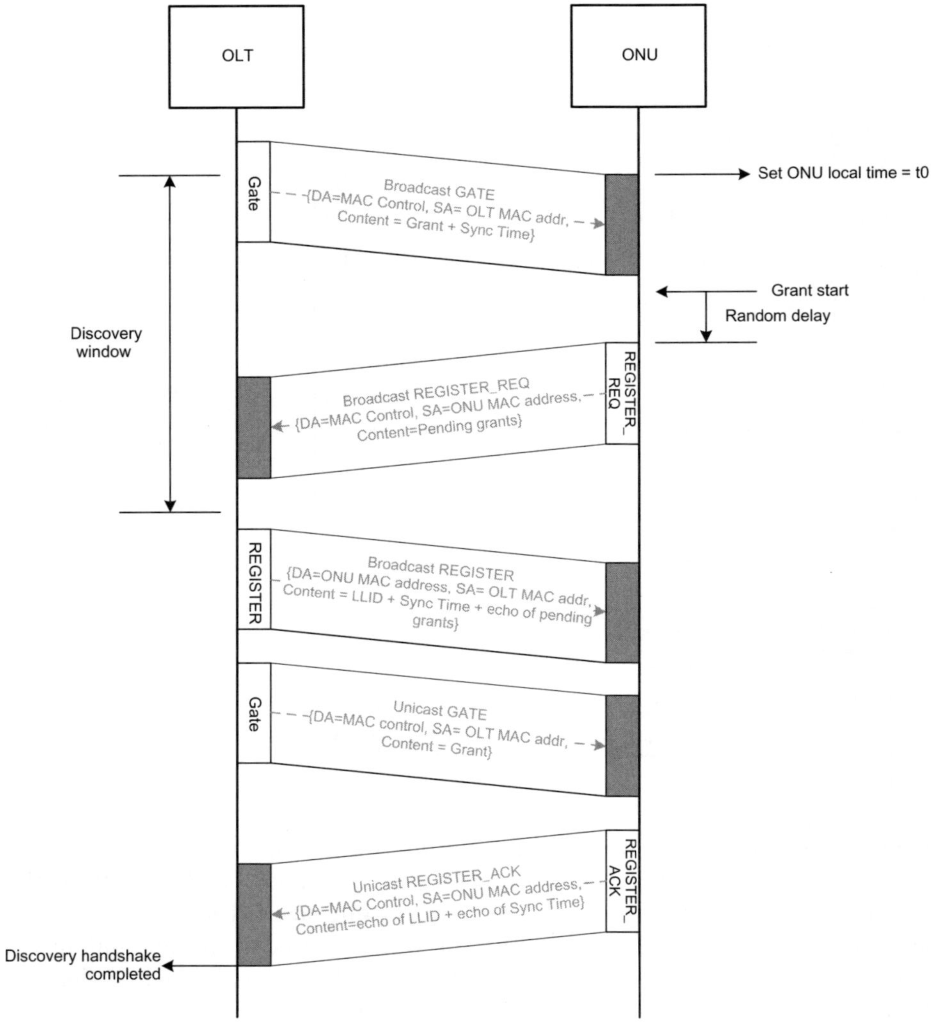

Figure 3.15. Discovery handshake.

port to the architecture, used for the logical broadcasting of frames to all downstream ONUs.

The LTE operates by assigning a logical link identifier (LLID) to each ONU during the auto-discovery process. Frames are transmitted with the 16-bit LLID embedded in a modified preamble (protected by a CRC). Frames matching the assigned LLID (or matching a broadcast LLID) are accepted by the ONU, and all other frames are discarded. In the upstream direction the ONU inserts the LLID into the preamble, allowing the OLT to "steer" the frame to a virtual MAC specific to the transmitting ONU, thereby emulating a full duplex point-to-point topology.

In order to improve physical reach of the optical distribution network, increase the splitting ratio, or simply improve the reliability of the communications channel, the IEEE included an optional forward error correction (FEC) mechanism. FEC provides the receiver with the ability to detect and even correct bit errors that occur during transmission. The encoder adds parity information to the frame to allow the decoder to detect the bit errors and reconstitute the correct information sequence. The FEC procedure selected by the 802.3ah taskforce is identical to that used in other PON technologies, most notably GPON, which is a block-based Reed–Solomon algorithm that adds 16 parity bytes for each 239-byte block frame data [22]. The IEEE standard defines a procedure to segment the 802.3 Ethernet frame into a sequence of 239 byte blocks. The final block is padded with zeros if necessary for the benefit of the FEC parity calculation. These padding bytes are not transmitted with the frame. The parity symbols are grouped together and transmitted at the end of the frame, leaving the 802.3 frame format unchanged. This decision represents a major advantage of the standardized frame-based technique, permitting interworking with non-FEC-enabled devices.

The final topic that deserves some discussion concerns privacy in EPON. Given that all frames are passively replicated to ONUs in the downstream direction and that ONUs rely on logical but not physical mechanisms for frame filtering, circumventing these logical restrictions would allow an ONU to eavesdrop on all downstream transmissions. Unlike local area networks, access networks consist of noncooperating users requiring privacy of communications; therefore it is surprising that the 802.3ah taskforce did not require an optional encryption component. However, many commercial network equipment manufacturers include encryption in both upstream and downstream directions enabled by the common availability of semiconductor products that integrate 128-bit advanced encryption standard (AES) encryption schemes [31, 32].

The rapid specification and commercialization of EPON technology is a testament to the flexibility and longevity of the 802.3 standard. Ethernet is the pervasive networking technology of choice for next-generation telecom networks, and EPON is one important step to bridging the domains of the subscriber's home LAN and the Ethernet domain of the metro and long-haul networks. The commercial future is exceedingly bright, due to the IEEE 802.3ah taskforces' efforts in maintaining backward compatibility with the billions of deployed Ethernet ports.

3.3.3 ITU-T Passive Optical Networks

Several PON technologies predate the IEEE EPON standardization efforts. These earlier PON technologies have been defined under the auspices of the International Telecommunications Union, Telecommunication Standardization Sector (ITU-T). In 1995 a group consisting of seven leading network service providers formed a special interest group called the Full Service Access Network (FSAN) group [33]. The FSAN had the worthy goal of creating a unified set of technical specifications covering a next-generation access network capable of delivering an expanded range of packet services in addition to the legacy transport of time division multiplex (TDM) and asynchronous transfer mode (ATM) services. The FSAN members developed a PON specification using ATM as the bearer protocol, known as asynchronous PON or APON. In 1997, the APON specification, now renamed broadband PON (BPON), was submitted to the ITU-T for standards consideration and became the G.983 series recommendations [19]. The original PON specification provided a symmetric 155-Mbit/s (155 Mbit/s downstream and 155 Mbit/s upstream) transport, which has been expanded to include an asymmetric 622-Mbit/s–155-Mbit/s and symmetric 622-Mbit/s data rates.

FSAN continued to define a minimum functional subset of the capabilities embodied in the recommendations called the Common Technical Specification (CTS). By proposing a common reduced feature set, FSAN expected lower equipment and operational costs due to lower equipment complexity and higher volumes. Although an admirable goal, the demands of efficient transport of legacy traffic (TDM and ATM) resulted in a scheme unsuited to the requirements of the predominant traffic type, namely, Ethernet. Addressing this limitation has been a focus of BPONs' successor, gigabit PON (GPON).

3.3.3.1 Broadband PON. The G.983 family of recommendations [19] specify a PON network with a 20-km reach, using either a single-fiber (two wavelengths) or a dual-fiber deployment, with a maximum splitting ratio of 1:32. As described earlier, BPON supports symmetric 155-Mbit/s, symmetric 622-Mbit/s, and asymmetric 622-Mbit/s–155-Mbit/s data rates. BPON differs from EPON in that it has an asymmetric framing structure: The frame formats used in the downstream and upstream differ, as shown in Figure 3.16.

The downstream (from OLT to ONU) is based on a pseudo-frame of 56 ATM cells. Each pseudo-frame consists of 54 "user" cells and two physical layer operations, administration and maintenance (PLOAM) cells. As with other PON

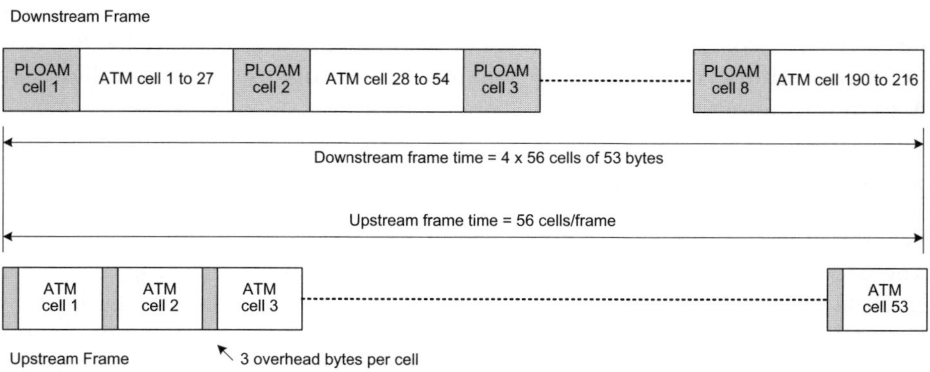

Figure 3.16. 622-Mbit/s–155-Mbit/s BPON frame format.

technologies, the downstream frame is passively replicated to all ONUs, with a particular ONU extracting cells with a matching VPI/VCI. The upstream frame format is a 56-byte cell consisting of a 3-byte preamble and a 53-byte ATM cell.

Broadband PONs arbitrate upstream bandwidth using a time division multiple access (TDMA) mechanism, where the OLT allocates timeslots to the ONU through a transmission of a GRANT PLOAM cell downstream. As part of the ONU discovery and calibration procedure, the OLT performs a "ranging" operation that estimates the round-trip time of each ONU. The OLT calculates an equalization delay that normalizes all ONUs to a delay equivalent of a uniform distance of 20 km from the OLT. The OLT provides the specific equalization delay to each ONU, thereby ensuring collision-free operation in the upstream direction.

Due to its early development, BPONs have been successfully deployed by NTT in Japan and by the former Bell South (now AT&T) in the United States, which has offered commercial services based on BPON since 1999 [34]; however, new PON deployments are based on the newer EPON and GPON technologies. Although surpassed by technology, BPONs represent an important contribution, providing many of the technical underpinnings for its successors: GPON and EPON.

3.3.3.2 Gigabit PON. FSAN and the ITU-T have continued to evolve the initial BPON to gigabit rates with the introduction of the G.984 series of recommendations standardizing gigabit PON (GPON). The general characteristics and the physical layer aspects of GPON are defined recommendations G.984.1 [20] and G.984.2 [21], respectively. These recommendations include the specification of 1.244-Gbit/s and 2.488-Gbit/s transmission rates for upstream and downstream. When included with the existing BPON data transmission rates, a total of seven rate options are available, providing the access network provider (ANP) considerable flexibility to engineer the access network. GPON continues the specification of dual or single-fiber (dual wavelength 1490 nm downstream, and 1310 nm upstream) systems. Common to other PON variants is an additional 1550-nm wavelength provided for overlay services. The GPON protocols can support splitting ratios up to 1:128 and a logical reach of 60 km, but these are physically constrained to 1:64 and 20 km due to the available optical power budget.

Recommendation G.984.3 [22] defines the transmission convergence (TC) function. Transmission convergence defines how user data are adapted into the PMD sublayer and includes network clock synchronization, framing formats, ranging procedures, and MAC layer functionality. This recommendation is also responsible for defining the basic management and control functions of the MAC, including auto-discovery, health and performance monitoring, and the configuration of optional features such as FEC and encryption. We will examine G.983.3 in more detail.

3.3.3.2.1 GPON Transmission Convergence. The transmission convergence layer specification of GPON defines the GPON TC (GTC), a mechanism

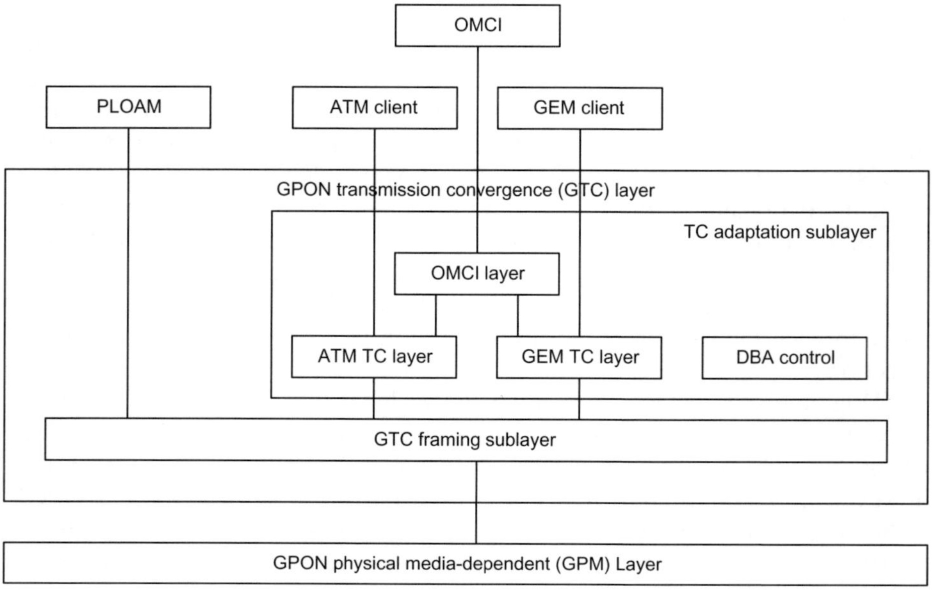

Figure 3.17. GPON transmission convergence sublayer.

for transporting user traffic across the GPON service (Figure 3.17). Currently, the standard provides both cell-based and frame-based techniques; and although both GPON and BPON support an ATM transmission convergence sublayer, these systems are not interoperable at common transmission rates. The inclusion of the Generic Encapsulation Method (GEM) addresses the prior inefficiencies with segmentation and reassembly of variable-length frames required by cell-based schemes. GEM borrows much from the Generic Framing Procedure (GFP) defined in ITU-T recommendation G.7041 [35]. GFP is a multiplexing technique that allows the mapping of variable-length user data into SONET/SDH or equivalent transport networks.

The GTC is further subdivided into two sublayers, the GTC framing sublayer and the TC adaptation sublayer. The framing sublayer is responsible for synchronization, encryption, FEC, MAC, and physical layer OAM (PLOAM). The adaptation sublayer specifies the two TC schemes, dynamic bandwidth allocation, and the definition of a management and control interface to the ONU (OMCI). Management aspects on the ONU are captured in G.984.4 [23].

GPON, similar to BPON, maintains an asymmetric framing structure, where the downstream frame format is different from the upstream format. In the downstream direction, the frame consists of a header portion [referred to as the physical control block downstream (PCBd)] and a user payload portion. The downstream frame is transmitted every 125 μs, independent of the transmission rate, resulting in a smaller user payload portion of the frame when transmitted with lower data rates. The 125-μs periodicity of the downstream transmissions

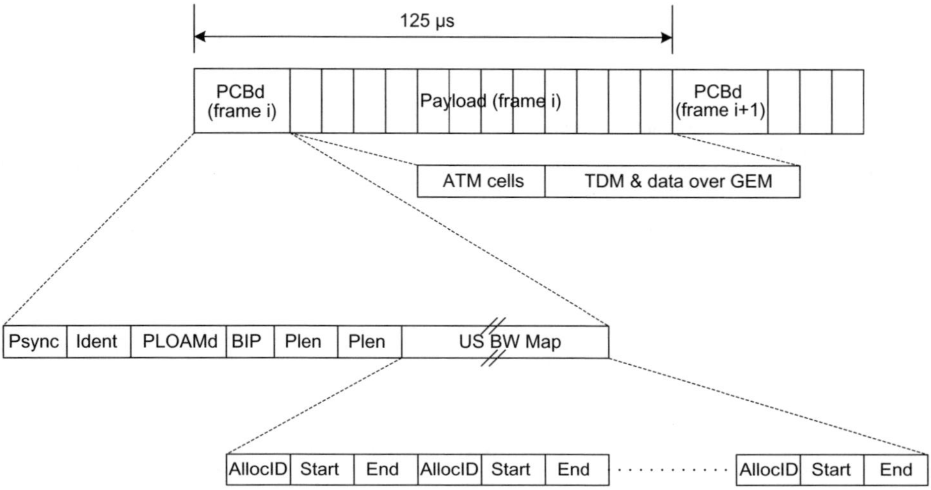

Figure 3.18. Downstream frame format.

allows the ONUs to derive an 8-kHz reference clock, which is particularly useful when interworking with TDM services. Recommendation G.984.3 includes a ranging procedure similar to BPON, where the OLT equalizes the ONU to a delay equivalent to a distance of 20 km. This equalization is provided to the ONU during the activation process.

As shown in Figure 3.18, the PCBd contains framing/preamble fields to enable physical layer synchronization and framing. This field is not scrambled. The 4-byte Ident field is used as part of the data encryption system. The PLOAM field contains a 13-byte embedded PLOAM message. Finally, the PCBd contains a scalar array of bandwidth allocations called the bandwidth map (BWmap). The BWmap contains an array of allocation structures. Each allocation structure consists of a queue identifier (alloc_ID) and pair of start and stop points. Each record is protected by a CRC that is capable of 2-bit error detection and single-bit error correction. The upstream framing structure consists of a variable number of header fields and a payload section (Figure 3.19). The upstream frame length is identical to the downstream length for all data rates. Each frame contains a number of transmissions from one or more ONUs coordinated by the downstream BWmap. The downstream payload section contains ATM cells, GEM frames, or both.

The upstream frame format provides for a variable number of headers to be included in the transmission. Flag fields contained in the allocation record of the BWmap specify the headers included in the frame. The four types of headers are:

- *Physical Layer Overhead (PLOu):* This header is mandatory and contains preamble, delimiters, and ONU identification information. The various fields in this header are protected by a bit interleaved parity (BIP).

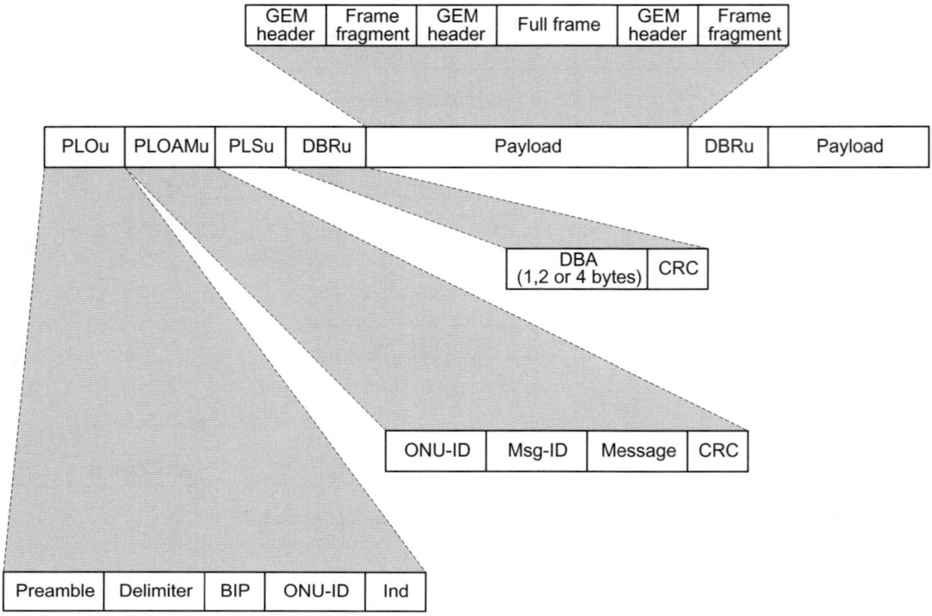

Figure 3.19. GPON upstream frame format.

- *Physical Layer Operations, Administration, and Maintenance (PLOAMu):* This header is optional, when included contains a 13-byte PLOAM message.
- *Power Level Sequence (PLSu):* This optional header is intended to allow for the adjustment of ONU power levels to reduce optical dynamic range as seen by the OLT.
- *Dynamic Bandwidth Report (DBRu):* An optional header that reports the ONUs bandwidth requirements. It may be formatted as a 1-, 2-, or 4-byte report, depending on the traffic types being reported. The reporting mode specifies a queue occupancy (in 48-byte quanta), a committed/peak rate tuple, or a combination of both.

The upstream payload section contains ATM cells, GEM frames, or DBA reports.

3.3.3.2.2 GEM Encapsulation. The Generic Encapsulation Method (GEM) is a frame-based mechanism capable of transporting Ethernet, TDM, and ATM in a single transport container using their native frame formats. A GEM frame consists of a 5-byte GEM header and user payload portion. The GEM header consists of a payload length field, a 12-bit port ID (which facilitates flow multiplexing), and a payload-type indicator (PTI). A 13-bit HEC field providing both error detection/correction capabilities and frame delineation is also included in the header.

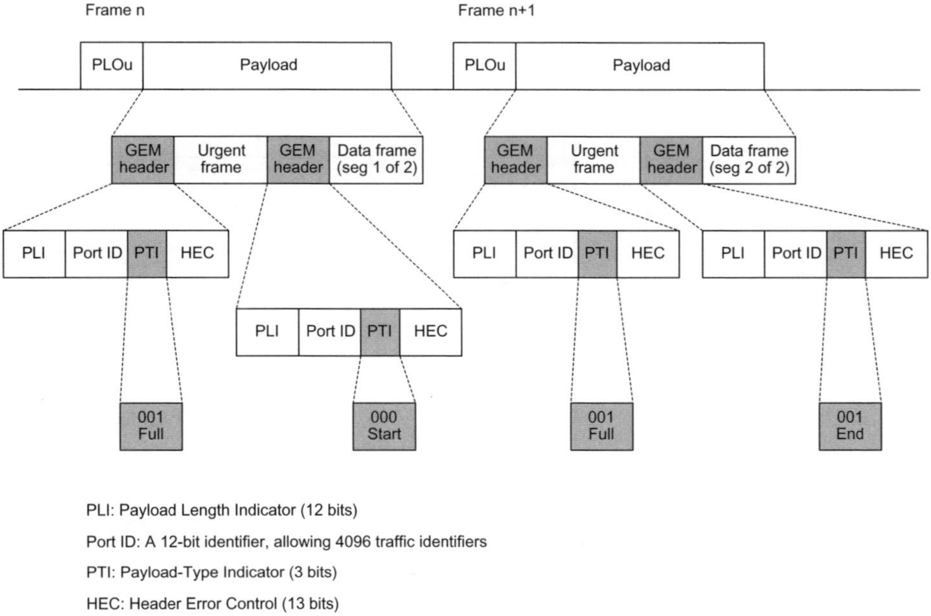

Figure 3.20. GEM header and fragmentation example.

An important distinction between EPON and GEM encapsulated Ethernet frames found in GPON is that GEM allows a client frame to be fragmented across multiple GEM payloads, whereas 802.3 precludes such fragmentation. This ability allows GEM to transport TDM while meeting the strict latency and jitter requirements by preempting non-urgent traffic with more time-sensitive TDM data. This fragmentation is shown in Figure 3.20.

GEM provides a connection oriented bidirectional communications channel identified by port ID. It is used to transport a user service flow between then ONU and OLT. GEM ports are aggregated into logical entities called traffic containers or T-Conts, which are identified by an Allocation ID (alloc-ID). Figure 3.21 represents the hierarchy of flows, ports, and containers.

As with other PON variants, the OLT is responsible for scheduling the upstream transmission in GPON with a scheme that provides for an extremely fine-grained traffic control. Rather than granting a transmission opportunity to a particular ONU, with the ONU selecting which frame to transmit from a set of pending frames, the OLT in GPON is responsible for scheduling each "traffic container" (or T-Cont) for each ONU. This fine-grained scheduling is facilitated by the ONU reporting the buffer occupancy of each T-Cont. A T-Cont is analogous to a class of service, which aggregates one or more physical queues containing traffic that requires similar treatment. The T-Cont treats the aggregate as a single entity with a specific requirement such as aggregated bandwidth, latency, or jitter. The benefits of such a fine-grained scheme are

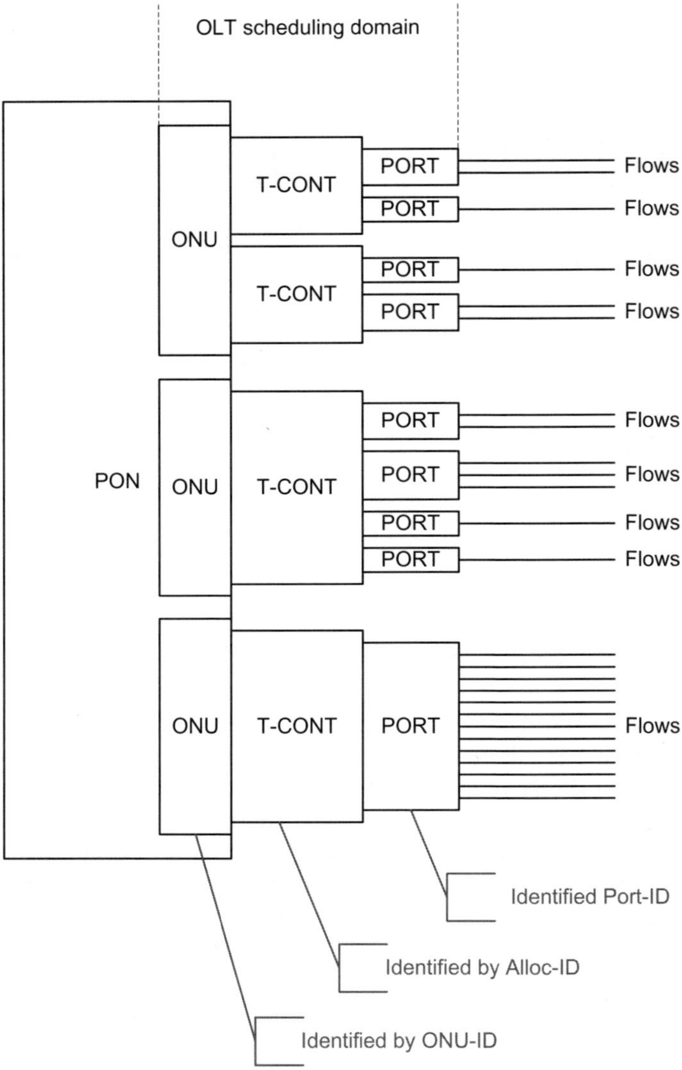

Figure 3.21. GEM traffic classification hierarchy.

improved network efficiency, reduced complexity (and therefore reduced cost) in the ONU, and the ability to provide service level agreements to all subscribers. The flipside of this approach is the potential for a significant amount of upstream and downstream bandwidth consumed for reporting and granting bandwidth for each T-Cont across all ONUs. Recommendation G.984.3 define five types of T-Conts corresponding to the different classes of service or scheduling modes:

- *TCONT1:* Legacy TDM emulation, providing unsolicited grants of bandwidth via a fixed payload size scheduled at periodic intervals. TCONT1 bandwidth is excluded from the dynamic bandwidth algorithm of the OLT.
- *TCONT2:* Intended for variable bit rate (VBR) traffic that has both a delay and throughput requirement. This TCONT is suitable for packetized voice and video.
- *TCONT3:* This traffic container is considered to be "better than best effort," meaning that it is opportunistically scheduled (it receives excess bandwidth when available) but receives a minimum bandwidth guarantee, preventing complete starvation.
- *TCONT4:* Considered as "best effort," opportunistically scheduled, with no minimum bandwidth guarantee.
- *TCONT5:* This is a superset of two or more of the other TCONTs. TCONT5 is intended to provide an aggregate bandwidth per ONU. The determination of which alloc-ID to transmit is at the discretion of the ONU.

ONUs report the amount of data waiting in a T-Cont buffer via the DBRu field in the upstream frame. The reporting mode is specific to the TCONT being reported. Mode 0 reports a single byte that represents the number of 48-byte blocks awaiting transmission. This mode is useful for best effort (TCONT4) traffic and is the only mandatory reporting mode specified by the recommendation. Reporting mode 1 is intended to report TCONT3 and TCONT5 bandwidth requirements. Mode 1 uses two bytes: The first byte represents the data requirement in terms of peak rate tokens, whereas the second byte reports the data requirements in sustained rate tokens. The ONU is required to police the incoming data rate with a token bucket policer. Mode 2 uses a 4-byte reporting record useful for TCONT5. Mode 2 encodes TCONT2 cells in the first byte. The second and third bytes contain TCONT3 peak and sustained transmission requirements, and the fourth byte contains the TCONT4 queue length. The OLT MAC performs the scheduling decision based on the upstream dynamic bandwidth reports received during the earlier epoch and long-term traffic limits applicable to each ONU, thus ensuring that all ONUs receive a minimum service even under heavily loaded operation.

3.3.4 Resource Management in PONs

The shared nature of upstream bandwidth in PONs results in challenges ensuring equitable allocation of bandwidth to ONUs while honoring service guarantees and simultaneously maintaining high network utilization. Proposing algorithms that address these challenges has been a rich research topic [27–30]. This section will introduce the concept of resource management in a PON network.

From the prior descriptions of EPON and GPON, resource management begins with resource negotiation. Both EPON and GPON allow an ONU to request upstream transmission bandwidth with the OLT granting the resource in accordance with some predefined policy. In an 802.3ah-compliant EPON, this is achieved through the REPORT/GATE MPCP message protocol exchange, and in a G.984 GPON the DBRu/BWmap mechanism accomplishes the same objective.

The second facet of resource management is selecting which ONU to grant the upstream bandwidth. This decision can be tightly coupled with the class of service that the ONU is requesting to send, as is the case with GPON, or it can be completely decoupled, as is the case with a simple round-robin scheduler. There is one aspect of resource management that is not completely specified in either EPON or GPON standards: the precise allocation of bandwidth to an ONU or class of service traffic of a particular ONU. This is referred to as dynamic bandwidth allocation (DBA). DBA algorithms take into consideration the amount and type of data buffered in the ONU awaiting transmission and allows the OLT to dynamically alter the amount of bandwidth granted to an ONU to service this buffered data. DBA algorithms are an important mechanism to improve upstream network utilization, improving throughput. Access networks are bursty in nature due to the relatively modest amount of traffic aggregation due to the limited number of subscribers. This is unlike metro and backbone networks that benefit from aggregating a large number of independent traffic sources, effectively "smoothing" the cumulative offered load. Examining conventional Ethernet traffic profiles [36] demonstrates that individual traffic sources are extremely bursty, with a self-similar nature, resulting in considerably variable bandwidth requirements over time. Static allocation of upstream bandwidths tends to underutilize links, resulting in poor throughput, increased packet latency, and potentially packet loss even at low utilizations. Employing DBA algorithms results in the network adapting to the instantaneous bandwidth requirements, allowing a greater degree of statistical multiplexing and hence higher network utilization resulting in higher throughput, lower latency, and packet loss.

3.3.4.1 IPACT. A common statistical multiplexing algorithm described in the literature [28, 30] is Interleaved Polling with Adaptive Cycle Time (IPACT). IPACT polls ONUs individually and issues transmission grants in a round-robin fashion. The grant window is equal to the backlog from the previous reporting epoch, thereby ensuring that the bandwidth is allocated dynamically based on queue occupancy. To prevent a single ONU from a monopolizing upstream bandwidth, an ONU is assigned a maximum transmission window (MTW). Once the ONU has exhausted its MTW, it is not granted any further bandwidth until the next polling cycle. There are a number of algorithmic variants that differ in the treatment of MTW:

- *Fixed:* The DBA ignores the requested window size, instead granting MTW with a constant polling interval.
- *Limited:* The DBA grants the requested window size, up to the MTW limit.

- *Credit:* The DBA grants the requested window size plus a constant credit that is proportional to the window size.
- *Elastic:* The DBA attempts to overcome the limitation of only granting an MTW per polling cycle. The OLT integrates over successive polling cycles; this ensures that over the last N grants, the assigned bandwidth does not exceed $N \times$ MTW, where N is equal to the number of ONUs.

3.3.4.2 Class of Service Schemes. Other scheduling algorithms attempt to take into consideration the type of traffic that the ONU is sending. In some circumstances, providing an aggregate bandwidth to an ONU may not be sufficient to ensure that all traffic receives the treatment required. For example, voice traffic imposes a strict delay budget on the access network: ITU-T Recommendation G.114, One-Way Transmission Time [37], specifies a 1.5-ms one-way propagation delay in the access network. Irrespective of the amount of bandwidth an ONU is granted, if it can be delayed for greater than 1.5 ms, voice quality could be impacted. Families of scheduling algorithms that consider not only bandwidth but also latency, jitter, and packet loss requirements, can be classified as providing either absolute Quality of Service (QoS) or relative QoS. Absolute assurances are quantitative and quantify the SLA requirements of the traffic in terms of bandwidth, latency, and packet loss ratio. Relative QoS assurances characterize the service guarantees in qualitative terms such as "low loss," "low latency," and whether bandwidth is assured. The various classes are scheduled in such a manner that their relative importance is preserved. In general terms, the absolute assurance schemes can be complex and may involve elements of admission control and per class rate control and shaping.

Although much valuable research has been conducted in this area, it is important to step back and analyze the need for such sophisticated techniques in the upstream direction, particularly in the context of residential broadband access networks. In the downstream direction, services for voice (low latency and low bandwidth), video (assured bandwidth with low loss), and a best-effort Internet access are clearly required. In the upstream direction, a simple high-priority service (for voice) and a best-effort Internet access service would seem sufficient for most subscriber services. Other often-cited classes of service for applications such as gaming and consumer-oriented video-conferencing offer marginal benefit and even more marginal revenue potential. Admittedly, business-oriented services may require more elaborate upstream bandwidth allocation to cater for TDM-based PBX traffic and a diversity in data traffic classes of service. These scenarios may warrant the additional complexity and cost associated with sophisticated upstream QoS mechanisms, allowing a PON to offer true service replacement.

3.3.5 Future Trends

Passive optical networks promise orders of magnitude more bandwidth than current day copper-based access networks; however, work is well underway examining future areas for expansion.

TDM-based PON networks can scale in multiple dimensions. Greater optical power budgets would yield higher splitting ratios and/or longer physical reach, greatly expanding the number of customers served by a single fiber. This is an important factor in determining the economic viability of a PON network, because it allows the cost to be amortized over a larger subscriber base, reducing the cost per subscriber. Amplification is another way of achieving the same goals, but with the loss of a valuable property of the optical distribution network (ODN)—its passive nature. Another scaling dimension is rate. International standards bodies are busy working on defining higher rate PONs, the IEEE is defining a 10-Gbit/s Ethernet PON standard through the P802.3av working group [25] with the goal of providing both a symmetric 10-Gbit/s and asymmetric 10-Gbit/s downstream–1-Gbit/s upstream network architecture. The ITU-T is considering similar extensions to the GPON recommendations to increase rates to 10 Gbit/s and beyond. Incrementally evolving the PON through rate increases preserves the investment in the current ODN and leverages its passive nature and rate independence, requiring only an upgrade to the optoelectronics of the OLT and ONUs.

Wave division multiplex (WDM) PONs [38] scale past the limits offered by the TDM PONs of today, as well as past the proposed higher-speed PONs currently in specification. WDM-PONs promise access to much greater bandwidths. In WDM-PONs, each ONU/OLT is assigned an individual wavelength pair to use, effectively creating a point-to-point link. Wavelength separation offers enhanced privacy and could provide splitting ratio's almost 10× what is achievable today with a reach in excess of 100 km. The challenges to be overcome involve developing cost efficient and reliable tunable optics.

PON networks have scaling potential in multiple dimensions, making them ideal candidates for next-generation access networks—networks that need to evolve with new and as yet unknown applications over a period of decades. This is a tough challenge indeed, but WDM-PONs and hybrid TDM/WDM architectures promise practically unlimited bandwidth. The future is bright indeed—optically speaking of course!

3.4 HYBRID FIBER COAXIAL NETWORKS

Community antenna television (CATV) operators have transitioned their core business from a supplier of video programming to a full service supplier of video, voice, and data telecommunication services. This has necessitated a technical evolution of the cable distribution network to include fiber distribution deeper into the network and the significant upgrade of what was originally a unidirectional broadcast network into a bidirectional multiservice network. Cable network operators [known as multiple system operators (MSOs)] have aggressively expanded their network to accommodate the distribution of multiple hundreds of HDTV channels, video on demand, personalized video services, and increased HSI access in response to competitive pressures from the telecom operators (telcos). In this section we will examine the fundamentals of the hybrid fiber coaxial (HFC) access network, with particular attention to DOCSIS 3.0 [39–42], the latest in a series of standards governing data transmission over cable.

3.4.1 Hybrid Fiber Coaxial Network Architecture

The CATV network was essentially a unidirectional network capable of broadcasting video. The network was tailored specifically to accommodate regional broadcast standards—in particular, radio-frequency (RF) transmission with channel characteristics that closely followed the regional terrestrial, over-the-air broadcasting technology; in North America, this translated to 6-MHz NTSC channels. In order to improve network reach and signal quality, fiber, with its low-loss properties was introduced to distribute the RF signal from the head end (HE). This architecture became known as the hybrid fiber coaxial (HFC) network and is shown in Figure 3.22.

In the above network, the HE acquires video content through standard satellite, terrestrial over-the-air, or other direct feeds. This is then mixed with local content and modulated in analog form to be transmitted via the fiber distribution network. Individual fibers terminate at fiber nodes that service a residential area compromising a few hundred to a few thousand residences. The fiber node performs the optical to electrical domain conversion and re-broadcasts the RF analog signal to the coaxial cable segments. Fiber is effectively lossless in the megahertz to gigahertz frequency range; however, coaxial is not, thus requiring amplifiers to ensure that all residences receive the signal with acceptable quality.

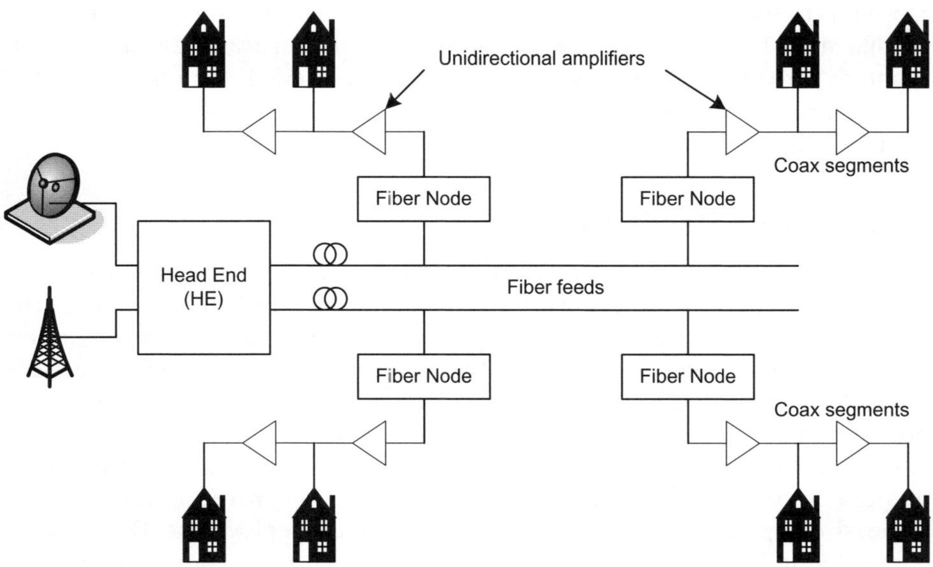

Figure 3.22. Unidirectional hybrid fiber coaxial network architecture.

The HFC network provided a total downstream spectrum of 54–550 MHz for video broadcast.

In order to provide full duplex communication, the HFC was upgraded to support bidirectional communications. The fiber portion of the network can use physically separate strands to support the upstream communication, but a single coaxial segment must accommodate both upstream and downstream signals. Frequency division duplexing separates a frequency band from 5 MHz to 42 MHz for upstream communication. In addition, the coaxial segment amplifiers are upgraded to support bidirectional amplification. In the downstream direction a 6-MHz channel is reserved for data. Using a 64-symbol QAM modulation technique, 30 Mbit/s of bandwidth is available for high-speed data. In the upstream direction, more robust (and lower bandwidth) modulation techniques, such as Quaternary Phase Shift Key (QPSK) and 16-symbol QAM, are used due to the noisy nature of the upstream 5- to 42-MHz frequency range.

Apart for network upgrades, additional equipment is required to enable full duplex communications in a HFC network. Cable modems (CMs) are customer premise equipment that convert digital information (typically Ethernet frames) into modulated RF signals in the upstream direction and convert the RF signal to digital information in the downstream direction. The Cable Modem Termination System (CMTS) located in the HE performs the converse operation. This involves upconverting the 6-MHz channel, combining with the other channels and converting to and from the optical domain. Figure 3.23 illustrates a HFC network that supports digital data transmission.

Given that all bandwidth in the HFC is shared, upstream transmissions must be coordinated. The bidirectional amplifiers in the coaxial segments prevent individual CMs from detecting collisions, precluding the use of carrier sense mechanisms such as CSMA/CD. A link layer TDMA mechanism similar in nature to that used in PONs provides collision-free upstream transmission. The HE regularly broadcasts solicitation messages. Newly added stations respond, a ranging procedure determines a latency offset, and the HE allocates a timeslot for the CM transmission.

3.4.2 DOCSIS Standards

The Data Over Cable Service Interface Specification (DOCSIS) standards specify the physical, MAC, and network layer operation for the CM and CMTS. The DOCSIS standards also include specification of security (protecting subscribers data in a shared bandwidth environment) and network management services. DOCSIS 1.0, released in 1997 [43], provided basic Internet connectivity. It supported 64 and 256 QAM downstream modulation schemes and QPSK and 16 QAM modulation for upstream channels. The value of the initial specification was the enforced equipment standardization, which improved interoperability, lowered equipment costs, and resulted in increased deployment. DOCSIS 1.1 completed in 2001 [44] improved security and added QOS capability, allowing MSOs to offer different services over DOCSIS such as voice and gaming. DOCSIS

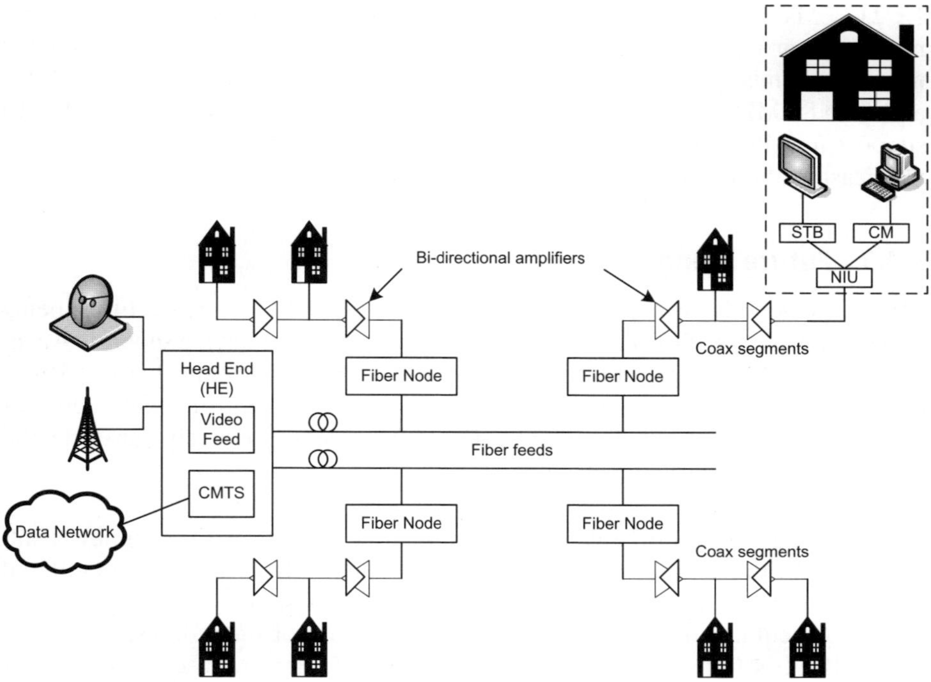

Figure 3.23. Bidirectional hybrid fiber coaxial network architecture.

2.0, introduced in 2002 [45], increased throughput (offering up to 30 Mbit/s upstream and 50 Mbit/s downstream) and provided a greater option of QAM modulation schemes for upstream channels. The latest version of DOCSIS (version 3.0) [39–42] provides channel bonding to increase upstream and downstream bandwidth in order to compete with VDSL/VDLS2 and FTTx networks. With bonding of four channels, 120-Mbit/s upstream and 160-Mbit/s downstream bandwidths are possible. Higher bandwidths are achievable with a greater number of channels. Channel bonding is a logical concatenation of multiple RF channels. In the downstream direction, the multiplexing is at the packet level, with individual packets contained within a single RF channel. In the upstream direction, the packet is "stripped" across the available channels in the group. For example, if a CM has a 1000-byte Ethernet packet to send, it requests a 1000-byte timeslot from the CMTS. In the bonded case, the CMTS responds with grants for segments of the packet across the bonded channels, which may grant a 500-byte timeslot on upstream channel #1, a 200-byte timeslot on upstream channel #2, and a 300-byte timeslot on upstream channel #3. The CM segments the packet into 500-, 200-, and 300-byte fragments and transmits the segments on the three upstream channels. Each segment contains a sequence number and a pointer to allow the CMTS to perform the necessary reassembly, reconstructing the original 1000-byte frame. Channel bonding in DOCSIS 3.0 is backwardly compatible with DOCSIS

1.x/2.0 CMs, allowing a seamless migration to the higher bandwidths. Other notable improvements included in DOCSIS 3.0 are the physically switchable upstream bands, including (a) the standard 5- to 42-MHz and the 5- to 65-MHz and 5- to 85-MHz frequency plans, providing enhanced upstream bandwidth, (b) support for IPv6, QOS support for IP multicast, and subscriber tracking of IP multicast in the CMTS, and (c) enhanced security.

3.4.3 Future Trends

MSOs are keenly aware of the threat posed by FTTH architectures being deployed by telcos. FTTH networks have the advantage of having separate wavelengths for video broadcast (1550 nm) and upstream/downstream (1310/1490 nm) data. The ability for the MSO to deliver similar bandwidth services requires further evolution of the HFC network. MSOs are aggressively pursuing the following evolutionary enhancements:

- Analog bandwidth reclamation reduces the number of TV channels available in the analog band. These are the channels that can be demodulated and received using the tuner built into most television sets. Analog reclamation requires digital encoding and compression of the video signal into a MPEG-4 transport. Decoding requires a customer premise set-top box or equivalent digital decoder function integrated into the television.
- Switched digital video (SDV) moves from a broadcast distribution to a selective multicast distribution model. In SDV, only channels that are being watched are delivered to a group of subscribers (known as a service group).
- Increasing the available spectrum of the plant. The spectrum available in the HFC has been progressively increased from 550 MHz to 750 MHz and presently 860 MHz. Activities are underway that increase this to 1002 MHz (1.002 GHz).
- Simplification of the frequency band plan by consolidating all specialty services, such as digital standard definition TV, video on demand, and high-speed Internet access into a single wide-band (>200 MHz) channel and delivering these services over a consolidated IP transport. Such an architecture will require sophisticated IP layer traffic management capabilities in the CMTS and CM to ensure that services are delivered with the required QOS.

Hybrid fiber coaxial networks and the MSOs that operate them have an incumbent position in the marketplace, delivering the video services of today into many households worldwide. Nevertheless, they are being forced to upgrade their networks in order to offer the raw bandwidths and services promised by next-generation telco networks. Given the financial imperative to survive and compete, HFC networks will continue to be an important aspect of wireline access networks, providing high bandwidth and rich services to many subscribers.

3.5 SUMMARY

Despite their technical differences, wireline access networks have evolved from a physical layer connection for a single service, to become a sophisticated multi-service network in which all traffic, including voice and video, use the Internet Protocol as a converged transport layer. In order to support this evolution, the wireline access networks have focused on increasing the physical layer bandwidth available to subscribers and increasing network intelligence, ensuring that each application or service, when transported over IP, receives the required quality of service necessary to meet the user's expectations. The next-generation broadband wireline access network will deliver hundreds of megabits and potentially gigabits of bandwidth into each home and business. The availability of this bandwidth will enable services and applications not possible or simply not even envisaged today.

REFERENCES

1. US Patent and Trade Mark Office, Patent # 244,426, A. G. Bell, July 19, 1881.
2. Chris Hellberg et al., Broadband Network Architectures: Designing and Deploying Triple Play Services, Prentice-Hall, 2007.
3. J. Cioffi et al., CuPON: The copper alternative to PON 100 Gb/s DSL networks, *IEEE Commun. Mag.*, Vol. 45, No. 6, pp. 132–139, June 2007.
4. D. Waring, J. Lechleider, and T. Hsing, Digital subscriber line technology facilitates a graceful transition from copper to fiber, *IEEE Commun. Mag.*, Vol. 29, No. 3, pp. 96–104, March 1991.
5. J. Cioffi and G. Ginis, Vectored transmission for digital subscriber line systems, *IEEE JSAC Special Issue on Twisted Pair Transmissions*, Vol. 20, No. 5, pp. 1085–1104, March 2001.
6. ITU-T, ITU-T Recommendation G.992.2: Single-Pair High-Speed Digital Subscriber Line (SHDSL) Transceivers, ITU-T, December 2003.
7. ITU-T, ITU-T Recommendation G.992.1: Asymmetric Digital Subscriber Line (ADSL) Transceivers, July 1999.
8. ITU-T, ITU-T Recommendation G.992.3: Asymmetric Digital Subscriber Line Transceivers 2 (ADSL2), January 2005.
9. ITU-T, ITU-T Recommendation G.992.5: Asymmetric Digital Subscriber Line (ADSL) Transceivers—Extended Bandwidth ADSL2 (ADSL2+), January 2009.
10. ITU-T, ITU-T Recommendation G.993.1: Very High Speed Digital Subscriber Line Transceivers, June 2004.
11. ITU-T, ITU-T Recommendation G.993.2: Very High Speed Digital Subscriber Line Transceivers 2 (VDSL2), February 2006.
12. Broadband Forum, DSL Technology Evolution—ADSL2/ADSL2plus/ADSL-RE/VDSL2, http://www.broadband-forum.org/downloads/About_DSL.pdf.
13. Broadband Forum, www.broadband-forum.com.
14. Broadband Forum, Technical Report TR-101, Migration to Ethernet-Based DSL Aggregation, April 2006.

15. IETF, RFC 2516—A Method for Transmitting PPP Over Ethernet (PPPoE), February 1999.
16. IETF, RFC 2661—Layer Two Tunneling Protocol "L2TP," August 1999.
17. Telenor Group, dynamic spectrum management—A methodology for providing significantly higher broadband capacity to the users, *Telektronikk*, No. 4, pp. 126–137, 2004.
18. L Hutcheson, FTTx: Current status and the future, *IEEE Commun. Mag.*, July 2008.
19. ITU-T, ITU-T Recommendation G.983.1: Broadband Optical Access Systems Based on Passive Optical Networks (PON), January 2005.
20. ITU-T, ITU-T Recommendation G. 984.1: Gigabit-Capable Passive Optical Networks (GPON): General Characteristics, March 2003.
21. ITU-T, ITU-T Recommendation G.984.2: Gigabit-Capable Passive Optical Networks (GPON): Physical Media Dependent (PMD) Layer Specification, March 2003.
22. ITU-T, ITU-T Recommendation G. 984.3: Gigabit-Capable Passive Optical Networks (GPON): Transmission Convergence Layer Specification, February 2004.
23. ITU-T, ITU-T Recommendation G. 984.4: Gigabit-Capable Passive Optical Networks (GPON): ONT Management and Control Interface Specification, June 2004.
24. IEEE, IEEE 802.3 Local and Metropolitan Area Networks—Specific Requirements, Part 3: Carrier Sense Multiple Access with Collision Detection (CSMA/CD) Access Method and Physical Layer Specifications, December 2005.
25. IEEE P802.3av task force (http://www.ieee802.org/3/av/).
26. IEEE, Ethernet PON (EPON) and the PAR + 5 Criteria, May 2001, www.ieee802.org/3/efm/public/may01/pesavento_1_0501.pdf.
27. G. Kramer, *Ethernet Passive Optical Networks*, McGraw-Hill, New York, 2005.
28. Y. Luo et al., Resource management for broadband access over time-division multiplexed passive optical networks, *IEEE Network*, Vol. 21, No. 5, pp. 20–27, September/October 2007.
29. J. Angelopoulos et al., Efficient transport of packets with QoS in an FSAN-aligned GPON, *IEEE Commun. Mag.*, Vol. 42, No. 2, pp. 92–98, February 2004.
30. M. McGarry, M. Maier, and M. Reisslein, Ethernet PONs: A survey of dynamic bandwidth allocation (DBA) algorithms, *IEEE Optical Commun. Mag., Special Supplement Optical Communications*, Vol. 42, No. 8, pp. 8–15, August 2004.
31. PMC-Sierra Inc. PAS6301 device, www.pmc-sierra.com.
32. Teknovus Inc. TK3701 device, www.teknovus.com.
33. Full Service Access Network (FSAN), http://www.fsanweb.org/.
34. H. Ueda et al., Deployment status and common technical specifications for a B-PON system, *IEEE Commun. Mag.*, Vol. 39, No. 12, pp. 134–141, December 2001.
35. ITU-T, ITU-T Recommendation, G.7041: Generic Framing Procedure (GFP), February 2003.
36. W. Leland, M. Taqqu, W. Willinger, and D. Wilson, On the self-similar nature of Ethernet traffic (extended version), *IEEE/ACM Trans. Networking*, Vol. 2, No. 1, pp. 1–15, February 1994.
37. ITU-T, ITU-T Recommendation, G.114: One-Way Transmission Time, May 2003.
38. K. Grobe and J.-P. Elbers, PON in adolescence: From TDMA to WDM-PON, *IEEE Commun. Mag.*, Vol. 46, No. 1, pp. 26–34, January 2008.

39. CableLabs, Data-over-cable service interface specifications, cable modem to customer premise equipment, interface specification, CM-SP-CMCIv3.0-I01-080320, March 2008.
40. CableLabs, Data over cable service interface specifications, DOCSIS 3.0, physical layer specification, CM-SP-PHYv3.0-I08-090121, January 2009.
41. CableLabs, Data-over-cable service interface specifications, DOCSIS 3.0, MAC and upper layer protocols interface specification, CM-SP-MULPIv3.0-I09-090121, January 2009.
42. CableLabs, Data-over-cable service interface specifications, DOCSIS 3.0, security specification CM-SP-SECv3.0-I09-090121, January 2009.
43. CableLabs, DOCSIS 1.0, http://www.cablemodem.com/specifications/specifications10.html.
44. CableLabs, DOCSIS 1.1, http://www.cablemodem.com/specifications/specifications11.html.
45. CableLabs, DOCSIS 2.0, http://www.cablemodem.com/specifications/specifications20.html.

4

FIBER–WIRELESS (FIWI) NETWORKS: TECHNOLOGIES, ARCHITECTURES, AND FUTURE CHALLENGES

Navid Ghazisaidi and Martin Maier

4.1 INTRODUCTION

We are currently witnessing a strong worldwide push toward bringing optical fiber closer to individual homes and businesses, leading to fiber to the home/fiber to the premises (FTTH/FTTP) networks [1]. In FTTx networks, fiber is brought close or all the way to the end user, where x denotes the discontinuity between optical fiber and some other, either wired or wireless, transmission medium. For instance, cable operators typically deploy hybrid fiber coaxial (HFC) networks where fiber is used to build the feeder network while the distribution network is realized with coaxial cables. Another good example for wired fiber-copper access networks are hybrid-fiber twisted-pair networks that are widely deployed by telephone companies to realize different variants of digital subscriber line (DSL) broadband access solutions.

From a capacity point of view, one might seriously argue that there is no techno-economic need and justification to replace hybrid-fiber twisted-pair-based DSL networks with all-optical solutions—for example, passive optical networks (PONs). According to Cioffi et al. [2], the so-called copper-PON (CuPON) multidropping DSL architecture is able to provide 50 Gbit/s of shared

Convergence of Mobile and Stationary Next-Generation Networks, edited by Krzysztof Iniewski

bandwidth in each direction on existing twisted pair of copper telephone lines through exploitation of all modes of crosstalk. Thus, CuPON is able to offer much higher data rates than state-of-the-art standardized access network solutions [e.g., IEEE 802.3ah Ethernet PON (EPON) and ITU-T G.984 Gigabit PON (GPON)] without requiring any costly replacement of widely installed twisted pairs by fiber. Note, however, that the speed of CuPON is higher than that of current fiber PONs not because copper has a wider bandwidth than fiber, but because current fiber PONs do not use their extra bandwidth. In fact, optical fiber provides an unprecedented bandwidth potential that is far in excess of any other known transmission medium. A single strand of fiber offers a total bandwidth of 25,000 GHz. To put this potential into perspective, it is worthwhile to note that the total bandwidth of radio on the planet Earth is not more than 25 GHz [3]. Besides huge bandwidth, optical fiber has some further advantageous properties such as low attenuation, longevity, and low maintenance costs that will eventually render fiber the medium of choice in wired first/last mile access networks. This trend can be observed in most of today's greenfield deployments where fiber rather than copper cables are installed for broadband access. On the other hand, in brownfield deployments it is important that installation costs, which largely contribute to overall costs of access networks, be reduced. A promising example for cutting installation costs is NTT's do-it-yourself (DIY) installation of FTTH optical network units (ONUs) deploying a user-friendly hole-assisted fiber that exhibits negligible loss increase and sufficient reliability, even when it is bent at right angles, clinched, or knotted, and can be mass produced economically [4]. Another interesting enabling technology is the so-called plastic optical fiber (POF), which is well-suited for simple wiring of low-cost optical home networks. POF provides consumers with user-friendly terminations, easy installation, and tolerance of dirty connections. Furthermore, POF's resistance to bending is comparable to that of twisted pair of copper telephone lines. An interesting application of POF-based networks is the concept of "Fiber to the Display," where POFs are directly connected to a large flat panel display to enable transmission rates of several gigabits per second in support of telemedicine or the emerging digital cinema standard for next-generation cinema [5].

FTTH networks are expected to become the next major success story for optical communications systems [6]. Future FTTH networks will not only enable the support of a wide range of new and emerging services and applications, but will also unleash their economic potential and societal benefits by opening up the first/last mile bandwidth bottleneck between bandwidth-hungry end users and high-speed backbone networks [7]. In this chapter, we assume that optical fiber paves all the way to and penetrates into the home and offices of residential and business customers. Arguing that due to its unique properties optical fiber is likely to entirely replace copper wires in the near- to mid-term, we will elaborate on the final frontier of optical networks, namely, the convergence with their wireless counterparts. Optical and wireless technologies can be thought of as quite complementary and will expectedly coexist over the next decades. Future broadband access networks will be bimodal, capitalizing on the respective strengths of both technologies and smartly merging them in order to realize future-proof fiber-wireless (FiWi) networks that strengthen our information society while avoiding its digital divide. By combining the capacity of optical fiber networks with the ubiquity and mobility of wireless networks, FiWi networks form a powerful platform for the support and creation of emerging as well as future unforeseen applications and services (e.g., telepresence). FiWi networks represent a powerful enabling technology for emerging fixed mobile converged (FMC) networks that enable seamless handoffs across (optical) wired and wireless networks [8]. Apart from their huge bandwidth, optical fibers provide transparency against modulation formats and protocols and are able to support a wide range of current and future wired and wireless standards. FiWi networks hold great promise to change the way we live and work by replacing commuting with teleworking. This not only provides more time for professional and personal activities for corporate and our own personal benefit, but also helps reduce fuel consumption and protect the environment, issues that are becoming increasingly important in our lives.

The remainder of this chapter is structured as follows. In Section 4.2, we set the stage by briefly reviewing radio-over-fiber (RoF) networks, a previously studied approach to integrate optical fiber networks and wireless networks, and explain their difference with regard to so-called radio-and-fiber (R&F) networks. Section 4.3 elaborates on enabling technologies of FiWi networks. In Section 4.4, we describe the state-of-the-art of FiWi network architectures. Section 4.5 covers the techno-economic comparison of two major optical and wireless enabling FiWi technologies. Finally, future challenges and imperatives of FiWi networks are discussed in Section 4.6. Section 4.7 concludes the chapter.

4.2 RoF VERSUS R&F FiWi NETWORKS

RoF networks have been studied for many years as an approach to integrate optical fiber and wireless networks. In RoF networks, radio frequencies (RFs) are carried over optical fiber links between a central station and multiple low-cost remote antenna units (RAUs) in support of a variety of wireless applications. For instance, a distributed antenna system connected to the base station of a micro-cellular radio system via optical fibers was proposed in Cha and Gans [9]. To efficiently support time-varying traffic between the central station and its attached base stations, a centralized dynamic channel assignment method is applied at the central station of the proposed fiber-optic microcellular radio system. To avoid having to equip each radio port in a fiber-optic microcellular radio network with a laser and its associated circuit to control the laser parameters such as temperature, output power, and linearity, a cost-effective radio port architecture deploying remote modulation can be used [10].

Apart from realizing low-cost microcellular radio networks, optical fibers can also be used to support a wide variety of other radio signals. RoF networks are

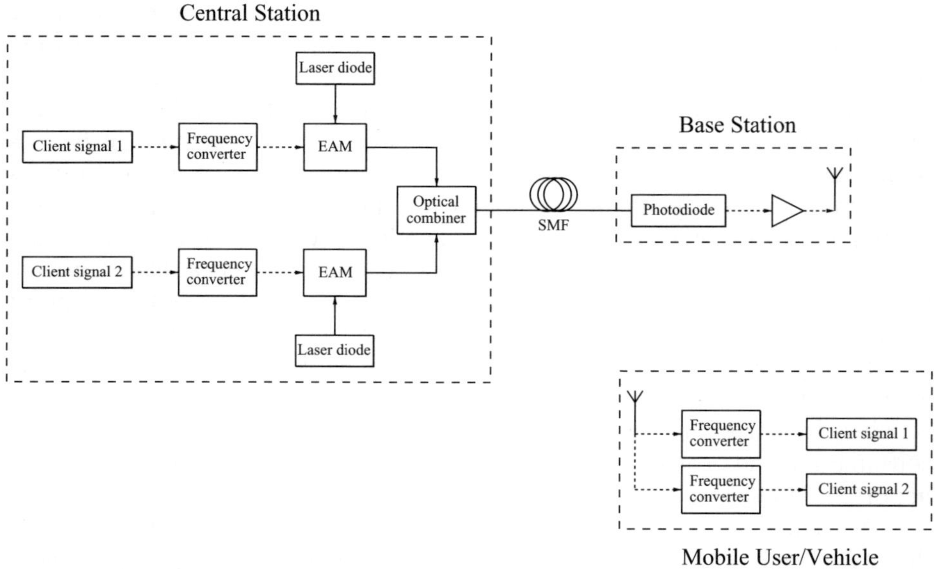

Figure 4.1. Radio-over-SMF network downlink using EAMs for different radio client signals [11].

attractive because they provide transparency against modulation techniques and are able to support various digital formats and wireless standards in a cost-effective manner. It was experimentally demonstrated in Tang et al. [11] that RoF networks are well-suited to simultaneously transmit wideband code division multiple access (WCDMA), IEEE 802.11a/g wireless local area network (WLAN), personal handyphone system (PHS), and global system for mobile communications (GSM) signals. Figure 4.1 illustrates the method investigated in Tang et al. [11] for two different radio client signals transmitted by the central station on a single-mode fiber (SMF) downlink to a base station and onward to a mobile user or vehicle. At the central station, both radio client signals are first upconverted to a higher frequency by using a frequency converter. Then the two RF signals go into two different electroabsorption modulators (EAMs) and modulate the optical carrier wavelength emitted by two separate laser diodes. An optical combiner combines the two optical signals onto the SMF downlink. At the base station, a photodiode converts the incoming optical signal to the electrical domain and radiates the amplified signal through an antenna to a mobile user or vehicle that uses two separate frequency converters to retrieve the two different radio client signals.

While SMFs are typically found in outdoor optical networks, many buildings have preinstalled multimode fiber (MMF) cables. Cost-effective MMF-based networks can be realized by deploying low-cost vertical cavity surface emitting lasers (VCSELs). In Lethien et al. [12], different kinds of MMF in conjunction with commercial off-the-shelf (COTS) components were experimentally tested

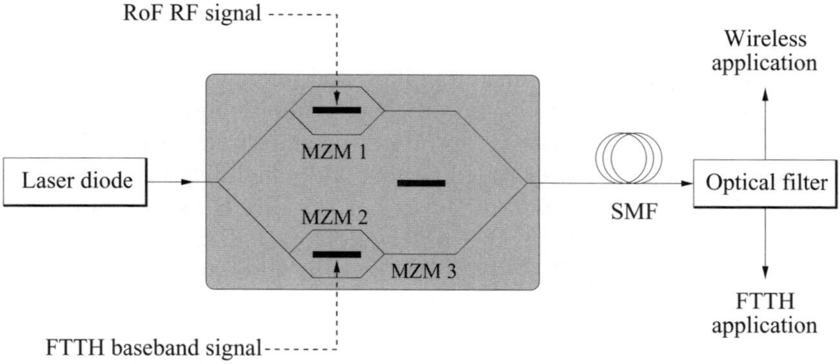

Figure 4.2. Simultaneous modulation and transmission of FTTH baseband signal and RoF RF signal using an external integrated modulator consisting of three Mach–Zehnder modulators (MZMs) [13].

to demonstrate the feasibility of indoor radio-over-MMF networks for the in-building coverage of second-generation (GSM) and third-generation cellular radio networks [universal mobile telecommunications system (UMTS)] as well as IEEE 802.11a/b/g WLAN and digital enhanced cordless telecommunication packet radio service (DECT PRS).

To realize future multiservice access networks, it is important to integrate RoF systems with existing optical access networks. In Lin et al. [13], a novel approach for simultaneous modulation and transmission of both RoF RF and FTTH baseband signals using a single external integrated modulator was experimentally demonstrated, as shown in Figure 4.2. The external integrated modulator consists of three different Mach–Zehnder modulators (MZMs) 1, 2, and 3. MZM 1 and MZM 2 are embedded in the two arms of MZM 3. The RoF RF and FTTH baseband signals independently modulate the optical carrier generated by a common laser diode by using MZM 1 and MZM 2, respectively. Subsequently, the optical wireless RF and wired-line baseband signals are combined at MZM 3. After propagation over an SMF downlink, an optical filter (e.g., fiber grating) is used to separate the two signals and forward them to the wireless and FTTH application, respectively. It was experimentally demonstrated that a 1.25-Gbit/s baseband signal and a 20-GHz 622-Mbit/s RF signal can be simultaneously modulated and transmitted over 50-km standard SMF with acceptable performance penalties.

The aforementioned research projects successfully demonstrated the feasibility and maturity of low-cost multiservice RoF networks. Their focus was on the investigation of RoF transmission characteristics and modulation techniques, considering primarily physical-layer-related performance metrics [e.g., power penalty, error vector magnitude (EVM)] and bit error rate (BER) measurements. It was shown that RoF networks can have an optical fiber range of up to 50 km. However, inserting an optical distribution system in wireless networks may have a major impact on the performance of medium access control (MAC) protocols

[14]. The additional propagation delay may exceed certain timeouts of wireless MAC protocols, resulting in a deteriorated network performance. More precisely, MAC protocols based on centralized polling and scheduling (e.g., IEEE 802.16 WiMAX) are less affected by increased propagation delays due to their ability to take longer walk times between central station and wireless subscriber stations into account by means of interleaved polling and scheduling of upstream transmissions originating from different subscriber stations. However, in distributed MAC protocols—for example, the widely deployed distributed coordination function (DCF) in IEEE 802.11a/b/g WLANs—the additional propagation delay between wireless stations and access point poses severe challenges. To see this, note that in WLANs a source station starts a timer after each frame transmission and waits for the acknowledgment (ACK) from the destination station. By default the ACK timeout value is set to 9 μs and 20 μs in 802.11a/g and 802.11b WLAN networks, respectively. If the source station does not receive the ACK before the ACK timeout, it will resend the frame for a certain number of retransmission attempts. Clearly, one solution to compensate for the additional fiber propagation delay is to increase the ACK timeout. Note, however, that in DCF the ACK timeout must not exceed the DCF interframe space (DIFS), which prevents other stations from accessing the wireless medium and thus avoiding collision with the ACK frame (in IEEE 802.11 WLAN specifications, DIFS is set to 50 μs). Due to the ACK timeout, optical fiber can be deployed in WLAN-based RoF networks only up to a maximum length. For instance, it was shown in Kalantarisabet and Mitchell [15] that in a standard 802.11b WLAN network the fiber length must be less than 1948 m to ensure the proper operation of DCF. In addition, it was shown that there is a tradeoff between fiber length and network throughput. As more fiber is deployed, the network throughput decreases gradually.

The aforementioned limitations of WLAN-based RoF networks can be avoided in so-called radio-and-fiber (R&F) networks [16]. While RoF networks use optical fiber as an analog transmission medium between a central control station and one or more RAUs with the central station being in charge of controlling access to both optical and wireless media, in R&F networks access to the optical and wireless media is controlled separately from each other by using in general two different MAC protocols in the optical and wireless media, with protocol translation taking place at their interface. As a consequence, wireless MAC frames do not have to travel along the optical fiber to be processed at the central control station, but simply traverse their associated access point and remain in the WLAN. In WLAN-based R&F networks, access control is done locally inside the WLAN without involving any central control station, thus avoiding the negative impact of fiber propagation delay on the network throughput. R&F networks are well-suited to build WLAN-based FiWi networks of extended coverage without imposing stringent limits on the size of the optical backhaul, as opposed to RoF networks that limit the length of deployed fibers to a couple of kilometers. Recall that this holds only for distributed MAC protocols such as DCF, but not for MAC protocols that deploy centralized polling and scheduling (e.g., WiMAX).

4.3 ENABLING FIWI TECHNOLOGIES

Both RoF and R&F technologies can be found in FiWi networks. In this section, we discuss enabling technologies of FiWi networks in greater detail.

4.3.1 RoF Technologies

Several RoF technologies have been emerging for the realization of low-cost FiWi networks. In the following, we briefly summarize some of the key enabling RoF technologies. For further details and a technically more profound discussion, we refer the interested reader to Jia et al. [17].

Optical RF Generation. To avoid the electronic bottleneck, the generation of RF signals is best done optically. The following novel optical RF generation techniques were experimentally studied and demonstrated in Jia et al. [17]:

- *FWM in HNL-DSF:* Four-wave mixing (FWM) in a highly nonlinear dispersion-shifted fiber (HNL-DSF) can be used to realize simultaneous all-optical upconversion of multiple wavelength channels by using optical carrier suppression (OCS) techniques. FWM is transparent to the bit rate and modulation format, which may be different on each wavelength. Due to the ultrafast response of HNL-DSF, Terahertz optical RF generation is possible.
- *XPM in HNL-DSF:* Cross-phase modulation (XPM) in a nonlinear optical loop mirror (NOLM) in conjunction with straight pass in HNL-DSF enables the all-optical up-conversion of multiple wavelength channels without any interference- and saturation-effect limitation.
- *XAM in EAM:* All-optical wavelength upconversion by means of cross-absorption modulation (XAM) in an electroabsorption modulator (EAM) has several advantages such as low power consumption, compact size, polarization insensitivity, and easy integration with other devices.
- *External IM:* External intensity modulation (IM) is another approach for optical RF generation, deploying one of three following modulation schemes: double-sideband (DSB), single-sideband (SSB), and OCS.
- *External PM:* Instead of external IM, external phase modulation (PM) can be used for optical RF generation.

According to Jia et al. [17], external intensity and phase modulation schemes are the most practical solutions for all-optical RF generation due to their low cost, simplicity, and long-distance transmission performance.

Remote Modulation. An interesting approach to build low-cost FiWi networks is the use of a single light source at the central office (CO) to generate a downlink wavelength that is reused at RAUs for upstream transmission by means of remote modulation, thereby avoiding the need for an additional light source at each RAU. The following remodulation schemes were experimentally studied in Jia et al. [17]:

- *DPSK for Downstream/OOK for Upstream:* PM is deployed to generate a differential phase-shift-keyed (DPSK) optical downstream signal. The DPSK is upconverted through OCS modulation. An optical splitter is used at each RAU to divide the arriving optical signal into two parts. One part is demodulated by a Mach–Zehnder interferometer and is subsequently detected by a photodetector. The other part is on–off-keyed (OOK) remodulated with upstream data using a Mach–Zehnder modulator and is sent to the CO.
- *OCS for Downstream/Reuse for Upstream:* At the CO, an optical carrier is split prior to optical RF generation by means of OCS and is then combined with the RF signal and sent downstream. Each RAU utilizes a fiber Bragg grating (FBG) to reflect the optical carrier while letting the RF signal pass to a photodetector. The reflected optical carrier is remodulated with upstream data and is then sent back to the CO.
- *PM for Downstream/Directly Modulated SOA for Upstream:* Similar to the aforementioned scheme, an optical carrier is combined with an RF signal, generated by means of PM, and sent downstream where an FBG is used at the RAU to reflect the optical carrier and pass the RF signal. The reflected optical carrier is amplified and directly modulated with upstream data using a semiconductor optical amplifier (SOA).

The use of a colorless (i.e., wavelength-independent) SOA as an amplifier and modulator for upstream transmission provides a promising low-cost RoF solution that is easy to maintain [17].

4.3.2 R&F Technologies

R&F-based FiWi access networks may deploy a number of enabling optical and wireless technologies.

Optical Technologies. Apart from PONs, the following optical technologies are expected to play an increasingly important role in the design of a flexible and cost-effective optical backhaul for FiWi networks [18].

- *Tunable Lasers:* Directly modulated external cavity lasers, multisection distributed feedback (DFB)/distributed Bragg reflector (DBR) lasers, and tunable VCSELs can be used as tunable lasers that render the network flexible and reconfigurable and help minimize production cost and reduce backup stock.

- *Tunable Receivers:* A tunable receiver can be realized by using a tunable optical filter and a broadband photodiode. Other, more involved implementations exist (see Kazovsky et al. [18]).
- *Colorless ONUs:* Reflective SOAs (RSOAs) can be used to build colorless ONUs that remotely modulate optical signals generated by centralized light sources.
- *Burst-Mode Laser Drivers:* Burst-mode transmitters are required for ONUs. They have to be equipped with laser drivers that provide fast burst on/off speed, sufficient power suppression during idle period, and stable, accurate power emission dur ing burst transmission.
- *Burst-Mode Receivers:* Burst-mode receivers are required at the central optical line terminal (OLT) of a PON and must exhibit a high sensitivity, wide dynamic range, and fast time response to arriving bursts. Among others, design challenges for burst-mode receivers include dynamic sensitivity recovery, fast level recovery, and fast clock recovery.

Wireless Technologies. A plethora a broadband wireless access technologies exist [19]. Currently, the two most important ones for the implementation of the wireless part of FiWi networks are WiFi and WiMAX.

WiFi. Due to the use of unlicensed frequency bands (2.4 GHz with 14 distinct channels) in IEEE 802.11b/g, providing up to 11/54 Mbit/s data rate, WLANs, also referred to as WiFi networks, have gained much attention. The initial IEEE 802.11 PHY layer includes (i) frequency hopping spread spectrum (FHSS), (ii) direct sequence spread spectrum (DSSS), and (iii) infrared (IR). IEEE 802.11b uses high-rate DSSS (HR-DSSS), while IEEE 802.11g deploys orthogonal frequency division multiplexing (OFDM). The IEEE 802.11 MAC layer deploys the above-mentioned DCF as a default access technique. In this contention-based scheme, subscriber stations (STAs) associated with the access point (AP) use their air interfaces for sensing channel availability. If the channel is idle, the source STA sends its data to the destination STA through the associated AP. If more than one STA try to access the channel simultaneously, a collision occurs. The standard uses the carrier sense multiple access/collision avoidance (CSMA/CA) mechanism to avoid collisions. Point coordination function (PCF) is another technique that may be used in the MAC layer. In PCF, the data transmission is arbitrated in two modes: (i) centralized mode, where the AP polls each STA in a round-robin fashion, and (ii) contention-based mode, which works similarly to DCF. In addition, the request to send (RTS)/clear to send (CTS) mechanism is applied to solve the hidden node problem. Next-generation WLANs (IEEE 802.11n) will offer a throughput of at least 100 Mbit/s measured at the MAC service access point (SAP) [20]. The IEEE 802.11n draft provides both PHY and MAC enhancements. By using multiple-input multiple-output (MIMO)-OFDM and channel bonding, 802.11n WLANs offer raw data rates of about 600 Mbit/s at the physical layer. To achieve a net MAC throughput of 100 Mbit/s and higher, 802.11n WLANs allow wireless stations for the truncation of transmission opportunities (TXOPs), reverse direction (i.e., bidirectional TXOP), and use of a reduced interframe space (RIFS) to decrease the dead time between frames (a TXOP, specified in IEEE 802.11e, is a time interval during which a wireless station following a single-channel access is allowed to send multiple data frames). The most important MAC enhancement of next-generation WLANs is frame aggregation. In 802.11n, the following two methods exist for frame aggregation: (i) aggregate MAC protocol data unit (A-MPDU) and (ii) aggregate MAC service data unit (A-MSDU). A-MPDU concatenates up to 64 MAC protocol data unit (MPDU) subframes into a single physical layer service data unit (SDU), provided that all constituent MPDUs are destined to the same receiver. A-MSDU concatenates multiple MAC service data unit (MSDU) subframes into a single MPDU, whereby all constituent MSDUs not only have to be destined to the same receiver but also must have the same traffic identifier (TID), that is, the same quality-of-service (QoS) level. A-MPDU and A-MSDU can be used separately or jointly to increase the MAC throughput of next-generation WLANs. Moreover, the emerging amendment IEEE 802.11s aims at specifying a wireless distribution system (WDS) among WLAN APs which can be used to realize municipal networks that provide public wireless access throughout cities, neighborhoods, and campuses. IEEE 802.11s introduces a new mesh frame format and radio-aware routing framework that uses the so-called hybrid wireless mesh protocol (HWMP) as default routing protocol [21]. HWMP works on layer 2, uses MAC addresses for path selection, and contains both reactive and proactive routing components.

WiMAX. The initial IEEE 802.16 WiMAX standard was established in the frequency band of 10–66 GHz, providing up to 75 Mbit/s data rate line-of-sight (LOS) connections in both point-to-multipoint (PMP) and mesh modes. IEEE 802.16a provides non-LOS connections in the frequency band of 2–11 GHz (licensed and unlicensed). The WiMAX PHY layer uses WirelessMAN-OFDMA (orthogonal frequency division multiple access) and transfers bidirectional data by means of time division duplex (TDD) or frequency division duplex (FDD). IEEE 802.16 is a connection-oriented standard; that is, prior to transmitting data between subscriber stations (SSs) and base station (BS), connections must be established. Each connection is identified by a 16-bit connection identifier (CID). The MAC layer is responsible for assigning CIDs as well as allocating bandwidth between SSs. It consists of the following three sublayers: (i) convergence sublayer (CS), whereby different higher-layer protocols are implemented in different CSs—for example, ATM CS and packet CS are used for ATM and Ethernet networks, respectively; (ii) common part sublayer (CPS), which is responsible for bandwidth allocation and generating MPDUs; and (iii) security sublayer. In the PMP mode, the requested services of each SS are first registered during the initialization phase and subsequently the connections are established. If a given SS changes its services, additional connections can be established in the network.

Each connection is associated with a service flow (SF). An SF is defined based on available scheduling services and includes a set of QoS parameters, an SF identifier (SFID), and a CID. To implement wireless mesh networks (WMNs), two scheduling types are used: (i) centralized and (ii) distributed. In the centralized scheduling mode, such as the PMP, each mesh-SS (MSS) sends its request to the mesh-BS (MBS) that manages the network. In the distributed scheduling mode, each MSS distributes its scheduling information and one-hop neighbors among all its adjacent MSSs. A three-way handshake mechanism is deployed for bandwidth allocation. Coordinated (collision-free) and uncoordinated (non-collision-free) methods are used for distributed scheduling. The two different mesh scheduling methods can be applied together by subdividing the data part of the frame into two parts, one for centralized scheduling and another one for distributed scheduling. The scalability and flexibility of the radio access technology and network architecture of the IEEE standard 802.16e, also known as mobile WiMAX, provide various services through broadband connections [22]. Mobile WiMAX is able to support multimedia transmissions with differentiated QoS requirements through the use of scheduling processes. The IEEE 802.16j [referred to as mobile multihop relay (MMR)] Working Group aims at extending network coverage and improving network throughput via multihop relay (MR) stations.

4.4 FiWi ARCHITECTURES

In this section, we present various state-of-the-art FiWi network architectures [23].

4.4.1 Integrated EPON and WiMAX

The integration of EPON and WiMAX access networks can be done in different ways; according to Shen et al. [24], the following four architectures can be used.

Independent Architecture. In this approach, WiMAX BSs serving mobile SSs are attached to an ONU just like any other wired subscriber. WiMAX and EPON networks are connected via a common standardized interface (e.g., Ethernet) and operate independently.

Hybrid Architecture. This approach introduces an ONU-BS that integrates the EPON ONU and WiMAX BS in both hardware and software. The integrated ONU-BS controls the dynamic bandwidth allocation of both the ONU and BS.

Unified Connection-Oriented Architecture. Similar to the hybrid architecture, this approach deploys an integrated ONU-BS. But instead of carrying Ethernet frames, WiMAX MPDUs containing multiple encapsulated Ethernet frames are used. By carrying WiMAX MPDUs, the unified architecture can be run like a WiMAX network with the ability to finely grant bandwidth using WiMAX's connection-oriented rather than EPON's queue-oriented bandwidth allocation.

Microwave-over-Fiber Architecture. In this approach, like RoF networks, the WiMAX signal is modulated on a carrier frequency and is then multiplexed and modulated together with the baseband EPON signal onto a common optical frequency (wavelength) at the ONU-BS. The central node consists of a conventional EPON OLT and a central WiMAX BS, called a macro-BS. The OLT processes the baseband EPON signal, while the macro-BS processes data packets originating from multiple WiMAX BS units.

4.4.2 Integrated Optical Unidirectional Fiber Ring and WiFi

This FiWi network, shown in Figure 4.3, interconnects the CO with multiple WiFi-based APs by means of an optical unidirectional fiber ring [25]. The CO is responsible for managing the transmission of information between mobile client nodes (i.e., STAs) and their associated APs as well as acting as a gateway to other networks. Each AP provides wireless access to STAs within its range. All STAs take part in the topology discovery, whereby each STA periodically sends the information about the beacon power received from its neighbors to its associated

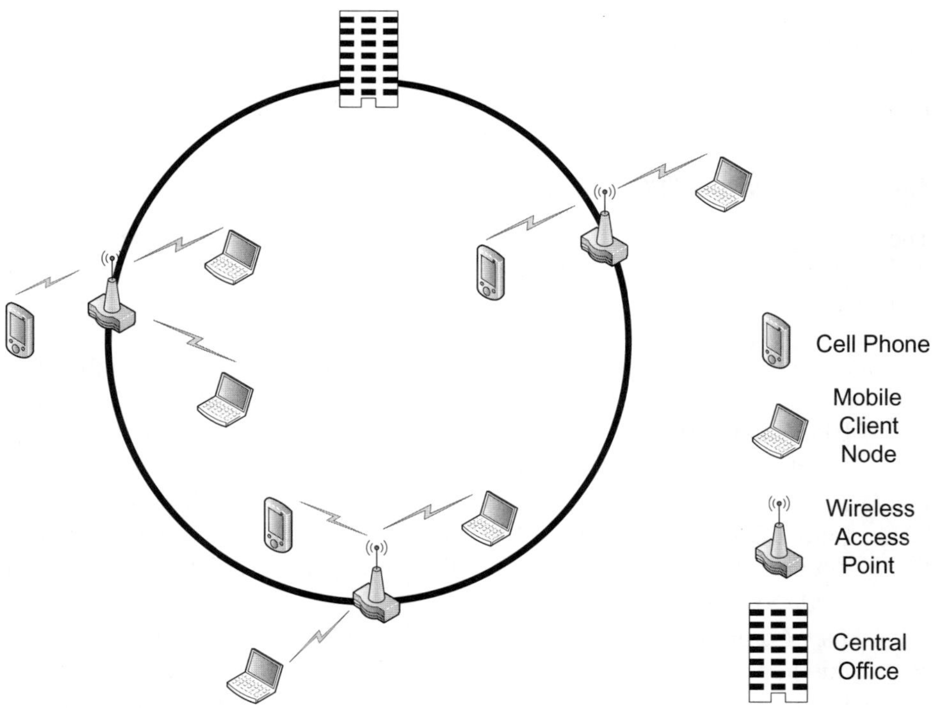

Figure 4.3. Optical unidirectional fiber ring interconnecting WiFi-based wireless access points [23].

AP. In doing so, APs are able to estimate the distances between STAs and compute routes. Multihop relaying is used to extend the range. To enhance the reliability of the wireless link, the CO sends information to two different APs (path diversity). The proposed implementation can support advanced path diversity techniques that use a combination of transmission via several APs and multihop relaying (e.g., cooperative diversity or multihop diversity). Consequently, the CO must be able to assign channels quickly and efficiently by using one or more wavelength channels on the fiber ring to accommodate multiple services such as WLAN and cellular radio network.

4.4.3 Integrated Optical Interconnected Bidirectional Fiber Rings and WiFi

Figure 4.4 shows a two-level bidirectional path protected ring (BPR) architecture for dense wavelength division multiplexing (DWDM)/subcarrier multiplexing (SCM) broadband FiWi networks [26]. In this architecture, the CO interconnects

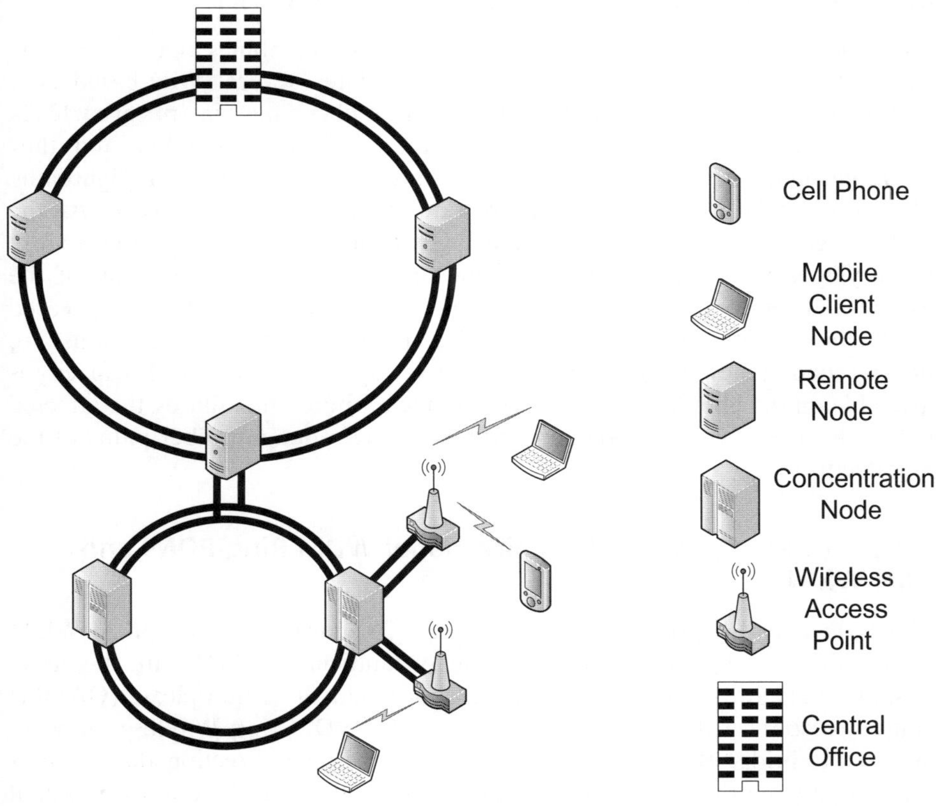

Figure 4.4. Optical interconnected bidirectional fiber rings integrated with WiFi-based wireless access points [23].

remote nodes (RNs) via a dual-fiber ring. Each RN cascades APs through concentration nodes (CNs), where each AP offers services to STAs. For protection, the CO is equipped with two sets of devices (normal and standby). Each RN consists of a protection unit and a bidirectional wavelength add–drop multiplexer based on a multilayer dielectric interference filter. Each CN contains a protection unit. The AP comprises an optical transceiver, a protection unit, up/down RF converters, and a sleeve antenna. Each AP provides channel bandwidth of at least 5 MHz and covers up to 16 STAs by means of frequency-division multiplexing (FDM). Under normal operating conditions, the CO transmits downstream signals in the counterclockwise direction via RNs and CNs to the APs. If a fiber cut occurs between two RNs or between two CNs, their associated controllers detect the failure by monitoring the received optical signal and then switch to the clockwise protection ring. If a failure happens at an AP, the retransmitted signals are protection switched through other optical paths by throwing an optical switch inside the affected AP. This architecture provides high reliability, flexibility, capacity, and self-healing properties.

4.4.4 Integrated Optical Hybrid Star-Ring and WiFi

Figure 4.5 depicts a hybrid FiWi architecture that combines optical star and ring networks [27]. Each fiber ring accommodates several WiFi-based APs, and is connected to the CO and two neighboring fiber rings via optical switches. The optical switches have full wavelength conversion capability, and they interconnect the APs and CO by means of shared point-to-point (P2P) lightpaths. The network is periodically monitored during prespecified intervals. At the end of each interval, the lightpaths may be dynamically reconfigured in response to varying traffic demands. When traffic increases and the utilization of the established lightpaths is low, the load on the existing lightpaths is increased by means of load balancing. Otherwise, if the established lightpaths are heavily loaded, new lightpaths need to be set up, provided enough capacity is available on the fiber links. In the event of one or more link failures, the affected lightpaths are dynamically reconfigured using the redundant fiber paths of the architecture.

4.4.5 Integrated Optical Unidirectional WDM Ring-PONs and WiFi-WMN

The FiWi network proposed in Shaw et al. [28] consists of an optical WDM backhaul ring with multiple single-channel or multichannel PONs attached to it, as shown in Figure 4.6. More precisely, an optical add–drop multiplexer (OADM) is used to connect the OLT of each PON to the WDM ring. Wireless gateways are used to bridge PONs and WMNs. In the downstream direction, data packets are routed from the CO to the wireless gateways through the optical backhaul and then forwarded to the STAs by wireless mesh routers. In the upstream direction, wireless mesh routers forward data packets to one of the wireless

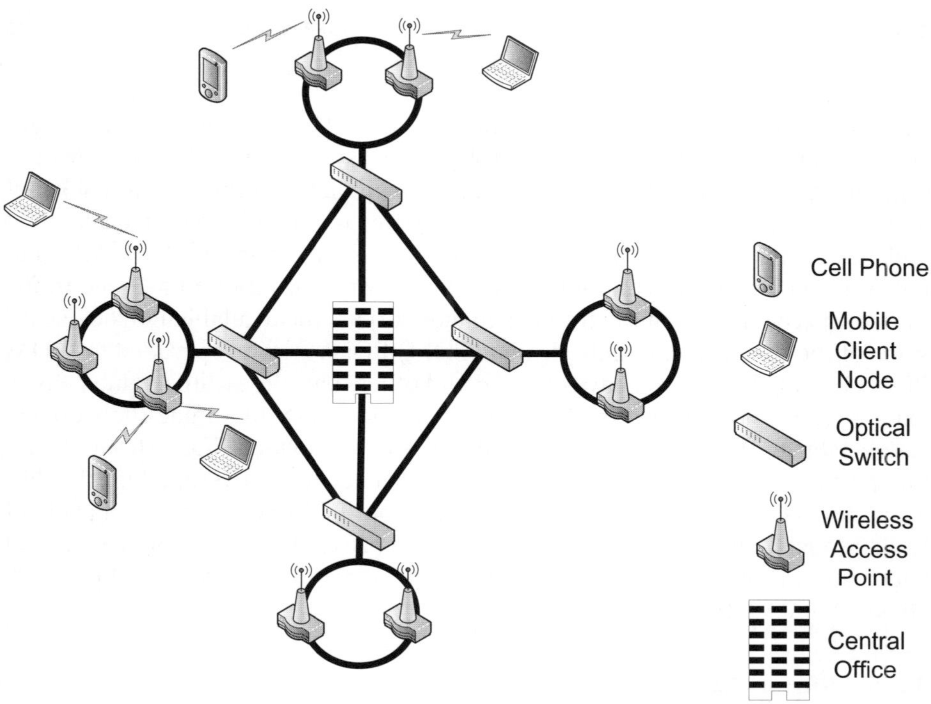

Figure 4.5. Optical hybrid star-ring network integrated with WiFi-based wireless access points [23].

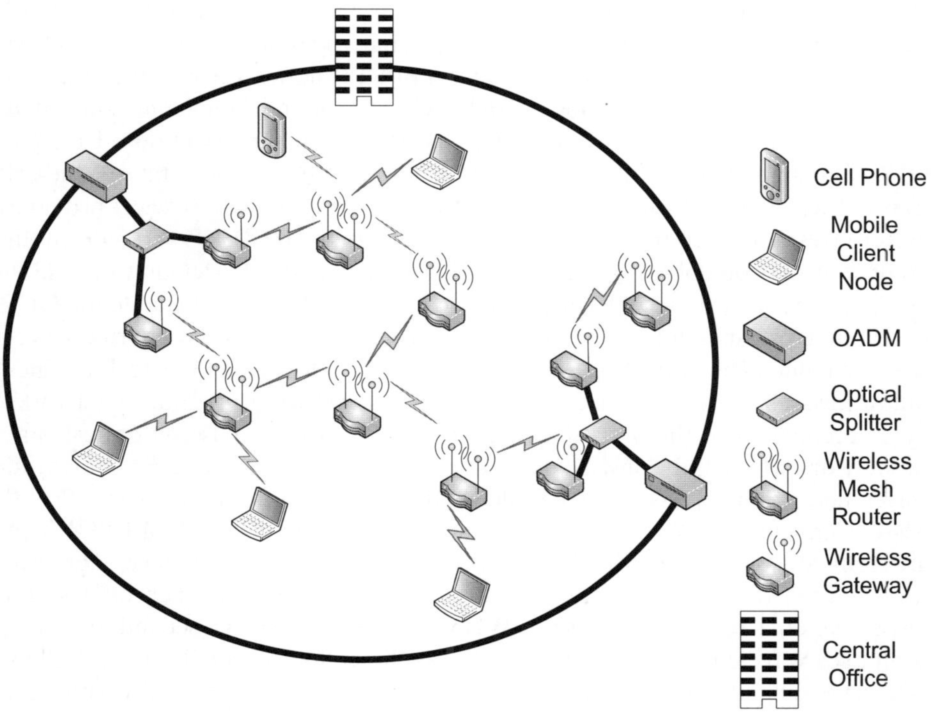

Figure 4.6. Optical unidirectional WDM ring interconnecting multiple PONs integrated with a WiFi-based wireless mesh network [23].

gateways, where they are then transmitted to the CO on one of the wavelength channels of the optical backhaul WDM ring, because each PON operates on a separate dynamically allocated wavelength channel. Since the optical backhaul and WMN use different technologies, an interface is defined between each ONU and the corresponding wireless gateway in order to monitor the WMN and perform route computation taking the state of wireless links and average traffic rates into account. When the traffic demands surpass the available PON capacity, some of the time division multiplexing (TDM) PONs may be upgraded to WDM PONs. If some PONs are heavily loaded and others have less traffic, some heavily loaded ONUs may be assigned to a lightly loaded PON by tuning their optical transceivers to the wavelength assigned to the lightly loaded PON. This architecture provides cost effectiveness, bandwidth efficiency, wide coverage, high flexibility, and scalability. In addition, the reconfigurable TDM/WDM optical backhaul helps reduce network congestion and average packet latency by means of load balancing. Moreover, the dynamic allocation of radio resources enables cost-effective and simple handovers.

4.4.6 SuperMAN

Figure 4.7 depicts the network architecture of SuperMAN. It builds on an all-optically integrated Ethernet-based access-metro network extended by optical-wireless interfaces with next-generation WiFi and WiMAX networks [29, 30]. More specifically, the optical part of SuperMAN consists of an IEEE 802.17 resilient packet ring (RPR) metro network that interconnects multiple WDM EPON access networks attached to a subset of RPR nodes. Each of the attached WDM EPONs has a tree topology with the OLT at the root tree being collocated with one of the P COs. No particular WDM architecture is imposed on the ONUs, thus allowing the decision to be dictated by economics, state-of-the-art transceiver manufacturing technology, traffic demands, and service provider preferences. The recommended WDM extensions to the IEEE 802.3ah multi-point control protocol (MPCP), described in greater detail in McGarry et al. [31], guarantee backward compatibility with legacy TDM EPONs and enable the OLT to schedule transmissions to and receptions from ONUs on any supported wavelength channel. The optical access-metro network lets low-cost PON technologies follow low-cost Ethernet technologies from access networks into metro networks by interconnecting the P collocated OLTs/COs with a passive optical star subnetwork whose hub consists of an athermal wavelength-routing $P \times P$ arrayed waveguide grating (AWG) in parallel with a wavelength-broadcasting $P \times P$ passive star coupler (PSC). It is important to note that in each WDM EPON two different sets of wavelengths, Λ_{OLT} and Λ_{AWG}, are used. The first wavelength set, Λ_{OLT}, is used for upstream and downstream transmissions between ONUs and respective OLT residing in the same WDM EPON, whereas the second set, Λ_{AWG}, comprises wavelengths that optically bypass the collocated OLT/CO and allow ONUs residing in different WDM EPONs to communicate all-optically with each other in a single hop across the AWG of the star subnetwork, provided that the

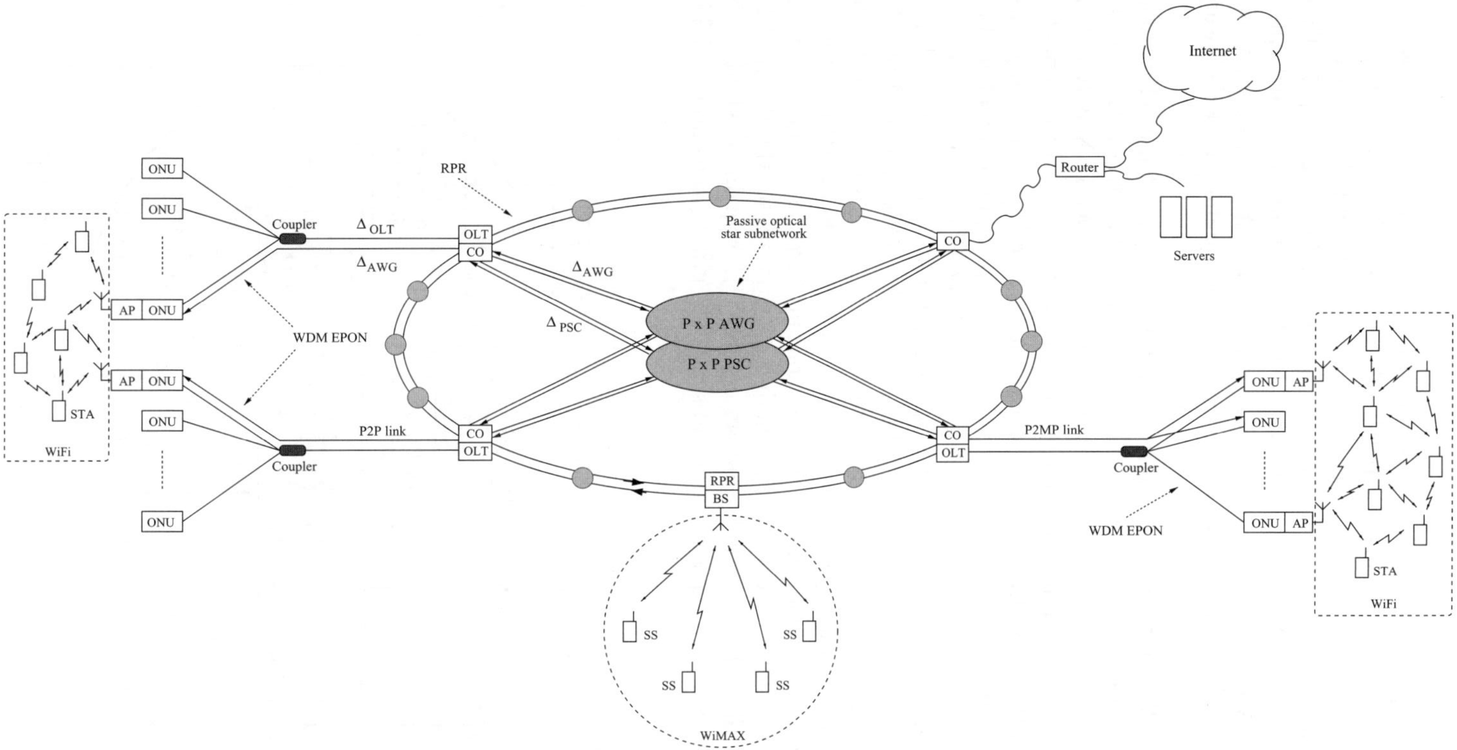

Figure 4.7. SuperMAN architecture integrating next-generation WiFi technologies with WDM EPON and next-generation WiMAX technologies with RPR [29].

ONUs are equipped with transceivers operating on these wavelengths. It is worthwhile to note that, similar to IEEE 802.3ah EPON, the optical part of SuperMAN is not restricted to any specific dynamic bandwidth allocation (DBA) algorithm. A plethora of DBA algorithms for WDM EPONs exist [32]. These DBA algorithms need to be adapted to SuperMAN. The aforementioned optical part of SuperMAN interfaces with next-generation WiFi and WiMAX networks. Both optical–wireless interfaces are described in greater detail in the following.

RPR–WiMAX Interface. As shown in Figure 4.7, some of the RPR nodes may interface with WiMAX rather than EPON access networks. Figure 4.8. depicts the optical–wireless interface between RPR and WiMAX networks in greater detail, where an integrated rate controller (IRC) is used to connect an RPR node to a WiMAX BS. In RPR, packets undergo optical–electrical–optical (OEO) conversion at each ring node. An RPR node deploys in general two separate electrical transit queues, one primary transit queue (PTQ) and one secondary transit queue (STQ), for service differentiation. In addition, an electrical stage queue is used to store traffic ready to be sent by the RPR station. The RPR scheduler gives priority to in-transit ring traffic over station traffic such that in-transit packets are not lost due to buffer overflow. Furthermore, RPR deploys a distributed fairness control protocol that dynamically throttles traffic in order to achieve network-wide fairness while maintaining spatial reuse. The WiMAX BS deploys a downlink (DL) scheduler and an uplink (UL) scheduler, whereby the latter one processes UL requests from and sends UL grants to its attached SSs.

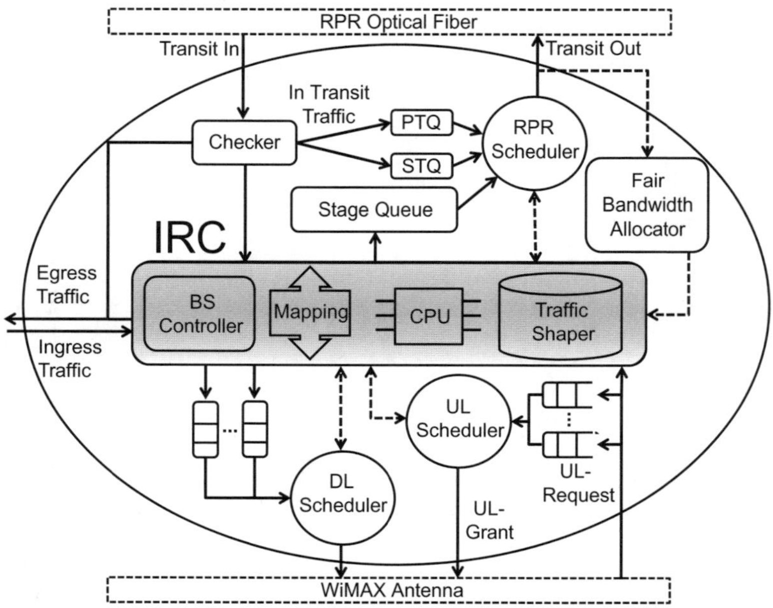

Figure 4.8. Optical–wireless interface between RPR and WiMAX networks [30].

In SuperMAN, IEEE 802.16e is considered as an enabling technology which adds mobility support to conventional IEEE 802.16d WiMAX networks. The IRC in Figure 4.8 comprises a BS controller, a traffic class mapping unit, a CPU, and a traffic shaper. It is used to seamlessly integrate both technologies and jointly optimize the RPR scheduler and WiMAX DL and UL schedulers. The BS controller is responsible for handling incoming and outgoing WiMAX traffic, besides providing handover for SSs between different RPR/WiMAX interface nodes. The traffic class mapping unit is able to translate the different WiMAX and RPR traffic classes bidirectionally. The traffic shaper checks the control rates of RPR traffic and performs traffic shaping according to the RPR's fairness policies. The role of the CPU is twofold: synchronizing all the operational processes occurring at different modules of the IRC, including alarm management, and monitoring and tuning shapers and schedulers dynamically in order to optimize QoS-aware packet delivery. The CPU monitors the RPR as well as the WiMAX DL and UL schedulers and their queues, in addition to the BS controller and traffic shaper. It supervises the traffic shaper in order to synchronize it with the BS controller and schedulers to avoid local congestion.

Figure 4.9 depicts the mean aggregate throughput of SuperMAN versus the speed of the SS (given in km/h) for RPR background traffic only (no SSs) and RPR background traffic in conjunction with WiMAX traffic coming from and going to 25 attached mobile SSs for different terrain types A, B, and C, where

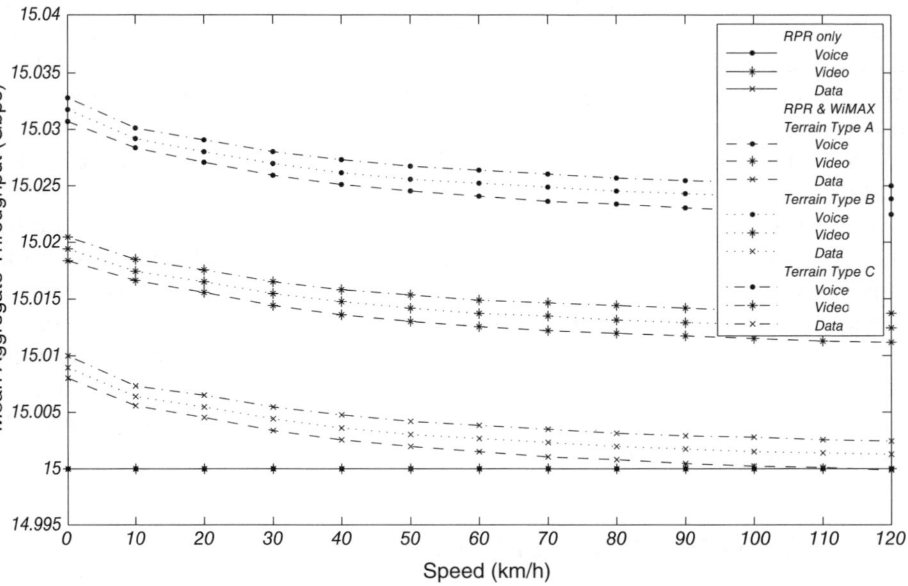

Figure 4.9. Mean aggregate throughput versus mobile SS speed with 25 attached mobile WiMAX SSs and 15-Gbit/s RPR background traffic for different terrain types using the hierarchical WiMAX scheduler [30].

type A describes an urban environment with maximum path loss, type B specifies a suburban environment with intermediate path loss, and type C describes a rural environment with minimum path loss. The considered RPR ring consists of eight RPR nodes with a 2.5-Gbit/s line rate for each ringlet with an RPR round-trip time (RTT) set to 0.5 ms. For RPR background traffic only, the mean aggregate throughput equals 15 Gbit/s for each of the three traffic classes (voice, video, and data) independent of speed and wireless channel conditions. As shown in Figure 4.9, for increasing speed the mean aggregate throughput decreases rapidly. This is due to the fact that for an increasing packet error rate, the number of required packet retransmissions increases, which in turn reduces the mean aggregate throughput of SuperMAN. Further results reported in Ghazisaidi et al. [30] show that deploying the proposed novel hierarchical scheduler at each RPR–WiMAX interface node improves the performance of SuperMAN in terms of mean aggregate throughput and mean delay for voice, video, and data traffic. The results prove that the proposed hierarchical scheduler with its multiple stages puts less backpressure on the RPR metro ring network and thereby achieves a higher mean aggregate throughput for all three traffic classes than a conventional weighted fair queuing (WFQ) scheduler for fixed users and mobile users with a speed of up to 120 km/h, under realistic wireless channel conditions.

WDM EPON–Next-Generation WiFi Interface. Recall from Section 2 that WiFi-based RoF networks can sustain acceptable throughput performance only if the inserted fiber does not exceed a certain maximum length. Because EPON can have a reach of up to 20 km, the WDM EPON tree networks with WiFi extensions are realized as R&F networks, where each WiFi-based network operates independently of its attached WDM EPON tree network. In the IEEE 802.11s WLAN mesh path selection algorithms of SuperMAN, proactive routing can be used to configure routing trees toward the collocated AP/ONU(s) that act as mesh portals bridging the WLAN mesh network to the optical (wired) WDM EPON access network. For intra-mesh communication between wireless stations, reactive routing may be applied to set a direct route between wireless stations, thereby eliminating the need to send intra-mesh traffic through the mesh portal. It is important to note that the routing framework of IEEE 802.11s is extensible. Thus, other routing protocols and routing metrics can be deployed in order to optimize network performance according to given traffic demands and usage scenarios. Moreover, frame aggregation as the most important MAC enhancement of next-generation WLANs may be considered to improve throughput-delay performance of SuperMAN.

4.5 TECHNO-ECONOMIC EVALUATION

As we saw in Section 4.4, different FiWi network architectures can be designed by using WiMAX and WiFi technologies. While low-cost WiFi is the technology of choice in home/office networks, it is somewhat unclear whether EPON or WiMAX provide the better solution in metro-access networks. Given the similarities of EPON and WiMAX, the two technologies are more likely to target the same network segment rather than being cascaded to cover different network segments. In other words, network operators are expected to make a choice between EPON and WiMAX, calling for a techno-economic comparison of the two technologies. During the last decade, the techno-economic evaluation of various network technologies has been an active research area. However, most of the previous techno-economic evaluations focused either on optical fiber-only (e.g., Weldon and Zane [33] and Tran et al. [34]) or wireless-only network architectures (e.g., Niyato and Hossain [35]). To date, only a few preliminary techno-economic evaluations of FiWi networks have been reported. A cost comparison of very-high-bit-rate DSL (VDSL) and a FiWi architecture consisting of cascaded EPON and WiMAX networks was carried out in Lin et al. [36]. The obtained results indicate the superior cost-efficiency of FiWi networks over conventional VDSL solutions. Different FiWi network design heuristics were investigated in terms of processing time, complexity, and installation cost in Sarkar et al. [37]. Despite these preliminary studies, a more thorough techno-economic evaluation of FiWi networks is necessary in order to gain deeper insights into the design, configuration, and performance optimization of emerging FiWi networks that are based on EPON and/or WiMAX technologies.

Figure 4.10 illustrates the proposed techno-economic model for the comparative analysis of EPON versus WiMAX. It consists of the following modules:

- *Scenario Description:* It defines various network deployment scenarios and terrain-type conditions (i.e., urban, suburban, and rural).
- *Technological Constraints:* This module determines the technological limitations of a given scenario, such as maximum distance between OLT and ONUs in EPON.
- *Initial Network Infrastructure:* It designs the initial network infrastructure of a given scenario with given constraints of the applied technology.
- *Cost-Modeling Techniques:* This module includes the cost-modeling methods used in the subsequent cost-efficient network design. The maximum cost-sharing approach as a cost-modeling technique is considered for EPON network. The maximum cost-sharing technique aims at minimizing the length of required distribution fibers (between optical splitter and ONUs). In this approach, the position of the OLT is fixed and the distance between OLT and ONUs is used as an important parameter in cost modeling. The costs of deploying EPON with ONUs being located at different ranges from the OLT are calculated. For WiMAX, the maximum QoS-coverage technique is considered. This approach aims at maximizing the range of a WiMAX network with QoS support for different traffic types.
- *Cost-Efficient Network Design:* This module modifies or redesigns the initial network infrastructure, making use of the cost-modeling techniques module.

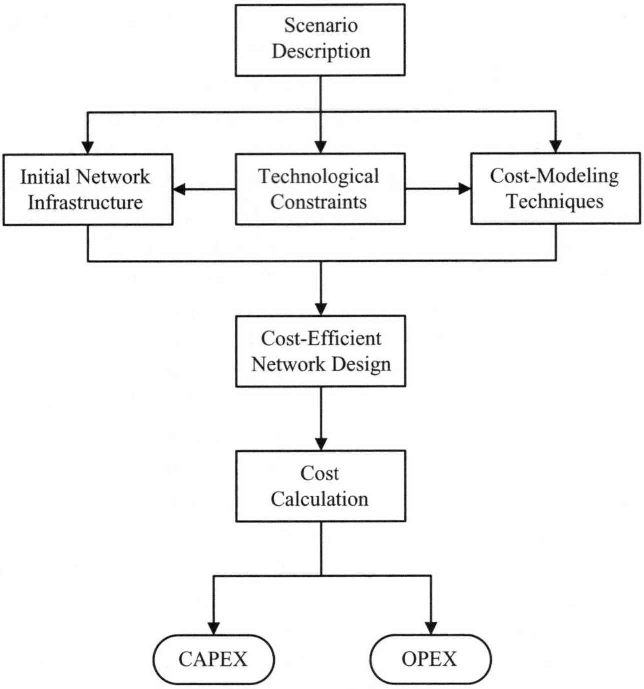

Figure 4.10. Proposed techno-economic model for the comparative analysis of EPON versus WiMAX.

- *Cost Calculation:* It calculates the network costs, which are categorized into (i) capital expenditures (CAPEX) that consist of initial network equipment and network installation costs and (ii) operational expenditures (OPEX) that comprise network operation, administration, and maintenance (OAM) costs. More specifically, CAPEX consist of equipment and installation costs for setting up the network, while OPEX comprise the OAM costs for running the network (e.g., power consumption, troubleshooting, and repairing).

Figure 4.11a shows the power consumption versus mean access data rate for EPON and WiMAX serving {16, 32, 64} ONUs/SSs at a range of 20 km for different terrain types. The power consumption increases for increasing mean access data rate, whereby EPON consumes less power than WiMAX. The power consumption of EPON is independent of the terrain type. The capacity of the BS in urban settings is smaller than that in suburban and rural settings, resulting in an increased power consumption. For both EPON and WiMAX, the power consumption grows for an increasing number of ONUs and SSs.

Figure 4.11b shows the total cost versus range for EPON and WiMAX for 32 ONUs/SSs with a fixed mean access data rate of 75 Mbit/s. The total cost of EPON increases for increasing range, while WiMAX total cost is largely

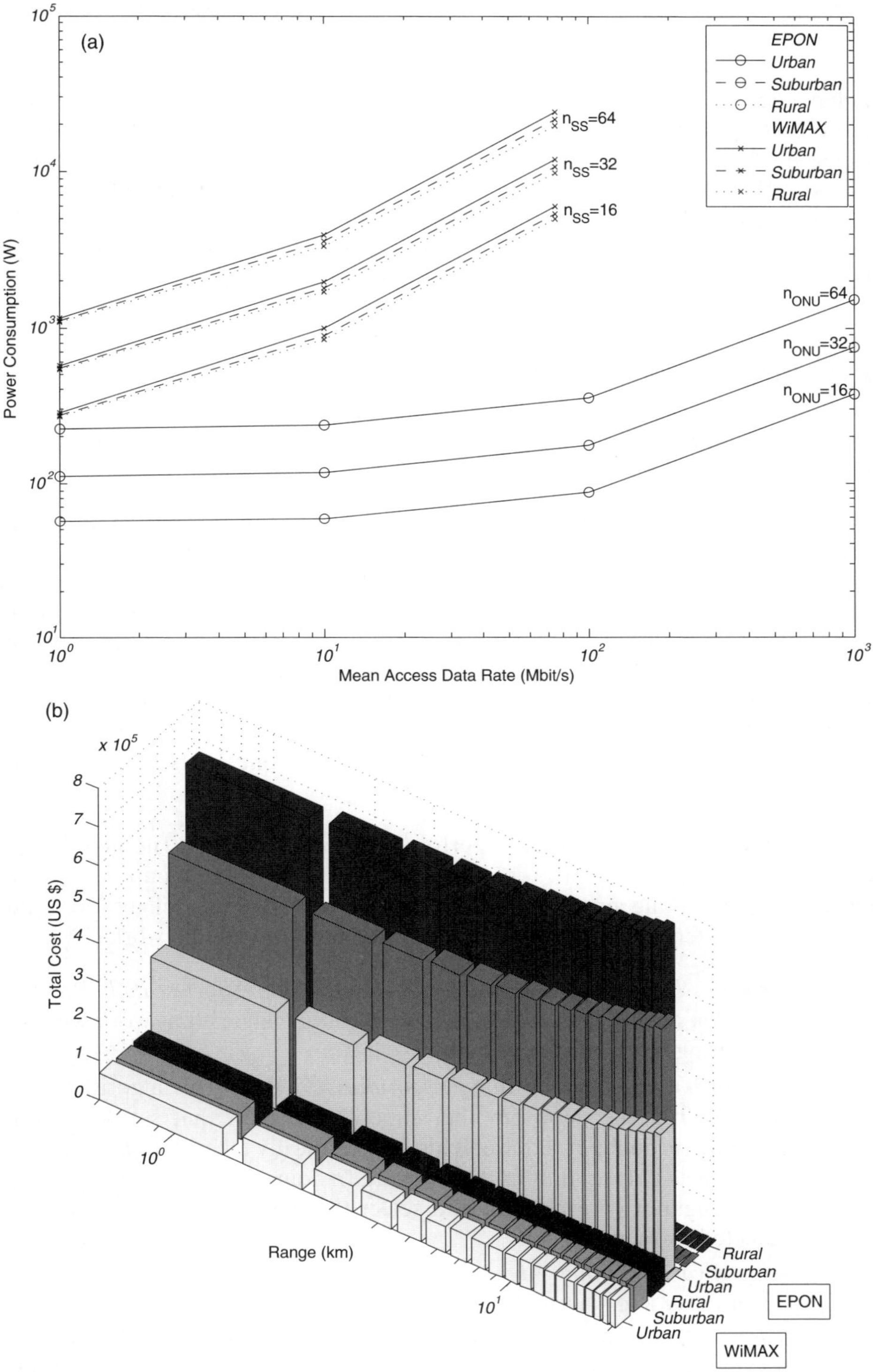

Figure 4.11. Techno-economic comparison of EPON versus WiMAX: **(a)** power consumption versus mean access data rate, **(b)** total cost versus range for 32 ONUs/SSs and a fixed mean access data rate of 75 Mbit/s.

independent of the range for a fixed number of SSs. We observe that WiMAX is more cost-efficient than EPON for a mean access data rate of up to 75 Mbit/s, especially for less populated suburban and rural terrain types. The cost difference between WiMAX and EPON becomes less pronounced for urban settings with high population densities. In fact, EPON might be a viable alternative to WiMAX in densely populated areas where the high installation costs of the required fiber infrastructure can be shared by a large number of subscribers.

4.6 FUTURE CHALLENGES AND IMPERATIVES

In this section, we elaborate on challenging and imperative issues that have to be addressed in future FiWi networks.

Architecture. The design of new FiWi network architectures is important in order to reduce their costs and increase their flexibility. All the aforementioned WiFi-based FiWi networks were implemented using an optical fiber ring. The combination of an optical fiber ring and WiMAX would be another interesting architecture where WiMAX SSs and WiFi STAs are able to access the network via integrated WiMAX and WiFi networks. Architectural upgrade paths are also important to improve the flexibility and cost-efficiency of already existing FiWi networks. A detailed feasibility and performance study of new FiWi network configurations in support of heterogeneous types of end users with different service requirements is desirable.

Routing. In WMNs, routing is performed by mesh routers. The mobility of end users affects the network topology and connectivity, which imposes severe challenges on routing protocols as well as on network (re)configuration and installation. Different routing algorithms can be used in FiWi networks [37]:

1. *Minimum-Hop Routing Algorithm (MHRA) and Shortest-Path Routing Algorithm (SPRA):* These shortest path routing algorithms work without considering given traffic demands.
2. *Predictive-Throughput Routing Algorithm (PTRA):* This algorithm is a link-state routing algorithm that chooses the path that satisfies given aggregate throughput requirements. It periodically takes link rate samples and predicts link conditions dynamically. In doing so, it is able to estimate the throughput of each path and select the path that gives the highest predicted throughput.
3. *Delay-Aware Routing Algorithm (DARA):* This algorithm focuses on packet delay and selects the path with the minimum predicted delay.
4. *Risk-and-Delay-Aware Routing Algorithm (RADAR):* This algorithm is an extension of DARA and can handle multiple-failure scenarios. In this algorithm, the path with the minimum estimated delay and packet loss is selected.

Among the aforementioned routing algorithms, RADAR shows the best performance in terms of delay, throughput, and load balancing under both high and low traffic loads, besides providing risk awareness. In addition to finding and maintaining routes for data flows, efficient routing protocols for future FiWi networks need to meet further requirements such as scalability, fast route (re)discovery, and QoS support.

Reconfigurability, Channel Assignment, and Bandwidth Allocation. The reconfigurability of FiWi networks is achieved using the following two mechanisms [38]:

1. *Dynamic Bandwidth Allocation (DBA):* DBA algorithms take full advantage of assigned channels and balance the traffic load among end users. They can be roughly categorized into statistical multiplexing algorithms and QoS-aware algorithms with absolute or relative QoS assurances.
2. *Dynamic Channel Assignment (DCA):* In fixed channel assignment (FCA)-based FiWi networks, only DBA algorithms can be used to accommodate real-time traffic demands efficiently. By contrast, DCA-based FiWi networks are able to reconfigure channel assignments according to current traffic loads and network conditions.

Performing load balancing periodically renders FiWi networks more robust. In RoF-based FiWi networks with centralized DBA and DCA systems, bandwidth demands in different access areas may vary over time. One approach to effectively achieve load balancing is the reallocation of bandwidth among regions, but a more profound study of alternative reconfigurability approaches is required.

QoS. Resource management and allocation mechanisms are crucial to provide QoS in wireless networks. A recent comparison of the aforementioned bandwidth allocation methods in PMP and mesh modes of WiMAX networks has shown that random access outperforms polling at low request rates but leads to a significantly decreased performance under high channel loads [39]. Thus, adaptive switching between random access and polling according to current traffic loads should be enabled to improve the performance of WiMAX-based FiWi networks. The mixing and mapping of different traffic classes used in PON, WiFi, and WiMAX networks is another important challenge in FiWi networks. While in EPON seven different traffic classes are specified by IEEE 802.1D, the vast majority of deployed WiFi networks support only one class of traffic and not more than four traffic classes are defined for WiMAX. Moreover, the design of QoS-aware routing protocols in WMNs is still an open issue and is not addressed within the emerging standard IEEE 802.11s.

Radio Interfaces. In WiMAX, selecting the TDD or FDD for RoF-based FiWi networks is a challenging issue. Using TDD in RoF-based FiWi networks seems to be the more cost-effective and simple option since only one set of devices (e.g., filters, oscillators, and amplifiers) is needed at RAUs, while FDD would require two separate sets of devices, one for upstream communication and another one for downstream communication. TDD seems more suitable than FDD for asymmetric traffic because of its ability to use wireless channel resources more efficiently. Furthermore, as the same frequency channel is applied for both directions in TDD, the current channel status is completely apparent for both sides [40]. Another important challenge is the ability of FiWi networks to mitigate the effect of different types of interference and delay in the wireless segment, especially in WMNs. Importantly, the following two types of delay must be taken into account in RoF-based FiWi networks [40]:

1. *Multipath Delay:* An end user located at the boundary of two different RAUs receives two identical signals with different delays. OFDMA renders WiMAX networks immune against multipath interference due to the long symbol duration and cyclic prefix. Because the delay added by the optical segment of FiWi networks can exceed the length of the cyclic prefix, received signals cannot maintain the orthogonality of subcarriers, resulting in a decreased throughput. Variable time-delay modules (VTDMs) can be used at the CO to generate different delays equal to the corresponding RoF link propagation delays prior to sending the signals to the RAUs.
2. *Propagation Delay:* The additional access waiting time in each transmitting period of TDD-based RoF reduces the channel utilization and capacity of FiWi networks due to the propagation delay between CO and RAUs. One approach to compensate for the propagation delay added by the transmission link is the use of fixed time-delay modules (FTDMs) at the CO. FTDMs are able to generate fixed delays equal to the corresponding propagation delays.

Despite recent progress in RoF-based FiWi networks, more research on physical-layer-related issues is needed given the high atmospheric absorption in high-frequency bands (e.g., millimeter wave band).

Scalability and Modularity. The capability of a network to increase the number of its elements (e.g., end users, gateways, and routers) without affecting the network performance is known as scalability. For an increasing number of end users in WMNs, the number of hops increases and the network throughput degrades significantly. Scalability is crucial to the successful deployment of WMNs and implies several challenging issues such as addressing and routing. As mentioned above, an RoF-based FiWi network is a combination of multiple simple RAUs and one complex CO. Extending the network by adding one or more RAUs should be possible in a cost-efficient manner. Modularity of FiWi network architectures in order to optimize network costs and capacity in a pay-as-you-grow manner is another open research issue.

Survivability. In FiWi networks, both the optical fiber and wireless segments should take part in the protection and failure recovery to improve network survivability. The cost-efficiency of different failure recovery techniques should be considered in the design of FiWi network architectures. Given the cost of installing spare fiber optics, the wireless segment of FiWi networks is expected to provide more cost-efficient failure recovery. WMNs offer various capabilities (such as multipath routing and high resiliency) that are subject to a number of different constraints. A more detailed study on the challenging issues of providing survivability and removing the limitations of WMNs seems necessary. Another property of reliable FiWi networks that requires more investigation is their ability to perform fast route (re)discovery.

Security. The wireless segment of FiWi networks can be affected by malicious nodes in several ways, such as passive and battery exhaustion attacks. One approach to eliminate malicious nodes is the use of authentication and encryption mechanisms. Routing protocols that use the enhanced authentication protocol (EAP) can perform authentication, whereby RAUs negotiate the session keys to be used with the encrypted data. The keys are used to control access to protected resources or services. To realize secure FiWi networks, two issues must be considered: (i) deploying a generic security management protocol in the range of each RAU and (ii) performing efficient resource monitoring and planning mechanisms to counteract denial-of-service attacks. The implementation of wireless authentication and security systems in FiWi networks needs to be studied in greater detail.

User-Friendliness. By offering various services, such as WLAN and cellular network, the cost-efficiency and flexibility of high-speed FiWi networks make them also attractive to both home and business clients. Despite recent developments (e.g., NTT's DIY installation of FTTH ONUs [4]), the installation of fiber cables for subscribers is not a negligible problem. On the other hand, wireless access networks offer an easier and more user-friendly installation.

Finding simplified methods to connect end users to the RAUs of FiWi networks requires more investigation. The idea of deploying autonomic networks is another interesting research avenue to enable self-configuring, self-optimizing, self-healing, and self-protecting FiWi networks.

Mobility and Bandwidth. In the wireless segment of FiWi networks, the mobility depends on the deployed wireless technology; for example, LOS requirements restrict end-user mobility. Advanced antennas that are able to perform fast and efficient handovers and work with adaptive routing protocols should be considered to decrease the restrictions on end-user mobility. Using WDM, such as in WDM PONs, takes full advantage of the huge capacity of optical fibers. Also, using wireless technologies that are able to provide higher-bandwidth connections in the wireless segment seems desirable. To provide QoS and reconfigurability in a cost-effective manner by means of centralized DBA and DCA algorithms, WDM RoF-based FiWi networks are an attractive design option, where RoF-based FiWi networks are used to avoid frequent handovers of fast-moving users in cellular networks [41]. Moreover, taking the frequency shift into account and adapting to the fast fading conditions in WMNs seems necessary. The investigation of possible methods to support high-speed transmissions with fast mobility is another open research issue.

Cost-Efficiency and Migration. Future-proofness of FiWi networks should be guaranteed by providing cautious pay-as-you-grow migration paths. The backward compatibility with implemented standards and technologies as well as the interoperability with future technologies in a cost-effective manner should be considered at all design and installation stages of future FiWi networks.

The above-discussed challenges of future FiWi access networks are summarized in Figure 4.12. In this figure, the different challenges are arranged in such a way that indicates whether they are better addressed in the wireless or optical segment of future FiWi networks. Their positions indicate the segment that seems more suitable to satisfy their requirements. For instance, mobility of end users can be easily addressed in the wireless segment, while their handovers may be better addressed in the optical segment. To realize reconfigurable FiWi networks, powerful load balancing and reconfiguration techniques must be developed for

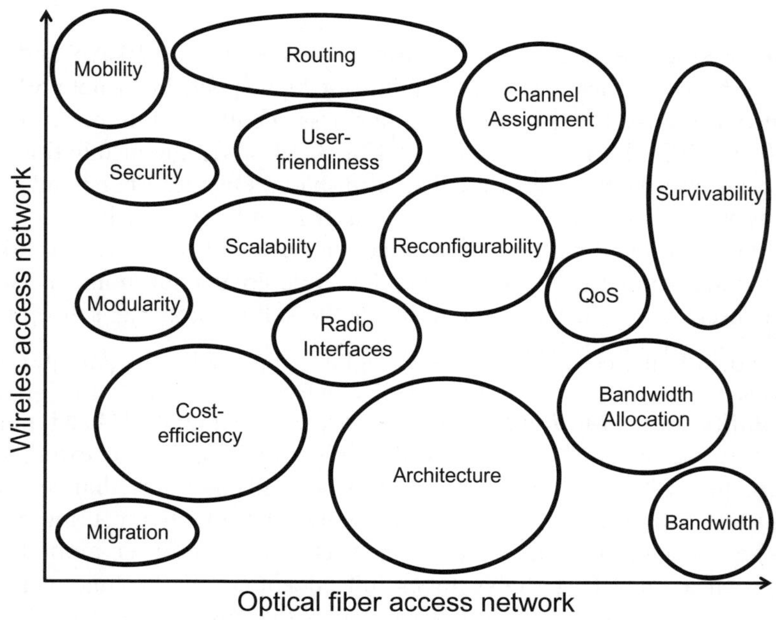

Figure 4.12. Challenges of future FiWi networks.

both wireless and optical segments. For instance, one promising approach might be that RAUs assign channels and bandwidth to end users, while the CO executes centralized DBA and DCA algorithms based on information periodically received from RAUs.

4.7 CONCLUSIONS

Hybrid optical–wireless FiWi networks form a powerful future-proof platform that provides a number of advantages. Introducing optical fiber into broadband wireless access networks helps relieve emerging bandwidth bottlenecks in today's wireless backhaul due to increasing traffic loads generated by new applications (e.g., iPhone). By simultaneously providing wired and wireless services over the same infrastructure, FiWi networks are able to consolidate (optical) wired and wireless access networks that are usually run independently of each other, thus potentially leading to major cost savings.

More interestingly, and certainly somewhat controversially, by paving all the way to and penetrating into homes and offices with high-capacity fiber and connecting wireless laptops and handhelds with high-throughput WiFi technologies to high-speed optical wired networks, FiWi networks give access to the ever-increasing processing and storage capabilities of memory and CPUs of widely used desktops, laptops, and other wireless handhelds (e.g., Wii). Note that nowadays desktop and laptop computers commonly operate at a clock rate of 1 GHz with a 32-bit-wide backplane, resulting in an internal flow of 2–8 Gbit/s with today's limited hard drive I/O, while future desktops and laptops are expected to reach 100 Gbit/s [7]. At present, these storage and processing capabilities are quite often utilized only in part. After bridging the notorious first/last mile bandwidth bottleneck, research focus might shift from bandwidth provisioning to the exploitation of distributed storage and processing capabilities available in widely used desktops and laptops, especially as we are about to enter the petabyte age with sensors everywhere collecting massive amounts of data [42]. An early example of this shift can be viewed in the design of P2P online game architectures that have begun to increasingly receive attention, where players' computing resources are utilized to improve the latency and scalability of networked online games, whose groundbreaking technologies might also be used to realize the future 3D Internet. On the other hand, in-house computer facilities might be replaced with computer utilities as in-house generators were replaced with electrical utilities [43]. Indeed, utility-supplied computing (e.g., Google) will continue to have an increasing impact on society and replace personal computer facilities unless new services and applications are developed that capitalize on them. Toward this end, it is important that FiWi networks are built using low-cost, simple, open, and ubiquitous technologies that allow all end users to have broadband access and to create unforeseen services and applications that help stimulate innovation, generate revenue, and improve the quality of our everyday lives, while at the same time minimizing the associated technical, economical, societal, and personal risks.

REFERENCES

1. T. Koonen, Fiber to the home/fiber to the premises: What, where, and when, *Proc. IEEE*, Vol. 94, No. 5, pp. 911–934, May 2006.
2. J. M. Cioffi, S. Jagannathan, M. Mohseni, and G. Ginis, CuPON: The copper alternative to PON 100 Gbit/s DSL networks, *IEEE Commun. Mag.*, Vol. 45, No. 6, pp. 132–139, June 2007.
3. P. E. Green, Optical networking update, *IEEE J. Selected Areas Commun.*, Vol. 14, No. 5, pp. 764–779, June 1996.
4. H. Shinohara, Broadband access in Japan: Rapidly growing FTTH market, *IEEE Commun. Mag.*, Vol. 43, No. 9, pp. 72–78, September 2005.
5. Y. Koike and T. Ishigure, High-bandwidth plastic optical fiber for fiber to the display, *IEEE/OSA J. Lightwave Technol.*, Vol. 24, No. 12, pp. 4541–4553, December 2006.
6. R. Ramaswami, Optical networking technologies: What worked and what didn't, *IEEE Commun. Mag.*, Vol. 44, No. 9, pp. 132–139, September 2006.
7. P. E. Green, *Fiber to the Home—The New Empowerment*, John Wiley & Sons, Hoboken, NJ, 2006.
8. J. Ghetie, *Fixed-Mobile Wireless Networks Convergence: Technologies, Solutions, Services*. Cambridge University Press, New York, 2008.
9. T.-S. Chu and M. J. Gans, Fiber optic microcellular radio, *IEEE Trans. Veh. Technol.*, Vol. 40, No. 3, pp. 599–606, August 1991.
10. J. Wu, J.-S. Wu, and H.-W. Tsao, A fiber distribution system for microcellular radio, *IEEE Photonics Technol. Lett.*, Vol. 6, No. 9, pp. 1150–1152, September 1994.
11. P. K. Tang, L. C. Ong, A. Alphones, B. Luo, and M. Fujise, PER and EVM measurements of a radio-over-fiber network for cellular and WLAN system applications, *IEEE/OSA J. Lightwave Technol.*, Vol. 22, No. 11, pp. 2370–2376, November 2004.
12. C. Lethien, C. Loyez, and J.-P. Vilcot, Potentials of radio over multimode fiber systems for the in-buildings coverage of mobile and wireless LAN applications, *IEEE Photonics Technol. Lett.*, Vol. 17, No. 12, pp. 2793–2795, December 2005.
13. C.-T. Lin, J. Chen, P.-C. Peng, C.-F. Peng, W.-R. Peng, B.-S. Chiou, and S. Chi, Hybrid optical access network integrating fiber-to-the-home and radio-over-fiber systems, *IEEE Photonics Technol. Lett.*, Vol. 19, No. 8, pp. 610–612, April 2007.
14. B. L. Dang and I. Niemegeers, Analysis of IEEE 802.11 in radio over fiber home Networks, in *Proceedings, IEEE Conference on Local Computer Networks*, Sydney, Australia, November 2005, pp. 744–747.
15. B. Kalantarisabet and J. E. Mitchell, MAC constraints on the distribution of 802.11 using optical fibre, in *Proceedings, European Conference on Wireless Technology*, Manchester, UK, September 2006, pp. 238–240.
16. P. S. Henry, Integrated optical/wireless alternatives for the metropolitan environment, *IEEE Communications Society Webinar*, April 2007.
17. Z. Jia, J. Yu, G. Ellinas, and G.-K. Chang, Key Enabling Technologies for optical–wireless networks: Optical millimeter-wave generation, wavelength reuse, and architecture, *IEEE/OSA J. Lightwave Technol.*, Vol. 25, No. 11, pp. 3452–3471, November 2007.

18. L. G. Kazovsky, W.-T. Shaw, D. Gutierrez, N. Cheng, and S.-W. Wong, Next-generation optical access networks, *IEEE/OSA J. Lightwave Technol.*, Vol. 25, No. 11, pp. 3428–3442, November 2007.
19. M. S. Kuran and T. Tugcu, A survey on emerging broadband wireless access technologies, *Computer Networks*, Vol. 51, No. 11, pp. 3013–3046, August 2007.
20. E. Perahia, IEEE 802.11n development: History, process, and technology, *IEEE Commun. Mag.*, Vol. 46, No. 7, pp. 48–55, July 2008.
21. G. R. Hiertz, S. Max, R. Zhao, D. Denteneer, and L. Berlemann, Principles of IEEE 802.11s, in *Proceedings, International Conference on Computer Communications and Networks (ICCCN)*, Honolulu, Hawaii, USA, August 2007, pp. 1002–1007.
22. B. Li, Y. Qin, C. P. Low, and C. L. Gwee, A survey on mobile WiMAX, *IEEE Commun. Mag.*, Vol. 45, No. 12, pp. 70–75, December 2007.
23. N. Ghazisaidi, M. Maier, and C. M. Assi, Fiber-wireless (FiWi) Access networks: A survey, *IEEE Commun. Mag.*, Vol. 47, No. 2, pp. 160–167, February 2009.
24. G. Shen, R. S. Tucker, and C.-J. Chae, Fixed mobile convergence architectures for broadband access: Integration of EPON and WiMAX, *IEEE Commun. Mag.*, Vol. 45, No. 8, pp. 44–50, August 2007.
25. V. Muralidharan, A. M. Wyglinski, and W. Wang, HiFi-WiN: Hybrid integrated fiber–wireless networking for broadband metropolitan area access, in *Proceedings, Virginia Tech Symposium on Wireless Personal Communications*, Blacksburg, VA, June 2007, pp. 1–8.
26. L. Wen-Piao, K. Ming-Seng, and C. Sien, A reliable architecture for broad-band fiber-wireless access networks, *IEEE Photonics Technol. Lett.*, Vol. 15, No. 2, pp. 344–346, February 2003.
27. S. Bhandari and E. K. Park, Hybrid Optical Wireless Networks, in *Proceedings, International Conference on Networking, International Conference on Systems, and International Conference on Mobile Communications and Learning Technologies (ICNICONSMCL)*, Mauritius, April 2006, pp. 113–117.
28. W.-T. Shaw, S.-W. Wong, N. Cheng, X. Zhu, M. Maier, and L. G. Kazovsky, Hybrid architecture and integrated routing in a scalable optical-wireless network, *IEEE/OSA J. Lightwave Technol.*, (Special Issue on Convergence of Optical Wireless Access Networks), Vol. 25, No. 11, pp. 3443–3451, November 2007.
29. M. Maier, N. Ghazisaidi, and M. Reisslein, The audacity of fiber-wireless (FiWi) networks (Invited Paper), in *Proceedings, ICST ACCESSNETS*, Las Vegas, NV, October 2008.
30. N. Ghazisaidi, F. Paolucci, and M. Maier, SuperMAN: Optical–wireless integration of RPR and WiMAX, *OSA J. Optical Networking*, Vol. 8, No. 3, pp. 249–271, March 2009.
31. M. P. McGarry, M. Maier, and M. Reisslein, WDM Ethernet passive optical networks, *IEEE Commun. Mag.*, Vol. 44, No. 2, pp. S18–S25, February 2006.
32. M. P. McGarry, M. Reisslein, and M. Maier, Ethernet passive optical network architectures and dynamic bandwidth allocation algorithms, *IEEE Commun. Surv. Tutorials*, Vol. 10, No. 3, pp. 46–60, 3rd Quarter 2008.
33. M. K. Weldon and F. Zane, The economics of fiber to the home revisited, *Bell Labs Tech. J.*, Vol. 8, No. 1, pp. 181–206, July 2003.

34. A. V. Tran, C. J. Chae, and R. S. Tucker, Ethernet PON or WDM PON: A comparison of cost and reliability, in *Proceedings, IEEE TENCON*, Melbourne, Australia, November 2005, pp. 1–6.
35. D. Niyato and E. Hossain, Integration of WiMAX and WiFi: Optimal pricing for bandwidth sharing, *IEEE Commun. Mag.*, Vol. 45, No. 5, pp. 140–146, May 2007.
36. P. Lin, T. Wang, J. Hu, Y. Suemura, S. Nakamura, and C. Qiao, Optical-to-wireless-integration cost modeling, in *Proceedings, High Performance Switching and Routing (HPSR)*, Brooklyn, NY, June 2007, pp. 1–6.
37. S. Sarkar, S. Dixit, and B. Mukherjee, Hybrid wireless–optical broadband-access network (WOBAN): A review of relevant challenges, *IEEE/OSA J. Lightwave Technol.*, Vol. 25, No. 11, pp. 3329–3340, November 2007.
38. J. J. V. Olmos, T. Kuri, and K. Kitayama, Dynamic reconfigurable WDM 60-GHz millimeter-waveband radio-over-fiber access network: Architectural considerations and experiment, *IEEE/OSA J. Lightwave Technol.*, (Special Issue on Convergence of Optical and Wireless Access Networks), Vol. 25, No. 11, pp. 3374–3380, November 2007.
39. Q. Ni, A. Vinel, Y. Xiao, A. Turlikov, and T. Jiang, Investigation of bandwidth request mechanisms under point-to-multipoint mode of WiMAX networks, *IEEE Commun. Mag.*, Vol. 45, No. 5, pp. 132–138, May 2007.
40. H. Kim, J. H. Cho, S. Kim, K.-U. Song, H. Lee, J. Lee, B. Kim, Y. Oh, J. Lee, and S. Hwang, Radio-over-fiber system for TDD-based OFDMA wireless communication systems, *IEEE/OSA J. Lightwave Technol.*, (Special Issue on Convergence of Optical and Wireless Access Networks), Vol. 25, No. 11, pp. 3419–3427, November 2007.
41. K. Kitayama, T. Kuri, J. J. V. Olmos, and H. Toda, Fiber-wireless networks and radio-over-fibre technique, in *Proceedings, Conference on Lasers and Electro-Optics/Quantum Electronics and Laser Science Conference (CLEO/QELS)*, San Jose, CA, May 2008, pp. 1–2.
42. The petabyte age, *WIRED*, p. 106ff, July 2008.
43. N. Carr, *The Big Switch: Rewiring the World, from Edison to Google*, W. W. Norton & Company, New York, 2008.

5

PACKET BACKHAUL NETWORK

Hao Long

5.1 INTRODUCTION

A backhaul network is usually responsible for transparently transmitting service from the front of an end user, such as a digital subscriber line access multiplexer (DSLAM) and a base station, to the point of presence (POP), and vice versa. The networking technique used in backhaul network is often called transport technology. In this chapter, we also use the word "transport" for the technology related description.

Figure 5.1 shows a typical transport network and its networking requirements:

- *Management/Control Plane:* A transport network must be a strictly controlled network. Generally, all the network resources are managed by the network management systems (NMS). A transport network is usually dependent on NMS, which provide the centralized control on the whole network. For a transport network, the control plane is optional, which means that the network should work well without any control plane. But in fact, most networks have a control plane because the control plane can facilitate the network management.

Convergence of Mobile and Stationary Next-Generation Networks, edited by Krzysztof Iniewski

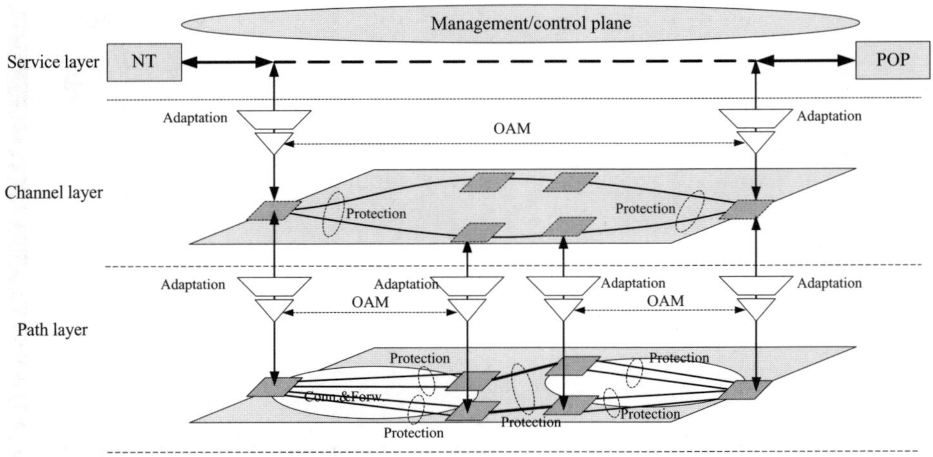

Figure 5.1. Transport network architecture.

- *Traffic Engineering:* A transport network is generally a traffic engineered network. The forwarding path, service access control, bandwidth allocation, and so on, are all under unified control. In most cases, the technology used in transport network is connection-oriented-based technology. Traffic engineering is an important means to satisfy the service level agreements (SLA).
- *Layered Network:* A typical transport network is usually divided into three layers that are functionally separated.
  - *Service Layer:* The layer of service traffic. The client service is available in this layer.
  - *Channel Layer:* This layer provides the separation between different service instances for service-level management. The channel layer label often can be viewed as a identification within the network.
  - *Path Layer:* This layer provides the transport connection between the ingress node and the egress node for each service. The transport network nodes forward the traffic based on the path layer label—for example, VC4 timeslot for SDH network.

For the three layers, the upper layer is carried on the lower layer. The adaptation function is used to encapsulate the upper-layer traffic into lower-layer traffic.

- *Operation, Administration, and Management (OAM):* Control-plane-independent OAM is a very important feature of transport network. OAM provides many tools for connectivity monitoring, performance monitoring, fault location, and so on. The connectivity monitoring tools is the basis for fast restoration in case of failure. The performance monitoring tools is critical for testing the satisfaction of the QoS requirement. The fault location can help the operator to quickly locate the failure point. Many alarm functions in the OAM tools can also play very important roles in practical network management. OAM is usually present in each layer of the network.

- *Fast Protection:* Fast protection is another very important feature of transport network. Fast restoration is a very popular service requirement. Many OAM tools are designed for helping the fast protection implementation. Typically, transport network should recover the service in 50 ms when a single failure occurs. Fast protection is usually present in the path layer, but also in the channel layer in some cases.

5.1.1 From SDH/SONET to Packet Transport

SDH/SONET has been proved to be a very successful technology used in transport/backhaul network. It has been widely deployed around the world. It has provided very good support for voice service and virtual private line service.

However, now the voice service is evolving from TDM traffic (i.e., 2G), to IP traffic (i.e., 3G and LTE); meanwhile data service is becoming popular. Most of the network traffic becomes data traffic with respect to the rapid growth of data service in IP form. As a TDM-based technology, SDH/SONET is hard to provide efficient support for this type of service.

- Packet service is a kind of burst service, which means that it is not even during its transmission. SDH/SONET can only provide fixed/hard bandwidth for each service, which means that it cannot efficiently transmit the packet service with statically multiplexing.
- SDH/SONET is designed with 10 Gbps as the highest rate. However, the bandwidth demands are growing so fast thanks to the data service. A rate of 10 Gbps might be hard to satisfy the bandwidth demands in future years.

Packet-based technology can solve the above problems. The statistically multiplexing feature of packet switching is good at transmitting the burst traffic. The bandwidth of packet switching is also easily increased by updating the switch fabric, line card port, and so on. IEEE is working on the 40GE/100GE standard, and until now the most technical problem has been resolved.

5.1.2 From Legacy Packet Technology to Packet Transport Technology

The legacy packet switching technologies are seldom used in a transport network. The Ethernet is always used as local network technology, while IP is used in a router core network. Thus, prior to its deployment in this area, many new functions should be integrated into packet switching to satisfy the requirements of transport network as described previously.

5.1.3 Packet Transport Technologies

The most acceptable way to build up a new packet transport technology is to stand on the basis of the current popular packet switching techniques. At present, there are three most popular packet switching techniques: IP, Ethernet, and MPLS.

IP switching is regarded as the most expensive solution because it uses of the variable-length match for forwarding. The Ethernet and MPLS are thought of as the most suitable techniques for extension to transport network. The following sections respectively describe the enhancement to the two techniques to become packet transport technologies. Since there are not many new things in management plane and layered network, the sections will focus on the other aspects shown in Figure 5.1, especially OAM and protection.

5.2 ETHERNET BACKHAUL NETWORK

5.2.1 Extending Ethernet into Carrier Network

The Ethernet is used as the most popular technology for local area network. It also began to be viewed as a promising access technology. However, since the Ethernet is designed for LAN application, it is hard to be applied in carrier transport network:

- Carrier-class OAM feature was not defined.
- Slow restoration: Ethernet restoration relied on the convergence of the spanning tree protocol, which needs at least hundreds of milliseconds, even several seconds.
- Traffic engineering is unsupported.
- Scalability is not good because the flooding and spanning tree protocol cannot support a big network with thousands of nodes.

In past years, IEEE and ITU-T put much effort on extending the Ethernet into the carrier network. IEEE 802.1ad defined provider bridges that support the demarcation of customer network and provider network by service VLAN tag. The function enables the flooding traffic to be limited in its owner domain. IEEE 802.1ah, published in 2008, defined provider backbone bridges that support of customer network and provider network by service instance tag and MAC-in-MAC encapsulation. This function separates the MAC address space of both the customer network and the provider network, both of which relieve the burden of MAC address table. These two standards provide much improvement of the scalability of the Ethernet network (Figure 5.2).

IEEE 802.1Q, 802.1ad, and 802.1ah also provide the support of a network management system that can be used to configure the Ethernet VLAN network. In a practical network, a point-to-point VLAN connection is a kind of typical deployment because in operator networks a point-to-point connection is the most commonly used connection type (Figure 5.3).

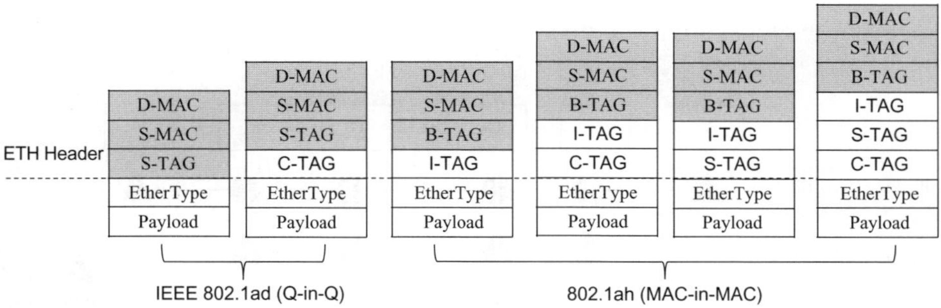

Figure 5.2. IEEE 802.1ad and 802.1ah.

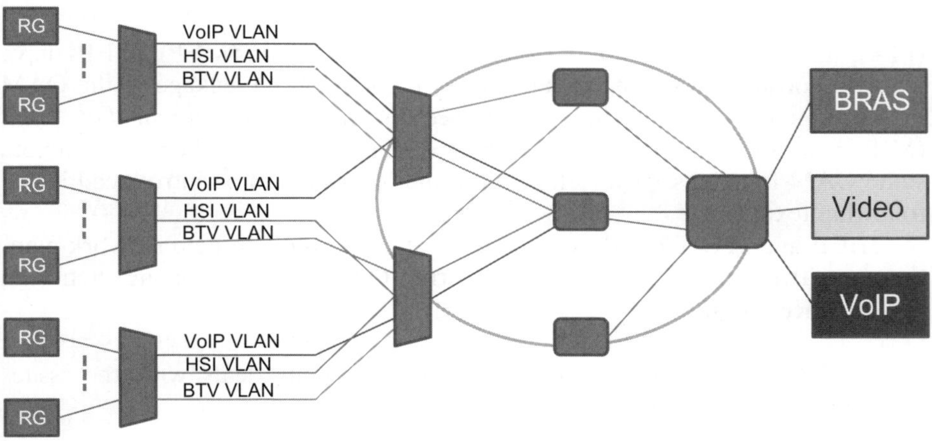

Figure 5.3. Ethernet in operator network.

IEEE and ITU-T also developed a very strong OAM and protection function for the Ethernet network. This is a key extension for the Ethernet to be used in an operator network.

5.2.2 Ethernet Operation, Administration, and Management (ETH OAM)

IEEE and ITU-T have defined some concepts for describing the OAM function. The *maintenance domain* (MD) is an enclosed domain that includes all the nodes connections to be maintained by the management system of the domain (Figure 5.4). Difference management systems have different maintenance scopes. The *MD level* is used to describe the scope of a maintenance domain. An MD with larger MD level may cover a smaller MD. Generally, the objects to be maintained are the connections contained in the maintenance domain. For a connection under maintenance, the group of its end points is called a *maintenance association*

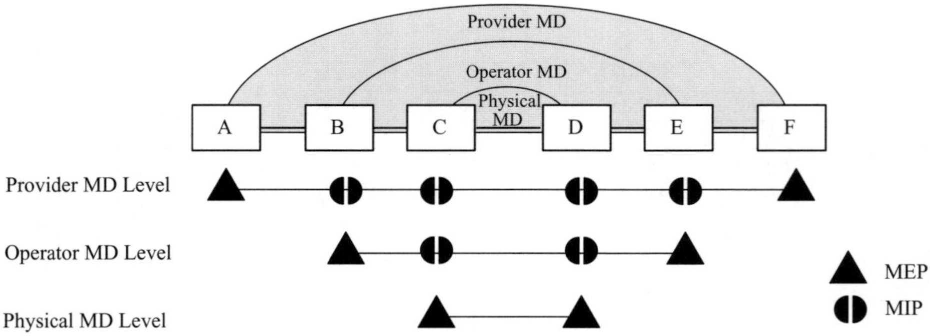

Figure 5.4. Maintenance domain.

(MA), and each end point is called an *MA end point* (MEP). MEPs have many important OAM functions including preventing the leakage of the OAM messages generated from the maintenance domain. An *MD intermediate point* (MIP) is a point inside the maintenance domain, and it can help to complete some OAM functions, such as link trace and loopback, to be introduced in the following text.

IEEE and ITU-T has developed a set of OAM tools to help network management. Two important aspects of Ethernet OAM are fault management and performance monitoring.

5.2.2.1 Fault Management. Fault management deals with the issues regarding link failure or node failure.

Continuity Check (CC) and Remote Defect Indication (RDI). Continuity check is the most used OAM function because it can detect link/node failure in milliseconds. The detection is between the two ends (MEPs), of a connection, typically bidirectional connection. The two ends send continuity check messages (CCM) to the remote peer(s) periodically (Figure 5.5). If one end doesn't receive any CCM in 3.5 times of the period, it will declare a fault detected on the direction from the remote side to the local side. To let the remote peer(s) know of this failure, it will set the *remote defect indication* (RDI) flag in the CCM messages sent by itself to the remote side (Figure 5.5).

Besides the fault detection, continuity check can also be used for connectivity verification. The CCM messages carry the MD name, the MA name and MEP ID for this purpose. If one end received an unexpected combination of the MD name, the MA name, and the MEP ID, it will declare a mismatch or misconnection defect. All the defects detected by a continuity check will be reported to a network management system to notify the operator.

Loopback. Loopback is the most used OAM tools for locating the failure point. Similar to continuity check, the idea of loopback function is also very easily

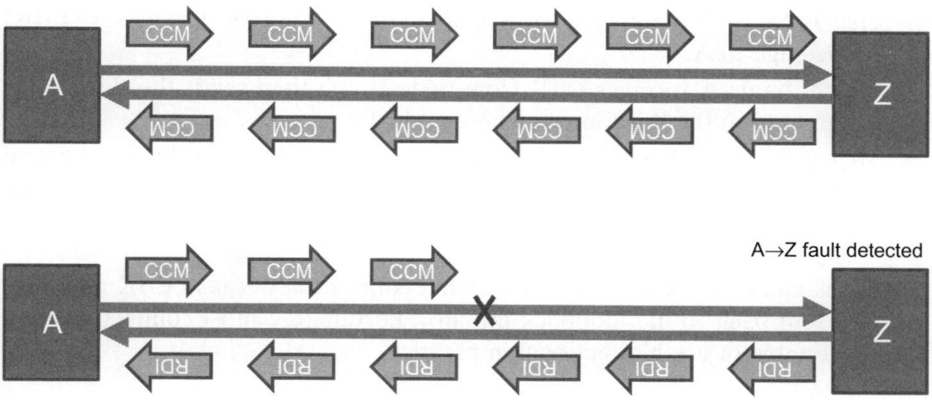

Figure 5.5. CCM and RDI.

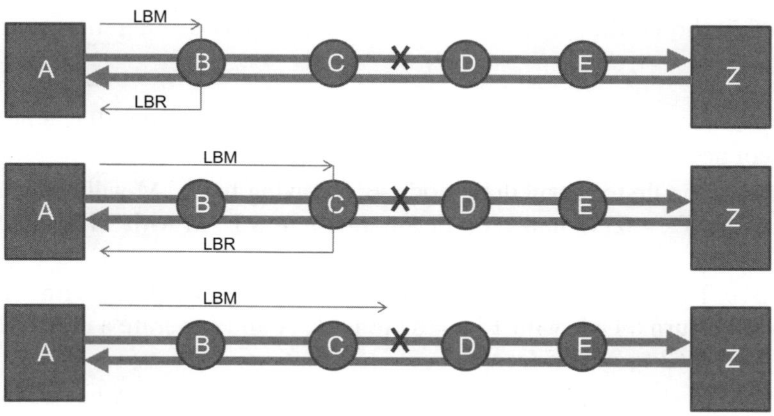

Figure 5.6. Loopback function.

understood. When a failure is detected on the connection, the NMS operator or the node itself will initiate a loopback function to locate the failure. Figure 5.6 shows an example:

Step 1: Node A learns a failure with respect to CCM loss or the RDI flag in the CCM received from Z, and it initiates loopback function and send a loopback message (LBM) to the first intermediate node, B.

Step 2: Node B receives the LBM and will return a loopback response (LBR) message to A.

Step 3: Node A receives LBR from node B, and it makes sure that the segment from A to B is OK. Node A then sends the second LBM targeted to the second intermediate node, C.

Step 4: Node C receives the LBM and will return a loopback response (LBR) message to A.

Step 5: Node A receives LBR from node B and then sends the third LBM targeted to the third intermediate node, D.

Step 6: Node A doesn't receive LBR from node D after a fixed time, and declare that the failure is located on node D or on the link between node C and node D.

Loopback can only work with bidirectional connection because LBR messages must be sent back to the loopback initiator. Fortunately, most connections in a transport network are bidirectional in practice.

Link Trace. Link trace is often used for topology discovery in the Ethernet network. In many cases, the head end of a connection doesn't know about the MAC addresses of the intermediate nodes. However, to know about this information is very important because it is useful when performing loopback function. Figure 5.7 shows the procedure for a head end collecting the connection topology:

Step 1: Node A initiates a link trace message (LTM) with TTL = 64 on the connection.

Step 2: The first intermediate node, B, receiving the LTM will return a LTR with TTL = 63 to node A and generate a new LTM with TTL = 63 to next nodes.

Step 3: The second intermediate node, C, receiving the LTM with TTL = 63 will return a LTR with TTL = 62 to node A and generate a new LTM with TTL = 62 to next nodes.

Step 4: Node D, E, Z will do the things that are similar to what node B and node C have done. Note: An exception is that node Z will not generate a new LTM because it is the end of the connection.

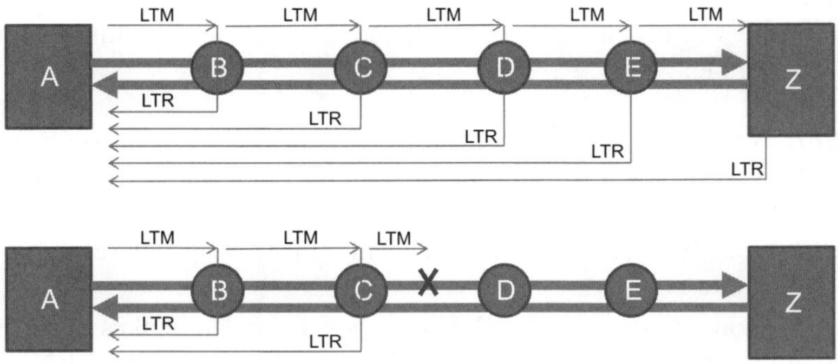

Figure 5.7. Link trace function.

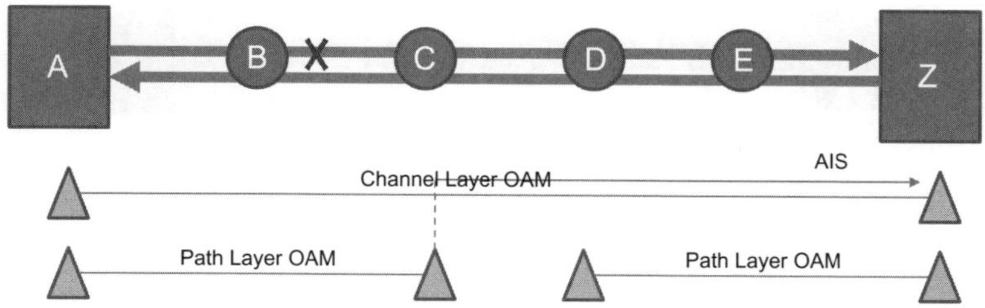

Figure 5.8. AIS function.

Step 5: Node A receives all the LTRs and sort the MAC addresses carried in the LTRs based on the TTL value. The less the TTL value, the closer the node to node A itself.

Besides topology discovery, link trace also can be used for fault location. In case of a failure on the connection, the head node will not receive LTR message as the lower part of Figure 5.7 shows from the downstream node of the failure point, and thus it can learn failure location. Similar to loopback function, a link trace also can only work with a bidirectional connection.

Alarm Indication Suppression/Signal. Alarm indication suppression/signal (AIS) is a very useful complementary tool for the network management. It is a common case that when a link/node failure occurs, many OAM instance will detect the failure because there should be many connection passes through the link/node. If all related OAM instance reports a failure to network management system, it is hard to find where the failure occurs on earth because NMS gets too many fault reports. AIS is used to suppress all other fault reports except for the OAM instance, which generates AIS.

Figure 5.8 shows an example of the AIS application. A path layer OAM instance is established between node A and node C and also between node D and node Z. A channel layer OAM instance is established between node A and node Z. When the link between node B and node C fails, node C and node Z will detect the failure. In this case, node C will generate AIS messages that are sent on channel layer OAM instance. When node Z receives the AIS message, it will suppress the fault report to NMS. The result is that only node C reports a fault to NMS, and it is easy for an operator to identify the failure source.

5.2.2.2 Performance Management. Performance management OAM is responsible for monitoring the service/path quality. It can be used to detect the signal degrade, partial fault, and so on. Two typical performance monitoring OAM tools are loss measure and delay measure.

Loss Measurement. Loss measurement is used to monitoring the packet loss ratio for a connection. Figure 5.9 shows the principle of loss measurement used in ETH OAM.

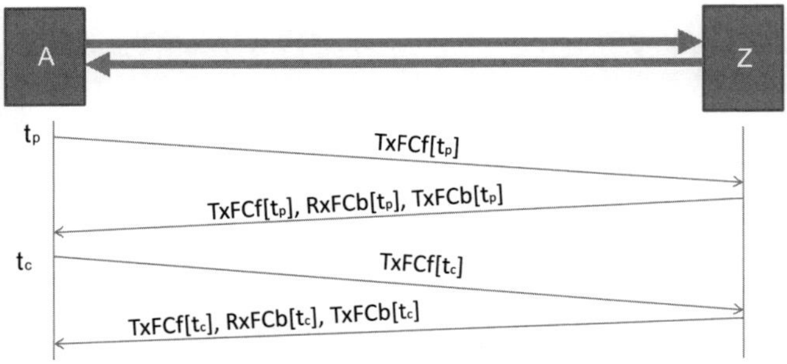

Figure 5.9. Loss measurement function.

Step 1: The head-end A sends the current transmitted frame count at time t_p, TxFCf[t_p], to the far end Z.

Step 2: The far-end Z returns the information of current received frame count, RxFCb[t_p], and current transmitted frame count, TxFCb[t_p], at time of receiving message from A. The return message also includes the receiving TxFCf[t_p].

Step 3: The head-end A sends the current transmitted frame count at time t_c, TxFCf[t_c], to the far end Z.

Step 4: The far-end Z returns the information of current received frame count, RxFCb[t_c], and current transmitted frame count, TxFCb[t_c], at time of receiving message from A. The return message also includes the receiving TxFCf[t_c].

Based on the information of the two rounds and its local received frame count at time t_p and t_c (i.e., RxFCl[t_p] and RxFCl[t_c]), the bidirectional packet loss ratios can be calculated out according to the following formulas:

$$\text{Frame Loss}_{\text{far-end}} = |\text{TxFCb}[t_c] - \text{TxFCb}[t_p]| - |\text{RxFCb}[t_c] - \text{RxFCb}[t_p]|$$

$$\text{Frame Loss}_{\text{near-end}} = |\text{TxFCf}[t_c] - \text{TxFCf}[t_p]| - |\text{RxFCl}[t_c] - \text{RxFCl}[t_p]|$$

The Frame $\text{Loss}_{\text{far-end}}$ represents the loss ratio on the far end, and the Frame $\text{Loss}_{\text{near-end}}$ represents the loss ratio on the local end.

In Ethernet OAM, there are two ways to carry the frame count information exchanged as specified in Figure 5.9. One is piggybacked on CCM, and the other is to define new messages for them—that is, loss measurement message (LMM) and loss measurement reply message (LMR).

Delay Measurement. Delay measurement is used to detect the packet transmission delay from one end to another end. Figure 5.10 shows the principle of delay measurement used in ETH OAM.

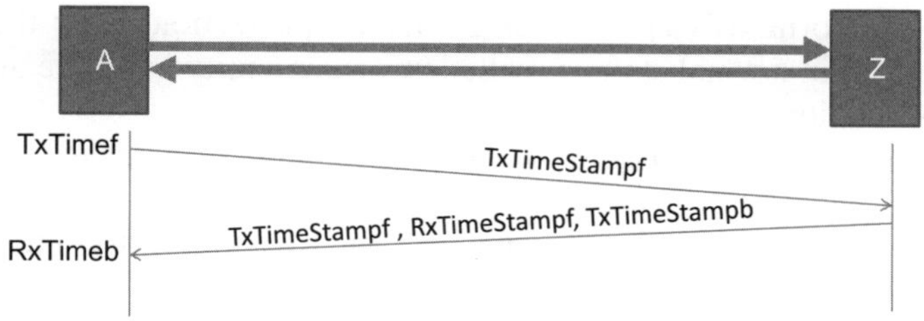

Figure 5.10. Delay measurement function.

Step 1: Head-end A sends a delay measure message (DMM) to the far-end Z. The DMM message carries a timestamp, TxTimeStampf, which represent the time sending the message.

Step 2: The far-end Z receiving the message will return a delay measure reply message (DMR) to node A. The DMR message will carry the timestamp for the time receiving the DMM message, RxTimerStampf, and the timestamp for the time transmitting the DMR message, TxTimeStampb. It also returns the receiving TxTimeStampf.

Step 3: Node A receiving the DMR will calculate out the bidirectional transmission delay according to the receiving time RxTimeb and timestamps carried in DMR with the following formula:

$$\text{Frame Delay} = (\text{RxTimeb} - \text{TxTimeStampf}) - (\text{TxTimeStampb} - \text{RxTimeStampf})$$

5.2.2.3 Additional OAM functions. This section shows the most frequently used OAM tools. In fact, there are many other OAM functions, such as client signal failure, lock, test, and so on. The reader can find the details in the recommendation ITU-T Y.1731.

5.2.3 Ethernet Protection Switching

End-to-end service recovery in 50 ms in the case of network failure is a typical requirement of transport network. Ethernet protection switching mechanism can switch the traffic from the failed working connection to a healthy backup connection. Ethernet gains the fast recovery capacity with the help of the fast failure detection with OAM function and the fast protection switching trigger mechanism. There are two typical protection switching mechanisms: linear protection switching and ring protection switching.

5.2.3.1 Ethernet Linear Protection. Ethernet linear protection switching is only applicable for point-to-point VLAN connection at present. In a linear protection switching mechanism, for each transport connection, a backup transport connection is preconfigured for protecting the service on it. The two connections are formed as a protection group. In a protection group, the transport connection used in normal status is called "working entity," and the another one is called "protection entity."

There are two typical linear protection switching mechanisms:

1+1 Protection: The traffic is sent both on working entity and protection entity. The far end takes one copy of the traffic from one of them.

1:1 Protection: The traffic is sent only on one of working entity and protection entity.

1:1 protection switching is most common in practice because the protection entity can be used for carrying other service traffic that is not as important. Ethernet linear protection switching utilizes an automatic protection switching (APS) message to coordinate the protection switching action on the both ends. Figure 5.11 shows the APS application in the 1:1 protection switching case.

Step 1: In normal status, the traffic is transmitted and received on working entity. The APS messages transmitted on the protection entity carry "No request (NR)" information, which means no protection switching request at present.

Step 2: A unidirectional failure occurs on the working entity in the direction from A to Z. Node Z detects the failure and switch to transmit and receive traffic from a protection entity. It also sends the APS(SF) message to the far end, A. SF means signal failure.

Step 3: When node A receives the APS(SF) message, it will also switch to transmit and receive traffic from protection entity.

After the three steps, the service traffic will be transmitted on and received from the protection entity on both directions. After the failure is repaired, there are two optional operations: return to work entity (revertive mode) or remain on

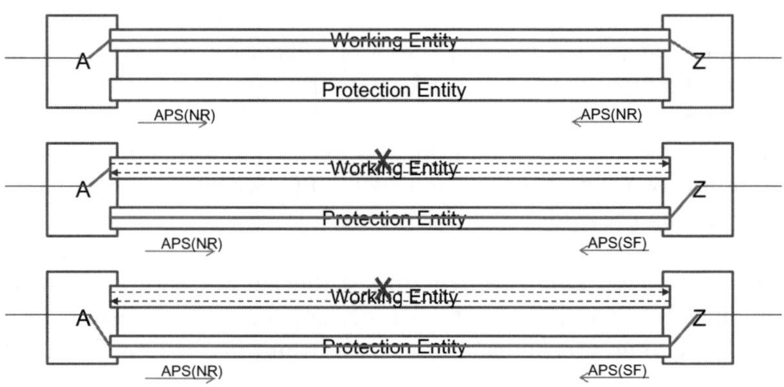

Figure 5.11. Ethernet 1:1 linear protection switching (failure case).

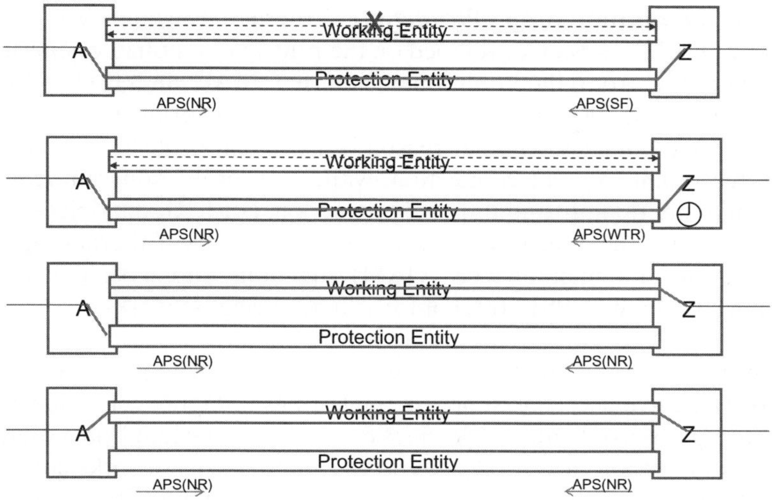

Figure 5.12. Ethernet 1:1 linear protection switching (recovery case).

protection entity (nonrevertive mode). In practice, the revertive mode is the most frequently used. Figure 5.12 shows the protection switching in the case of link recovery with revertive mode:

Step 1: In failure status, the traffic is transmitted and received on a protection entity. The APS messages transmitted on a protection entity carry a "signal failure (SF)" indication.

Step 2: The failed link gets repaired. Node Z detects the recovery by the OAM function. It will start a *wait to restore* (WTR) timer. During the WTR timer running, node Z also sends an APS(WTR) message to notify node A current status. Note that the WTR timer is very useful in carrier network. It can prevent the protection switching flapping, which is caused by the instability when repairing a failure. The WTR timer is usually between 5 and 12 minutes.

Step 3: When the WTR timer expires, node Z will switch back to a working entity. It will also transmit an APS(NR) message to node A.

Step 4: When node A receives an APS(NR) message from node Z, it will also switch back to a working entity. To this step, the client traffic has been switched back to a working entity in both directions.

Ethernet linear protection switching also provides many other APS events to cover the cases that would happen in practice as much as possible. For example, manual switch and force switch are defined for operator's maintenance activity. In fact, the whole protection switching logic is a very complicated system, although its principle is very easily understood. People who have great interest can find the details in ITU-T G.8031.

5.2.3.2 Ethernet Ring Protection. Ethernet ring protection (ERP) defined by G.8032 has been developed on the principle of utilizing generic mechanisms inherited from the traditional Ethernet MAC and bridge functions. The requirement of ERP comes from the fact that many of the present networks have ring-link topologies. Another reason is that linear protection switching can only protect point-to-point connection, while ring protection can achieve the protection of multipoint-to-mulitpoint connection. For ERP, the objective of fast protection switching is achieved by integrating mature Ethernet operations, administration, and maintenance (OAM) functions and a simple automatic protection switching (APS) protocol for Ethernet ring networks.

Ring Topology. Figure 5.13 shows the possible variants of Ethernet ring topology. The current G.8032 supports single rings in Figure 5.13a and 5.13b. In the multi-ring cases of Figure 5.13c, rings can be interconnected via a shared link. In the second version of G.8032,[1] multi-rings and ladder networks consisting of conjoined Ethernet rings are also supported by the Ethernet ring protection.

Client Channel, APS Channel and Channel Loop Avoidance. ERP separates a ring APS channel from client channel since one APS channel can be used to carry APS information for many client channels. ERP uses a separate VLAN within the scope of the ring as a ring APS channel (R-APS channel).

As Figure 5.14 shows, both the client channel and the R-APS channel will form a ring loop in the ring topology. The ring loop will destroy the network in the Ethernet because the Ethernet has no information as basis of terminating an unknown frame in a data forwarding plane. To break the loop, the ERP mechanism will ensure that there is at least one blocked ring port onthe ring.

In the normal state, one ring link is designated as the *ring protection link* (RPL), which blocks Ethernet traffic to guarantee the loop avoidance. An *RPL owner*, which is attached to one end of the RPL, is designated to perform traffic blocking. The RPL owner sets the port on the RPL as blocked, and thus it drops any client traffic received from or sent to the RPL. The RPL owner plays a very important role in G.8032, because it is responsible for use of the RPL for ring

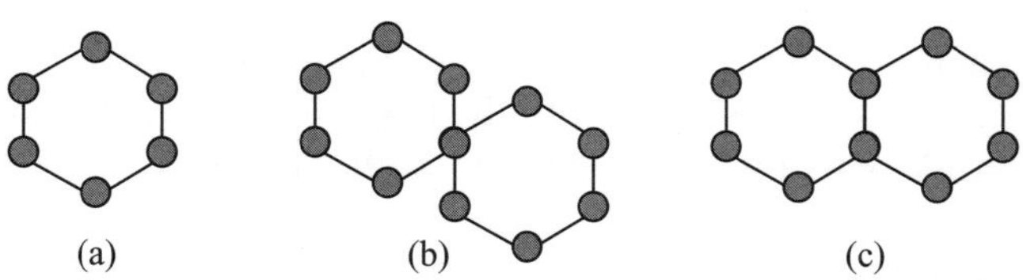

Figure 5.13. Possible ring topologies: **(a)** Single ring. **(b)** Two single rings with a shared mode. **(c)** Multi-ring with shared link and nodes.

[1]G.8032 v2 have developed some mechanisms to support multi-ring topology.

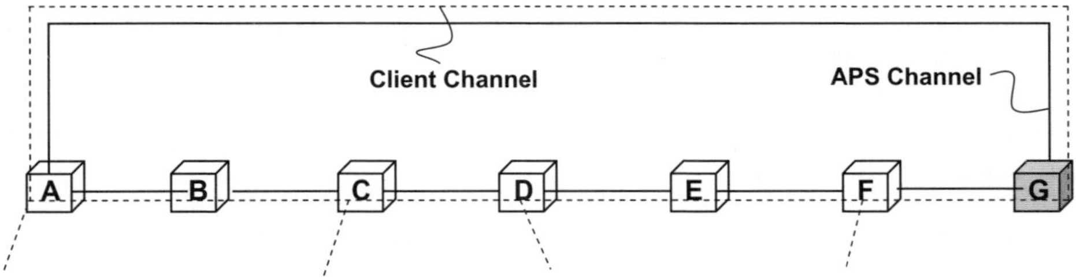

Figure 5.14. Client channel and ring APS channel.

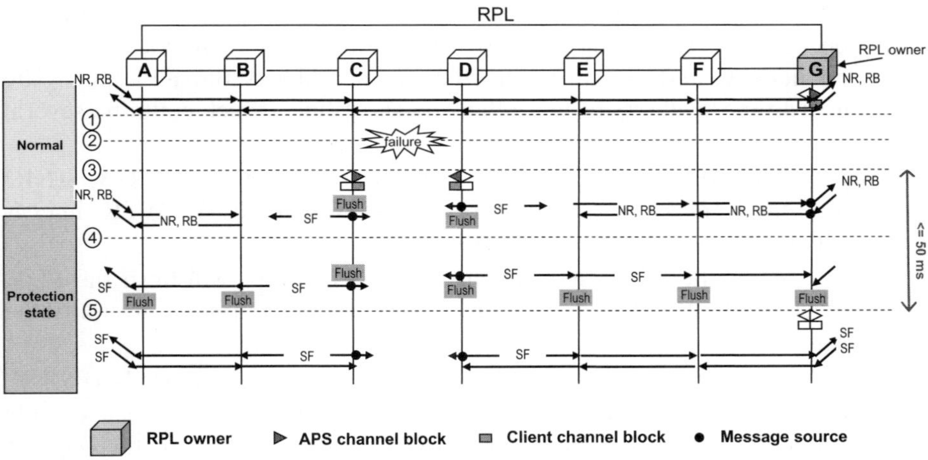

Figure 5.15. Example sequence diagram of link failure.

protection switching. When a failure occurs, the RPL owner will open the RPL port to resume the connectivity of all ring nodes.

Ethernet Ring Protection Switching. Figure 5.15 illustrates a scenario in the case of link failures which shows ERP processes as follows:

Step 1: In the normal state, the RPL owner blocks its port connected to the RPL. The RPL owner is also responsible for transmitting APS(NR, RB) messages on the ring to indicate that the ring is OK and the RPL is blocked.

Step 2: A link failure occurs between node C and node D.

Step 3: Nodes C and D detect the local SF condition, and each node flushes its FDB, blocks the failed port, and transmits an R-APS(SF) message on both ring ports, followed by periodic transmission of the same messages, while the SF condition persists.

Step 4: Other nodes flush FDBs on when receiving the R-APS(SF) message from node C or node D. In addition, the RPL owner will unblock its port on the RPL.

Step 5: The following R-APS(SF) will not trigger FDB flush and the ring will enter the protection state.

Figure 5.16 illustrates a scenario in the case of link failures which shows ERP processes as follows:

Step 1: The link between node C and node D is repaired.

Step 2: Nodes C and D detect the local SF, which is cleared. The two nodes will start the guard timer, and transmit an R-APS(NR) message on both ring ports. Note that the node running guard timer will discard R-APS messages. Here the guard timer is used to prevent the full loop because the remaining R-APS(SF) messages would unblock the ports on node C and D before the RPL owner blocks the RPL port. The guard timer will expire in a relatively short time, which could be several seconds.

Step 3: The RPL starts the WTR timer while receiving an R-APS(NR) message to make sure the link repair has completed. The WTR timer is usually between 5 and 12 minutes.

Step 4: After the WTR timer expires, the RPL owner will flush the FDB, block the RPL link, and send an R-APS(NR, RB) message to notify the whole ring.

Step 5: All other ring nodes will flush the FDB when receiving an R-APS(NR, RB) message. In addition, node C and D will unblock the blocked ports and stop transmitting the R-APS(NR) message. To this step, the ring has switch back to the normal state.

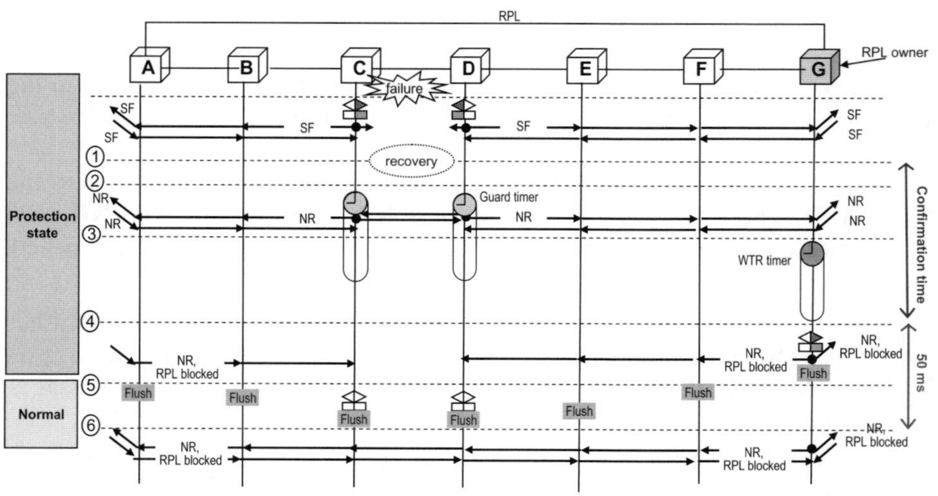

Figure 5.16. Example sequence diagram of link recovery.

Here, this section only provides a general description of the procedure in case of the simplest scenario. The ITU-T has defined much more, such as manual switch, nonrevertive, force switch, ring interconnection, and so on. The readers who have an interest can find the details from ITU-T G.8032.

5.2.4 Provider Backbone Bridge—Traffic Engineering (PBB-TE).

The PBB-TE is an extension of provider backbone bridge technology which is specified by IEEE 802.1ah, and it defines a kind of connection-oriented network based on the Ethernet frame format. The PBB-TE inherited the forwarding rules from the Ethernet legacy—that is, forwarding based on destination address (DA) and VLAN TAG. In a network, usually only a subset of Ethernet VLAN is used as PBB-TE. The PBB-TE VLAN has the following features:

- It forbids self-learning and flooding unknown frames.
- It does not attend the process of spanning tree protocol.
- The forwarding table is statically configured by a network management system (NMS).

A PBB-TE uses the triple <DA, SA, VID> as a connection identifier. For point-to-point PBB-TE connection, DA is the unicast address of the connection end. For point-to-multipoint PBB-TE connection, DA is the multicast address. The triple must be specified in the packet header as the forwarding information as shown in Figure 5.17.

Extension of ETH OAM for a PBB-TE Network. It is almost the same for the end-to-end OAM function (such as continuity check) between the Ethernet legacy and the PBB-TE. The only difference is that the header of the OAM packet should be replaced by the PBB-TE connection identifier. But for the OAM function involving intermediate nodes, since the destination address field has been occupied by a PBB-TE connection identifier, it cannot be used for

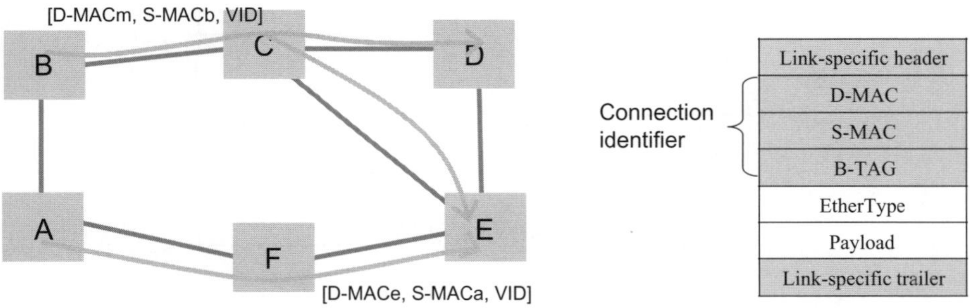

Figure 5.17. PBB-TE connection.

targeting the intermediate nodes. In PBB-TE OAM, for loopback function, the target intermediated node information is carried in the LBM PDU. In addition, since the PBB-TE connection allows the different VLAN IDs to be used in two directions, for loopback and link trace function, the reverse VLAN ID information is also needed to be carried in LBM and LTM. The reverse VLAN ID information is used to indicate the correct reverse connection identifier, which is encapsulated on LBR and LTR.

PBB-TE Linear Protection. PBB-TE also specified its own protection switching mechanism. It is a kind of 1:1 linear protection switching, but it does not utilize the APS mechanism used in the VLAN network. The difference is that it uses RDI to trigger protection switching in case of unidirectional failure as shown in Figure 5.18. In addition, PBB-TE makes a large simplification according to Ethernet linear protection switching by assuming that PBB-TE connection is always under the same management system. The reduction includes the negotiation of manual switch, force switch, and so on.

At present, the IEEE has completed the standardization of PBB-TE. People can find the details from IEEE 802.1Qay.

5.3 MPLS BACKHAUL NETWORK

5.3.1 Extending MPLS into Backhaul Network

At present, MPLS is usually used as a core network technology. Most core routers have MPLS capacity. MPLS has a short, fixed label for efficient forwarding. In addition, the greatest advantage of MPLS technology is that it can support the most used service types, such as L2VPN, L3VPN, TDM service, and so on. However, since MPLS is designed for core network, it has the following shortcomings when it is applied in a carrier transport network:

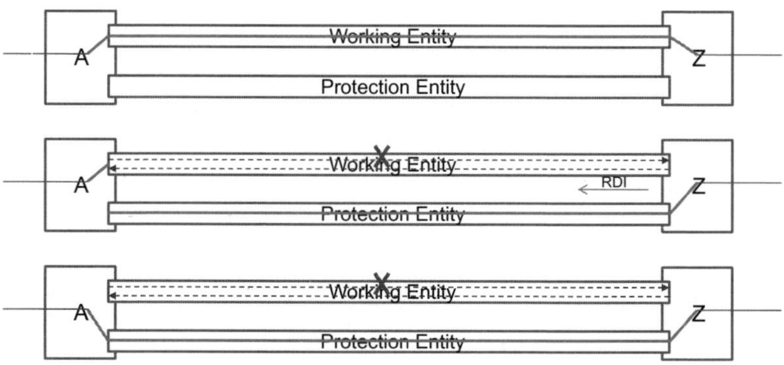

Figure 5.18. PBB-TE protection switching.

- MPLS OAM is dependent on a control plane and an IP layer, both of which are not necessary in a transport network.
- MPLS fast reroute is hard to use when there is a large number of LSPs.
- MPLS LSP is defined as unidirectional LSP, while in a transport network the most case is bidirectional connection.
- Some MPLS forwarding behavior is dependent on an IP layer.
- MPLS cannot work without a control plane.

As a promising technology, many vendors and carriers did much work on how to make it a transport technology.

5.3.2 MPLS-TP: A Transport Profile of MPLS Technology

IETF has started to define a transport profile of MPLS technology for the application in a transport network. Figure 5.19 shows the evolution from MPLS to MPLS-TP. Actually as the figures shows, ITU-T had specified a technology transport MPLS (T-MPLS), which removed the layer 3 dependence and added the function of data-plane-based OAM and protection switching. However, T-MPLS definition cannot interoperate well with current MPLS network become it changes some definition of MPLS packet format. IETF refused the T-MPLS work and started the MPLS-TP work. The MPLS-TP will produce the following enhancement with regard to MPLS:

- It will have centralized NMS, and it must be independent of the control plane.
- It must support OAM and protection switching in the data plane.
- Forwarding must be based on an MPLS label. PHP and ECMP are disabled.
- OAM must be carried in an ACH channel.

Figure 5.19 presents a general summary of the evolution from MPLS to MPLS-TP.

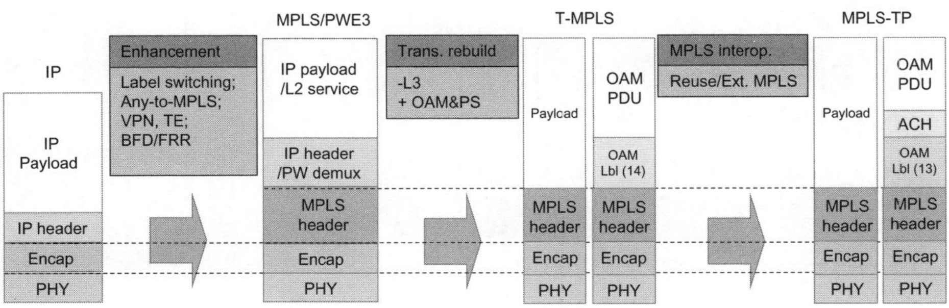

Figure 5.19. From MPLS to MPLS-TP.

5.3.3 MPLS-TP Standardization Progress

The MPLS-TP standardization work was started from the first half of 2008. The standardization work is under the MPLS-TP Joint Working Team, which is formed by the IETF and the ITU-T as shown in Figure 5.20. Most of the transport network vendors are its members. It is one of the most active working groups in IETF.

According to the opinions of most members, the MPLS-TP should have a function similar to that of present transport technology—that is, strong OAM and protection, strong central management, and so on. However, most of the mechanisms haven't been specified at present.

5.3.4 A Comparison Between MPLS-TP and Ethernet Transport

It is like a drama for the battle between MPLS-based and Ethernet-based transport technologies. The battle can be traced to 2005 or even before. Ethernet transport technologies, especially the PBB-TE, had a very good position at the beginning. More than 10 vendors had declared to support the PBB-TE at that time. It also started to be deployed in some parts of the carrier network from 2005 to 2007. However, entering 2008, several carriers made a declaration to abandon the PBB-TE one by one. The MPLS-TP gets much support from carriers, and thus the MPLS-TP seems to be nearly winning the battle at present. The background is that the router camp is much stronger in the market decision because data communication is dominated by routers in carrier networks.

With respect to the technical view, the MPLS-TP has the following advantages compared with Ethernet technologies:

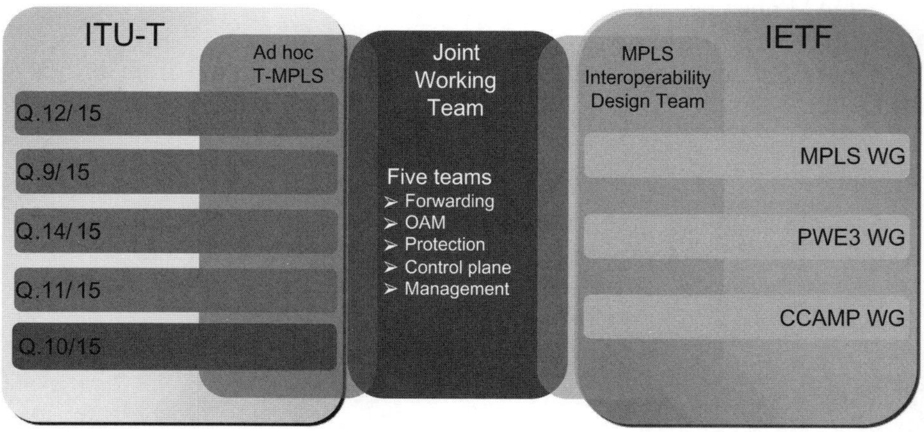

Figure 5.20. MPLS-TP standard body.

- It can be seamlessly integrated into routers as per present-day design specifications.
- It inherits the pseudowire feature from MPLS and thus has more mature support of different services, such as E1, STM, fiber channel, ATM, and so on.
- It can perfectly support L3VPN, which also inherits from MPLS technology.

In fact, Ethernet-based transport technology also has its own advantages, such as the perfect merge of connection-oriented and connectionless technique, and thus have good support on multipoint-to-multipoint connection. Their standardization also goes further than MPLS-TP. However, the market will decide everything. This is not to say that Ethernet transport has been replaced. The battle is still going on. The Ethernet is still getting the market share because the market is different everywhere. For example, Ethernet ring protection is getting more and more attention from carriers.

5.4 SUMMARY

This chapter provides a general overview of current packet transport technology progress at present. Packet transport technology is very promising in the application of a backhaul network from the market point of view. There are two routes that are used to introduce the packet transport technology into the real world: from Ethernet or from MPLS. This chapter gives a general description of progress of the two routes. Both the technologies have good support for transport network application, and they will find their own positions in the world market.

BIBLIOGRAPHY

1. IEEE Std. 802.1Q, Virtual Bridged Local Area Networks, 2005.
2. IEEE Std. 802.1ad, Virtual Bridged Local Area Networks—Amendment 4: Provider Bridges, 2005.
3. IEEE Std. 802.1ah, Virtual Bridged Local Area Networks—Amendment 6: Provider Backbone Bridges, 2008.
4. IEEE Std. 802.1Qay, Virtual Bridged Local Area Networks—Amendment: Provider Backbone Bridge Traffic Engineering, 2009.
5. ITU-T Recommendation Y.1731, OAM Functions and Mechanisms for Ethernet Based Networks, 2008.
6. IEEE Std. 802.1ag, Virtual Bridged Local Area Networks—Amendment 5: Connectivity Fault Management, 2007.
7. ITU-T Recommendation G.8032/Y.1344, Ethernet Ring Protection Switching, 2008.

8. ITU-T Draft Recommendation G.8032/Y.1344 version 2, Ethernet Ring Protection Switching, 2009.
9. ITU-T Rec. G.8031/Y.1342, Ethernet Protection Switching, 2006.
10. J.-D. Ryoo et al., Ethernet ring protection for carrier ethernet networks, *IEEE Commun. Mag.*, September 2008.
11. D. Ward and M. Betts, eds., MPLS-TP Joint Working Team, MPLS architectural considerations for a transport profile, April 18, 2008.
12. ITU-T Recommendation G.8011.1, Architecture of Transport MPLS (T-MPLS) Layer Network, 2006.
13. ITU-T Recommendation G.8011.1 Amendment 1, Architecture of Transport MPLS (T-MPLS) Layer Network Amendment 1, 2007.

6

MICROWAVE BACKHAUL NETWORKS

Ron Nadiv

6.1 INTRODUCTION

Microwave radios play a key role in today's telecommunication networks and are much more common than many may suspect. Though sometimes hidden from the eye, point-to-point (PtP) microwave radios make up almost half of the cellular backhaul connections worldwide and comprise more than two-thirds of worldwide connections outside of the United States.

Radios are also common in carriers' long-haul connections and metropolitan networks serving as fiber replacements. Private enterprise networks, public safety and military networks, and utility companies are also utilizing microwave systems. Still, the most common deployment scenario of PtP radios remains the backhauling of cellular networks.

When talking about mobile backhaul, let us have a look on the market size and its growth: The number of worldwide mobile subscribers was 3.3 billion in 2007 and will grow to 5.2 billion by 2011. According to a report by Infonetics Research, the number of backhaul connections is expected to grow from 2.6 million in 2007 to 4.3 million by 2011.

Convergence of Mobile and Stationary Next-Generation Networks, edited by Krzysztof Iniewski

The microwave radio is so widely deployed mainly because fiber cannot be made available everywhere. In addition, microwave makes for fast and cost-effective deployment compared to fiber. In many cases, self-build and ownership of microwave is cheaper than leasing fiber or copper T/E lines from the incumbent carrier. The new offering of microwave radios supporting carrier Ethernet specifically suits the needs of the growing mobile backhaul, which is pushed into Ethernet technology nowadays.

New mobile technologies such as 3G, WiMAX, and LTE deliver high data rates and lead to a rapid increase in bandwidth-consuming applications. This requires operators to significantly expand their networks or to risk not being able to provide an adequate service quality. The main problem facing operators today is that the demand for bandwidth is not accompanied by a proportional revenue growth. Users demand data bandwidth but are not willing to pay a premium for it as they have for voice services. Instead, they expect a flat-rate model like they get for their residential high-speed data services like DSL or cable. Considering this problem, the cost of expansion of legacy networks becomes unacceptable and operators are forced to move toward new solutions based on the more cost-efficient carrier Ethernet.

The carrier Ethernet represents the only viable solution because it brings the cost model of Ethernet to mobile backhaul. Certainly the technology also has some risks, for instance, supporting legacy networks and the key element of clock synchronization required by mobile applications. Be that as it may, many analysts agree that the number of cell sites served by an Ethernet backhaul will continue to expand rapidly. Analyst group Heavy Reading [2,3] claims that in 2008 only 23,000 cell sites worldwide were served with Ethernet, about 1% of the total ~2 million. This figure is expected reach to 713,000 by the end of 2012, or 24% of the total ~3 million.

As mobile services become more and more sophisticated, and considering a widely spreading trend of convergence with fixed networks, operators need to do more than simply deliver basic voice and best-effort data. For this they need the tools to differentiate their services, and they need to integrate those tools into the transport layer. The new microwave radios are much more than transport "pipes." Integrating advanced functionality such as service differentiation and policy enforcement tools is needed in order to achieve successful and profitable backhaul deployment.

Microwave radios offered today include a wide selection of legacy transport radios, such as (a) carrier Ethernet radios and (b) hybrid radios supporting both legacy and Ethernet technologies that allow for smooth migration from legacy to next-generation networks [4].

In this chapter we will review the microwave radio technology and focus specifically on (a) point-to-point radio applications for mobile backhaul and (b) the way it evolves today for supporting the rapidly growing cellular networks.

The first part of the chapter will go through some fundamental radio techniques and will compare various types of this technology, including point-to-point versus point-to-multipoint radios. The second part of the chapter will focus on networking features and attributes of modern radios, primarily explaining Ethernet radio technology and what is required for such radios for carrier Ethernet backhaul. The final section will be dedicated to mobile backhaul evolution challenges and possible solution alternatives.

6.2 MICROWAVE RADIO FUNDAMENTAL TECHNIQUES

6.2.1 Radio Link Budget, System Gain, and Availability

Radio link budget is a basic calculation that is essential for any radio deployment [5]. In order to maintain a radio link, the received signal level at each receiver should be above a threshold level that ensures the minimum signal-to-noise ratio (SNR) and enables the receiver to lock on the signal and demodulate it successfully. This threshold level is typically measured at a bit error rate (BER) of 10^{-6}, denoted as P_{th}. BER level depends on the radio modulation scheme, modem implementation (such as coding strength and modem quality), and radio implementation factors such as noise figure, branching losses from the antenna, and other losses.

The received signal strength can be calculated as follows:

$$P_{RX} = P_{TX} + G_{TX} - \text{FSL} - L + G_{RX}$$

where P_{RX} is the received signal level, P_{TX} is the transmitted power level, G_{TX} and G_{RX} are the transmitting and receiving antenna gain, FSL is the free-space propagation loss (proportional to $1/R^2$, where R is the link distance, as well as to $1/f^2$, where f is the link frequency), and L represents additional losses caused by many factors such as atmosphere gas absorption, rain fading, or obstacles.

The *fade margin* of a given link, with a given threshold, refers to the maximum loss (L) at which the link still delivers data at the specified BER:

$$\text{Fade Margin} = \text{P}_{\text{RX}} - \text{P}_{\text{TH}} = \text{P}_{\text{TX}} + \text{G}_{\text{TX}} - \text{FSL} + G_{RX} - P_{TH}$$

The term $P_{TX} - P_{TH}$ is usually referred to as *system gain.*

$$\text{SG} = P_{TX} - P_{TH}$$

The system gain is a commonly used figure of merit for comparing the performance of different competing radio systems. A radio with higher system gain will have a higher fade margin, allowing it to be used over a longer link distance or in combination with smaller (less expensive) antennas compared with other lower system gain radios.

Radio Availability. Availability is defined in terms of percentage of yearly time of service and is calculated as 100% of the time minus the annual outage time in which traffic loss occurs. For example, availability of 99.999%, sometimes referred to as "five 9's," means that the expected traffic loss is not more than 0.001% of the time in a year, which is approximately 5 minutes a year.

Unlike wireline applications, radio networks are designed to deal with expected outages. While in wireline networks a link breakdown (such as fiber cut) is obviously a failure mode, in wireless radio links, a link breakdown due to channel fades is simply one more scenario that needs to be planned for. Correct planning for specific service level objective enables operators to rely on their microwave network with just the same confidence as with wired network.

Rain Fade Margin and Availability. The main cause for radio link fades is rain attenuation. The more intense the rain is (as measured in millimeters/hour), the more it attenuates the propagating signal, with exponential dependency. Other precipitation forms such as snow or fog do not cause significant propagation loss at the microwave frequencies (unlike optical systems for example, which are severely affected by fog). Rain attenuation increases significantly in frequency and also varies in polarization. Since falling rain drops are not shaped as spheres, the horizontal polarization is attenuated more than vertical polarization. It should be noted that at low frequencies (below 8 GHz), rain attenuation may not be the dominant outage factor, but rather multipath effects resulting from reflections. This will be described further in the chapter when we discuss diversity.

Link fade margin planning is based on statistical distribution models that measure rain intensity. These models rely on historical rain measurements, and they divide the globe into "rain zones."

The two commonly used models for calculating rain-fade margin are the ITU-R and Crane models. Both allow planners to compute the required fade margin for a given radio link frequency, link length, and rain zone, in order to reach the desired availability. It should be noted that rain effects result from the actual blockage of the radio path by heavy rain cells. Once this happens, neither equipment nor antenna redundancy can help to overcome the blockage. On the other hand, because heavy rain usually occurs within relatively small areas at a given time, path diversity may help. By using redundant path topologies such as a ring, operators can maintain their services on a network level even when a particular radio path is temporarily blocked.

6.2.2 Multipath Fading and Radio Diversity

As mentioned above, rain fading is less severe in lower frequencies, but multipath fading becomes more dominant. Multipath fading results from the reflected waves reaching the receiver and mixing with the desired signal at random phases, as shown in Figure 6.1.

Multipath fading depends mainly on the link geometry and is more severe at long-distance links and over flat surfaces or water. Multipath fading is

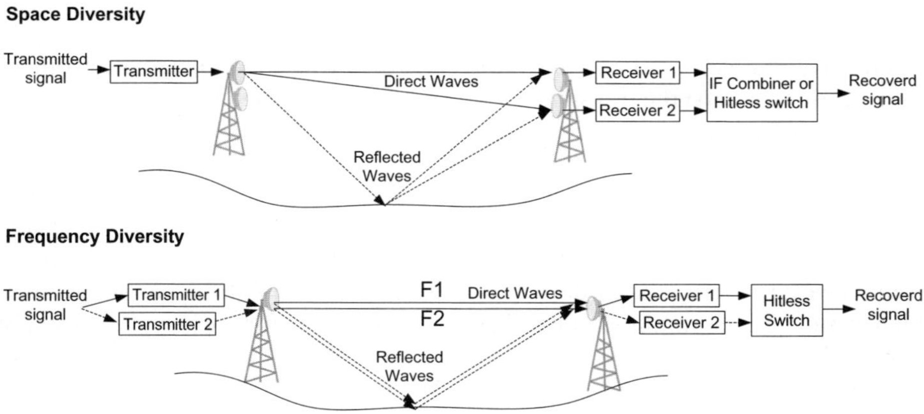

Figure 6.1. Radio diversity systems.

also affected by air turbulence and water vapor, and it can vary quickly during temperature changes due to rapid changes in the reflections phase.

It is common to distinguish between *flat fading*, which equally affects the entire channel bandwidth, and *selective fading*, which is frequency-dependent over the given channel bandwidth.

The common way to handle severe multipath conditions is to use diversity techniques, mainly space or frequency diversity, as shown at Figure 6.1.

Multipath fading depends on geometry. So if we place two receivers (with separate antennas) at a sufficient distance from one another, multipath fading will become uncorrelated between both receivers. Thus, statistically, the chance for destructive fading at both ends decreases significantly. Frequency diversity uses a single antenna and two simultaneous frequency channels. While both diversity solutions require a second receiver, space diversity also requires a second antenna. On the other hand, frequency diversity requires double the spectrum usage (this is generally less favored by regulators).

With space diversity there are two options to recover the signal. One is through *hitless switching*, where each receiver detects its signal independently, but the output is selected dynamically from the best one. Switching from one receiver to the other must be errorless. Another signal recovery option is to combine both signals in phase with each other to maximize the signal-to-noise ratio. *IF combining*, as it is commonly referred to, is usually handled by analog circuitry at the intermediate frequency (IF). This way is much better because the in-phase combining of both signals can cancel the notches at both received signals and thus increase significantly the dispersive fade margin, as well as increase the system gain up to 3 dB with flat fading.

With frequency diversity, as the different signals are carried over different carrier waves, IF combining is not possible but only hitless switching is relevant.

6.2.3 Modulation, Coding, and Spectral Efficiency

Microwave radio is based on transmitting and receiving a modulated electromagnetic carrier wave. At the transmit side, the modulator is responsible for conveying the data on the carrier wave. This is done by varying its amplitude, frequency, or phase. At the receive side, the demodulator is responsible for synchronizing on the receive signal and reconstructing the data. Legacy radio systems implemented simple analog modulation schemes. One such scheme is frequency shift keying (FSK), where for each time period T, the *symbol time*, one of several carrier frequencies was selected. Modern radios use digital modulators that allow much higher numbers of amplitude and phase possible states for each symbol, usually from 4 QAM (QPSK) up to 256 QAM. The more *states* the modulation has, the more data it can carry over one symbol. For example, QPSK has four possible states and thus carries 2 bits/symbol, whereas 256 QAM carries 8 bits/symbol.

The channel bandwidth that is occupied by the transmitted signal depends on the *symbol rate*, which is $1/T$ (where T is the symbol time), and on the signal filtering following the modulator. Hence, for a given channel bandwidth, there is a limited symbol rate that the radio can use, and consequently there is a limited data rate that it can transmit.

Let us now look at the following example. ETSI [11,12] defines for most frequency bands a channel separation of 1.75 MHz to 56 MHz. ETSI also defines a spectrum mask for each channel and modulation to guarantee a certain performance of spectral emissions and interferences of adjacent and second adjacent channels. A modern digital radio system can usually utilize a symbol rate of 50 MHz over a 56-MHZ channel, and thus it can deliver 400 Mbit/s by using 256 QAM modulation.

Modern digital radios also use error correction coding that requires parity bits to be added to the information bits. The term *coding rate* is used to define the portion of net data bits out of the total number of bits. For instance, a coding rate of 0.95 means that out of each 100 bits transmitted, 95 are the "useful" information bits. A stronger code will use more parity bits, and thus it will have a lower rate. In addition, part of the symbols may be used for synchronization purposes and not for carrying data.

Now that we can compute the net data rate provided by a radio system, we can go on to monitor its spectral efficiency. The term *spectral efficiency* refers to the number of information bits that are carried over 1 Hz of spectrum. If we now take again the above example of 256 QAM over 56 MHz and assume a coding rate of 0.93, we get the net data rate of 372 Mbps and obtain a spectral efficiency of 6.64 bits/Hz, as shown in Table 6.1.

Since bandwidth is a precious resource, with massive demand and growth in high capacity radio deployments, regulators today encourage the deployment of high spectral efficiency systems and sometimes even requires a certain minimal efficiency.

The Tradeoffs of Spectral Efficiency. When calculating spectral efficiency, we need to consider the required system gain and the system's complexity and cost.

TABLE 6.1. System Gain and Availability Examples

		QPSK, 56-MHz Channel	256-QAM, 56-MHz Channel
Symbol rate (Msymbol/s)		50	50
QAM (bits/symbol)		2	8
Coding rate		0.95	0.93
Data rate (Mbit/s)		95	372
Spectral efficiency (bits/Hz)		1.69	6.64
Required SNR (dB)		7	28
Practical P_{TH} (dBm)		−85	−67
Practical P_{TX} (dBm)		22	19
SG (dB)		107	83
Link availability at 23 GHz, 5 km, 1 ft antennas	ITU-R rain zone K (NY-USA)	99.9991 (~5 min/year)	99.985 (~80 min/year)
	ITU-R rain zone N (tropical areas)	99.993 (~37 min/year)	99.93 (~6 h/year)

System Gain. As a general rule, higher spectral efficiency means lower system gain and therefore poor availability. The information theory bounds the information rate that can be delivered over a given channel with a given *signal-to-noise ratio* (SNR). This is defined by the well-known Shannon formula: $C = W \log 2(1 + \text{SNR})$, where C is channel capacity in bits/hertz and W is channel bandwidth in hertz. Note that C/W just equals spectral efficiency. From the Shannon formula we can clearly understand that in order to have higher spectral efficiency, we need a better SNR to demodulate the signal and restore the information successfully. The formula shows that at higher SNRs each increase of 1 bit/Hz will require doubling the SNR—that is, improve it by 3 dB. The need for higher SNR means higher thresholds and results in lower system gain, both of which translate into lower availability (under the same other link conditions).

System Complexity. Implementation of higher efficiency modulations involves an increasing cost and complexity for both the transmitter and the receiver. Higher modulations require lowering the quantization noise. This in turn will require higher-precision analog-to-digital and digital-to-analog converters, along with much more digital hardware for performing all digital processing at a higher precision. As higher modulation schemes become much more sensitive to signal distortion, they require better linearity and phase noise of the analog circuitry, as well as additional complex digital processing (such as longer equalizers) to overcome such distortions. Additionally, the requirement for linearity typically means that by using the same power amplifier, a system can obtain less transmit power at a higher modulation, so system gain is decreased twice: one time due to higher SNR and another time due to lower P_{TX}.

The selection of one system or another depends on the specific requirements of the operator and restrictions (or lack thereof) of the regulator. Luckily, many modern radio systems are software-defined, so one piece of equipment can be configured to different working modes according to user preferences.

The example in Table 6.1 compares two modern radio systems and shows the tradeoff between system gain and spectral efficiency. As can be seen in the table, the 256 QAM system delivers almost four times the capacity of the QPSK system, but has a SG lower by 24 dB! Given the same link conditions, we can also see the difference in availability between the systems, which is very significant especially at high rain rate zones such as ITU-R zone N.

A Word About Error Correction Codes. We have mentioned briefly the option to use some transmitted bits as parity bits to allow error correction and robust delivery of information bits. We've also mentioned the Shannon formula, which sets the theoretical limit for channel capacity within a given SNR. When Shannon's formula was published in 1948, there was no implementation foreseen that was, or would be, able to get near its limit. Over the years, implementations of codes such as Reed–Solomon became very common in radio design, enabling systems to reach within a few decibels of Shannon's limit. More recent techniques, such as turbo codes or low-density-parity-check (LDPC) codes, come even closer to reaching the theoretical Shannon limit—at times as close as only a few tenths of a decibel. Such codes require a very high computational complexity, but the advances in silicon processes have made them practical (even at very high bit rates) and allow them to be implemented in state-of-the-art radios today.

Enhancing Spectral Efficiency with XPIC. One way to break the barriers of spectral efficiency is to use dual-polarization radio over a single-frequency channel. A dual polarization radio transmits two separate carrier waves over the same frequency, but using alternating polarities. Despite its obvious advantages, one must also keep in mind that typical antennas cannot completely isolate the two polarizations, and isolation better than 30 dB is hard to achieve. In addition, propagation effects such as rain can cause polarization rotation, making cross-polarization interferences unavoidable. The relative level of interference is referred to as cross-polarization discrimination (XPD). While lower spectral efficiency systems (with low SNR requirements such as QPSK) can easily tolerate such interferences, higher modulation schemes cannot and require *cross-polarization interference canceler* (XPIC). The XPIC algorithm allows detection of both streams even under the worst levels of XPD such as 10 dB. This is done by adaptively subtracting from each carrier the interfering cross carrier, at the right phase and level. For high-modulation schemes such as 256 QAM, an improvement factor of more than 20 dB is required so that cross-interference does not limit performance anymore. XPIC implementation involves system complexity and cost since the XPIC system requires each demodulator to cancel the other channel interference.

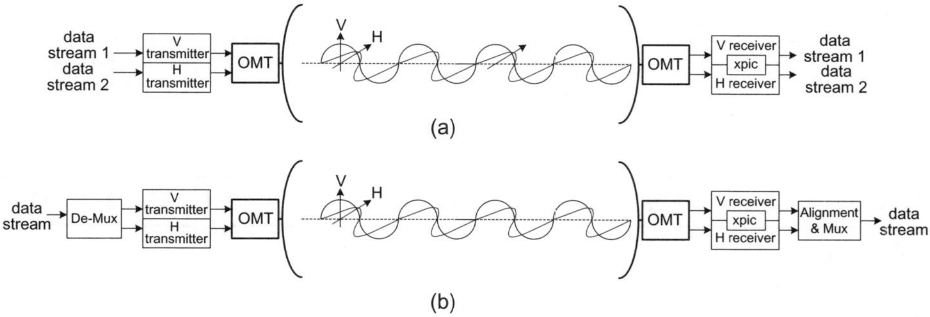

Figure 6.2. **(a)** XPIC system delivering two independent data streams. **(b)** XPIC system delivering a single data stream (multi-radio).

XPIC radio may be used to deliver two separate data streams, such as 2xSTM1 or 2xFE, as shown at Figure 6.2a. But it can also deliver a single stream of information such as gigabit Ethernet, or STM-4, as shown at Figure 6.2b. The latest case requires a demultiplexer to split the stream into two transmitters, and it also needs a multiplexer to join it again in the right timing because the different channels may experience a different delay. This system block is called "multi-radio," and it adds additional complexity and cost to the system.

It should be noted that there are different techniques regarding how to split the traffic between radios. For example, Ethernet traffic can be divided simply by using standard Ethernet link aggregation (LAG), but this way is not optimal. Since LAG divides traffic-based flow basis, it will not be divided evenly, and it can even be that all traffic is on a single radio, thus not utilizing the second radio channel. A more complex but optimal way is to split the data at the physical layer, taking each other bit or byte to a different radio, so all radio resources are utilized.

Adaptive Modulation. Adaptive modulation means dynamically varying the modulation in an errorless manner in order to maximize throughput under momentary propagation conditions. In other words, a system can operate at its maximum throughput under clear sky conditions and can decrease it gradually under rain fade [26].

Legacy microwave networks based on PDH or SDH/SONET deliver a fixed, "all or nothing" pipe of information and cannot accommodate for any reduction in the link's capacity. This can be further explained by the following example. An STM-1/OC-3 radio link with 155.52-Mbit/s capacity is carried over a radio link using a 28-MHz channel. In this case, a modulation of 128 QAM is required, with a typical SNR threshold of ~25 dB. Let us now suppose that a short fade occurs, causing the SNR to drop by 2 dB below this threshold. Should this happen, the connection fails entirely, since the SDH equipment will not be able to tolerate either the errors or any decrease in throughput. Unlike legacy networks, emerging packet networks allow the link to simply lower the modulation scheme under

TABLE 6.2. ACM Example

	1 Highest Availability, Lowest Throughput	2	3	4	5 Highest Throughput, Lowest Availability
ACM Point	QPSK	16 QAM	64 QAM	128 QAM	256 QAM
SNR_{TH} (dB)	10	16	22	25	28
TX power (dBm)	22	21	20	19	18
System gain (dB)	104	97	90	86	82
Throughput (Mbit/s) (at 28 MHz channel)	40	75	135	160	180
Availability (38 GHz, K rain zone, 2.5 km, 1 ft antennas)	99.9992 (~5 × 9's)				99.994 (~4 × 9's)

the conditions described above. In this particular case, dropping the modulation to 64 QAM will decrease the capacity to approximately 130 Mbps but will not cause the link to fail and will continue to support the service.

An adaptive QAM radio can provide a set of working points at steps of approximately 3 dB. This results from the modulation scheme change, since the SNR threshold changes by 3 dB for each additional bit per hertz. For "finer" steps, it is also possible to change the coding scheme [adaptive coding and modulation (ACM)]. This may further increase throughput by as much as 10%. System performance can be further enhanced by combining adaptive modulation with adaptive power control. As we have already seen, a lower modulation scheme usually allows higher power levels due to relaxed linearity requirements. Hence an additional increase in system gain can be achieved when reducing the modulation scheme and simultaneously increasing the transmitted power.

It should be noted that adaptive radios are subject to regulatory authorization. ETSI standards were recently adopted to allow such systems, and a similar process has begun by the FCC. Still, not all local regulators allow it.

An example for ACM system with five working points is shown in Table 6.2, demonstrating the range of system gain versus throughput and also showing a test case of availability difference over this range. Figure 6.3 shows the behavior over time of such system under a severe fade situation.

6.3. DIFFERENT TYPES OF RADIO TECHNOLOGIES

Modern telecommunication networks employ a variety of radio technologies. All of these technologies involve transmitting radio waves between two or more locations, but they also differ a lot from one another both in the method of transmission and in the application they are best suited for. The lion's share

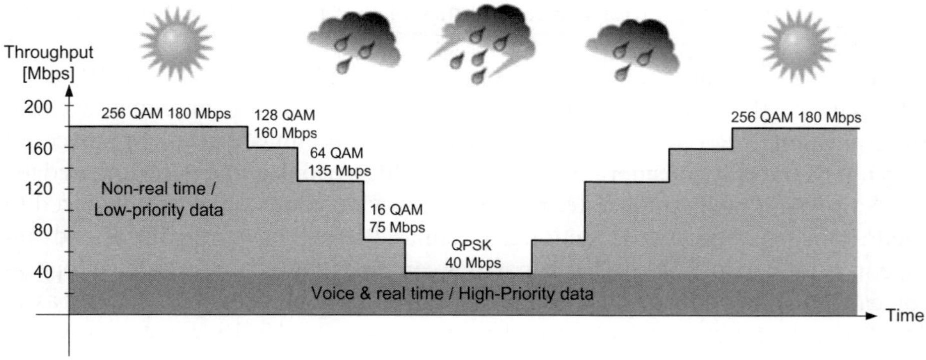

Figure 6.3. ACM example.

of this chapter will focus on point-to-point (PtP) radios over licensed frequencies that make up for more than 90% of wireless backhaul solutions today. In addition, we will describe several other technologies, considering the pros and cons of each one.

6.3.1 Point-to-Point and Point-to-Multipoint Radios

A point-to-point (PtP) radio link consists of two symmetrical terminals transmitting and receiving signals between the two sites. A point-to-multipoint (PtMP) system, on the other hand, contains a central site (sometimes called "hub") that is connected to number of remote terminals.

Let us now consider the advantages and disadvantages of PtP and PtMP architectures. The example depicted in Figure 6.4 shows a hub site connected to

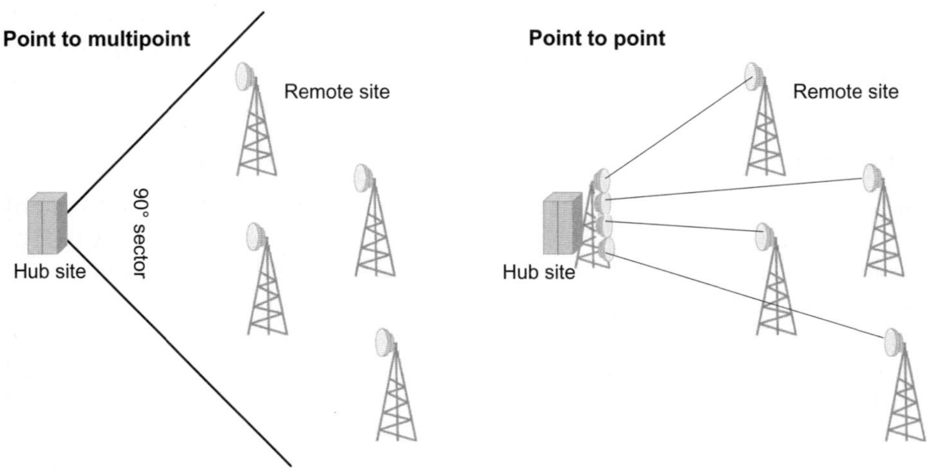

Figure 6.4. PtP and PtMP.

four edge sites by using four PtP radio links, or one PtMP 90° sector serving the four users.

Using a pure equipment-count comparison, it's plain to see that PtMP has an edge. The PtMP sector will require 4 + 1 equipment units and antennas, whereas the equivalent PtP solution requires 2 × 4 units and antennas. But equipment count may not be the only consideration, and we should also look at the radio link budget.

Since the PtMP sector antenna has a very wide beam width as compared to the PtP link that uses two directional antennas, PtP will have a significant advantage in system gain. Thus higher capacities can be delivered, or, in turn, an operator can use lower-power and smaller antennas to achieve similar results as in PtMP. For example, a typical sector antenna gain is 10–15 dBi, whereas a typical gain of a PtP antenna in use today is around 40 dBi and even higher. So, assuming that the remote radio antenna is identical, the difference in system gain can be around 30 dB in favor of PtP. This big difference is a major limit in reaching high capacities with high availably using PtMP radios.

Another major difference between the PtP and PtMP is the use of the radio-frequency spectrum as a shared media versus dedicated media. The PtMP sector shares the same frequency channel among several users to allow more flexibility and enable statistical multiplexing between users. The use of a dedicated channel for each PtP link, however, does not allow using excess bandwidth of one user and pass it to another. Shared media PtMP also offers the benefit of statistical multiplexing; but in order to enjoy it, operators must deploy a wide sector antenna that has a downside of interferences between other links at the same channel, thus poor spectrum reuse. In contrast, PtP antennas make it easier to reuse frequency channels while avoiding interferences from one radio link to another due to its pencil beam antennas.

The wide-angle coverage of PtMP makes frequency planning more complex and limited, and frequency reuse is not trivial and hard to regulate with interferences between different hubs, especially with large coverage deployments such as mobile backhaul. Therefore it is uncommon to find PtMP architecture in these applications, unlike radio access networks asWiMAX where PtMP is more commonplace.

6.3.2 Line of Sight and Near/No Line of Sight Radios

The segmentation of line of sight (LOS) and near/no line of sight (NLOS) radios goes hand in hand with the frequency segmentation of above and below 6 GHz. For high-frequency radios, an LOS between the radios is generally required and the accepted criteria for LOS is usually defined as keeping the first Fresnel zone clear of any obstacles. Such clearance allows radio waves to propagate across an uninterrupted path, while any obstacle on the way (which can be any physical object of conducting or absorbing materials like trees, buildings, and ground terrain) will cause attenuations, reflections, and diffractions of the signal. The ability to maintain a radio connection under such conditions decreases as the frequency increases.

NLOS communication typically refers to frequencies up to 6 GHz. Unlike LOS propagation in which signal level decrease over distance R is proportional to $(1/R)^2$, NLOS signal level decreases faster, proportional to a higher order of $1/R$. Hence the connection distance is decreased significantly, depending on link obstacles and terrain. In addition, NLOS connections require radio capability to handle signal distortions and interferences resulting from the NLOS propagation (such as OFDM modems), which are not trivial to implement—especially with high-bandwidth signals and high spectral efficiency modulation.

Most access technologies today including cellular, WiFi, and Wimax handle NLOS connections as a matter of definition. NLOS backhaul can make for easier and cheaper deployment, since LOS is not always possible or available and can require high-cost construction of antenna towers.

Still, NLOS has a number of major drawbacks when compared to LOS:

- *Limited Available Spectrum.* Sub-6-GHz bands are limited. Most licensed bands in this range are assigned to access applications (e.g., cellular, Wimax) and are therefore expensive to purchase. Unlicensed bands (such as WiFi) are congested and unreliable for backhaul usage.
- *Limited Capacity.* NLOS propagation and low-gain antennas limit traffic capacities that cannot compete with those of high frequencies.
- *Complicated Planning.* Complicated planning and the inability to accurately predict propagation and interferences make it very difficult to guarantee a robust backhaul network with carrier-grade availability.

6.3.3 Licensed and Unlicensed Radios

This is a regulative aspect of microwave radios that needs to be mentioned. Most countries consider spectrum a national resource that needs to be managed and planned. Spectrum is managed at a global level by the UN's International Telecommunication Union (ITU), at a regional level by bodies such as the European Conference of Postal and Telecommunications Administrations, and at national levels by agencies such as the Office of Communication (OFCOM) in the United Kingdom or the Federal Communication Commission (FCC) in the United States.

The implications of using a licensed spectrum radio are simple to understand: The operator must pay the toll, and in return he is guaranteed that no interferences from other licensed operators will harm the radio performance. For the backhaul network designer, such a guarantee is crucial for guaranteeing his service level.

The nature of the license differs from country to country and between frequency bands. For example, some licenses cover specific channels (usually paired as "go" and "return" channels), while others cover a bulk frequency (such as LMDS); licenses can be given for nationwide transmission, whereas others are given for a single radio link only; and regulators might license the channel at both polarizations separately, or they might license them both together.

The benefit of using unlicensed frequency is of course cost savings, as well as speed of deployment because there is no need for any coordination; the drawbacks are service reliability due to interferences and inability to guarantee service level as desired for carrier class backhaul.

6.3.4 Frequency Division Duplex and Time Division Duplex Radios

Another, more technical comparison of radios refers to the way the radio performs two-way communication. With frequency division duplex (FDD), the radio uses a pair of symmetric frequency channels, one for transmitting and one for receiving. The radio transmits and receives simultaneously, thus requiring a very good isolation between the two frequencies which therefore must be at a certain frequency space from each other. With time division duplex (TDD), the radio uses a single-frequency channel that is allocated part of the time for transmitting and part of the time for receiving.

Because TDD shares the same frequency channel for both directions, the transmission is not continuous, and there is a need to synchronize both terminals on switchover times, and to keep a guard time between transmission. Additionally, synchronization time overhead is needed to allow the receiver to lock before the meaningful data are transmitted.

These overheads make TDD less preferable with regard to bandwidth efficiency and also latency.

On the other hand, TDD has an advantage in cases where asymmetry is present because bandwidth can be divided dynamically between the two ways.

TDD also has some potential for lower-cost systems. Because the radio does not need to transmit and receive at the same time, isolation is not an issue. Furthermore, radio resources can be reused both at the transmitter and at the receiver.

A major advantage of FDD is that it makes radio planning easier and more efficient. This is because different frequency channels are used for the different radio link directions.

In reality, most microwave bands allocated for PtP radios are regulated only for FDD systems, with a determined allocation of paired channels. TDD is allowed in very few licensed bands above 6 GHz, specifically in LMDS bands (26, 28, 31 GHz) and in the high 70 to 80-GHz bands.

To sum up this part of our discussion: While some radio systems use TDD, FDD is far more common in most access systems and is certainly dominant in the PtP radios for backhaul applications.

6.4 MICROWAVE RADIO NETWORKS

In this chapter we will describe the networking aspects of microwave radio, mainly focusing on mobile backhaul applications. Typical carrier networks are usually built out of a mix of technologies. While there are almost no microwave-only networks in existence, certain segments of a telecommunications network can be based extensively on microwave radios. Legacy SDH/PDH radios are used to support only the transport function of a telecom network, requiring additional equipment to perform networking functions such as cross-connecting or switching. Modern Ethernet-based radios, on the other hand, are much more than mere transport devices and can support most of the functionality required of a mobile backhaul network.

6.4.1 Terms and Definitions

Backhaul networks employ a range of different technologies designed in a variety of architectures and use a wide set of terminology and jargon. In this chapter, we will refer to mobile backhaul as the radio access network (RAN) between a cellular base station at the cell site and the network controller site or the edge core node. We will use the terms RAN-BS and RAN-NC as defined by the Metro Ethernet Forum (MEF) Mobile Backhaul Implementation Agreement (MBH-IA [19]). RAN-BS, the RAN Base Station, stands for any mobile technology base station such as GSM, CDMA, or Wimax base stations (BTS), a 3G Node-B, or LTE eNB (evolved Node-B). RAN-NC, the RAN network controller, stands for GSM base station controller (BSC), 3G RNC, Wimax access service network (ASN) gateway, or LTE serving gateway (SGW).

As an example, Figure 6.5 describes 3G architecture; the mobile backhaul in this case includes primarily the interface called "Iub" between node-B's and RNC, and the interface called "Iur" between RNCs. The first part of the RAN connecting the RAN-BSs is often referred to as the access RAN, whereas the

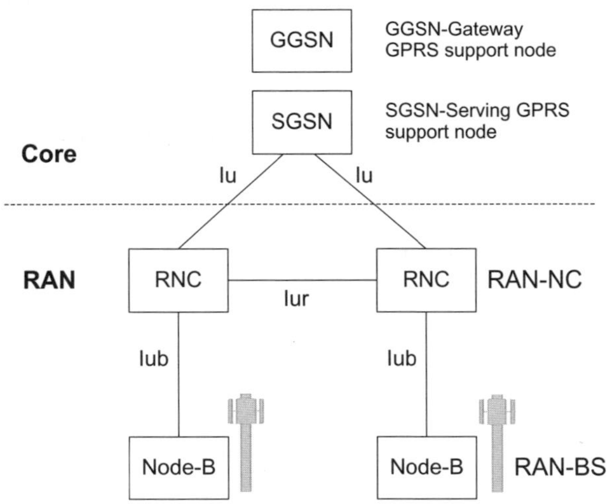

Figure 6.5. 3G architecture.

part connecting the RAN-NCs is referred to as the aggregation RAN. The access RAN may include thousands of base station connections typically aggregated over several steps. Therefore the emphasis in access RAN systems is on efficiency and low cost. The aggregation RAN, on the other hand, is more limited in scale, but with much more strict resiliency requirements.

Mobile backhaul networks have traditionally been realized using TDM and ATM technologies. These technologies are transported over the PDH radios in the access RAN, and SDH/SONET radios in the aggregation RAN.

Next-generation mobile equipment and networks, however, will be based on Ethernet. Carrier Ethernet services will provide the connectivity in the mobile backhaul network, either in dedicated Ethernet networks or in a converged network together with fixed services. We will examine the technical aspects of microwave radios both in legacy networks and in Ethernet networks, and examine the evolution paths from legacy to next-generation radio networks.

6.4.2 PDH and SDH Radios

The most common legacy interfaces in mobile access backhaul are E1 and T1 used to multiplex a number of 64-Kbps voice PCM channels (32 at E1, 24 at T1), and they became the common interface for legacy 2G GSM and CDMA base stations. Newer technologies, such as 3G UMTS, use ATM over the same TDM E1 interfaces or over a bundle of such interfaces [referred to as IMA (inverse multiplexing)], and CDMA2000 uses HDLC over T1s or DS-3.

Legacy mobile backhaul networks are typically deployed in either a tree or a ring topology, or a combination of these, as shown in Figure 6.6. The backhaul network collects multiple E1/T1s from the cell sites and transports them toward the core. Tree topologies require low-capacity links at the access edge, and they gradually increase capacity requirements as the network advances toward the core. PDH radios offering bandwidth in increments of either E1 (2 Mbit/s) or T1

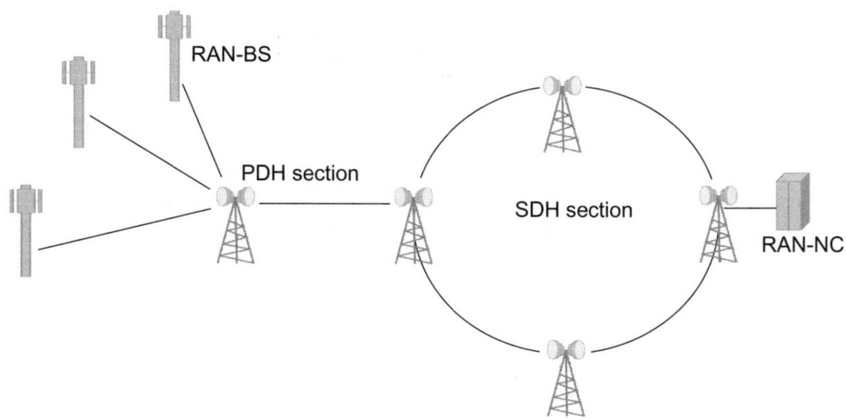

Figure 6.6. Legacy PDH/SDH backhaul example.

(1.55 Mbit/s) are widely used in such deployments. Ranging from low-capacity/low-cost 4xE1/T1s radio links at the tail site (the last base station in a network) PDH radios can reach 64xE1/T1 or more connections at higher aggregation levels.

At the aggregation RAN, connections have higher capacity and resiliency is more important, and thus SDH/SONET radios are often used. Such radios were developed to replace the PDH system for transporting larger amounts of traffic without synchronization problems and with supported resiliency over ring deployments. Since the RAN-NCs connect several hundreds of cell sites, often its interface is STM-1/OC-3, which supports 155 Mbit/s of throughput; and the aggregating rings, built of fiber optics or microwave SDH/SONET radios, can be of multiple STM-1, STM-4, or even more. Modern SDH radios support bandwidth in increments of STM-1, up to 2xSTM-1 over single carrier, or 4xSTM-1 with dual carrier (possibly over same channel using XPIC).

An E1/T1 path from the RAN-NC to the individual cell site may consist of many "hops" across the metro ring and over the access tree. To ensure ongoing and uninterrupted communication, the radio system must maintain the signal performance parameters over this multihop path. A radio delivering TDM signals should not only reconstruct the E1/T1 data at its remote side, but also perform two major requirements: fault management and clock delivery.

TDM signals use fault management indications such as AIS (alarm indication signal) and RDI (remote defect indication) that are propagated in responses to signal loss or signal out of frame. AIS and RDI serve to alert the rest of the network when a problem is detected. A TDM radio should initiate such indications at its line ports, also in response to radio link failures.

Clocking performance is a major issue for some legacy base stations that use the TDM signals not only for delivering traffic but also for synchronization. Degraded synchronization may result in poor performance of the cellular network, and the critical handover procedure in particular. The ITU G.823/G.824 [7,8] standards define two kinds of TDM interfaces: traffic interface and synchronization interface. Each has its requirements for short-term phase variations, referred to as *jitter*, and long-term phase variation, referred to as *wander* (defined as slow phase variations, at a rate of less than 10 Hz). Note that some base stations (such as CDMA) use external synchronization signals (usually GPS). In these cases, TDM synchronization over the radio network is not an issue.

A radio link that delivers TDM signals must maintain the clock for each delivered signal over the radio link. Since typically the radio itself uses an independent clock for transmission which is not related to the line clock, it utilizes justification bits to align the clocks at the transmit side. This technique also restores the line clock at the receive side using a clock unit. In the case of PDH, this is performed separately for each E1/T1 signal.

It is important to note that clock performance is degraded as the number of radio hops between two end points increases. Hence, there is a limit to the number of radio hops that can be supported without aligning back to a common clock from a reliable external source.

Cross-Connecting. Traditional radio network implementations used PDH radios for transport and connected them to each other using a manual wiring device called a digital distribution frame (DDF). Modern radio systems may already include a multiradio platform with embedded cross-connect function. This provides a one-box solution for radio aggregation sites and adds advantages such as better reliability, improved maintenance with a common platform to manage, and the ability of remote provisioning. Not the least bit important, a single-box solution makes for smaller system footprint and will most likely have a better cost position.

In aggregation sites closer to the core, a large quantity of signals is connected, and it becomes cumbersome to use separate connections for each E1/T1. Hence radio systems with cross-connect are usually groomed for capacity interfaces such as channelized STM-1.

Resiliency. Resiliency is another important aspect in radio network deployment. Resiliency ensures the network's reliability and overall service quality and should be treated at the element level and the network level. Radio networks need to be resilient from both equipment failures as well we from radio link failure due to propagation effects as heavy rain fade. Resiliency is an important consideration when designing the network's topology. In a network built in a tree or chain topology, the closer a failure is to the core, the more cell sites it can put out of service; therefore operators typically employ protected radio links in aggregation sites where a single link failure can affect a significant portion of the network. Such an example is depicted in Figure 6.7a below where each radio link affecting more than one cell site is protected. It is important to note that this protection scheme shields against equipment failures, but not against propagation effects that will influence both radios in a protection scheme, nor against a complete site disaster.

In ring deployments, resiliency is inherent in the topology because traffic can reach each and every node from both directions, as shown in Figure 6.7b. Thus, there is no need to protect the radio links, so typical ring nodes feature unprotected radio links in both east and west directions. One additional advantage that rings have over tree topologies is that ring protection not only covers equipment failures, but also provides radio path diversity that can help in dealing with propagation fading.

SONET/SDH rings offer standardized resiliency at the path level and can be deployed in several architectures, such as UPSR or BLSR [9]. When deploying PDH rings, path protection may be handled over the entire ring (similar to SDH but proprietary because there is no available standard) or on a connection-by-connection basis, providing end-to-end trail protection using subnetwork connection protection (SNCP). The common way to implement SNCP would be 1 + 1, which means that traffic is routed over two diverse paths, and a *selector* at the egress point selects the best path of the two. SNCP can be used in other topologies as well. Figure 6.7c depicts an SNCP implementation for a partial mesh case. The SNCP can be defined at different levels and is usually applied at the E1/T1 level.

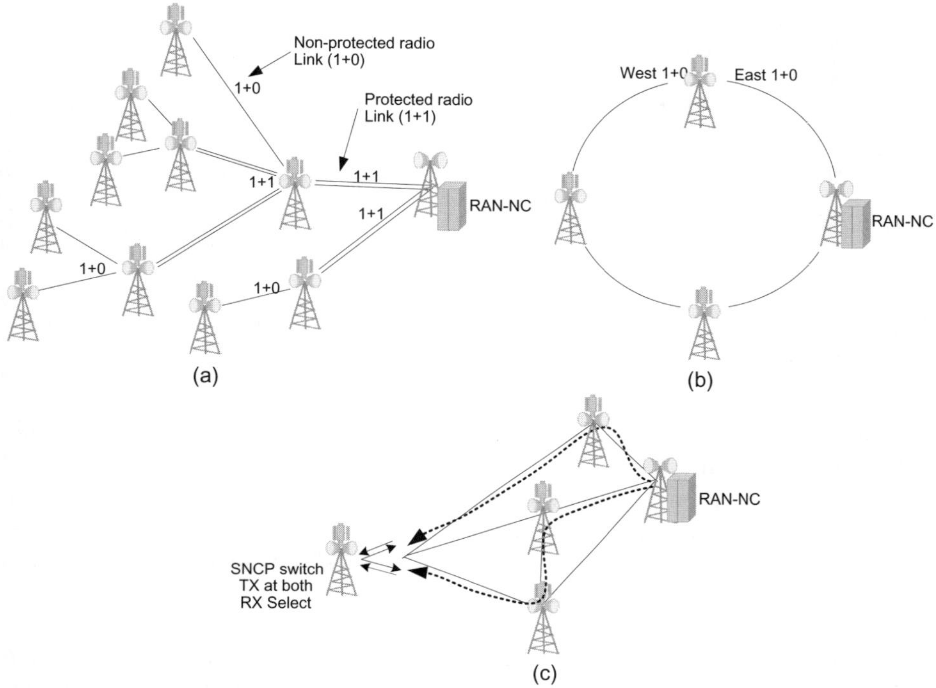

Figure 6.7. Backhaul resiliency. (a) Protected links in a tree backhaul. (b) Resiliency in a ring backhaul. (c) SNCP resiliency in a general (mesh) backhaul.

6.4.3 Ethernet Radio

Types of Ethernet Radio. Ethernet radio can be described as an Ethernet bridge in which one (or more) of its ports is not an Ethernet physical layer (PHY), but a radio link. The unique characteristics of the radio introduce a certain level of complexity and therefore require special planning of the bridge.

Simple radios typically feature two ports: a line port and a radio port (Figure 6.8). Such systems are sometimes referred to simply as "pipe," since traffic flows only from the line to the radio and vice versa. Yet, even such "simple" systems have a certain level of sophistication which varies from basic "transparent" devices that feature no traffic processing, to sophisticated two-port devices, capable of a variety of classifications and VLAN encapsulations, bridging for inband management and sophisticated traffic management along with features such as policing and shaping, OA&M support, and more.

Some very basic radios, for example, are designed to directly modulate the carrier with the physical Ethernet signal. Such radios would deliver even faulty signals, with error frames, since they lack an Ethernet MAC function. Other radios may include Ethernet MAC and drop errored checksum frames. In order to allow radio inband management, a bridging capability is needed to bridge the

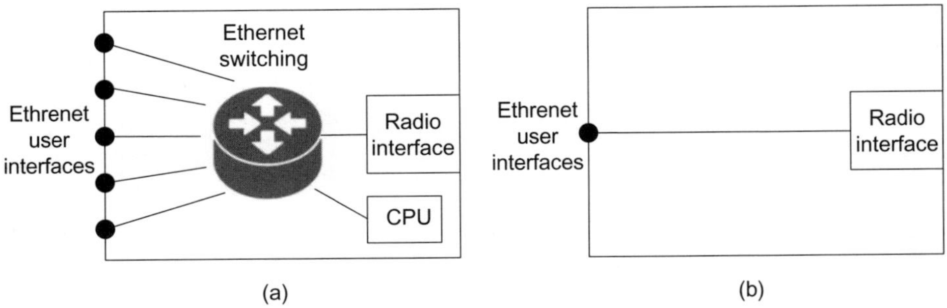

Figure 6.8. Types of Ethernet radios. (**a**) Multiport switched Ethernet radio. (**b**) "Pipe" Ethernet radio.

management frames to the CPU and vice versa (usually tagged by a unique defined management VLAN).

Multiport devices, on the other hand, may include several line ports and several radio ports to provide an integrated radio node solution. Such systems also vary from very basic systems such as "multipipes" lacking bridging functionality between ports, to full-blown sophisticated bridges.

Framing. Native packet radios differ from legacy PDH/SDH radios in that traffic is delivered in bursts and in different-sized frames rather than a continuous fixed rate of traffic. On the other hand, the radio link still delivers a continuous bit rate that may be either fixed or dynamic with ACM. In order to carry frames on such a continuous transmission, the radio system marks the beginning of each packet and also fills in the "silent" gaps between frames. Common techniques for this task are HDLC (High-Level Data Link Control) and GFP (Generic Framing Protocol) defined by the ITU-T to allow mapping of variable-length signals over a transport network such as SDH/SONET.

Rate Gap Between Line and Radio. A challenging issue of packet radio is line rate and radio rate differences. One of Ethernet's greatest advantages is its scalability. A fast Ethernet (FE) line can deliver every rate up to 100 Mbit/s, while a gigabit Ethernet (GbE) line can deliver up to 1000 Mbit/s.[1] A GbE radio will usually not support the full line rate, creating a congestion problem. Congestion happens not only when the average bit rate of the radio line exceeds the radio link rate, but also when long bursts occur in the line.

Let us consider the following example: a GbE interface supports an average traffic rate of 200 Mbit/s and utilizes a radio link of 200 Mbit/s. Since each frame enters at the line rate of 1 Gbit/s and is transmitted at 200 Mbit/s, it takes five times longer to egress than ingress, thus requiring a buffer to hold the burst of data until it is transmitted. As long as the bursts are not bigger than the buffer

[1]These are physical layer rates. For net traffic rate, one should also consider the Ethernet line overheads.

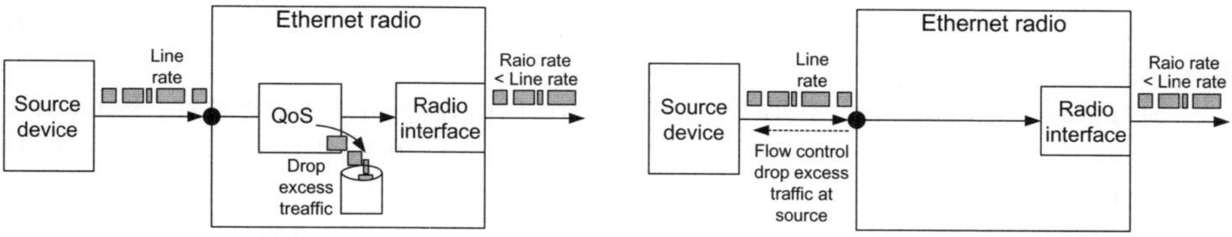

Figure 6.9. Handling the rate difference between line and radio.

size, no congestion will occur. But take, for example, an entire minute of full GbE traffic followed by 4 minutes of silence, which is still 200 Mbit/s of traffic on average. In this case, the burst size is much bigger than any practical buffer, making traffic loss unavoidable.

A radio system can handle congestions in one of the following two alternatives, as depicted in Figure 6.9: Drop the excess traffic frames in the radio, or avoid any excess traffic in the radio by signaling backwards to the source so it can drop it (also known as flow control). Dropping frames in a way that least affects the end-to-end service requires a Quality of Service (QoS) mechanism with buffers large enough to accommodate the desired burst size. Large buffers obviously increase latency and cause a problem to real-time services. Still, good QoS planning can solve such issues by classifying the traffic into different classes, allowing real-time delay-sensitive flows (such as voice and video calls) to receive higher priority with small buffering, whereas non-real-time traffic (as web browsing or files transfer) will be classified to lower priority with larger buffering.

It should be noted that the majority of data traffic today consists of "elastic flows" such as TCP-IP or FTP which are non-real time and not delay-sensitive. Only small portions are actually "streaming flows" like voice or real-time video which are delay-sensitive and require a constant bit rate.

The second option for handling the difference between line rate and radio rate is to signal back to the feeder in case of congestion. The common mechanism for doing so is by employing Ethernet flow control (802.3x). This mechanism, however, has a number of drawbacks. First, it stops all the data without any consideration to priority. Second it can introduce latency that may lead to packet loss if no sufficient buffering exists. This may happen because flow control messages are carried "inband" as Ethernet frames and as such may be delayed behind other large data frames. In such case, buffers may reach congestion before the flow control message reaches the feeder. Lastly, careless use of flow control can spread the back pressure in the network and turn a local problem into network-wide one.

Ethernet Radio QoS. As mentioned, QoS is a key performance issue in Ethernet radio due to the difference between (a) consists line rate and radio rate and (b) the varying rate of ACM radio. A basic Ethernet QoS mechanism is of (a) a classifier that sorts the traffic frames into different priority queues and (b) a scheduler that selects the order at which traffic frames are transmitted from the queues

Simple classifications can be based on different markings such as the Ethernet frame VLAN ID, VLAN priority bits (the 802.1p standard defines 8 classes of service) or the layer 3 (IP) priority bits marking (such as IPv4 TOS/DSCP that defines 64 classes of service). MPLS EXP bits may also be used in the case of IP/MPLS carried over Ethernet. Such classifications will typically use 4 or 8 queues, whereas more sophisticated classifications can be based on traffic flows and utilize a much larger number of queues.

Common scheduling methods are Strict Priority (SP), Weighted Round Robin (WRR), Weighted Fair Queuing (WFQ), or a combination of these. Assigning a queue to higher SP means that as long as there is even a single frame in this queue, it will be transmitted first. In case other queues are overloaded at this time, frames of lower SP queues will be discarded.

Unlike SP, WRR and WFQ select frames from all non-empty queues according to assigned weights. Often a combination of the methods is used, for example to give strict priority to extremely important traffic such as synchronization where other queues are scheduled using WRR.

Bandwidth management and congestion avoidance are two additional mechanisms that are commonly used for Ethernet QoS. The bandwidth management mechanism includes rate limiters (policers) and traffic shapers. These may be implemented at the port level and at the queue level, as well as per VLAN tag or per traffic type such as unicast/broadcast. For example, introducing a policer at each queue allows enforcing a bandwidth profile per class of service. Introducing an egress shaper allows the conformation of the egress rate of the line (or radio) and smoothes traffic bursts to better utilize its bandwidth.

Congestion avoidance mechanisms include such algorithms as Random Early Detection (RED) and Weighted RED (WRED). These mechanisms aim to improve the performance of TCP flows in the network by early discarding frames already before the queue is congested. The random dropping eliminates the process known as TCP synchronization, where multiple TCP flows are simultaneously slowing, resulting in an underutilized network.

Class of Service (CoS) Levels. For effective QoS operation, one should define the network service classes. As an example, Table 6.3 describes the 3GPP definition for traffic classes and their mapping into the mobile backhaul network service classes. The example uses the 3 CoS model, which is one of the models defined by the MEF CoS IA (draft). The model uses three classes: high, medium, and low. Each level is associated with its identifier (such as VLAN tag priority bits or IPv4 DSCP) and has performance attributes such as bandwidth, frame loss, frame delay, and frame delay variation. Other models can map the same traffic types into other CoS levels.

Ethernet Adaptive Modulation (ACM) Radio. One of the advantages Ethernet radio has over PDH/SDH radios is the ability to transport only part of

TABLE 6.3. 3GPP Traffic Classes and Mapping to 3 CoS Model

Traffic Type	Traffic Class	Example Application	Service Class Mapping 3 CoS Model
User plane	Conversational	Voice	High
	Streaming	Streaming video	Medium
	Interactive	Web browsing	Low
	Background	Background download of emails	Low
Management plane	Synchronization	IEEE 1588 timing packets	High
	High-priority management	Configurations, alarms	High
	Low-priority management	FTP download	Low
Control plane	Signaling		High
	Control	Layer 2 Control Protocols (L2CP)	High

the data using ACM, without dropping all of it when propagation conditions do not allow the radio to work at its maximum capacity. In order to handle the reduced bandwidth, QoS must be implemented, so that frames are not dropped randomly but rather in order of importance, according to the operator service policy.

ACM radios have another advantage when delivering TCP/IP data. In a non-ACM radio, even a short interruption of 50 ms can cause TCP/IP timeouts. As a result, drastic throughput decrease can occur until the TCP sessions are recovered, a process that may even take as long as several seconds. With ACM radio, complete traffic interruption is avoided, but instead there is only a momentary drop in throughput. This is not likely to cause any TCP timeouts, and the TCP flow can handle any loss by adjusting and dropping the rate.

One Ethernet transport aspect that is intensified by ACM is delay variation. As ACM radio rate drops to lower rate, it introduces more latency. Thus delay variation is worse with ACM than it is with fixed modulation. This introduces challenges in some applications, such as synchronization over packet, as will be explained further in this chapter.

Ethernet Radio Optimizations. The Ethernet physical layers defined by the IEEE 802.1 standard, uses some non-payload fields such as Preamble and Start Frame Delimiter (SFD), as well as Inter frame Gap (IFG) following each frame. These fields are not required by the radio media and can be easily reconstructed at the far-end radio, saving up to 25% of the bandwidth with short frame length.

Further bandwidth savings can be achieved by omitting the CRC and recomputing it at the far end. In order to avoid fault frames, there is a need to monitor radio performance and drop radio payload frames in case of errors. Thus this may have penalty of error multiplication, as radio code words are not synchronized to the Ethernet frames, and one errored word can cause the drop of several frames even in cases where not all of them were indeed errored.

More sophisticated optimizations techniques include compressions, such as MAC header compression, or even higher layers header compression (IP, UDP, TCP, etc). Since such headers may repeat themselves many times during a stream of traffic, it is possible to deliver a frequent "headers map" to the remote side and then transmit a short ID instead of the long header. The saving in such a case is dependent on the nature of traffic. If traffic actually turns out to be random, there may be even a loss of bandwidth resulting from transmitting the key maps. In many mobile applications, a limited number of connections may exist so such optimizations can yield a significant saving (results up to 25% were demonstrated by leading vendors).

Compression at higher layers and payload compression should be carefully considered according to traffic nature, since it can bring benefit, but also may become useless. For example, IP header compression (as defined by IETF RFC 2507/2508/3095) may be useless in UMTS networks, since 3G already employs it in itself. Payload compression may also be useless if encryption is used in the network, due to the random nature of encrypted data.

Carrier Ethernet Radio. Traditional Ethernet LAN is based on frames forwarding according to dynamic learning tables, as well as on frames broadcasting in case the target address is unknown. Such networks are also based on Spanning Tree Protocols (STP) for preventing loops in the network. When deploying carrier networks, the use of learning and STP yields poor performance; it is not scalable and does not utilize the network resources in an efficient way, nor does it allow traffic engineering or fast resiliency.

The MEF, under the mission of accelerating the worldwide adoption of carrier-class Ethernet networks and services, sets the requirements and attributes that distinguish carrier Ethernet from familiar LAN-based Ethernet. These include primarily a definition of standardized services and requirements of QoS, reliability, scalability, and service management.

The defined services include a few types of Ethernet connections (EVC) [18]: a point-to-point Ethernet connection called E-line, a multipoint-to-multipoint connection called E-LAN, and a rooted point-to-multipoint connection called E-tree. Reliability means demanding availability and rapid (sub-50 ms) recovery time. QoS, setting, and provisioning service level agreements (SLAs) that deliver end-to-end performance based on bandwidth profile, frame loss, delay, and delay variation are additional requirements. Service management requires carrier-class OAM and an ability to monitor diagnose and manage the network.

Carriers evolving to Ethernet backhaul are not likely to rely on LAN-based Ethernet, but rather demand tools to engineer the traffic and to provision and manage connections with defined SLAs. In the following sections we will describe some aspects of carrier Ethernet required in today's modern radios for carrier applications as mobile backhaul.

It should be noted that different technologies can be utilized to implement these capabilities over Ethernet, from simple (and limited scale) VLAN tagging (802.1Q) [17], to provider-bridging (802.1ad), provider backbone bridges (PBB—802.1ah) [22], provider backbone bridge traffic engineering (PBB-TE) [23], and VPLS/VPWS over MPLS (multiprotocol label switching)-based technologies. The MEF does not define the "how" but rather the "what." The right choice of technology depends on many variables. While at the fiber core of the networks MPLS is the dominant solution, for more limited-scale segments of the network including mobile backhaul, simpler solutions can suit better. This is true in particular when it comes to radio networks in which bandwidth is not unlimited, and the overhead introduced by more sophisticated technologies can be a burden.

OAM (Operations, Administration, and Maintenance). OAM refers to tools needed to detect and isolate faults, as well as to monitor the performance of the network connection. Such tools are a crucial requirement in any carrier network because they enable the detection of link failures, verification of end-to-end connectivity, and monitoring the network's performance.

Originally, Ethernet had no OAM tools; however, in recent years, new standards were introduced to address the need for Ethernet OAM over carrier and service provider network. Such standards include: IEEE 802.3ah (Ethernet in the First Mile), which deals with link level monitoring; IEEE 802.1ag [16] (and also ITU Y1731), which deals end-to-end service connectivity fault management (CFM); and ITU Y1731 [15], which, in addition to 802.1ag CFM, also defines Ethernet performance monitoring (PM).

Ethernet OAM defines monitoring points known as maintenance end points (MEPs) and maintenance intermediate points (MIPs). MEPs are located at the edge of the Ethernet service, at the user–network interface (UNI), or at the network-to-network interface (NNI), where MIPs are located within the Ethernet network at intermediate points of the connection. MEPs can generate various OAM test frames in the Ethernet data stream and respond to OAM requests, whereas MIPs can only respond to OAM requests.

The basic CFM operation of a MEP is to send periodic connectivity check messages (CCMs) to other MEPs belonging to the same service. In a normal operation, each MEP periodically receives such messages from all other MEPs. If CCMs fail to arrive, the MEP detects that a failure occurred in the connection. MIPs are used in the process of fault isolation, responding to loopback requests (similar to IP ping) or link-trace requests used to trace the path of the service. Ethernet OAM defines eight maintenance levels of operation to allow seamless interworking between network operators, service providers, and customers.

For the Ethernet radio being part of carrier Ethernet networks, implementing OAM is a crucial requirement, as is the case for any other bridge in the network.

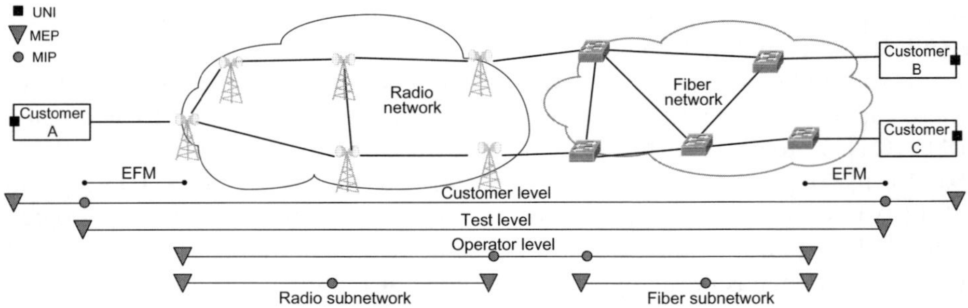

Figure 6.10. Ethernet OAM.

Figure 6.10 shows an example of a network connection between three customer sites, across two networks. Both the operators and the customer are using OAM to monitor the connection by locating MEPs and MIPS along the connection, each at different maintenance levels. In addition, EFM is used at both connections to monitor the line link between operator and customer.

Resiliency. As Ethernet evolves to carrier networks, resiliency requirements become just as important as it is for legacy TDM networks. As with OAM, originally Ethernet had no such tools, but new standards were developed to provide rapid service restoration that deliver SDH/SONET grade resiliency. These include (a) ITU-T G.8031 [14] for Ethernet linear protection switching and (b) G.8032 [13] for Ethernet ring protection switching (ERPS). The objective of fast protection switching is achieved by integrating Ethernet functions and a simple automatic protection switching (APS) protocol.

When considering tree or chain topologies in which path redundancy cannot be found, Ethernet networks employ the same methodology as presented in the TDM case; that is, radio protection handles equipment failures. It is important to mention in this respect the use of a link aggregation group (LAG) in order to connect protected equipments. When a logical link consists of a LAG and one or more members of the LAG fails, the LAG continues to deliver traffic with reduced capacity; thus LAG is a simple common way to connect the protected radio to the fiber switch feeding the wireless network.

When considering a general (mesh) topology network, spanning tree protocols (STP) are often used. However, since convergence time is very long, typically >30 s for STP and 1 s for rapid STP (RSTP), it is not suitable for carrier networks where 50 ms is the industry benchmark traditionally achieved by SONET/SDH.

Ring STP and G.8032 ERPS. Though not standardized, many vendors have created ring-optimized enhancements of RSTP, allowing fast restoration that approaches the 50-ms goal. STP uses bridge protocol data units (BPDUs) that propagate in the network and need to be processed at every node with complex tree computation. The ring STP takes advantage of the known and simple topology of the ring, whereas the tree computing simply results in the

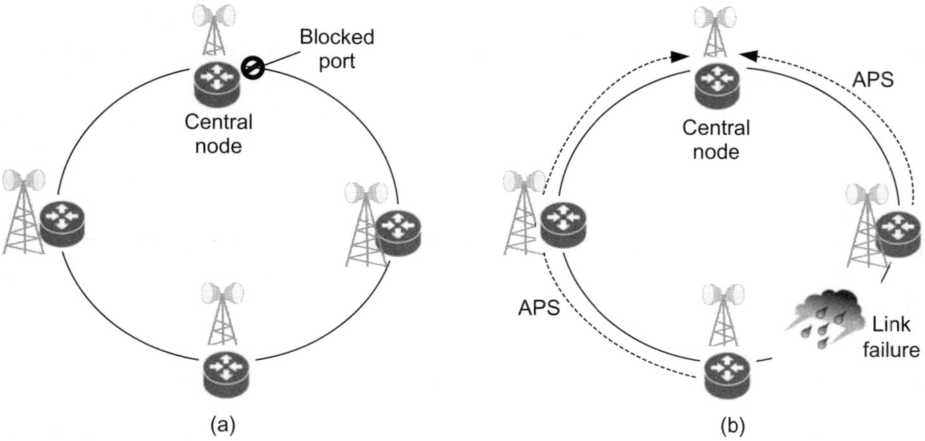

Figure 6.11. Ethernet ring protection. (**a**) Initial ring state (central node blocks one port). (**b**) After a link failure (blocked port is now open).

selection of one port to be blocked. Thus it is able to process and propagate the BPDUs much faster and achieve the required fast convergence time.

A similar approach is taken by the recent ITU-T standard protocol G.8032, which is designed to achieve sub-50-ms ring resiliency in a ring topology. Also dubbed ERP (Ethernet ring protection), G.8032 defines a central node that blocks one of its ports, thereby blocking one link in the ring and not allowing a loop to be formed for the Ethernet traffic. OAM CCMs are then used to detect a link failure in the ring, while automatic protection switching messages (APS) are used to announce such a failure and trigger the opening of the blocked link, as shown in Figure 6.11. For sub-50-ms protection, fast CCM rate of 3.3 or 10 ms is needed.

For Ethernet radios in a ring, the implementation of ring STP or ERPS is a key issue in supporting carrier ethernet service. The implementation should not differ from any wire-line device, but rather it has to be able to initiate an APS request when radio failure events occur.

G.8031 Ethernet Linear Protection Switching. As mentioned, a carrier Ethernet network will not likely be based on LAN Ethernet concepts (learning and STP), but rather it will based on an engineered connection. With general mesh networks, it is possible to define more than one path for E-line EVC and have a redundant path protection. Just as SNCP described for protecting TDM trails, G.8031 specifies point-to-point connection protection schemes for subnetworks constructed from point-to-point Ethernet VLANs.

Protection switching will occur based on the detection of certain defects on the transport entities within the protected domain, based on OAM CCMs and using APS protocol. For sub-50-ms protection, a fast CCM rate of 3.3 or 10 ms is required.

G.8031 defines linear 1 + 1 and 1:1 protection switching architectures with unidirectional and bidirectional switching. Using 1:1 has the benefit of utilizing the standby path bandwidth as long as it is not active (but just running CCMs), whereas 1 + 1 consumes both paths bandwidth constantly. In this respect we can highlight the benefit of Ethernet over TDM trails with SNCP, where the standby path bandwidth cannot be used by other services because there is no statistical multiplexing.

G.8031can fit any general topology; it allows utilizing the network in an optimal way by engineering the connections in the optimal working path and diverse path.

6.4.4 The Hybrid Microwave Radio

The hybrid radio combines both TDM traffic and Ethernet frames simultaneously while keeping the important characteristics of both. It has the capability of keeping the TDM clock attributes and allowing it as traffic and synchronization interface, as well as the capability to transport the Ethernet frames natively and being able to drop frames under ACM conditions according to QoS policy. The radio bandwidth can be dynamically allocated between the TDM fixed bandwidth and variable Ethernet bandwidth to optimally combine both kinds of traffic. Unused E1/T1 bandwidth should be automatically allocated for additional Ethernet traffic.

In early deployments, in which Ethernet is primarily added for lower-priority "best-effort" data services, it is assumed that under fading situation with ACM, the hybrid radio will first drop only Ethernet traffic and then keep the TDM voice and real-time traffic connections. With network convergence and new deployments of Ethernet-based services, this does not always have to be true. Thus, more sophisticated solutions should be configured to drop some lower-priority Ethernet traffic first, but also allow the higher-priority Ethernet to be the last to drop—even after lower-class TDM connections have been.

6.5 MOBILE BACKHAUL EVOLUTION ALTERNATIVES

The evolution from 2G to 3G to LTE and onwards goes hand in hand with the shift from voice (2G) to data (3G) and to fully mobile broadband connectivity (LTE/4G), together with the shift from circuit to packet connectivity. Another migration is the one toward convergence of fixed and mobile networks, in which the same backhaul network serves not only the mobile network but also that of fixed services such as business or residential broadband access and VPNs.

The growing demand for mobile data services requires enormous capacity growth in the backhaul segment, which makes TDM backhaul neither scalable nor affordable anymore. For example, LTE eNB backhaul requires 50xE1s (supporting 100 Mbit/s), compared to 2xE1s used at typical 2G BTS. When multiplying this number by the number of cell sites in a network, it's plain to see that simply adding E1 connections makes no sense.

In addition, the latest mobile technologies introduced like WCDMA-R5 and onwards—EVDO, LTE and WiMAX—are "All-IP" technologies and are based on pure packet architecture. This makes Ethernet the natural first choice for backhaul growth.

Up to this point, we have counted a number of advantages that Ethernet has over TDM. But in order for Ethernet to completely displace TDM, it must provide the necessary carrier Ethernet requirements of QoS and OAM. Additionally, next-generation Ethernet systems must ensure synchronization delivery and fast resiliency while supporting legacy TDM. Though Greenfield and long-term deployments may be Ethernet only, existing legacy networks will not quickly disappear, but will continue to coexist with new technologies for many years to come.

6.5.1 The Migration: Evolving to Ethernet While Supporting Legacy TDM

Figure 6.12 depicts the evolution of mobile networks and backhaul technology. For each technology, the diagram indicates the maximal user throughput, and describes a typical RAN-BS deployment case. In the following sections we will discuss three different alternatives for backhauling next-generation mobile traffic over radio networks:

1. Pure TDM backhaul
2. Pure packet backhaul
3. Hybrid backhaul, with overlay of both networks

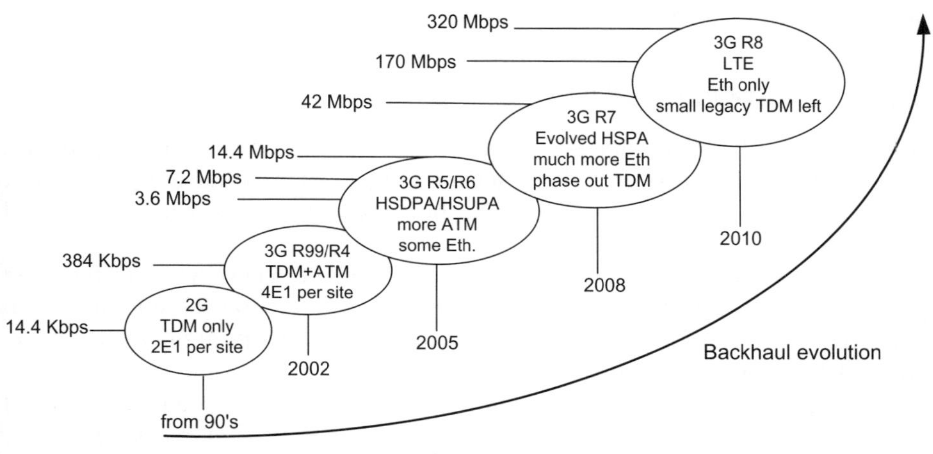

Figure 6.12. Mobile networks evolution.

TDM, hybrid, and all-IP backhaul can be served by copper, fiber, and radio transport systems. Each alternative has its strong and weak points, and there is no single "right way." In the long run, telecom operators hope to implement all-IP 4G architectures for reasons of simplicity and cost-cutting. However, until reaching maturity of pure packet backhaul and considering the fact that 3G and even 2G networks will not disappear overnight, we can assume that all of these alternatives will coexist together for a long time.

6.5.2 Pure TDM Backhaul

Keeping the legacy TDM (PDH/SDH) network in the long term is a valid option; but over time, legacy technology will be outmatched by the advancement of mobile technologies. While HSDPA initial deployments still used TDM interfaces and required only few additional E1/T1s to carry data services, next-generation cellular base stations are expected to have Ethernet interfaces and fit into all-IP architectures.

Mapping Ethernet frames over TDM is not a new concept, but operators who chose this solution will quickly run into scalability and performance issues. Maintaining RANs based on costly E1/T1 connections to support data rates in the tens or even hundreds of Mbps per cell site is not likely to generate a profitable business case.

While this book is being written, some microwave radios still do not support native packet transport. Such systems implement Ethernet transport by mapping frames on groups of E1/T1 connections that are carried all over the network as fixed-rate circuits. This solution obviously has several drawbacks including: granularity, as bandwidth is allocated at multiples of E1/T1; scalability, as a very large number of E1/T1s are gathered along the network; and, most important, no statistical multiplexing is available at the network's aggregation points. An exception for this is the use of ATM aggregation which allows statistical multiplexing for ATM-based traffic as with 3G early releases, yet this option is not future-proofed for IP-based technologies.

6.5.3 Packet-Only Backhaul

A pure packet backhaul appeals primarily as it is future-proofed and provides the complete solution over a single technology. As such, it has the potential of saving in both CAPEX and OPEX, but also maintains some risks and drawbacks.

A pure packet backhaul requires us to map the TDM/ATM traffic over packet using TDM to Packet Generic Inter-Working Function (GIWF), as shown in Figure 6.13a, in contrast to maintaining the legacy network in the hybrid backhaul concept shown in Figure 6.13b.

Several standards define the emulation of TDM/ATM over a packet-switched network (PSN). These include IETF pseudo-wires (PW) [24,25], MEF circuit emulation services over Ethernet (MEF8) [20,21] and some ITU-T recommen-

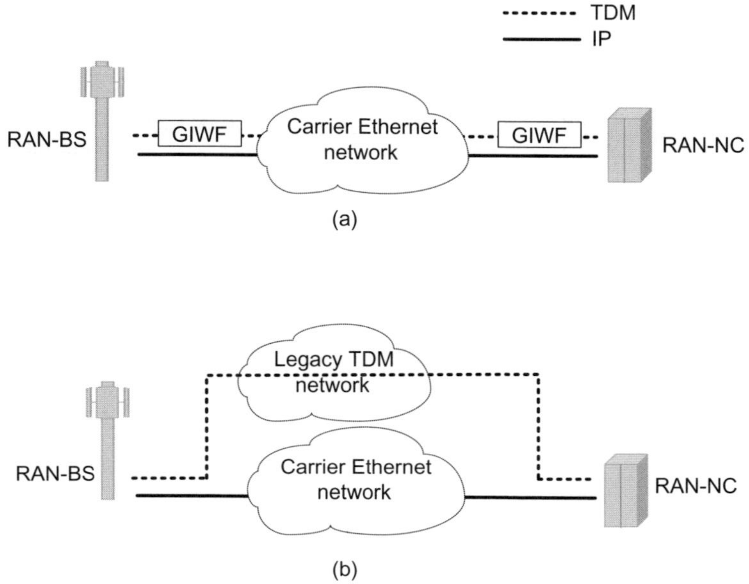

Figure 6.13. Hybrid backhaul and pure packet backhaul. (**a**) Pure packet backhauling. (**b**) Hybird backhaul concept.

dations. The IETF long list of RFCs and ITU-T recommendations define PWs for most traditional services as TDM, ATM, frame relay, Ethernet, and others. Thus it allows the use of a single-packet infrastructure for all. The major benefit of using emulated solutions is the ability to map everything over the same network and thus install, manage, and maintain a single network for all service types. A major drawback is the overhead introduced by the encapsulation of frames. While such overhead may be insignificant at the fiber core, it can become a major problem over copper and microwave in the access and metro. The encapsulation overhead depends on the exact configuration and can be reduced by using larger frames that encapsulate more TDM/ATM traffic at a time. Yet this approach results in longer delays and error multiplication.

It should be noted that at some implementations, such as IP/MPLS PW, the overhead will be introduced not only to the TDM traffic, but also to the Ethernet traffic, as several PWs are assigned to all types of traffic coming from the cell site. Thus in such implementations, Ethernet packets are encapsulated by IP/MPLS PW headers, and then transported again over Ethernet, resulting in a significant bandwidth overhead and low radio utilization.

Another major issue with pure packet backhaul is delivering synchronization over the packet network while meeting the requirement for TDM services (G.823/G.824) [7,8] under many network conditions. Synchronization between RAN-BSs is critical to the mobile network performance. A requirement of ±50-ppb clock accuracy is defined for GSM, 3G, and LTE FDD systems, and

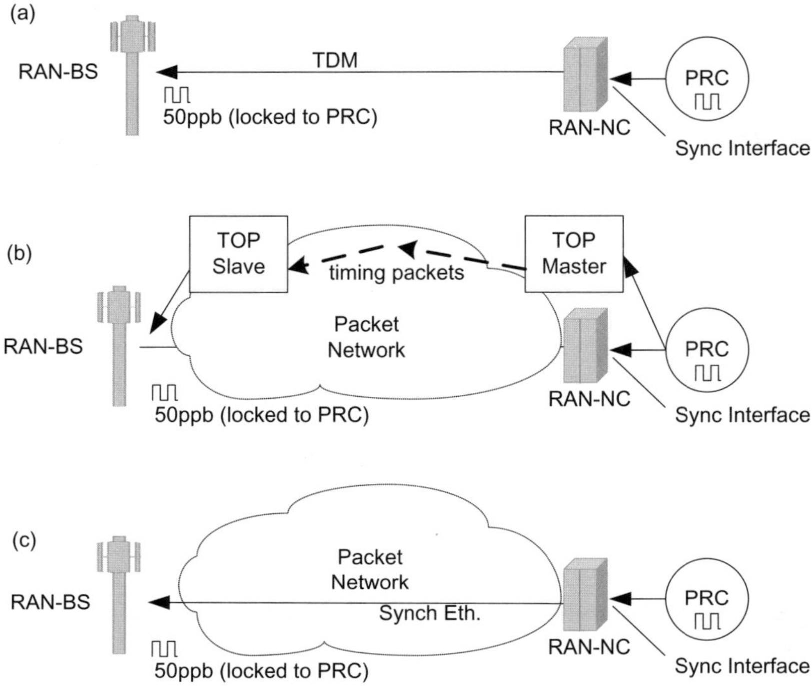

Figure 6.14. RAN-BS synchronization methods at the backhaul. (**a**) TDM synchronization. (**b**) Timing over packet synchronization. (**c**) Synchronous Ethernet synchronization.

systems need to comply with strict jitter and wander specifications. With the exception of a few networks that utilize an external source for the clock delivery (usually a GPS receiver, which is commonly used in IS-95/CDMA2000 networks), most legacy networks use SDH/PDH backhaul to deliver synchronization.

As shown in Figure 6.14a, the TDM generated in a central site (MSC) is locked to a primary reference clock (PRC) via a synchronization interface (such as E1 or STM-N). The RAN-BS is synchronized on the backhaul incoming TDM traffic signal, so eventually it locks the RAN-BS to the same PRC. When using PW/CES backhaul, the RAN-BS is now synchronized on the TDM recovered from CES frames.

Standards that define the delivery of timing over packet (TOP) are IEEE Precision Time Protocol (PTP) 1588v2 and IETF NTP. Both are based on exchanging time information via dedicated timing packets (using time-stamps) and restoring the clock using this information, as shown in Figure 6.14b. Such techniques are ubiquitous and work over any transport technology, but the restored clock accuracy is highly dependent on the network performance in terms of packet delay variation (PDV). Whatever sophisticated algorithms are employed at the clock restoration, there are limits to the accuracy that can be achieved as the PDV gets too high.

When observing the Ethernet backhaul network, PDV becomes more of an issue as the radio links are narrower in bandwidth. We can observe a simple test case of a cell site with a backhaul connection of 5 Mbit/s versus a 50-Mbit/s radio link. At the cell site, TOP frames are transmitted at a fixed rate, but also large data frames of 1500B may be transmitted in between. Obviously the TOP frames should be classified to higher-priority queue and scheduled with strict priority, but it may happen that such a frame is scheduled just after a large data frame (known as "head-of-line blocking") and delayed longer. Since the transmission time of a 1500B frame is only ~0.24 ms at the 50-Mbit/s link versus ~2.4 ms at the 5-Mbit/s link, the PDV will be 10 times higher at the narrow bandwidth link. To calculate the total end-to-end PDV, one should consider several connection hops. Still, because usually the closer we are to the core, the larger the connection's bandwidth becomes, the dominant contributor to PDV is the access network.

There are proprietary methods to mitigate the problems discussed above in a radio link; however, it is important to keep it mind that with standard Ethernet transport, time over packet is risky when it comes to access networks, because it is highly dependent on PDV performance.

A different technique to deliver synchronization over Ethernet backhaul is synchronous Ethernet (ITU-T G.8261) [10], shown at Figure 6.14c. Synchronous Ethernet, or Sync-Eth, is a physical layer frequency distribution mechanism similar to SDH/PDH that uses the actual Ethernet bit stream to deliver the clock. Its biggest advantage is that the clock accuracy is independent from network load. Additionally, it has no demands on bandwidth resources and is not affected by any congestion or PDV. Thus Sync-Eth represents an excellent SDH/PDH replacement option.

Sync-Eth drawbacks are that it requires special hardware at every node, it is limited to a single clock domain, and, unlike ToP, it can only deliver frequency and not phase.

A radio supporting Sync-Eth should have the capability to deliver the output Ethernet traffic at the remote side, locked to input clock at the near side and maintaining its quality (jitter and wander as specified by G.823/G.824). As PtP radio naturally delivers the clock of the transmitted signal, supporting Sync-Eth is relatively easy. Additional functionality is required to (a) select timing from several inputs for network protection and (b) provide timing traceability via SSM (Synchronous Status Messages).

6.5.4 Hybrid Backhaul

Hybrid backhaul means overlaying existing circuit backhaul networks with a carrier Ethernet transport layer used for data traffic and for connecting base stations with IP interfaces, as shown in Figure 6.13b. This solution allows keeping the legacy connections and synchronization over TDM, while adding flexibility, increasing capacity and enjoying additional benefits using a packet network.

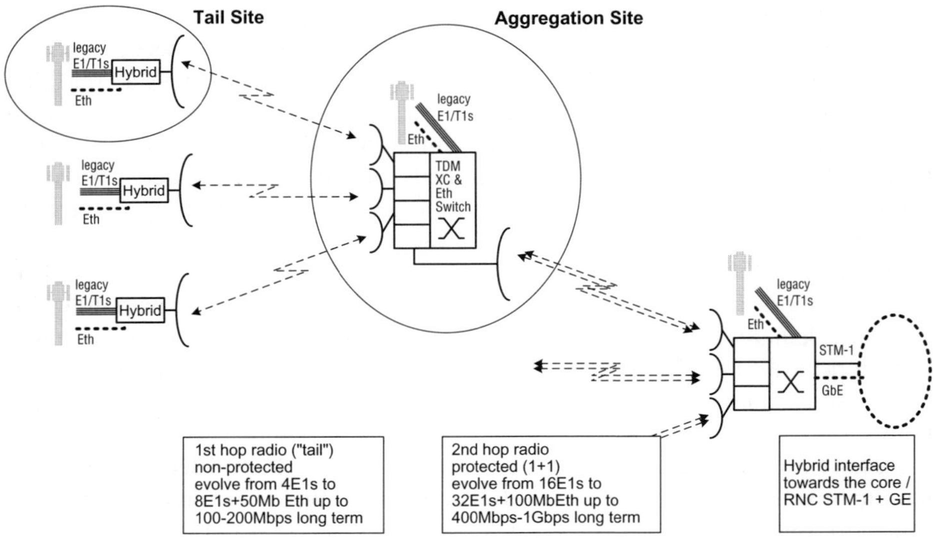

Figure 6.15. Hybrid radio backhaul network example.

The concept of hybrid is sometimes referred to in the wireline backhaul segment as "offload." In the wireline world, hybrid systems sometimes use a different media for each kind of network—for example, maintaining the TDM over leased lines and delivering Ethernet over DSL. This method has an upside of more cost-effective delivery of data using DSL and has the obvious downside of having to manage and maintain two separate networks. In the wireless world, hybrid means having two logical networks on a single physical network. Figure 6.15 demonstrates a hybrid aggregation network, where each tail site is accessed with both Ethernet and several E1/T1 interfaces. Each aggregation site includes a node with TDM cross-connect and an Ethernet switch. All traffic is gathered finally toward the core: Ethernet on a GE interface, and all TDM on a channelized STM-1 interface.

Enabling smooth migration of the backhaul is a complicated task. Still, in many cases the hybrid approach, employing radios with ACM, carries the lowest risk. Figure 6.16 shows a typical migration case for legacy TDM network based on microwave radios carrying 4xE1s per site over 7 MHz or 14 MHz channels. Adding the new Ethernet traffic for data services is possible while maintaining the same radio planning and antennas. Enabled data rates are up to 50 Mbit/s over 7 MHz channels or 100 Mbit/s over 14 MHz channels, while the TDM traffic stays the same, thus without any synchronization issues.

We can see the importance of ACM, being the enabler for such migration. Without ACM, legacy services could not be maintained with the same service level; thus radio planning would have needed to redesign (such as using bigger antennas), and migration cost would have been significantly higher.

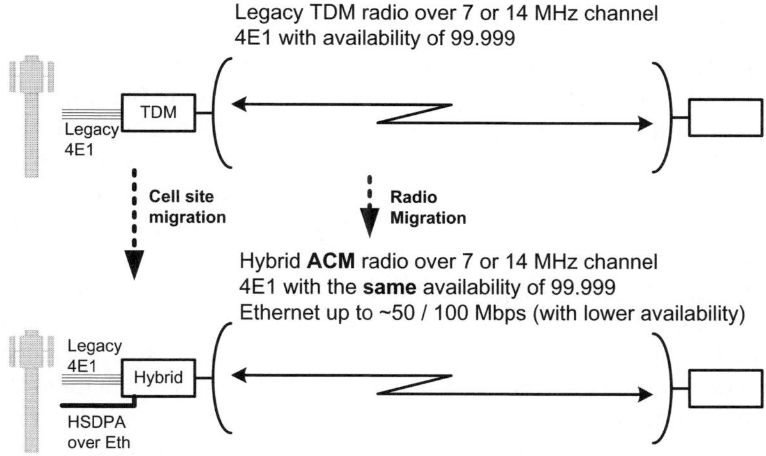

Figure 6.16. Migration of RAN-BS connection from TDM to hybrid.

6.5.5 Hybrid Access with Packet Aggregation

We mentioned the difficulties and risks of pure packet backhaul, especially when it comes to the access portion, where there is narrower available bandwidth. In the access segment the PDV is higher, and the overhead of solutions like PW is more significant. On the other hand, a pure packet backhaul has its advantage of convergence and simplicity, managing all kinds of traffic over the same network.

A solution that combines a hybrid access network and pure packet with PWs at the aggregation network can benefit from all advantages and avoid some risks. It also has the potential of cost saving as PW equipment is not needed at every site, but only at the aggregation sites, as depicted in Figure 6.17.

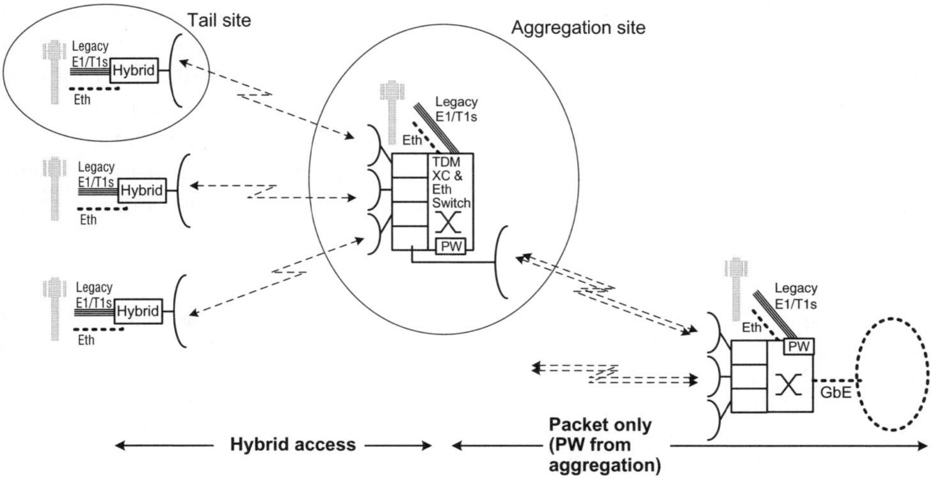

Figure 6.17. Combining the hybrid and the packet backhaul.

6.5.6 Radio Backhaul Topologies

There are many parameters to be considered when selecting the topology of every network, and even more so when it comes to radio networks, where LOS, rain zone, and other propagation factors should be taken into account—as should aspects such as antenna sites and towers considerations. The common topology choices for radio networks are usually a tree and a ring, or often a combination of both. The tree topology in itself is a combination of two additional basic topologies: the chain and the star, as shown in Figure 6.18.

A star topology uses a separate link for each site. This is very simple but not efficient with microwave, because it requires longer radio links and a LOS from origin point to each site, which is unlikely to happen. The star topology also makes for very poor frequency reuse. This is because all the links originate at the same place and the interferences are more likely to happen between links at the same frequency. With chain topology all sites reside on a single path overcoming the star microwave aspects, but resulting in a very sensitive topology where the first link failure causes a complete network failure. Thus almost all links should be protected. Combining the chain and the star yields the tree, at which fewer links can cause major network failures, and only those need to be protected. On the other hand, closing the chain yields the ring, which is the most efficient in terms of protection.

Focusing on the ring and the tree, we will highlight some of these considerations using a simple test case, shown in Figure 6.19. The test case describes a typical radio cluster with one fiber site and 10 cell sites requiring 50 Mbit/s each and aggregated to a total of 400 Mbit/s. Also, it is assumed that every link supporting more than one site needs to be protected. Several aggregation topologies are suggested: tree, single ring, and "tree of rings" (two smaller rings). The tree uses protected links wherever a link failure affects more than a single site.

Comparing the network fixed assets cost (CAPEX), we can see that the ring requires fewer radios. On the other hand, more high-capacity radios are needed in a ring, usually at a higher cost and consuming more spectrum. The ring also requires a few more antennas; thus cost comparison is not straightforward and can vary, depending on the particular case. Another factor that can influence cost is spectrum reuse. Since rings have no more than two links at every node, better frequency reuse is usually achieved and often rings are implemented

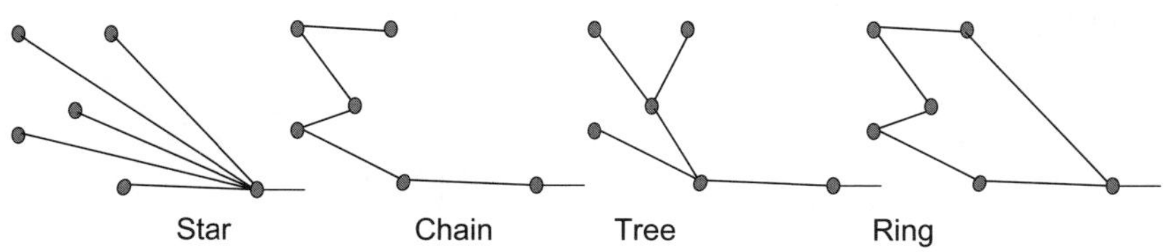

Figure 6.18. Network topologies.

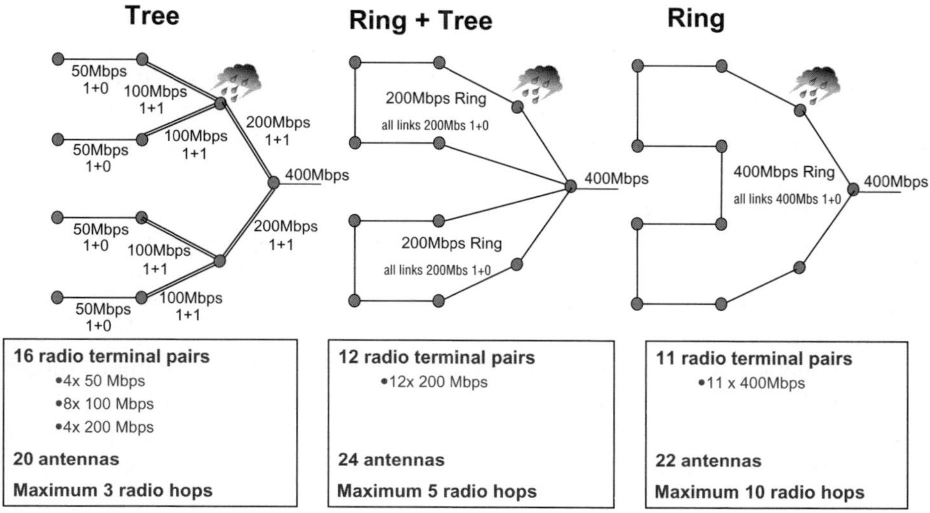

Figure 6.19. Aggregation topologies test case.

by only a single pair of frequency channels (this depends on the geography of course).

A clear-cut advantage for ring topology is its superior resiliency. The protected tree is indeed protected against equipment failures, but does not provide any path redundancy. Thus it is more vulnerable to heavy fade conditions, as well as to complete site failure (as electricity failure, or weather disaster happens).

Consider the rain situation shown in Figure 6.19; in the case where this site is in complete failure (as a reason of rain, electricity breakdown, or any other failure), it causes the other four sites to fail in the tree, but no other sites in the ring.

The ring also gives a much better availability thanks to its path diversity, because it is necessary for both paths to fail in order to create a service failure. Thus, for achieving the same end-to-end target availability within a tree and a ring, each ring link can be designed for lower availability compared to the tree link. This can save costs through the usage of smaller antennas and lower power at the ring links.

A disadvantage of the ring is that it takes more radio hops to reach the distanced cells. If designed properly, the shortest path can be selected for each traffic flow, but in case of protection where the ring is cut, traffic can flow over $N - 1$ hops (with N nodes ring). The number of hops can be an issue when considering latency and delay variation, affecting particularly synchronization delivery. Still, when comparing the smaller number of hops in the tree, one should remember that some of them are also with smaller bandwidth than the ring and thus can be with worse performance of delay variation, so this disadvantage is definitely arguable. Limiting the number of maximum hops can be a drive to move into a ring-tree combination, with several smaller rings, as shown in Figure 6.19.

Statistical multiplexing is another advantage of ring topology as derived from the fact that more links share traffic to several sites. On the other hand, since all links are of the same capacity, it may create some disadvantage when adding capacity because all links should be upgraded. This is in contrast to tree topology, where only the tree trunk can be upgraded when adding another branch.

One way to increase ring capacity without upgrading its entire links is to evolve it into a mesh by adding crossing links and breaking it into smaller rings, but this introduces another complexity of managing the connections and protection schemes.

To conclude, there is no single "right" topology. Network planners should consider the particular environmental conditions, business conditions (such as spectrum, radio and antennas costs), reliability requirement, and application characteristics, in order to determine the best solution for their needs.

Backhaul Topology at LTE. Unlike all prior cellular technologies which are "hub and spoke" networks at which traffic from all RAN-BSs at the cellular sites is aggregated together to the RAN-NC, the LTE defines a flat architecture with direct connectivity between one RAN-BS (called eNB) and its neighbors. Figure 6.20 describes the LTE architecture in comparison to 3G architecture, showing the S1 interface that is toward the core and also showing the X2 interface defined between neighboring eNBs.

The LTE flat architecture defines actually a mesh mobile backhaul network, but since the X2 interface is used primarily for handoffs (move of a mobile user from one BS to another) with relatively little traffic, it is not likely that the X2 will justify a dedicated physical link. Thus, the mesh connectivity can remain logical but not physical. A logical mesh can be defined over every existing tree or ring topology by creating Ethernet connections such as E-Line or E-LAN.

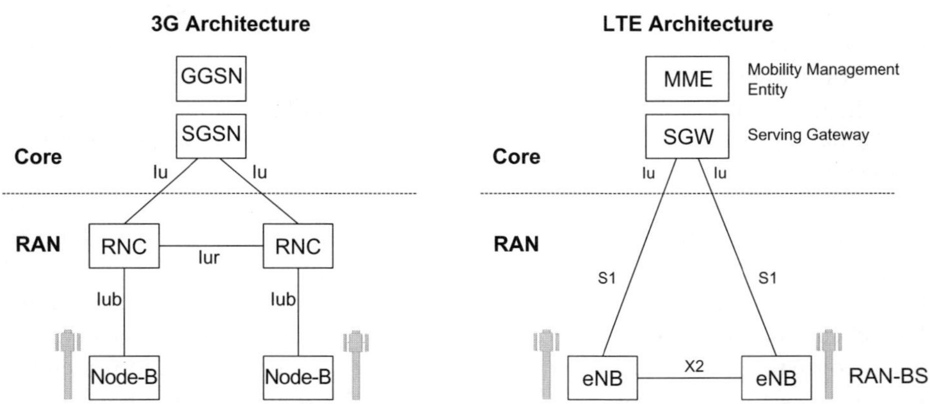

Figure 6.20. LTE flat architecture in comparison to 3G architecture.

REFERENCES

1. Infonetics Research, *Research Mobile Backhaul Equipment, Installed Base & Services*, Copyright 2008, Infonetics Research, Inc.
2. Heavy Reading, *Ethernet Backhaul Quarterly Market Tracker*, Copyright 2008, Heavy Reading.
3. Heavy Reading, *Ethernet Backhaul: Mobile Operator Strategies & Market Opportunities*, Copyright 2007, Heavy Reading.
4. Unstrung Insider, Ethernet Microwave: Backhaul & Beyond, Vol. 6, No. 4, April 2007, Copyright 2007, Light Reading.
5. Harvey Lehpamer, *Microwave Transmission Networks—Planning, Design and Deployment*, McGraw-Hill, New York, 2004.
6. Ethernet ring protection for carrier Ethernet networks, *IEEE Commun. Mag.*, September 2008.
7. ITU-T Rec. G.823, The Control of Jitter and Wander Within Digital Networks which Are Based on the 2048 kbit/s Hierarchy, 2000.
8. ITU-T Rec. G.824, The Control of Jitter and Wander Within Digital Networks which Are Based on the 1544 kbit/s hierarchy, 2000.
9. ITU-T Rec. G.841, Types and Characteristics of SDH Network Protection Architectures, 1998.
10. ITU-T Rec. G.8261/Y.1361, Timing and Synchronization Aspects in Packet Networks, 2006.
11. ETSI EN 302 217-2-1 V1.2.1, Fixed Radio Systems; Characteristics and Requirements for Point-to-Point Equipment and Antennas, Part 2-1: System-Dependent Requirements for Digital Systems Operating in Frequency Bands where Frequency Coordination, 2007.
12. ETSI EN 302 217-2-2 V1.2.3, Fixed Radio Systems; Characteristics and Requirements for Point-to-Point Equipment and Antennas, Part 2-2: Harmonized EN Covering Essential Requirements of Article 3.2 of R&TTE Directive for Digital Systems Operating in Frequency Bands where Frequency Co-ordination Is Applied, 2007.
13. ITU-T Rec. G.8032/Y.1344, Ethernet Ring Protection Switching, 2008.
14. ITU-T Rec. G.8031/Y.1342, Ethernet Linear Protection Switching, 2006.
15. ITU-T Draft Rec. Y.1731, OAM Functions and Mechanisms for Ethernet Based Networks, 2006.
16. IEEE Std. 802.1ag, Local and Metropolitan Area Networks, Virtual Bridged Local Area Networks, Amendment 5: Connectivity Fault Management.
17. IEEE Std. 802.1Q, Virtual Bridged Local Area Networks, 2005.
18. MEF 6.1, Ethernet Services Definitions—Phase 2, 2008.
19. MEF 22, Mobile Backhaul Implementation Agreement—Phase 1, 2009.
20. MEF 3, Circuit Emulation Service Definition, Framework and Requirements in Metro Ethernet Networks, 2004.
21. MEF 8, Implementation Agreement for the Emulation of PDH Circuits over Metro Ethernet Networks, 2004.
22. IEEE P802.1ah-2008, IEEE Standard for Local and Metropolitan Area Networks—Virtual Bridged Local Area Networks—Amendment 6: Provider Backbone bridges.

23. IEEE P802.1Qay/D3.0, IEEE Standard for Local and Metropolitan area Networks—Virtual Bridged Local Area Networks—Amendment: Provider Backbone Bridge Traffic Engineering, 18 April 2008.
24. IETF RFC 5086, Structure-Aware Time Division Multiplexed (TDM) Circuit Emulation Service over Packet Switched Network (CESoPSN), 2007.
25. IETF RFC 4553, Structure-Agnostic Time Division Multiplexing (TDM) over Packet (SAToP), 2006.
26. Ceragon Networks, White paper, Flex your backhaul network with adaptive coding & modulation, 2008.

PART II

WIRELINE TECHNOLOGIES

7

PAVING THE ROAD TO Gbit/s BROADBAND ACCESS WITH COPPER

Thomas Magesacher, Per Ödling, Miguel Berg, Stefan Höst, Enrique Areizaga, Per Ola Börjesson, and Eduardo Jacob

7.1 INTRODUCTION

Ubiquitous low-cost broadband access is a key enabler of quality of life and modern economy. The demand on end-user data rates keeps increasing, which in turn fuels the development and deployment of new systems. Fixed broadband access technology is evolving from exclusively copper-based solutions to hybrid fiber/copper architectures. A recent analysis of this evolutionary process has revealed that there is a gap—a missing, not foreseen system generation [1]. This chapter is devoted to this expected next step in the evolution of broadband systems, here named the 4th-Generation Broadband concept. It identifies a technical, infrastructural, and economical niche and describes how the fiber access network is extended and forked to feed a last and ultimate generation of DSL systems, shown to have gigabit potential.

Our classification of broadband systems into "generations" contains only broadband systems operating on the twisted copper pairs of the public telephony network and optical fiber—that is, DSL systems and fiber access systems. The future, as well as the present, will certainly see also other technologies such as coaxial cable access systems (using cable TV infrastructure) or fixed wireless access systems, but we leave these outside the scope of this presentation.

Convergence of Mobile and Stationary Next-Generation Networks, edited by Krzysztof Iniewski

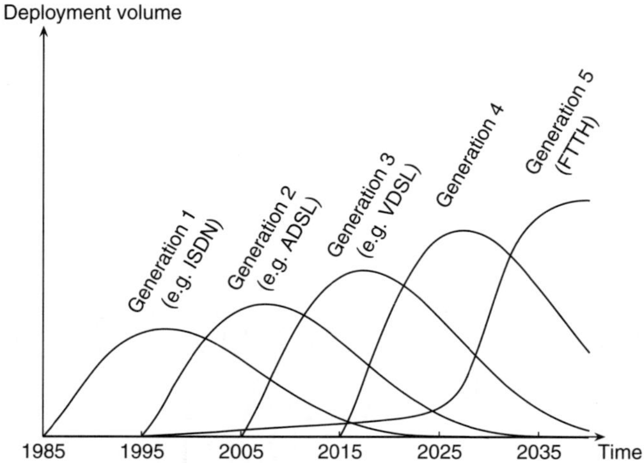

Figure 7.1. A sketch of deployment volumes of broadband access techniques (number of new installations or upgrades per time unit). The *x* axis is based on historical data (up to the present time), while the *y* axis is no more than an illustration of trends.

In Figure 7.1 the principal deployment history for broadband access equipment is sketched. Note that the classification of systems as generations in Figure 7.1 is introduced to define and emphasize a gap in the foreseen broadband evolution and is not a generally accepted terminology. The term broadband access equipment loosely denotes communications equipment intended, for example, for Internet access with a permanent connection—that is, post-dial-up systems. During the last two decades, two generations of broadband access systems for telephone loops have been rolled out: Generation 1, which is mainly based on ISDN (cf., e.g., Stallings [2]) and Generation 2, which is mainly based on ADSL (cf., e.g., Golden et al. [3]). Both generations are characterized by systems deployed from the Central Office. Generation 1 marked the start of data communication beyond dial-up modems, while Generation 2 added a "real" transport network and user bandwidths comfortably greater than voiceband modems.

Today we are seeing the launch of the Generation 3 broadband access system, the VDSL family [3], which will provide customer data rates of up to 100 Mbit/s. While ADSL operates from the central office, often over cables that are several kilometers long, VDSL is designed to operate over shorter loops. Therefore, the VDSL equipment is normally placed in cabinets, resulting in a typical loop length that is below 1 km. The backhaul solution—that is, the technology to bring data between the transport network and the cabinet—is today almost exclusively based on optical fiber technology. The transition from Generation 2 to Generation 3 thus implies an extension of the fiber network from the Central Offices to the cabinets. This is a first and fundamental step toward spawning a large-scale fiber-to-the-home (FTTH) infrastructure.

Generation 4, presented and discussed here, is nothing more than the logical extension of the thinking behind Generation 3. The communication needs of the future are assumed to require data rates an order of magnitude higher than those of Generation 3—that is, a step from around 100 Mbit/s to around 1 Gbit/s. To deliver these data rates using the in-place copper architecture then requires even shorter loops. The key question is whether or not there exists a natural place to deploy the new transmission equipment in an economical fashion.

With the 4th-Generation Broadband concept, we would like to bring out the "Last Distribution Point," hereinafter referred to as Last DP, as a candidate from which broadband services could be delivered in a technically and economically feasible fashion. The copper plant is a star network, forking out into finer and finer segments (fewer and fewer lines running together) until eventually individual twisted pairs reach their respective user premises. The Last DP can be found by following the lines from the users' homes and backwards into the network, where normally after 20–200 m you find a point in which a number of lines merge together and form a bundle. This is the most outward point at which a modem pack can be installed serving several (say 10–30) customers. The Last DP was touched upon as early as 1990, in the form of fiber-to-the-building (FTTB) and fiber-to-the-curb (FTTC) discussions but at the time not associated with a corresponding new generation of copper access (DSL) equipment making full use of the greater bandwidth offered by the shorter loops. The earlier FTTB and FTTC discussion left little mark in the standardization processes and were essentially abandoned.[1] We believe that it may be time to awaken the idea of moving to the Last DP, but now dressed in modern technology and based on 20 years of experience from the development of the broadband market.

A natural question is, of course, how much this infrastructural quantum leap will cost, especially in comparison with installing optical fibers all the way out to the customer (FTTH). We return to this in the next chapter.

Figure 7.1 contains historical data for Generations 1 to 3, along with predictions of the deployment timescale for the 4th- and the 5th-Generation broadbands.[2] So far, the transition between any consecutive pair of earlier generations has taken about 10 years. This suggests that the process that leads to the creation of a new-generation broadband has a period of 10 years, based on the lead time in standardization and product realization.

The supported data rate increases by roughly an order of magnitude from generation to generation. This also applies to the step from voiceband modems, which can be viewed as Generation 0, to Generation 1. The step to 4th-Generation broadband (4GBB), using the Last DP and, possibly, vectoring technology [4], will provide data rates on the order of 1 Gbit/s—that is, 10 times the data rate of Generation 3 (e.g., VDSL2 with up to 100 Mbit/s).

[1]It was recently pointed out by British Telecom researchers that technology developments will soon make it feasible to exploit the Last DP to deliver broadband services, spawning the work presented here.

[2]*Source*: T. Magesacher, P. Ödling, S. Höst, E. Areizaga, P. O. Börjesson, and E. Jacob.

Applying the above argument to 20 years from now, the bandwidth demand should then have increased another order of magnitude to 10 Gbit/s per household, serving as an outlook toward the technical specifications of Generation 5 (FTTH). According to the prediction in Figure 7.1, the 5th Generation's deployment volume will gradually increase, exhibiting a peak around 2035.

7.2 HYBRID FIBER/COPPER-BASED BROADBAND ACCESS NETWORK

In most operational telecom networks, the topology of the access loop looks like the example network situation depicted in Figure 7.2, where one primary cable connects the Central Office (CO) to various street cabinets (in the figure labeled "DP in Cab" meaning distribution point in cabinet), and from there we have stepwise forking out to reach the users' premises.

This structure is typical to all major telecom networks, although the distances and the number of lines per cable vary both between countries and between central offices. A typical CO has on the order of 15 primary cables, each with about 1500 pairs. Each cable, in turn, serves around half a dozen street cabinets, making normally between 50 and 100 cabinets per central office, serving some 20,000 households and other customer locations.

The average length of a copper pair connecting the customer with the CO is ranging from 1.5 km to 3 km, depending on country and area. This distance is the main obstacle to increasing the bandwidth from Generation 2 systems, where best-in-class is ADSL2+ in practice normally providing between 10 and 20 Mbit/s, to the higher bit-rates offered by Generation 3, today VDSL2 with up to 100-Mbit/s per copper pair. By instead placing the transmission equipment in cabinets, it is possible to reduce the average length to less than 1 km (cf. Figure 7.2). The cabinets will then typically be connected to the CO using optical fiber, as well as to the users with VDSL2. However, and mainly in urban areas,

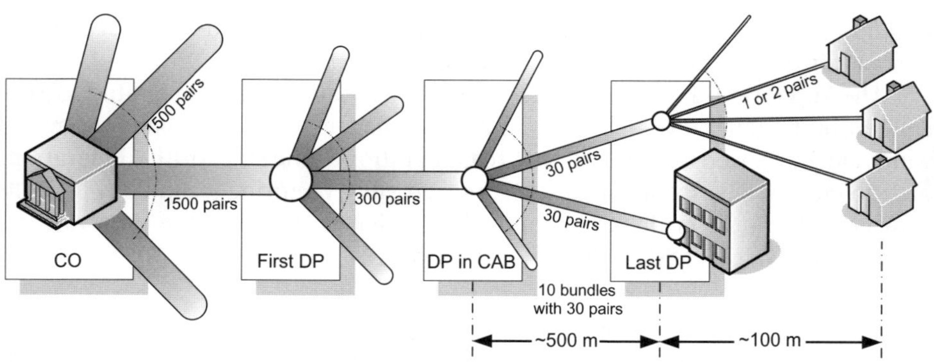

Figure 7.2. Access network topology and deployment scenario. This particular cable serves the outskirts of a small city in Sweden.

the deployment of cabinets is unfeasible due to the difficulties in their installation and in the obtention of the required permissions.

The Generation 4 broadband is the next logical step to shorten the loops, increase the bandwidth, and extend the optical fiber access network. The transmission equipment would then be placed in the Last DP (cf. Figure 7.2) and typically connected to the CO with newly deployed fiber. The user will still be linked to the Last DP by means of a copper pair.

The alternative to the Generation 4 concept described here is to deliver the fiber all the way out to every customer—that is, FTTH or Generation 5 systems in our terminology. The problem with all deep-fiber strategies (see Figure 7.3), and the reason why the technique is detained, is the cost of deploying the fiber (i.e., civil works), as well as the cost for the fiber itself. The question is then how far it should be extended. Considering that the most costly part of the connection is from the Last DP to the customer, since it means digging and ducting for each residence, it is of prime importance to take into account the dominant type of dwelling house per country (Figure 7.4). The Last DP is installed where a balance is found between the amount of homes passed and their distance to the DP. Therefore, in Spain, where almost 60% of its population lives in apartment buildings or towers, the Last DP is usually located in basements (FTTB). On the opposite side, Ireland, where most people live in detached/semi-detached houses, the Last DP is mostly present within street cabinets (FTTN). In the vast majority of cases, the distance between the Last DP and the customers' homes should be less than 100 m. In Copper and Faulkner [5], the drop wire distribution for an access network is shown; here only a small fraction of the lines are longer than 60 m.

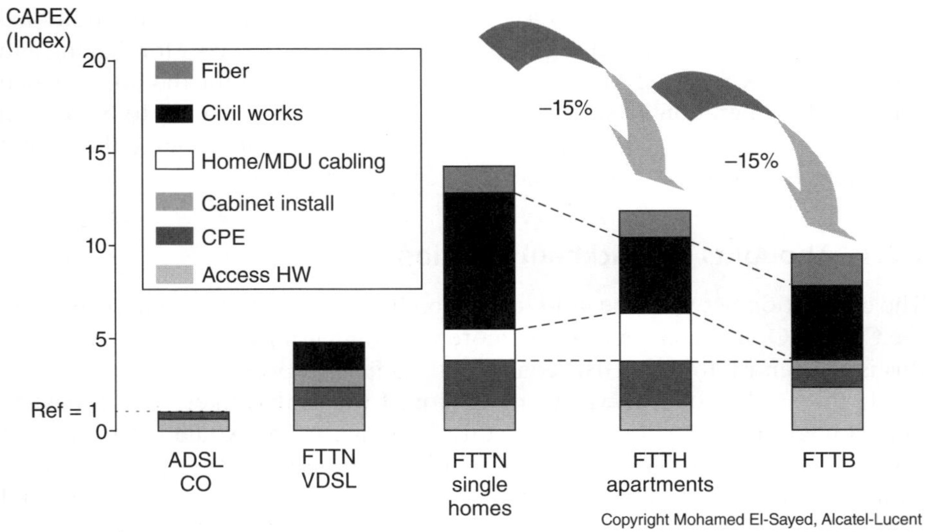

Figure 7.3. CAPEX for the different FTTx flavors.

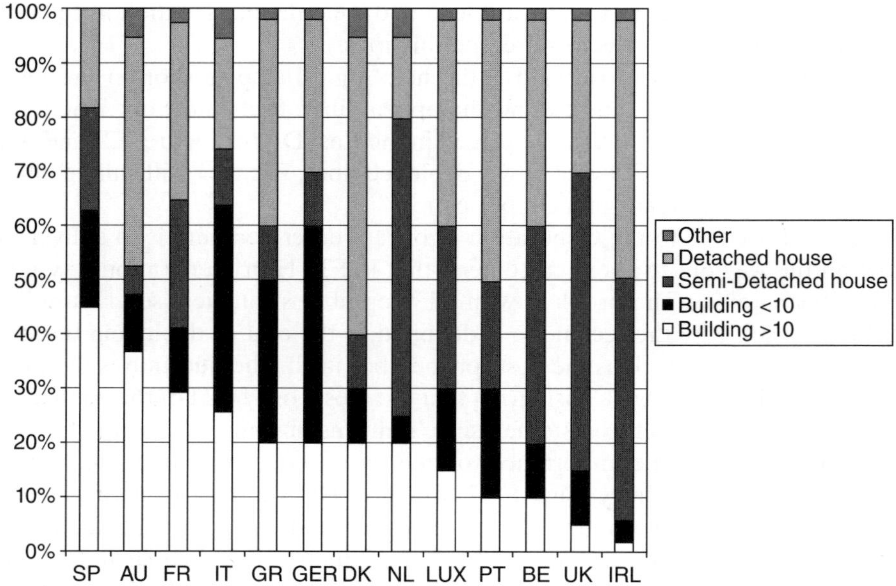

Figure 7.4. Dwelling-house type per country in Europe.

According to the techno-economic investment evaluations in, for example Olsen et al. [6], the deployment of FTTH can mostly be justified in particularly dense urban areas, while the cost of deploying fiber to the Last DP is moderate. As a rough estimation, using the example of Figure 7.2, replacing the copper from the cabinet to the Last DP will imply digging 5 km (500 m × 10 bundles), while replacing the cabling from the Last DP to each house will mean an additional 30 km per cabinet (100 m × 30 × 10). Considering an average cost for digging and ducting of 105 kEuro per km, then the cost is 0.515 MEuro for the fiber needed for the 4th-Generation broadband, while an extra 3.15 MEuro has to be thrown in to go to FTTH. This cost difference is the key realization bringing the insight that there is a niche for a 4th-Generation broadband system.

7.2.1 Thoughts on Backhaul Solutions

The connection between the 4th-Generation broadband (4GBB) equipment and the Central Office can be realized in more ways than using optical fiber. Although this is not central to the 4GBB concept, it is a field of possible innovation.

If the 4GBB concept was to be deployed today, it is likely that a passive optical fiber network (PON) architecture (see Figure 7.5), would offer the most priceworthy solution. This solution could be reasonably "future proof" in that, without additional investments in fiber, the optical transmission equipment could be upgraded—for example, from G-PON (gigabit-capable PON) to 10G-GPON or WDM-PON—when such new technologies become available.

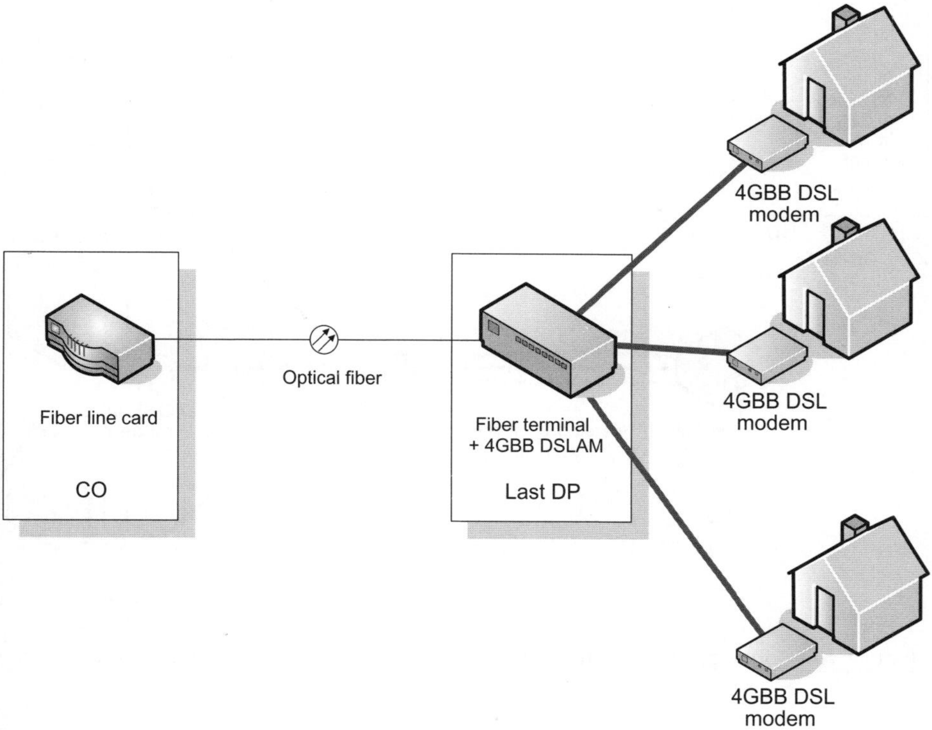

Figure 7.5. Topology with fiber to the Last DP.

Staying with technology available today, a principal alternative to a new, extended, fiber infrastructure would be to utilize the copper that is already in the ground. A group of copper pairs, preferably a whole cable binder (bundle), could be allocated to a DSL technology, creating a large shared bit-pipe (see Figure 7.6). This shared bit-pipe would use bonding [7, 8] to provide trunking gain thanks to statistical multiplexing of user traffic, and vectoring (defined by the ITU standards project "G.vector") to cancel crosstalk between the pairs, increasing the attainable bit rate (techniques discussed in more detail in Section 7.3). In reference 9, 0.5 Gbit/s was achieved with a prototype bonding and vectoring system, over a distance of 0.5 km using six pairs. Extrapolating this result to a binder with 30 pairs then gives 2.5 Gbit/s, which is on par with the performance of a GPON link (2.448 Gbit/s). It is expected that further refinements of the technology could achieve such performance over longer distances than 0.5 km, perhaps up to 1 km depending on cable gauge.

A backhaul reach of 0.5–1 km is of course much shorter than the reach of a GPON link (typically 20 km). However, it could become an option to use the copper binder as a backhauling solution between the cabinet and the Last DP where the distance is typically less than 1 km (and where there often is no ducting; see Figure 7.6), thereby avoiding or postponing the cost of digging.

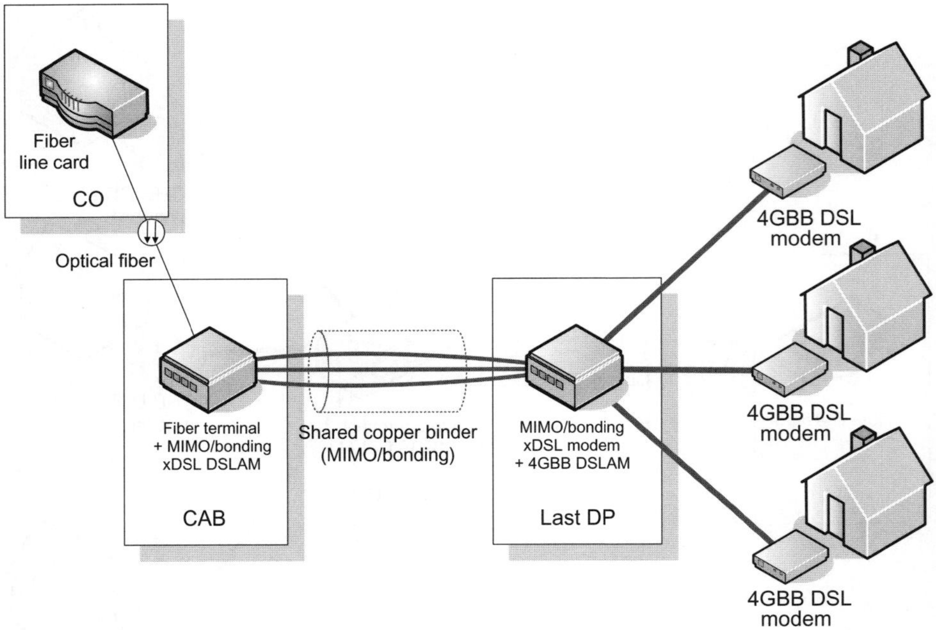

Figure 7.6. Topology where a shared copper binder is used between cabinet and Last DP.

Further capacity enhancements can be achieved if the copper binder is not restricted to use only differential signaling (see Section 7.3.2). Using Multiple-Input Multiple-Output (MIMO) schemes [10] applied to cancel crosstalk and spatially correlated noise, the 30-pair binder could be converted to a 60 × 60 or 59 × 59 MIMO channel depending on whether or not the binder shield can be used as a conductor. The copper backhaul solution discussed above could, as discussed, be suitable for shorter ranges, supporting several Gbit/s from the cabinet to the Last DP [11]. This scheme would then be similar to the CuPON concept proposed by Cioffi et al. [12] in the sense that the copper is shared but different in the sense that the shared DSL system is only used for the backhaul (e.g., from the cabinet to the Last DP).

It is also important to understand how the DSLAMs at the Last DP will, as any other active equipment, be powered. There are essentially three alternatives to power equipment in the Last DP:

- Local power from utility poles, lamp posts, or similar sources. This could be very expensive for the small number of lines considered here since there is often a metering fee incurred. However, if the Last DP is located in a building basement, it could be possible to get power from the building owner without a metering fee.
- Forward powering, using the available twisted-pair copper to feed power from the CO or cabinet to the Last DP. Commercial solutions for forward

powering exist today but are often too powerful, bulky, and expensive for the small number of lines considered here. If suitable equipment for forward powering to the Last DP becomes available, it would be natural to feed power over the same copper lines as used for the backhaul discussed above.

- Reverse powering, where power is fed from the subscriber equipment to the DSLAMs [5]. Since the copper lines between the Last DP and customers' homes are very short, the resistive losses will typically be much lower than for forward powering. This is perhaps the most promising solution, but there are some issues that need to be solved before this can become successful—for example, a model for how to compensate the subscribers for paying the DSLAM power consumption.

The three powering solutions above are likely the most realistic ones. The results of Copper and Faulkner [5] show that reverse powering is the cheapest when there are few subscribers per node while local powering becomes cheaper for large number of subscribers. In certain environments it may also be possible to utilize battery-backed solar power or wind power, but it remains to be seen whether such a powering solution will be feasible.

7.3 PHYSICAL-LAYER TECHNIQUES FOR THE LAST DROP

This section discusses the technical ingredients that are available to enable data rates on the order of Gbit/s over the copper cable connecting the Last DP with the customer's premises. The data rate that can be achieved with an arbitrarily low bit error rate is limited by the channel capacity:

$$C = \sum_{m=1}^{M} \int_{f=f_\ell}^{f_u} \log_2(1+\text{SINR}_m(f))df$$

in bit/s, where M is the number of independent "channels" (or modes), $B = f_u - f_\ell$ is the available bandwidth in hertz, and $\text{SINR}_m(f)$ is the frequency-dependent signal-to-interference-plus-noise-power ratio of the receive channel mode m. Consequently, there are the following ways to increase the throughput:

- Increase bandwidth B.
- Increase the number of independent modes M.
- Increase the signal-to-interference-plus-noise-power ratio $\text{SINR}_m(f)$ of the receive signal for all (or some) frequencies $f \in [f_\ell, f_u]$.

In general, the potential gain in throughput when increasing the number of channels M or the available bandwidth B is larger compared to increasing the signal-to-noise-plus-interference-power ratio $\text{SINR}_m(f)$ since C grows linearly with M and B but only logarithmically with $\text{SINR}_m(f)$. While increasing the bandwidth is technically straightforward, there are many options for both increasing the number of modes and increasing the signal-to-noise-plus-interference-power ratio. In the following, a survey of these (mostly) physical-layer techniques is presented.

7.3.1 Bonding

In many countries, there are cables with two (or even more) twisted pairs connecting the Last DP with a customer's home. Only two of these wires (one pair) are actually used for telephony and DSL. Without much effort, a second pair (or even more, if available) can be exploited (cf. Figure 7.7a). In essence, this does not require any physical-layer processing, but only a simple interleaving of the data streams to be received or sent over the two (or more) lines—a technique that is already in place in various forms and referred to as bonding [7, 8]. As simple as it is, adding channels via bonding does not exploit the potential of a cable to its full: While the number of independent channels M increases, the signal-to-noise-plus-interference-power ratio $\mathrm{SINR}_m(f)$ of all channels $m \in \{1, \ldots M\}$ decays due to increased crosstalk among the bonded lines.

7.3.2 Alternative-Mode Signaling

Traditionally, signaling over copper cables is realized via loops formed by twisted-wire pairs. The information is represented as the voltage applied (at the transmitting end) or measured (at the receiving end) between the two wires—a way of signaling that is referred to as differential-mode signaling. The main advantage of differential-mode signaling is its high immunity with respect to surrounding electromagnetic fields.

A way to increase the number of channels M is to exploit alternative ways of signaling which result in alternative modes. Recent research suggests that we entirely abandon the twisted-pair concept and adopt a multi conductor view: Instead of using the K differential modes of a K-pair cable, $2K - 1$ independent transmission modes can be exploited using alternative modes. In case the shield or earth is exploited serving as a conductor available at both ends of the cable, the number of exploitable modes is $2K$.

From a technical perspective, there are various possibilities to obtain these modes. Figure 7.7 depicts a few examples. For the sake of simplicity, the number of wire pairs (loops) is chosen to be only $K = 2$. Common-mode signaling excites both wires of a loop with respect to a common reference, which can be the common potential of another loop yielding the configuration depicted in Figure 7.7b with three (in general $2K - 1$) independent modes [13–15]. In case earth or the shield is used as common reference [15], the common mode of each loop in the cable can be exploited, yielding four (in general $2K$) independent modes as depicted in Figure 7.7c. Another setup, referred to as split-pair signaling [11], uses one wire as reference yielding three (in general $2K - 1$) independent modes for a two-pair cable (cf. Figure 7.7d). Extending this idea to a cable with a shield, yields four (in general $2K$) modes (cf. Figure 7.7e). In general, alternative

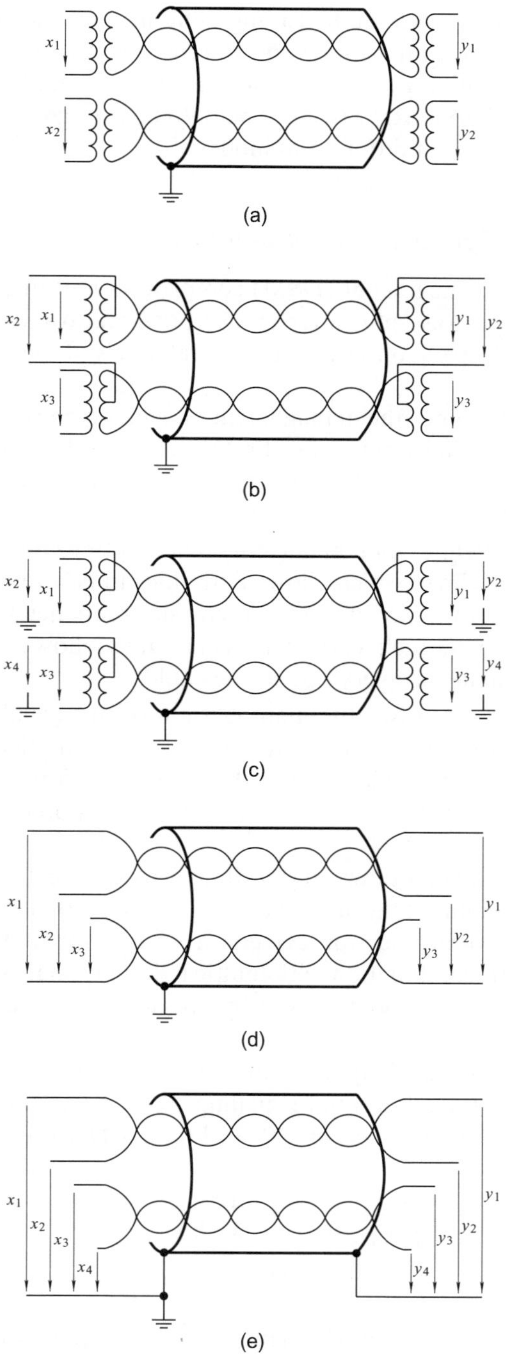

Figure 7.7. Various ways to increase the number of channels (modes): **(a)** Bonding. **(b)** Common-mode signaling. **(c)** Common-mode signaling exploiting the shield or earth. **(d)** Split-pair signaling. **(e)** Split-pair signaling exploiting the shield.

modes may be more susceptible to surrounding electromagnetic fields and thus yield lower signal-to-noise-plus-interference-power ratio values $\mathrm{SINR}_m(f)$ compared to differential-modes signaling. In order to exploit the potential of alternative modes, the $\mathrm{SINR}_m(f)$-values need to be brought to the same order of magnitude as available on differential modes.

7.3.3 Dynamic Spectrum Management

Dynamic spectrum management (DSM) comprises a multitude of techniques to improve the signal-to-noise-plus-interference-power ratio and is widely embraced within the DSL industry [16]. There are several levels of management:

- DSM Level 1 defines the management of average power values on a single-line basis. DSL lines practicing DSM Level 1 behave more "politely" to other lines by, for example, reducing the average power to the level that is needed instead of transmitting with the level that is actually permitted.
- DSM Level 2 defines joint management and optimization of average power of several DSL lines, which allows DSL lines to be even "more polite" and avoid the generation of crosstalk in certain frequency bands. The philosophy is simple but effective: If all lines in a cable follow a "politeness policy," there is a benefit for every line in the cable.
- DSM Level 3 comprises the manipulation of the signals itself (instead of just their power), which allows for signal processing that eliminates (or at least mitigates) crosstalk either at the receiver (referred to as interference cancellation) or at the transmitter (referred to as precoding).

DSM Level 3 is sometimes also referred to as "vectoring" or "vectored transmission"—a terminology motivated by the fact that it is convenient to arrange signals of co-located transceivers in vectors for joint processing using linear algebra [4]. In combination with multicarrier modulation, vectoring allows us to eliminate the impairment caused by crosstalk at the cost of some signal processing.

A little example should help to illustrate the idea. Consider a two-pair system (four wires) used with differential signaling. For one subcarrier, the resulting spatial channel can be described by a 2×2 matrix $\mathbf{H}$; for example,

$$\begin{bmatrix} y_1 \\ y_2 \end{bmatrix} = \underbrace{\begin{bmatrix} 1 & 0.12 \\ -0.08 & 1 \end{bmatrix}}_{\mathbf{H}} \begin{bmatrix} x_1 \\ x_2 \end{bmatrix} + \begin{bmatrix} n_1 \\ n_2 \end{bmatrix}$$

The direct paths have unit gain. The signal x_2 transmitted on line No. 2 multiplied by 0.12 is the far-end crosstalk (FEXT) seen on line No. 1. The signal x_1 transmitted on line No.1 multiplied by –0.08 is the FEXT seen on line No. 2. These values correspond to crosstalk-coupling functions' magnitudes of around

−18.42 dB ($10\log_{10}(0.12^2)$) and −21.94 dB ($10\log_{10}(0.08^2)$), respectively. With a signal-to-noise-power ratio of 30 dB, the resulting signal-to-interference-plus-noise-power ratios on the two lines are about 18.12 dB ($10\log_{10}(1/(0.12^2+0.001))$) and 21.31 dB ($10\log_{10}(1/(0.08^2+0.001))$), respectively. These are the values that can be achieved with bonding.

An example of vectoring with co-located receivers (interference cancellation) is based on the QR-decomposition of the channel matrix **H** yielding

$$\mathbf{H} = \underbrace{\begin{bmatrix} -0.9968 & 0.0797 \\ 0.0797 & 0.9968 \end{bmatrix}}_{\mathbf{Q}} \underbrace{\begin{bmatrix} -1.0032 & -0.0399 \\ 0 & 1.0064 \end{bmatrix}}_{\mathbf{R}}$$

where **R** is upper-triangular and **Q** is a unitary matrix. Post-processing the receive signals with $\mathbf{Q}^H$ yields the resulting channel $\mathbf{Q}^H\mathbf{H} = \mathbf{R}$, which has an upper-triangular structure. Consequently, x_2 can be detected first. Assuming correct detection, the FEXT generated from line No. 2 can be reconstructed and subtracted before detecting x_1. Since **Q** is a unitary matrix, post-processing does not change the noise power. The resulting detection signal-to-interference-plus-noise-power ratios are about 30.03 dB ($10\log_{10}(1.0032^2/0.001)$) and 30.06 dB ($10\log_{10}(1.0064^2/0.001)$), respectively. Note that these values in fact exceed the signal-to-noise-power ratio of 30 dB. In this sense, vectoring has turned the impairment caused by FEXT into an advantage.

The same decomposition can be utilized for vectoring with co-located transmitters. Direct application of this precoding idea, however, results in a transmit-power increase. Nonlinear precoding can be used to amend this problem [17, 18].

7.3.4 Multiple-Input Multiple-Output Techniques

In contrast to vectoring, which requires co-location of wire pairs on only one of the two sides, dedicated Multiple-Input Multiple-Output (MIMO) techniques require co-location of wire pairs on both sides. Continuing the example from the previous section, a simple MIMO-processing technique [19] could, for example, evaluate the singular value decomposition of the channel yielding

$$\mathbf{H} = \underbrace{\begin{bmatrix} 0.7415 & 0.6710 \\ 0.6710 & -0.7415 \end{bmatrix}}_{\mathbf{V}} \underbrace{\begin{bmatrix} 1.025 & 0 \\ 0 & 0.985 \end{bmatrix}}_{\mathbf{S}} \underbrace{\begin{bmatrix} 0.6710 & 0.7415 \\ 0.7415 & -0.6710 \end{bmatrix}}_{\mathbf{U}}$$

Pre-processing the transmit signals with $\mathbf{U}^H$ and post-processing the receive signals with $\mathbf{V}^H$ yields the resulting channel $\mathbf{V}^H\mathbf{H}\mathbf{U}^H = \mathbf{S}$, whose off-diagonal elements are zero. Since **V** and **U** are unitary matrices, pre- and post-processing does not change signal power nor noise power. The processing yields signal-to-interference-plus-noise-power ratios on line No. 1 and line No. 2 of around 30.21 dB($10\log_{10}(1.025^2/0.001)$) and 29.87 dB ($10\log_{10}(0.985^2/0.001)$), respectively.

Although the gains depend on the actual coupling values, simple processing of this kind can yield notable improvements in signal-to-interference-plus-noise power ratio. Nevertheless, crosstalk caused by neighboring lines that are not part of the MIMO system (or of the vectoring system) remains and can be tackled using the ideas discussed next.

7.3.5 Extrinsic Interference Cancellation

Besides crosstalk, which originates inside the cable or the wire system consisting of a number of lines, there may be interference originating from radio sources. Examples include AM radio stations, TV stations, and electrical household appliances. In fact, the major part of the 30 to 200-MHz frequency band is occupied by broadcast TV and radio stations. Furthermore, crosstalk originating from lines that do (for whatever reason) not participate in DSM practices constitutes extrinsic interference. In contrast to crosstalk originating from a line belonging to the system (and thus practicing DSM), there is no reference (a strongly correlated signal) available for interference cancellation. However, a reference can be obtained by exploiting an unused line to "listen." In essence, this line functions like a receive antenna providing a signal that is strongly correlated with the interference and can thus be used for interference cancellation.

It is reasonable to assume that the susceptibility with respect to extrinsic interference of alternative modes is higher compared to differential modes. It may thus be beneficial to employ interference cancellation together with alternative-mode signaling. As pointed out in Lee et al. [11], the ratio of achievable data rates with alternative-mode signaling and with standard differential-mode signaling can be roughly estimated as follows. Without extrinsic interference, the gain is proportional to the number of modes: $(2K - 1) / K$. In the presence of extrinsic interference, one alternative mode can be used to acquire a reference signal of the interference for subsequent cancellation. The resulting ratio of data rates is thus roughly $(2K - 2) / K$. Consequently, for a two-pair drop cable without shield operated in the presence of extrinsic interference, alternative-mode signaling may yield no note worthy improvement. Clearly, the ingress/egress issue is critical since alternative-mode loops are not twisted. However, cable shields could mitigate the problem.

7.4 REGULATORY AND LEGAL ASPECTS

Copper cables used for data transmission act like antennas and thus both pick up unwanted interference (a process referred to as ingress) and emit electromagnetic waves (a process referred to as egress). While the former impairs the performance of high-speed data transmission over the cable, the latter may create conflicts with other services.

Ingress and egress mechanisms, described in Foster and Cook [20], can be roughly sketched as follows. Transmission of data over a wire pair is carried out by differential excitation of the pair (i.e., excitation of the circuit formed by the two wires of a pair). Due to imperfection of geometrical and consequently also of electrical symmetry of each wire pair with respect to earth, the differential signal causes a corresponding common-mode excitation of the wire pair (i.e., excitation of the circuit formed by the wire pair constituting a single conductor and earth). The pair of wires then behaves like a transmit antenna and causes unwanted egress. The degree of symmetry (or asymmetry) causing the differential-mode to common-mode conversion, an important property of a wire pair or a cable, is referred to as *balance* and is quantified by the ratio of the corresponding voltages or currents. While the balance can reach values around 70 dB in the voiceband (i.e., in the kilohertz-range), it decreases significantly with increasing frequency [10]. Extrapolating measurement results collected for frequencies up to 30 MHz, the conservative assumption that the balance decays linearly from 35 dB in the voiceband to 25 dB at 100 MHz is adopted hereinafter.

Conversely to egress, a time-varying electromagnetic field in the vicinity of a wire pair causes a common-mode excitation of both wires with respect to earth. The wire pair simply behaves like a receive antenna. The balance, which is a reciprocal property, determines the amount of resulting differential-mode ingress caused by common-mode to differential-mode conversion.

As a principal assumption, Foster and Cook [20] suggests that an electromagnetic field with electric field strength x V/m causes an induced worst-case common-mode voltage of x V, an observation that is mainly based on experience gained through measurements both in the laboratory and in the field. Independent theoretical and experimental work [21, 22] supports this observation to an extent large enough to warrant application for throughput predictions.

7.4.1 Egress

For the frequency range 0–30 MHz, the invoked egress limits orientate themselves on the standardized VDSL band plans, adopting the transmit power spectral density (PSD) limit of –60 dBm/Hz. The standard[1] [23], hereinafter referred to as CISPR-22, suggests a quasi-peak limit for radiation caused by an electric appliance measured at a distance of 10 m. The limit, which is specified in terms of electric field strength, is 30 dBμV/m measured in any band of 9 kHz width in the frequency range 30–230 MHz. A transmit PSD below –63 dBm/Hz ensures that this limit is on average not violated assuming a balance of 30 dB.

It is pointed out in Foster and Cook [20] that radiated emission with a field strength as low as 0 dBμV/m might be detected by, and thus cause disturbance for, radio receivers. Consequently, it is reasonable to assume a transmit PSD that decays linearly with frequency from –60 dBm/Hz at 30 MHz to a value at 100 MHz that ensures a field strength below 12 dBμV/m, which is the level caused by today's modems in the amateur radio bands. Standardized band plans limit the PSD within these bands to –80 dBm/Hz. The impact of these bands on the

[1] Comité international spécial des perturbations radioélectriques.

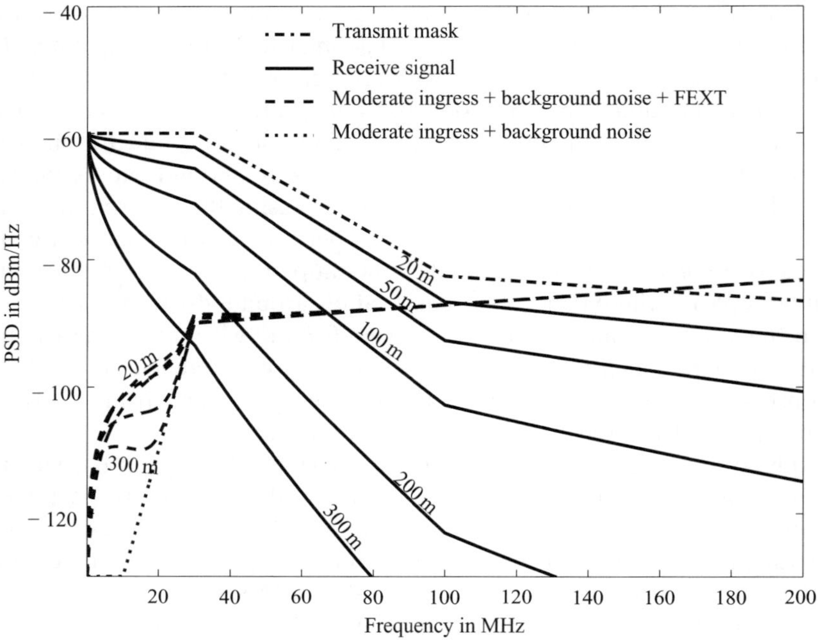

Figure 7.8. Transmit and receive power spectral densities (PSDs): The uppermost line (dashed–dotted) is the transmit PSD mask, followed by receive PSDs (solid lines) for loop lengths 20 m, 50 m, 100 m, 200 m, and 300 m (AWG24/0.5 mm). Transmission is impaired by moderate ingress (–110 dBm/Hz), FEXT from one equal-length crosstalker, and background noise (–130 dBm/Hz).

throughput analysis is insignificant and thus neglected. Figure 7.8 depicts the resulting transmit PSD mask (dashed–dotted line) and the corresponding receive PSDs (solid lines) for different loop lengths.

7.4.2 Ingress

Reversely to the egress mechanism, the wires will pick up radiation caused by devices operating in close vicinity. Assuming that these devices operate at the radiation limits suggested by the CISPR-22 standard, the resulting ingress PSD[1] is roughly –133 dBm/Hz for a balance of 30 dB—a level comparable to background noise. Apart from the radiation-induced interference, there is disturbance caused by conducted common-mode interference. Assuming that the wire pair obeys the limits suggested by CISPR-22, the resulting ingress PSD[2] is roughly

[1]An electric field-strength of 47 dBμV/m causes a differential-mode voltage of 7.08 μV (balance 30 dB), which corresponds to a PSD of roughly –133 dBm/Hz in 100 Ω over a measurement bandwidth of 9 kHz.

[2]A conducted common-mode voltage of 90 dBμV causes a differential-mode voltage of 1 mV (balance 30 dB), which corresponds to a PSD of roughly –90 dBm/Hz in 100 Ω over a measurement bandwidth of 9 kHz.

–90 dBm/Hz for a balance of 30 dB. Instead of an ingress level of –133 dBm/Hz, which corresponds to an ingress-free environment and is a rather unrealistic scenario, –110 dBm/Hz is assumed to characterize the situation of "moderate ingress." A level of –90 dBm/Hz, on the other hand, characterizes "strong ingress" and appears to be a rather pessimistic assumption: Although radio interference is an issue for currently deployed DSL systems that operate mainly at frequencies below 10 MHz, ingress levels observed in the field are way below the worst-case CISPR-22 level mentioned above. To summarize, the two levels should embrace ingress levels encountered in practice and serve as a basis for throughput predictions.

In a FEXT-free environment, a background-noise PSD of –130 dBm/Hz is a widely accepted, though conservative, value for frequencies up to 30 MHz. Aiming at assumptions that can be referred to as *realistic till conservative*, a linear transition (in log domain) from the background-noise level at lower frequencies (ca. 10 MHz) to the CISPR-22 ingress level at higher frequencies (ca. 30 MHz) is assumed. The resulting noise PSD for a FEXT-free environment is shown in Figure 7.8 (dotted line).

As discussed in the previous section, the cable segments at the "customer end" of the access network exhibit short lengths and a low number of pairs. Consequently, it is reasonable to assume that the number of expected crosstalkers is low. Apart from the FEXT-free case, a scenario with both ingress and one equal-length FEXT disturber is considered.

7.5 A THROUGHPUT PREDICTION

The previous section established the feasibility of the new 4GBB hybrid fiber–copper topology at an affordable investment-cost per customer. This section presents a projection of the achievable throughput connecting the Last DP and the customer.

It turns out that electromagnetic compatibility of interacting equipment and services sets major limitations for the achievable data rates. Data transmission over wires causes radiation and potentially disturbs nearby equipment. This undesired effect is referred to as *egress* and limits the applicable transmit PSDs. Reversely, cables—in particular aerial drop wires—pick up extrinsic disturbances (generated outside the cable) referred to as *ingress*. Lacking dedicated ingress and egress regulations, we derive realistic ingress levels and transmit PSD masks from ingress and egress limits defined in the existing international standard on radio interference CISPR-22 [23]. Together with wideband cable models [24], these transmit PSDs and ingress levels provide the basis for throughput predictions.

Once the constraints in the form of transmit PSD, ingress PSD, and background-noise level are found, the computation of the achievable data rate is straightforward (cf., e.g., Golden et al. [3]). A signal-to-noise ratio gap of 9.5 dB is assumed, which yields the throughput achieved with uncoded QAM transmission at a bit error rate of 10^{-7}, from which state-of-the-art channel coding would reduce the bit error rate to low enough levels for all common services. The results presented in the following represent aggregate downstream and upstream data rates.

The following technology options for the link from the Last DP to the customer are compared:

- State-of-the-art (differential-mode) signaling over one twisted pair (existing solution).
- State-of-the-art signaling over one twisted pair in combination with vectoring, yielding one FEXT-free channel (or mode).
- State-of-the-art MIMO signaling over two twisted pairs in combination with vectoring (which eliminates crosstalk generated by lines outside the MIMO system), yielding two FEXT-free modes.
- Alternative-mode MIMO signaling over two twisted pairs in combination with vectoring (which eliminates crosstalk generated by lines outside the MIMO system), yielding three FEXT-free modes.

The achievable data rate versus exploited bandwidth is depicted in Figure 7.9 (top plot) for a 50-m drop wire. In general, the data rate increases rapidly with frequency for low frequencies and flattens out for high frequencies. Exploiting the bandwidth up to frequencies where the data rate flattens out will yield the throughput versus loop length depicted in Figure 7.9 (bottom plot). Exploiting the available bandwidth with state-of-the-art signaling is not sufficient to approach the Gbit/s limit. Very short loops in combination with vectoring, however, support data rates around 1 Gbit/s. MIMO techniques and signaling via alternative modes exceed the Gbit/s limit in the presence of strong ingress and boost the throughput to several gigabits per second in the presence of moderate ingress.

7.6 CONCLUSIONS

In many countries, more than one-third of the population regularly uses the Internet. In addition to the tremendous residential usage, the market for small and medium-size enterprises is growing. The capacity within the backbone networks is virtually unlimited (at least in the sense that it is economically feasible to upgrade it to match virtually any need), leaving the transmission bottlenecks located within the first mile. The development of broadband services markets worldwide is thus dependent on having access networks that live up to the vision of the future society. In this area, a key enabler for the development of society, few new concepts have been presented. Notable exceptions are the work of Cioffi et al. [12], to which the 4th-Generation Broadband concept presented in this chapter is a natural complement.

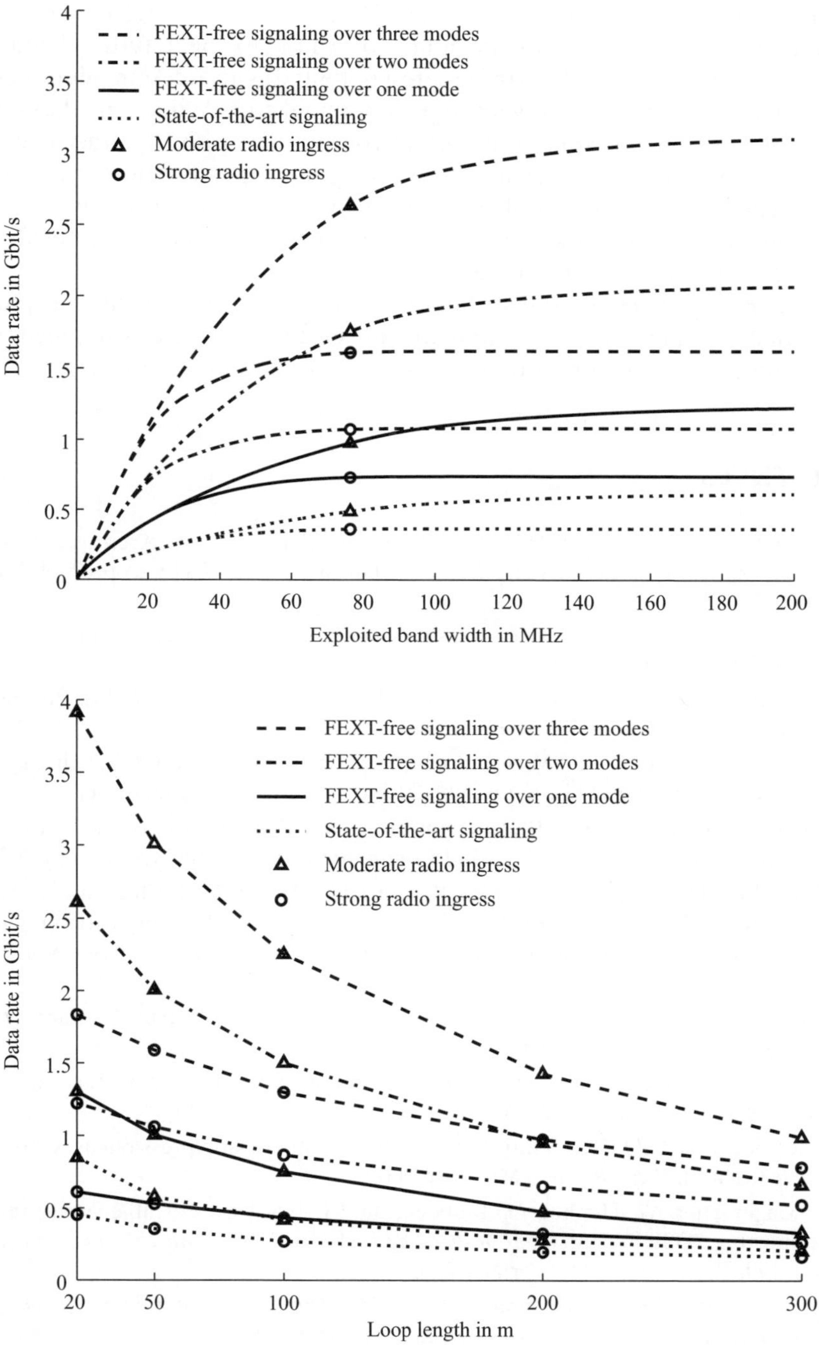

Figure 7.9. **Top**: Throughput over a 50-m drop wire versus exploited bandwidth for different technology options and both strong ingress (–90 dBm/Hz, marked by circles) and moderate ingress (–110 dBm/Hz, marked by triangles). **Bottom**: Throughput versus loop length.

Today, broadband strategies beyond VDSL2 are based on telecom operators eventually deploying fiber to the homes (FTTH) to meet future bandwidth demands. Although fiber offers the greatest potential as an access medium, deployment is hampered by prohibitively large investment costs. While remedies for this are sought, copper has still an important role to play, with the 4th-Generation broadband systems as a candidate for bridging the gap between today's VDSL2 and FTTH. New fiber–copper-based systems that carry a partial investment in extending the fiber network manage a smooth migration from legacy networks to an all-fiber-optic network of the future.

Taking advantage of last distribution points close to the customer, exploiting the available bandwidth, and employing advanced signal processing techniques bring data rates on the order of gigabits per second to the customer.

REFERENCES

1. P. Ödling, T. Magesacher, S. Höst, P. O. Börjesson, M. Berg, and E. Areizaga, The fourth generation broadband concept, *IEEE Commun. Mag.*, Vol. 47, No. 1, pp. 62–69, January 2009.
2. W. Stallings, *Integrated Services Digital Networks (ISDN)*, IEEE Computer Society Press, Washington, DC, 1985.
3. P. Golden, H. Dedieu, and K. Jacobsen (editors). *Fundamentals of DSL Technology*, Auerbach, Boca Raton, FL, 2005.
4. G. Ginis and J. M. Cioffi. Vectored transmission for digital subscriber line systems. *IEEE J. Selected Areas Commun.*, Vol. 20, No. 5, pp. 1085–103, June 2002.
5. I. Copper and D. Faulkner, Reverse powering over DSL, in *Proceedings of European Conference on Networks and Optical Communications*, Krems, Austria, July 2007.
6. B. T. Olsen, D. Katsianis, D. Varoutas, K. Stordahl, J. Harno, N. K. Elnegaard, I. Welling, F. Loizillon, T. Monath, and P. Cadro, Technoeconomic evaluation of the major telecommunication investment options for European players, *IEEE Network*, Vol. 20, No. 4, pp. 6–15, 2006.
7. International Telecommunication Union, Atm-based multi-pair bonding. *ITU-T Recommendation G.998.1*, 2005.
8. International Telecommunication Union, Ethernet-based multi-pair bonding, *ITU-T Recommendation G.998.2*, 2005.
9. Ericsson (NASDAQ:ERIC), http://www.ericsson.com/ericsson/press/releases/20090316-1297846.shtml, *Press Release*, March 16 2009.
10. T. Magesacher, W. Henkel, G. Tauböck, and T. Nordström. Cable Measurements Supporting xDSL Technologies. *Journal e&i Elektrotechnik und Informationstechnik*, Vol. 199, No. 2, pp. 37–43, February 2002.
11. B. Lee, J. M. Cioffi, S. Jagannathan, and M. Mohseni, Gigabit DSL, *IEEE Trans. Commun.*, Vol. 55, No. 9, pp. 1689–1692, September. 2007.
12. J. M. Cioffi, S. Jagannathan, M. Mohseni, and G. Ginis, CuPON: The copper alternative to PON 100 Gb/s DSL networks, *IEEE Commun. Mag.*, June:132–139, 2007.

13. T. Magesacher, P. Ödling, P. O. Börjesson, W. Henkel, T. Nordström, R. Zukunft, and S. Haar. On the capacity of the copper cable channel using the common mode, in *Proceedings, Globecom 2002*, Taipei, Taiwan, November 2002.
14. T. Magesacher, P. Ödling, P. O. Börjesson, and S. Shamai (Shitz), Information rate bounds in common-mode aided wireline communications, *Eur. Trans. Telecommun. (ETT)*, Vol. 17, No. 2, pp. 533–545, 2006.
15. S. Jagannathan, V. Pourahmad, K. Seong, J. Cioffi, M. Ouzzif, and R. Tarafi, Common-mode data transmission using the binder sheath in digital subscriber lines, *IEEE Trans. Commun.*,Vol. 57, No. 3, pp. 831–840, March 2009.
16. K. B. Song, S. T. Chung, G. Ginis, and J. M. Cioffi, Dynamic spectrum management for next-generation dsl systems, *Commun. Mag. IEEE*, Vol. 40, No. 10, pp.101–109, October 2002.
17. M. Tomlinson. New automatic equaliser employing modulo arithmetic, *Electron. Lett.*, Vol. 7, pp. 138–139, March 1971.
18. H. Harashima and H. Miyakawa, Matched-transmission technique for channels with intersymbol interference. *IEEE Trans. Commun.*, Vol. 20, pp. 774–780, August 1972.
19. G. Tauböck and W. Henkel, MIMO systems in the subscriber-line network, in *Proceedings of Fifth International OFDM Workshop*, pp. 18.1–18.3, Hamburg, Germany, September 2000.
20. K. T. Foster and J. W. Cook, The radio frequency interference (RFI) environment for very high-rate transmission over metallic access wire-pairs, *ANSI Contribution T1E1.4/95-020*, 1995.
21. R. Stolle. Electromagnetic coupling of twisted pair cables. *IEEE J. Selected Areas Commun.*, Vol. 20, No. 5, pp. 883–889, June 2002.
22. R. B. Armenta and C. D. Sarris, Modeling the terminal response of a bundle of twisted-wire pairs excited by a plane wave, *IEEE Trans. Electromagnetic Compatibility*, Vol. 49, No. 4, pp. 901–913, November 2007.
23. CENELEC, Information Technology Equipment—Radio Disturbance Characteristics, Limits and Methods of Measurement, *European Standard EN55022:1998 (CISPR 22:1997, modified)*, September 1998.
24. T. Magesacher, J. Rius, I. Riu, M. Jakovljević, M. Loiola, P. Ödling, and P. O. Börjesson, Modeling and measurement of short copper cables for ultra-wideband communications, in *Proceedings of SPIE OpticsEast Broadband Access Communication Technologies*, Boston, October 2006.

8

DYNAMIC BANDWIDTH ALLOCATION IN EPON AND GPON

Björn Skubic, Jiajia Chen, Jawwad Ahmed,
Biao Chen, and Lena Wosinska

8.1 INTRODUCTION

Dynamic bandwidth allocation (DBA) in passive optical networks (PON) presents a key issue for providing efficient and fair utilization of the PON upstream bandwidth while supporting the quality of service (QoS) requirements for different traffic classes. A PON consists of an optical line terminal (OLT) located at the provider central office and a number of optical network units (ONUs) or optical network terminals (ONTs) at the customer premises. In time-division multiplexing (TDM), PON downstream traffic is handled by broadcasts from the OLT to all connected ONUs, while in the upstream direction an arbitration mechanism is required so that only a single ONU is allowed to transmit data at a given point in time because of the shared upstream channel (see Figure 8.1). The start time and length of a transmission timeslot for each ONU are scheduled using a bandwidth allocation scheme. A merit of TDM-PON is the possibility to exploit the statistical multiplexing of network traffic in access networks by oversubscribing the shared optical distribution network in order to achieve high network utilization. A prerequisite for exploiting these multiplexing gains in the upstream is a well-designed and efficient DBA algorithm.

Convergence of Mobile and Stationary Next-Generation Networks, edited by Krzysztof Iniewski

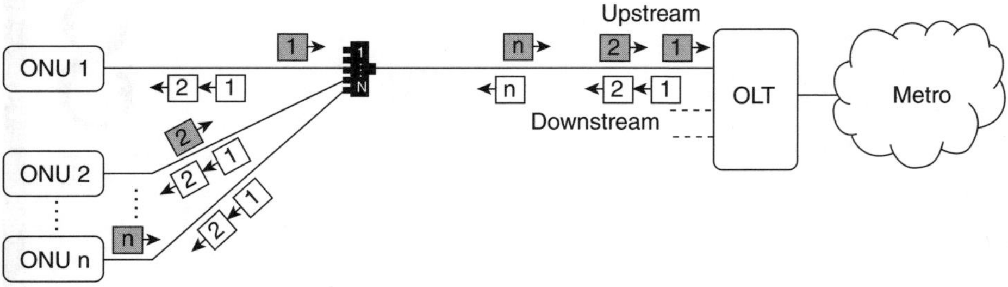

Figure 8.1. Schematic view of TDM-PON.

In this chapter we present an overview of DBA schemes for the two major standards for TDM-PON, namely, Ethernet PON (EPON) and gigabit-capable PON (GPON). The particular PON standard sets the framework for the design and operation of the DBA. As we will see, the challenges for designing DBA are quite different within the two standards. We illustrate the differences between EPON and GPON and how they are overcome. Furthermore, we consider the evolution toward next-generation TDM-PON with higher upstream bit rates and how this affects the design of the DBA.

8.2 STANDARDS

The EPON and GPON standards are said to embrace different philosophies, with EPON focusing on simplicity and looser hardware requirements, while GPON focuses on tighter hardware requirements and a fulfillment of telecom operator requirements. On a more detailed level, the two philosophies boil down to differences in guard times, overheads, and other forms of parameters influencing bandwidth allocation. The implementation of bandwidth allocation is outside the scope of both the EPON and GPON standards, although in GPON several aspects of the DBA are specified in the standard, introducing some constraints to the DBA implementation. These underlying design choices for the PON govern how DBA should be designed in order to cope with imposed traffic requirements and fairness policies while still maintaining efficient utilization of the PON's shared upstream channel.

The design of the logical layer is crucial for the DBA implementation. One important parameter is burst overhead—that is, overhead related to the transmission of an optical burst from one ONU to the OLT. Burst overhead consists of the guard band between transmission of two bursts and, depending on implementation, either unused slot remainders (USR) or fragmentation overhead for treating allocation slot remainders which are too small to fit complete Ethernet frames (see Table 8.1). For systems with an overall large burst overhead, the DBA must provide a more coarse grained scheduling in order to maintain a large average burst size and minimize efficiency loss due to burst overhead. The EPON standard allows a relatively large burst overhead and consequently requires more coarse-grained scheduling.

TABLE 8.1. Protocol Differences in the EPON and GPON Upstream that Affect Bandwidth Allocation

	EPON	10/10	10/1	GPON	2.488/1.244
Raw upstream bit rate		10.3125 Gbit/s	1.250 Gbit/s		1.244 Gbit/s
Line coding		64B/66B	8B/10B		NRZ-scrambling
Upstream bit rate		10 Gbit/s	1.000 Gbit/s		1.244 Gbit/s
Frame overhead	IPG + preamble	20 bytes	20 bytes	GEM	5 bytes
Guard band	Ton	512 ns	512 ns		
	Trec_settling	800 ns	400 ns	Guard time	4 bytes
	Tcdr	400 ns	400 ns	Preamble + delimiter	8 bytes
	Burst delimiter	8 bytes	8 bytes	Burst header	3 bytes
	Burst terminator	24 bytes	24 bytes		
DBA overhead	REPORT	64 bytes	64 bytes	SR	2 bytes
Slot remainders	USR	0–1537 bytes	0–1537 bytes	Fragmentation overhead	5 bytes

Another important part of the standard is the DBA communication messages. Both the EPON and GPON standards define (a) report messages used by the ONUs to communicate instantaneous buffer occupancy information to the OLT and (b) grant messages used by the OLT to communicate scheduling information back to the ONUs. This communication mechanism is vital for the operation of the DBA. The details of how these messages are defined will affect the design of the DBA in several ways. They govern the overhead associated with the ONU-OLT communication, which in turn affects the potential frequency of communication messages exchange. As shown in Table 8.1, EPON messages have a larger overhead associated with them compared to GPON messages. The definition of the DBA messages may also constrict at which points in time the DBA information exchange between the OLT and ONUs can occur. Regarding this issue, the EPON standard provides more flexibility in the design of the DBA algorithm than GPON. Within GPON, several status reporting modes are defined. For our comparison with EPON we consider what is most commonly used and referred to as status reporting mode 0. Our comparison of EPON and GPON is therefore based on a comparable type of communication between the OLT and ONUs.

8.2.1 EPON

EPON for 1 Gbit/s data transfer is defined in IEEE 802.3-2008 [1]. The standard defines raw downstream and upstream bit rates of 1.25 Gbit/s, which, by using 8B/10B line encoding, provides symmetric bit rate data transmission of 1 Gbit/s. There is an Ethernet frame overhead of 20 bytes—that is, 12 bytes interpacket gap (IPG) and 8 bytes preamble—which affects overall efficiency but is independent of DBA implementation. The DBA-dependent penalties are burst overhead and DBA communication overhead.

A large part of the burst overhead is guard band between two consecutive upstream bursts (Figure 8.2). For conventional 1G EPON, guard band consists of laser on–off time, receiver settling, and clock and data recovery (CDR). IEEE 802.3 presents maximum values for these overhead gaps. There is no fragmentation of Ethernet frames in EPON. As a result, there is typically burst overhead related to unused slot remainders (USR) that are too small to fit the consecutive Ethernet frame. The size of this USR overhead can be considerable but decreases with fewer ONUs, and higher bit rates as in 10G EPON, and it can even be eliminated by means of advanced DBA schemes.

The Multipoint Control Protocol (MPCP) was designed in order to facilitate the discovery and registration of ONUs as well as medium access control in EPON. The MPCP consists of five messages, namely, REGISTER REQ, REGISTER, REGISTER ACK, GATE and REPORT. The first three messages are used for the discovery and registration of new ONUs, while the last two (REPORT and GATE) are used for bandwidth allocation and constitute the DBA overhead. GATE messages are sent by the OLT to grant nonoverlapping transmission windows to different ONUs in EPON. Usually, the information contained in GATE includes the start time and size of the granted transmission windows. REPORT messages are used to report the buffer occupancy of up to

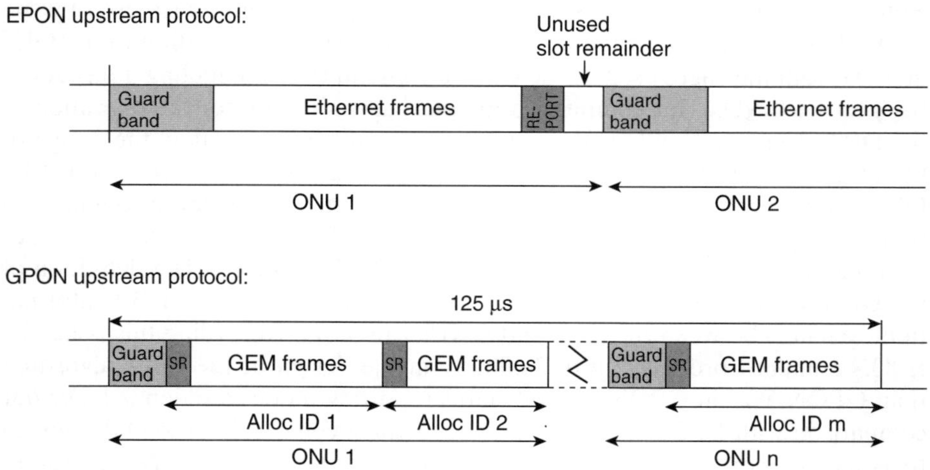

Figure 8.2. Schematic diagram of EPON and GPON upstream transmission [6].

eight internal queues at an ONU to the OLT. Their exchange allows the time slots to be assigned according to the traffic demand of the individual ONUs and the available bandwidth. The MPCP allows for a very flexible communication mechanism between the OLT and ONUs. The drawback of MPCP is the rather large DBA overhead due to the large size of the REPORT and GATE messages, which are defined as the smallest size Ethernet frame (64 bytes).

In order to cater for the ever-increasing bandwidth requirements from end subscribers, the 10G EPON Task Force was formed, known as IEEE 802.3av [2], with an initiative to standardize requirements for the next-generation 10G EPON in 2006. The IEEE 802.3av draft focuses on a new physical layer standard while still keeping changes to the logical layer at a minimum, such as maintaining all the MPCP and operations, administration, and maintenance (OAM) specifications from the IEEE 802.3 standard. 10G EPON will consist of both a symmetric 10/10 Gbit/s and a nonsymmetric 10/1 Gbit/s solution (see Table 8.1). Considering the symmetric solution with 10 Gbit/s upstream, there will be a 64B/66B line coding using a raw bit rate of 10.3125 Gbit/s. The maximum value for receiver settling is increased and burst delimiters and terminators are introduced, effectively increasing the guard band overhead, while the DBA overhead and USRs are less significant because of the increased data rate.

8.2.2 GPON

The GPON standard is defined in the International Telecommunication Union–Telecommunication Standardization Sector (ITU-T) G.984.x series of Recommendations [3] sponsored by the full service access network (FSAN). The GPON standard is based on operator requirements and provides a management layer based on the GPON physical and logical layers. Upstream and downstream rates up to 2.48832 Gbit/s are specified in the standard, although upstream rates of 1.24416 Gbit/s are conventionally used in products. The GPON protocol strips off the IPG and preamble and introduces a 5-byte GEM (GPON Encapsulation Method) header to each Ethernet frame, improving link efficiency.

The requirements on the physical layer are tighter than those for EPON, providing significantly reduced guard band between bursts. The GPON protocol is based on the standard 125-μs periodicity used in the telecommunications industry. This periodicity provides certain efficiency advantages, because messages (control, buffer report, and grant messages) can efficiently be integrated into the header of each 125-μs frame, implying reduced DBA overhead. In order to efficiently pack Ethernet frames into the 125-μs frame, GEM encapsulation has been designed to support Ethernet frame fragmentation, which means that USRs are avoided at the cost of an extra GEM frame header and also at the expense of increased protocol complexity. Frame fragmentation allows the system to transport fragments of Ethernet frame in order to utilize bandwidth associated with slots that are too small for complete frames. The benefit of the increased protocol complexity of implementing frame fragmentation depends on the average burst size in bytes and is reduced as higher bit rates are introduced.

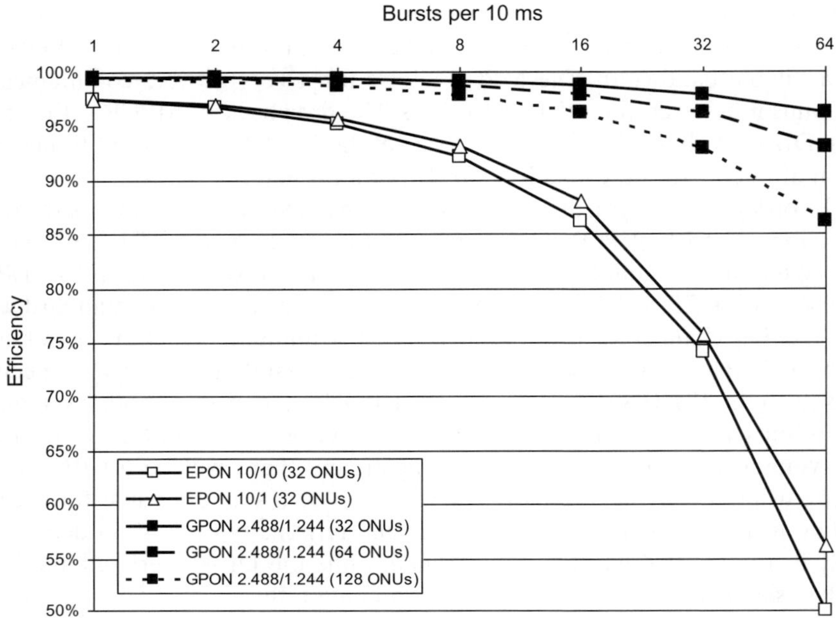

Figure 8.3. PON efficiency dependence on burst rate, protocol (EPON, GPON), and number of ONUs.

With the introduction of frame fragmentation and the reduced burst overhead compared to EPON, GPON efficiency is not as dependent on the DBA design as EPON. As illustrated in Figure 8.3, the average burst size is a crucial parameter for EPON DBA performance. The figure shows how the efficiency of the PON system depends on the burst rate (burst frequency) for an average ONU. For a given burst rate, split ratio is an additional parameter that further affects average burst size and thereby efficiency. The evaluation in Figure 8.3 was made based on a simple traffic model (30% 64-byte frames, 70% 1518-byte frames) and with worst-case parameters for EPON from Table 8.1. Note that efficiency is plotted after line-coding and FEC and does not include efficiency loss due to these operations. Increased burst rate and increased split ratio will significantly degrade EPON efficiency through increased overhead. Decreased burst rate will, on the other hand, increase delay by increasing the response time of the DBA. This tradeoff presents a key challenge for the design of an efficient EPON DBA algorithm. On the other hand, for GPON it is the set of constraints imposed by the DBA messaging protocol (Figure 8.2) that present the key challenge for the design of an efficient DBA algorithm.

In order to support more bandwidth demanding services, FSAN is currently working toward the next-generation 10G GPON (XG-PON, G.987.x) standard. The solution will most likely have a ~10-Gbit/s downstream rate and 2.48832-Gbit/s upstream line rate.

8.3 TRAFFIC REQUIREMENTS

The implementation and performance of a DBA algorithm is related to the nature and the requirements of different types of network traffic. Traffic is mainly characterized by frame size distribution and burstiness. Several measurements of frame size distribution show that a large portion of network traffic consists of either minimum-sized Ethernet frames or maximum sized Ethernet frames. An older measurement is presented in [4]. The Hurst parameter (H), described in more detail in Sala and Gummalla [5], is a measurement of the degree of self-similarity or burstiness of traffic. Network traffic with $H = 1$ exhibits burstiness on all timescales.

Traffic requirements are given in terms of throughput, delay, jitter, and packet loss ratio, and they differ depending on service. Traffic requirements are typically given with respect to end-to-end requirements that need to be broken down to requirements on the access part of the network. We will now summarize the important aspects of network traffic that a PON must cater for.

- *Voice:* Voice traffic in access networks is carried through either legacy telephony or voice over IP (VoIP). EPON and GPON are designed for VoIP. The nature of VoIP traffic depends on codec and how voice is encapsulated in IP packets. Common speech codecs are G.711, G.723, and G.729. These convert speech to a bit stream of 2.15 to 64 kbit/s, depending on codec used. Instead of transmitting an 8-bit sample every 125 μs as in legacy telephony, several 8-bit samples are collected and packaged in an IP packet. The size of the IP packets is limited by coding latency. For G.711 and a 64-kbit/s stream, voice payloads are typically 160 bytes (20 ms of speech) or 240 bytes (30 ms of speech). Including packaging overhead, bit rates of 83 kbit/s and 76 kbit/s are required for the two schemes, respectively. Voice activity detection is used to eliminate packets of silence to be transmitted, reducing the average data rate by one-third. These variations can be modeled by the on–off model [7]. For acceptable service it is required that one-way end-to-end delay be less than 100 ms. VoIP is typically jitter sensitive although usually there is some buffer capacity for coping with a small amount of jitter. Some encoding schemes can handle single-packet losses maintaining acceptable voice quality. Voice is transported in accordance with the most stringent QoS requirements in Table 8.2.
- *Video:* The largest growth in network traffic during the coming years is expected to be because of video applications. However, for access networks this is primarily true for the downstream, although some amount of video traffic will still occupy the upstream. Video traffic is typically bursty and depends on the compression algorithm used in the codec. Common codecs are H.263, H.263+, V6, MPEG-2, and H.264/MPEG-4. Videos encoded without rate control show strong correlations over long time periods in contrast to videos encoded with rate control. It is the strong temporal correlations in video content that is exploited by the compression

TABLE 8.2. QoS Parameters for Different Traffic Types for the Access Part of the Network [6]

	Low	Medium	High
Delay	5 ms	100 ms	500 ms
Jitter	2/5 ms (DS/US)	15 ms	40 ms
Packet loss ratio (PLR)	10^{-6}	10^{-5}	10^{-4}

algorithms. In MPEG-2 every 15th frame is a larger I-frame. Between the I-frames a sequence of smaller P- and B-frames are transmitted. There have been measurements on frame size distribution and Hurst parameter for different video streams. Hurst parameters have been shown to be in the range of $H = 0.7$–0.8 [8]. Interactive video applications such as video conferencing impose high QoS requirements regarding jitter and delay. Video traffic within video on demand (VoD) applications is transmitted with some degree of QoS assurance, whereas a large amount of internet video content is transported as best effort Internet traffic.

- *Data:* Data traffic primarily consists of file sharing and Web browsing. Web browsing possesses self-similar properties and is commonly modeled as a sum of on–off sources of Pareto-distribution with Hurst parameter 0.8. In access networks data traffic is usually managed in a best effort manner. Data are transported in accordance with the most relaxed QoS requirements in Table 8.2.

Traffic requirements are crucial for the implementation of the DBA. There is often a tradeoff between supporting delay-sensitive traffic classes and providing overall high bandwidth utilization. This tradeoff will be discussed in Section 8.5 for the different DBA algorithms.

8.4 PERFORMANCE PARAMETERS

There are a number of performance parameters used for evaluating and comparing the operation of different DBA algorithms. These parameters represent important aspects of a DBA algorithm, and several are typically connected to traffic requirements.

8.4.1 Delay

An important performance parameter for any access technology is the delay imposed on data transported through the system. This delay is a combination of waiting time in a receiving buffer (at the ONU) and transmission time through the system. The delay of a certain traffic class typically depends on the priority it is given by the scheduling mechanism in the PON. The mechanisms responsible for delay differ between a congested traffic class and a noncongested traffic class. In a noncongested traffic class, delay is determined by the execution delay of the scheduling mechanism, the average cycle time of the scheduling algorithm, and the propagation delay through the PON. In a congested traffic class, delay is mainly decided by the magnitude of the congestion, which in turn depends on traffic characteristics such as burstiness and amount of traffic with the same or higher priority, buffer sizes, and PON efficiency.

It is primarily the noncongested traffic delay that is of interest when comparing delay of different algorithms. Traffic classes that are delay-sensitive are treated with higher priority by the scheduling algorithm and to a smaller extent overbooked. In a noncongested traffic class there is available bandwidth for all incoming traffic, and delay is determined by the delay of the scheduling algorithm.

Delay for a congested traffic class is of less interest because it refers to best-effort traffic with loose delay requirements. Here, delay can instead be seen as an inverse measure of PON efficiency as more efficient DBA algorithms are capable of more effectively utilizing the available bandwidth.

8.4.2 Jitter

Jitter is the standard deviation of delay. In analog to delay it is the jitter of non-congested traffic classes that are of importance. Some Internet applications are sensitive to jitter such as online multiplayer games and many video conferencing applications. It is determined by the operation of the DBA algorithm rather than traffic characteristics.

8.4.3 Efficiency

DBA algorithm efficiency is a critical performance metric. One way of defining efficiency is

$$\text{Efficiency} = \frac{\text{Useful data bit rate}}{\text{Bit rate}}$$

For DBA efficiency (rather than PON efficiency), one would typically include Ethernet frame headers in the useful data rate and use the post line-code, post FEC bit rate denominator as in Figure 8.3. The efficiency metric is only meaningful for congested systems.

There are a number of sources of efficiency loss in a PON upstream that can be controlled by the DBA algorithm:

- *PON Guard Band.* The size of the guard band between two bursts is defined in the standard. The efficiency loss related to guard band depends on the defined size of the guard band and on the average burst size which is controlled by the DBA algorithm. Small bursts lead to increased guard band overhead.

- *DBA Messages.* Each PON standard defines the size of the DBA messages used for reporting buffer occupancy and granting bandwidth. The efficiency loss related to DBA messages depends on the size of these messages and on how often these messages are exchanged. The latter is controlled by the DBA algorithm.
- *Unused Slot Remainders (USR).* USR is an inherent problem for EPON systems which do not support Ethernet frame fragmentation. The contribution of USR to efficiency loss is determined by traffic profile and average burst size. There exist DBA algorithms for EPON that eliminate USR completely, but these typically introduce some additional DBA delay.
- *Fragmentation Overhead.* For systems that support Ethernet frame fragmentation, a small overhead is introduced due to fragmentation. Efficiency loss due to fragmentation depends on average burst size and average number of queues per ONU.
- *DBA Waiting Time.* For some nonoptimal DBA or computational expensive algorithms the PON system must wait for the DBA algorithm to complete its execution.
- *Over-granting.* Efficiency loss due to bandwidth over-granting typically occurs in DBA algorithms where bandwidth demand is estimated or predicted. This type of overhead may be reduced by using more conservative DBA algorithms.

8.4.5 Fairness

It is often required that the DBA algorithm should allocate bandwidth fairly among the queues. Fairness is a concept that becomes relevant first for congested traffic classes, where multiple queues compete for a limited amount of resources. We will now briefly examine the concept of fairness.

Fairness is related to some set of pre-decided principles for fair bandwidth allotment. For noncongested traffic classes, demand dictates the desired bandwidth distribution. For congested traffic classes, it is usually desired that a combination of demand and weights decide the bandwidth distribution. A bandwidth allocation algorithm may be designed to achieve the desired fair bandwidth distribution or an approximation of this fair distribution. Usually fairness is considered with respect to queues. Fairness with respect to queues on the same ONU (sibling fairness) is often regarded as important. Fairness with respect to queues on different ONUs (cousin-fairness) can be difficult to achieve within certain DBA schemes (hierarchical scheduling).

Fairness is also related to timescale. For example, an algorithm can be regarded as very unfair on a short timescale but fair on a longer timescale. Many times it is long-term fairness that is desired. Simple DBA algorithms are implemented to be fair on the timescale of the DBA cycle. Reducing the DBA cycle may therefore implicitly lead to a fairness on a shorter timescale at the expense of efficiency. For this reason it may be useful to extend the timescale for possible unfair allotment for the sake of overall efficiency. For bursty traffic, the tradeoff between fairness and efficiency becomes more accentuated.

8.5 DBA SCHEMES

The design of an efficient DBA scheme depends on the PON protocol (Section 8.2), the nature of network traffic (Section 8.3) in the system, and the service requirements (Section 8.4). As we have seen in previous sections, the GPON standard, with the 125-μs framing and standardized DBA output, imposes more stringent constraints to the implementation of the DBA. Flexibility is larger in EPON, which partly explains why there has been a significant amount of academic work on EPON DBA [9–23], whereas research on GPON DBA has been limited to a few number of system and chip vendors and research institutes [24–28]. Designing EPON DBA is in several aspects more challenging than designing GPON DBA. Large burst overhead and absence of Ethernet frame fragmentation in the EPON standard, delegates some of the complexity to the EPON DBA, which in GPON is handled by the GPON framework.

There are several ways of categorizing DBA algorithms. One important distinction relates to whether the DBA algorithm is interleaved with respect to ONU bursts (burst interleaved) or with respect to time intervals (time frame interleaved). Since the execution time of the DBA (including OLT-ONU communication) is non-negligible, it is necessary to interleave several DBA processes in order to avoid ONU idle time. In burst interleaved schemes the DBA is executed per ONU burst upon reception of an ONU report at the OLT (Figure 8.4a). In time frame interleaved schemes the DBA is executed at regular time intervals (Figure 8.4b). EPON DBA can be implemented either way but the most well-known implementations are burst interleaved. GPON DBA is restricted by the logical layer to a time-frame-interleaved implementation. The effect of this difference has profound implications and is discussed more in detail in Section 8.5.3.

Another way of categorizing algorithms is whether the DBA is implemented with a fixed or a variable cycle length. Due to the absence of Ethernet frame fragmentation and flexibility of the protocol, EPON algorithms and burst interleaved schemes are more prone to have a variable DBA cycle. Due to the fixed protocol structure, GPON algorithms and time-frame-interleaved schemes are more prone to have a fixed DBA cycle. In general, a variable cycle length is advantageous with respect to bandwidth efficiency, whereas a fixed cycle simplifies QoS assurance.

DBA algorithms can also be categorized depending on if they are centralized or distributed. Figure 8.5 shows the taxonomy for general DBA schemes. We use this taxonomy as a framework for discussing DBA schemes.

The remainder of this section is organized as follows. First we discuss the two main scheduling categories for DBA, namely, single-level (Section 8.5.1) and

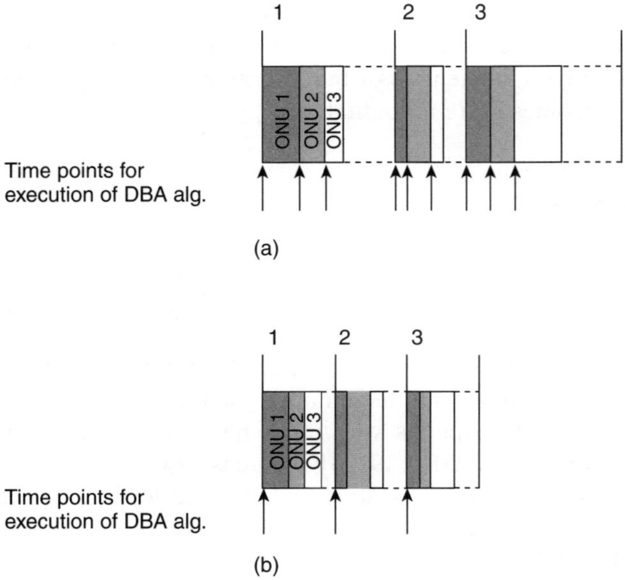

Figure 8.4. Burst interleaved and time frame interleaved DBA algorithms. (**a**) Burst interleaved scheme. (**b**) Time-frame interleaved scheme.

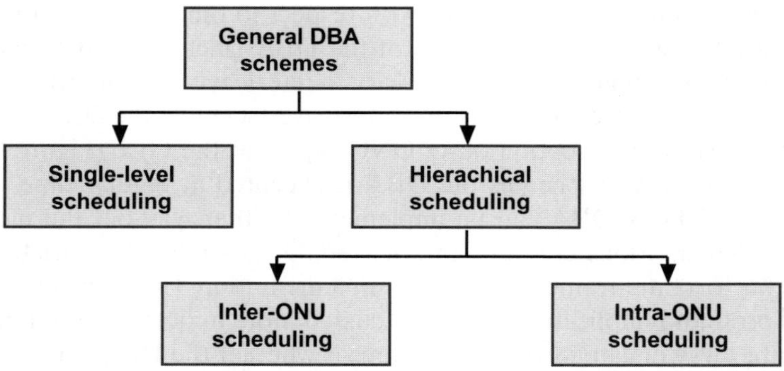

Figure 8.5. Taxonomy for general DBA schemes.

hierarchical (Section 8.5.2) scheduling. These schemes are discussed in the context of EPON, where the DBA communication protocol is more flexible. In Section 8.5.3, we discuss DBA within the GPON context. Finally, we illustrate DBA schemes for some alternative PON architectures.

8.5.1 Single-Level Scheduling

Single-level scheduling is used to denote centralized scheduling where a single scheduler at the OLT manages bandwidth allocation to the ONUs on a per queue basis. In EPON, single-level scheduling for ONUs with multiple internal queues is typically arranged by introducing an individual LLID (logical link ID) for each internal queue. The internal queue is then regarded as a virtual ONU. One of the most well-known single-level scheduling algorithms that can be implemented in EPON is the interleaved polling with adaptive cycle time (IPACT) algorithm [9]. In IPACT, the OLT polls and issues transmission grants to the ONUs cyclically in an interleaved fashion. The polling cycle is defined as the time between two consecutive report messages sent from the same ONU to the OLT. In IPACT the polling cycle is variable and adapts to the instantaneous bandwidth requirements of the ONUs.

The interleaved polling of ONUs entails that the OLT must inform the $(i+1)$ st ONU of grant information, including the start time and the size of the granted window, during or before the time that the ith ONU is transmitting Ethernet frames in the upstream. For efficient operation and high bandwidth utilization, the grants for the $(i+1)$st ONU must be received before the data transmission of the ith ONU is completed and the transmission slots must be scheduled in such a way that the first bit from the $(i+1)$st ONU arrives at the OLT right after the OLT receives the last bit from the ith ONU.

There are mainly three ways of allocating timeslot sizes to the ONUs in IPACT, namely, gated, limited, and limited with excess distribution. The most straightforward service discipline is the gated scheme, where the grants from the OLT are directly equal to the buffer occupancy reported by the ONUs. Here the polling cycle is here determined by the overall PON load. The scheme is efficient in terms of bandwidth utilization but inadequate in terms of fairness and QoS support. For example, it may lead to a situation where an ONU with heavy traffic load monopolizes the upstream channel so that frames from the other ONUs are delayed. To solve this problem, the limited service discipline was proposed [9] where a minimum guaranteed bandwidth is predefined for each ONU. The full bandwidth request by the ith ONU is granted if it is smaller than a predefined value $B_i^{\max}$. If the request is larger, then the ONU is granted bandwidth corresponding to $B_i^{\max}$ which sets an upper bound to the bandwidth allocated to the ith ONU in a given cycle. This scheme introduces some element of QoS control and fairness among the ONUs at the cost of reduced efficiency. A drawback with the limited service algorithm is that there can be a shrinking of the polling cycle due to burstiness of traffic arriving at the ONUs, degrading bandwidth utilization. The third service discipline, limited with excess distribution [10], was proposed to alleviate this shrinking of the polling cycle. In this discipline the ONUs are partitioned into two groups, namely, underloaded and overloaded. The underloaded ONUs are those with a bandwidth request below the guaranteed minimum. The unused capacity associated with these ONUs is shared in a weighted manner amongst the overloaded ONUs.

Additional variations of IPACT have been proposed with the aim of enhancing performance, in particular with respect to delay [11–13]. References 11 and 12 propose estimation-based schemes for effective upstream channel sharing among multiple ONUs. By estimating the amount of new packets arriving between two consecutive polling cycles and granting ONUs with excess bandwidth based on these estimations, proposed schemes can achieve reduced delays at light load compared to the limited service IPACT. Reference 13 proposes a heuristic where the OLT grants bandwidth to the ONUs in order of ONU buffer occupancy. In this way, packet delay can be reduced. In order to support differentiated services, some advanced algorithms are proposed for QoS diversification (e.g., in references 14 and 15). In Reference 14, a scheme is proposed where an ONU requests bandwidth for all of its queued traffic, and all traffic classes proportionally share the bandwidth based on their instantaneous demands and predefined threshold. Reference 15 shows that queuing delay using a strict priority algorithm results in an unexpectedly long delay for lower-priority traffic classes (light-load penalty), and they suggest the use of DBA with appropriate queue management to resolve the problem.

The main drawback of IPACT-based schemes is that the complexity of the algorithm increases dramatically with introduction of QoS support. IPACT is otherwise considered very efficient when it comes to best-effort traffic. For efficiency, the size of the polling cycle is a crucial parameter [16]. The polling cycle is determined by a combination of the total traffic load and the type of service discipline. For smaller polling cycles there may be severe efficiency penalties in terms of guard band, USRs, and ONU idle time due to non-optimal interleaving. In IPACT, efficiency increases with increased polling cycle time. However, increased polling cycle also leads to increased delay and jitter.

A general problem with single-level schemes concerns the scheduling of a large number of queues. Scheduling a large number of queues requires plenty of control messages between the OLT and ONUs. For example, an EPON system with 32 ONUs, 128 subscribers per ONU, and three queues (for service differentiation) per subscriber will be required to handle a total of 12,288 queues. This adds a considerable amount of GATE and REPORT messages, and hence the important performance metrics (bandwidth utilization, delay, jitter, etc.) may be significantly degraded. Thus, single-level schedulers are not scalable with respect to the number of queues. This scalability problem can be resolved by hierarchical scheduling.

8.5.2 Hierarchical Scheduling

Hierarchical scheduling is a type of distributed scheduling where the scheduler is divided into an intra-ONU and an inter-ONU scheduler. The intra-ONU scheduler manages bandwidth allocation to queues within each ONU, while the inter-ONU scheduler takes care of bandwidth allocation to the ONUs. The concept of hierarchical scheduling is shown in Figure 8.6. The inter-ONU scheduler treats each ONU as one aggregated queue and does not need information on the internal bandwidth requirements of the queues. The control messages exchanged between the OLT and ONUs only contain grant and report information related to the aggregated bandwidth requirements of the ONU (i.e., a large allocated timeslot that can be internally shared among the queues within an

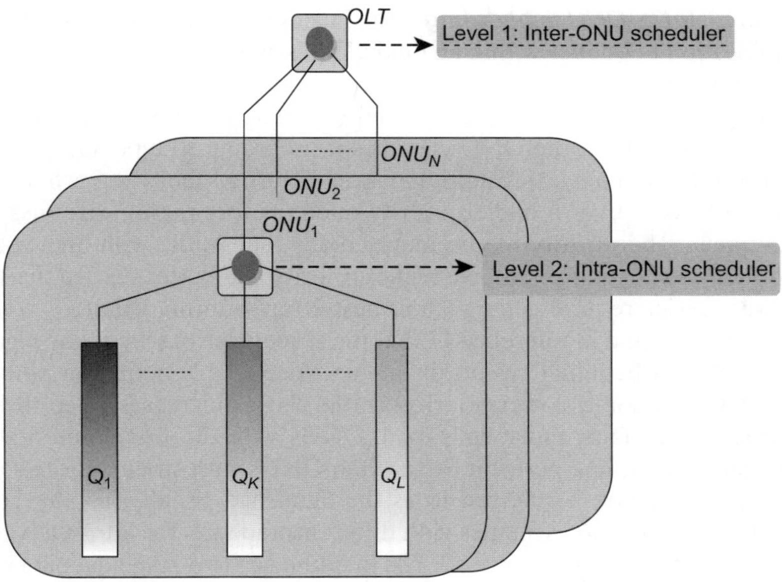

Figure 8.6. Framework for the hierarchical scheduling algorithm.

ONU). Compared to single-level scheduling, the complexity of both the intra-ONU and inter-ONU scheduler is relatively low due to the smaller number of queues that needs to be handled at each level. The reduced number of queues handled at the OLT also reduces the amount of control messages compared to single-level scheduling. There is furthermore a potential performance gain in using hierarchical scheduling as the intra-ONU and inter-ONU algorithms are run in parallel. Hence, the main advantage of hierarchical scheduling is scaling for a large number of queues. The nature of how the scalability problem will develop for next-generation TDM-PON depends on to what extent the increased bandwidth in next-generation TDM-PON is used for increased bandwidth per subscriber and to what extent it is used for aggregating more subscribers in the PON system.

The most challenging problem in hierarchical scheduling is supporting global QoS characteristics (such as global fairness, global priority, etc.) of resource distribution among queues of different ONUs. Hierarchical scheduling algorithms in references 17–21 allow fairness and/or priority only among queues within the same ONU. Failure to provide global fairness and priority may imply poor distribution of the available bandwidth among subscribers of different ONUs. The fact that concepts such as fairness and priority only have local meaning within each ONU has the consequence that high-priority traffic at one ONU may not receive sufficient bandwidth due to lower priority traffic at another ONU. The following sections describe recent work in references 17–23 related to the intra-ONU and inter-ONU scheduling.

8.5.2.1 Intra-ONU Scheduling. The intra-ONU scheduler manages bandwidth allocation to queues within an ONU. We assume that an ONU is equipped with L queues serving L priority traffic classes (denoted $Q_1, Q_2, \ldots, Q_L$ in Figure 8.6) with Q_1 being the highest priority and Q_L being the lowest. When a packet is received from a user, the ONU classifies it according to type and places it in the corresponding queue. In traditional strict priority scheduling, when a grant arrives, the ONU serves a higher-priority queue before taking care of a lower-priority queue. The priority requirement entails that traffic with higher service requirements, such as voice, receive higher priority and better service than traffic with lower service requirements such as best-effort Internet traffic.

It has been found in reference 17 that the strict priority scheduling algorithm for intra-ONU scheduling results in an unexpected phenomenon where the average delay for some (lower-priority) traffic classes increases when the traffic load decreases. In fact, under light load, ONUs with the first come first serve (FCFS) queue discipline perform better than ONUs with strict priority scheduling. This phenomenon is referred to as the light-load penalty. To alleviate this penalty, two optimization schemes with different tradeoffs for intra-ONU scheduling were proposed in reference 17. The first one is a two-stage queuing scheme that totally eliminates the light-load penalty at the expense of increased packet delay for all types of traffic. The second scheme attempts to predict high-priority packet arrivals. This scheme eliminates the light-load penalty for most of the packets. Some low-priority packets are delayed excessively, but the number of such packets is small and does not affect the average packet delay. The drawback of this second scheme is the increased complexity due to the estimation of the traffic-arrival process.

Another consequence of strict priority scheduling is that non-greedy queues of lower priority are mistreated as the system is overloaded. The urgency fair queuing (UFQ) [18] scheme for intra-ONU scheduling is proposed in order to achieve a better balance in bandwidth assignment among different traffic classes within the same ONU. In the UFQ scheme, packets are scheduled based on urgency regarding delay requirements; that is, packets with higher priority which are not so urgent for transmission can give way to the ones with lower priority. Other schemes have been proposed where bandwidth is assigned to each traffic class within an ONU based on its load-based weight. One of these, the modified start-time fair queuing (MSFQ) algorithm, tracks aggregate ONU service via a global virtual time. Variables are also maintained for tracking local per-queue start and finish time which are related to the global virtual time and the weight of different traffic classes. The packet in the queue with minimal start time is selected to be transmitted first. In this way, load-based fairness can be achieved. The MSFQ algorithm can provide fairness for differentiated services even when the network is overloaded. Simulations of MSFQ [19] show that when a traffic class of priority i is greedy (i.e., it requires more bandwidth than guaranteed), the other traffic classes with higher priority have stable delay while the classes of low-volume traffic with the lower priority maintain throughput performance, but with slightly increased average delay. Conversely, the strict priority scheduler yields unacceptably high delay and throughput degradation for the classes of traffic with lower priority than i.

The MSFQ algorithm does not adequately fulfill the priority requirement when the load distribution of different traffic classes is changed. To solve this problem, the modified token bucket (MTB) [20] algorithm for the intra-ONU scheduling was proposed. The MTB algorithm assigns bandwidth in two stages. In the first stage, bandwidth is allocated to each queue according to the size of a token that is related to a load-based weight. This first stage limits greedy traffic classes from monopolizing all bandwidths. In the second stage, the remaining bandwidth is allocated according to the strict priority policy. MTB provides a method of obtaining a compromise between complying with strict priority and preventing single queues from monopolizing all bandwidths, which holds also when the load distribution of different traffic classes is changed. The computational complexity of the proposed MTB algorithm is $O(k)$ where k is the total number of packets that can be sent in one grant window. This can be compared to the strict priority and the MSFQ algorithms which have a complexity of $O(k)$ and $O(k \log L)$, respectively.

Each ONU may contain multiple queues for multiple subscribers and multiple services. In reference 21 a hierarchical intra-ONU scheduling scheme was designed in order to handle scalability issues at the ONU with respect to a large number of queues. As shown in Figure 8.7, there may be several queues for each priority class for different users. There are two levels of scheduling: One is the interclass scheduling (to serve L classes of traffic with differentiated priorities) and the second one is the intraclass scheduling (to allocate fairly the bandwidth

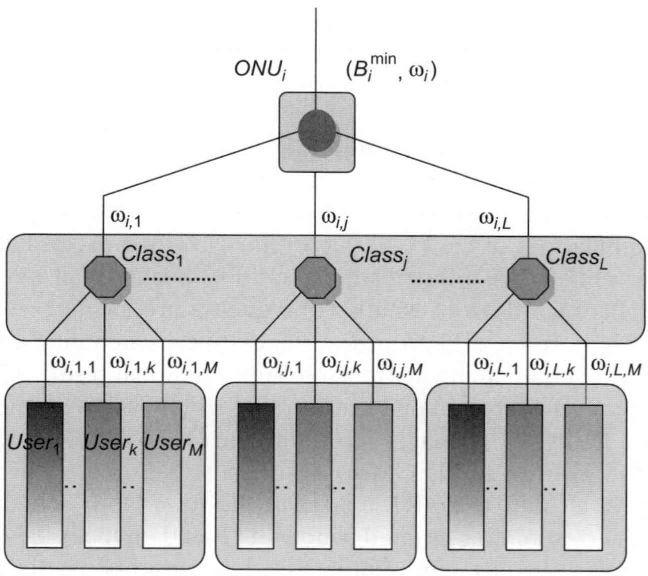

Figure 8.7. Framework of the hierarchical intra-ONU scheduling algorithm in reference 21.

among M users within the same class). The proposed hierarchical intra-ONU scheduler realizes fine granularity scheduling to support QoS for traffic of each individual user by combining the MTB algorithm [20] and the MSFQ algorithm [19], where MTB is used for interclass scheduling and MSFQ is used for intraclass scheduling.

8.5.2.2 Inter-ONU Scheduling. The inter-ONU scheduler allocates bandwidth to the ONUs. It treats each ONU as one aggregated queue and does not need information on the internal queues of each ONU [17–22]. Using EPON as an example, the GATE and REPORT messages would grant and request aggregated bandwidth per ONU rather than to individual queues.

Most single-level scheduling algorithms can also be applied to inter-ONU scheduling. For instance, the limited and limited with excess distribution schemes are widely employed in inter-ONU scheduling to allocate aggregated bandwidth to each ONU. In reference 17, IPACT with the limited service discipline algorithm described in references 9 and 10 is used for inter-ONU scheduling. This service discipline has the same problems with shrinking of polling cycle as the single-level version. The limited with excess distribution service discipline from reference 10 adopted in references 19 and 20 can alleviate this shrinking of the polling cycle. However, by utilizing this weighted inter-ONU scheduling, an overloaded ONU may get more bandwidth than requested, and thus some bandwidth may be wasted on over-granting. With this in mind, a novel inter-ONU scheduler based on recursive calculation is proposed in reference 21 to guarantee that no ONU gets more bandwidth than requested. For an ONU where assigned bandwidth is less than the requested, the bandwidth actually used may be less than the assigned, if packet fragmentation is not supported such as in EPON. This causes an increase in USRs and a decrease in bandwidth utilization. In reference 21 a novel GATE/REPORT approach for EPON to eliminate unused timeslot reminders is introduced in order to further improve the bandwidth utilization.

There has also been work on adjusting hierarchical scheduling to support global QoS characteristics. The fair queuing with service envelopes (FQSE) [22] algorithm was developed in order to realize global fair scheduling in hierarchical scheduling. Simulations of the FQSE algorithm show that excess bandwidth due to idle queues can be redistributed among multiple queues in proportion to their assigned weights regardless of whether the queues are located on the same or different ONUs. Reference 23 proposes a hierarchical scheduling algorithm with a novel inter-ONU scheduling approach to support global traffic priority among multiple service providers and end users. Using EPON as an example, an ONU needs to issue REPORT messages informing the OLT of the aggregated queue sizes of all the priority traffic classes destined to different service providers. After collecting all the REPORT messages, the inter-ONU scheduler at OLT calculates the corresponding granted bandwidth based on the weight and priority of the aggregated queues representing different priority traffic classes from various ONUs.

8.5.3 Schemes for GPON

Research on GPON DBA is limited to a few number of research institutes and vendors [24–28], and there is limited public domain information available on the topic. Resource scheduling in GPON differs significantly from EPON. In GPON the scheduling of bandwidth is inherently connected to the 125-μs periodicity of the GTC superframes and effective DBA algorithms must be tailored for GPON. The connection between the DBA and the frame structure is due to the way DBA messages, status reports (SRs), and upstream bandwidth maps (BW maps) are integrated into the GTC (GPON transmission convergence) frames. Figure 8.8 illustrates an example of the DBA process in GPON. Upstream bandwidth maps, each specifying the bandwidth allocation for a specific upstream GTC frame, are generated at the OLT and broadcasted every 125 μs (not shown in Figure 8.8) to the ONUs integrated in the GTC downstream headers. Upon request, each Alloc ID (logical queue) prepends a 2-byte SR message to the upstream data transmission specifying its current buffer occupancy. Status request could be collected during a single GTC frame as in Figure 8.8 or during several GTC frames. Once status reports from all Alloc IDs have been received, bandwidth allocation can be calculated. The resulting bandwidth allocation may be scheduled over a series of upstream bandwidth maps.

As a consequence of the GPON framework, GPON DBA is most naturally interleaved with respect to GTC frames rather than ONU bursts as in IPACT. Because of the way complete bandwidth maps are broadcasted to the ONUs, it is nontrivial to implement an IPACT-type algorithm in GPON. The GPON DBA

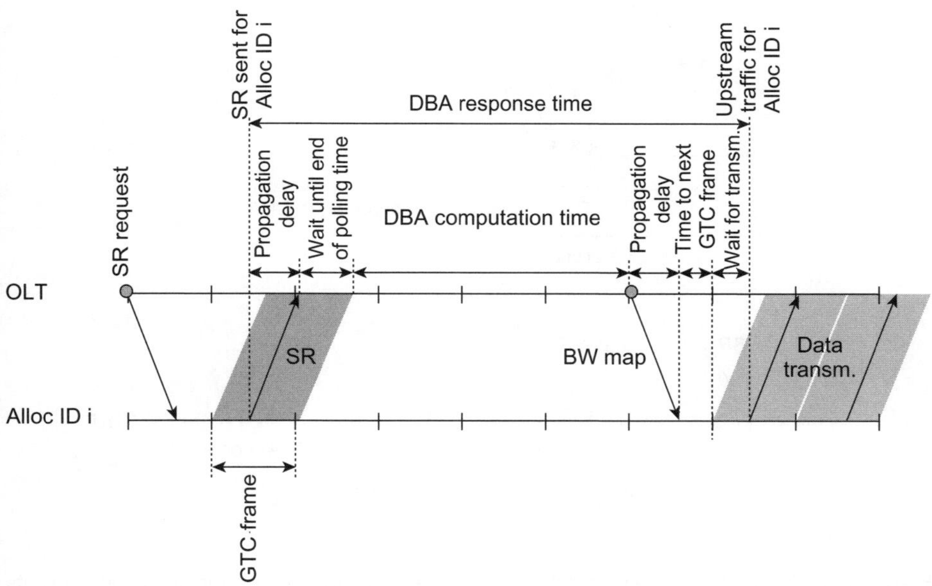

Figure 8.8. Schematic overview of GPON DBA.

must produce a complete prescription for bandwidth allocation at the end of each DBA cycle. This prescription is then used as input toward generating a sequence of bandwidth maps for transmission to the ONUs. The GPON procedure stands in contrast to EPON where GATE messages are transmitted flexibly to individual ONUs at any point in time. The two ways of interleaving DBA processes provide different DBA challenges. Interleaving bandwidth allocation with respect to different ONUs is more challenging with respect to fairness and QoS provisioning, because these concepts typically have an inter-ONU nature. Interleaving bandwidth allocation with respect to time frames is more challenging with respect to accurately predicting bandwidth demand, because bandwidth demand depends on bandwidth allocation in subsequent time frames. Because of the framing structure used in GPON, the DBA typically uses a fixed cycle.

GTC frame interleaving is illustrated in Figure 8.9. Figure 8.9a shows a schematic view of the single DBA process described in Figure 8.8. This process updates the bandwidth allocation to the ONUs through the upstream bandwidth map embedded in the header of each downstream GTC frame. We refer to the DBA delay as the delay from the GTC frame of the issuing of a status report at the ONU to the GTC frame of the transmission of data according to the updated bandwidth map (Figure 8.9a). With DBA process we refer to the DBA mechanism executed during the DBA delay as well as the data transmission period during which the updated bandwidth allocation is used (Figure 8.9a). In order to continuously update the bandwidth allocation, we execute multiple DBA

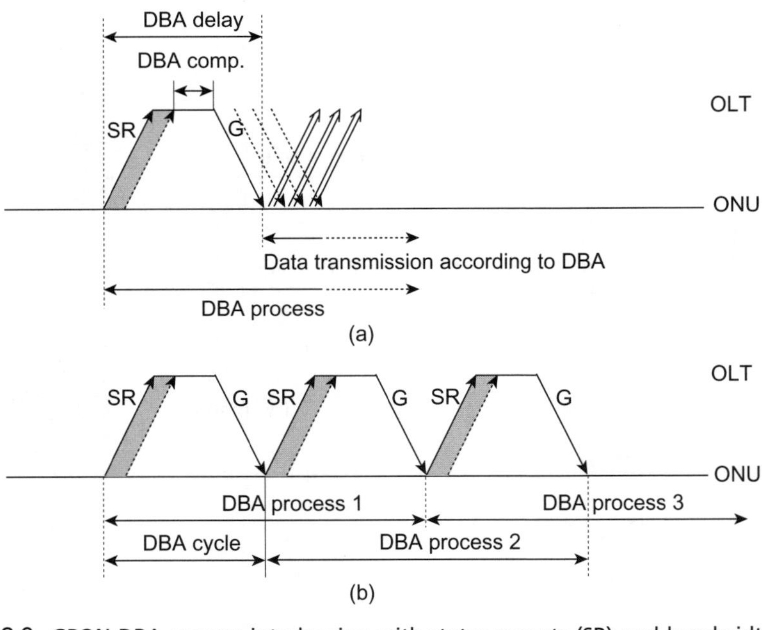

Figure 8.9. GPON DBA process interleaving with status reports (SR) and bandwidth grants (G). (a) Single DBA process. (b) Multiple DBA processes (DBA cycle = DBA delay).

processes (Figure 8.9b). The frequency defines the DBA cycle. In Figure 8.9b we set the DBA cycle equal to the DBA delay. As a result, the DBA process is here twice as long as the DBA delay with the first half of the DBA process used for updating the bandwidth allocation and the second half for transmitting data according to the updated allocation. Note that in this example the DBA processes are interleaved in the sense that there are always two DBA processes active at a given time. While one process is in a data transmission phase, the other process is in the phase of updating the bandwidth allocation map. With this example structure of the GPON DBA, we find that the average DBA response time is given by the sum of the DBA cycle and the DBA delay as shown in Figure 8.10. This response time is valid for noncongested traffic classes.

The GTC framing implies that DBA messages are handled compactly, resulting in small DBA overhead. Combined with the tight requirements on the physical layer overhead and Ethernet frame fragmentation, GPON provides high efficiency also for very small upstream ONU bursts. In principle, bursts from all ONUs could be collected every 125 μs without significant overhead, although typically the DBA is executed less frequently. Another consequence of the small overheads in GPON there is that is no scalability problem with respect to number of queues per PON.

Because of the structure provided by the GPON framework, the bandwidth allocation task in GPON DBA algorithm can be subdivided into three tasks: (1) prediction of bandwidth demand, (2) bandwidth assignment, and (3) scheduling. This makes it possible to isolate tasks related to different performance metrics. In IPACT these three tasks are integrated and interdependent, making it more difficult to understand the effect of small changes in the DBA algorithm on different performance metrics.

8.5.3.1 Predicting Bandwidth Demand. Due to the type of interleaving used in GPON DBA, predicting bandwidth demand for an Alloc ID presents a greater challenge in GPON than in EPON. In practice, the status report information from an Alloc ID may be outdated once the resulting bandwidth allocation will be used for upstream transmission. Since the point in time when the status report was issued, data have both entered and exited the Alloc ID buffer. Poor estimation of bandwidth demand leads to problems with over-granting, which in

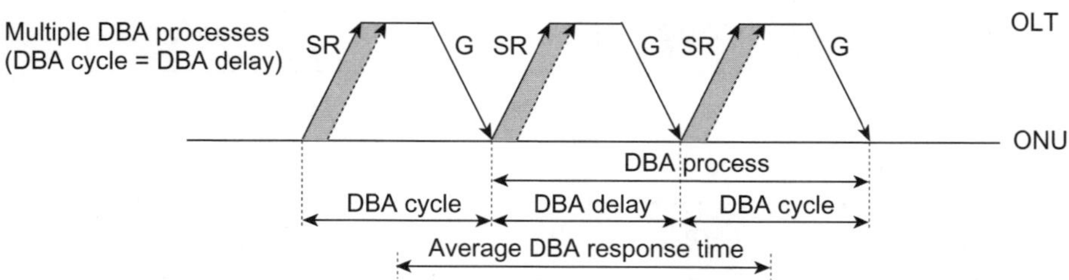

Figure 8.10. GPON DBA response time.

turn leads to serious problems at high load as capacity is used in a suboptimal manner. High-priority traffic is typically unaffected by over-granting, whereas low-priority traffic experiences reduced bandwidth. The minimum guaranteed T-CONT content scheme (MGTC) [25] was designed to eliminate the problem of over-granting. MGTC provides a more conservative estimation of bandwidth demand. The minimum guaranteed Alloc ID content is extracted by subtracting an estimate of the total outgoing traffic from the Alloc ID since the status report was issued and assuming there is no incoming traffic to the Alloc ID. The method eliminates the over-granting problem at the expense of slightly higher DBA response time. This leads to slightly increased delay for high-priority traffic, but with substantially increased efficiency as a result.

8.5.3.2 Bandwidth Assignment. The bandwidth assignment task consists in partitioning available bandwidth to Alloc IDs based on predicted bandwidth demand and on requirements such as priority and fairness. Several descriptions of possible GPON bandwidth assignment algorithms are available [26, 27]. Bandwidth assignment is relatively straightforward in GPON as bandwidth demand for all Alloc IDs from one polling period is used to produce a global bandwidth assignment. GPON frame fragmentation also simplifies bandwidth assignment because the DBA algorithm does not need to consider USRs. For EPON DBA at high load, in order to avoid a large frame from completely blocking a queue, a minimum slot size might be required which introduces extra complexity to the EPON DBA.

8.5.3.3 Scheduling. The scheduling task refers to the problem of scheduling assigned bandwidth over a sequence of GTC frames through a series of upstream bandwidth maps. A partial solution to the scheduling problem is described in reference 28. There are many ways of implementing the scheduling. In general, the scheduling can be made more or less granular (Figure 8.11). A more fine-grained scheduling leads to lower delay and jitter at the expense of

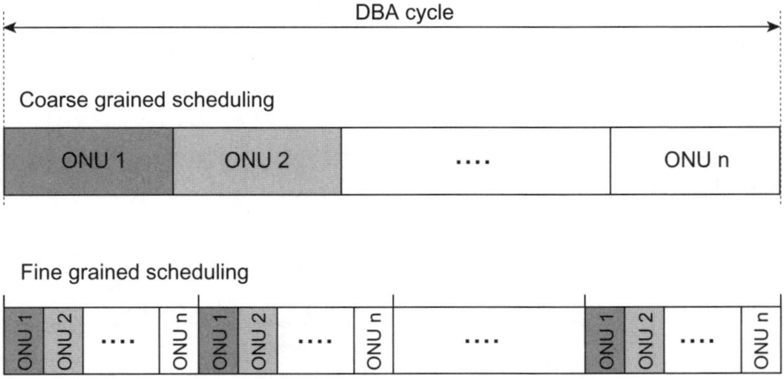

Figure 8.11. Fine-grained and coarse-grained scheduling.

more burst overhead and reduced efficiency. Ultimately, the scheduling depends on traffic requirements; and as these differ for different traffic classes, one could schedule different traffic classes in different ways.

8.5.4 DBA for Alternative PON Architectures

All the DBA schemes discussed above are for the standard PON architecture, where downstream traffic is broadcasted while in the upstream direction the ONUs cannot detect signals from the other ONUs. In the following subsections we consider DBA for two alternative architectures (i.e., broadcast PON and two-stage PON).

8.5.4.1 Broadcast PON. The broadcast PON architecture proposed in references 29 and 30 supports reflection of the upstream signal back to each ONU, as illustrated in Figure 8.12a. This architecture can employ decentralized DBA schemes, in which the OLT is excluded from the implementation of the resource scheduling. In reference 30, the full-utilization local-loop request contention multiple access (FULL-RCMA) scheme is proposed for dynamically allocating bandwidth to ONUs in a broadcast PON. In FULL-RCMA, one of the ONUs is designated as master and generates a token list. According to the order determined by the master, all the ONUs pick up the token to access the medium for their data transmission. Performance analysis [30] shows that FULL-RCMA can outperform IPACT with the limited service discipline in terms of upstream bandwidth utilization. In the DBA scheme proposed in reference 31 the ONUs independently compute the transmission schedule of all ONUs. This scheme can also integrate inter- and intra-ONU scheduling at the ONU in order to provide better QoS support. Compared to standard PON, ONUs that support broadcast PON would be more expensive, since they must (1) contain more expensive hardware to support the implementation of the medium access control, (2) have an extra receiver for the reflected upstream signal, and (3) need higher-power lasers to compensate for the loss caused by splitting upstream signals which are reflected back to other the ONUs.

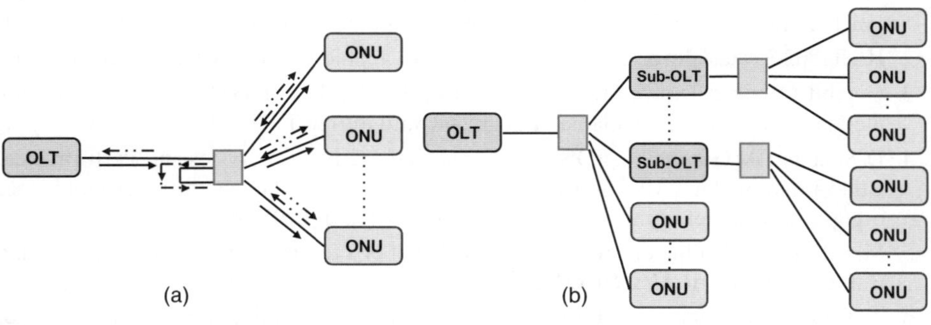

Figure 8.12. Two representative alternative architecture: (**a**) Broadcast PON [29, 30]. (**b**) Two-stage PON [31].

8.5.4.2 Two-Stage PON. A two-stage architecture [31] allows more end-users to share the bandwidth in a single PON and enables longer access reach (beyond the usual 20 km; e.g., defined in references 1–3). As illustrated in Figure 8.4b, in the first stage of this architecture there are several ONUs and sub-OLTs, which can regenerate the optical signals in the upstream and downstream as well as aggregate the traffic to/from their child ONUs in the second stage. Reference 31 proposes a DBA scheme that can take advantage of some of the predictability of the aggregated traffic at the sub-OLTs. Compared with the traffic from a single ONU, the aggregated traffic from several child ONUs of the sub-OLT tends to be less bursty and more predictable if traffic is not fully self-similar.

8.5 CONCLUSIONS

Dynamic bandwidth allocation presents a key issue for providing efficient utilization of the upstream bandwidth in TDM-PON. As we have shown, the design of an efficient DBA scheme depends on the PON standard, the nature of network traffic in the system, and the service requirements of different traffic classes. The requirements on the DBA algorithm are to provide efficient and fair utilization of the upstream bandwidth while still satisfying the minimum service requirements for the different traffic classes. We have described how differences in the GPON and EPON standards result in a series of distinct challenges for the design of the DBA. We have furthermore shown how these differences result in different design choices and have discussed some of these choices. With the evolution of TDM-PON toward higher bit rates, there will be a shift in the parameters governing the design of the DBA. As a result, DBA solutions will have to be tailored for future generations of TDM-PON, depending on parameters in the standard, the nature of network traffic, and new service requirements.

REFERENCES

1. IEEE Standard for Information technology-Specific requirements—Part 3 [Online], IEEE Standard 802.3, 2008. Available at http://standards.ieee.org/getieee802/802.3.html.
2. IEEE 802.3av task force home page [Online]. Available at :http://www.ieee802.org/3/av.
3. Gigabit-Capable Passive Optical Networks (GPON), ITU-T G.984.x series of recommendations [Online]. Available at http://www.itu.int/rec/T-REC-G/e.
4. D. Sala and A. Gummalla, PON Functional Requirements: Services and Performance, Ethernet in the First Mile Study Group 2001 [online]. Available at http://grouper.ieee.org/groups/802/3/efm/public/jul01/presentations/sala_1_0701.pdf.
5. M. S. Taqqu, W. Willinger, and R. Sherman, Proof of a fundamental result in self-similar traffic modeling, *ACM/SIGCOMM Computer Commun. Rev.*, Vol. 27, pp. 5–23, 1997.
6. A. Cauvin, A. Tofanelli, J. Lorentzen, J. Brannan, A. Templin, T. Park, and K. Saito, Common technical specification of the G-PON system among major worldwide access carriers, *IEEE Commun. Mag.*, Vol. 44, pp. 34–40, October 2006.

7. P. Seeling, M. Reisslein, and B. Kulapala, Network performance evaluation using frame size and quality traces of single layer and two layer video: A tutorial, *IEEE Commun. Surv. Tutorials*, Vol. 6, pp. 58–78, 2004.
8. S. H. Hong, R.-H. Park, and C. B. Lee, Hurst parameter estimation of long-range dependent VBR MPEG video traffic in ATM networks, *J. Visual Commun. Image Representation*, Vol. 12, pp. 44–65, June 2001.
9. G. Kramer, B. Mukherjee, and G. Pesavento, IPACT: A dynamic protocol for an Ethernet PON (EPON), *IEEE Commun. Mag.*, Vol. 40, pp. 74–80, February 2002.
10. C. M. Assi, Y. Ye, S. Dixit, and M. A. Ali, Dynamic bandwidth allocation for quality-of-service over Ethernet PONs, *IEEE J. Selected Areas in Commun.*, Vol. 21, pp. 1467–1477, November 2003.
11. H. Byun, J. Nho, and J. Lim, Dynamic bandwidth allocation algorithm in Ethernet passive optical networks, *Electron. Lett.*, Vol. 39, pp. 1001–1002, June 2003.
12. Y. Zhu and M. Ma, IPACT with grant estimation (IPACT-GE) scheme for Ethernet passive optical networks, *IEEE/OSA J. Lightwave Technol.*, Vol. 26, pp. 2055–2063, July 2008.
13. S. Bhatia and R. Bartos, IPACT with smallest available report first: A new DBA algorithm for EPON, in *Proceedings of the IEEE International Conference on Communications (ICC'07)*, Glasgow, UK, June 2007.
14. J. Xie, S. Jiang, and Y. Jiang, A dynamic bandwidth allocation scheme for differentiated services in EPONs, *IEEE Commun. Mag.*, Vol. 42, pp. S32–S39, August 2004.
15. Y. Luo and N. Ansari, Bandwidth allocation for multiservice access on EPONs, *IEEE Commun. Mag.*, Vol. 43, pp. S16–S21, February 2005.
16. B. Skubic, J. Chen, J. Ahmed, L. Wosinska, and B. Mukherjee, A comparison of dynamic bandwidth allocation for EPON, GPON, and next-generation TDM PON, *IEEE Commun. Mag.*, Vol. 47, pp. S40–S48, March 2009.
17. G. Kramer, B. Mukherjee, S. Dixit, Y. Ye, and R. Hirth, On supporting differentiated classes of service in EPON-based access network, *OSA J. Optical Networking*, pp. 280–298, August 2002.
18. Y. Zhu and M. Ma, Supporting differentiated services with fairness by an urgent queuing scheduling scheme in EPONs, *Photonic Network Commun.*, Vol. 12, pp. 99–110, July 2006.
19. N. Ghani, A. Shami, C. Assi, and M. Y. A. Raja, Intra-ONU bandwidth scheduling in ethernet passive optical networks, *IEEE Commun. Lett.*, Vol. 8, pp. 683–685, August 2004.
20. J. Chen, B. Chen, and S. He, A novel algorithm for Intra-ONU bandwidth allocation in Ethernet passive optical networks, *IEEE Commun. Lett.*, Vol. 9, pp. 850–852, September 2005.
21. B. Chen, J. Chen, and S. He, Efficient and fine scheduling algorithm for bandwidth allocation in Ethernet passive optical networks, *IEEE J. Selected Topics Quantum Electron.*, Vol. 12, pp. 653–660, July–August 2006.
22. G. Kramer, A. Banerjee, N. Singhal, B. Mukherjee, S. Dixit, and Y. Ye, Fair queueing with service envelopes (FQSE): A cousin-fair hierarchical scheduler for subscriber access networks, *IEEE J. Selected Areas Commun.*, Vol. 22, No. 8, pp. 1497–1513, October 2004.

23. J. Chen, B. Chen, and L. Wosinska, A novel joint scheduling algorithm for multiple services in 10G EPON, in *Proceedings of Asia-Pacific Optical Communications Conference (APOC'08)*, Vol. 7137, pp. 71370L–71370L-6, October 2008.
24. H. Yoo, B.-Y. Yoon, K.-H. Doo, K.-O. Kim, M.-S. Lee, B.-T. Kim, and M.-S. Han, Dynamic bandwidth allocation device for an optical network and method thereof, WO 2008/039014 A1, September 28, 2007.
25. B. Skubic, B. Chen, J. Chen, J. Ahmed, and L. Wosinska, Improved scheme for estimating T-CONT bandwidth demand in status reporting DBA for NG-PON, *Asia Communications and Photonics Conference and Exhibition* (ACP '09), Vol. 2009, Supplement, pp. 1–6, November 2–6, 2009.
26. Y.-G. Kim, B.-H. Kim, T.-S. Park, J.-W. Park, J.-Y. Park, J.-K. Kim, D.-K. Kim, S.-H. Kim, J.-Y. Lee, J.-H. Kim, and H.-J. Yeon, GPON system and method for bandwidth allocation in GPON system, US 2007/0133989 A1, November 3, 2006.
27. Y.-G. Kim, B.-H. Kim, T.-S. Park, J.-W. Park, J.-Y. Park, J.-K. Kim, D.-K. Kim, S.-H. Kim, J.-Y. Lee, J.-H. Kim, and H.-J. Yeon, GPON system and method for bandwidth allocation in GPON system, US 2007/0133988 A1, November 3, 2006.
28. E. Elmoalem, Y. Angel, and D. A. Vishai, Method and grant scheduler for cyclically allocating time slots to optical network units, US 2006/0233197 A1, April 18, 2005.
29. C. Foh, L. Andrew, E. Wong, and M. Zukerman, FULL-RCMA: A high utilization EPON, *IEEE J. Selected Areas Commun.*, Vol. 22, pp. 1514–1524, October 2004.
30. S. R. Sherif, A. Hadjiantonis, G. Ellinas, C. Assi, and M. A. Ali, A novel decentralized Ethernet-based PON access architecture for provisioning differentiated QoS, *IEEE/OSA J. Lightwave Technol.*, Vol. 22, pp. 2483–2497, November 2004.
31. A. Shami, X. Bai, N. Ghani, C. M. Assi, and H. T. Mouftah, QoS control schemes for two-stage Ethernet passive optical access networks, *IEEE J. Selected Areas Commun.*, Vol. 23, pp. 1467–1478, August 2005.

9

NEXT-GENERATION ETHERNET PASSIVE OPTICAL NETWORKS: 10G-EPON

Marek Hajduczenia and Henrique J. A. da Silva

1G-EPON, which is part of IEEE 802.3-2008 [1–3], is considered to have sufficient capacity for the next few years [4], provided that current bandwidth demand growth is maintained [5, 6]. Proprietary nature of higher-speed EPON solutions [7] meant that there was a limited supplier base and restricted interoperability between system integrators, which initially caused some concerns about deployment of such systems. This is what the market situation was at the time of 10G-EPON Call for Interest, presented during one of the IEEE plenary meetings in 2006 [8]. However, recent adoption of 2G-EPON specifications by CCSA (http://www.ccsa.org.cn/english) indicates a growing popularity of this solution at least on the Chinese market.

Taking into consideration that more than 30 million active ports of 1G-EPON have already been deployed up to date (2010), 10G-EPON seems like a natural step in the evolution toward more multimedia-rich, bandwidth-intensive applications of the future, where high-definition, distributed contents and file sharing as well as networked hardware play increasingly important roles. Given the successful history of IEEE equipment and a number of identified market applications, 10G-EPON will certainly enjoy deployment scales beyond anything

Convergence of Mobile and Stationary Next-Generation Networks, edited by Krzysztof Iniewski

that competitive PON architectures have ever seen. Providing 10 times more raw bandwidth than current 1G-EPON (approximately 8.9 Gbit/s is available for subscriber data due to mandatory FEC for all 10 Gbit/s links), it is poised to deliver the bandwidth required for next-generation applications following an evolutionary scenario rather than forcing operators to completely replace legacy 1G-EPON equipment. Assuming that market models conceived for 10G-EPON P802.3av project become reality in the near future, this new addition to EPON architecture will enjoy deployment costs per subscriber comparable to current 1G-EPON equipment, while allowing for much higher subscriber density at the CO, securing ROI for already deployed hardware.

9.1 ACRONYMS

APD	Avalanche photodiode
BER	Bit error rate
ChIL	Channel insertion loss
CRC	Cyclic redundancy check
CSMA/CD	Carrier sense multiple access/collision detection
DBA	Dynamic bandwidth allocation
DML	Directly modulated laser
EML	Externally modulated laser
EPON	Ethernet passive optical network
FEC	Forward error correction
FIFO	First in, first out
GMII	Gigabit MII
IP	Internet protocol
IPG	Interpacket gap
IPTV	Internet protocol TV
LA	Limiting amplifier
LLC	Logical link control
LLID	Logical link identificator
MAC	Medium access control
MACC	MAC client
MDI	Medium-dependent interface
MII	Media interface independent
MPCP	Multipoint control protocol
MPCPDU	MPCP data unit
ODN	Outside distribution network
OLT	Optical line termination
ONU	Optical network unit
P2MP	Point to multipoint
P2P	Point to point
PCS	Physical coding sublayer
PIN	Positive, intrinsic, negative diode

PMA	Physical medium attachment
PMD	Physical medium dependent
PON	Passive optical network
QoS	Quality of service
ROI	Return on investment
RS	Reconciliation sublayer
SLD	Start of LLID delimiter
TDMA	Time division multiple access
TF	Task force
TIA	Transimpedance amplifier
TQ	Time quantum
WDM	Wavelength division multiplexing
XGMII	10 Gbit/s MII

9.2 10G-EPON ARCHITECTURE

The IEEE 802.3 Working Group is historically focused only on the two bottom layers of the layered Open Systems Interconnection (OSI) reference model, namely the physical and data link layers, leaving network architecture and higher level management to the IEEE 802.1 Working Group and other Standard Development Organizations. To facilitate reuse of individual elements comprising 802.3 layers, they are further divided into sublayers, connected by standardized interfaces. This enables projects like P802.3av to build on specifications from earlier projects (e.g., P802.3ae 10GE or P802.3ah Ethernet in the First Mile), by introducing extensions necessary to support new functionalities. Such a modular construction of 802.3 specifications translates directly into lower cost and faster development cycles for new equipment, because experience and design from the previous generation of devices (or even other product lines) can be applied directly to new products.

The physical sublayer is subdivided into six blocks:

1. MDI specifies characteristics of the electrical signals which are received from/transmitted to the underlying medium. Additionally, it also contains definitions of mechanical and electrical interfaces used to exchange data between PMD and medium.
2. PMD specifies the basic mechanisms for exchange of data streams between medium and PCS sublayer. The bottom part of PMD contains physical devices, like receiver and transmitter.
3. PMA sublayer specifies functions responsible for transmission, reception, clock recovery, and phase alignment.
4. PCS defines a set of functions, which are responsible for converting a data stream received from xMII into codewords, which can then be passed through PMA and PMD and finally transmitted into the medium. In the receive path, PCS performs the reverse function—that is, decodes the

received data and recreates the original data stream. PCS houses such critical functionalities as data decoder/encoder, FEC encoder/decoder, data detector (ONU transmit path only), and scrambler/descrambler as well as gearbox, adjusting the data rates between PCS (bursty transmission) and PMA (continuous data stream).

5. xMII specifies a standardized interface between the MAC and PHY layers. This is one of the major interfaces in the 802.3 stack, allowing for modular interconnection of various PHY layers to MAC.
6. RS maps MAC service primitives into xMII signals, effectively transferring data into PHY and vice versa. In the EPON architecture, RS plays also one more critical role: It is responsible for LLID insertion and filtering for all data passing from MAC or PHY.

The data link layer is further subdivided into three major sublayers:

1. The MAC sublayer defines a set of medium independent functions, enabling MAC clients to exchange data with their link peers. MAC supports general data encapsulation (including framing, addressing, and error detection) and medium access (collision detection and deferral process for shared medium environment—not used in case of P2P and P2MP systems).
2. The MAC control sublayer is optional and performs real-time control and manipulation of MAC operation. MAC control sublayer specifications are open, which allows new projects to extend MAC functionality while leaving MAC itself intact.
3. LLC is already considered out of the scope of 802.3, which means that all underlying sublayers (MAC and MAC control) are specified in such a way that LLC is completely transparent to them.

P802.3av, as per its Project Authorization Request approved in September 2006, was focused exclusively on extending the EPON network architecture by adding a set of new PHYs, capable of supporting higher data rates (10 Gbit/s effective) and power budgets with higher ChIL. Minor modifications to management and MPCP sublayer were identified also as part of the project, though their scope was limited to new PHYs. There were two reasons why P802.3av made no changes to 1G-EPON specifications, even if that could potentially facilitate coexistence: (i) the P802.3av TF had a mandate to introduce changes into existing 1G-EPON specifications, and (ii) introduction of changes to a standard describing mass-deployed equipment might potentially cause compliance issues, thus was discouraged. The approval of IEEE 802.3av 10G-EPON standard on September 11, 2009, the numerous and geographically varied attendance of the meetings, the deep involvement of many companies, and commercial availability of 10G-EPON equipment from the first system suppliers come as a result of more than three years' worth of continuous work of a dedicated group of experts. It is anticipated that the first commercial deployments of 10G-EPON systems will become a reality by Q4 2010, further fueling market success of this technology.

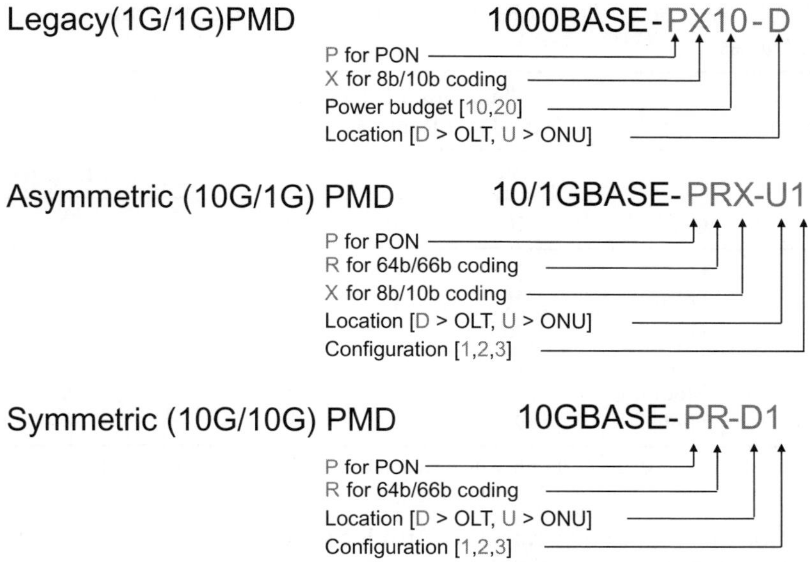

Figure 9.1. PMD naming conventions for (a) 1G-EPON, (b) 10/1G-EPON, and (c) 10/10G-EPONversions.

9.2.1 Physical-Medium-Dependent (PMD) Sublayer

PMD specifications are included in Clause 75 of IEEE 802.3av, representing the result of more than 2 years of technical discussions and a number of compromises reached between major parties involved in this process. A number of ad hoc activities were carried out, focusing on high-power budget and high split systems, link channel model, nonlinear effects in the fiber, and so on, the conclusions and recommendations of which are available on the official website of P802.3av TF (http://www.ieee802.org/3/av/).

9.2.1.1 PMD Naming Convention. PMD naming generated a long and very hot discussion, even though it seems like a minor issue compared with technical topics, which needed to be closed at the time. Symmetric, 10 Gb/s PMDs were quickly stabilized in the form presented in Figure 9.1c, while asymmetric PMDs (10 Gb/s downstream, 1 Gb/s upstream) proved to be more tricky and took time to reach the final version presented in Figure 9.1b. The legacy 1G-EPON PMD naming convention is presented in Figure 9.1a, for reference. Note also that there is no 10GBASE-PR-U2 PMD, since 10GBASE-PR-U1 is shared between PR10 and PR20 power budget classes.

9.2.1.2 Power Budgets. Power budgets in the 10G-EPON specification describe P2MP media supported by the given PMD, similarly to their definition in 1G-EPON. However, due to the existence of symmetric and asymmetric data rate PMDs, when referring to (for example) a low-power budget, it is not clear whether symmetric or asymmetric PMD is meant. For that purpose, a new

TABLE 9.1. Major Power Budget Parameters and Their Mapping into Power Budget Classes

Power Budget		Supported Effective Data Rate [Gb/s][a]					
		DS	US		ChIL (dB)		
Name	Class	10	10	1	20	24	29
PR10	Low	X	X		X		
PR20	Medium	X	X			X	
PR30	High	X	X				X
PRX10	Low	X		X	X		
PRX20	Medium	X		X		X	
PRX30	High	X		X			X

[a]Effective data rate at MAC level rather than channel data rate observed at the PHY.

designation of *power budget class* was introduced, which can be seen as a superset comprising both PR-type and PRX-type power budgets, characterized by the given ChIL. Therefore, a PRX-type power budget describes an asymmetric rate PHY, operating at 10 Gb/s downstream and 1 Gb/s upstream over a single single-mode fiber, while a PR-type power budget describes a symmetric rate PHY, operating at 10 Gb/s downstream and 10 Gb/s upstream, also over a single single-mode fiber. Furthermore, each power budget is identified with a number designating its *class*, where "10" represents low power budget, "20" represents a medium-power budget, and "30" represents high-power budget. Table 9.1 provides an overview of 10G-EPON power budgets and their association with power budget classes.

9.2.1.3 Implementation Choices for Individual Power Budget Classes. The following discussion presents the outcome of technical decisions, included in the standard in the form of PMD parameter tables in Clauses 75.4 and 75.5. Note that the standard provides specific numbers for individual parameters, while not describing the motivation behind them. Additionally, specific technical solutions supporting particular power budgets are also not indicated in the standard, leaving it up to implementers to decide what hardware solution to choose. It is worth noting that during the design process, special attention was paid to PMD sharing across at least two power budget classes to minimize the number of PMD elements to be developed.

Several PMDs share either transmit or receive path parameters; for example, PR(X)–D1 and PR(X)-D3 PMDs share a common transmitter design, and PR(X)–U1 and PR-U2 PMDs as well as PR-D2 and PR-D3 share a common receiver design. Additionally, in the case of symmetric 10/10G-EPON systems, PR-U1 (ONU) PMD is shared between two power budgets, namely PR10 and PR20, meaning that PR10 and PR20 ONUs are exactly the same. This means that the most cost-sensitive devices (ONUs) can be produced in larger quantities, leading to a faster decrease of their price.

Table 9.2 presents a typical ONU and OLT PMD design for all target power budget classes, with the indication of individual design choices as well as target transmit power and receive sensitivity levels.

9.2.1.4 Dual-Rate Burst-Mode Receivers—Architecture and Technical Challenges. The 10G-EPON OLT receiver must support burst-mode operation, resulting from the TDMA channel access mechanism used in 1G-EPON and 10G-EPON. In the case of a single data rate OLT receiver, supporting only one 1.25 Gbit/s or 10.3125 Gbit/s signal, the receiver can be optimized to handle the target upstream data rate and line code. However, in the case of a dual-rate device, supporting both 1.25 Gbit/s and 10.3125 Gbit/s upstream data rates, the OLT receiver becomes more complicated. It must perform gain adjustment as well as data rate sensing, without additional information from MPCP, at the expected data rate of incoming data bursts.

A dual-rate OLT receiver must thus be capable of receiving both 1G-EPON and 10G-EPON ONU burst transmissions. The single optical interface must receive optical signals in the 1260–1360 nm band, while rejecting anything else [the rejection function is typically achieved by optical filters in the receiver optical sub-assembly (ROSA)]. Two electrical interfaces carry the signals detected at 1.25 Gbit/s and 10.3125 Gbit/s. Therefore, from a topological point of view, the OLT receiver must split the incoming signal into two independent paths, which will then be fed through the stack and reach either the MAC or MAC client sublayers. The location of such a signal split is arbitrary and is not prescribed in the standard:

1. The signal can be split in the optical domain (via a regular 1:2 power splitter) and then fed into two independent photodetectors, as presented in Figure 9.2a, or
2. The signal can be detected with the use of a single (optimized) photodetector and then split in the electrical domain after the TIA, as shown in Figure 9.2b.

Option 1 has a much simpler electronic block, since both 1.25 Gbit/s and 10.3125 Gbit/s receivers can be optimized to provide maximum sensitivity for the respective signals. However, the extra 1:2 split used in the upstream channel introduces an extra loss of ≈3.5 dB, which deteriorates the signal power level at the receiver, degrading the BER and potentially breaking the power budget requirements. More stringent power budgets (e.g., PR20 or PR30) may become technically challenging to implement under such conditions, if at all possible.

Option 2, on the other hand, uses only a single optical module, though its photodetector and TIA unit must be dynamically adjusted to the data rate of the incoming signal, to prevent sensitivity deterioration resulting from incompatibility between signal data rate and receiver operating conditions. The electronic block must therefore switch rapidly between 1.25 Gbit/s and 10.3125 Gbit/s bursts. Such an operation is hard to achieve, because there is no available signal indicating what is the data rate of the next incoming burst, and thus data rate detection must be done on the fly.

TABLE 9.2. Target Implementation of 10G-EPON PMDs

PMD Location	PMD Type	Tx Technology	Rx Technology
OLT	PR30	10.3125-Gbit/s EML Launch power +2÷+5 dBm Mandatory FEC RS(255,223)	10.3125-Gbit/s APD –28 dBm for BER ≤ 10^{-3} Mandatory FEC RS(255,223)
	PRX30		1.25-Gbit/s PIN + FEC/APD[a] –29.78 dBm for BER ≤ 10^{-12} Optional FEC RS(255,239)
	PR20	10.3125-GBd EML + OA[b] Launch power +5÷+9 dBm Mandatory FEC RS(255,223)	10.3125-Gbit/s APD –28 dBm for BER ≤ 10^{-3} Mandatory FEC RS(255,223)
	PRX20		1.25-Gbit/s PIN + FEC/APD[c] –27 dBm for BER ≤ 10^{-12} No FEC requirement
	PR10	10.3125-Gbit/s EML Launch power +2÷+5 dBm Mandatory FEC RS(255,223)	10.3125-Gbit/s PIN –24 dBm for BER ≤ 10^{-3} Mandatory FEC RS(255,223)
	PRX10		1.25-Gbit/s PIN –24 dBm for BER ≤ 10^{-12} No FEC requirement
ONU	PR30	10.3125 DML/EML + OA Launch power +4÷+9 dBm Mandatory FEC RS(255,223)	10.3125-Gbit/s APD –28.5 dBm for BER ≤ 10^{-3} Mandatory FEC RS(255,223)
	PRX30	1.25-Gbit/s DML Launch power +0.6÷+5.6 dBm Optional FEC RS(255,239)	
	PR20	10.3125 DML Launch power –1÷+4 dBm Mandatory FEC RS(255,223)	10.3125-Gbit/s PIN –20.5 dBm for BER ≤ 10^{-3} Mandatory FEC RS(255,223)
	PRX20	1.25-Gbit/s DML Launch power –1÷+4 dBm No FEC requirement	
	PR10	10.3125 DML Launch power –1÷+4 dBm Mandatory FEC RS(255,223)	10.3125-Gbit/s PIN –20.5 dBm for BER ≤ 10^{-3} Mandatory FEC RS(255,223)
	PRX10	1.25-Gbit/s DML Launch power –1÷+4 dBm No FEC requirement	

[a]APD is typically used to avoid the use of FEC in upstream channel.
[b]Currently, such transmitters are under development and their commercial viability is yet to be proven.
[c]APD is typically used to avoid the use of FEC in upstream channel.

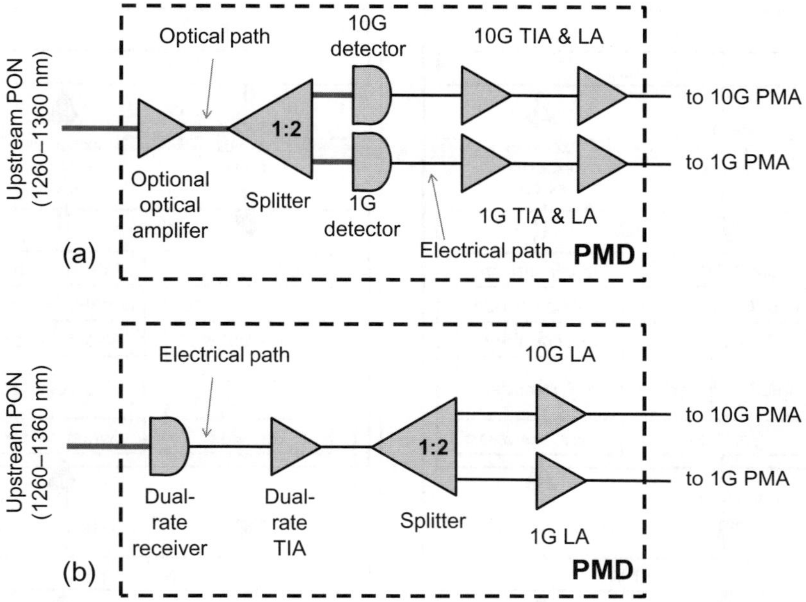

Figure 9.2. Implementations of the dual-rate PMD with signal split in **(a)** optical or **(b)** electrical domain.

9.2.2 Physical Coding Sublayer (PCS)

The PCS sublayer is responsible mainly for converting the data stream received from xMII into codewords, which can then be passed through PMA and PMD and finally transmitted into the medium. In the receive path, PCS performs the reverse function, that is, it decodes the received data and recreates the original data stream. PCS houses such critical functionalities as data decoder/encoder, FEC encoder/decoder, Data Detector (ONU transmit path only), scrambler/descrambler, and gearbox, adjusting the data rates between PCS (bursty transmission) and PMA (continuous data stream).

Figure 9.3 shows the functionalities included in the PCS sublayer for the downstream and upstream data paths. In the upstream data path, the ONU uses the data detector located below the FEC encoder to drive the laser on and off, as described in Section 9.2.2.2.

9.2.2.1 Idle Deletion. The Idle Deletion process in the 10G-EPON transmit path, implemented at the top of the PCS sublayer, is responsible for the removal of a number of excess IDLE characters, inserted by the MAC between subsequent frames. MACC enforces larger spacing between consecutive frames to prepare the data stream for insertion of the FEC parity at the PCS sublayer. At the output of the Idle Deletion function, the data stream is bursty and contains

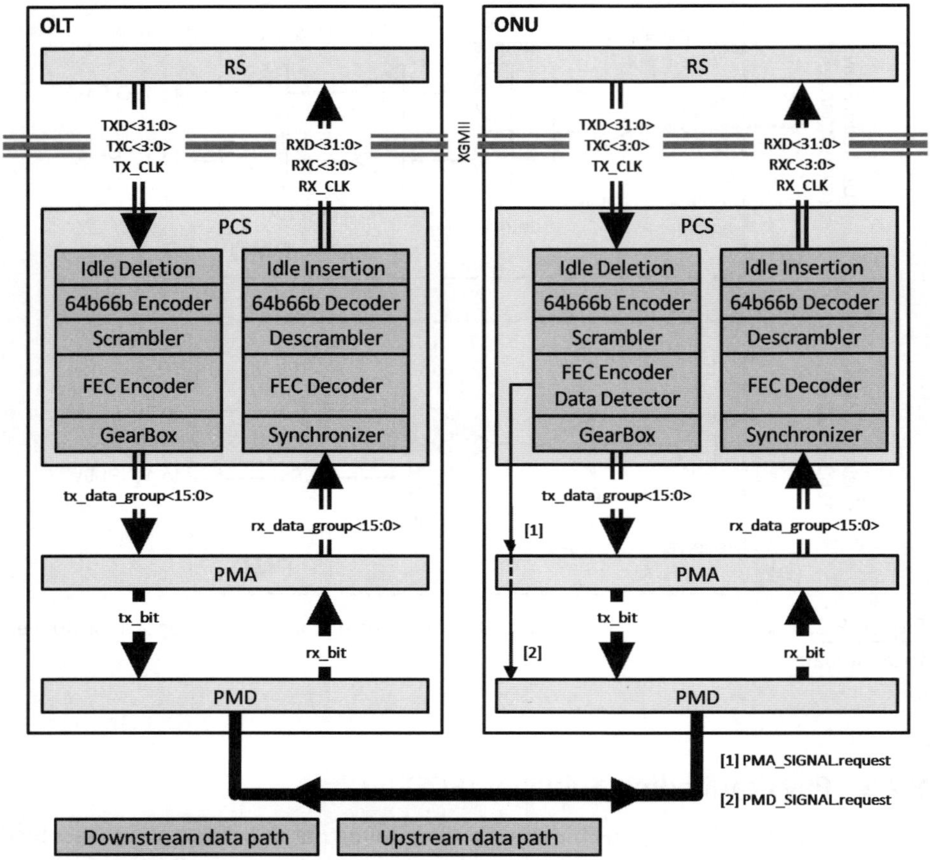

Figure 9.3. Functional block diagram of the 10/10G-EPON PCS: Downstream and upstream data paths.

gaps, which will be used by the FEC encoder to insert calculated parity without increasing the PMD data rate. Effectively, the Idle Deletion process will delete 4 IDLE vectors of 72 bits (64 bits of data and 8 bits of control) for every 31 vectors of 72 bits received from the XGMII. Apart from deleting excess IDLE characters from the data stream, the Idle Deletion function must also guarantee that the minimum IPG is preserved between two subsequent Ethernet frames.

Operation of the Idle Deletion process, as well as the existence of gaps in the data stream at the output of the Idle Deletion process, are illustrated in Figure 9.4.

Due to the operation of the DIC as described in Clause 46.3.1.4 in IEEE 802.3-2008, the IPG may vary in size. DIC sometimes adds or deletes IDLE characters between subsequent frames, to ensure alignment of the /Start/ control character with the lane 0 on the XGMII interface. This introduces another variable to the already complex task of the MACC entity, attempting to calculate

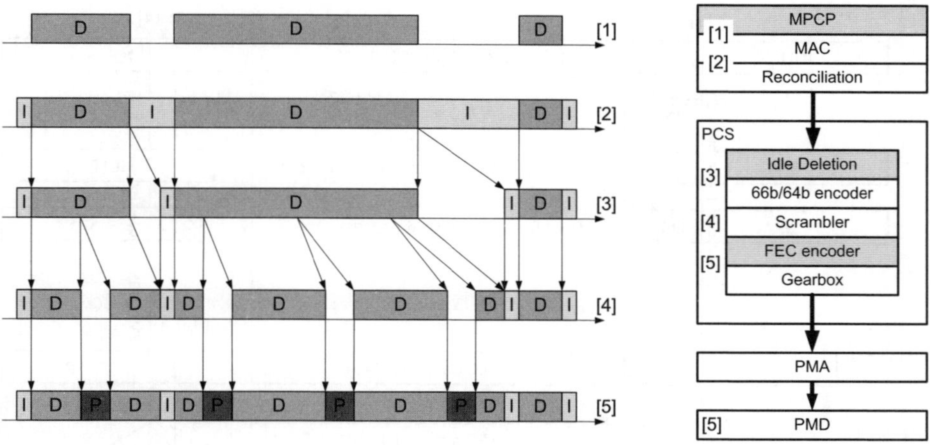

Figure 9.4. Transfer of data between the MACC and the output of the FEC encoder, when a data stream is ready for transmission. Special attention is paid to gaps in the data stream at the output of Idle Deletion process.

how much to delay the transmission of the next frame in order to allow the FEC encoder to insert calculated parity at the end of each codeword.

9.2.2.2 Data Detector. In the upstream channel, a number of ONUs contend for accessing a single OLT interface, requiring a TDMA mechanism in place to negotiate access of individual subscriber stations. In order to avoid the so-called *capture effect*, where spontaneous noise generated by one or more ONU located close to the OLT could mask data transmission from more distant ONUs and prevent them effectively from delivering subscriber data, ONUs must switch their lasers off completely between transmissions. For that purpose, the ONU PCS was equipped with a Data Detector mechanism, which detects the presence of transmitted data and generates a laser control signal (setting the PMA_SIGNAL.request(tx_enable) primitive to ON or OFF, as necessary).

The Data Detector is designed in the form of a FIFO buffer, operating as a delay line and storing a sequence of codewords to be transmitted next toward the PHY. The 10G-EPON Data Detector is composed of an input process loading data into the FIFO buffer and an output process retrieving data from the FIFO buffer. In this way, both processes can operate in an asynchronous manner, as long as the input process does not allow for the FIFO buffer to be emptied by the output process.

Figure 9.5 shows the relationship between the condition of the Data Detector delay line and the generation of the laser control signal. The depth (or length) of the Data Detector is chosen in such a way that the introduced delay is sufficient to switch the laser on and transmit the necessary data burst elements:

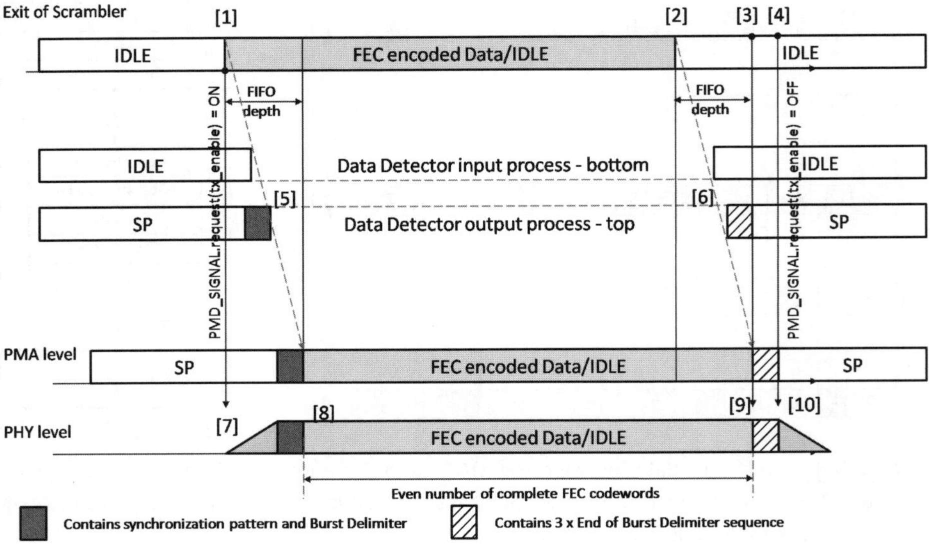

Figure 9.5. Operation of the ONU Data Detector and resulting shape of a data burst. The FIFO depth in vertical and horizontal directions is not up to scale.

- Synchronization pattern, which guarantees that the OLT burst-mode receiver has sufficient time to adjust its gain ($T_{\text{receiver\_settling}}$) and synchronize its receive clock (T_{CDR}).
- Burst Delimiter, a single 66-bit sequence that allows the OLT receiver to identify the beginning of the FEC protected data stream. Burst Delimiter is selected in such a way that the pattern can be reliably detected even in the presence of bit errors.
- A certain number (currently defined as 2) of IDLE characters, the first of which is used to synchronize the OLT data descrambler while the second one is used to provide the necessary initial IPG before the data stream can be fully synchronized.

Operation of the data detector is relatively simple. At the start, the Data Detector FIFO buffer is filled with IDLE characters and the laser is disabled. Effectively, no data are transmitted into the PHY layer.

Upon arrival of the first data (non-IDLE) character (see point [1] in Figure 9.5), the Data Detector enables the laser by setting the PMA_SIGNAL.request(tx_enable) primitive to ON (point [7]). The laser takes a certain time to switch on, a period during which the MAC is transmitting IDLE characters, which in the Data Detector are replaced with SP sequence. After a complete series of SP pattern is transmitted (number of transmitted 66-bit blocks equal to SyncLength), the Data Detector output process at point [5] substitutes the last IDLE character by the Burst Delimiter sequence, which indicates the start of the

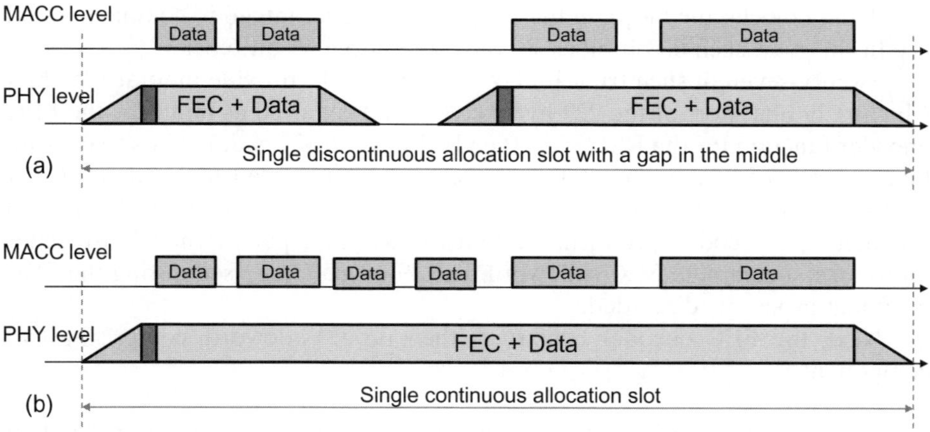

Figure 9.6. Two possible layouts of the upstream allocation slot: (**a**) with a long run of IDLE characters causing laser to switch off in the middle and (**b**) a continuous upstream slot.

FEC protected data stream—see point [8]. Data are delivered from the output of the scrambler until the last data character—see point [2]. Next, the FIFO buffer starts filling with IDLE characters, until it is completely full at point [3]. At that time, the Data Detector output process replaces three consecutive IDLE characters by the End of Burst delimiter pattern—see point [6]. The laser is still transmitting at this time. Only when the last of the End of Burst delimiter sequences is transmitted—see point [10]—the Data Detector sets the PMA_SIGNAL.request(tx_enable) primitive to OFF—see point [4]. This effectively starts switching the ONU laser off.

Longer sequences of IDLE characters can be received between data frames. However, if the FIFO buffer in the Data Detector is not emptied, the laser will not be switched off, though it is possible that, during a longer burst, the laser is switched off due to a very long run of IDLE characters. Such a situation (among others) is shown in Figure 9.6a. Note that the delay between the MACC and the PHY level, introduced by the Data Detector operation, was neglected to simplify the diagram.

9.2.2.3 FEC in 10G-EPON. All the 10.3125 Gbit/s links in the 10G-EPON architecture use stream-based FEC employing the Reed–Solomon code RS(255, 223). In 10/10G-EPON, this FEC code is used for both downstream and upstream links, while in 10/1G-EPON it is used only for the downstream link.

9.2.2.3.1 FEC Encoding Process. The 64b/66b encoder produces a stream of 66-bit data blocks, 27 of which are aggregated at the FEC encoder to allow for generation of a single FEC codeword. Prior to FEC encoding, each of the said 66-bit bits data blocks is preprocessed by removing the redundant first bit from the sync header (bit [0]). This process does not impact data integrity because, for all data blocks, the bit [0] in the sync header is guaranteed to complement bit [1]. In this way, each 66-bit block is converted into a 65-bit block.

Twenty-seven of such truncated 65-bit data blocks provide in total 1755 bits, which is 29 bits short of the 223 bytes (1784 bits) required as input for the FEC encoder function for the RS(255,223) code. Therefore, 29 padding bits (each with binary value of 0) are prepended to the said 27 truncated 65-bit data blocks, forming the complete 223-byte payload portion of a FEC codeword. These data are then FEC-encoded, producing 32-byte bytes of FEC parity, which is used later on to form a complete FEC codeword. The 29-bit padding used during the FEC encoding process is discarded.

Next, the FEC encoder constructs the output codeword, comprising two components:

- Original 27 data blocks, each containing 66 bits (including the redundant bit [0] in the sync header), which were used to calculate the FEC parity.
- FEC parity, where each of the four data blocks of 64 bits obtained from the FEC encoder is prepended with a 2-bit sync header, resulting in a properly formed 66-bit block resembling a regular output word produced by the 64b/66b encoder. The FEC parity is distinguished from regular data blocks through the use of a specific sequence of sync headers. P802.3av TF selected 00 11 11 00 for this purpose; that is, the resulting FEC parity sequence looks like this: [00 P1] [11 P2] [11 P3] [00 P4], where P1 … P4 are subsequent 64-bit FEC parity blocks.

After this process is complete, the FEC encoder outputs 31 data blocks of 66 bits toward the PMA, and then it aggregates another sequence of 27 data blocks of 66 bits from the output of the scrambler.

9.2.2.3.2 FEC Decoding Process. The 10G-EPON FEC decoder has the ability to correct or at least confirm the correctness of each of the 27 data blocks carried within an FEC codeword, based on the information carried within four FEC parity blocks of 66 bits each. The FEC code used in 10G-EPON is capable of correcting up to 16 errored symbols (a single symbol is 8 bits wide) per FEC codeword and detecting uncorrectable FEC codewords. Once this step is complete, the FEC decoder forwards the processed 66-bit data blocks to the descrambler and discards the parity blocks.

Additionally, the FEC decoder is also responsible for setting bit 0 of the sync header to the inverse of bit 1 of the sync header, thus making sure that the recovered bit stream is properly marked as data blocks.

9.2.2.3.3 Stream-Based FEC Versus Frame-Based FEC. 1G-EPON adopted optional frame-based FEC (used for all 10 Gb/s links), while 10G-EPON uses mandatory stream-based FEC. Both mechanisms are poised to provide extended protection against bit errors occurring during transmission in the optical channel. However, both are also quite different in many ways.

A stream-based FEC mechanism processes Ethernet frames and IDLEs as a stream of data symbols, resulting in a much simpler implementation, which is critical for high data-rate systems. This particular FEC-encoding method requires both transmitter and receiver communicating over a physical medium to use the very same framing structure. A device not supporting FEC encoding will not be able to retrieve data and separate it from parity. This means that all ONUs in an 10G-EPON must use FEC. In the stream-based method, the parity symbols generated after each data block are inserted immediately after the FEC parity codeword that they are protecting, resulting in an interleaving pattern of data blocks and parity blocks.

In the frame-based method, the parity symbols generated for each block are grouped together and are appended at the end of a frame. This leaves the data frame itself unaltered, representing a major advantage of this particular encoding method. Any device not supporting FEC encoding may still receive the data, though will not take advantage of the enhanced FEC bit protection. In 1G-EPON, adoption of this particular FEC coding method allows for mixing ONUs with enabled and disabled FEC on the same ODN.

9.2.3 Media-Independent Interface (GMII/XGMII)

An xMII (the first interface of this type was used in IEEE P802.3u Fast Ethernet and could operate at 100 MBd or 10 MBd) is a generic purpose interface connecting various types of PHYs to one and the same speed-agnostic MAC, through the RS sublayer. This means that a network device is capable of interacting with any type of underlying PHY over one and the same hardware interface, independently of the transmission medium this PHY is connected to. Effectively, the xMII interface shields the upper layers of the stack from having to interface with a plethora of different PHY types.

The success of the initial xMII included in P802.3u Fast Ethernet specifications led to the development of extended versions of this interface, capable of operating at gigabit data rates (GMII: 1 Gb/s MII) and then even 10 Gb/s data rates (XGMII: 10 Gb/s MII).

In terms of its physical structure, each xMII interface is composed of data lanes and control lanes. The number of data lanes and operating frequency predetermine the target data rate at which the xMII interface can transfer data:

- GMII (see Figure 9.7, as defined in Clause 35 of IEEE 802.3-2008) has a slightly different structure when compared with XGMII described below. It is composed of two 8-bit-wide data path, one 1-bit-wide clock path, and two 2-bit-wide control paths, the use of which depends on whether the transmit or the receive direction is considered. The transmit and receive data paths are unidirectional and independent, allowing for full duplex operation. In case of the transmit direction, TX_EN (transmit

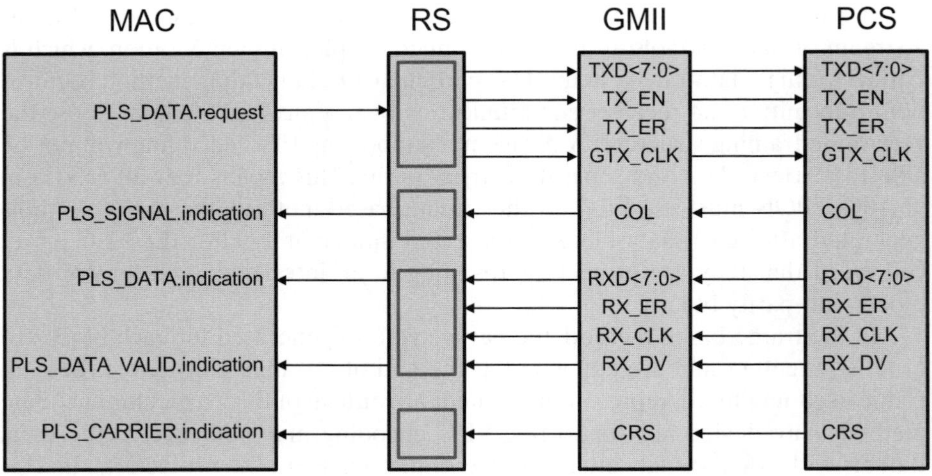

Figure 9.7. Internal structure of the GMII interface and interconnection between RS and PCS sublayers. MAC service primitives are also depicted for a complete picture of signal cross-relations.

enable) and TX_ER (transmit error) signals are delivered. In case of the receive direction, RX_DV (receive data valid) and RX_ER (receive error) signals are delivered. Additionally, a 2-bit-wide data path is provided in the receiver direction, namely CRS (carrier sense) and COL (collision detected).

- In the case of XGMII (as defined in Clause 46 of IEEE 802.3-2008), this interface is composed of two 32-bit-wide data paths (capable of carrying 4 bytes of data at the same time with the clock rate of 312.5 MHz, providing an effective throughput of 10 Gb/s), two 4-bit-wide control paths (used to indicate whether data or control character is carried on the 8-bit-wide data path), and two 1-bit-wide clock paths. All the transmit and receive data paths are unidirectional and independent, allowing for full duplex operation. No Carrier Sense signal is transmitted through XGMII, and it may be generated only in the RS, if needed, as shown in Figure 9.8.

The symmetric 10/10G-EPON devices use exclusively the XGMII interface between RS and PCS sublayers, as shown in Figure 9.9, while the asymmetric 10/1G-EPON devices must have both XGMII and GMII interfaces implemented, as presented in Figure 9.10. A 10/1G-EPON OLT will have the transmit path from XGMII and the receive path from the GMII interface implemented, while at the 10/1G-ONU the situation is reversed; that is, there is only receive path from XGMII and transmit path from GMII interfaces. Such a mixed xMII interface use is only one of the examples of inventive steps that were taken during the development of the 10G-EPON system.

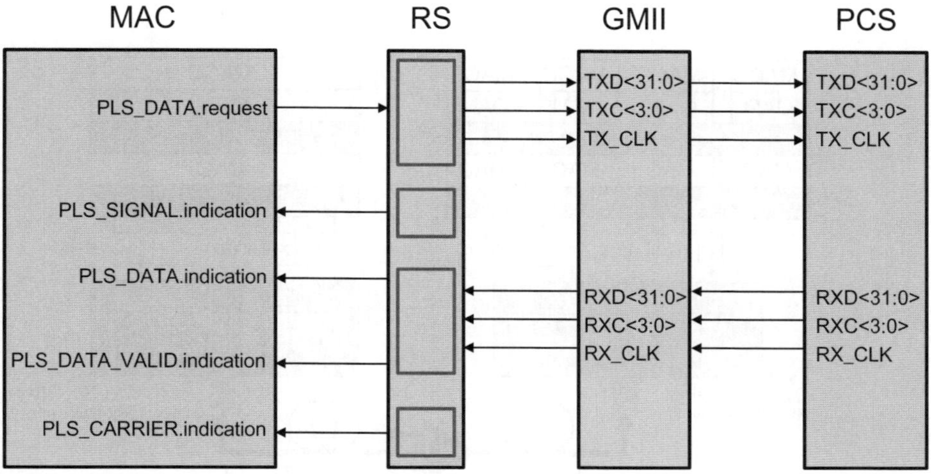

Figure 9.8. Internal structure of the XGMII interface and interconnection between RS and PCS sublayers. MAC service primitives are also depicted for a complete picture of signal cross-relations.

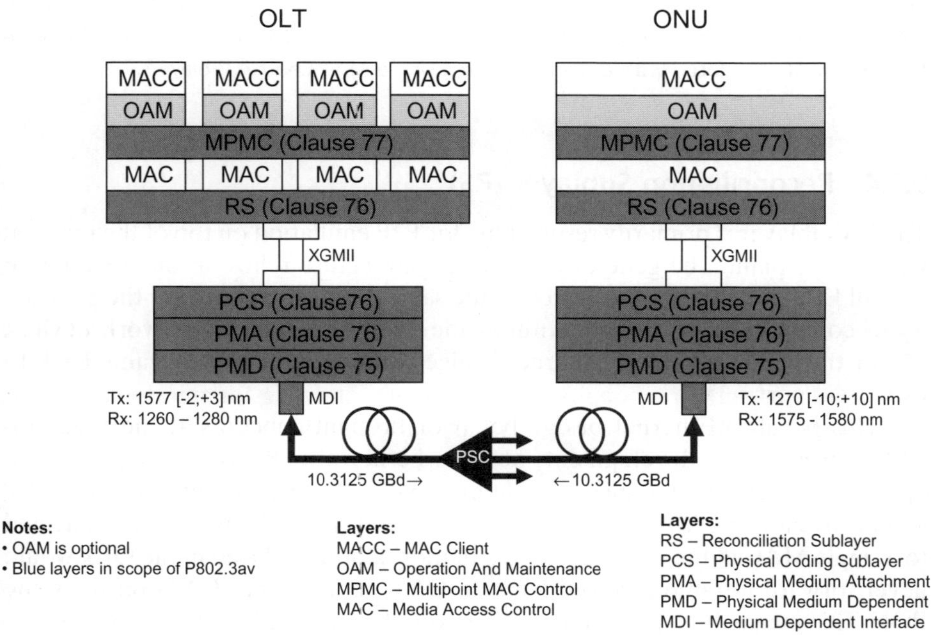

Figure 9.9. Symmetric, 10-Gbit/s-downstream and 10-Gbit/s-upstream EPON system architecture, with reference to specific clauses in IEEE 802.3av™-2009. *Note*: There can be more than one MAC interfaced with a single RS and a single MAC Control sublayer.

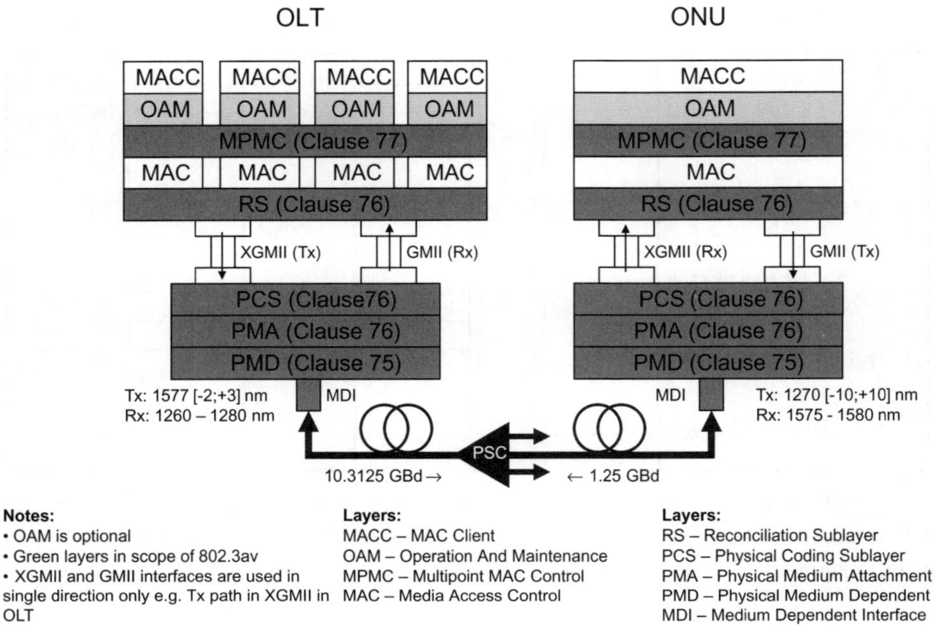

Figure 9.10. Asymmetric, 10-Gbit/s-downstream and 1-Gbit/s-upstream EPON system architecture, with reference to specific clauses in IEEE 802.3av™-2009. *Note*: There can be more than one MAC interfaced with a single RS and a single MAC Control sublayers.

9.2.4 Reconciliation Sublayer (RS)

The RS sublayer is primarily responsible for P2P emulation on top of the physical P2MP fiber plant. The general IEEE 802 architecture relies on the assumption that all Ethernet devices connected to the same physical media have the possibility of communicating directly, without the need of any extra network devices. Under this assumption, an Ethernet bridge will never forward a frame back to its ingress port.

This peculiar Ethernet bridge behavior brought concerns at the time 1G-EPON was under development, about whether the P2MP architecture can operate correctly under these conditions. Without a P2P emulation, the OLT would have a single MAC instance; thus effectively all ONUs would be connected to a single MAC at the OLT. In such a situation, a bridge placed in the OLT would never forward a frame received from an ONU to any other ONU on the same PON. This means that, physically, ONUs would require L3 connectivity in order to exchange data, which contradicts the requirement for L2 Ethernet connectivity between these devices.

Effectively, in order to overcome this problem, EPON systems require each downstream and upstream frame to be tagged with a network unique LLID, identifying the given target/source entity in an unambiguous manner. The number of LLIDs instantiated in a particular ONU has a significant impact on the system's performance and is one of the most vital design choices for a fully functional EPON system with inherent tri-play support. Typically, two solutions are considered, namely one LLID per ONU or one LLID per queue (multiple LLIDs per ONU). Considering an ONU as a functional rather than physical entity, both LLID assignment policies remain compliant with the IEEE 802.3-2008 standard (see Clause 64.1.1). In the case of the latter approach, a single physical ONU (in the form of customer premises equipment) may have a number of virtual (logical) ONUs instantiated, each with a single LLID assigned to it. In order to ensure high QoS, the multiple LLID per ONU approach allows for traffic prioritization and better bandwidth management (polling) via the MPCP control plane.

To keep in line with the standard definitions and simultaneously assure that it is possible to support multiple LLIDs per ONU, system integrators and specifically EPON chip vendors typically develop chipsets capable of instantiating several logical (functional) ONUs per single physical ONU device. In this way, from an architectural point of view, a single optical port is connected to several ONUs. The OLT in this case recognizes each logical ONU in the physical ONU as an independent entity with only a single LLID. The RS includes two major functions related with P2P emulation on the P2MP environment, namely LLID insertion and LLID extraction coupled with LLID based filtering.

9.2.4.1 LLID Structure and LLID Subranges in 10G-EPON. An LLID is composed of the mode bit (most significant bit) and logical_link_id partition, as defined in Clause 65.1.3.1 and reused in 10G-EPON. The mode bit was introduced to EPON architecture to guarantee compliance with shared LAN architecture, where a single station can communicate with any other station on the network segment. However, considering that access networks are completely different from corporate LANs in terms of data security and service models, very few of the existing EPON deployments actually utilize the mode bit. Its presence halves the range of available LLID addresses to 0x0000–0x7FFF, where 0x7FFF is reserved in 1G-EPON for broadcast transmissions and the range of 0x0000–0x7FFE can be used for unicast LLID assignment. In 10G-EPON, a new broadcast LLID was needed, and thus 0x7FFE was reserved for this purpose. Additionally, to ensure more future-proof definition of system specifications, a block of LLIDs was reserved for future use (0x7FFD–0x7F00 range). This leaves the range of 0x000–0x7EFF for unicast LLID assignment.

9.2.4.2 LLID Insertion Process. The LLID insertion process is used in the transmit path of the P2MP extended RS sublayer, where each frame transmitted by the MAC layer toward the PHY layer is processed by inserting a number of EPON-specific extension fields, namely SLD and LLID. The original CRC8 calculated by the MAC entity is also replaced in this process. The SLD byte is inserted on position 3, while the LLID is inserted on positions 6 and 7, in an 8-byte-long preamble. Once the insertion of new fields is complete, the CRC8 is recalculated to guarantee integrity of such an extended preamble structure.

Details of this process are described in Clause 76.2.6.1 of IEEE 802.3av™-2009 for 10G-EPON and in Clause 65.1.3.2 of IEEE 802.3-2008.

In terms of functionality, the LLID insertion process performs in exactly the same way in 1G-EPON and 10G-EPON, the only difference being the range of available LLIDs, as discussed in section 9.2.4.1. There is, however, a difference between the LLID insertion processes at the ONU and at the OLT.

In an ONU, a single MAC is connected via RS to the underlying PHY. This means that, in the 1 LLID per ONU environment, a frame leaving an ONU MAC can be tagged with one of two possible LLIDs: a broadcast LLID (either 0x7FFF, for 1.25 Gbit/s channel, or 0x7FFE, for 10.3125 Gbit/s channel) or a unicast LLID assigned to the ONU during the discovery process (following the registration handshake).

The situation is quite different on the OLT side, where a number of MAC instances (see Figure 9.9 or Figure 9.10) are connected via the RS sublayer to the underlying PHY. At any time, only one MAC is active and transmitting downstream. The LLID insertion process must therefore insert the LLID in the function of target ONU (unicast channel) assigned to it during the discovery and following registration processes, or insert a broadcast LLID (0x7FFE in 10G-EPON—assuming 10/1G-EPON or 10/10G-EPON OLT).

9.2.4.3 LLID Extraction Process and LLID-Based Routing. In the receive data path, a frame received with the LLID in the preamble is passed through the RS sublayer, where the LLID tag is parsed and the routing information is extracted, reconstituting the original Ethernet preamble format. The extracted LLID is compared with the broadcast LLID as well as with the local unicast LLID assigned to the given ONU during the discovery and following registration processes. If the comparison criteria defined for 10G-EPON in Clause 76.2.6.1.3.2 are met, a frame is passed to the proper MAC (*LLID-based routing*) entity. Otherwise, a frame is dropped. This functionality is commonly referred to as *LLID filtering*, and it allows for logical isolation of subscriber channels transmitted over the P2MP shared media of the ODN plant.

The LLID routing function in ONUs is relatively simple, since all frames passing through the LLID filtering function are then directed to a single MAC. In the OLT, however, LLID routing is responsible for directing the filtered frame to one of multiple MAC instances connected to the RS.

Since the LLID insertion and LLID extraction functions are coupled for each data link, LLID tags exist only in the data path between two RS sublayers, and the MAC entities on the transmitting and receiving sides are not even aware of their existence. In this way, the MAC entities on both sides of the link are operating in the standard P2P manner, while the P2P emulation for P2MP environment is located in the RS sublayer.

9.2.4.4 Operation of the RS Sublayer with XGMII and GMII in a 10/1G-EPON. As indicated before, in the case of a 10/1G-EPON device (both ONU and OLT) the RS sublayer is directly connected to two xMII interfaces types, namely XGMII and GMII. Such architecture is novel in IEEE PMD layers, and generated long discussions related with its technical feasibility as well as with the description to be included in the draft. Operation of the so-called dual rate Media-Independent Interface (drMII, acronym used only in this text) is described in Clause 76.2.2.

The 10/1G-EPON has to support different data rates in the transmit and receive paths, due to its inherent data rate asymmetry. In such a case, a combination of XGMII and GMII data paths is used for transmission and reception in a full duplex manner, while only specific halves of individual xMII interfaces are enabled at any time. This means that, at the 10/1G-EPON OLT, the transmit path in XGMII and the receive path in GMII interfaces are enabled. The situation is reversed in the 10/1G-EPON ONU, where the transmit path in GMII and the receive path in XGMII interfaces are enabled. For practical reasons, implementations are expected to include full GMII/XGMII interfaces (if implemented at all, i.e., some integrated chip designs do not need such structured interfaces at all), where unnecessary data paths are disabled. The mapping between XGMII/GMII service primitives and the PLS_DATA.request and PLS_DATA.indication (service primitives of the RS sublayer) is described in Clause 76.2.2.4. Figure 9.10 depicts the 10/1G-EPON architecture with drMII interface.

9.2.5 Media Access Control (MAC)

The Ethernet MAC specification describes a medium-independent entity responsible for a number of data delivery functions, including among others:

a. Data encapsulation (transmit path) and de-encapsulation (receive path), which has further a number of specific functions, that is:
   i. Delimitation of frame boundaries, by adding framing information to the payload information provided by the upper layer MAC client entities, as well as frame synchronization.
   ii. General-purpose address handling, by insertion of source and destination addresses in the transmit path as well as their paring in the receive path; this function is responsible for directing received frames to the proper MAC clients based on the target address, thereby providing selective frame routing.
   iii. Error detection, based on the Frame Check Sequence field attached to the end of the assembled frame. The transmitting MAC attaches a CRC32 Frame Check Sequence to the end of the assembled frame, and the receiving MAC utilizes this sequence to guarantee data integrity and lack of bit errors.
b. Media Access Management, responsible for controlling media access and guaranteeing that a frame leaving the MAC service interface will be transmitted through PHY with a minimum delay. There are two very specific functions in this group, namely:

i. Medium allocation, which is responsible for controlling the time when the frame can be transmitted, by observing the current state of the underlying medium (collision avoidance).

ii. Contention resolution, which is responsible for handling data collision situations, when retransmission of the previous frame is necessary.

9.2.6 Multipoint MAC Control (MPMC)

The MAC Control (MACC) sublayer provides real-time control and manipulation of the MAC sublayer, allowing to customize operation and behavior of this PHY agnostic sublayer. Examples of MACC clients include the Bridge Relay Entity, LLC, or other applications characteristic of the particular IEEE network device. In the case of EPON, MACC entities include, for example, Discovery client, DBA client, and so on, and as such it is out of scope of the IEEE 802.3 standard to prescribe their exact behavior. However, MPMC provides a generic framework in the form of MPCP, providing a generalized mechanism for operation of MPMC clients. MPCP remains largely unchanged in 10G-EPON, as compared with 1G-EPON, and controls the discovery and registration processes as well as scheduling of the upstream bandwidth.

9.2.7 Extended Discovery Process in 10G-EPON

The Discovery process allows all previously inactive, deregistered, or powered-off ONUs to register in the EPON system, effectively gaining access to the shared upstream medium. This process is driven by the OLT Discovery agent, which periodically opens a discovery window in the upstream channel, during which no registered ONUs are allowed to transmit. During the same window, all unregistered ONUs are given the opportunity to announce their presence to the OLT by sending REGISTER_REQ MPCPDUs. The frequency of such discovery windows is not defined by the standard, and it depends only on implementation.

Due to the potential coexistence of the 1G-EPON, 10/1G-EPON, and 10/10G-EPON ONUs on the same ODN, the P802.3av TF found it necessary to extend the Discovery process defined currently in IEEE 802.3-2008, Clause 64. The principal requirement was to enable proper operation in the multi-rate environment, where a single OLT can support all three types of ONUs, with dual rate burst-mode transmission in the upstream channel. To ensure proper operation of the DBA clients located in the OLT and ONUs (out of scope of the standard), and utilizing a common time unit of TQ (equal to 16 ns), it is necessary to assure the existence of the following functionalities, which were added in IEEE 802.3av™-2009 to the Discovery Process previously used in 1G-EPON:

- Identification of the upstream/downstream channel data rate for a given target ONU. The information on the data rate used by the given ONU in the US channel is crucial, since the OLT DBA client must know in advance at what data rate the given ONU will be transmitting, to allocate the proper size of the transmission slot. Identification of the upstream channel data rate is based on the LLID address carried in REGISTER_REQ MPCPDU.

- Identification of the upstream channel ONU capabilities, determining whether the given ONU is 1G, 10G, or dual-rate-capable. Such information is required for proper registration of the particular types of ONUs, especially in the case of dual-rate-capable devices, which may choose to register in either 1G or 10G Discovery Windows opened by the OLT.
- Proper operation over two independent DS data channels (1G and 10G, separated using WDM) as well as over the TDMA shared, dual rate, upstream channel.

9.2.7.1.1 Initiation of Discovery Process at the OLT and ONU. The Discovery process is initiated by the OLT discovery agent, resulting in the transmission of a discovery GATE MPCPDU, which carries such information as the starting time and length of the discovery window, together with the Discovery Information flag field, as defined in Clause 77.3.6.1. Individual flags contained in the Discovery Information field are used to notify all ONUs about the upstream and downstream channel transmission capabilities of the given OLT. This flag field is defined in such a way that the OLT can potentially support more than one data rate in each transmission direction, if such devices were deemed economically justified.

Upon receiving a broadcast Discovery GATE MPCPDU, ONUs parse it and retrieve information carried within. To ensure proper RTT measurement and time slot alignment, each ONU resets its local clock on reception of a downstream time-stamped MPCPDU. The Discovery GATE MPCPDU is an example of such a time-stamped MAC Control frame.

Next, an ONU with unregistered LLID(s) will wait for the start of the Discovery Window and then transmit upstream a REGISTER_REQ MPCPDU. Any other ONUs with unregistered LLID(s) will perform likewise, which means that during the Discovery Window multiple ONUs can access the PON medium simultaneously, potentially resulting in transmission overlap between data bursts from individual ONUs. The EPON system lowers the probability of burst overlap by operating a contention algorithm at all the ONUs, where each ONU waits a random amount of time (typically shorter than the length of the Discovery Window itself) before transmitting the REGISTER_REQ MPCPDU. The length of such an additional delay time is randomly selected for each ONU, resulting in a Random Delay Mechanism. In this way, if the Random Delay Mechanism is successful, the OLT can receive multiple valid REGISTER_REQ MPCPDUs during a single Discovery Window.

Each REGISTER_REQ MPCPDU delivers two pieces of vital information about the source ONU, namely its MAC address and the depth of the grant queue, which in turns defines how many grants can be assigned to the given ONU in advance. Additionally, it also carries the Discovery Information field, characterizing transmission capabilities of the given ONU in the upstream and downstream channels, as specified in Clause 77.3.6.3. Moreover, in order to optimize upstream channel utilization and minimize the size of guard bands between data bursts from individual ONUs, the REGISTER_REQ MPCPDU also carries laser on/off parameters, providing the OLT with information about the quality of ONU hardware.

9.2.7.1.2 Initial ONU Registration at the OLT. After successful reception of a REGISTER_REQ MPCPDU, the OLT has sufficient information to start the registration process. A new LLID is created at the OLT and associated with the MAC address of the registering ONU. As a follow-up, the OLT transmits downstream a unicast REGISTER MPCPDU to the newly discovered ONU (MAC unicast channel, with broadcast LLID since the ONU does not have an associated LLID at this moment). This message carries the newly assigned LLID as well as information on the synchronization time required by the OLT. For confirmation purposes, the OLT echoes the maximum number of pending grants (though the purpose of this echo is not defined in the standard). The OLT also transmits the target laser on/off parameter values, which are to be used by the ONU during the following operation. It is assumed that the parameter values transmitted by the OLT may be different than what the ONU indicated in the REGISTER_REQ MPCPDU, though they must not be smaller than the ONU advertised values, which would prevent the ONU from proper operation.

9.2.7.1.3 ONU Confirmation Scheduling. Once the REGISTER MPCPDU is transmitted and the LLID association is created at the OLT side, the OLT has sufficient information to allow the given ONU to access the PON medium. The DBA Client operating at the OLT side selects the upstream channel transmission window and schedules it by sending downstream a (LLID) unicast message to the ONU in the form of a GATE MPCPDU.

The transmission slot carried in this GATE MPCPDU will allow the registering ONU to transmit upstream a REGISTER_ACK MPCPDU and thus complete successfully the registration process. After this stage, the ONU is considered as completely activated and bidirectional traffic flow may commence.

9.2.7.1.4 Repeated ONU Registration. Under certain circumstances (e.g., excessive BER, problems with bidirectional connectivity, timeout, signal loss, etc.), an ONU must go through repeated discovery and registration processes, trying to remedy the existing connectivity problems. Additionally, there may be also situations where an ONU needs to inform the OLT of its desire to deregister, due to (for example) a request from higher management layers.

In the first case, the OLT may request the ONU to deregister by sending a REGISTER MPCPDU to this particular ONU with the Deregister flag enabled. In the latter case, an ONU may deregister by sending upstream a REGISTER_REQ MPCPDU with the Deregister flag set, indicating that the OLT should release the LLID association for the given MAC address and allow

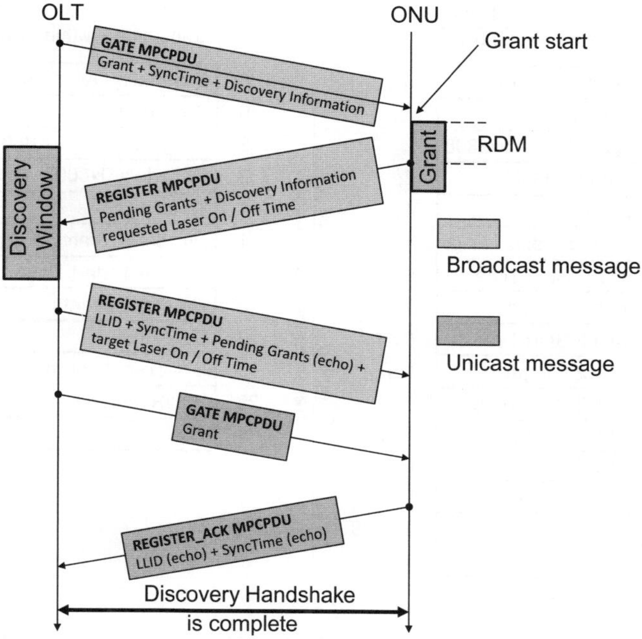

Figure 9.11. Exchange of MPCPDUs during the Discovery Handshake process.

the given ONU to go through the discovery and registration processes once more, as presented in Figure 9.11.

9.2.7.2 Changes to MPCPDUs. In order to support dual-rate operation, as well as optimize the use of the upstream channel by allowing laser on/off time negotiation between ONU and OLT, several changes to the MPCPDUs were introduced in IEEE 802.3av™-2009.

9.2.7.2.1 GATE MPCPDU. The regular granting GATE MPCPDU was not changed and maintains its internal structure, as defined in Clause 64 for 1G-EPON (see also Clause 77.3.6.1 in IEEE 802.3av™-2009). The Discovery GATE MPCPDU was, however, subject to some changes (see Figure 9.12).

The most visible change in the GATE MPCPDU structure is the addition of the Discovery Information field, which contains information on the OLT transmission capabilities for both downstream and upstream channels. This field was deliberately defined with 16 bits, to allow for future extensions of this message and the scope of information carried in this field, without the need to redefine the message itself. The internal structure of the Discovery Information field is presented in Table 9.3.

Additionally, there are minor changes in the definition of the "Grant #n Length" field, which is still expressed in TQ units but must account for all necessary

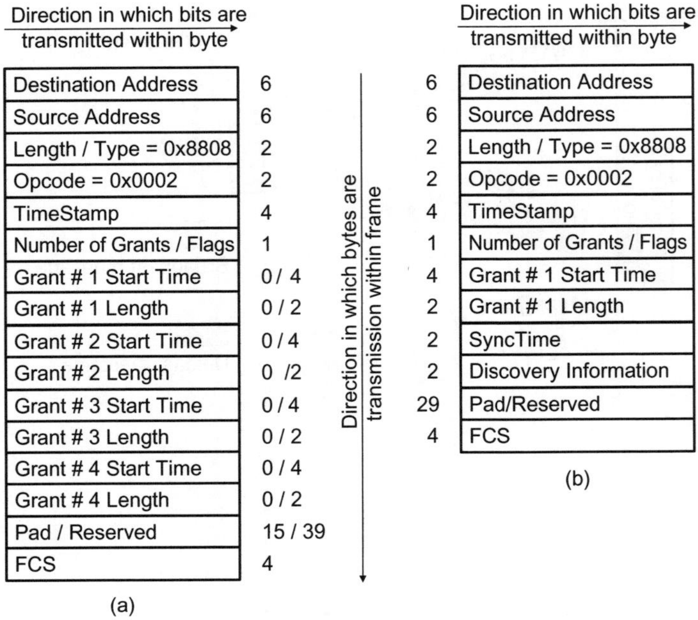

Figure 9.12. Internal structure of the GATE MPCPDU in **(a)** granting and **(b)** Discovery versions.

TABLE 9.3. Internal Structure of the Discovery Information Field in the GATE MPCPDU

Bit	Field Name	Description / Values
0	1.25 Gbit/s upstream OLT capability	0—No 1—Yes
1	10.3125 Gbit/s upstream OLT capability	0—No 1—Yes
2–3	Reserved/undefined	Ignored upon reception
4	OLT opens 1.25 Gbit/s Discovery Window	0—No 1—Yes
5	OLT opens 10.3125 Gbit/s Discovery Window	0—No 1—Yes
6–15	Reserved/undefined	Ignored upon reception

elements of transmission overhead, namely laserOnTime, syncTime, laserOffTime, BURST_DELIMITER, two initial IDLE blocks, FEC parity overhead, and burst terminator sequence (composed of three END_BURST_DELIMITER blocks), which consume part of the allocated bandwidth slot.

9.2.7.2.2 REGISTER_REQ MPCPDU. REGISTER_REQ MPCPDU suffered changes in its internal structure (see Figure 9.13), due to the addition of a

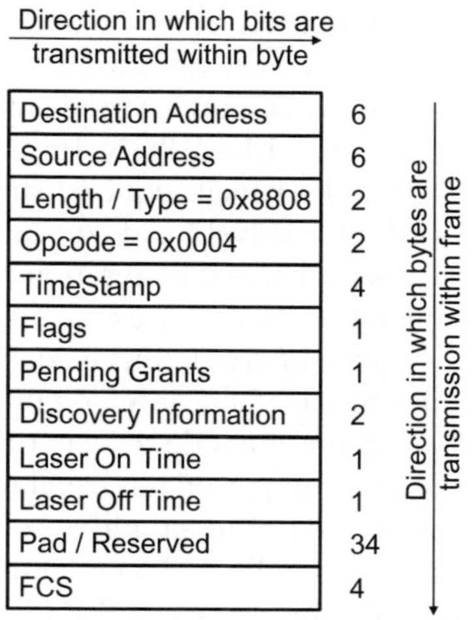

Figure 9.13. Internal structure of the modified REGISTER_REQ MPCPDU.

TABLE 9.4. Internal Structure of the Discovery Information Field in the REGISTER_REQ MPCPDU

Bit	Field Name	Description / Values
0	1.25 Gbit/s upstream ONU capability	0—No 1—Yes
1	10.3125 Gbit/s upstream ONU capability	0—No 1—Yes
2–3	Reserved/undefined	Ignored upon reception
4	ONU attempts registration in 1.25 Gbit/s Discovery Window	0—No 1—Yes
5	ONU attempts registration in 10.3125 Gbit/s Discovery Window	0—No 1—Yes
6–15	Reserved/undefined	Ignored upon reception

number of new data fields, namely Discovery Information, Laser On Time, and Laser Off Time.

The Discovery Information field (see its internal structure in Table 9.4) is a counterpart of the field with the same name added to the Discovery GATE MPCPDU. It has a similar function; that is, it indicates the transceiver data rate capability in the transmit and receive paths, though this time for the ONU rather than for the OLT. The current definition of this field ensures its future extensibility to higher data rates, as well as transmission of any other necessary information related with the ONU downstream/upstream capabilities, which are currently not covered in the standard.

The addition of the Laser On/Off Time fields to the REGISTER_REQ MPCPDU was dictated by the necessary optimization of the upstream channel utilization, where the large 512 ns laser on/off period length defined in 1G-EPON standard was deemed excessive for 10G-EPON. Initial deployments relied on laser drivers designed for P2P links, which required such long on/off times. The introduction of dedicated BM laser drivers and constant improvements in their design brought the on/off times for current generation of 1.25 Gbit/s lasers down to the level of several dozen nanoseconds. It is expected that 10.3125 Gbit/s lasers will not exhibit worse performance figures in this field. Furthermore, to eliminate once and for all a static allocation of the laser on/off times, the P802.3av TF decided to allow for negotiated guard band size, where an ONU would indicate the minimum value of the laser on/off period it is capable to support and the OLT would adjust the value (upwards only) to simplify the DBA operation and use the same laser on/off times for all ONUs, independently from the manufacturer. Note that the laser on/off period value was capped at 512 ns, which was considered as the maximum necessary value, even with very low quality laser drivers.

The Laser On Time/Laser Off Time field has the form of an 8-bit-wide field, where the value of the laser on/off time is expressed in TQ units. This allows for simple coverage of the complete 0- to 512 ns range (with 16 ns increments) in a single 8-bit value.

9.2.7.2.3 REGISTER MPCPDU. The REGISTER MPCPDU has changes complementary (see Figure 9.14) to those of the REGISTER_REQ MPCPDU, representing feedback received from the OLT in response to the registration request transmitted during the Discovery Window. As such, the REGISTER MPCPDU (as compared to Clause 64 for 1G-EPON) was added only two new

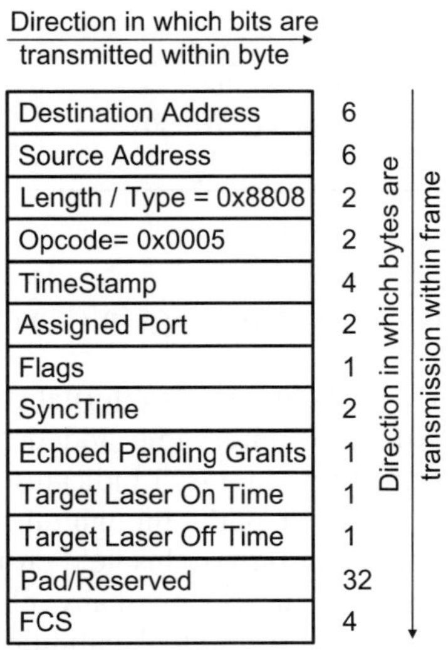

Figure 9.14. Internal structure of the modified REGISTER MPCPDU.

data fields, namely Laser On Time and Laser Off Time. Definition and internal structure of these two fields is identical to those included in the REGISTER_REQ MPCPDU, and thus further elaboration is not deemed necessary.

However, it is worth noting a slightly different interpretation of these two fields when compared with the REGISTER_REQ MPCPDU laser on/off time parameters. In case of the REGISTER_REQ MPCPDU, the ONU would indicate what minimum laser on/off period length can be supported by the existing transceiver. Therefore, this defines the minimum length of the interburst gap. The OLT, once the laser on/off period length characteristic for the given ONU is received during the registration process, may decide to increase these values to any value higher than the value indicated by the ONU, while taking into consideration that neither of these two parameters can exceed 512 ns. Next, the value of laser on/off time selected by the OLT is transmitted downstream, for the ONU to comply with. The ONU adjusts the depth of its Data Detector in the PCS to accommodate the changes requested by the OLT. Note also that, physically, ONU lasers may switch on/off faster, though the size of the guard band is still defined by the depth of the Data Detector in the PCS.

9.2.7.2.4 OTHER MPCPDUs. Other MPCPDUs, namely REPORT and REGISTER_ACK, were not modified in IEEE 802.3av™-2009, apart from minor editorial changes targeting clarification of the already existing specifications as well as elimination of any doubts, which the P802.3av TF had during the balloting stage.

9.2.7.3 Impact of Stream FEC on Operation of the MPCP Sublayer. 10G-EPON operates with mandatory FEC for all 10.3125 Gbit/s links. This means that part of the usable bandwidth is occupied by parity code, which increases the robustness of the data channel to bit errors, providing the means for achieving the target BER = 10^{-12} which would be impossible otherwise, considering the loss in receiver sensitivity due to the increase of the transmission data rate.

For this reason, transmission in 10G-EPON is substantially different from that in 1G-EPON, where FEC was frame-based rather than stream-based. On one hand, in 1G-EPON a station with no FEC support could still delineate data and try to receive it without using FEC gain. This is not possible in 10G-EPON, since data and FEC parity are interleaved. On the other hand, parity insertion at the PCS requires additional space between the frames to maintain a constant data rate at the PHY interface. This means that the MAC data rate and the PCS data rate must meet exactly the ratio of 223/255, corresponding to the FEC RS code designation; that is, 255 bytes of data are transmitted at PCS layer, containing only 223 bytes of subscriber data and 32 bytes of FEC parity. Effectively, data transmitted from the MACC toward the PHY, through the MAC, must have a larger IPG between individual frames, sufficient in size to accommodate the FEC parity later on (see Figure 9.4).

This means that the MACC must ensure that a sufficient IPG is left between consecutive frames, to allow the Idle Deletion process in the upper PCS to remove the extra IDLE characters and separate the given frame (including necessary IDLE characters) into FEC codewords, each of which is then extended with FEC parity. This new functionality required substantial changes in the definition of the FEC_Overhead and CheckGrantSize functions, which have to estimate the quantity of extra IDLE characters to be inserted after the end of the given frame. To fully appreciate the complexity of the problem, consider that the MACC does not know in what condition the PCS-based FEC encoder is and where the given FEC codeword boundary is located. It does not know exactly how many IDLE characters will be deleted in PCS (see Figure 9.4).

If too many IDLEs are inserted, part of the useful bandwidth will be wasted. On the other hand, if too few IDLEs are inserted, data will be overwritten in the FEC encoder, resulting in data loss. The P802.3av TF took therefore a number of meetings to arrive at the solution, in which the MACC can track (very) precisely the location and quantity of FEC parity which is inserted at the PCS sublayer. That means that the quantity of inserted IDLEs can be calculated precisely, optimizing the bandwidth utilization especially in the upstream channel.

9.3 COEXISTENCE OF 1G-EPON AND 10G-EPON

The gradual evolution toward 10G-EPON systems requires replacement of the minimum amount of active equipment up-front, leaving the underlying fiber infrastructure intact. In this way, Service Providers can have a rare opportunity of maximizing the ROI for the systems they already heavily invested in when deploying 1G-EPON. Simultaneously, introduction of next-generation equipment into the network structure allows for delivering more bandwidth demanding services to (premium) customers willing to pay a slightly higher connection cost per port, representing early adopters of higher capacity 10G-EPONs. Thus, it comes as no surprise that the issues related to the coexistence with the legacy equipment on the same PON plant have been considered critical from the very beginning of the project, warranting the investigation of the wavelength allocation schemes, dual rate operation, and the necessary changes to the MPCP framework resulting from coexistence of various data rate devices in the same infrastructure.

9.3.1 Downstream Channel

Due to the requirement of complete backward compatibility, in 10G-EPON the downstream 10.3125 Gbit/s and 1.25 Gbit/s channels will be WDM multiplexed, thus creating in effect two independent P2MP domains. The guard band between data channels ought to be sufficiently large to allow for their uninterrupted operation under any temperature conditions provided for in the technical specifications of the emerging hardware. The 1.25 Gbit/s downstream link will therefore remain centered at 1490 [±10] nm (in accordance with the IEEE 802.3-2008 standard, Clause 60), while the new 10.3125 Gbit/s downstream link is centered around 1577

[+3, −2] nm, creating a significantly narrower band, limited at the lower end by the RF video overlay and by the OTDR filter cutoff band at the upper end (see Figure 9.15). After long debates, the P802.3av TF agreed on aligning all power classes in terms of wavelength allocation, allowing optical subassembly vendors to develop a single ONU optical filter design. A comparison of 10G-EPON and 1G-EPON wavelength allocation plans is depicted in Figure 9.15.

9.3.2 Upstream Channel

The upstream channel in any PON system is always considered technically critical, mainly because of the need to balance technical complexity and cost of the resulting ONU hardware. In the case of 10G-EPON, WDM multiplexing for 10.3125 Gbit/s and 1.25 Gbit/s channels in the upstream is discouraged, mainly because of lack of available wavelength bands. The minimum dispersion window (around 1310 nm) is already in use by 1G-EPON, thus leaving apparently no space for introduction of a new, 10.3125 Gbit/s channel, as shown in Figure 9.15.

Accepting the fact that existing 1G-EPON specifications must not be modified in any way, causing potentially some of the deployed equipment to become standard incompliant, only dual rate burst-mode multiplexing remains as a viable option. Therefore, both 10.3125 Gbit/s and 1.25 Gbit/s upstream transmissions will partially overlap in frequency domain (though remain separated in time domain via TDMA), with the 1G-EPON ONUs remaining centered at 1310 [±50] nm, while the 10G-EPON ONUs will transmit at 1270 [±10] nm, taking advantage of the fact that only narrower-band DFB LDs must be used at 10.3125 rates [9],

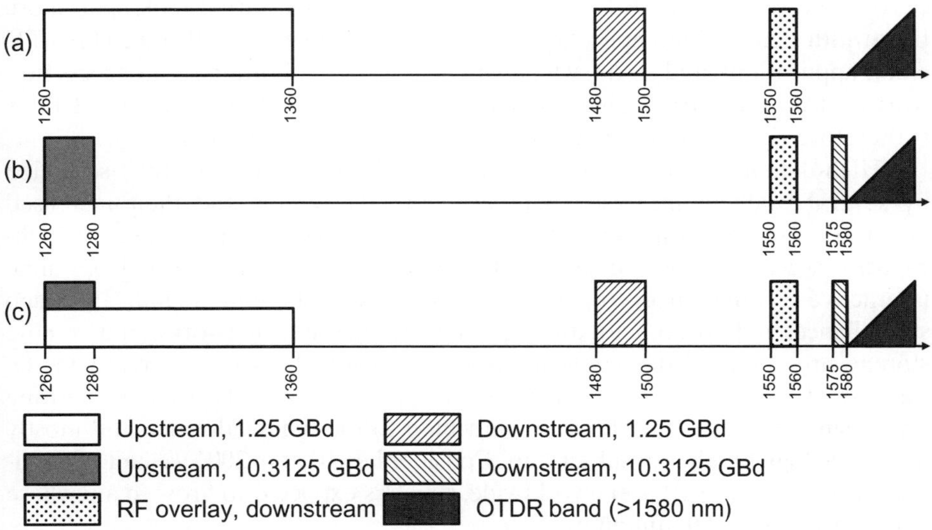

Figure 9.15. Wavelength allocation plan for (a) 1G-EPON, (b) 10G-EPON, and (c) combined, 1G-EPON and 10G-EPON. All options account for the RF video overlay and OTDR service bands.

which are inherently more narrowband. In this way, the upstream transmission will become not only bursty but also dual-rate, representing a new level of technical complexity for the OLT receiver.

In such a system configuration, the OLT receiver will have to be equipped with a set of new functionalities. The AGC required for burst-mode reception, currently considered as a state-of-the-art technical achievement, will be surpassed by a dual-rate burst-mode device. Such a component will have not only to ensure proper power level adjustment, but also identify the incoming data rate and perform receiver adjustment in order to maximize its sensitivity for each particular burst.

However, developing an OLT dual rate burst-mode receiver and implementing it in accordance with the IEEE specifications may prove to be a nontrivial task, requiring significant research to be conducted by electronics and receiver manufacturers. Initial prototypes of such dual-rate burst-mode receivers were already presented publicly, though their commercial availability remains at this time undefined.

9.4 TARGET APPLICATIONS FOR 10G-EPON SYSTEMS

Given the virtually constant increase in bandwidth requirements from end-subscribers as well as changes in commonly utilized networked services, 1G-EPON networks will soon (some estimate that in the next 3–5 years) become inadequate to support the next generation of multimedia rich digital contents. When it was first introduced, 1G-EPON represented a substantial step in the evolution of access networking, creating a platform for delivery of bandwidth intensive applications like IPTV or VoIP. Now, a few years later, users started looking at more bandwidth demanding applications, like HD-IPTV, Picture in Picture (PiP) TV, cloud applications, and so on, which will again cause a bottleneck in access networks if transition to a higher-capacity platform is not carried out. 10G-EPON is therefore right on time to meet growing customer expectations.

High-definition, video-centric, multimedia-rich services are on the rise, fueled by several years of increased popularity of HDTV sets and their increased penetration in typical households. Other rapidly growing applications for high-capacity access networks include video-conferencing, interactive video, online interactive gaming, peer-to-peer networking, karaoke-on-demand, IP video surveillance, and so on. Cloud computing, storage area networks, and remote storage are some of the applications that evolved thanks to increase in the capacity of the access loop. Even such applications as VoIP, typically generating rather small data streams, tend to contribute to the bandwidth demand mostly due to the high number of subscribers. Consider here that in 2006 we had approximately 35 million VoIP users worldwide, which is expected to grow to anywhere between 120 and 150 million by 2011.

10G-EPON was also designed with other two target applications in mind. The MDU market is the first of them, focusing on residential areas with high population density, where a large fraction of home/apartment owners also subscribe to digital services. Such markets exist mainly in certain regions of Europe, as well as in Asia. MDU development is not very popular in the United States, and thus this application does not fit the needs of the American market. The 10G-EPON is also naturally suited for deployment in such areas as hospitals, schools, and business campuses, as well as governmental and educational institutions, where a large number of wired/wireless users generate a substantial quantity of data traffic which then needs to be delivered to aggregation networks. Currently existing solutions based on DSL access are simply not future-proof, given the constant increase in the number of connected computers, PDAs, and other equipment with data interfaces.

The last target application considered at the time of conception of the 10G-EPON SG was mobile backhauling, which has received recently substantial attention due to the ongoing transition to 3G and 4G mobile networks. Base stations implementing these new standards provide subscribers with substantially more bandwidth than 2G devices, which again puts more stress on the data uplink to the nearest aggregation point. Existing ATM solutions are already limiting data rates provided for 2G devices, not to mention newer base stations or even wireless access points operating under IEEE 802.11b/g/n or 802.16. Such access points must be connected to high-capacity backhaul links, typically with symmetric transmission capacity. The 10G-EPON fits perfectly this application, and the first 10G-EPON ONUs integrated into Base Station devices are expected by the end of 2009.

Last but not least, the 10G-EPON can find its applications anywhere the 1G-EPON was once so successful, providing higher transport capacity and lower cost per subscriber and, ultimately, providing subscribers with more choices for their digital entertainment. Initial deployment plans for 10G-EPON systems indeed target overbuilt existing 1G-EPON networks, by migrating at least a fraction of premium subscribers to newer equipment. Recent announcements of first demonstration versions of 10G-EPON equipment [10, 11] proves also continued support for this technology among equipment vendors.

9.5 CONCLUSIONS

It is expected that 10G-EPON equipment will follow the path of 10-fold capacity increase at three times the port price, so characteristic of all Ethernet equipment: Some of the technical challenges the new system will be faced with include (a) backward compatibility with legacy EPONs (including RF video overlay) and (b) support for asymmetric data rates (10 Gb/s downstream and 1 Gb/s upstream). PHY layer challenges include, among others, (a) dispersion penalties and decreased receiver sensitivity due to the 10-fold increase of the data rate, (b) nonlinear effects in the fiber plant (ODN) due to high launch powers in 29-dB ChIL compatible PMDs, and (c) inherent jitter and clocking problems due to dual rate operation.

Dual-rate MAC stack and dual-rate burst-mode reception represents the next level of technical complexity, but were resolved by Q1 2010 with successful demonstration of commercial dual-rate OLTs. The economic feasibility of dual-rate EPONs is currently questionable from a practical standpoint, though overbuild (brown-field) scenarios can potentially benefit from such a solution, providing extended ROI on relatively new 1G-EPON equipment.

The development process of 10G-EPONs will keep on driving state-of -the-art engineering in the area of burst-mode receivers, high-power laser sources, and ultrasensitive high-data-rate photodetectors. Chip integration as well as protocol implementation will also present several challenges yet to be surmounted, mainly in the form of a reliable Discovery Process, data rate negotiation, and so on.

It is also anticipated that the rapid stabilization of 10G-EPON PMD parameters may benefit cooperation between FSAN/ITU-T and IEEE PON groups, resulting in convergence of 10G-EPON and Next-Generation PON (NG-PON) systems for some of power budgets, at least at the PHY level. This will allow hardware manufacturers to achieve higher production volumes and cut equipment costs, making both PON systems far more economically attractive than when considered separately.

ACKNOWLEDGMENTS

The authors would like to thank Glen Kramer for a careful review and many insightful comments.

REFERENCES

1. G. Kramer, *Ethernet Passive Optical Networks*, 1st edition, Communications Engineering Series, McGraw-Hill Professional, New York, 2005.
2. G. Kramer and G. Pesavento, EPON: Challenges in building a next generation access network, in *1st International Workshop on Community Networks and FTTH/P/x*, Dallas, 2003, pp. 66–73.
3. A. Kasim, P. Adhikari, N. Chen, N. Finn, N. Ghani, M. Hajduczenia, P. Havala, G. Heron, M. Howard, L. Martini, R. Metcalfe, M. O'Connor, M. Squire, W. Szeto, and G. White, *Delivering Carrier Ethernet: Extending Ethernet Beyond the LAN*, 1 edition, McGraw-Hill Osborne Media, New York, 2007.
4. G. Kramer, What is next for Ethernet PON?, in *5th International Conference on Optical Internet (COIN 2006)*, Jeju, South Korea, 2006.
5. S. Swanson, Ethernet standards evolve to meet high-bandwidth networking needs, *Lightwave*, Vol. 12, 2006. Available at http://www.lightwaveonline.com/about-us/lightwave-issue-archives/issue/ethernet-standards-evolve-to-meet-high-bandwidth-networking-needs-53434312.html.
6. H. Steenman, End User Perspective on Higher Speed Ethernet, AMS-IX, online report, available at: http://www.ieee802.org/3/hssg/public/sep06/steenman_01_0906.pdf, 2006.

7. Teknovus Ltd., Teknovus and Fiberxon Cooperate on "Turbo" EPON, Teknovus Press Release, available online at http://teknovus.com/Page.cfm?PageID=140&CategoryID=14, 2007.
8. IEEE 802.3, Call For Interest: 10 Gbps PHY for EPON, online report, available at http://www.ieee802.org/3/cfi/0306_1/cfi_0306_1.pdf, 2006.
9. IEEE 802.3av TF, "Baseline Proposals," electronic report, available at http://www.ieee802.org/3/av/public/baseline.html, 2007.
10. fibresystems.org, ZTE unveils world's first next-generation PON equipment, fibresystems.org, online article, available for download at http://fibresystems.org/cws/article/newsfeed/36940, 03.12.2008.
11. Lightwave, Teknovus to demo 10G EPON at FOE, Lightwave, online article, available for download at: http://lw.pennnet.com/Articles/Article_Display.cfm?Section=ARTCL&SubSection=Display&PUBLICATION_ID=13&ARTICLE_ID=350716&pc=ENL, 19.01.2009.

10

BROADBAND POWER-LINE COMMUNICATIONS

Lars Torsten Berger

10.1 INTRODUCTION

The idea of using power lines also for communication purposes was already around at the beginning of the last century [1, 2]. It is now broadly referred to as *power line communications* (PLC). The obvious advantage is the widespread availability of electrical infrastructure, so that theoretically deployment costs are confined to connecting modems to the existing electrical grid.

Today, applications include the provisioning of Internet to end customers, referred to as *Access* PLC, or *broadband over power line* (BPL) [3, 4]. Besides, PLC technology is successfully being used for the distribution of audio, video, voice, and data services within the users' homes, also referred to as *In-Home* PLC. Furthermore, utility companies are becoming more and more interested in *automatic meter reading infrastructure* (AMI) and *smart grid*, allowing a more efficient electrical network management [5–7].

Early PLC systems made use of narrow bandwidths on high-voltage lines. The operating frequency range went up to a couple of hundred kilohertz, and data rates were on the order of hundreds of bits per second (bit/s) [1, 2]. Up to

Convergence of Mobile and Stationary Next-Generation Networks, edited by Krzysztof Iniewski

the present day, narrowband power-line systems are in operation mainly for control services that require data rates below 2 Mbit/s. Popular narrowband PLC systems are, for example, *X10*, *KNX*, *INSTEON*, *BACnet*, and *LonWorks*. However, they are not at the focus of this contribution. More on past and present narrowband PLC systems may be found in references 8 and 9.

In line with the advances of digital communications in general, also PLC systems were able to enormously boost their data rates. In the last decade, PLC chips by semiconductor vendors, such as Intellon [10], and DS2 [11], came to market that operate in the band from around 1 to 30 MHz. The chips are mainly based on two consortia-backed specifications developed within the frameworks of the *HomePlug Powerline Alliance* (HomePlug) [12] and the *Universal Powerline Association* (UPA) [13]. The HomePlug specification comes in two main releases, *HomePlug 1.0* and its evolution *HomePlug AV*, with *physical layer* (PHY) peak data rates of 14 Mbit/s, and 200 Mbit/s, respectively [14–20]. Both releases target the In-Home market. An Access specification called *HomePlug Access BPL* is currently under development. The rivaling UPA specification was selected by the European Union IST research project OPERA as baseline technology [21] and provides an Access as well as an In-Home solution. The In-Home solution, called *Digital Home Standard* (DHS), has a peak PHY data rate of 240 Mbit/s, while the Access solution provides at best 205 Mbit/s [22–25]. More recently, a third specification named *High-Definition Power Line Communications* (HD-PLC), a trademark of Panasonic [26] that is promoted within the *HD-PLC Alliance*, was released. It is designed to distribute multimedia content In-Home and has a theoretical PHY peak data rate of 210 Mbit/s [27].

Departing from all these specifications, continuous research and development efforts have led to "next-generation" prototypes, enabling data rates in excess of 400 Mbit/s [11, 28]. Besides advances in digital signal processing, these systems also owe their throughput boosts to an increased spectrum usage, ranging from 1 MHz up to around 200 MHz. With the ability to fulfill additionally high *Quality of Service* (QoS) and coverage expectations, it becomes clear that PLC can, in some cases, be not only a complement, but even a superior alternative to state-of-the-art wire line as well as wireless systems like xDSL, Wi-Fi, WiMAX, UMTS/HSPA and CDMA2000 EV-DV [29].

Nevertheless, to make PLC systems an even broader success, an internationally adopted PLC standard is essential. The International Telecommunications Union (ITU), as well as the Institute of Electrical and Electronics Engineers (IEEE) commenced work on such next-generation standards, namely ITU-T G.hn and IEEE P1901. ITU-T G.hn focuses on home networking and smart grid applications and has recently been identified by the U.S. National Institute of Standards and Technology (NIST) as an important standard for smart grid interoperability [30]. ITU-T G.hn is applicable not only to power lines but also to phone lines and coaxial cables, thus for the first time defining a single standard for all major wire line communications media. At the end of 2008, the PHY layer and the overall architecture were consented in ITU-T Recommendation G.9960 [31]. In the same year, the HomeGrid Forum was founded to promote the ITU-T G.hn standard and to address certification and interoperability issues [32]. In parallel the IEEE P1901 is working on the "Draft Standard for Broadband over Power Line Networks: Medium Access Control and Physical Layer Specifications." It will cover the aspects Access and In-Home, as well as coexistence of Access-In-Home and In-Home-In-Home networks [33]. However, to get sufficient industry support, the IEEE P1901 standard might have included two incompatible PHY and *medium access control* (MAC) substandards. They are based on HD-PLC and HomePlug AV. This inherent fragmentation of IEEE P1901 makes some analysts believe that ITU-T G.hn will emerge as the dominant next-generation solution [34, 35].

All PLC-systems have to tackle the problem of PLC-generated interference, also referred to as *electromagnetic compatibility* (EMC). The problem has been addressed by limiting the used *power spectral density* (PSD), as well as adaptively notching selected frequencies that are in use, for example, by Amateur Radio or television broadcasting services. However, achievable PLC data rates are primarily related to the available *signal-to-noise ratio* (SNR) over a certain frequency range. With a cap on the PSD, the remaining outer factors in the struggle for higher PHY data rates are the attenuation imposed by the power-line channel, as well as noise at the receiver side.

Issues of EMC, channel, and noise characteristics, as well as the expected mean SNR, are dealt with in Sections 10.3 to Section 10.6, respectively. In the sequel, Section 10.7 provides a PHY- and MAC-centered overview of the current specifications HomePlug AV, UPA (Access and DHS), and the prospective international ITU-T G.hn standard. First, however, different PLC deployment scenarios are introduced in Section 10.2.

10.2 POWER-LINE SCENARIOS

Power-line communications can make use of high-voltage, medium-voltage, and low-voltage grids as shown in Figure 10.1.

High-voltage (HV) lines, with voltages in the range from 110 kV to 380 kV, are used for nationwide power distribution and consist of long overhead lines with little or no branches. Theoretically, these lines could be used for communication purposes. However, their potential for broadband power-line services is limited. High-voltage arcing noise is a problem, and signal attenuation in decibels increases approximately linearly with distance. Furthermore, fiber-optic backhaul networks are frequently running alongside HV lines, providing a more attractive communication alternative [4].

Medium-voltage (MV) lines, with voltages in the range from 10 kV to 30 kV, are connected to the HV lines via primary transformer substations. The MV lines are used for power distribution between cities, towns, and

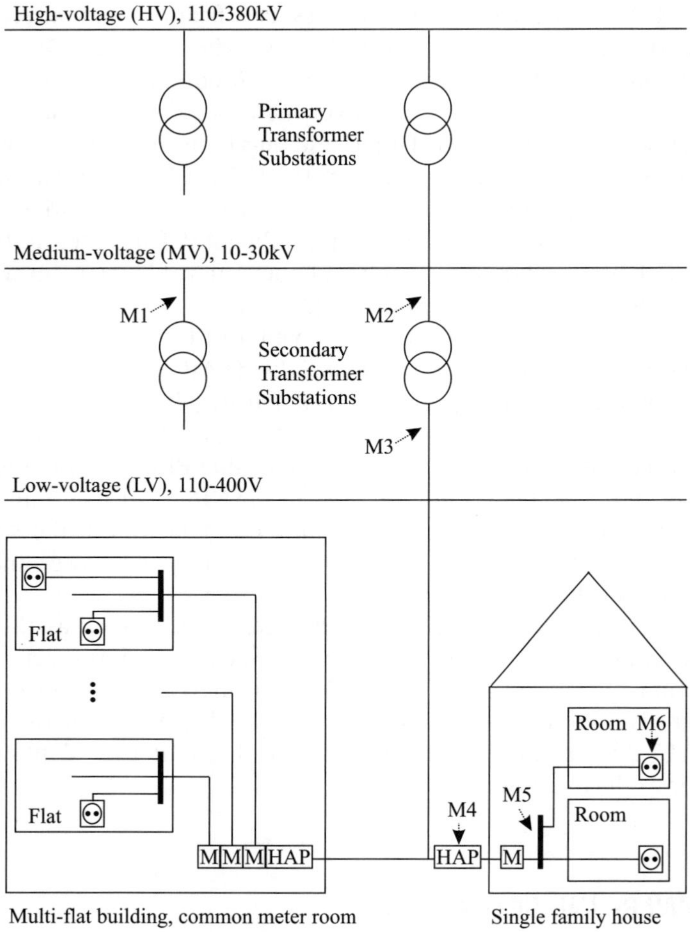

Figure 10.1. Power-line deployment scenarios.

larger industrial customers. They can be realized as overhead or underground lines. Furthermore, they exhibit a low level of branches. From a communications point of view, their potential to serve as backhaul for Access networks, especially in rural areas, is much higher than that of HV lines.

Low-voltage (LV) lines, with voltages in the range from 110V to 400V, are connected to the MV lines via secondary transformer substations. A communication signal on an MV line can pass through the secondary transformer onto the LV line, but with a heavy attenuation on the order of 55–75 dB [36]. Hence, a special coupling device or a PLC repeater are required if one wants to establish a communications path. As indicated in Figure 10.1, the LV lines lead directly or over street cabinets to the end customers' premises. LV lines are therefore at the heart of the power-line Access network. Note that considerable regional topology difference exits. For example, in the United States, one smaller secondary transformer on a utility pole might service a single house or a small number of houses. In Europe, however, it is more common that up to 100 households get served from a single secondary transformer substation. Furthermore, as pointed out in reference 37, significant differences exist between building types. They may be categorized as *multi-flat buildings with riser*, *multi-flat buildings with common meter room*, *single-family houses*, and *high-rise buildings*. Their different electrical wiring topologies influence signal attenuation as well as interference between neighboring PLC networks [38]. In most cases the electrical grid enters the customers' premises over a *house access point* (HAP) followed by an *electric meter* (M) and a distribution board (fuse box). From the distribution board the LV lines run up to the different power sockets in every room.

Besides the depicted *Access* and *In-Home* scenarios, there are cases of PLC deployments within vehicles such as cars, trucks, ships, airplanes, or even space crafts. However, such *In-Vehicle* PLC is not the focus of this chapter. Instead, the interested reader may refer to references 39–41 and references therein.

10.3 ELECTROMAGNETIC COMPATIBILITY REGULATIONS

Power-line cables were not designed to carry communication signals. In fact, in most cases, power cables are unshielded and far less homogeneous than, for example, twisted-pair telephone wiring. Hence, the deployment of PLC equipment gives rise to not only conducted emission, but also to radiated emission that does not stay confined to the power grid and can therefore interfere with radio receivers (such as Amateur Radio) and television broadcast receivers.

The main source of radiated emission is the *common mode* current [38]. Considering a two-port with a phase and a neutral conductor, the common mode is defined as the current flowing in both conductors in the same direction. In this case, the return path is closed over an undefined earth plane. The phase-aligned currents in both conductors generate two in-phase electric fields. This can lead to a considerable amount of electromagnetic interference. To avoid interference, PLC modem manufacturers aim at injecting the signal as symmetrically as possible. This way, two 180° out-of-phase electric fields are generated that neutralize each other, resulting in little radiated emission. This desired symmetrical way of propagation is also known as *differential mode*. Nevertheless, even if the PLC transmitter would be able to inject the signals in a fully symmetric manner, inhomogeneities and asymmetries, especially in In-Home wiring, always lead to differential-to-common-mode conversion and in the sequel to unintended radiated emission.

In the *Comité International Spécial des Perturbations Radioélectriques* (CISPR), founded in 1934 and now part of the *International Electrotechnical Commission* (IEC), efforts are currently been made to regulate PLC-generated interference. Two main topics of ongoing discussion are (i) how electromagnetic interference from PLC equipment is to be measured and (ii) how much electromagnetic interference can be tolerated by other equipments. It must be said that international EMC standardization is a slow process. It becomes even more difficult by the fact that the electrical grid topologies and power-line deployment scenarios are very diverse. Furthermore, the wireless spectrum usage varies from country to country. Hence, PLC-generated interference affects existing services differently. The consequence is that the CISPR/I/PT PLT working group, in charge of PLC standardization, has not yet been able to agree on a standardization proposal [42, 43]. It could, however, be that an amendment to the existing CISPR 22 standard [44] will in the future regulate PLC emissions. In the current testing procedure an *equipment under test* (EUT) is with its mains port connected to a testing device called an *artificial mains network* (AMN). Then the voltages between phase ground and neutral ground are measured and compared against emission limits. Furthermore, a communications EUT is connected with its communications port to a testing device called an *impedance stabilization network* (ISN). Then common mode currents are measured and compared to specified emission limits. In the past there was a clear distinction between a device's *mains port* and its *communications port*. However, for PLC equipment, both ports fall together. If the strict CISPR 22 mains port regulations would be used to limit the PLC injected signal power, commercially viable PLC deployments would hardly be possible. Therefore, a special PLC amendment to CISPR 22, currently under discussion, could include:

- An emission measurement procedure at the mains-communications port while no communication takes place.
- A second emission measurement procedure at the mains-communications port when normal PLC communication takes place.
- A general cap on the injected PSD, for example, of –55 dBm/Hz.
- A procedure for adaptive notching, meaning that the PLC equipment senses the presence of radio services and notches the affected frequencies for its own operation.
- A procedure of adaptive power management, meaning that the transmitting equipment limits its transmit power as a function of channel attenuation and noise to a level below the allowed maximum, which is just sufficient to achieve the required data rate.

Once an amended CISPR 22 standard is in place, there is a good chance that it will become part of European Union legislation. Responsible here is the *European Committee for Electrotechnical Standardization* (CENELEC), which through liaison groups maintains a close collaboration with CISPR.

In the United States the *Federal Communications Commission* (FCC) is in charge of regulating electromagnetic emissions. In general, all digital equipment has to comply with the FCC part 15 standard (47 CFR §15) [45]. Specifically, Access PLC systems over medium- and low-voltage power lines and for a frequency range from 1.705 to 80 MHz are treated in the standard's Section G. Conducted emission limits are explicitly not applicable, but radiated emission limits are imposed through a transmit power spectral density mask. Additionally, PLC systems have to be able to notch certain frequencies that might be used by other services. Furthermore, the FCC defines excluded bands where no PLC signal shall be injected, as well as geographical exclusion zones close to which no Access PLC systems may be deployed. Furthermore, procedures in which service providers inform about prospective PLC Access deployments as well as complaint handling procedures are a requirement.

Looking at the developments in CISPR 22, as well as at FCC part 15, it becomes clear that next-generation PLC equipment has to be highly configurable to apply a power spectral density shaping mask, as well as adaptive notching.

10.4 CHANNEL CHARACTERISTICS

The PLC channel exhibits frequency-selective multipath fading and a low-pass behavior. Furthermore, cyclic short-term variations and abrupt long-term variations can be observed. Below we look at these different channel characteristics in more detail.

To understand the effects that lead to frequency selective fading consider, for example, the open stub-line schematic in Figure 10.2a adapted from references 46 and 47. An impedance-matched transmitter is placed at A. B marks the point of a branch, also called an electrical T-junction. An impedance-matched receiver is placed at C. Assume for now that a 70-Ω parallel load is connected at D. l_x and Z_x represent the line lengths and characteristic impedances. More specifically, lines 1 to 3 are characterized by (20 Ω, 10 m), (50 Ω, 20 m), and (20 Ω, 30 m), respectively. At any impedance discontinuity (e.g., from impedance Z_a to Z_b), an injected signal undergoes reflection and transmission, described by the power reflection coefficient [48]

$$r_{ab} = \frac{Z_b - Z_a}{Z_b + Z_a} \tag{10.1}$$

and the power transmission coefficient

$$t_{ab} = 1 + r_{ab}. \tag{10.2}$$

Specifically, for the situation in Figure 10.2, r_{1B} is, for example, given by

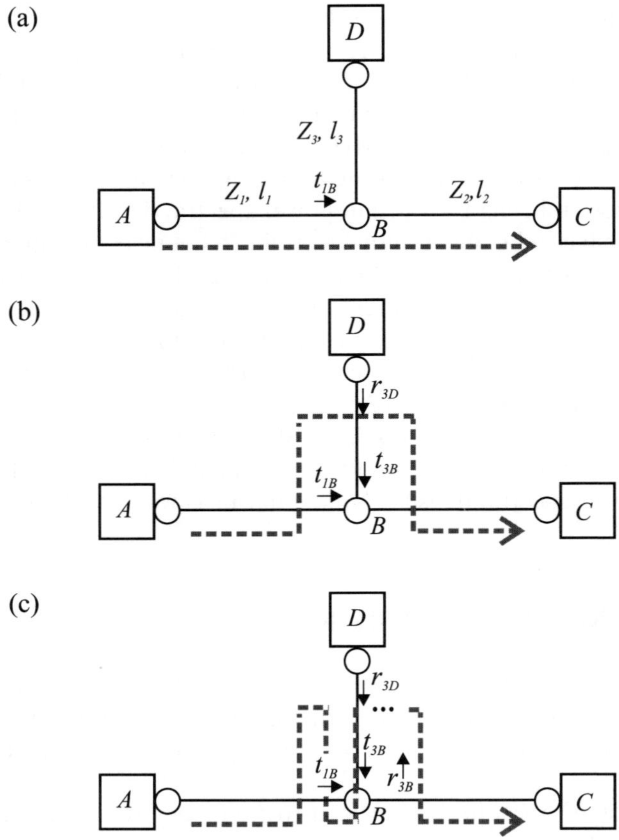

Figure 10.2. Multipath propagation in stub line [46]. Copyright 2009 Academy Publisher.

$$r_{1B} = \frac{(Z_2 \parallel Z_3) - Z_1}{(Z_2 \parallel Z_3) + Z_1} \tag{10.3}$$

where $(Z_2 \| Z_3)$ represents the impedance of Z_2 and Z_3 when connected in parallel. The other coefficients can be derived by inspection in a similar manner. Transmissions and reflections lead to a situation where a PLC signal travels in the form of a direct wave from A over B to C as displayed in Figure 10.2. Another PLC signal travels from A over B to D, bounces back to B, and reaches C, as depicted in Figure 10.2. All further signals travel from A to B, and undergo multiple bounces between B and D before they finally reach C, as can be seen in Figure 10.2. The result is a classical multipath situation, where frequency selective fading is caused by in-phase and anti-phase combinations of the arriving signal components. In reference 47 it is shown how the stub-line example from Figure 10.2 can be represented by an infinite impulse response filter. Its frequency transfer function is plotted in Figure 10.3.

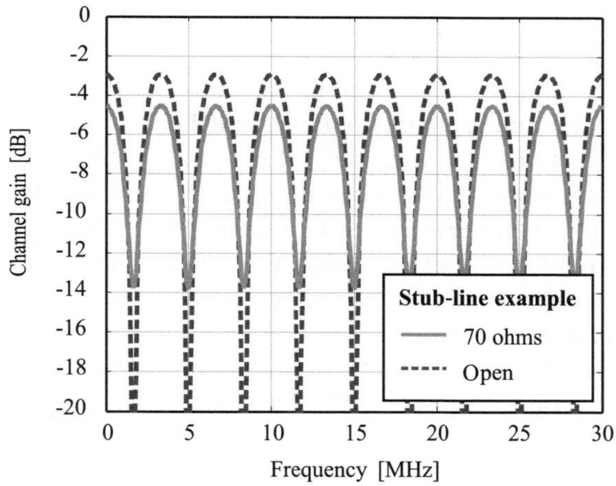

Figure 10.3. Frequency transfer function of stub line example. 70 ohms, or open termination at point D.

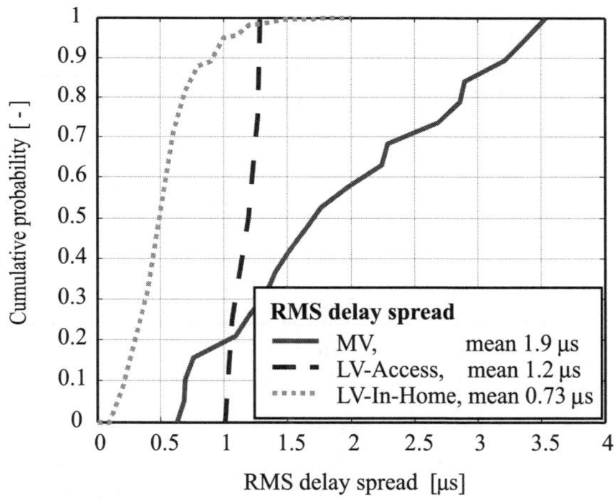

Figure 10.4. Root mean square delay spread statistics.

One important parameter capturing the frequency selectivity characteristics is the *root mean square* (rms) *delay spread* (DS). For example, when designing *orthogonal frequency-division multiplexing* (OFDM) systems [49], the guard interval might be chosen as two to three times the rms DS to deliver good system performance [50]. MV, LV-Access and LV-In-Home DS statistics extracted from references 36 and 50 are presented in Figure 10.4. The displayed rms DS statistics correspond to bands up to 30 MHz. Note that the rms DS may be obtained following various procedures. Dependent on the procedure, the results vary by up to 15% [36]. Furthermore, only a small measurement set was available for the LV-Access case. Hence, it is lacking statistical relevance. Nevertheless, Figure 10.4 gives a good indication regarding which order of rms DS to expect in the different scenarios.

Besides multipath fading, the PLC channel exhibits time variation. It is long known that the channel changes when loads are connected or disconnected [51]. To see this, consider that the 70-Ω load in the stub-line example from Figure 10.2 is unplugged and the line remains open at point D. In this case the channel frequency transfer function changes as displayed in Figure 10.3. It is easy to imagine that similar long-term variations occur if entire line segments are connected or disconnected. In between these rather long-term switching intervals, many early PLC channel characterizations regarded the channel as stationary [52]. Only through synchronizing channel measurements with the electrical grid mains cycle, Cañete et al. were able to show that the In-Home channel changes in a cyclostationary manner [53–55]. As an example of this cyclic short-term variation, consider the measured reflection coefficient, Γ, of a halogen lamp in Figure 10.5a. It is

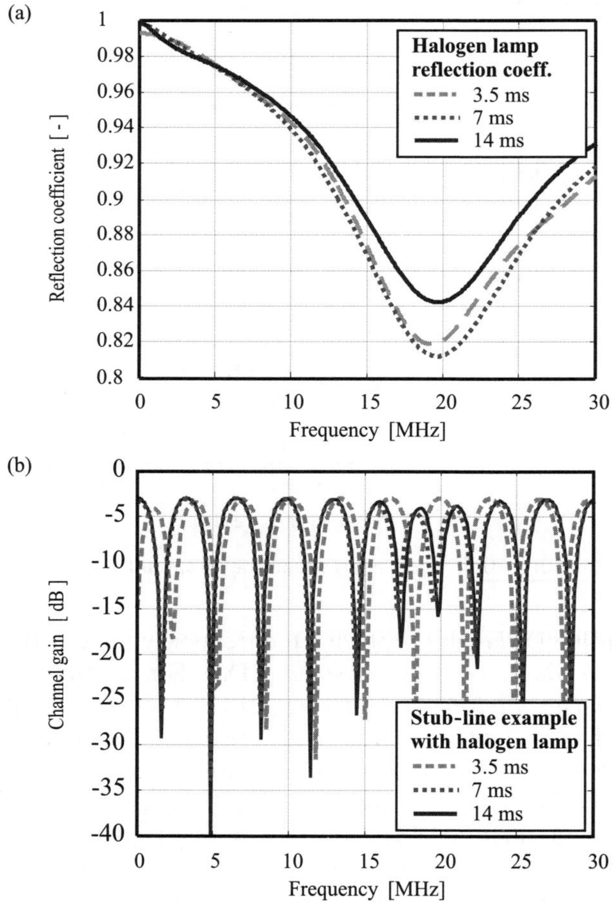

Figure 10.5. (a) Halogen lamp reflection coefficient. (b) Time variant frequency response for stub line example [75]. Copyright 2008 IEEE.

plotted at 3.5, 7, and 14 ms after the positive zero crossing of a 50-Hz AC mains cycle. The reflection coefficient relates to the load impedance via

$$Z_{in} = Z_0 \cdot \frac{1-\Gamma}{1+\Gamma} \tag{10.4}$$

where Z_0 is the reference impendence of the network analyzer. With the help of (10.4), the reflection coefficient measurements can be included into the stub-line example. The resulting cyclically varying channel transfer functions are displayed in Figure 10.5b. Models for long-term and cyclic short-term variation can, for example, be found in reference 56.

Until now the low-pass behavior of PLC channels has not been considered. It results from dielectric losses in the insulation between the conductors, and it is more pronounced in long cable segments such as outdoor underground caballing. Transfer function measurements on different cable types and for different length can be found in references 57 and 58. Using a large set of field trials, low-pass mean gain models are derived in reference 36. Over the range from 1 to 30 MHz, the mean gain in decibels is approximated by linear models. Consider again the PLC scenarios from Figure 10.1. The mean gain from the secondary transformer to the HAP, M3 to M4, is expressed as [36]

$$\bar{g}_{\text{LV-Access}} = -(a_1 \cdot f \cdot d + a_2 \cdot f + a_3 \cdot d + a_4) \tag{10.5}$$

where f is frequency in megahertz, d is distance in meters, and the coefficients a_1 to a_4 are 0.0034 dB/(MHz m), 1.0893 dB/MHz, 0.1295 dB/m, and 17.3481 dB, respectively.

The mean gain model in dB for MV lines, as well as for LV-In-Home situations, is given by [36]

$$\bar{g}_{\text{MV or LV-In-Home}} = -(b_1 \cdot f + b_2). \tag{10.6}$$

For the LV-In-Home situation the mean gain is given from the mains distribution board to a socket in a room, labeled M5 and M6 in Figure 10.1. The coefficients are b_1 = 0.596 dB/MHz and b_2 = 45.325 dB. The MV gain describes the channel between two primary transformers on the MV side, indicated by M1 and M2 in Figure 10.1. Its coefficients are b_1 = 1.77 dB/MHz and b_2=37.9 dB. In both situations the model is not distant-dependent. For the MV situation, this is due to the fact that not enough measurement results were available to construct a distant-dependent model. Hence, in this case the model is limited to situations where the distance between M1 and M2 is around 510 m. Nevertheless, correction factors are proposed in reference 36 to determine the mean gain at other distances. For the LV-In-Home situation the model is not distance-dependent because "distance" in an In-Home situation is a hard-to-define term. Power-line networks in such situation exhibit usually a large amount of branches, and a detailed floor plan to determine cable length cannot always be obtained.

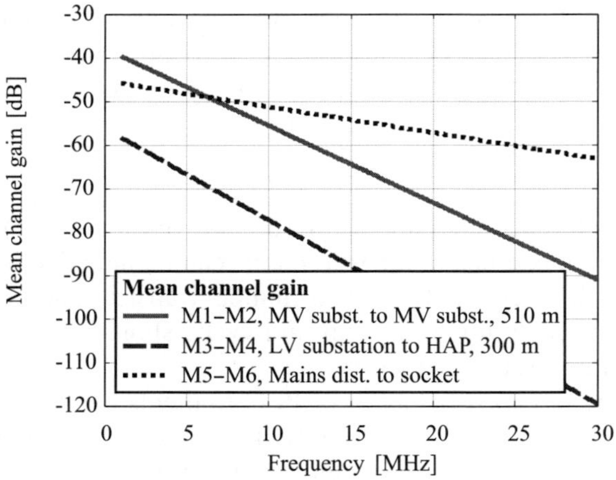

Figure 10.6. Mean channel gains.

Using these linear models, the mean gains for the three cases are plotted in Figure 10.6. A distance of 300 m is used in the LV-Access graph. It can be seen that the low-pass behavior is less pronounced in the In-Home case. It can further be seen that in the MV and the LV-Access situation the attenuation drastically increases with frequency. This goes well in line with the findings in reference 59 and is one of the reasons why Access networks are frequently operated in the lower frequency range (e.g., between 1 and 10 MHz), while In-Home networks might operate at frequencies above 10 MHz.

10.5 NOISE CHARACTERISTICS

Power-line noises can be grouped based on temporal as well as spectral characteristics. Following, for example, references 58 and 60, one can distinguish among the following:

Colored Background Noise. Its PSD exhibits only long-term time variations on the scale of minutes or even hours. It is caused by the superposition of many noise sources with little power. It is often observed to decrease with increasing frequency.

Narrowband Noise. Like the colored noise, narrowband noise only exhibits long-term time variation. It is mainly caused by other radio services (Amateur Radio, television broadcasting, etc.) that get coupled into the power grid. Thus, it consists mainly of modulated sinusoidal signals with power levels clearly above the background noise.

Periodic Impulsive Noise Asynchronous to the AC Frequency. These noises often have repetition frequencies between 50 kHz and 2 MHz and are attributed to switching power supplies. They have PSDs that decrease with frequency, and they are sometimes considered as part of the colored background noise.

Periodic Impulsive Noise Synchronous to the AC Frequency. They occur with repetition frequencies that are a multiple of the AC frequency. For example, in a 50-Hz network they would occur at 50 or 100 Hz. They are attributed to net-synchronous power converters used in dimmers or, more generally, to all kinds of diode-containing rectifier circuitry. The impulses are usually very narrow in the time domain and can have a considerable amplitude, with the consequence that they can have an adverse effect over a wide frequency range.

Aperiodic Impulsive Noise. This noise occurs at random intervals and may consist of several consecutive spikes in the time domain also called bursts. Aperiodic impulsive noise is attributed to all kinds of switching effects (e.g., within electrical motors or in condenser discharge lamps). Its amplitude can be significant (e.g., 50 dB above the colored noise floor), and its randomness in time makes it specifically difficult to deal with from a communication system point of view. In environments only lightly affected by this noise, one might observe 1 impulse in 10 seconds, while in heavily affected environments there might be around 100 impulses per second.

In reference 58, all these noises are modeled directly at the receiver. Time-invariant *additive white Gaussian noise* (AWGN) plus an exponentially decaying spectral filtering process is used for the conglomerate of colored background noise and the periodic impulsive noise asynchronous to the mains. Further, narrowband noise is modeled by a superposition of sinusoidal signals with different amplitudes and random phases. Periodic impulsive noise synchronous to the AC mains is modeled by filtering AWGN and adding it at synchronous periodic intervals. Finally, the aperiodic noise is also generated from filtering AWGN, but by adding it at random intervals that are determined by an underlying Markov process. A special procedure based on two interconnected Markov processes is used to implement noise bursts.

Instead of modeling the noise directly at the receiver, Cañete et al. proposed to model the noise at its origin and to filter it by the channel transfer function [53, 61]. Advantages are that temporal correlation effects between channel changes and noise variations, as well as correlated noise events as seen by different receivers, could be modeled. Disadvantages are that in many cases the channel transfer functions to the different receivers might not be known, or complex channel modeling might be required to obtain them.

A statistical approach to average colored background noise modeling is presented in reference 36 based on a large amount of noise measurements in MV as well as LV-Access and LV-In-Home situations. Although a lot of the details get lost by averaging, the results can still deliver some interesting rule of thumb when one wants to determine a likely average noise level. One general finding is that the mean noise power falls off exponentially with frequency. Derived from reference 36, the mean noise PSD in dBm/Hz is given by

TABLE 10.1. Mean Noise Model Coefficients [36]

Location	c_1 (dB)	c_2 (1/MHz)	c_3 (dBm/Hz)
M1 and 2, secondary transformer, MV	37	0.17	−105
M3, secondary transformer, LV	24.613	0.105	−116.721
M4, house access point, LV	29.282	0.12	−114.941
M5, main distribution board, LV	39.794	0.07	−118.076
M6, socket in private home, LV	17.327	0.074	−115.172

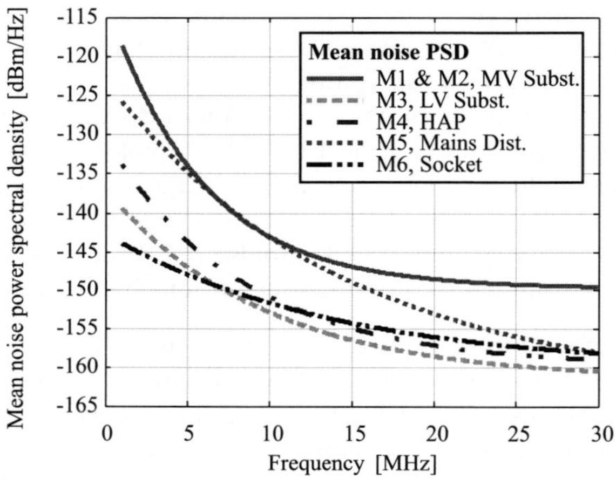

Figure 10.7. Mean noise power spectral densities.

$$\overline{P}_N = c_1 \cdot e^{(-c_2 \cdot f)} + c_3 - 10 \cdot \log_{10}(30000) \quad (10.7)$$

where the last term normalizes out the 30-kHz bandwidth used in the noise measurement process. The coefficients c_1 to c_3 are given in Table 10.1. The resulting noise models correspond to the measurement points M1 to M6 in Figure 10.1 and are plotted in Figure 10.7.

10.6 MEAN SIGNAL-TO-NOISE RATIO

Assume that a power-line signal with $\overline{P}_S = -55\,\text{dBm/Hz}$ may be injected. Using the gain and noise models from (10.5) to (10.7) the mean SNR can be approximated as

$$\overline{SNR} = \overline{g} + \overline{P}_S - \overline{P}_N. \quad (10.8)$$

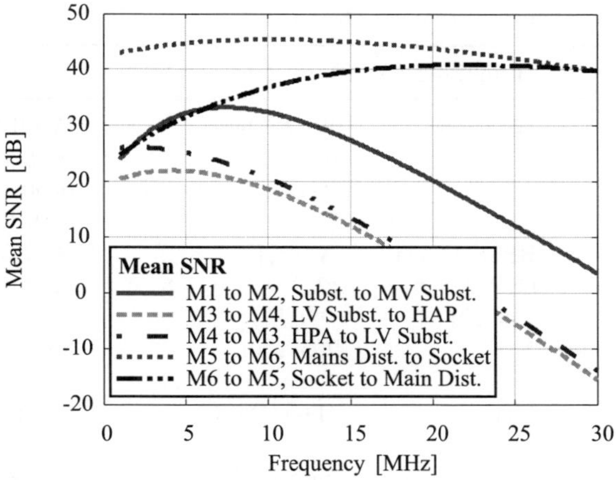

Figure 10.8. Mean SNRs for the various connections between the measurement points M1 to M6.

The mean SNRs for the various connections between the measurement points M1 to M6 are plotted in Figure 10.8. One should note that although the channel gain between two measurement points is symmetric, the noise at the measurement points differs. Hence, five different graphs are produced.

It can be seen that especially the lower part of the spectrum, up to 10 MHz, is very well suited for Access and Backhaul applications. Furthermore, for In-Home applications the entire spectrum from 1 to 30 MHz promises high mean SNRs on the order of 40 dB, which also goes well in line with the findings in reference 62. The results show that there is a high potential for PLC if the estimated mean SNRs can be exploited in PLC modems. However, the presented results have to be handled with care. One should bear in mind that the mean SNR models from reference 36 exhibit a significant standard deviation. With respect to the individual link SNRs, the standard deviation ranges from 13.5 to 23.4 dB. Furthermore, effects due to frequency selectivity, narrowband interference, impulsive noise, and time variation were not considered. Whether the estimated mean SNRs translate into high PLC data rates depends on the PLC modem's signal processing algorithms and its component quality.

10.7 PLC TECHNOLOGY OVERVIEW

We will now look at the PHY and MAC of the consortia-backed specifications HomePlug AV and UPA (Access/DHS), as well as at the upcoming international standard ITU-T G.hn. An overview of the main parameters is provided in

TABLE 10.2. PLC Systems—Parameter Overview

	HomePlug AV	UPA (Access; DHS)	ITU-T G.hn
Modulation	W-OFDM	W-OFDM	W-OFDM
Spectrum (MHz)	2–28	0–30	0–200
Bandwidth modes (MHz)	—	Access: 30, 20, 10 DHS: 30, 20	0–50, 0–100 100–200
Constellation mapping (bpc)[a]	BPSK/QAM (1, 2, 3, 4, 6, 8, 10)	ADPSK (2, 3, 4, 5, 6, 7, 8, 9, 10)	BPSK/QAM (1, 2, 3, 4, 6, 8, 10, 12), rx optional: (5, 7, 9, 11)
Maximum PHY data rate (Mbit/s)	200	205; 240	Target 1000
Forward error correction	CTC	4D-TCM, RS	QC-LDPC-BC
Retransmission	SACK	go-back-N ACK, or No ACK	SACK, or No ACK
Medium access method	TDMA-CSMA/CA	ADTDM	TDMA-CSMA/CARP
Network admission	NMK	RADIUS server	ITU-T X.1035
Data encryption	AES-128	3DES, or AES-128/256	AES-128
Neighboring network coexistence	TDM	TDM, FDM	TDM, FDM

[a] bpc, bits per carrier.

Table 10.2. The HD-PLC specification and the upcoming IEEE P1901 standard are not considered, to confine the overview to a concise and manageable level. Nevertheless, the interested reader may refer to reference 63 for a comparison that includes HD-PLC. Besides, one should note that IEEE P1901 basically includes the PHY of HD-PLC and HomePlug AV. Note further that ITU-T G.hn is applicable not only to power lines but also to phone lines and coaxial cables. However, only the power-line-specific parameters are presented here. Finally, note that all reviewed PLC systems have a robust communication mode. However, for the sake of simplicity, the following subsections will only deal with the modulation and coding of the data payload in normal operation. Similarly, for the sake of simplicity, details on interleaving and scrambling are omitted. The interested reader may refer to the actual documents [16, 22–24, 31] instead.

10.7.1 Windowed OFDM Modulation

All PLC systems in Table 10.2 use *windowed orthogonal frequency division multiplexing* (W-OFDM) [49], a multicarrier technology that is, for example, also used in Wi-Fi, WiMAX, and xDSL. The data bits to be transmitted are first mapped onto a constellation point in the complex plain. A set of constellation points is grouped to form a so-called OFDM symbol. The so assembled *frequency domain* symbol is passed through an *inverse fast Fourier transform* (IFFT) to obtain a *time domain* OFDM symbol. A *cyclic prefix* (CP) is added, by copying some of the time domain samples from the end of the symbol to its beginning. Afterwards, windowing and possibly up-conversion are performed before the samples are sent to the *digital-to-analog converter* (DAC). From there, the analog signal is fed to a power amplifier before it is coupled onto the power line by capacitive or inductive coupling devices.

At the receiver the signal is decoupled from the power line and is sent to an *automatic gain control* (AGC) stage. This AGC ensures that the received signal strength, which after propagating through the PLC channel exhibits a considerable dynamic range, falls well into the limited input conversion range of the *analog-to-digital converter* (ADC). Afterwards, a carefully selected subset of the obtained digital time domain samples is passed through a *fast Fourier transform* (FFT). The subset selection, better known as synchronization, removes the CP that had been inserted by the transmitter. After the FFT, the signal consists of a set of raw soft symbol points in the frequency domain. Dependent on the used constellation mapping, which can be *coherent* or *differential*, the raw soft symbols might have to be phase-rotated and scaled before they can be related to the originally transmitted constellation points and in the sequel to the transmitted bits.

Clearly, the exact procedure within the PLC transmitter and the PLC receiver is vendor discretionary, and performance can differ significantly even if two vendors implement the same communication specification. The fact that all PLC systems in Table 10.2 deploy OFDM is explained by the manifold advantages of OFDM when used over power-line channels:

- OFDM allows tight carrier spacing and therefore enjoys high spectral efficiency.
- As pointed out earlier, the PLC channel is frequency-selective with sometimes long multipath echoes. Nevertheless, dependent on the CP length and the quality of the synchronization process, OFDM makes it possible to avoid *intersymbol interference* (ISI).
- Simple frequency domain equalization may be deployed on a per carrier basis. Furthermore, the underlying IFFT and FFT signal processing operations are well understood, with the consequence that highly efficient algorithms exist to implement the required operations on a semiconductor chip.
- The fact that OFDM symbols consist of a set of orthogonal carriers in the frequency domain makes it easy not to load some carriers with signal energy. This way, adaptive notching and power mask requirements, imposed by EMC regulations, can be flexibly implemented. Furthermore, through the option to avoid some "bad" carriers, OFDM exhibits an inherent resilience to jammers such as narrowband radio stations.

10.7.2 Constellation Mapping

As already mentioned, to form an OFDM symbol in the frequency domain, a complex constellation point is assigned to every carrier. All PLC technologies in Table 10.2 have the option to select among various constellations with different size and therefore with a different number of *bits per carrier* (bpc). This task is performed by vendor-specific bit-loading algorithms. However, a standard approach would be to select a constellation that together with the deployed *forward error correction* (FEC) scheme leads to a certain expected *block error rate* (BLER) after decoding.

Coherent and differential constellation mapping mechanisms exist. Coherent methods require that the receiver compensate for amplitude and phase variations introduced by the channel. Channel estimation and channel tracking mechanisms are required. On the other hand, differential mechanisms encode the information in the difference between two symbols. Hence, to decode the symbol information the receiver uses the previous symbol as reference. Direct channel estimation and tracking is not required, which makes differential constellation mapping robust against abrupt channel changes. Nevertheless, the spectral efficiency of the differential schemes is usually reduced [64].

Also hybrids between coherent and differential mapping exist. An example is the *amplitude differential phase shift keying* (ADPSK) scheme used by the UPA specification. Here information is differential encoded in phase, allowing quick recovery after abrupt channel changes. In good situations—that is, a stable channel and low noise—information is additionally coherently encoded in various amplitude levels. The constellation points of 5 bpc ADPSK are plotted in Figure 10.9a as an example.

The other technologies in Table 10.2, HomePlug AV and ITU-T G.hn, use purely coherent mappings. In *binary phase shift keying* (BPSK) 1 bpc is mapped to two antipodal points. Furthermore, *quadrature amplitude modulation* (QAM) can be used where information is encoded through amplitude variations in the complex plain. HomePlug AV supports the bpc (1, 2, 3, 4, 6, 8, 10). ITU-T

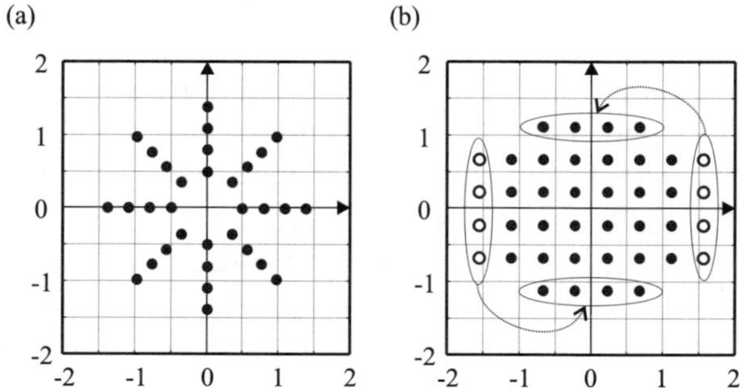

Figure 10.9. Odd constellation examples. (a) UPA 32-ADPSK. (b) ITU-T G.hn 32-QAM.

G.hn additionally supports 12 bpc and makes the support of odd constellation QAM schemes with bpc (5, 7, 9, 11) mandatory for the transmitter but optional for the receiver. An example of the 5 bpc ITU-T G.hn 32-QAM constellation points can be found in Figure 10.9b. As indicated, odd constellations can be constructed by first setting a fat rectangle of points and then transforming the outer columns into rows.

10.7.3 Forward Error Correction

Forward error correction—that is, adding redundant bits to the transmitted data that help the receiver to detect the original information bits even in a harsh communication channel—is a standard procedure in modern digital communications. Two main classes of FEC codes, *convolutional codes* and *block codes*, exist. Convolution codes work on continuous streams of data, while block codes work on data blocks [65, 66]. It has also been common practice to concatenate an inner convolutional code with an outer block code, used for example in DVB-T or ADSL.

As seen in Table 10.2, convolutional *four-dimensional trellis-coded modulation* (4D-TCM) and a *Reed–Solomon* (RS) block code are concatenated in the UPA specification. The other two technologies use only a single code. HomePlug AV uses a *convolutional Turbo code* (CTC), which consists of two interleaved convolutional coders. Similar CTCs are, for example, also used in the third-generation wireless standards UMTS and CDMA2000. The ITU-T G.hn/G9960 standard uses a *quasi-cyclic low-density parity-check block code* (QC-LDPC-BC), which is also used, for example, in WiMAX (IEEE 802.16e), 10GBase-T Ethernet (IEEE 802.3an), and DVB-S2. Both CTC and QC-LDPC-BC have a higher spectral efficiency than the 4D-TCM RS solution. Comparing CTC and QC-LDPC-BC, it turns out that at BLER > 10^{-3} they have similar coding gain, albeit QC-LDPC-BCs allow higher throughputs. At BLER < 10^{-3}, QC-LDPC-BCs outperform CTCs in coding gain and throughput when implementing similarly complex decoding structures [67].

10.7.4 Retransmission Schemes

Any FEC scheme can only correct a limited amount of errors. When, for example, a strong impulsive noise is experienced at the PLC receiver, it could be that the deployed FEC cannot decode the incoming signal correctly. In such an event, the PHY layer would inform the higher layers about the incorrectly received data. Dependent on the transfer control protocol, the receiver might then request the transmitter to resend the data. Different retransmission schemes exist. A general overview can, for example, be found in reference 68.

UPA has two transfer control protocol modes. In *ACK mode* a *sliding window go-back-N acknowledgement* procedure is implemented. The receiver keeps track of the received data packets and sends the packet identifier of the last correctly received packet back to the transmitter. The transmitter keeps track of transmitted packets and triggers a retransmission based on the ACK information if deemed necessary. In the second mode, called *No ACK*, the receiver only keeps track of the correctly received sequence of data packets; however, it cannot request a retransmission.

HomePlug AV uses a *selective repeat acknowledgment* (SACK) transfer control protocol. The receiver acknowledges the correct reception of PHY blocks. Only those blocks that are not positively acknowledged will be retransmitted. While SACK avoids duplicate transmissions of correct packets, the implementation complexity is higher, due to increased buffering requirements at the receiver.

ITU-T G.hn supports, like UPA, two modes of operation: ACK mode and No ACK mode. However, in ACK mode it uses a selective repeat ARQ scheme, similar to the one used in HomePlug AV. Only the packets that are not positively acknowledged are retransmitted. The receiver can still accept and acknowledge packets that it receives after the reception of an erroneous packet. The No ACK operation can be used in situations with very stable and high SNRs, as well as for services where occasional packet errors may be tolerated.

10.7.5 Medium Access Control

Until now only a simple PLC network topology with a single transmitter and a single receiver was considered. However, many PLC modems might be connected to the same power-line network. In this case, *medium access control* (MAC) protocols have to arbitrate access to the shared power-line medium. When QoS has to be guaranteed, centrally coordinated MAC protocols have advantages over probabilistic ones like *carrier sense multiple access* (CSMA). Therefore, although all PLC technologies in Table 10.2 use some kind of CSMA, QoS demanding data is scheduled in all of them using a centrally coordinated *master/slave* topology.

In HomePlug AV, the master is called *Central Coordinator* (CCo). The CCo assigns timeslots to the other AV nodes and broadcast these assignments within a so-called *beacon region*. The entire beacon is repeated every two AC mains cycles, leading to a repetition period of 33.33 ms or 40 ms in a 60- and a 50-Hz mains grid, respectively. The beacon consists of the *beacon region*, a *CSMA region*, and a *contention-free region*. Usually, applications that have strict QoS requirements are scheduled in the contention-free region. Delay-tolerant applications contend for access in the CSMA region. The overall MAC protocol is also referred to as *time division multiple access, carrier sense multiple access with collision avoidance* (TDMA-CSMA/CA).

The MAC protocol of the UPA specification uses a control signal called a *token*. The token is passed from a master to its slaves. With the token, a slave receives the right to put data onto the shared power-line medium. After the slave has no more data to send, or once its allocated maximum time is up, it returns the token to the master. The UPA Access part specifies three entities: The *head end* (HE), the *time division repeater* (TDR), and the *customer premises equipment*

(CPE). Equivalently, UPA DHS specifies the *access point*, the *repeater*, and the *end point*. Considering an Access network, the HE is the master to all nodes in its direct reach, be it CPEs or TDRs. The TDRs function as masters to nodes that are hidden to the HE. The UPA DHS MAC works along the same lines. The MAC protocol is also said to implement *advanced dynamic time division multiplexing* (ADTDM).

ITU-T G.hn uses a mix of the MAC features of HomePlug AV and UPA. The important network entities are called *domain master*, *relay node*, and *regular node*. The domain master assigns *transmission opportunities* (TXOPs) to all nodes to facilitate QoS support. The information is conveyed in a so-called *medium access plan* (MAP). TXOPs exist in two flavors, *contention free* (CFTXOP) and *shared* (STXOP). During CFTXOP, only a preassigned transmitter is allowed to send data to ensure QoS. The STXOPs are themselves subdivided into *contention-free timeslots* (CFTS) and *contention-based timesslots* (CBTS). During CFTS the nodes perform an implicit token passing, which means that, following an order, only one node is allowed to transmit at a time. In the CBTS, all nodes contend for medium access using a *carrier sense multiple access with collision avoidance and resolution protocol* (CSMA/CARP) [68]. Furthermore, ITU-T G.hn knows three modes of domain operation called *peer-to-peer mode* (PM), *centralized mode* (CM), and *unified mode* (UM). In PM, data packets may directly be exchanged between any node A and any node B as long as both are within reach of each other. Relaying is not allowed. In CM, all data packets are first transmitted to the so-called *domain access point* (DAP), which often coincides with the domain master. From the DAP the data packets are retransmitted to the destination node. In CM the DAP is the only allowed relay node. Finally, in UM, connections are of a peer-to-peer type. However, multiple relay nodes may exist to connect two far-away nodes that cannot directly see each other. Which of the three modes is used is decided and signaled by the domain master.

10.7.6 Security

Although PLC signals are mainly confined to the physical line, it is possible that they are received by a third party. Thus, all PLC technologies in Table 10.2 specify mechanisms to restrict the access to a network. Furthermore, data encryption techniques are used.

More specifically, the UPA specifications use a *remote authentication dial in user service* (RADIUS) [69], to allow the master to manage access of slaves to a network. Once a node is admitted, data are encrypted using the *triple data encryption standard* (3DES) [68]. Alternatively, UPA DHS may use the 128-bit or the 256-bit *advanced encryption standard* (AES) [70]. AES is commonly regarded as a successor of 3DES [68].

To connect to a HomePlug AV network, a node must have knowledge of the so-called *network membership key* (NMK)—for example, through the insertion of the right password. A node with correct NMK is passed the *network encryption key* (NEK). This NEK is then used to encrypt data transmissions with the 128-bit AES.

ITU-T G.hn uses an advanced authentication and secure admission procedure based on ITU-T X.1035 [14]. Moreover, ITU-T G.hn uses unique encryption keys for each pair of communicating nodes or per multicast group. The encryption mechanism is based on AES-128.

10.7.7 Coexistence and Neighboring Network Support

With an increasing penetration of PLC networks, the risk that these networks interfere with each other also increases. In the worst case, interference can be so significant that neither of the interfering networks may operate satisfactory. Hence, if not tackled, this so-called *neighboring network interference* could cause very negative consumer experience and limit consumer acceptance of PLC technology as a whole. Therefore, UPA and HomePlug have developed mechanisms to make networks of the same specification coexist. These will be outlined in the following:

The UPA coexistence specification [22] demands that UPA networks detect their mutual presence by transmitting and sensing a preamble signal. If neighboring UPA networks are detected, all nodes follow a fixed set of rules to avoided interference. The avoidance process is based on dynamic *frequency division multiplexing* (FDM) as well as dynamic *time division multiplexing* (TDM). For this purpose the spectrum is subdivided into three bands called *FB1* to *FB3*. FB1 refers to frequencies below 12 MHz, and FB2 refers to frequencies above 13 MHz. Finally, FB3 refers to the entire spectrum from 1 to 30 MHz. Besides, the time interval between transmitted preamble signals is subdivided into 12 timeslots. The result is a time-frequency grid over which transmissions of neighboring UPA networks are scheduled. Consider, for example, the situation where an UPA In-Home network operates over the entire frequency spectrum, that is, FB3. At some point the presence of an UPA Access network is detected in FB1. In this case the In-Home network switches its data transmission to band FB2. However, it continues to listen to the preamble signals over FB1 and FB2. Should it detect that the Access network has become inactive, it may dynamically expand its operation using the full spectrum, that is, FB3, again. Consider another example where one UPA In-Home network is using FB3 while a second UPA In-Home network is powered up that would like to use FB3 as well. Listening to the preamble both networks detect each other and start to use preassigned timeslots. More precisely, the first In-Home network starts to use timeslot 1 and 7. The second network uses timeslots 3 and 9. The remaining 8 timeslots may be used by either of the two based on a controlled contention process. In case of congestion, these slots are shared evenly and collision-free. Besides, the UPA coexistence specification supports *spatial reuse*. To understand the idea behind spatial reuse, imagine a network with five nodes, labeled A, B, C, D, and E. Imagine further that all nodes are within reach of node C. However, nodes A and B are not within reach of node D and E, and vice versa. In such situations, C may operate as a relay, for example, to establish a connection between A and E or between B and D. More importantly, nodes A and B may communicate with each other using the same timeslot and the same frequency band as E and F. Having several peer-to-peer connections operating in parallel, while exploiting merely spatial separation, can boost overall network throughput significantly.

Looking at HomePlug AV, several neighboring AV and HomePlug 1.0 networks can detect each other. Similar to the UPA networks, they transmit and sense preamble signals. Once a neighboring network situation is detected, the different CCos negotiate who may use which timeslots in the beacon's contention-free region. Only one network is allowed to transmit in every timeslot, this way implementing a TDM system. Furthermore, all the nodes may contend for medium access in the beacon's CSMA region.

It was early recognized that merely coexistence between same-technology networks was not sufficient to gain widespread consumer acceptance. Hence, in 2005 the *Consumer Electronics Powerline Communication Alliance* (CEPCA) [71] was brought to live. It was founded by Panasonic, Mitsubishi, and Sony with the objective to develop specifications that enable the coexistence between different In-Home technologies as well as between different In-Home and Access technologies. A tight cooperation with UPA was quickly established, which resulted in the submission of a coexistence standard proposal to the IEEE P1901 standardization process. The proposal was several times modified and amended until a proposal including HomePlug and ITU-T G.hn coexistence mechanisms was finally confirmed at the end of 2008. Around the same time, ITU-T G.cx was formed to tackle coexistence of ITU-G.hn with existing home networking technologies. Although not finally consented, it is likely that IEEE P1901 as well as ITU-T G.hn devices will incorporate an *intersystem protocol* (ISP) containing technical elements from the UPA and the HomePlug coexistence protocols described earlier [72]; that is, neighboring networks detect each others' presence with the help of preamble signals. Afterwards, TDM and FDM mechanism are applied to avoid destructive intersystem interference when operating on the same medium [73, 74].

10.8 CONCLUSIONS

An overview of past, present, and upcoming broadband power-line technologies was provided. It can be concluded that current-generation PLC technologies complement other wire line and wireless technologies. The main advantages of PLC are that no new wires are required and that the coverage is in many cases higher than that of wireless solutions.

A large amount of PLC deployment scenarios exists, which makes it difficult to come up with a "one fits all" channel and noise characterization. Thus, only average channel and noise models have been presented throughout this text. However, many references have been provided, pointing the interested reader to parametric, as well as physical, models. Based on the presented mean SNR models, it can be concluded that In-Home scenarios exhibit usually higher SNRs than Access scenarios. Furthermore, Access scenarios suffer severe SNR degradation with increasing operating frequency.

To make PLC a widespread success, it is essential to come to a single broadly adopted standard. The upcoming ITU-T G.hn standard, with its single PHY and single MAC, can fill this gap, and it becomes even more attractive because it is applicable not only to PLC, but also to coaxial cables and phone lines. Thus, for the first time a standard has the potential to unify the entire wire line industry. This is good news for consumers, as well as for consumer equipment manufacturers, because they are no longer forced to choose among non-coexisting proprietary technologies.

REFERENCES

1. K. Dostert, Telecommunications over the power distribution grid—Possibilities and limitations, in *International Symposium on Power Line Comms and Its Applications (ISPLC)*, pp. 1–9. Essen, Germany, April 1997.
2. P. A. Brown. Power line communications—past present and future, in *International Symposium on Power Line Communicatons and Its Applications (ISPLC)*, September 1999, pp. 1–8.
3. J. Abad, A. Badenes, J. Blasco, J. Carreras, V. Dominguez, C. Gomez, S. Iranzo, J. C. Riveiro, D. Ruiz, L. M. Torres, and J. Comabella, Extending the power line LAN up to the neighborhood transformer, *IEEE Commun. Mag.*, Vol. 41, No. 4, pp. 64–70, April 2003.
4. G. Held, *Understanding Broadband over Power Line*, CRC Press, Boca Raton, FL, 2006.
5. A. Sendí Escalona, R. Saorín Adán, J. Hartmann, and J. A. Garrigosa, White paper AMR. IST Integrated Project Deliverable D41v1.1, The OPERA Consortium, December 2004. EC/IST FP6 Project No 507667.
6. M. Koch, H. Hirsch, and M. Ianoz, Powerline communication for the coordination of distributed energy generation, in *7th International Symposium on Electromagnetic Compatibility and Electromagnetic Ecology*, Saint-Petersburg, Russia, June 2007, pp. 40–43.
7. Q. Liu, B. Zhao, Y. Wang, and J. Hu, Experience of AMR systems based on BPL in China, in *IEEE International Symposium on Power Line Communications and Its Applications (ISPLC)*, Dresden, Germany, April 2009, pp. 280–284.
8. A. Dutta-Roy, Networks for home, *IEEE Spectrum*, Vol. 36, pp. 26–33, December 1999.
9. K. Dostert, *Powerline Communications*, Prentice-Hall, Upper Saddle River, NJ 2001.
10. Intellon, Home page: http://www.intellon.com/, September 2009.
11. DS2, Design of Systems on Silicon, Home page: http://www.ds2.es/, September 2009.
12. HomePlug Powerline Alliance, Home page: http://www.homeplug.org/home, September 2009.
13. Universal Powerline Association (UPA), Home page: http://www.upaplc.org/, September 2009.
14. Telecommunication Standardization Sector, Study Group 17, Password-authenticated key exchange (PAK) protocol. ITU-T Recommendation X.1035, International Telecommunication Union (ITU), February 2007.

15. M. K. Lee, R. E. Newman, H. A. Latchman, S. Katar, and L. Yonge, HomePlug 1.0 powerline communication LANs-protocol—Description and performance results, *Int. J. Commun. Syst.*, Vol. 16, pp. 447–473, 2003.
16. HomePlug Powerline Alliance, HomePlug AV baseline specification, May 2007. Version 1.1.
17. HomePlug Powerline Alliance, HomePlug AV white paper. Technical Report HPAVWP-050818, HomePlug Powerline Alliance, 2005.
18. K. H. Afkhamie, S. Katar, L. Yonge, and R. Newman, An overview of the upcoming HomePlug AV standard, in *International Symposium on Power Line Communications and Its Applications (ISPLC)*, Vancouver, Canada, April 2005, pp. 400–404.
19. S. Gavette, HomePlug AV technology overview. White paper, Microsoft, April 2006.
20. Telecommunications Industry Association (TIA), Committee: TR-30.1. Medium-speed (up to 14 Mbps) power line communications (PLC) modems using windowed OFDM, May 2008. TIA-1113.
21. Open PLC European Research Alliance (OPERA), Home page: http://www.istopera.org/, September 2009.
22. OPERA Consortium, Opera technology specification—Part 1. European Union Project Deliverable, SSWG v1.0, January 2006. IST Integrated Project No 507667.
23. OPERA Consortium, Opera technology specification—Part 2. European Union Project Deliverable, SSWG v1.0, January 2006. IST Integrated Project No 507667.
24. Universal Powerline Association (UPA), UPA—Digital Home Specifications (UPA DHS), v 1.0., February 2006.
25. S. Arroyo, Digital home specification white-paper, Technical Report 1.0, Universal Powerline Association (UPA), May 2006.
26. Panasonic Communications Co., Ltd., Home page: http://panasonic.net/corporate/segments/pcc/, September 2009.
27. High Definition Power Line Communication Alliance (HD-PLC), Home page: http://www.hd-plc.org, September 2009.
28. Gigle home page: http://www.gigle.biz/, September 2009.
29. J. Simón, J. P. Faure, and R. Martínez, White paper: Comparison of access technologies. IST Integrated Project Deliverable D7v2.1, The OPERA Consortium, March 2009. IST Integrated Project No 026920.
30. National Institute of Standards and Technology (NIST), U.S. Department of Commerce. NIST framework and roadmap for smart grid interoperability standards. NIST Draft Publication, Release 1.0, September 2009.
31. International Telecommunications Union (ITU). ITU-T Recommendation G.9960, Unified High-Speed Wire-Line Based Home Networking Transceivers—Foundation, August 2009.
32. HomeGrid Forum, Home page: http://www.homegridforum.org/home, September 2009.
33. Institute of Electrical and Electronic Engineers (IEEE), Standards Association, Working group P1901. Home page: http://grouper.ieee.org/groups/1901/, September 2009.
34. J. Byrne, Fruitless battle of proprietary home-networking tech nears end. Blog, posted by the Linley Group Silicon Valley, http://blog.linleygroup.com/2009/06/fruitless-battle-of-proprietary-home.html, June 2009.

35. J. C. Dvorak, HDMI: Who needs this aggravation? *PC Magazine*, Davis Media Inc., http://www.pcmag.com/article2/0,2817,2349468,00.asp, July 2009.
36. P. Meier, M. Bittner, H. Widmer, J.-L. Bermudez, A. Vukicevic, M. Rubinstein, F. Rachidi, M. Babic, and J. Simon Miravalles, Pathloss as a function of frequency, distance and network topology for various LV and MV European powerline networks. Project Deliverable, EC/IST FP6 Project No 507667 D5v0.9, The OPERA Consortium, April 2005.
37. A. Rubinstein, F. Rachidi, M. Rubinstein, A. Vukicevic, K. Sheshyekani, W. Bäschelin, and C. Rodríguez-Morcillo, EMC guidelines. IST Integrated Project Deliverable D9v1.1, The OPERA Consortium, October 2008. IST Integrated Project No 026920.
38. A. Vukicevic, *Electromagnetic Compatibility of Power Line Communication Systems*. Dissertation, École Polytechnique Fédérale de Lausanne, Lausanne, Switzerland, June 2008, No 4094.
39. S. Galli, T. Banwell, and D. Waring, Power line based LAN on board the NASA space shuttle, In *IEEE 59th Vehicular Technology Conference*, Vol. 2, May 2004, pp. 970–974.
40. S. Tsuzuki, M. Yoshida, and Y. Yamada, Characteristics of power-line channels in cargo ships, in *International Symposium on Power Line Communications and Its Applications (ISPLC)*, Pisa, Italy, March 2007, pp. 324–329.
41. V. Degardin, M. Lienard, P. Degauque, E. Simon, and P. Laly, Impulsive noise characterization of in-vehicle power line, *IEEE Trans. Electromagn. Compat.*, Vol. 50, No. 4, pp. 861–868, November 2008.
42. W. Baeschlin, S. Arroyo, L. Feltin, M. Koch, B. Wirth, J.-P. Faure, and M. Heina, First report on the status of PLC standardisation activities in CISPR, European Union Project Deliverable D30, OPERA Consortium, February 2008, IST Integrated Project No. 026920.
43. W. Baeschlin, L. Feltin, J.-P. Faure, M. Rindchen, and H. Hirsch. Second report on the status of PLC standardisation activities in CISPR, European Union Project Deliverable D31, OPERA Consortium, January 2009, IST Integrated Project No. 026920.
44. Comité International Spécial des Perturbations Radioélectriques, Information technology equipment; radio disturbance characteristics; limits and methods of measurement, International Standard Norme CISPR 22, Edition 6.0, ICS CISPR, September 2008.
45. FCC, Title 47 of the code of federal regulations (CFR), Technical Report 47 CFR §15, Federal Communications Commission, July 2008.
46. M. Zimmermann and K. Dostert, A multi-path signal propagation model for the power line channel in the high frequency range, in *International Symposium on Power-line Communications and Its Applications (ISPLC)*, Lancaster, UK, April 1999, pp. 45–51.
47. L. T. Berger and G. Moreno-Rodríguez, Power line communication channel modelling through concatenated IIR-filter elements, *Acad. Publisher J. Commun.*, Vol. 4, No. 1, pp. 41–51, February 2009.
48. D. M. Pozar, *Microwave Engineering*, 3rd edition, John Wiley & Sons, Hoboken, NJ, 2005.

49. R. van Nee and R. Prasad, *OFDM for Wireless Multimedia Communications*. Universal personal communication. Artech House Publishers, Norwood, MA, 2000.

50. S. Galli, A simplified model for the indoor power line channel, in *IEEE International Symposium on Power Line Communications and Its Applications (ISPLC)*, Dresden, Germany, March 2009, pp. 13–19.

51. F. J. Cañete Corripio, L. Díez del Río, and J. T. Entrambasaguas Muñoz, A time variant model for indoor power-line channels, in *International Symposium on Power Line Communications (ISPLC)*, Malmö, Sweden, March 2001, pp. 85–90.

52. A. J. Han Vinck and G. Lindell, Summary of contributions at ISPLC 1997–2001, in *International Symposium on Power Line Communications (ISPLC)*, Malmö, Sweden, March 2001, pp. 383–413.

53. F. J. Cañete, L. Díez, J. A. Cortés, and J. T. Entrambasaguas, Broadband modelling of indoor power-line channels, *IEEE Trans. Consumer Electron.*, Vol. 48, No. 1, pp. 175–183, February 2002.

54. J. A. Cortés, F. J. Cañete, L. Díez, and J. T. Entrambasaguas, Characterization of the cyclic short-time variation of indoor power-line channels response, in *International Symposium on Power Line Communications and Its Applications (ISPLC)*, Vancouver, Canada, April 2005, pp. 326–330.

55. F. J. Cañete Corripio, J. A. Cortés Arrabal, L. Díez del Río, and J. T. Entrambasaguas Muñoz, Analysis of the cyclic short-term variation of indoor power line channels, *IEEE J. Sel. Areas Commun.*, Vol. 24, No. 7, pp. 1327–1338, July 2006.

56. S. Sancha, F. J. Cañete, L. Díez, and J. T. Entrambasaguas, A channel simulator for indoor power-line communications, in *IEEE International Symposium on Power Line Communications and Its Applications (ISPLC)*, Pisa, Italy, March 2007, pp. 104–109.

57. M. Zimmermann and K. Dostert, A multipath model for the powerline channel. *IEEE Trans. Commun.*, Vol. 50, No. 4, pp. 553–559, April 2002.

58. M. Babic, M. Hagenau, K. Dostert, and J. Bausch, Theoretical postulation of PLC channel model, IST Integrated Project Deliverable D4v2.0, The OPERA Consortium, March 2005.

59. H. Liu, J. Song, B. Zhao, and X. Li, Channel study for medium-voltage power networks, in *IEEE International Symposium on Power Line Communications (ISPLC)*, Orlando, FL, March 2006, pp. 245–250.

60. M. Zimmermann and K. Dostert, An analysis of the broadband noise scenario in power-line networks, in *International Symposium on Power Line Commun. and Its Applications (ISPLC)*, Limerick, Ireland, April 2000, pp. 131–138.

61. F. J. Cañete, J. A. Cortés, L. Díez, and J. T. Entrambasaguas, Modeling and evaluation of the indoor power line transmission medium, *IEEE Commun. Mag.*, Vol. 41, No. 4, pp. 41–47, April 2003.

62. A. Schwager, L. Stadelmeier, and M. Zumkeller, Potential of broadband power line home networking, in *Second IEEE Consumer Communications and Networking Conference*, January 2005, pp. 359–363.

63. P. Siohan, A. Zeddam, G. Avril, P. Pagani, S. Person, M. Le Bot, E. Chevreau, O. Isson, F. Onado, X. Mongaboure, F. Pecile, A. Tonello, S. D'Alessador, S. Drakul, M. Vuksic, J. Baudais, A. Maiga, and J. Herald, State of the art, application scenario and specific requirements for PLC. European Union Project Deliverable D3.1 v1.0, OMEGA, 2008. IST Integrated Project No ICT-213311.

64. J. G. Proakis, *Digital Communications*, 4th edition, McGraw-Hill International Series in Electrical Engineering: Communications and Signal Processing, McGraw-Hill, 2001.

65. T. M. Cover and J. A. Thomas, *Elements of Information Theory*, John Wiley & Sons, New York, 1991.

66. T. K. Monn, *Error Correction Coding, Mathematical Methods and Algorithms*, John Wiley & Sons, Hoboken, NJ, 2005.

67. S. Galli, PLC standardization progress and some PHY considerations. *Keynote Speech at IEEE International Symposium on Power Line Communications (ISPLC)*, Dresden, Germany, March 2009. Digital Object Identifier 10.1109/ISPLC.2009.4913389.

68. A. S. Tannenbaum, *Computer Networks*, 4nd edition, Prentice-Hall International, Englewood Cliffs, NJ, 2003.

69. J. Hassell, *RADIUS—Securing Public Access to Private Resources*, O'Reilly & Associates, Sebastopol, CA, 2002.

70. National Institute of Standards and Technology (NIST), U.S. Department of Commerce. Specification for the advanced encryption standard (AES). Federal Information Processing Standards Publication 197, November 2001.

71. Consumer Electronics Powerline Communication Alliance (CEPCA), Home page: http://www.cepca.org/, September 2009.

72. B. O'Mahony, C. Gómez, J. Egan, V. Oksman, M. Finocchiaro, and S. Galli, G.hn compatibility with existing home networking technologies. HomeGrid Forum White Paper, April 2009, Revision 1.0.

73. S. Galli, A. Kurobe, and M. Ohura, The inter-PHY protocol (IPP): A simple coexistence protocol for shared media, in *IEEE International Symposium on Power Line Communications and Its Applications (ISPLC)*, Dresden, Germany, March 2009, pp. 194–200.

74. Institute of Electrical and Electronics Engineers (IEEE) Standards Association, Working group P1901. IEEE P1901 Draft standard for broadband over power line networks: Medium access control and physical layer specifications, July 2009.

75 G. Moreno-Rodríguez and L. T. Berger. An IIR-filter approach to time variant PLC-channel modelling, in *IEEE International Symposium on Power Line Communications and Its Applications (ISPLC)*, Jeju, South Korea, April 2008, pp. 87–92.

11

POWER-LINE COMMUNICATIONS AND SMART GRID

Tae Eung Sung and Adam Bojanczyk

With the ever-increasing demand for high-speed data communication and its quality of service (QoS), broadband connectivity to and within the home has been available to consumers through various technologies. Among those technologies, power-line communications (PLC) is becoming an excellent candidate for providing broadband connectivity as it exploits an already existing infrastructure. This infrastructure is much more pervasive than any other wired alternatives, and it allows virtually every line-powered device to take advantage of value added services that are being developed. Therefore, PLC may be considered as the technological enabler of a variety of future applications that probably would not be available otherwise [1].

PLC is not new. At a very early stage of its development, the first reported applications of PLC were remote voltage monitoring in telegraph systems and remote meter readings. Today the interest in PLC spans several important applications: broadband Internet access, indoor wired local area networks (LANs) for residential and business premises, in-vehicle data communications, smart grid applications (advanced metering and control, real-time energy pricing, peak shaving, mains monitoring, distributed energy generation, etc.), and other municipal applications, such as traffic light and street lighting control [2].

Convergence of Mobile and Stationary Next-Generation Networks, edited by Krzysztof Iniewski

In this chapter, an overview of power-line communications (PLC) and smart grid is provided together with a description of recent works on modeling of time-varying PLC channels. Historical overview, standards, practical issues, and future potentials for smart grid system are also presented.

11.1 INTRODUCTION

Originally, power-line networks were designed for distribution of power at 50 Hz or 60 Hz. The use of this medium for data communication at higher frequencies presents several technical challenges. The structure of the mains grid, as well as indoor wiring and grounding practices, differ from country to country and even within a country. Additionally, the power-line channel is a harsh and noisy transmission medium that is very difficult to model, is frequency-selective, is impaired by colored background noise, and also is affected by periodic and aperiodic impulsive noise. The power-line channel is also time-varying [3]. The channel transfer function of the power-line channel may vary abruptly when the topology changes—that is, when devices are plugged in or out or switched on or off [4]. However, the power-line channel also exhibits a short-term variation because the high-frequency parameters of electrical appliances depend on the instantaneous amplitude of the mains voltage [5]. A fundamental property of the power-line channel is that the time-varying behavior mentioned previously is actually a periodically time-varying behavior, where the frequency of the variation is typically twice the mains frequency (50 or 60 Hz). Additional challenges are due to the fact that power-line cables are often unshielded and thus become both a source and a recipient of electromagnetic interference (EMI). As a consequence, PLC technology must include mechanisms to ensure successful coexistence with wireless and telecommunication systems, as well as be robust with respect to impulse noise and narrowband interference.

11.2 POWER LINE COMMUNICATIONS (PLC)

Power-line communications basically means any technology that enables data transfer at narrowband (NB) or broadband (BB) speeds through power lines by using advanced modulation and coding strategies [6]. It has been around for quite some time, but its use has been limited to narrowband tele-remote relay applications, public lighting, and home automation. Broadband communication over power lines (sometimes called BPL) was introduced at the end of the 1990s.

Electrical power is normally transmitted over high-voltage (HV) networks (110–380 kV) at a considerably long distance within a continent, distributed over medium-voltage (MV) (10–30 kV) networks at a size of large cities and big commercial vendors, and used at low-voltage (LV) (220 V in Europe, 110 V in the United States) for the end user supply inside buildings or private homes [7]. Most PLC technologies limit themselves to a set of wires such as premises wiring, but sometimes cross-leveled technology between the distribution network and premises wiring is also realizable. PLC technology has the ability of communicating data or information signals via the electrical supply network (ESN), and therefore it can extend an existing local area network (LAN) or share an existing Internet connection through electric plugs with the installation of adapter units.

The principle of PLC consists in superimposing a high-frequency (HF) signal (1.6 to 30 MHz) at low energy levels over the 50-Hz (Europe) or 60-Hz (United States) electrical signal. The combined signal is transmitted via the power infrastructure and is decoded at remote locations. An integrated coupler at the PLC receiver entry points eliminates low-frequency components before the signal is post-processed [8].

From the viewpoint of customer side applications, PLC is becoming an alternative to existing wireless technology for a seamless in-home network environment where the wireless applications cannot supply consistently stable, high-throughput service. As shown in Figure 11.1a and 11.1b, every PC or peripheral device is attached to PLC connecting outlets or adapters that behave as modems. Figure 11.1c illustrates an integrated power-line MAC/PHY transceiver that requires no new wiring to support transmission at speeds of up to 14 Mbit/s. It provides the ability to interconnect multiple interfaces to the external MAC controller. Figure 11.1d shows a smart coffeemaker that communicates via PLC when it is plugged to an AC outlet and communicates wirelessly via radio frequency (RF) when operated on batteries.

11.2.1 Narrowband (NB) PLC

11.2.1.1 Home Control. Typically, home-control PLC transmitters operate by modulating a carrier wave of between 20 and 200 kHz into the household wiring. The carrier is modulated by digital signals. Receivers may be either plugged into regular power outlets or permanently wired in place. Since the carrier signal may propagate to nearby homes or apartments on the same distribution system, receivers are assigned individual house internet protocol (IP) addresses that identify their owners. Thus receivers can be individually controlled by the signals transmitted over the household wiring.

11.2.1.2 Low-Speed Narrowband Communications. Narrowband applications of mains communications vary enormously, as would be expected of such a widely available medium. One simple application of narrowband power-line communication is the control and telemetry of electrical equipment such as meters, switches, heaters, and domestic appliances. A number of active developments are considering such applications from a systems point of view, such as demand-side management. Domestic appliances would intelligently coordinate their use of resources like limiting peak loads. Such applications are being developed for the emerging smart grid systems that embrace the PLC technologies.

Meanwhile, control and telemetry applications include both (a) utility-side applications, which utilize equipment belonging to the utility company (i.e., from

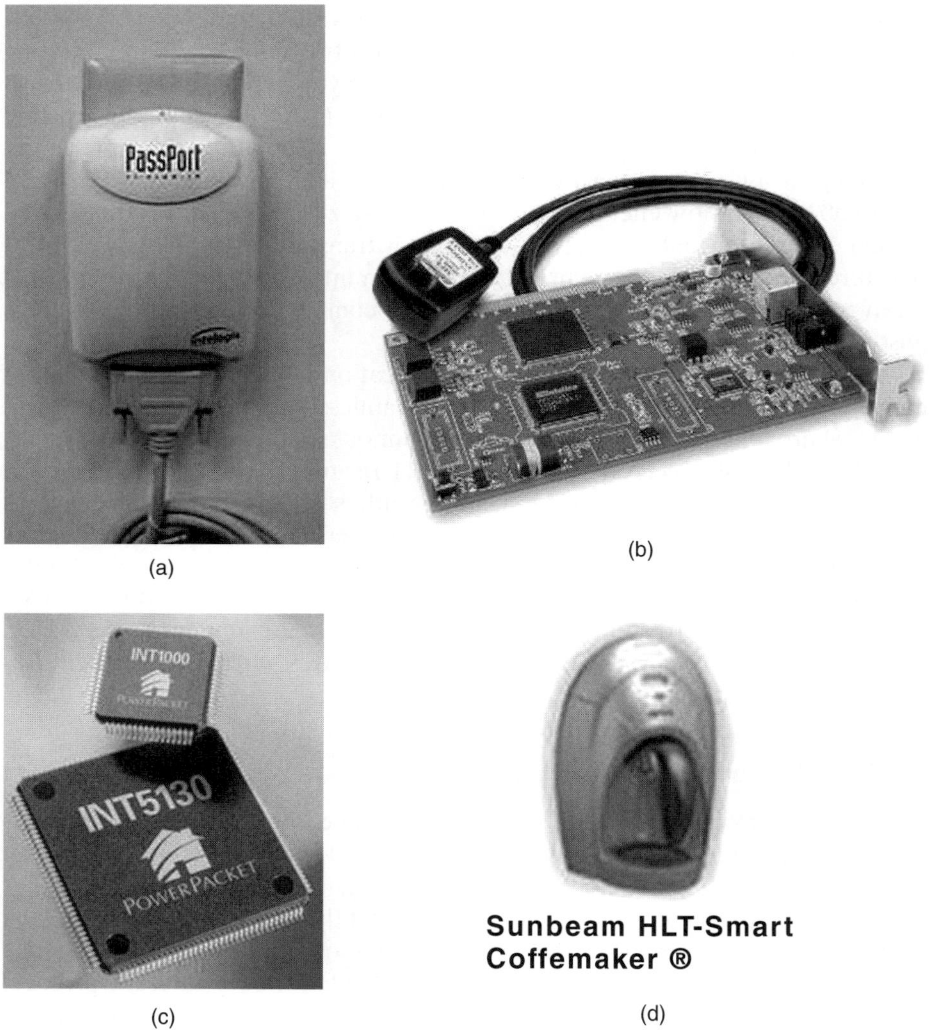

Figure 11.1. Various PLC products: **(a)** Power-line PC networking connectors. **(b)** Power-line network adapter. **(c)** integrated powerline MAC/PHY transceiver. **(d)** HLT-smart coffeemaker.

the supply transformer substation up to the domestic meter), and (b) consumer-side applications, which utilize equipment in the consumer's premises. Possible utility-side applications include automatic meter reading, dynamic pricing, load management, load profile recording, financial credit control, pre-payment, remote connection, fault detection, and network management, and they could be extended to gas and water control.

It is known that the most robust low-speed power-line technology uses differential code shift keying (DCSK) technology available from Yitran

Communications [9]. Renesas Technology licenses this technology from Yitran and incorporates it in the single-chip microcontroller unit (MCU) PLC family of devices known as M16C/6S. Renesas also licenses a state-of-the-art network layer for Automatic Meter Reading/Automated Meter Management (AMR/AMM) applications that can run on these devices [9].

11.2.1.3 High-Speed Narrowband Communications. Distribution line carrier (DLC) uses existing electrical distribution network mainly in the medium voltage (MV). It is very similar to the powerline carrier. DLC uses NB PLC technology in frequency range of 9–500 kHz with data rate up to 576 kbit/s [10]. DLC is suitable for multiple real-time energy management applications. It can be implemented under the Real-Time Energy Management over Power Lines and Internet (REMPLI) System as well as the Supervisory Control and Data Acquisition (SCADA), AMR, and Power Quality Monitoring System (PQMS) [11]. DLC complies with the standards such as EN 50065 (CENELEC), IEC 61000-3 and FCC Part 15 Subpart B [10].

Apparently, there are no interference issues between DLC and radio users or electromagnetic radiation. With external inductive or capacitive coupling, a distance of 15 km or further can be reached over a medium-voltage network. On low-voltage networks, a direct connection can be made because the DLC has a built-in capacitive coupler. This allows end-to-end communications from substation to the customer premises without repeaters.

Recent DLC systems significantly improve upon and differ from other PLC segments. DLC is mainly useful for a backhaul infrastructure that can be integrated with corporate wide-area networks (WANs) via TCP/IP, serial communication or leased-line modem to support for multiservices real-time energy management systems.

11.2.1.4 Utility Applications. Most utility companies mainly adopt special coupling capacitors to connect medium frequency (MF) radio transmitters to the power-frequency AC conductors. The active frequency region lies in the range of 24–500 kHz, with transmitter power levels up to hundreds of watts. These signals may be superimposed on one or more conductors of a high-voltage AC transmission line. Multiple PLC channels may be coupled onto one HV line. Filtering devices are applied at substations to prevent the carrier frequency current from being suppressed by the station equipment and to ensure that distant faults do not affect the isolated segments of the PLC system. These circuits are contrived to control switchgear and to protect transmission lines. For example, a protection relay can make use of a PLC reference channel to stay on a line if a fault is detected between its two terminals, but to leave the line in operation if the fault is elsewhere on the system.

While utility companies use microwave and fiber-optic cables for their primary system communication needs, the power-line carrier module may still be useful as a backup channel or for relatively very simple low-cost installations that do not require installing fiber-optic lines.

11.2.2 Broadband PLC

HF communication may reuse large portions of the radio spectrum or may use selected bands, depending on the technology.

11.2.2.1 In-Home Networking. PLC can be used to interconnect home or office computers, peripheral devices or other networked consumer electronics. As shown in Figure 11.2, a typical application of home networking is the streaming of high-definition (HD) video content through consumer premise equipment (CPE) such as a PC or home server to a HD-TV in the living room.

Various wired (e.g., Ethernet, Coax) and wireless networks (e.g., WiFi) exist to establish these home networking functions. However, there might be some drawbacks of these solutions, especially for "room-to-room" connectivity and long distances within the house. The data throughput of wireless connections decreases if the signal is attenuated by walls or ceilings. Wired networks may require inconvenient installation efforts. For mains-powered devices, PLC technology enables new and highly convenient networking functions without any additional cables. An in-home backbone connecting all devices or clusters in the house is provided by PLC, as can be seen in Figure 11.2. Wireless devices can communicate via an access point to the PLC network [12]. Although there is not yet a universal standard for this type of ubiquitous application, standards for power-line home networking are being regulated by numerous companies within the framework of HomePlug Power-Line Alliance (HPA) and the Universal Power-Line Association (UPA) [13].

Today's PLC solutions theoretically promise data rates up to 200 Mbit/s on physical layer. Measurements in buildings show significant lower bit rates due to

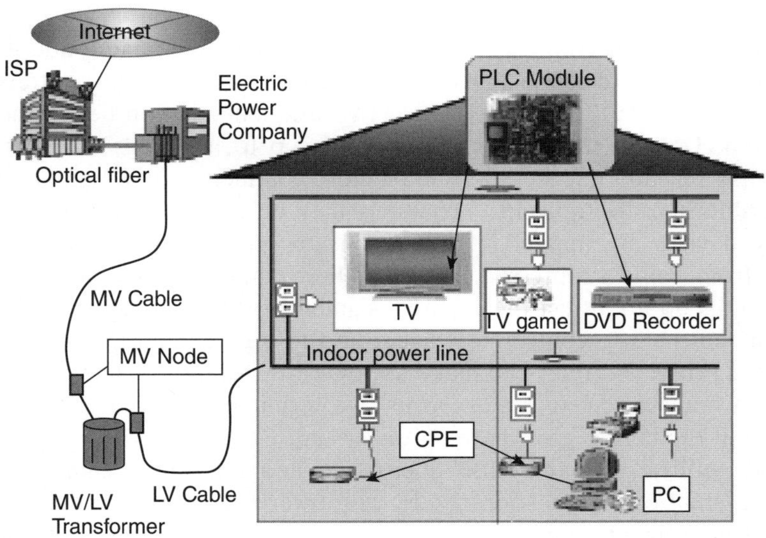

Figure 11.2. In-door power-line applications. (Reproduced by courtesy of reference 16.)

high attenuation, frequency-selective transfer functions, and noise in many cases. Typically, today's PLC systems use one transmit and one receive port for data communication. However, in most parts of the world, three-wire installations allow more feeding and receiving options. In the presence of multiple feeding and receiving ports, MIMO principles can be exploited.

The in-home power-line technology focuses on delivering a short-distance solution. Services like power-line intranet solutions and power-line music distribution (that follow the EN 55022 directive [14]) belong to the in-home networking category. These kinds of services compete against other in-home interconnection technologies such as wireless, even so though the bit rates in wireless connections are definitely lower [15]. In-home power-line technology can also provide services like sending a small amount of data with low bit rate (for example, to open an automatic door or to control the switching on and off of a light).

Besides in-door PLC applications, we also have outdoor utility-side applications that have intermediate MV nodes and repeaters (REP) to connect between power-line backbone infrastructure and supply transformer substations (see Figure 11.3). To accommodate multiple consumer premise equipments (CPE) and guarantee reliable transmission of the data or information, we need to insert REP between transformers and CPE.

11.2.2.2 In-Vehicle Networking. In a similar way as in-home PLC networking, power-line transmission can be applied to in-vehicle networks. In any

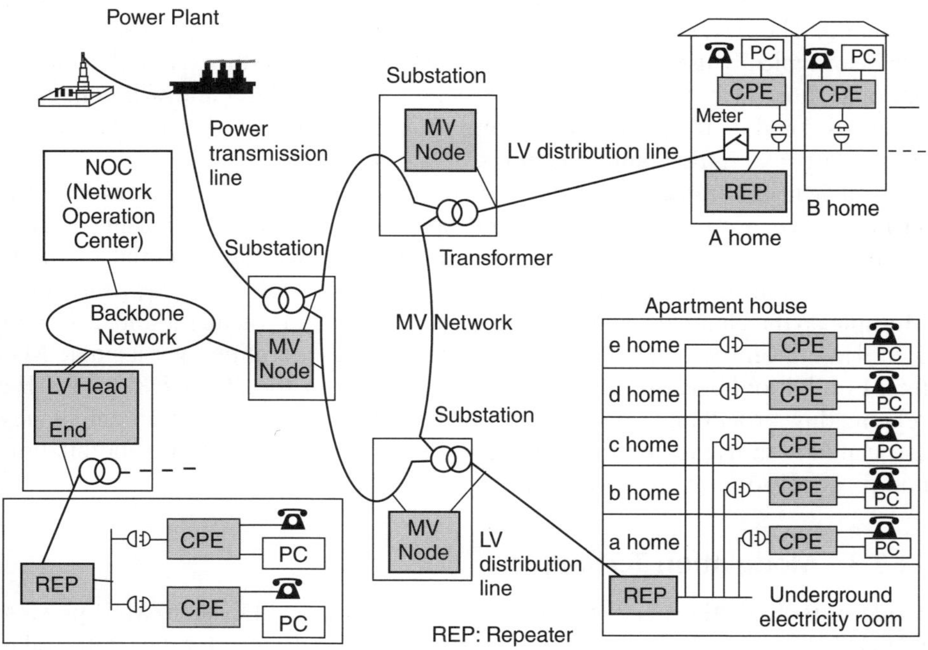

Figure 11.3. Outdoor power-line transmission. (Reproduced by courtesy of reference 16.)

vehicle (from automobiles to ships, from aircraft to space vehicles), separate cabling is used to establish a broadband physical layer for a local command and control network. The in-vehicle power distribution network may serve as an infrastructure supporting both power delivery and broadband digital connectivity [16].

11.2.2.3 Broadband Internet Access. Broadband over power lines (BPL) literally provides broadband Internet access through ordinary power lines. A computer or workstation only needs to plug a BPL modem into any outlet within equipped homes or offices to have high-speed Internet access. BPL may offer benefits over regular cable modem or digital subscriber line (DSL) connections. The extensive infrastructure already pre-installed enables people to access the Internet in remote locations without additional equipment costs. Also, such ubiquitous availability would make it easier for hooking up televisions, sound systems, and so on.

PLC modems transmit data signals onto HF electric carriers. The asymmetric speed in the modem is generally from 256 kbit/s (uplink) to 2.7 Mbit/s (downlink). In the repeater, it can be increased up to 45 Mbit/s. Meanwhile, the speed from the head ends to the Internet rises up to 135 Mbit/s, relatively a favorable speed to the end customers.

The PLC system faces a number of challenges. The primary one is that power lines are inherently very noisy. Every time a device turns on or off, it introduces a pop or click into the line. Energy-saving devices often introduce noisy harmonics into the line. The second major issue is the signal strength and operating frequency. The system is expected to operate in frequencies of 10–30 MHz, which has been typically used for decades by amateur radio operators as well as international shortwave broadcasters. Power lines are unshielded and will behave as antennas for the signals they carry, and thus they will experience interference from shortwave radio communications.

Much faster transmissions using microwave frequencies transmitted via a surface wave propagation mechanism have been demonstrated using only a single power-line conductor. These systems have shown the potential for symmetric and full duplex communications in excess of 1 Gbit/s in each direction. Multiple WiFi channels with simultaneous analog television in the 2.4- and 5.3-GHz unlicensed bands have been demonstrated operating over a single MV line. In addition, because it can operate anywhere in the 100-MHz to 10-GHz region, this broadband technology can completely avoid interference issues associated with use of shared spectrum while offering flexibility for modulation and protocols of a microwave system [17].

11.2.3 Modulation Techniques

Modern BPL systems use the Orthogonal Frequency Division Multiplexing (OFDM) technique, which allows customers to mitigate interference with radio waves by removing specific frequencies [17]. The most commonly used trans-

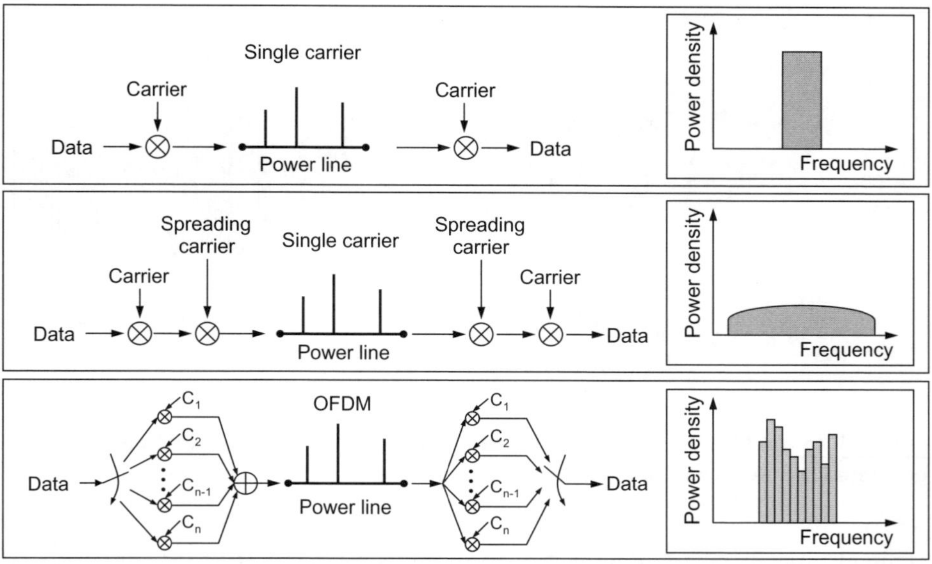

Figure 11.4. Transmission methods for PLC. (Reproduced by courtesy of reference 18.)

mission methods are single-carrier, spread-spectrum, and OFDM modulation schemes (see Figure 11.4). OFDM is preferred over the two other methods because due to limited spectral resources PLC technology must achieve maximum spectral efficiency. Moreover, implementing high data rates results in the generation of contiguous wideband transmission signals. While spread-spectrum modulation additionally adopts spreading carriers to obtain a widely spread flat spectrum, OFDM achieves multiple narrowband subchannels that are mutually orthogonal.

In OFDM techniques, due to the subchannels' narrowband property, attenuation and group delay are constant within each channel. Thus, equalization is easy and can be performed by only a single tap. Orthogonality of all carriers leads to outstanding spectral efficiency, which has been identified as a key element for the success of high-speed PLC.

11.3 CHANNEL CHARACTERISTICS OF POWER LINES

The development of power-line communication systems requires detailed knowledge of the channel properties, such as transfer function, interference scenario, and channel capacity for choosing suitable transmission methods. This section presents appropriate power-line channel models, which form the basis of a channel emulator that was developed in references 19 and 20. The emulator proved to be helpful for various tests and the comparison of performance of different communication systems. In particular, numerical simulations of power-line

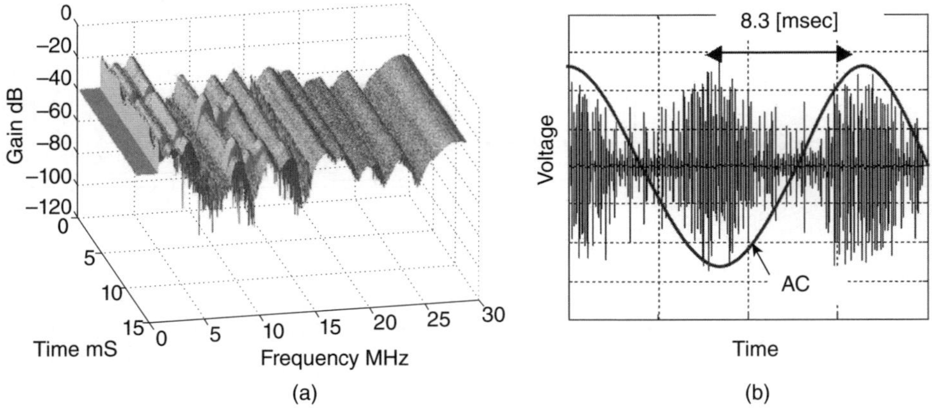

Figure 11.5. **(a)** Measured time variation of indoor power-line channel. **(b)** Noise waveform generated by the dimmer of a halogen light.

channel capacity clearly demonstrate enormous potential of PLC for high-speed communication purposes [18].

According to measurements, a fundamental property of the power-line channel is that the time-varying behavior is periodic, where the period is typically half the AC mains period (50 or 60 Hz). Parallel to the channel I/O response behavior, the noise statistics exhibit a cyclo-stationary component with the same period. An example of this behavior unique to the power-line channel is shown in Figure 11.5 [19], where we recognize that the power-line channel is time-varying and frequency-selective, and the noise variation has a spectral component of envelope with the period of 8.3 (ms).

11.3.1 Multipath Channel Model

From a multipath point of view, the signal components of the N individual paths are combined by superposition, and thus the frequency response can be represented in a simplified model as follows [21]:

$$H(f) = \sum_{i=1}^{N} g_i \cdot e^{-\left(a_0 + a_1 f^k\right) d_i} \cdot e^{-j2\pi f(\tau_i)} \tag{11.1}$$

where the first, second, and third multipliers are weighting factor, attenuation portion, and delay portion, respectively. The parameters for the multipath echo model above can be obtained from measurements of the complex channel transfer function. The attenuation parameters a_0 (offset of attenuation), a_1 (increase of attenuation), and k (exponent of attenuation) can be obtained from the magnitude of the frequency response. To determine the path parameters d_i and g_i, the impulse response is necessary. The impulse response gives information about the time delay τ_i of each path, which is proportional to d_i. The weighting factors

g_i can be obtained from the amplitude of each impulse. Typical values for the number of paths N are in the range of 5–50.

11.3.2 Noise

In general, we consider four types of noise components, usually present in the power distribution network: background noise, random impulse noise, synchronous noise, and continuous noise.

Background noise has a smooth spectrum. The most important sources of noise with a smooth spectrum are universal motors—that is, small motors with serial windings. The power spectral density $N(f)$ of this noise type was found to be a decreasing function of the frequency f in the frequency band of interest and on average can be described as follows [22]:

$$N(f) = 10^{K-3.95\times10^{-5} f} \qquad (\mathrm{W/Hz}) \tag{11.2}$$

The value of K in (11.2) changes with time and transmitter/receiver locations. However, when compared to the targeted bit rates of 256 kbit/s (uplink) and 2.7 Mbit/s (downlink), K is known to be constant for long periods of time: During the daytime, K normally remains unchanged for many seconds to minutes, whereas at night, often no real changes occur for hours. Although the distribution of measured values for K differs somewhat between locations, a specific type of noise is present more often than in another environment.

Synchronous noise is mainly generated at frequencies synchronous to the power-line base frequency mostly by light dimmers. Silicon-controlled rectifiers triggered by power voltage cause a very short break in current flow. The length of the break determines the intensity of light. Because switching is synchronous to the power frequency, a series of harmonics with various amplitude is generated. The setting of the dimmer and the characteristics of the lamp (bulb) dictate which harmonics carry most power. Usually, harmonics are small compared to fundamental frequency, but the fundamental frequency can also be far below their level. All switching devices operating on a similar principle tend to produce noise spikes synchronous to 50 or 60 Hz.

Another type of noise encountered in the power network is the periodic nonsynchronous noise that is continuous over time. It is generated by television sets and computer monitors. This kind of noise is mainly caused by silicon-controlled rectifiers (SCR), which switch a certain number of times every mains frequency [23]. What results is a train of noise impulses in the time domain, or noise at higher harmonics of the power system frequency in the frequency domain.

The only type of noise that spread spectrum modulation cannot deal with efficiently is white background noise, which is a hybrid of both background and random impulse noise. Its frequency spectrum occupies all the communication bandwidth; therefore extending the signal spectrum does not provide any gain. On the contrary, the noise power increases while (due to numerous factors) the signal power usually has to stay constant. The problem becomes more serious as the main source of white noise is believed to be universal motors, which can be found in many household appliances (electric drill, mixer, coffee grinder, hair dryer, fan, etc.).

11.3.3 Attenuation

The total signal attenuation of the channel consists of two parts: coupling losses and line losses. Coupling losses depend entirely on the design of the transmitter and can in principle be made arbitrarily small. The main subject in this section will, therefore, be line losses—that is, the amount of signal power lost into unwanted directions as a function of distance between transmitter and receiver.

In principle, two receivers are needed to accurately measure the line losses as a function of time and frequency: One receiver measures the received signal power close to the transmitter, and the other at the same time measuries the received signal power at a distance *d* from the transmitter. Subtraction of the two then cancels possible coupling losses and renders the line losses. However, since only one receiver setup is available, usually the measurements are taken at two well-separated time instances. Although both the received signal power at a distance *d* from the transmitter and the signal power close to the receiver show large frequency dependencies, the subtraction of the two measurements gives a value of the line losses that is almost frequency-independent. This implies that the frequency dependencies in the received signal are mainly due to variations in the coupling losses.

11.3.4 Electromagnetic Interference (EMI)

Electromagnetic fields radiated from power lines may cause disruption to existing critical radio communication services. To address this issue, research groups have been collaborating with Agilent Technologies in the development of PLC systems with low EMI radiation. Recently, the potential use of optical inductive couplers for broadband power line modems to reduce EMI radiation has been demonstrated. Also, mathematical models are being developed to predict radiated electromagnetic fields from power lines under different cabling arrangements. We expect to develop new signal-injection technologies for suppressing EMI radiation from PLC systems to guard against harmful effect of the radiation [24].

To minimize possible electromagnetic interferences (EMI), PLC access networks have to operate with limited signal power, which in turn may reduce the system data rate. In order to compete with other access technologies, PLC systems have to ensure higher levels of network utilization. A medium access control (MAC) protocol that maximizes network utilization is thus necessary. Both mathematical and simulation models have been successfully developed for MAC protocol performance analysis under impulsive noise interference.

Currently, a new protocol is being designed by the IEEE Standardization Group; it is called P1775, which would optimize packet length to achieve maximum network utilization in different PLC environments [25].

11.4 COUPLING TECHNIQUES

The communication signal is modulated on RF signals between an electrical power distribution line and the electrical power wiring at a customer's premises. This includes (a) a first transceiver for transmitting and receiving electromagnetic energy, which is inductively coupled to the distribution line, and (b) a transceiver for receiving electromagnetic energy from, and transmitting electromagnetic energy to, the first transceiver. The second transceiver is coupled to the customer's power wiring to receive communication signals from, and to supply communication signals to, the customer's power wiring [26].

11.4.1 Capacitive Coupling and Inductive Coupling

Once the data signal has been generated, it needs to be placed on the power line by some kind of coupling network. The idea is to superimpose the data signal onto the 240-V, 50-Hz (or 110-V, 60-Hz) power waveform and extract it afterwards at the receiving end. There are three possible combinations of lines on which to couple the signal: live to ground, neutral to live, and neutral to ground. Differential mode coupling is a scheme where the live wire is used as one terminal which the neutral is used as the other. In the case where a neutral line is not present, the ground line acts as the second terminal. Common mode coupling involves the live and neutral being treated as one terminal, with the ground being treated as the other. This kind of coupling is potentially not safe, and hence it is not used.

For coupler implementation, differential mode coupling is often used. The basic component used for the coupling may be capacitive or inductive [27].

An inductive coupler is a kernel component that transmits only PLC signals between the power line and the PLC data transceiver. It is quick and easy to install without connecting the communication cable to the power line because it is clamped around power line. The PLC signal can be transmitted a to power line with low signal loss for a wide frequency range. In fact, an inductive coupler provides more benefits when compared with a capacitive coupler. It can be applied to low- and high-speed PLC application regardless of voltage, and it can be applied to PLC application such as an automation of electric power distribution system, power line monitoring system, home network, automatic meter recording (AMR), and ship network, among others [28].

For a prototype implementation of the system we are trying to establish simple communication between two data communication equipments. Thus we need two couplers of different frequency responses, one in the transmit direction and the other in the receive direction. At the receiver side the coupling device should possess a band-pass characteristic, blocking the 50-Hz mains voltage and passing signal at the carrier frequency. At the transmitter side, the coupler should possess high-pass properties, passing the communication signal unattenuated. The coupler should also be impedance-matched to the power line for maximum power transfer [29].

11.4.2 Mitigation of Inductive Interference

Electric power systems, like almost everything run by electricity, depend on internal electric and magnetic fields; some of these fields are affected by the surrounding environment. The strongest of these fields can then induce voltages and currents in nearby devices and equipment and, in some cases, can interfere with the internal fields being used by electrical equipment in the vicinity. These induced voltages and currents, which are due to the coupling between the energized source and the electrical equipment, are called inductive interference [30].

Overhead power lines cause practically all of the problems due to inductive coupling. For this reason and for safety considerations, power lines are built in special corridors far from inhabited areas. Spacing between them and the requirements of their surroundings are considered and carefully calculated to minimize possible interference. These are often shared by telephone lines, communication circuits, railroads, and sometimes trolley buses, each of which must be considered for possible inductive coupling.

Modern telephone and communication circuits are well-shielded and rarely encounter interference from nearby power lines. However, where a long parallel exposure exists, inductive coupling can be reduced by balancing the operation of the power line—that is, by simultaneously transporting power and data signals. Fences, long irrigation pipes, and large underground objects within the corridors may experience considerable inductive coupling and must be grounded for safety [31].

11.5 STANDARDS

Despite numerous advantages in broadband over power lines (BPL) connectivity to and within the home, the fundamental obstacle for adopting this technology is the lack of an international technical standard issued by a credible and globally recognized standards-setting body [32]. The first standard for PLC applications is the European CENELEC EN 50065 regulated in 1991, which mandates the use of the frequency range between 3 kHz and 148.5 kHz [33]. Since then, much effort has been made to regulate the standards and enlarge the practices of consumer premise equipments. We expect that most of technical challenge will be overcome soon through the work of the IEEE P1901 Corporate Standards Working Group [34].

11.6 PRACTICAL EXAMPLE

The power-line channel is time-varying due to the appliances connected to outlets. Appliances often exhibit input impedance that is a function of the instantaneous amplitude of the mains voltage which results in a periodically time-varying channel response. Here we explore a discrete time-varying block channel model in the time domain [35], which allows for more realistic evaluation of modulation and coding strategies as well as for the design of bit loading strategies that can be resilient to the particular type of distortion in this harsh medium. Block transmission is a very efficient scheme to combat intersymbol interference (ISI) caused by frequency-selective time-dispersive channels at the cost of inter-block interference (IBI) occurring during block transmission. In particular, in order to suppress the IBI in the power-line communication channel, a novel method called the lifting-trailing-zeros (LTZ) technique is introduced in Section 11.6.1.2. We also propose a power-line network simulator and demonstrate its effectiveness by numerical verification.

11.6.1 System Model

11.6.1.1 Discrete-Time Block Model for Time-Varying System. Let us define the output samples $y[k] := y(kT_s)$ with symbol duration T_s. Then, a single-input single-output (SISO) time-varying system can be represented as

$$y[k] = \sum_{n=-\infty}^{\infty} x[n]h[k,k-n] \tag{11.3}$$

where the time-varying impulse response is $h[k,k-n] := \int_{-\infty}^{\infty} h(kT_s,\tau)p((k-n)T_s-\tau)\, d\tau$ and $p(t)$ is a pulse shaping filter with a Nyquist characteristic.

In (11.3) we find the well-known fact that for a linear time-invariant (LTI) system, the effective discrete-time impulse response is

$$h[k,k-n] = \sum_{m=-\infty}^{\infty} e^{j\frac{2\pi T_s}{T_0}mk} h_m[k-n] \tag{11.4}$$

where $h_m[k]$ is the discrete-time impulse response of the mth lag at time k. This leads to writing the output as well as a sum of components:

$$y[k] = \sum_{m=-\infty}^{\infty} e^{j\frac{2\pi T_s}{T_0}mk} y_m[k] \tag{11.5}$$

where $y_m[k]$ is the discrete-time output response of the mth lag at time k [35].

11.6.1.2 Lifting. The technique we use parses the data in blocks of size P bigger than the maximum memory of the discrete-time equivalent system L. The objective is to account explicitly for interblock (IBI) interference while hiding the parameters that cause the intersymbol interference (ISI) inside the mixing matrices that map input blocks into output blocks. The rationale behind the assumptions that the blocks are large enough and the system is causal is that only the previous block will interfere with the current block, allowing us to describe the input–output relationship in a compact form.

For a length-P block, let us define the ith block as follows:

$$\underline{\mathbf{x}}[i] = \begin{bmatrix} x[iP] \\ x[iP+1] \\ \vdots \\ x[iP+P-1] \end{bmatrix}, \quad \underline{\mathbf{y}}[i] = \begin{bmatrix} y[iP] \\ y[iP+1] \\ \vdots \\ y[iP+P-1] \end{bmatrix} \tag{11.6}$$

Then, (11.3) becomes

$$\underline{\mathbf{y}}[i] = \sum_{j=-\infty}^{\infty} H_{i,i-j}\underline{\mathbf{x}}[j] \tag{11.7}$$

where the channel transformation matrices $H_{i,i-j}$ are of size $p \times p$ and their (k,n)th element is defined as

$$\{H_{i,i-j}\}_{k,n} = h[iP+k,(i-j)P+k-n] \qquad (k,n = 0,\dots,P-1) \tag{11.8}$$

If the system is linear time-invariant (LTI), then (11.7) can be represented as

$$\underline{\mathbf{y}}[i] = \sum_{j=-\infty}^{\infty} H_{i-j}\underline{\mathbf{x}}[j] \tag{11.9}$$

where the (k,n)th element of H_{i-j} is defined as follows:

$$\{H_{i-j}\}_{k,n} = h[(i-j)P+k-n] \qquad (k,n = 0,\dots,P-1) \tag{11.10}$$

The following lemma is proved in reference 35.

LEMMA 11.1

If the channel memory L (i.e., L + 1 taps) is finite and $L < P$, then $H_{i,i-j}$ is nonzero only for $i - j = 0,1$, that is,

$$H_{i,i-j} = \begin{cases} H_{i,0} & \text{iff} \quad i = j \\ H_{i,1} & \text{iff} \quad i = j+1 \\ 0 & \text{otherwise} \end{cases} \tag{11.11}$$

Thus (11.7) becomes

$$\underline{y}[i] = H_{i,0}\underline{x}[i] + H_{i,1}\underline{x}[i-1] \tag{11.12}$$

while (11.9) becomes

$$\underline{y}[i] = H_0\underline{x}[i] + H_1\underline{x}[i-1] \tag{11.13}$$

where $H_{i,0}$ and $H_{i,1}$ are band lower-triangular and upper-triangular, respectively, and H_0 and H_1 are both *Toeplitz*.

From Lemma 11.1, a useful corollary follows.

COROLLARY 11.2

If $\boldsymbol{x}[i]$ has L trailing zeros, where L is the channel memory, then for all i *we obtain* $H_{i,1}\boldsymbol{x}[i-1] = 0$ and hence (11.12) and (11.13) are simplified to

$$\underline{y}[i] = H_{i,0}\underline{x}[i] \tag{11.14}$$

$$\underline{y}[i] = H_0\underline{x}[i] \tag{11.15}$$

The corollary allows us to describe the cascade of N linear systems recursively, as described by the following Theorem.

THEOREM 11.3

Let us consider a cascade of N linear systems. If the memory of the ith system is finite and equal to L_i, then the memory of the cascade of all N linear systems is equal to $L^{(1,\ldots,N)} = \sum_{j=1}^{N} L_j$.

Now, choosing $P \geq L^{(1,\ldots,N)}$, we can apply Lemma 11.1 and obtain

$$H_{i,i-j}^{(1,\ldots,N)} = \begin{cases} \Pi_{j=N}^{1} H_{i,0}^{(j)} & \text{iff} \quad i = j \\ H_{i,0}^{(N)} H_{i,1}^{(1,\ldots,N-1)} + H_{i,1}^{(N)} H_{i-1,0}^{(1,\ldots,N-1)} & \text{iff} \quad i = j+1 \end{cases} \tag{11.16}$$

where $H_{i,0}^{(1,\ldots,N)}$ is still band lower-triangular while $H_{i,1}^{(1,\ldots,N)}$ is upper-triangular.

11.6.1.3 Transmission Lines (TL) as Two-Port Networks (2PN). In TL theory, a common way to represent a two-port network (see Figure 11.6) is to use the transmission matrix, also known as the ABCD matrix [36].

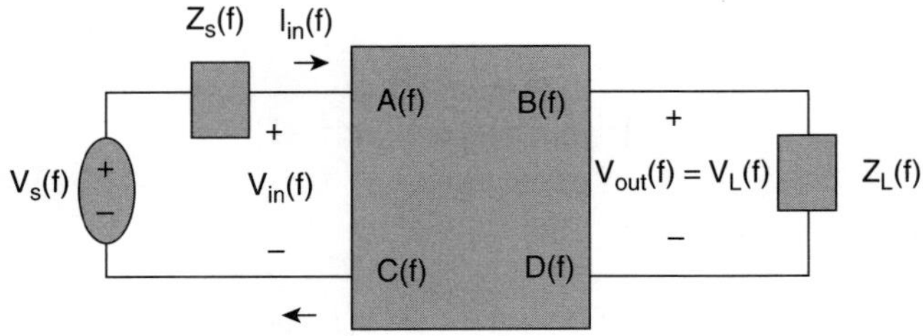

Figure 11.6. Frequency domain representation of a two-port network (2PN).

In TL theory the relationship between current and voltage in frequency domain at the two ports of a 2PN is given by

$$\begin{bmatrix} V_{\text{in}}(f) \\ I_{\text{in}}(f) \end{bmatrix} = \begin{bmatrix} A(f) & B(f) \\ C(f) & D(f) \end{bmatrix} \begin{bmatrix} V_{\text{out}}(f) \\ I_{\text{out}}(f) \end{bmatrix} \tag{11.17}$$

where the quantities above are all complex phasors. This description can be easily mapped into an overall system transfer function. Let us consider load impedance $Z_L(f) = z_0^{(L)}$ constant in time and frequency with a closed output port, a generator transmitting a signal with Fourier transform $V_s(f)$, and source impedance $Z_S(f) = z_0^{(s)}$ constant in time and frequency. Then we can represent (11.17) as follows:

$$V_{\text{out}}(f) = \frac{1}{A(f)} V_{\text{in}}(f) - \frac{B(f)}{A(f)} I_{\text{out}}(f) \tag{11.18}$$

$$V_{\text{out}}(f) = \frac{1}{C(f)} I_{\text{in}}(f) - \frac{D(f)}{C(f)} I_{\text{out}}(f) \tag{11.19}$$

The above expressions show that the effects of the *ABCD* parameters in (11.17) for voltage and current phasors can also be interpreted as a filtered version of the input voltage and current signals.

In general, a power line point-to-point link consists of several sections of power cables including bridged taps. The end-to-end system can be decomposed in the cascade of several subsystems, and each subsystem can be modeled with an appropriate 2PN. By a well-known Chain Rule, we can obtain the overall channel transmission matrix by simply multiplying the transmission matrices of each subsystem.

In the case of a time-varying system, expressions (11.18) and (11.19) do not hold. Nevertheless, the relations between current and voltage remain linear and, as such, one can still express the dependency of the output voltage as functions of the input voltage, the input current, and the output current via the following integral relationships:

$$v_{\text{out}}(t) = \int_{-\infty}^{\infty} v_{\text{in}}(\tau)\alpha(t, t-\tau)\,d\tau + \int_{-\infty}^{\infty} i_{\text{out}}(\tau)\beta(t, t-\tau)\,d\tau \tag{11.20}$$

$$v_{out}(t) = \int_{-\infty}^{\infty} i_{in}(\tau)\gamma(t,t-\tau)d\tau + \int_{-\infty}^{\infty} i_{out}(\tau)\sigma(t,t-\tau)d\tau \qquad (11.21)$$

where $\alpha[t, t - \tau]$, $\beta[t, t - \tau]$, $\gamma[t, t - \tau]$, $\sigma[t, t - \tau]$ are defined as the time-varying inverse Fourier transform pairs of $\frac{1}{A(f)}, \frac{-B(f)}{A(f)}, \frac{1}{C(f)}, \frac{-D(f)}{C(f)}$, respectively. We can represent each of the above continuous time convolutions by their discrete-time equivalent similarly as done in Section 11.6.1.1 and write the following discrete-time model:

$$v_{out}[k] = \sum_{n=-\infty}^{\infty} v_{in}[n]\alpha[k,k-n] + \sum_{n=-\infty}^{\infty} i_{out}[n]\beta[k,k-n] \qquad (11.22)$$

$$v_{out}[k] = \sum_{n=-\infty}^{\infty} i_{in}[n]\gamma[k,k-n] + \sum_{n=-\infty}^{\infty} i_{out}[n]\sigma[k,k-n] \qquad (11.23)$$

where $\alpha[k, k - n]$, $\beta[k, k - n]$, $\gamma[k, k - n]$, and $\sigma[k, k - n]$ are defined as in (11.4).

The relationship between source and load voltages and the output voltage and current can also be easily expressed in the discrete time:

$$v_{in}[k] = v_s[k] - z_0^{(s)} i_{in}[k] \qquad (11.24)$$

$$v_{out}[k] = z_0^{(L)} i_{out}[k] \qquad (11.25)$$

Let us define $\mathrm{v}_s[i]$, $\mathrm{v}_{out}[i]$, $\mathrm{v}_{in}[i]$, $\mathrm{i}_{out}[i]$, and $\mathrm{i}_{in}[i]$ in an analogous manner as in (11.6). Then if the system memory is finite and $L < P$, then (11.22) and (11.23) can be cast in their lifted form as follows [20]:

$$\underline{\boldsymbol{v}}_{out}[i] = A_{i,0}\underline{\boldsymbol{v}}_{in}[i] + A_{i,1}\underline{\boldsymbol{v}}_{in}[i-1] + B_{i,0}\underline{\boldsymbol{i}}_{out}[i] + B_{i,1}\underline{\boldsymbol{i}}_{out}[i-1] \qquad (11.26)$$

$$\underline{\boldsymbol{v}}_{out}[i] = C_{i,0}\underline{\boldsymbol{i}}_{in}[i] + C_{i,1}\underline{\boldsymbol{i}}_{in}[i-1] + D_{i,0}\underline{\boldsymbol{i}}_{out}[i] + D_{i,1}\underline{\boldsymbol{i}}_{out}[i-1] \qquad (11.27)$$

$$\underline{\boldsymbol{v}}_{in}[i] = \underline{\boldsymbol{v}}_s[i] - z_0^{(s)}\underline{\boldsymbol{i}}_{in}[i] \qquad (11.28)$$

$$\underline{\boldsymbol{v}}_{out}[i] = z_0^{(L)}\underline{\boldsymbol{i}}_{out}[i] \qquad (11.29)$$

where $A_{i,0}$ (and $B_{i,0}$, $C_{i,0}$, $D_{i,0}$) is defined similarly as in (11.8),

$$\{A_{i,i-j}\}_{k,n} := \alpha[iP+k,(i-j)P+k-n] \qquad (k,n=0,\ldots,P-1) \qquad (11.30)$$

Using these block matrices which are band lower-triangular for the current input blocks and upper-triangular for the previous blocks, we are able to remove the interblock interference (IBI) terms via the lifted-trailing-zeros (LTZ) technique and find a simplified input–output (I/O) relationship in two-port network topology. Once we obtain a closed form of I/O relationship for adjacent front-end pairs, we apply it to multiple segments of power-line cable in a cascaded manner and find the overall I/O relationship.

11.6.1.4 Transmission Line I/O Relationship in the DT Case. For the continuous-time channel impulse response $h^{(1,\ldots,N)}(t, \tau)$ of the 2PN, we can write the following input–output relationship

$$v_{\text{out}}(t) = \int_{-\infty}^{\infty} v_s(\tau) h^{(1,\ldots,N)}(t,\tau)\, d\tau \tag{11.31}$$

and this maps to the discrete-time equivalent relationship

$$v_{\text{out}}[k] = \sum_{n=-\infty}^{\infty} v_s[n] h^{(1,\ldots,N)}[k, k-n] \tag{11.32}$$

Our objective is to calculate the lifted form $H_{i,i-j}^{(1,\ldots,N)}$ corresponding to $h^{(1,\ldots,N)}[k, k-n]$

$$\underline{\mathbf{v}}_{\text{out}}[i] = H_{i,0}^{(1,\ldots,N)} \underline{\mathbf{v}}_s[i] + H_{i,1}^{(1,\ldots,N)} \underline{\mathbf{v}}_s[i-1] \tag{11.33}$$

In order to do this, we first tackle the single system case and then we extend our results to the cascade of multiple systems.

There are various ways to find the input–output (I/O) relationship in the single system case. First, based on *Lemma 11.1*, we can prove the following theorem.

THEOREM 11.4

Given a system with finite memory $L < P$, the input–output relationship in terms of $\mathrm{v}_{out}[i]$, $\mathrm{v}_s[i]$ and $\mathrm{v}_{in}[i]$ is given by the expression

$$G_{i,0} \underline{\mathbf{v}}_{\text{out}}[i] + G_{i,1} \underline{\mathbf{v}}_{\text{out}}[i-1] = \underline{\mathbf{v}}_s[i] + J_{i,1} \underline{\mathbf{v}}_s[i-1] + Q_{i,1} \underline{\mathbf{v}}_{\text{in}}[i-1] \tag{11.34}$$

where we introduce newlydefined block matrices as follows:

$$G_{i,0} = A_{i,0}^{-1}\left(I - 1/z_0^{(L)} B_{i,0}\right) + z_0^{(s)} C_{i,0}^{-1}\left(I - 1/z_0^{(L)} D_{i,0}\right) \tag{11.35}$$

$$G_{i,1} = -1/z_0^{(L)} A_{i,0}^{-1} B_{i,1} - z_0^{(s)}/z_0^{(L)} C_{i,0}^{-1} D_{i,1} \tag{11.36}$$

$$J_{i,1} = C_{i,0}^{-1} C_{i,1} \tag{11.37}$$

$$Q_{i,1} = A_{i,0}^{-1} A_{i,1} - C_{i,0}^{-1} C_{i,1} \tag{11.38}$$

Since block matrices $A_{i,0}$ and $C_{i,0}$ are band lower-triangular and full rank, thus they are invertible. Hence we only need to check the invertibility of $G_{i,0}$ in (11.35).

COROLLARY 11.5

Under the assumption of trailing zeros, i.e. forcing the last L input symbols to be zeros, then the input–output relationship between $v_s[i]$ and $v_{out}[i]$ is

$$\underline{\boldsymbol{v}}_s[i] = G_{i,0}\underline{\boldsymbol{v}}_{\text{out}}[i] \quad (11.39)$$

If $G_{i,0}$ is invertible, then the channel transformation matrix $H_{i,0}$ is $G_{i,0}^{-1}$, otherwise $H_{i,0}$ is given by the pseudo-inverse matrix $G_{i,0}^{+}$.

11.6.2 Implementation of PLC Network

11.6.2.1 Overall Power Line Network Simulator. Using the lifted forms for the relationship between source and load voltages and the output voltage and current in (11.26)–(11.29), we find a generalized network simulator which is suitable for cascading time-varying transmission line segments.

For a P-length source voltage vector $v_s[i]$ in (11.28), we introduce an M-length symbol vector $s[i]$ via a selected precoding matrix F_l (l is channel memory) as follows:

$$\underline{\boldsymbol{v}}_s[i] = \sum_{l=-\infty}^{\infty} F_l \underline{\mathbf{s}}[i-l] \quad (11.40)$$

Let us construct $\bar{s}[i] = [s^T[i] \cdots [s^T[i+K-1]]^T$ and $\bar{v}_s[i] = \left[\underline{\boldsymbol{v}}_s^T[i] \cdots \underline{\boldsymbol{v}}_s^T[i+K-1]\right]^T$ for K sufficiently large, which result from lifting $s[i]$ and $v_s[i]$ twice. Assuming that the channel memory is finite, similarly to (11.12) the equation (11.40) simplifies to two terms.

$$\bar{v}_s[i] = \bar{F}_0\bar{s}[i] + \bar{F}_1\bar{s}[i-1] \quad (11.41)$$

where $\bar{F}_0$ multiplies the current input vector and $\bar{F}_1$ multiplies the past (IBI) term.

$\bar{F}_0$ and $\bar{F}_1$ are block matrices of size KP-by-KM. Blocks of $\bar{F}_0$ are band lower-triangular matrices. Blocks of $\bar{F}_1$ are upper-triangular matrices with few nonzero elements in their upper-right corners. Then, (11.26)–(11.29) can be rewritten with IBI terms in doubly lifted forms as follows:

$$\bar{v}_{\text{out}}[i] = \bar{A}_{i,0}\bar{v}_{\text{in}}[i] + \bar{B}_{i,0}\bar{i}_{\text{in}}[i] + \text{IBI}_{\bar{v}_{\text{out}}}[i-1] \quad (11.42)$$

$$\bar{i}_{\text{out}}[i] = \bar{C}_{i,0}\bar{v}_{\text{in}}[i] + \bar{D}_{i,0}\bar{i}_{\text{in}}[i] + \text{IBI}_{\bar{i}_{\text{out}}}[i-1] \quad (11.43)$$

$$\bar{v}_{\text{in}}[i] = \bar{F}_0\bar{s}[i] - Z_0^{(s)}\bar{i}_{\text{in}}[i] + \text{IBI}_{\bar{v}_{\text{in}}}[i-1] \quad (11.44)$$

$$\bar{v}_{\text{out}}[i] = Z_0^{(L)}\bar{i}_{\text{out}}[i] + \text{IBI}'_{\bar{v}_{\text{out}}}[i-1] \quad (11.45)$$

where we have defined

$$\mathrm{IBI}_{\bar{v}_{\mathrm{out}}}[i-1] := \bar{A}_{i,1}\bar{v}_{\mathrm{in}}[i-1] + \bar{B}_{i,1}\bar{i}_{\mathrm{in}}[i-1] \tag{11.46}$$

$$\mathrm{IBI}_{\bar{i}_{\mathrm{out}}}[i-1] := \bar{C}_{i,1}\bar{v}_{\mathrm{in}}[i-1] + \bar{D}_{i,1}\bar{i}_{\mathrm{in}}[i-1] \tag{11.47}$$

$$\mathrm{IBI}_{\bar{v}_{\mathrm{in}}}[i-1] := \bar{F}_1\bar{s}[i-1] - Z_0^{(s)}\bar{i}_{\mathrm{in}}[i-1] \tag{11.48}$$

$$\mathrm{IBI}'_{\bar{v}_{\mathrm{out}}}[i-1] := Z_0^{(L)}\bar{i}_{\mathrm{out}}[i-1] \tag{11.49}$$

In addition, $Z_0^{(s)} := z_0^{(s)} I_{KP\times KP}$, $Z_0^{(L)} := z_0^{(L)} I_{KP\times KP}$ are block matrices and $\bar{A}_{i,k}$, $\bar{B}_{i,k}$, $\bar{C}_{i,k}$, $\bar{D}_{i,k}$ (k = 0,1) have similar structures as $\bar{F}_0$ and $\bar{F}_1$, except that they are of size KP-by-KP.

Now we consider designing a two-port power line network simulator via iterative updates. As shown in (11.46)–(11.49), all IBI terms are determined by the past values. If the two-port network is activated at rest, we can arrange the system equation (or equivalently, its symbolic representation) as follows:

$$\underbrace{\begin{bmatrix} \mathrm{IBI}_{\bar{v}_{\mathrm{out}}}[i-1] \\ \mathrm{IBI}_{\bar{i}_{\mathrm{out}}}[i-1] \\ \mathrm{IBI}_{\bar{v}_{\mathrm{in}}}[i-1] \\ \mathrm{IBI}'_{\bar{v}_{\mathrm{out}}}[i-1] \end{bmatrix}}_{\bar{W}_{i-1}} + \underbrace{\begin{bmatrix} 0 \\ 0 \\ \bar{F}_0\bar{s}[i] \\ 0 \end{bmatrix}}_{\bar{S}_i} = \underbrace{\begin{bmatrix} -\bar{A}_{i,0} & -\bar{B}_{i,0} & 0 & I \\ -\bar{C}_{i,0} & -\bar{D}_{i,0} & I & 0 \\ I & -Z_0^{(s)} & 0 & 0 \\ 0 & 0 & -Z_0^{(L)} & I \end{bmatrix}}_{\bar{\Psi}_i} \underbrace{\begin{bmatrix} \bar{v}_{\mathrm{in}}[i] \\ \bar{i}_{\mathrm{in}}[i] \\ \bar{i}_{\mathrm{out}}[i] \\ \bar{v}_{\mathrm{out}}[i] \end{bmatrix}}_{\bar{Y}_i} \tag{11.50}$$

where $\bar{W}_{i-1}$ is determined from the input current and voltage vectors in the previous block at the first stage and the current vector in the previous block at the last stage.

In particular, since $\bar{\Psi}_i$ is nonsingular in (11.50), we obtain the generalized I/O relationship between $\bar{Y}_i$ and $\bar{S}_i$ as follows:

$$\bar{Y}_i = \bar{\Psi}_i^{-1}[\bar{S}_i + \bar{W}_{i-1}] \tag{11.51}$$

where $\bar{\Psi}_i$ is invertible if block matrices $\bar{A}_{i,0}$, $\bar{B}_{i,0}$, $\bar{C}_{i,0}$, and $\bar{D}_{i,0}$ are full rank. The relation (11.51) forms a basis for the PLC network simulations used in Section 11.6.2.3.

11.6.2.2 Bistatic Load Impedances and Channel Capacity. Let us now consider the case of switched impedances. We can approximate the transition between the two responses as instantaneous and decompose the time-varying load impedance $Z(t,f)$ into the sum of two alternating contributions characterized by the locally time-invariant base-band equivalent impedances $Z_1(f)$, and $Z_2(f)$ as follows:

$$Z(t,f) \approx S_A(t)Z_1(f) + (1 - S_A(t))Z_2(f) \tag{11.52}$$

where

$$S_A(t) = \begin{cases} 1 & \text{if } t \in A \\ 0 & \text{otherwise} \end{cases}$$

We also have that if the eigenvalue decomposition (EVD) of input covariance matrix is $R_{\underline{v}_s}[i] = U[i]\Lambda[i]U^H[i]$, the maximum information rate per ith block can be expressed as

$$I(\underline{\mathbf{v}}_{\text{out}}[i];\underline{\mathbf{v}}_s[i]) = \frac{1}{P}\sum_{m=1}^{\bar{P}}\log(1+\lambda_{mm}(i)\phi_{mm}(i)) \tag{11.53}$$

where $\lambda_{mm}(i)$ are eigenvalues with $\{\Lambda[i]\}_{m,k} = \delta[m-k]\lambda_{mm}(i)$, $\phi_{mm}(i)$ are power constraints, and $\bar{P}$, $\bar{P} \le P$, is the number of nonzero eigenvalues.

In addition, when $\phi_{mm}(i)$ equal to the FCC constraints, $\phi_{mm}(i) = \text{FCC}_m$, by averaging over the two channel frequency responses $T_1(f_m)$ and $T_2(f_m)$, we obtain the total achievable average channel capacity $C^{(\text{TOT})}$ as

$$C^{(\text{TOT})} = \frac{0.5}{PT}\sum_{j=1}^{2}\sum_{m=1}^{P}\log\left(1+\frac{|T_j(f_m)|^2}{N_0}\text{FCC}_m\right) \tag{11.54}$$

where $T_j(f_m)$ ($j = 1, 2$) is one of the two transfer functions for the mth symbol block.

11.6.2.3 Numerical Results. Figure 11.7 illustrates a simple model with two appliances, where each of them has a time-varying characteristic of high impedance for half of the AC cycle and low impedance for the other half. The actual values of the transfer functions are calculated using the PLC model given in reference 4. We want to compare $C^{(\text{TOT})}$ for two schemes; one is the bistatic

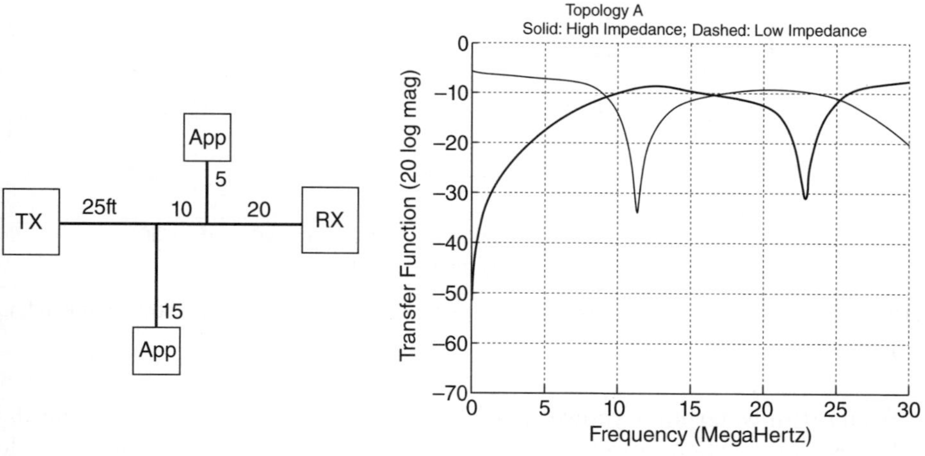

(a) (b)

Figure 11.7. Schematic diagram of **(a)** a simple model and **(b)** its corresponding transfer functions.

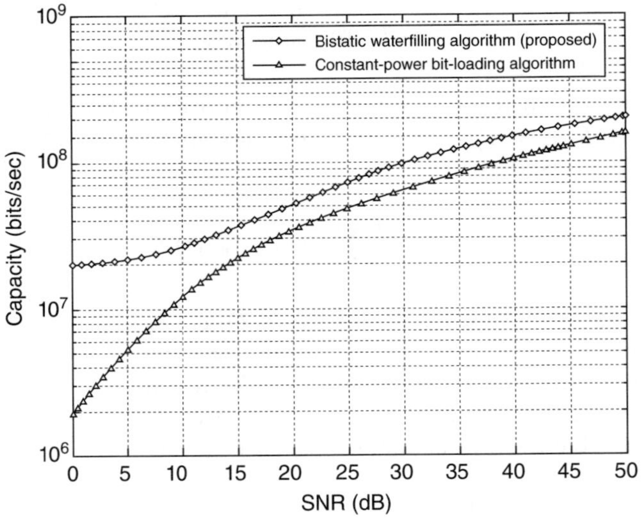

Figure 11.8. Comparison between bistatic channel capacity and constant power loading capacity over various SNR (dB).

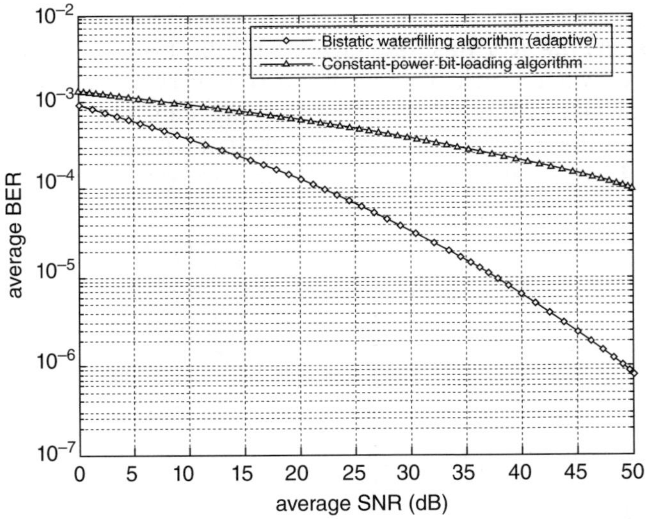

Figure 11.9. Comparison between constant-power bit-loading and adaptive water-filling algorithms with target Pe = 10^{-3}.

capacity with waterfilling (power loading [37]) proposed in this section, and the other is the constant-power bit-loading capacity (as commercial modems do now). The achievable capacity in the band 2–30 MHz for bistatic channels is obtained by (11.54). In Figure 11.8, the bistatic waterfilling capacity is compared with constant-power bit-loading capacity. Figure 11.9 illustrates the behavior of the two schemes in terms of average bit error rate (BER) versus average received SNR for a fixed target BER of $Pe = 10^{-3}$. The uniform-power bit-loading scheme has a simpler structure but slow convergence, while the adaptive waterfilling scheme exhibits relatively faster convergence. It is seen from Figure 11.8 and 11.9 that the bistatic algorithm exhibits better performance than the constant power bit-loading algorithm.

The example we have discussed shows that a power-line block transmission model over time-varying channels is useful to find the characteristics of the entire PLC network topology. The examples also show that the PLC technology can provide high-capacity and low-BER performances. Thus the PLC technology is a good candidate for inclusion in the smart grids that are described next.

11.7 SMART GRID SYSTEMS

A smart grid can achieve the increased utilization of capital assets while minimizing operations and maintenance costs. Optimized power flows reduce waste of high-cost generation resources and maximize use of low-cost generation resources. Harmonizing local distribution with cross-regional energy flows and transmission traffic improves the use of existing grid resources and reduces grid congestion and bottlenecks, which can ultimately produce consumer savings.

11.7.1 Features

Today existing and planned implementations of smart grids provide a wide range of features including the PLC technology. As illustrated in Figure 11.10 [16], the PLC technology can be directly applied to smart grid system. For example, automated meter reading (AMR) and automated meter management (AMM) via power lines can be easily incorporated. In cooperation with consumer premise equipments (CPE) and repeaters (REP), WebHost Manager (WHM) controls the flow of raw MV level powers before reaching customer premises. It also feeds back the measurement information to MV node and eventually electric power company, which operates daytime fare setting, power quality check, and remote switching.

11.7.1.1 Load Adjustment. Total load connected to the power grid can vary significantly over time. Total load is the aggregate of individual loads, thus it is not guaranteed to be stable or slow-varying. Using mathematical prediction algorithms, it is possible to predict how many standby generators need to be used, and hence to overcome a certain failure rate. In a smart grid, the load reduction by even a small portion of the clients may resolve the problem of fast-varying loads without the need for large number of additional standby generators.

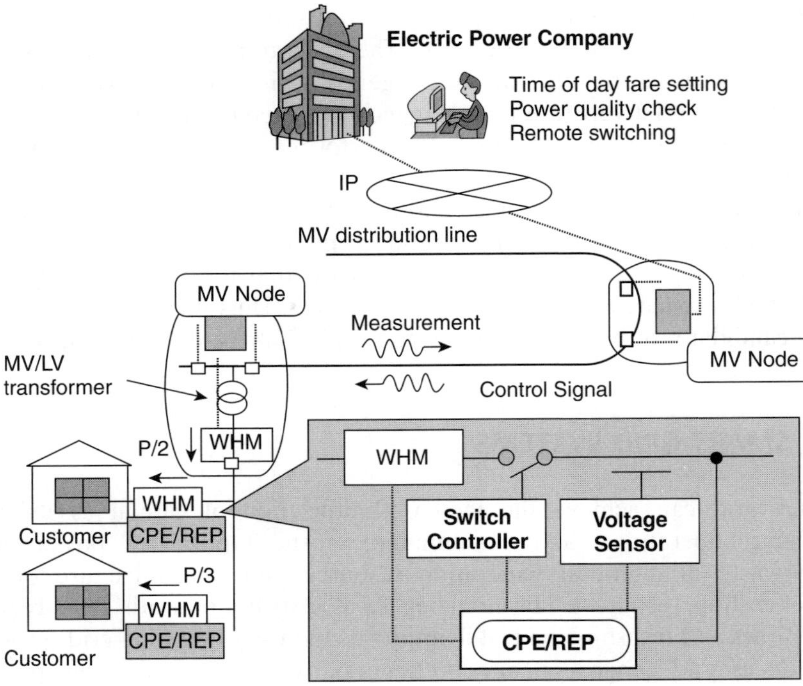

Figure 11.10. Convergence of PLC and smart grid. (Reproduced by courtesy of reference 16.)

11.7.1.2 Demand Response Support. Demand response support [38], enables generators and loads to automatically interact in real time, coordinating the demand to flatten power demand spikes. Eliminating the fraction of demand that occurs in these spikes lowers the cost for additional reserve generators, extends the life of equipment, and allows users to cut their energy bills by notifying low-priority devices to use energy only when it is cheapest. Current power grid systems have varying degrees of communication within control systems for their high value resources, such as generating plants, transmission lines, substations, and major energy users.

11.7.1.3 Distributed Power Generalization. Distributed power generation in terms of fault tolerance of smart grids allows individual consumers to generate power on the spot, using whatever generation method they find appropriate. This allows individual loads to fit their generation directly to their load, making them independent from grid power failures. Early grids were designed for one-directional flow of electricity, but if a local subnetwork generates more power than it is consuming, the reverse flow can cause safety and reliability issues. Smart grid solutions are being developed to cope with these issues.

11.7.2 Technology

Many smart grid technologies are already used in other applications such as manufacturing and telecommunications and are being adapted for use in grid operations. In general, smart grid technology can be grouped into five key categories [39].

11.7.2.1 Integration of Communications. Plenty of communication technologies have been developed over time but have not been fully integrated. In most cases, data are being collected via modem rather than direct network connection. Room for improvement lies in substation automation, demand response, distribution automation, supervisory control and data acquisition, energy management systems, power-line carrier communications, fiber-optics, and so on [40]. Integrating several types of communications technologies will allow for the following: real-time control; information and data exchange to optimize system reliability; resource utilization; and security issue.

11.7.2.2 Sensing and Measurement. In the area of sensing and measurement, important tasks are evaluating congestion and grid stability, monitoring equipment health, preventing energy theft, and supporting control strategies. Such technologies include: advanced microprocessor meters (smart meter) and meter reading equipment, wide-area monitoring systems, online readings by distributed temperature sensing, real-time thermal rating (RTTR) systems, electromagnetic signature measurement/analysis, real-time pricing tools, advanced switches and cables, backscatter radio technology, and digital protective relays [41].

11.7.2.3 Smart Meters. In a smart grid system, digital meters that record usage in real time are substituting analog mechanical meters. Smart meters are similar to advanced metering infrastructure meters and provide a communication path extending from generation plants to electrical and other smart grid-embedded devices. Such devices can be shut down according to customer preference during times of peak demand [42].

11.7.2.4 Phasor Measurement Units (PMU). It has been believed that high-speed sensors distributed throughout the electrical supply network can be used to monitor power quality and automatically control the state of the network. Phasors represent the waveforms of alternating current, which are identical everywhere on the network. In the 1980s, the clock pulses from Global Positioning System (GPS) satellites were used for precise time measurements in the grid. Thanks to the ability to record phases of alternating current everywhere on the grid, automated systems are expected to facilitate the management of power systems by responding to system conditions in a rapid, dynamic manner.

11.7.2.5 Wide-Area Measurement System (WAMS). A Wide-Area Measurement System (WAMS) is a smart network of PMUs that can perform real-time monitoring on a regional and national scale. Most research scientists in the power systems areas believe that the Northeast blackout of 2003 would have been limited to the smaller region if a wide-area phasor measurement network was in place [39].

11.7.3 Applications

Smart grid communications solutions are crucial to building an integrated intelligent grid. High-performance smart grid applications require two-way communications in real-time between centralized fusion center and a number of smart devices throughout the electrical network.

As a practical example, BPL Global offers a variety of smart grid communications solutions such as fiber, wireless, broadband over power line, WiMax, GPRS, Ethernet, radio, and other communication technologies [43]. The broadband over power-line network technology can be utilized in combination with a utility fiber or Ethernet communications infrastructure. In addition to designing and building smart grid communications networks, BPL Global provides full communications and electrical network monitoring and management tools to ensure high quality of service (QoS), cost-effective operations, and effective monitoring management of the smart grid.

Open architecture described in Figure 11.10 enables all smart grid applications to be supported through one common communications network. The reliability of the communications network is essential because this is the backbone for smart grid applications deployed by the utility company [38,43].

11.7.4 From Smart Grids to Energy Internet

Achieving secure and reliable delivery of energy is essential to modern society, but is very challenging due to increasing demand and declining resources. The ongoing effort to restructure the current delivery infrastructure is to improve its performance so that energy can be utilized with higher efficiency. Smart grids have a number of unique features compared to their predecessors: (a) detecting and correcting power flow problems at their very early stage, (b) receiving and responding to broader range of information, (c) possessing rapid recovery capability, (d) adapting to changes and self-reconfiguring accordingly, (e) building in reliability and security from design, and (f) providing operators advanced visualization tools [44–46].

Emerging smart grids start to resemble the Internet system and hence have become known as Energy Internet. Apparent benefits from energy internet are its openness, robustness and reliability. The availability of resources determines that massive generation of energy, such as electricity, has to be centralized. While customers are highly distributed, an extremely sophisticated transmission and distribution network is needed for energy delivery. The challenge is that our current knowledge about complex systems like the electric power grid does not enable us to regulate it efficiently and reliably. Often a compromise has to be made between efficiency and reliability. Consumers will have higher expectations for the service, both for quality and quantity. On the other hand, resources are limited. Hence, generating and saving the energy at the same time will be the main target to satisfy both efficiency and reliability.

11.7.5 Interoperability of Smart Grids through Standards

While there are existing standards for Smart Grid technology that should be used (e.g., distributed generation, metering standards, communications standards), new standards will have to be developed for new interfaces [47]. These include interfaces from the generation sources to the equipment back at the homes and businesses that will need to communicate through the grid, to real-time systems for energy transmission, storage, billing, load balancing, and so on. Therein, interoperability through standards will be the key to making plug-and-play capabilities and to driving down the costs of the various hardware and software systems.

11.7.6 Electric Vehicle Interconnection with Smart Grids

Renewable energy sources have a problem of variability or intermittency, and means of storage are needed to ensure the stability of the grid [47]. Batteries are one appropriate solution. Another possibility that could address both transportation requirements and energy management would be the advent of widespread deployment of plug-in electric or plug-in hybrid electric vehicles.

Electric vehicles, if connected to smart grid, could provide additional non-transportation functionality (e.g., a distributed energy storage medium), which could help in load regulation. A variety of activities need to be done so that vehicle to grid interconnection becomes a reality. First of all, hardware requirements need to be defined for the vehicle and grid interfaces. Secondly, requirements for communications, metering, and billing need to be identified. Finally, utility contracts have to be in place to make use of any capabilities for optimizing two-way electricity flows.

11.7.7 Promises of Future Smart Grids

Future electric power grid is expected to be very different from what it is today. It has to accommodate a large number of renewable generators whose outputs are intermittent and stochastic. Consumer demand will also be stochastic due to demand response programs, smart meters, and intelligent appliances. Extensive deployment of high-resolution sensors and high-speed sensor networks will provide time-synchronous measurements in milliseconds, thus enabling better control of the power grid. However, more research will be needed to understand a new paradigm for the power system operation of future smart grid.

11.8 CONCLUSIONS

Today we have a better understanding of the power-line channel. For practical, promising PLC-smart grid systems, grounding and wiring practices should be further exploited for transceiver optimization. Harmonization of standards and regulations can make analysis of signal transmission more focused.

The innovation potential for PLC is enormous, creating considerable economic values, from which, due to the nature of the powerline medium, everybody may benefit. Toward this goal, it is a primary intention of this chapter to help evaluate the promises and limitations of PLC with respect to everyone's individual needs. For a clear and complete illustration of the various facets of PLC, recent results and challenges in terms of industry practices and theoretical analysis are presented in order to prepare the potential user for this emerging technology.

Smart grid technology is being recognized as a key solution to challenges such as increasing electric demand and the environmental impact of greenhouse gases produced during electric generation. Integrated smart grid solutions combine advanced sensing technology, two-way high-speed communications and home energy management solutions to provide enhanced services for the endusers.

In addition, the energy Internet is emerging as an implementation of smart grids. It will provide openness, robustness, and reliability. Building an Internet type of energy network for the future will help to resolve some of the pressing energy challenges. Advances in information technology and ongoing research on power infrastructure and complex system will make this goal achievable.

REFERENCES

1. K. Dostert, *Powerline Communications, Prentice-Hall*, Upper Saddle River, NJ, 2001.
2. H. Hrasnica, A. Haidine, and R. Lehnert, *Broadband Powerline Communications: Network Design*, John Wiley & Sons, Hoboken; NJ, 2004.
3. S. Barmada, A. Musolino, and M. Raugi, Innovative model for time-varying power line communication channel response evaluation, *IEEE J. Sel. Areas Commun*, Vol. 24, No. 7, pp. 1317–1326, July 2006.
4. S. Galli and T. Banwell, A novel approach to the modeling of the indoor power line channel—Part II: Transfer function and its properties, *IEEE Trans. Power Delivery*, Vol. 20, No. 3, pp. 1869–1878, July 2005.
5. F. Corripio, J. Arrabal, L. Del Rio, and J. Munoz, Analysis of the cyclic short-term variation of indoor power line channels, *IEEE J. Sel. Areas Commun*, Vol. 24, No. 7, pp. 1327–1338, July 2006.
6. K. Dostert, Telecommunications over the power distribution grid—possibilities and limitations, in *IEEE International Symposium on Power Line Communications and Its Applications (ISPLC'97)*, Germany, 1997.

7. http://en.wikipedia.org/wiki/Power_line_communication.
8. http://en.kioskea.net/contents/cpl/cpl-intro.php3.
9. http://www.yitran.com/index.aspx?id=3387.
10. http://www.powerq.com.my/telecommunication/distribution-line-carrier-system.
11. G. Hamoud, R. L. Chen, and I. Bradley, Risk assessment of power systems SCADA, in *IEEE Power Engineering Society General Meeting*, July 2003.
12. D. Schneider, L. Stadelmeier, J. Speidel, and D. Schill, Precoded spatial multiplexing MIMO for inhome power line communications, in *IEEE Global Telecommunications Conference (GLOBECOM'08)*, New Orleans, 2008.
13. http://www.tgdaily.com/content/view/24994/103/.
14. E. Mainardi and M. Bonfè Powerline communication in home-building automation systems, *J. Robotics Automation in Construction*, 2008.
15. Y. Lin, A. Latchman, and R. E. Newman, A comparative performance study of wireless and power line networks, *IEEE Communi. Mag.*, Vol. 41, No. 4, pp. 54–63, April 2003.
16. www.argreenhouse.com/papers/sgalli/Sapienza06_PLC.pps.
17. Z. Mingyue, L. Chunying, and B. Haiying, Study of Channel Characteristics of Power Line Communications Networks, in *Proceeding Parallel and Distributed Computing Applications and Technologies (PDCAT'05)*, 2005.
18. M. Gotz, M. Rapp, and K. Dostert, Power line channel characteristics and their effect on communications system design, *IEEE Communi. Mag.*, Vol. 42, No. 4, pp. 78–86, April 2004.
19. T. Sung, Innovative PLC network design for bit loading algorithm and bistatic channel capacity, *IEEE International Symposium on Consumer Electronics (ISCE'09)*, Kyoto, Japan, May 2009.
20. T. Sung, Weighted OFDM and MDFB over time-Varying power line block transmission models, in *IEEE International Conference on Signal Processing System (ICSPS'09)*, Singapore, May 2009.
21. M. Zimmermann and K. Dostert, A multipath model for the powerline channel, *IEEE Trans. Commun.*, Vol. 50, No. 4, pp. 553–559, April 2002.
22. S. Jung, A channel model for power line communication in home network, in *Proceeding CISL*, February 2002.
23. http://training.corinex.com/corinex/company/all-news/11-press-releases/39-access-broadband-over-powerline.
24. http://www3.ntu.edu.sg/ntrc/Research_2.htm.
25. http://grouper.ieee.org/groups/bpl/index.html.
26. www.patentstorm.us/patents/7286812/description.html.
27. http://www.freshpatents.com/Power-line-communication-system-and-capacitive-signal-coupling-unit-dt20081211ptan20080303609.php.
28. http://www.mattrone.com/eng/inductive%20coupler.html.
29. http://gomponent.hobbyist.de/ac-coupling/plcc.pdf.
30. M. Yuichiro and K. Toru, Inductive coupling unit and bypass tool for power line communications, *J. Mitsubishi Electricity Adv.*, Vol. 109, pp. 18–20, 2005.
31. http://www.answers.com/topic/inductive-coordination.
32. S. Galli and O. Logvinov, Recent developments in the standardization of power line communications within the IEEE, *IEEE Communi. Mag.*, Vol. 46, No. 7, pp. 64–71, July 2008.

33. www.echelon.com/support/documentation/datashts/153x0.pdf.
34. http://grouper.ieee.org/groups/1901/index.html.
35. T. Sung, A. Scaglione, and S. Galli, Time-varying power line block transmission models over doubly selective channels, in *IEEE International Symposium on Power Line Communications and Its Applications (ISPLC'08)*, Jeju island, Korea, April 2008.
36. E. Biglieri, S. Galli, Y. Lee, H. Poor, and A. Han Vinck, Power line communications, Guest Editorial for the Special Issue on PLC, *IEEE J. Sel. Areas Commun*, Vol. 24, No. 7, pp. 1261–1266, July 2006.
37. T. Cover and J. Thomas, *Elements of Information Theory*, John Wiley & Sons, Hoboken, NJ, 2006.
38. http://www.grid-net.com.
39. http://en.wikipedia.org/wiki/Smart_grid.
40. http://thegreenbutton.com/blogs/chris_blog/archive/2008/05/13/262489.aspx.
41. http://tdworld.com/test_monitor_control/highlights/lios-middle-east-technology-0109.
42. http://www.smarthomeusa.com/info/UPB/about/.
43. http://www.bplglobal.net/eng/markets/index.aspx.
44. L. H. Tsoukalas and R. Gao, From smart grids to an energy internet: assumptions, architectures and requirements, *DRPT 2008 IEEE International Conference*, Nanjing, China, April 2008.
45. European Commission, Vision for Europe's electricity networks of the future, *European Smart-Grids Technology Platform*, EUR 22040, 2006.
46. San Diego Smart Grid Study Final Report, Energy Policy Initiatives Center (EPIC), October 2006.
47. R. DeBlasio and C. Tom, Standards for the smart grid, in *IEEE Energy 2030*, Atlanta, GA, November 2008.

PART III

WIRELESS TECHNOLOGIES AND SPECTRUM MANAGEMENT

12

SIGNALING FOR MULTIMEDIA CONFERENCING IN 4G: ARCHITECTURE, EVALUATION, AND ISSUES

Chunyan Fu, Ferhat Khendek, and Roch Glitho

The fourth-generation wireless system (4G) is seen as an integration and an evolution of existing wireless network architecture such as 2G and 3G, with new ones such as mobile ad hoc networks (MANETs). One major issue in 4G is the provisioning of ubiquitous and seamless service access with different underlying wireless technologies. Multimedia conferencing is seen as a service that can enable many "killer" applications such as audio/video conferencing, gaming, and public debating in 4G. In this chapter, we discuss an important technical aspect of conferencing: signaling. It refers to session establishment, modification, and termination. It is indispensable and critical for each phase of conferencing. We will focus on the signaling architectures in 4G (including 3G network, MANETs, and integrated MANETs/3G), and we analyze the signaling performance, the issues, and some solutions.

This chapter consists of three sections. In Section 12.1, we introduce the background information and the state of the art. The concepts of 4G, MANETs, conferencing, and signaling are presented in that section. Section 12.2 is devoted to signaling architectures in 4G. In Section 12.3, we will discuss some signaling performance issues and present a solution that is based on cross-layer design.

Convergence of Mobile and Stationary Next-Generation Networks, edited by Krzysztof Iniewski

12.1 BACKGROUND: 4G, MOBILE AD HOC NETWORKS, AND CONFERENCING

In this section, background information that helps to understand the signaling for conferencing in 4G is provided. We start by introducing the definition and research challenges of 4G. As the new network component in 4G, the concept, classification, and technologies of mobile ad hoc network (MANET) are also presented. The integration of MANETs and 3G is discussed afterward. We introduce the concept, the technical components, and the classification of conferencing in the last subsection.

12.1.1 A Brief Overview of 4G

12.1.1.1 The Concept of 4G. The driving forces of research and development of the fourth-generation wireless system (4G or beyond 3G) are the increasing number of mobile service subscribers, the increasing needs for data and multimedia services, and the demand for high-speed and ubiquitous communications. There are different visions of 4G from different organizations at different stages [1]. However, a broad consensus has been reached around the vision of the Wireless World Research Forum (WWRF) [2]. The WWRF foresees 4G wireless systems as an integration of both legacy wireless systems such as 2G and 3G, and it also foresees new networks such as mobile ad hoc networks (MANETs) and sensor networks. The purpose of 4G system is to provide "a high-data rate transmissions and highly sophisticated services, comparable to those offered by wired networks and even going beyond them." In order to achieve this purpose, IP comes in vision and is to be supported all over the system. This makes the ubiquitous and seamless service access possible. Handoffs between networks and systems are considered to ensure a global roaming across multiple wireless and mobile networks. High bandwidth is critical to guarantee the end-to-end Quality of Service (QoS). In addition, the concepts of context awareness, service composition, mobility, and adaptation are involved in order to provide comprehensive applications and services.

We use the definition from Kim and Prasad [1] to summarize the 4G concept:

> The 4G will be a fully IP-based integrated system of systems and network of networks achieved after the convergence of wired and wireless networks as well as computer, consumer electronics, communication technology, and several other convergences that will be capable of providing 100 Mbps and 1 Gbps, respectively, in outdoor and indoor environments with end-to-end QoS and high security, offering any kind of services anytime, anywhere, at affordable cost and one billing.

12.1.1.2 Main Research Issues in 4G. 4G research issues can be classified into issues related to high-speed wireless access, issues involving network heterogeneity, and issues associated with service and application provisioning. The issues related to high-speed wireless access focus on an upgrade of existing wireless systems and applying new air-interface technologies. For examples, the researches related to multiantenna and multidimensional channel modeling (e.g., multiple-input and multiple-output [MIMO]), and the researches related to short-range high-speed networks fall into this category. Research issues related to network heterogeneity have been identified in reference 3. Examples are network interoperation, handoff, mobility, location coordination, resource coordination and network failure and backup. These issues are quite related to what network and terminal technologies will be involved in the future 4G system. The issues associated to service provisioning are identified by WWRF in reference 2. Examples are service adaptation, context awareness, service overlay, charging and billing, and security and authentication.

In the next subsection, we will introduce one of the new components of 4G: the mobile ad hoc networks, which not only provide high bandwidth wireless access, but also can increase the coverage of cellular networks.

12.1.2 Mobile Ad Hoc Networks

Mobile ad hoc networks (MANETs) can be defined as transient networks formed dynamically by a collection of arbitrarily located wireless mobile nodes, without the use of existing network infrastructure or centralized administration [4]. They rely on wireless technologies such as IEEE 802.11 and Bluetooth. An important assumption for MANETs is the multihop routing. Each node in MANETs may play the roles of both a router and host. Devices in a MANET can be heterogeneous, such as personal digital assistants (PDAs), laptops, palm PCs, and cell phones.

12.1.2.1 Classifications of MANETs. We introduce two classifications for MANETs according to two different criteria. The first one is related to the coverage area. The second one focuses on the relationships with other networks. In terms of coverage area, ad hoc networks can be classified into four types: body, personal, local, and wide-area networks [5]. Wide-area ad hoc networks (WANs) are large-scaled mobile multihop wireless networks. They generally are of interest to military users. On smaller scales, body area network (BAN) is strongly correlated with wearable computers. The components of a wearable computer are distributed over the body (e.g., head-mounted displays, microphones, earphones, etc.), and BAN provides the connectivity among these devices. The communication range of a BAN is about 1 m. Personal area network (PAN) is a network in the environment around the person. PAN connects mobile devices to other mobile or stationary devices. The typical communication range of a PAN is 10 m. Wireless LAN has communication range of 100–500 m, so it is the solution for home and office automation. In relation to other networks, mobile ad hoc networks can be classified into standalone ad hoc networks or integrated ad hoc networks [6]. A standalone ad hoc network is a network in which every node only communicates with other nodes in the same networking area. It does not have a connection with other networks, such as Internet. An integrated mobile ad hoc network is a MANET that connects with some infrastructure-based networks, such as 3G networks and Internet. The integration of MANETs and 3G networks is of special interest for us because it is one of the important scenarios in 4G wireless system. The benefit of it includes extending the coverage of the 3G wireless cells and balancing the load between these cells. We introduce the integration technology in Section 12.1.3.

12.1.2.2 Technologies and Standards. Technologies and standards for MANETs are emerging. In the physical and data link layers, there are several standards for wireless PAN and wireless BAN. One such example is IEEE 802.15.3 [7], which provides high data-rate personal area ad hoc networks. Within the same working group, a very high data rate wireless PAN is being investigated in IEEE 802.15.3a, which supports ultra-wideband (UWB) devices (500 Mbit/s). The other example is Bluetooth [8] (IEEE 802.15.1), which intends to serve as universal low-cost air interfaces that will replace the plethora of proprietary interconnecting cables between personal devices.

A major WLAN standard is the IEEE 802.11 (or WiFi as marketing term) [9], which is commercially successful and widely used in enterprises and educational organizations. There are two possible settings in the IEEE 802.11. One is infrastructure-based setting in which the access point (AP) is defined. The AP is normally connected to a wired network, thus providing Internet access. The other is the ad hoc setting in which nodes are dynamically configured to set up a temporary network. MANETs are only related to this setting. Other standards of WLANs are Digital Enhanced Cordless Telecommunications (DECT), HiperLAN, and Infrared WLANs. Further details about these technologies can be found in reference 10.

The standards for WANs are under development. Three Wireless Metropolitan Area Networks (WMANs) are emerging: IEEE 802.16 (or WiMax as marketing term), ETSI HiperMAN, and WiBro (from South Korea). There is no ad hoc network setting in the current versions of WMAN. However, since a MANET built using a WMAN technology is of special interest to military users [11], a MANET extension of IEEE 802.16 has been proposed in reference 12.

At the network layer, IETF MANET working group has standardized four IP routing protocols. They are Ad Hoc On-Demand Distance Vector (AODV) routing [13], Optimized Link State Routing Protocol (OLSR) [14], Topology Dissemination Based on Reverse-Path Forwarding (TBRPF) [15], and Dynamic Source Routing Protocol (DSR) [16]. OLSR and TBRPF are proactive link state routing protocols, while AODV and DSR are reactive routing protocols.

12.1.2.3 Research Issues and Challenges for MANETs. Because of the unique characteristics of MANETs—that is, the absence of an infrastructure, the unreliable network links, the scarce network resources, and the mobile, transient, and heterogeneous network nodes—research issues and challenges can be found in each of the network layers. Liu and Chlamtac [4] summarize these challenges and highlight research issues for each layer as shown in Figure 12.1.

Network Layers	Research issues	Cross-Layer issues
Application layer Presentation layer Session layer	→ New/killer applications Network auto-configuration Location services Security (authentication, encryption)	Energy conservation, QoS Reliability Scalability Network Simulation Performance Optimization
Transport layer	→ TCP adaptation, Backoff window	
Network layer	→ IP routing, addressing, optimization, multicasting	
Data link layer	→ Media access control, error correction, optimization	
Physical layer	→ Spectrum usage/allocation	

Figure 12.1. Research issues in MANETs.

12.1.3 Integrated MANETs/3G Networks

An integrated MANETs/3G network is also recognized as a type of Multihop Cellular Network (MCN) in the wireless network domain. MCN is a concept contrasting with the traditional Single-hop Cellular Network (SCN) such as 2G, 2.5G, and 3G wireless networks. Lin and Hsu [17] first posed the concept. After that, many other researchers have contributed to the lower-layer connection techniques [18–23], routing strategies [24–26] and mobility management [27].

Cavalcanti et al. [28] summarizes the connection alternatives of a MANET and a 3G cellular network. In these alternatives, a general assumption is that users have two wireless interfaces: one to MANET and the other one to 3G. The users may communicate directly through 3G interfaces. They may also communicate directly through MANET interfaces. Furthermore, a user in a MANET may connect to a gateway or a relay, which can establish a connection with another user in 3G network. Another assumption is that MANETs and 3G networks are tightly connected; that is, all the users share the same 3G core network.

To illustrate how the integration is concretely done at the lower layers, we use two examples, iCAR [18] and UCAN [19]. They take different advantages of the integrated 3G/MANETs and they use different methods for integration. According to Fu et al. [56].

> In iCAR, MANETs are used to balance the traffic load between the wireless cells. An entity, ad hoc relaying station (ARS), is defined to divert the traffic from congested cells to lightly loaded ones. ARSs are wireless devices deployed by the network operator. They have two air interfaces, one to communicate with the cellular base transceiver stations (BTSs) and the other to communicate with mobile host (MH) and other ARSs. Three strategies are defined for traffic relaying. First, an ARS directly relays new calls from a congested cell to a neighboring cell. This is

called primary relaying. However, if an MH is not close to an ARS, the system will re-sort the traffics and follow the second strategy—that is, release the channel from the MHs that are close to the ARSs, relay their traffic to neighboring cells, and allocate the channel to the MH in need. In this case, an MH-to-MH call via ARSs only (i.e., without BTSs involved) is defined. The third relaying strategy, called cascaded relaying, is the double uses of the second strategy. It covers the situation when both cells, where the calling party and the called party are located, are congested.

In UCAN, MANETs are used to extend the coverage of the wireless cells. The system aims at improving the throughput using multihop routing when the quality of the signal in the downlink channel between the BTS and the MH is poor. Dissimilar to iCAR, it does not define the deployed entity. Instead, it uses proxy clients and relay clients to relay packets toward the destination MH. Proxy clients are the MHs that have better downlink signals with the BTS and act as the interface between the MANET and the cellular network. Relay clients are hops to relay the traffic between proxies and destination MHs. In order to find a proxy client, two discovery protocols have been proposed, a proactive greedy scheme and a reactive on-demand protocol. These protocols use the pilot burst information (that reflects the downlink channel condition) collected by the MHs to discover a proper proxy MH.

12.1.4 Multimedia Conferencing

Multimedia conferencing (also known as multimedia multiparty sessions) can be defined as the conversational exchange of multimedia content between several parties. It consists of three components: conference control, signaling, and media-handling. Conference control is related to conference policies, admission control, floor controls, and voting. Signaling is used to set up, modify, and tear down sessions. It is required before, during, and after the media transmission. Media handling is concerned with the transmission and the mixing of the media streams. It also performs the possible trans-coding between different types of media streams.

12.1.4.1 Classifications. There are many classification criteria for conferencing. The most commonly used are presented in this section. According to Fu et al. [56].

Conferences can be with or without floor control. Floor control is a technology that deals with conflicts in shared work spaces [29]. It coordinates the concurrent usage of shared resources and data among participants of a conference. A typical example of floor control is the management of the turn of speaking in a conference—that is, how and when to allocate the audio channels to involved parties in order to ensure fairness and avoid collisions. Conferences can be prearranged or ad hoc. A prearranged conference starts at a predetermined time and is sponsored by specific parties. The duration of a conference may also be predefined. An ad hoc conference starts when the first two parties decide to create a session. Parties may join and leave during the conference, and it ends when the last two parties leave. Another criterion is whether the conference is private (closed) or public (open). A closed or private conference does not welcome parties to join freely. Only the parties who are invited by the conference participants can join. An open or public conference on the other hand publishes its information to all parties in a network. Any party can join the conference if and when it wishes. Yet another criterion is whether the conference has subconferencing capabilities. The subconferencing capability simulates a conference with different rooms as in the real world. In each room, called a subconference, parties can hear/see each other, but they cannot hear/see others that are in different subconferences.

The remaining commonly used criterion is the topology used for signaling and media handling. Schulzrinne and Rosenberg [30] has discussed four main topologies for signaling and media handling: end-system mixing, full mesh, multicast, and centralized. In end-system mixing, one of the participants in the conference does the mixing for all the other participants. In general, due to the limited capability of participants, very few participants can be supported in this type of conferences. In full mesh, every end-system does its own mixing and has a signaling link with every other end-system. Multicast is an enhanced form of full mesh. Every end-system still does its own mixing. However, packets are sent to a multicast address instead of being sent point-to-point. In centralized conferences, a conference bridge is defined to do mixing for all the end systems. Each end-system has a direct signaling connection with the bridge. In this model, a participant may either call the bridge to join a conference (dial-in) or be called by the bridge to join (dial-out). A similar but more recent classification has been presented in IETF RFC 4353 [31]. Three models are defined and different names are used: loosely coupled conference (use of multicast), tightly coupled conference (centralized model), and fully distributed conference (full mesh model).

12.1.4.2 Techniques and Standards. IETF and ITU-T have developed standards for each aspect of conferencing. We present them in this subsection. We also present the related work on signaling for conferencing outside of the standard bodies.

12.1.4.2.1 STANDARDS FOR CONFERENCE CONTROL. Conference control has been defined by ITU-T in T.120 Series [32]. The control policies are mainly focused on centralized conferences. From a historical point of view, early conference control contributions in IETF are based on loosely coupled, multicast conferences. Protocols such as Multimedia Conference Control (MMCC) [33], Agreement Protocol [34], and Conference Control Channel Protocol (CCCP) [35] were defined. Simple Conference Control Protocol (SCCP) [36] was the first draft that tried to map loosely coupled conference into ITU-T T.120 series. However, it still relies on multicast. Recently, IETF has changed its focus from loosely coupled to tightly coupled conference, known as a Common Conference Information Data Model for Centralized Conferencing (XCON [37]). It also defines Conference Policy Control Protocol (CPCP) [38] for centralized conference control. The IETF Binary Floor Control Protocol (BFCP [39]) is defined for floor control.

12.1.4.2.2 STANDARDS FOR MEDIA HANDLING AND MEDIA CONTROL. The most widely used media transmission protocols are Real-time Transport Protocol (RTP) and RTP Control Protocol (RTCP), which are defined by IETF RFC 3550 [40] and RFC 3551 [41], respectively. They are also included in H.323 series as real-time media transport protocols. RTP provides end-to-end network transport functions that are suitable for applications that transmit real-time data, such as audio, video, or data, over multicast or unicast network services. It supports the use of translators and mixers. RTCP allows monitoring of the data delivery of RTP.

The Media Gateway Control Protocol (or Megaco) is a signaling protocol between conference signaling and media handling. It is defined by IETF RFC 3525 [42] and ITU-T H.248.1 [43]. The idea is a separation of call control and media handling logics, adding the control commands between them. This separation introduces more flexibility for deploying multimedia conference services. The 3GPP has also used this protocol for conferencing. The Media Server Control Markup Language (MSCML [44]) is a new emerging IETF standard for media control. It embeds media control command as an XML format in SIP message body. Comparing to Megaco, it is much more lightweight but is with less functionality.

12.1.4.2.3 Standards for Signaling. Signaling protocols have also been defined by both ITU-T and IETF. The most widely applied signaling protocols are H.323 [45] from ITU-T and Session Initiation Protocol (SIP) [46] from IETF.

H.323 is a set of specifications. It is actually the very first set of signaling standards created after Signaling System 7 (SS7 that is used for circuit switching networks). Reference 47 provides an overview. H.323 defines four entities: terminal, gateway, gatekeeper, and Multipoint Control Unit (MCU). Basic signaling and media handling functions of H.323 are located in terminals. A gateway is a component that bridges H.323 sessions to other types of networks. A gatekeeper, although it is not a mandatory entity, may have many functions such as user admission, zone management, bandwidth control, and address translation. The specifications of H.323 cover more than signaling. The H.323 protocols are binary encoded and include three different signaling protocols: Registration Admission and Status (RAS), H.225, and H.245. RAS is used between end-points and gatekeepers. It enables gatekeeper discovery and registration/un-registration of end-points with gatekeepers. H.225 is the protocol for call establishment and teardown. H.245 enables media capability negotiation.

Multimedia conference control in H.323 is done via MCU. An MCU can be further divided up into two entities: multipoint controller (MC) and multipoint processor (MP). MC handles signaling while MP handles media. MP is an optional entity. It is not required in a decentralized conference model in which media are distributed through multicast. MC is mandatory. It is a central control point for both centralized (i.e., where media mixing is done in a central MP) and decentralized models. The conference models defined in H.323 are shown in Figure 12.2. H.323 has been applied in 2G wireless system for voice over IP (VoIP) services.

SIP is specifications including a baseline protocol and a set of extensions. In baseline protocol, it defines four entities: user agent (UA), proxy server, location

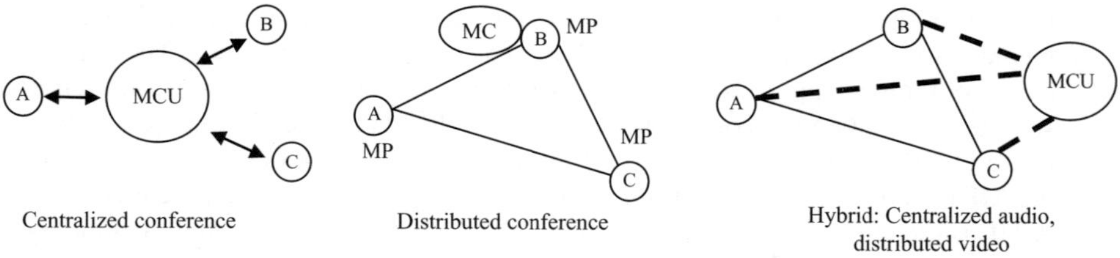

Figure 12.2. Conference models for H.323.

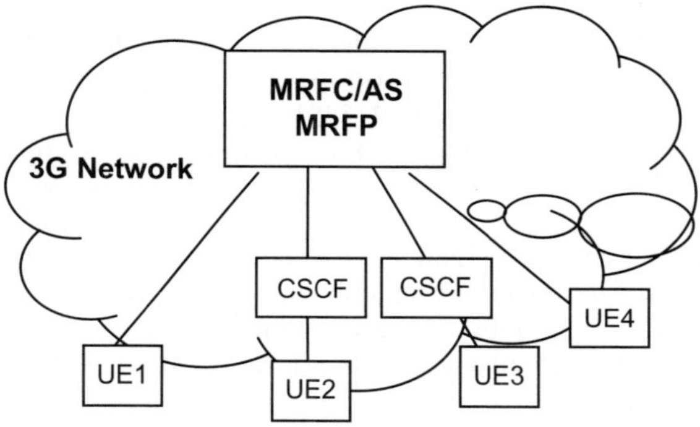

Figure 12.3. 3GPP 3G conference architecture (Copyright IEEE 2006 [56].)

server, and registrar. Session control functions are located in UAs. SIP servers are nonmandatory entities that help to route SIP messages and to locate SIP user agents. Reference 48 gives an overview of SIP. SIP is lightweight and extendable, and it has been adopted by the two main standards bodies for 3G networks (i.e., 3GPP and 3GPP2) as the sole signaling system for multiparty sessions. It is a text-based request/reply protocol.

IETF has been working on SIP as conferencing signaling protocol since year 2000. SIP has been used for two conference models—loosely coupled and tightly coupled. A loosely coupled conference is based on multicast. IETF draft [36] describes SCCP, a loosely coupled conference control protocol that uses SIP as the signaling protocol. The signaling architecture is centralized. Signaling messages are exchanged between a controller and a participant through multicast.

A tightly coupled conference is central-server-based. SIP usage in this sort of conference model is defined in reference 31. SIP creates sessions between each participant and a conference focus (i.e., the central server). This conference model is with more interest because it is also applied by 3GPP [49]. Figure 12.3 shows the simplified 3GPP 3G conferencing architecture. According to Fu et al. [56].

> In the architecture, the conference focus can be implemented in the media resource function controller (MRFC) and/or in the conferencing application server

(AS). The MRFC is the functional entity that handles signaling. It considers the media resource function processor (MRFP) as a default media mixer. The AS hosts conference applications. Any party that wants to participate in a conference should either invite the conference focus (i.e., following the dial-in model) or be invited by the conference focus (i.e., following the dial-out model). In addition to the MRFC and the AS, there is another functional entity, user equipment (UE). A UE is a conference participant that has the required conferencing functionality in the end-users' terminals.

12.1.4.2.4 Signaling Approaches from Outside of the Standard Bodies. Several approaches have been proposed from outside of the standards bodies. Some of them target specific issues that are not solved by the standard approaches, while others propose more comprehensive solutions. The works proposed in Koskelainen et al. [50] are examples of the former, while GlobalMMCS [51] and ICEBERG [52] are examples of the latter. According to Fu et al. [55].

The SIP-based conference defined in Koskelainen et al. [50] is another example of work for using SIP for conferencing. It extends the tightly coupled conference model of SIP in order to improve the scalability. Multiple conference focuses are proposed, and each focus manages a set of local participants. The conference focuses are interconnected and form a tree structure.

GlobalMMCS [51] is designed to bridge H.323, SIP, access grid clients, and 2.5G/3G cellular phones in audio–visual collaborative applications. The system makes use of a publication/subscription-based message delivery middleware, the NaradaBroking overlay network. As far as multimedia conferencing is concerned, the system borrows the ideas of MCU, MC, and MP from H.323. However, unlike H.323, the MCs can be distributed. There can be several in the same conference, each one managing a subset of participants.

ICEBERG signaling [52] proposes a signaling system for the management of dynamic and multidevice multiparty sessions. Unlike the other signaling protocols such as SIP, it is a signaling system that is directly designed for multimedia conferencing. Two entities are defined: call agent and call session. They are both dynamic entities created during call session establishment. The call session entity is the control center that manages all the information related to that session. There is one call agent per party. It manages the information related to that party. Changes related to the session are propagated as follows. A designated serving call agent periodically receives a report from each party in the session, and it forwards the report to the call session entity. The call session entity maintains the states of all of the parties in a table, and it updates the table when it receives the reports. It also propagates the information to each of the call agents.

12.2 SIGNALING FOR CONFERENCING IN 4G

In this section we introduce signaling architectures and protocols for conferencing in 4G. We present the signaling architectures for MANETs and integrated MANETs/3G networks.

12.2.1 Signaling for Conferencing in MANETs

The signaling for multimedia conferencing in MANETs is very challenging. A signaling scheme not only needs to establish, modify, and terminate sessions, but also has to take into consideration the network statuses such as the lack of infrastructure, the frequently changing participants, and the limited resources. A very basic requirement for conferencing signaling in MANETs is that none of the signaling entities can be a permanent or static central-control point. The other functional requirement is that the system should be able to dynamically propagate conference-related information (e.g., who joins, who leaves) to all the involved parties. This is not an easy task, because conferences are normally very dynamic in MANETs. Parties can join and leave at any time and very frequently. A party may leave the conference when it decides to do so or when it is forced to because it has moved out of the coverage area or its battery power is used up. We term the first case (which is general to all networks) "voluntary departure" and call the second (which is specific to MANETs) "unintentional departure." If a party in a conference temporally moves out of range or if its link breaks for a very short time, the sessions that it has maintained should be recovered after its connections are recovered.

Signaling for conferencing in MANETs has not yet been standardized. IETF discussed some issues related to distributed SIP in Kelley [53], which is applied later in the proposal: SIP framework for MANET [54]. These early investigations do not comprehensively cover the signaling requirements for MANETs. The cluster-based signaling protocol [55] is so far a much-detailed proposal for conference signaling in MANETs. It discusses different signaling issues and provides simulation results. We will first discuss early investigations and then provide more detailed information on our cluster-based signaling solution.

12.2.1.1 IETF Distributed SIP and SIP Framework for MANETs. SIP has been addressed for a fully distributed model. In the model, each participant maintains a SIP session with other participants. Reference 53 describes this approach in detail. This is of special interest to MANETs because it only involves SIP end systems (UAs) and no central server is required. However, this approach has several limitations. A first drawback is the way the session-related information is dynamically propagated to parties. There is a problem when two (or more) parties are invited to join an ongoing session at the same time. There is no general solution to ensure that each of the invited parties is made aware of the other invited parties. This problem is identified as the "coincident joins problem," and no solution is provided.

The framework defined in Khlifi et al. [54] applies the architecture defined in Kelley [53] for MANETs, but it resolved the "coincident joins problem." It proposes a conference leader that propagates session-related information to all participants. Any participant change should report to the conference leader. This work cannot support a large number of participants due to the full-meshed

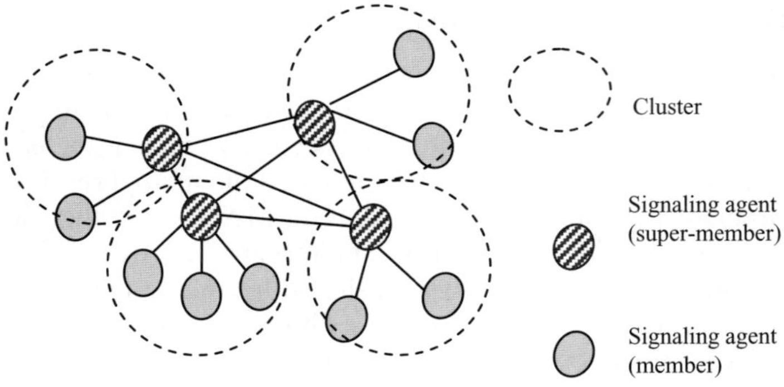

Figure 12.4. Signaling architecture for standalone MANETs.

signaling connections among participants. In addition, it does not consider the issues such as session recovery.

12.2.1.2 Cluster-Based Signaling in MANETs. Clusters enable scalability without centralization, and they can help in solving the signaling issue in MANETs.[1] A cluster-based signaling architecture for conferencing in standalone MANETs is proposed in Fu et al. [55]. The clusters are formed in the application layer and only when there is a conference. We first present the architectural principles, followed by a description of the clusters' operational procedures. We then discuss two critical issues related to the operational procedures: how to exchange node capabilities and how to handle unintentional departure and session recovery.

12.2.1.2.1 The Architecture and General Principles. Figure 12.4 gives an overall view of the cluster-based signaling architecture. The only functional entity is the signaling agent (SA). There is one per party, or more generally, one per node in a MANET. They are grouped in clusters that we call signaling clusters. These clusters are application-level clusters and are independent of lower layer clusters such as routing clusters. In each cluster, at any given time, there is one and only one cluster head (i.e., super-member), and all the other members of the cluster are connected to it. A super-member has direct signaling links to the super-members of the neighboring clusters.

There are two general parameters of a cluster: split value (Sv) and merge value (Mv). Every node in a conference maintains the same Sv and Mv. If the size of a cluster reaches Sv, the cluster will split into two clusters. If it reaches Mv, the cluster will find another cluster to merge with.

[1]Sections 12.2.1.2–12.2.1.4 are taken from reference 55. Copyright © IEEE 2009.

A super-member is responsible for keeping track of the information of its members and its neighboring super-members. It also propagates the information when there is a change in membership. In addition, it detects the eventual unintentional departures of the nodes connected to it by sending periodic heartbeat messages.

In this architecture, it is the node with the most capabilities that is elected as the super-member. A participant that initiates a conference is responsible for collecting the capabilities of the called party before the conference is initiated. Super-members keep track of the capability changes of their members and neighboring super-members during the conference.

12.2.1.2.2 OPERATIONAL PROCEDURE OF CLUSTERS. Clusters are dynamically created and deleted for conferencing. The signaling system is responsible for maintaining the state of conference and clusters. Each signaling cluster has a life cycle. The first phase is its creation. A super-member is elected in this phase. After its creation, the cluster moves to an active phase. The membership of the cluster evolves (parties join and leave). These changes may lead a cluster to split into two, or to merge with another cluster. Ongoing activity may also lead to the election of a new super-member, triggered by the departure of the super-member, for example. The life cycle ends with the deletion of the cluster. In this section, we describe the signaling procedures related to each of the phases of the cluster life cycle.

(a) Cluster Creation and Deletion. The first cluster is created when a conference starts. The creation procedure is as follows: First, the party (called the initiator) that wishes to establish a session collects the capability of the called party. It compares its own capability to the capability of the called party and designates the one with more capability as the super-member. Second, it requests the super-member (itself or the called party) to create a session. The initiator needs to set the Sv and Mv and passes the parameters to the called party. After the first session is set up, the super-member starts to periodically collect the capabilities of its members.

The last cluster is deleted when the last two parties leave the session. All the states and parameters of the cluster are cleared.

(b) Super-member Election. An election algorithm is used whenever there is a need to select a new super-member among several candidates. This happens when a new cluster is created or when a cluster merges with another cluster or when a super-member leaves. The basic rule is that the candidate with the most capability is selected as a super-member. The election algorithm is quite straightforward. The capability of each super-member candidate is compared to other capabilities, and the one with most capability wins.

(c) Member Joining and Leaving. Both members and super-members can invite parties to join a conference. If it is a super-member that is inviting and it is capable of handling more members, the super-member directly establishes a

session with the party. If the super-member cannot handle more members, it may ask a neighboring super-member to do so. If a member invites a party, that member will ask its super-member to establish the session. A new member is then added to the cluster. The super-member of the cluster propagates the membership change to neighboring clusters.

Any participants, including members and super-members, may leave a conference whenever they want to. In the case of a member departure, the member terminates its connection with its super-member and the super-member propagates the membership change to the neighboring clusters. With the departure of a super-member, that super-member designates a new super-member (choosing the member with most capability among the member list) before leaving. It passes its member-list and neighboring super-member list to the new super-member. The new super-member sets up a session with each member and each neighboring super-member and forms a new cluster. After this procedure, the old super-member terminates all its connected sessions. In the case where there is no member in a cluster, the super-member that wishes to leave simply terminates all its connected sessions.

(d) Splitting. When a new member is added to a cluster, the super-member initiates a split procedure if the size of the cluster reaches Sv or if the super-member does not have enough capability to handle more members. A cluster may also split when its super-member does not have enough capability to handle its existing members. This happens, for instance, when the battery power of the super-member decreases. First, the super-member selects a new super-member, based on capabilities. It also selects half of its members that are to become members of the new cluster. The selection may be by random or according to some rules, such as the sequence number of members. The super-member that wishes to split the cluster then asks the new super-member to form a new cluster that contains the selected members, and it passes the selected member list and neighboring super-member list to the new super-member. The new super-member creates a new cluster by establishing sessions. The super-member then terminates sessions with the selected members. Figure 12.5 shows signaling architectures before and after splitting.

(e) Merging. If the size of the cluster diminishes to Mv, the super-member initiates a merger procedure. This procedure starts by a searching for an existing cluster with which to merge, with the constraint that the size after the merger will be less than Sv. A new super-member is elected as soon as the merger begins. The new super-member will be one of the two super-members (the one with more capability) of the two clusters. The procedure continues as follows: The elected super-member establishes sessions with the members of the cluster to merge with. The un-elected super-member then terminates sessions with its members and sets the elected super-member as its super-member, and it becomes a regular member. The merger information will then be propagated to the neighboring super-members.

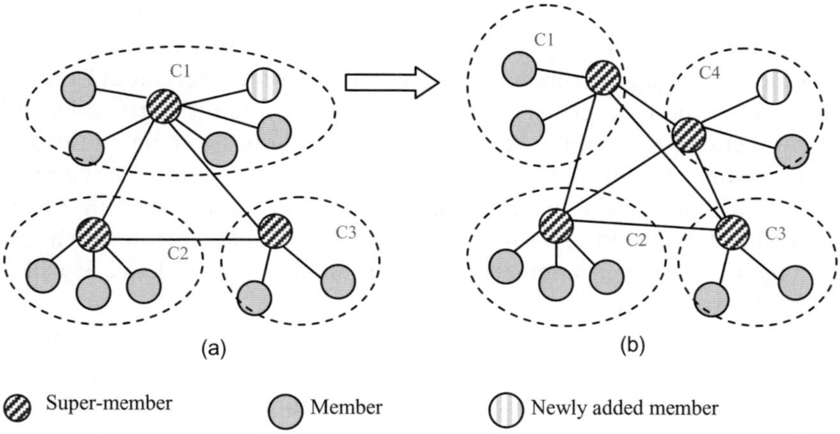

Figure 12.5. Cluster splitting **(a)** Before splitting. **(b)** After splitting (Copyright IEEE 2009 [55].)

(f) Information Propagation. In order to maintain a signaling cluster system, efficient information propagation is required—that is, rapid propagation with as little introduced overhead as possible. In this architecture, super-members are responsible for propagating membership and capability information whenever there is a change. The information can be propagated to all the signaling agents in no more than two hops.

(g) The Issue of Coincident Behavior of Participants. One issue of a distributed signaling architecture is the state synchronization when there are coincident behaviors of participants in the conference. Such behaviors may cause inconsistent states among participants; for example, with a coincident joining (defined in Kelley [53]), two newly joined parties have no way to know each other and no session will be established between them, and thus the fully distributed signaling architecture cannot be maintained. However, with the information propagation procedure, the cluster scheme defined in this proposal can handle most coincident behaviors. In some cases, protection mechanisms are used to prevent inconsistencies. We present this issue case by case:

Coincident Join. Two or more parties join a conference at the same time. They may join the same cluster or different clusters. The cluster scheme can handle this case because the coincidently joined parties do not have a direct session with each other. Instead, they establish sessions with the super-members that are already in the cluster, and later they can "know" each other from their super-member(s).

Coincident Departure. Two or more participants leave a conference at the same time. They may leave the same cluster or different clusters. Similar to the first case, the cluster scheme can handle the coincident departure of members and less than two super-members. The scheme does not support coincident departure of super-members, so it defines a protection phase when a super-member leaves. A super-member should reject any session establishment or termination request when it starts to leave. The protection phase ends when the super-member has completed the leaving procedure. Within this protection phase, a super-member leaving procedure will fail if another super-member is leaving at the same time, because the newly selected super-member cannot establish a session with a leaving neighboring super-member. If a super-member fails to leave, it will retry after a random period of time.

Coincident Splitting. Two or more clusters split at the same time. With the mesh structure of super-members, the super-members in older clusters maintain a session with every newly split super-member. After a run of the information propagation procedure, a newly split super-member will have knowledge of the other new super-members. The logic added in order to handle this case is that if a super-member finds that it has not established a session with a neighboring super-member and if it has a higher address, it will establish a session with that super-member.

Coincident Merging. Two or more pairs of clusters merge at the same time. The scheme can handle this case because there is no new super-member elected. The cluster state can be propagated to all neighboring super-members.

There are two other critical issues related to the signaling procedures. The first is the participant capability discovery that is critical to super-member election, and the second is the detection of unintentional departure and session recovery. We present how the cluster scheme handles the two issues.

12.2.1.2.3 Capability Exchange Mechanism. A simple application-level protocol is presented as part of the cluster-based signaling scheme for handling capability discovery. The entity involved is the signaling agent. There are three types of messages defined: Cap_subscribe, Cap_notify, and Cap_publish. Cap_subscribe is a request message containing a subscription interval. Cap_notify is a response message containing a sequence number and the current capability level, and Cap_publish is a message containing the current capability level.

Cap_subscribe is used in the following scenarios:

- When the initiator of a conference establishes the first session, it sends the message to the called party and sets the subscription interval to zero. In this case, the called party sends only one Cap_notify back to the initiator with its current capability loaded in the message.
- A super-member sends a Cap_subscribe message to a member when a session is established. In this case the subscription interval is set to a nonzero value, and in each interval period the member sends back a Cap_notify message loaded with its current capability.

- A super-member sends a Cap_subscribe to a member with a zero subscription interval value. The member sends back a Cap_notify response and stops the periodic Cap_notify messages.

Cap_publish is sent between super-members. When its capability is changed, a super-member sends a Cap_publish to every neighboring super-member.

12.2.1.2.4 Unintentional Departure Detection and Session Recovery. A failure-detection and recovery mechanism is used by the signaling architecture for handling the unintentional departures. The basic idea is that each session in a conference maintains a heartbeat—a periodical exchange of a request and a reply. There are three timers defined: heartbeat rate Th, transaction timer Tt, and recovery timer Tr. A heartbeat sender sends a request to a heartbeat receiver in each period of Th. A backup super-member list is contained in the request. If the sender does not receive a response in Tt, it will resend the request. If it does not receive a response after n resent requests, it will determine that the session is inactive. If a receiver does not receive a heartbeat request in Th+Tt*n, it will determine that the session is inactive. An inactive session will be removed from the conference, and a recovery procedure will be activated.

The idea of session recovery is that if a sender or a receiver detects that the last session has terminated, it does not remove the conference state immediately, but keeps the state for a time (Tr) and tries to establish a session with one of the backup super-members in each period of Tt until one session is created.

12.2.2 Signaling for Conferencing in Integrated MANETs/3G

An important scenario in 4G is the integrated MANETs/3G networks. We present the conferencing signaling for the scenario in this section. It is an integration of 3GPP 3G conferencing architecture and the clustering architecture presented in Fu et al. [55]. In the rest of this section, we will present the proposal in detail.

12.2.2.1 Network Assumptions. The integrated MANETs/3G network considered in the proposal [56] is shown in Figure 12.6. According Fu et al. [58].

> In the figure a *multihop routing area* is defined as an area in which all the nodes are working on a MANET interface, and the nodes can reach each other by direct wireless connections or multihop routing. It also shows that there is more than one multihop routing area in the system. The assumed network supports three types of devices: devices with only MANET interfaces, devices with both MANET and 3G interfaces, and devices with only 3G interfaces. The first two types are called the *multihop mobile station (MS)*, and the third is called the *single-hop MS*.

12.2.2.2 Architectural Principles[2]

12.2.2.2.1 The Architecture. The architecture is depicted in Figure 12.7. It includes three entities: the signaling agent (SA), the conference focus (MRFC/

[2]Sections 12.2.2.2.1, 12.2.2.2.2, and 12.2.2.3 are taken from reference 56. Copyright © IEEE 2006.

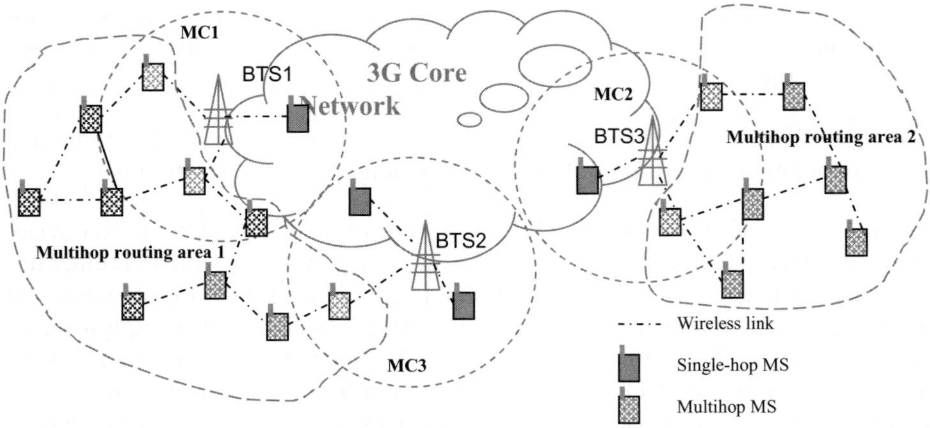

Figure 12.6. Considered integrated network. (Copyright IEEE 2008 [58].)

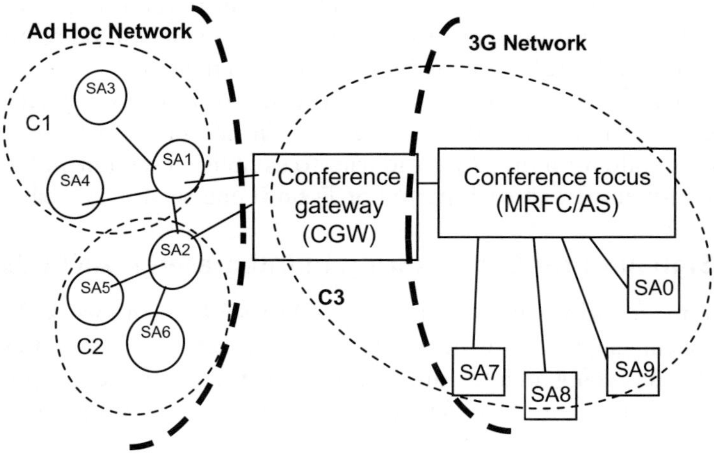

Figure 12.7. Integrated conferencing architecture. (Copyright IEEE 2006 [56].)

AS), and the conference gateway (CGW). The MRFC/AS is the entity defined in the 3GPP standard [49]. In this architecture, its functionality as per the 3GPP standard is enhanced. The enhancement is a CGW discovery functionality—that is, the ability to find a suitable CGW that can handle sessions with the participants in MANETs. The SAs are conference participants. They are either 3GSAs (i.e., participants in 3G), or MSAs (i.e. participants in MANETs). 3GSAs are the signaling parts of the 3G User Equipment defined in reference 49. They can establish sessions with the MRFC/AS. MSAs are the same as the SAs defined in Section 12.2.1.2. An MSA can be either a super-member or a simple member. Here also the CGW discovery functionality is needed in addition to the functionality of the MSA defined in Fu et al. [55]. It should be noted that 3GSAs and MSAs may run different signaling protocols.

The CGW is a new entity introduced. It is a mediator deployed by the 3G network operator (or a trusted third party). It has an infrastructure-based interface that is connected to the 3G conference focus. It also has a MANET interface that is connected to MANET SAs. Unlike the client proxy defined in UCAN or the ARS defined in iCAR, the CGW is an application layer entity. Its two interfaces are not physical air interfaces, but instead they are application layer interfaces to signaling components. A CGW has six major functions. First, it has the functionality of a signaling agent that is capable of establishing sessions with MSAs and with the conference focus. Second, it understands the signaling protocols for multimedia conferencing in both MANET and 3G, and it performs the translation (if required). Third, it understands the conferencing signaling architectures (e.g., centralized versus distributed) used in both MANET and 3G, and it does the mapping (if required). Fourth, it collects the membership information in both networks, converts it (if required), and distributes it. Fifth, it provides the functionality of publication and discovery so that MSAs and the conference focus can find and use its services. Sixth, it provides registration functions and manages the repository of MANET participants.

The architecture relies on three main principles. First, participants in MANET see the CGW as a special super-member that never leaves, splits, or merges with other super-members. Second, participants in 3G (i.e., 3GSAs) see the MRFC/AS as a centralized control point to which every participant, in 3G networks or in MANETs, is connected. Third, the MRFC/AS sees the CGW as a sub-focus that aggregates and manages sessions for MANET participants. The basic assumption is that participants that are not 3G users are implicitly seen by the 3G conference focus as MANET participants, to which sessions are created through a CGW. The same assumption is made for MSAs.

12.2.2.2.2 CGW Discovery. The CGW can be discovered by the 3G conference focus and by the MANET participants. The architecture proposed to reuse any of the MANETs service discovery protocols such as Konark [57]. Two basic scenarios are presented. First, a CGW periodically publishes its location and capability to MANET nodes. The MANET node caches the CGW information when it first receives it, and it registers with the CGW. Second, a MANET node sends a CGW request message that contains CGW capability requirements to the network. A MANET node that has the proper CGW capability information or a CGW that matches the capability requirements can respond. The MANET node that receives the responses then registers with the CGW.

CGW location information contains the CGW's address and listening port. CGW capability information includes parameters such as the "signaling protocols supported," "conference type supported," "network information," "architectures supported," and "encoding supported."

12.2.2.3 Conferencing Scenarios. Four different conferencing scenarios are enabled using the signaling architecture: conferencing with 3GSA parties,

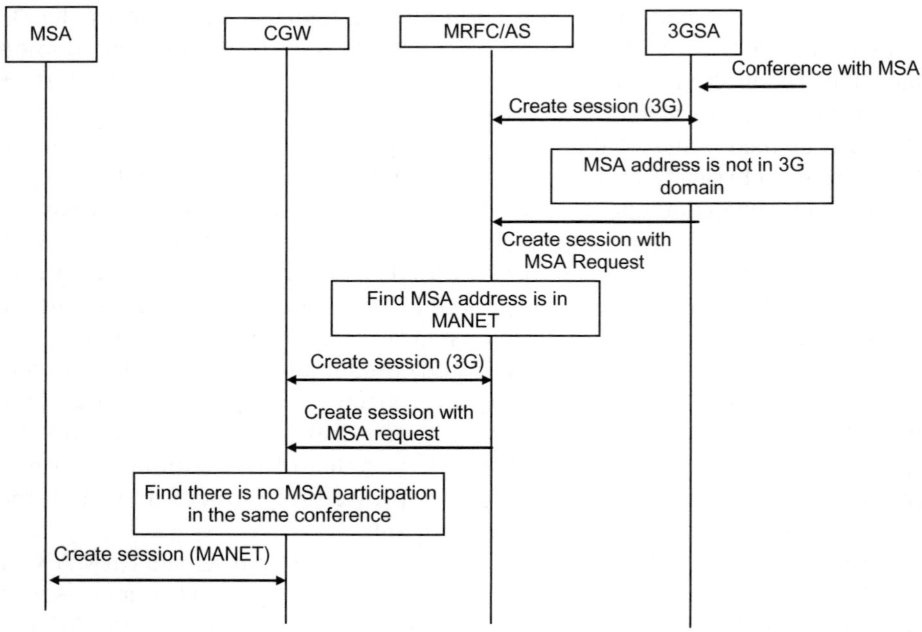

Figure 12.8. Conference initiations: A 3G participant initiates a session with a MANET party. (Copyright IEEE 2006 [56].)

conferencing with both 3GSA and MSA parties, conferencing with MSA parties in the same multihop routing area, and conferencing with MSA parties in different routing areas. The first and the third scenario do not require the use of a CGW. The second scenario requires CGWs to perform protocol translations and signaling routing. The last scenario does not require a protocol translation, but it uses CGWs and the 3G conference focus as signaling routing mediators. Here, we show one example scenario of conference initiation.

Figure 12.8 depicts sequence that a 3GSA initiates a conference with an MSA. The 3GSA first creates a session with its MRFC/AS using the 3G signaling protocol. Then it requests the conference focus to create a session with the MSA. The conference focus finds that the MSA is in a MANET. It then creates a session with the CGW and asks the CGW to create a session with the MSA. Finally, the CGW creates a session with the MSA using the signaling protocol of MANET and designates the MSA as a super-member.

12.3 OPTIMIZATION OF SIGNALING SYSTEMS: USING CROSS-LAYER DESIGN

In this section we study the performance and related issues of the conferencing signaling proposals for MANETs and integrated MANETs/3G networks. We present a solution [58] to optimize the signaling performance using a cross-layer design.

12.3.1 Performance Issues of Signaling Architectures[3]

Four issues have been identified for the signaling proposals: the overhead introduced by the heartbeat message, the overhead introduced by capability exchange protocol, the CGW deployment issues, and the suboptimal routing issues.

Application-Layer Heartbeat. The heartbeat mechanism proposed for the signaling architecture in MANETs deals with the unintentional departures of nodes. Although a reduction of the overhead introduced by the heartbeat messages have been considered (e.g., a one-way request/replay scheme used and the heartbeat messages are simple and lightweight), it can be found that the heartbeat does introduce significant overhead. Especially when the heartbeat rate is high, the overhead can exceed the session establishment overhead [55]. On the other hand, this overhead cannot be removed because the detection and handling of unintentional departures is important for conferencing in MANETs.

Application-Layer Capability Exchange. The application-layer capability exchange scheme is used for optimal use of node capabilities. The capability information is maintained by each super-member by periodically exchanging node capabilities between a super-member and its members. Similar to the heartbeat message, even though the protocol is lightweight, the overhead introduced is nontrivial because of periodical messages.

Both heartbeat and capability exchange mechanisms help to meet the signaling requirements that are not related to session management, but are necessary for handling the particularities of MANETs. For example, the unintentional departure is often caused by link break, node moving out of range, or node crash. These issues are common to every layer of MANETs, and they have been discussed frequently in relation to lower layers. A proof is that the AODV [13] routing protocol has used three nonresponded "Hello" messages to detect a link break. The same thing happens to the optimal use of node capabilities. The optimal use of capabilities is a common target for all MANET layers because resources in MANETs are scarce. Actually, some of the lower layer protocols have considered this requirement in their design. For example, WCA [59] uses the capability of nodes as one of the criteria to decide cluster heads. Thus, we doubt that the use of heartbeat and capability exchange mechanisms in the application layer is not very efficient, but cross-layer design helps to share information among layers and to improve the overall performance.

[3]Sections 12.3.1 to 12.3.4.4 are taken from reference 58. Copyright © IEEE 2008.

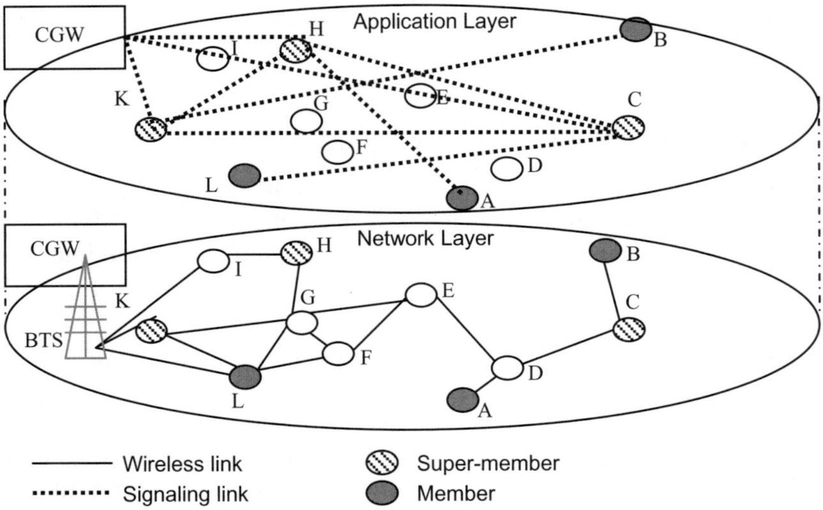

Figure 12.9. Issue of suboptimal routing. (Copyright IEEE 2008 [58].)

CGW Deployment and CGW Publication/Discovery. CGW is a key entity in the integrated signaling architecture. The proper deployment of a CGW can improve performance. There are two questions to be answered. The first is where to deploy a conference gateway. The second is how many conference gateways should be deployed. These questions are not answered in Fu et al. [56]. Furthermore, the CGW publication/discovery is yet a task that requires periodical exchange of messages at the application layer. Are there opportunities to reduce this overhead?

Suboptimal Routing. This is an issue caused by application-layer clusters. A new cluster member joins a cluster when the super-member or a member of the cluster invites it. The scheme has not considered whether the joiner is physically close to the cluster. This may introduce serious performance problems. In Figure 12.9, for instance, the shortest path between party A and party B for the routing layer is 3 hops. If signaling clusters in the application layer have been formed, the real path distance between party A and party B will be 11 hops: 4 hops between A and its super-member H, 5 hops between B and its super-member K and 2 more hops between the two super-members H and K. This issue can be somehow avoided if the application layer "knows" physical location of nodes; for example, if super-member K knows that party B is close to super-member C, it may not invite B but asks super-member C to do so. We believe that cross-layer design can help in this situation.

From the above analysis, a need for cross-layer design is identified. Two of the performance issues are directly related to lower-layer problems such as routing and link-break detection. The issue of optimal resource usage is common to every layer and the CGW deployment is related to lower-layer network architectures. In the next section, we will introduce the concept of cross-layer design and the requirement of cross-layer design for signaling optimization. We also analyze some existing cross-layer design proposals with respect to the requirements.

12.3.2 Cross-Layer Design for MANETs

Cross-layer design refers to protocol design done by actively exploiting the dependence between protocol layers to obtain performance gain. This is unlike layered design, where the protocols at different layers are designed independently [60]. Layered design has obtained great successes in today's infrastructure-based networks such as telephony networks and Internet, while cross-layer design is mainly motivated in wireless networks. Wireless networks have their distinctive characteristics such as user mobility, frequent link failure, limited link capability, and limited battery power of mobile devices. The argument is that applying layered design in wireless networks usually causes suboptimal performances, but a careful cross-layer design helps to solve the issue. A cross-layer design may remove the duplicate functionality or data in layers. It may also optimize the parameters in each layer so that the performance in a single layer or in a whole system is enhanced.

The cross-layer designs can be classified into four types [60]: interface breaking, merging of adjacent layers, design couplings, and vertical calibrations. The interface breaking can further be sorted into three types: upward information flow, downward information flow, and back-and-forth. The design coupling and merging of adjacent layers are self-explainable from their name. The purpose of vertical calibration is to perform adjustment in each layer in order to achieve a global performance gain. A method of implementing vertical calibration is using a shared database for all layers.

12.3.2.1 Cross-Layer Design Requirements for Signaling Optimization. The cross-layer design for signaling optimization should meet the following requirements. First, it should be easy to implement and does not introduce too much network overhead. Second, it should respect some cautionary aspects of cross-layer design—for example, no design loop and an ease of upgrade. Third, the cross-layer information should be time-sensitive. This is important for the signaling system because conferencing is a real-time application. Fourth, the interoperability should be considered for the integrated signaling system. For example, an SA with a cross-layer design should be capable of interacting with an SA designed in traditional manner.

12.3.2.2 Related Work in MANETs. We divide the existing cross-layer design proposals in MANETs into two categories: global versus local solutions. The former usually presents new cross-layer design methods that are applicable for different architectures and benefit different layers. The latter consists of local solutions for specific architectures and requirements. For example, the work defined in Setton et al. [61] uses cross-layer design for the optimization of real-time video transmission. Local solutions are difficult to adapt for other applications or situations. Therefore, we will not discuss them further. Four global cross-layer design solutions are examined. CLASS [62] follows the method of direct communication between layers. It proposes a general mobility management architecture that allows for direct signaling flows between non-neighboring layers. This signaling interaction enables efficient mobility management for the whole system. However, it does not meet the requirements. For example, it requires logical changes in each layer, which is complicated to implement and upgrade. In addition, it can lead to conflicts and loops if the signaling has not been designed very carefully.

MobileMAN [63] considers the cautionary aspects of cross-layer design, and it makes use of a shared database. It defines a repository called *Network Status* from which each layer can write and read information. It provides optimization for all the network functions and improves local and global adaptations. However, it is not easy to implement because it requires the redesign of protocols in each layer. Furthermore, the expiration of the data in the repository have not been discussed, so the optimization scheme may not be efficient for time-sensitive applications. References 64 and 65 present some other drawbacks of MobileMAN; for example, it may be cumbersome for protocols that neither write nor read the Network Status.

CrossTalk [64] extends the vision of MobileMAN. It introduces a global view of the network status while specifying the *Network Status* defined in MobileMAN as a local view. It is capable of providing real-time information for the optimization processes. However, the global view is collected through a data dissemination procedure that incurs a significant overhead.

With the design goals of rapid prototyping, minimum intrusion, portability, and efficiency in mind, ECLAIR [65] is proposed for the optimization of mobile device protocol stacks. It uses an approach similar to that of MobileMAN and CrossTalk. The difference is that it not only collects data from layers and stores them in a repository, but also develops the optimization processes outside of the protocol stack. This abstraction makes the design more flexible and ensures a fast deployment. Similar to MobileMAN, it does not consider expiration of data, so the real-timeliness requirement is not fulfilled.

12.3.2.3 Related Work in Integrated MANETs. In the context of integrated MANETs (or MCNs), cross-layer design has been considered recently (e.g., references 66 and 67). The work in reference 66 provides a cross-layer design for BTS routing discovery. It adjusts the functionality of physical, MAC, and network layers. It also collects information from the three layers so that an efficient route can be discovered. The performance evaluation shows that the proposed schemes can achieve faster route discovery and more reliable route setup. However, the proposal is a specific method that cannot be used by our

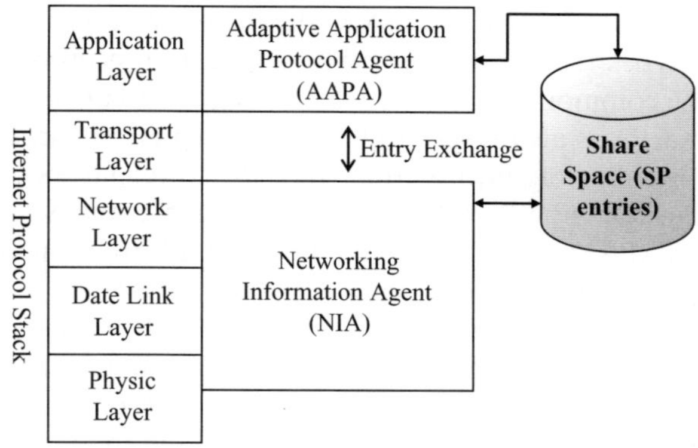

Figure 12.10. Cross-layer optimization architecture. (Copyright IEEE 2008 [58].)

signaling system. Furthermore, it has not considered interoperability issue. Similar to what is presented in reference 66, reference 67 presents another MCN routing protocol using cross-layer design. It considers a set of constraints (e.g., interference level) when discovering a route. These constraints are collected from physical or MAC layer. Like reference 66, this work cannot meet our requirements. To the best of our knowledge, there is no global cross-layer design solution proposed for integrated MANETs.

In next section we present the cross-layer architecture [58] that is designed for signaling optimization and is also generally suitable for application-layer optimizations.

12.3.3 A Cross-Layer Optimization Architecture

The cross-layer optimization architecture is shown in Figure 12.10. In this architecture, the shared database method is used and the cautionary aspects of cross-layer design are considered. The general principles are as follows: An application protocol defines optimization schemes and specifies the types of information that it wants to acquire from lower layers. The lower layers provide the information, which is stored in and retrieved from a shared database.

12.3.3.1 Entities. The architecture involves three entities: Share sPace (SP), Adaptive Application Protocol Agent (AAPA), and Networking Information Agent (NIA). SP is a repository from which application protocols can retrieve lower-layer information. A share space contains a set of entries. Each entry represents one type of information. It contains an entry type, an entry value, and a keep-fresh timestamp, which is used to ensure the freshness of the information. If the information has not been updated for a given period of time, it will expire.

An entry is initially defined by an AAPA, which is responsible for translating the information types that are requested by application protocols into common entry types. These common entry types are exchanged and understood by AAPAs and NIAs. An AAPA is also responsible for creating entries in the SP. An NIA is responsible for collecting information from the lower layers and updates the relative entries in the SP. When there is a new update, the SP will inform the AAPA and the AAPA will then cache the updated information, reformat it, and send it to the application protocol.

The purpose of using AAPA and NIA is to mask the complexity of the cross-layer design so that existing protocols in single layers do not need to have major upgrades to support cross-layer optimization. The protocols in the lower layers only need to open their data structures to the NIA, and the application protocol simply sends requests to and gets responses from its AAPA. Any functional upgrade or version change in a layer is still independent of other layers. However, the AAPA and NIA should be upgraded accordingly.

12.3.3.2 Benefits of the Optimization Architecture. This architecture benefits from the advantages of both layered design and cross-layer design. Compared to the existing proposals, the architecture is more application-layer oriented. Although shared database-based proposals such as MobileMAN and ECLAIR have defined some standard parameters for each layer, the shared data cannot really adapt to the different requirements of diverse applications. For example, routing cluster information is useful for optimizing a clustering-based application but it may not be considered as a standard parameter. This architecture makes it possible to exchange the data entries, and the new entries can be negotiated before performing an optimization. This allows space for customization of cross-layer parameters.

In the solution, the lower-layer information is retrieved locally. Timestamps are used to ensure the freshness of the information. There is no extra spreading overhead introduced in the network. On the other hand, the SP may not include as much of the information as that included in the global view defined in CrossTalk, but it includes necessary information required by application. Similar to ECLAIR, our architecture is flexible and quickly deployable. The application optimization schemes are designed independent of basic application logics. Thus, it can simply return to the basic logic if an optimization is not performed.

12.3.4 Optimization Schemes

There are six optimization schemes that can be applied in application layer: link-break handling, capability usage, suboptimal routing, super-member election based on clustering, super-member election based on signal power, and CGW deployment. The last two schemes are specific to the signaling for an integrated MANET/3G network. As a cross-layer design approach, each scheme uses a particular type of information from lower layers. There is no specific lower layer protocol required for the architecture. The general principle is that an optimization scheme is only performed when the lower-layer protocol(s) can provide the related information. This principle ensures that the optimized signaling scheme can run on different lower-layer routing protocols. It should be noted that the six optimization schemes are independent of each other.

12.3.4.1 Common Entries. Six common entries are defined using the format <type, value, timestamp>. These entries are the *routing table* <routing_table, route_list, timestamp>, the *node capability list* <node_capability, capability_info_list, timestamp>, the *neighbor list* <neighbor_hop_list, address_list, timestamp>, the *routing-layer clustering* <routing_cluster, cluster_info, timestamp>, the *routing topology* <routing_topo, route_list, timestamp>, and the *signal power* <signal_power, power_value, timestamp>.

In the *routing table* entry, each route in the route_list consists of a source address, a destination address, a path, a hop count, an active status, and a timestamp. In the *node capability list* entry, each capability_info consists of a node address, capability types, and corresponding capability values. In the *neighbor list* entry, a list of neighboring nodes' addresses is stored. In the *routing layer clustering* entry, cluster_info contains a cluster head. If the node is a cluster head, it also contains a list of cluster members. This entry may not contain all of the nodes' capabilities. The *routing topology* entry stores the routes to reach every MANET node in an integrated architecture. In the *signal power* entry, a signal power value is stored. The signal power of a node is its wireless signal strength to the BTS.

12.3.4.2 General Optimization Schemes for Signaling in MANETs. *Link Break Handling Optimization Scheme.* This is performed when the routing table entry is updated. Every routing protocol provides this information. A super-member checks the route status for each of its connected members and super-members. If it discovers a route failure, it will terminate the session. This scheme helps the signaling system to handle an unintentional departure gracefully without using a heartbeat mechanism.

Capability Usage Optimization Scheme. When the first super-member is elected, participants check their local node capability list. If there is a fresh capability list, the participant will copy and use this list for super-member election; that is, the participant with the highest level of capability becomes the super-member. After the establishment of the first session, the super-member checks its local capability list only when it needs to elect a new super-member. This scheme can only be invoked when the lower layer protocol takes node capabilities into consideration. Using this scheme, it can avoid invoking the application-layer capability exchange mechanism.

Suboptimal Routing Optimization Scheme. This uses the entry of the neighbor list. Some of the routing protocols (e.g., proactive routing protocols) can provide this information. The scheme can be described as follows.

- When a new member joins the conference, all the super-members are informed.
- Each super-member checks its neighbor list. If the information is fresh and if it contains the address of the new joiner, it will ask the super-member that has invited the joiner (which we call the *original super-member*) to switch the new joiner to its cluster. As there may be more than one switching request, the original super-member will choose the one with the most available capabilities to switch the session.
- Periodically, a super-member checks its neighbor list. If there is a conference member that is its neighbor but is not its member, it will ask the super-member that has connected to the member to switch the session. The super-member may initiate a switch or refuse the switch request if it is also adjacent to the member.

This scheme ensures that a member joins a cluster when the cluster head has a direct link with the member. However, it does not ensure the shortest-path. It is not recommended for an obligatory shortest-path optimization because it may lead to very frequent session switches. In addition, it may seriously increase the network overhead. For example, super-members would be required to exchange their routing information frequently.

Super-member Election Based on Clustering Scheme. It is performed when the routing cluster entry is updated. Node capability is still the major criterion for super-member election. However, if all the candidates have a similar capability level, the cluster heads in the routing layer will have priority to be chosen as super-members in the application layer. This helps to further optimize the signaling route. This scheme can only be invoked when the routing protocol uses a clustering scheme.

12.3.4.3 Specific Optimization Schemes for Integrated MANETs/3G Networks. *CGW Deployment Optimization Scheme.* This uses the entry of route topology. In most of the routing protocols of MCNs, there is a route topology stored in an entity of the infrastructure side of the network. The entity may be a BTS (e.g., in reference 66) or a Mobile Switching Center (MSC) (e.g., in reference 21), depending on what type of lower-layer routing protocol is deployed. The idea is to collocate a CGW with the entity where there is a routing topology. In this case, the CGW can periodically acquire the MANET nodes' information from the routing topology and update its participant location repository. The overhead related to the frequent application-layer location update can then be avoided. For the question of how many CGWs should be deployed, in the conditions of considered environment, one CGW can be deployed per multihop routing area. This will facilitate the signaling routing procedure.

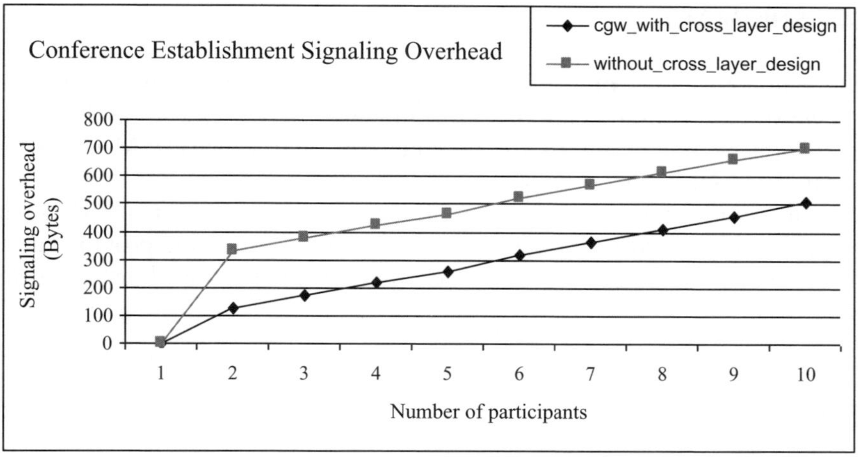

Figure 12.11. Simulation results for CGW deployment optimization. (Copyright IEEE 2008 [58].)

Figure 12.11 depicts performance results of comparing the signaling overheads using and without using CGW optimization. It shows that with the CGW optimization scheme, system introduces much less overhead than the usage of a CGW discovery protocol in application layer.

Super-member Selection Based on Signal Power Scheme. The signal power entry is used. A node with the highest signal power has a higher priority to be a super-member. The signal power value is provided by the physical layer. In most cases, signal power is influenced by the distance between a BTS and an MS, and thus a higher signal power may reflect a short path between a super-member and the CGW. Also, it is a general case that when a node is physically closer to a BTS, it is less prone to move out of range. This helps to optimize the signaling route and improve cluster stability.

12.3.4.4 Interoperability Analysis. The interaction between parties with and without cross-layer optimizations is a complex issue. Some of the optimizations may cause *interoperability* problems while others may not, depending on local versus global effect of an optimization scheme. A local effect does not cause an *interoperability* issue, while the global effect causes an issue if both parties make a decision at the same time and their decisions are in conflict. Within the optimization schemes introduced thus far, those that may cause serious *interoperability* problems are the super-member election optimization schemes. In a cluster, more than one party may be selected as a super-member based on different rules. This is not allowed in the cluster scheme. One solution is that whenever a super-member detects another super-member in the same cluster, it checks if its cross-layer information is active. The super-member without active cross-layer information changes itself to a cluster member.

12.4 SUMMARY

In this chapter, we have presented signaling architectures for conferencing in 4G. We reviewed the conferencing protocols in legacy networks such as 3G, and we also introduced the proposals for new network and scenarios: MANETs and integrated MANETs/3G networks.

Another important content of this chapter is the optimization of the signaling schema. We characterized and discussed the signaling issues and provided a solution, optimization architecture, and schema, which are based on cross-layer design.

REFERENCES

1. Y. K. Kim and R. Prasad, *4G Roadmap and Emerging Communication Technologies*, Artech House, Norwood, MA, 2005.
2. Wireless World Research Forum (WWRF), *Technologies for the Wireless Future*, Vol. 3, K. David, (ed.), John Wiley & Sons, Hoboken, NJ, 2008.
3. U. Varshney and R. Jain, Issues in emerging 4G wireless networks, *IEEE Computer*, Vol. 34, No. 6, pp. 94–96, June 2001.
4. Jennifer J. –N. Liu and I. Chlamtac, Mobile ad hoc networking with a view of 4G wireless: Imperatives and challenges, *Mobile Ad Hoc Networking*, Wiley-IEEE Press, Hoboken, NJ, 2004.
5. M. Conti, Body, personal, and local ad hoc wireless networks, in *Handbook of Ad Hoc Networks*, M. Ilyas (ed.), Chapter I, CRC Press, New York, 2003.
6. B. Xu, S. Hischks, and B. Walke, The role of ad hoc networking in future wireless communications, in *Proceedings of ICCT*, 2003.
7. IEEE 802.15.3 TM, *Part 15.3: Wireless Medium Access Control (MAC) and Physical Layer (PHY) Specifications for High Rate Wireless Personal Area Networks (WPANs)*, IEEE Computer Society, Washington, DC, 2003.
8. C. Bisdikian, An overview of bluetooth wireless technology, *IEEE Commun. Mag.* pp. 86–94, December 2001.
9. IEEE std 802.11, *Wireless LAN Medium Access Control (MAC) and Physical Layer (PHY) Specifications*, IEEE Communications Society, New York, 1999.
10. G. V. Zaruba and S. K. Das, Off-the-shelf enablers of ad hoc networks, in *Mobile Ad Hoc Networking*, Wiley-IEEE Press, Hoboken, NJ, 2004.
11. J. L. Burbank and W. T. Kash, IEEE 802.16 broadband wireless technology and its application to the military problem space, in IEEE MILCOM 2005, Vol. 3, 17–20 October 2005, pp. 1905–1911.
12. M. Sherman et al., A PMP-friendly MANET networking approach for WiMAX/IEEE 802.16, in *IEEE MILCOM 2006*, 23–25 October 2006, pp. 1–7.
13. C. Perkins, E. Belding-Royer, and S. Das, Ad Hoc On-Demand Distance Vector (AODV) Routing, RFC 3561, July 2003.
14. T. Clausen and P. Jacquet (eds.), Optimized Link State Protocol (OLSR), RFC 3626, October 2003.

15. R. Ofier, F. Templin, and M. Lewis, Topology Dissemination Based on Reserve-Path Forwarding (TBRPF), RFC 3684, February 2004.
16. D. Johnson, Y. Hu, and D. Maltz, The Dynamic Source Routing Protocol (DSR) for Mobile Ad Hoc Networks for IPv4, RFC 4728, February 2007.
17. Y. Lin and Y. Hsu, Multihop cellular: A new architecture for wireless communication, *IEEE INFOCOM 2002*, Vol. 3, pp. 1273–1282, 2002.
18. H. Wu, C. Qiao, S. De, and O. Tonguz, Integrated cellular and ad hoc relaying systems: iCAR, *IEEE JSAC*, Vol. 19, pp. 2105–2115, 2001.
19. H. Luo et al., UCAN: A unified cellular and ad-hoc network architecture, in *Procedings, ACM Mobicom*, September 2003.
20. Z. Dawy, S. Davidovic, and I. Oikonomidis, Coverage and capacity enhancement of CDMA cellular system via multihop transmission, *IEEE Globecom*, pp. 1147–1151, 2003.
21. H. Lee and C. Lee, An integrated multi-hop cellular data network, in *IEEE Vehicle Technology Conference*, Vol. 4, pp. 2232–2236, 2003.
22. Y. Liu et al., Integrated radio resource allocation for multihop cellular networks with fixed relay stations, *IEEE JSAC*, Vol. 24, pp. 2137–2146, November 2006.
23. G. Kannan, S. N. Merchant, and U. B. Desai, Access mechanism for multihop cellular networks, *IEEE VTC-2007*, pp. 279–283, 2007.
24. A. Kusuma, L. Andrew, and S. V. Hanly, On routing in CDMA multihop cellular networks, *IEEE Globecom*, pp. 3063–3067, 2004.
25. Y. Wu, K. Yang, and J. Zhang, An adaptive routing protocol for an integrated cellular and ad-hoc network with flexible access, in *ACM IWCMC'06*, Vancouver, pp. 263–268, 2006.
26. A. R. Wolff and C. Lee, Large scale routing in a multi-hop cellular network using a radial geometric approach, *IEEE WCNC*, pp. 4446–4451, 2007.
27. P. P. Lam and S. C. Liew, Nested network mobility on the multihop cellular network, *IEEE Commun. Mag.*, Vol. 45, No. 9, pp. 100–104, September 2007.
28. D. Cavalcanti, D. Agrawal, C. Cordeiro, B. Xie, and A. Kumar, Issues in Integrating Cellular Networks, WLANs and MANETs: A Futuristic heterogeneous Wireless Network, IEEE Wireless Communications, June 2005.
29. H. Dommel and J. Aceves, Floor control for multimedia conferencing and collaboration, *ACM Multimedia Systems Mag.*, Vol. 5, No. 1, pp. 23–38, 1997.
30. H. Schulzrinne and J. Rosenberg, Signaling for internet telephony, in *Procedings, 6th International Conference on Network Protocols*, IEEE Computer Society, Washington, DC, pp. 298–307, 1998.
31. J. Rosenberg, A Framework for Conferencing with the Session Initiation Protocol (SIP), IETF RFC 4353, February 2006.
32. ITU-T Recommendation T.120, Data Protocols for Multimedia Conferencing, ITU-T, July 1997.
33. E. M. Schooler and S. L. Casner, An architecture for multimedia connection management, in *Proceedings, IEEE 4th Comsoc International Workshop on Multimedia Communications*, MM'92, pp. 271–274, Monterey.
34. S. Shenker, A. Weinrib, and E. Schooler, Managing shared ephemeral teleconferencing state: Policy and mechanism, IETF Internet draft, July 1995.

35. M. Handley, I. Wakeman, and J. Crowcroft, The conference control channel protocol (CCCP): A scalable base for building conference control applications, in *Proceedings of ACM SIGCOMM'95*, Boston, August 1995.
36. C. Bormann, J. Ott, and C. Reichert, Simple conference control protocol, IETF Internet draft, December 1996.
37. O. Novo, G. Gamarillo, D. Morgan, and R. Even, A common conference information data model for centralized conferencing (XCON), IETF Internet draft, April 2006.
38. H. Khartabil, P. Koskelainen, and A. Niemi, The conference policy control protocol, IETF Internet draft < draft-ietf-xcon-cpcp-01 >, 2004.
39. G. Gamarillo, J. Ott, and K. Drage, The Binary Floor Control Protocol (BFCP), IETF RFC 4582, November 2006.
40. H. Schulzrinne et. al., RTP: A Transport Protocol for Real-Time Applications, IETF RFC3550, July 2003.
41. H. Schulzrinne and S. Casner, RTP Profile for Audio and Video Conferences with Minimal Control, IETF RFC 3551, July 2003.
42. C. Groves el al., Gateway Control Protocol Version 1, IETF RFC 3525, June 2003.
43. H.248.1, Gateway Control Protocol: Version 3, ITU-T, September 2005.
44. J. Van Dyke, E. Burger, and A. Spitzer, Media Server Control Markup Language (MSCML) and Protocol, IETF RFC 5022, September 2007.
45. H.323 series, ITU-T Recommendations, Geneva 2003.
46. Rosenberg et al., SIP: Session Initiation Protocol, RFC 3261, June 2002.
47. H. Liu and P. Mouchtaris, Voice Over IP Signaling: H.323 and Beyond, *IEEE Commun. Mag.*, Vol. 38, No. 10, pp. 142–148, October 2000.
48. H. Schulzrinne and J. Rosenberg, The session initiation protocol: internet centric signaling, *IEEE Commun. Mag.*, Vol. 38, No. 10, pp. 134–141, October 2000.
49. 3GPP TS 24.147, V8.2.0, Conferencing using the IP multimedia (IM) core network (CN) subsystem, Stage 3, Release 8, December 2008.
50. P. Koskelainen, H. Schulzrinne, and X. Wu, A SIP-based conference control framework, in *ACM NOSSDAV '02*, May 2002.
51. A. Uyar, W. Wu, H. Bulut, and G. Fox, Service-oriented architecture for a scalable videoconferencing system, in *ICPS '05*, 11–14 July 2005, pp. 445–448.
52. H. J. Wang et al., Iceberg: An Internet core network architecture for integrated communications, *IEEE Personal Commun.*, Vol. 7, No. 4, pp. 10–19, August 2000.
53. M. Kelley, Distributed Multipoint Conferences Using SIP, IETF Internet draft, March 8, 2000.
54. H. Khlifi, A. Agarwal, and J.-C. Gregoire, A framework to use SIP in ad-hoc networks, in *Electrical and Computer Engineering, IEEE CCECE 2003*, Canadian Conference on 4–7 May 2003, pp. 985–988 Vol. 2.
55. C. Fu, R. Glitho, and F. Khendek, Signaling for multimedia conferencing in standalone mobile ad hoc network, *IEEE Trans. Mobile Computing*, Vol. 8, issue. 7, pp. 991–1005, 2009.
56. C. Fu, F. Khendek, and R. Glitho, Signaling for multi-media conferencing in 4G: The case of integrated 3G/MANETs, *IEEE Commun. Mag.*, Vol. 44, No. 8, pp. 90–99, August 2006.

57. S. Helal, N. Desai, V. Verma, and C. Lee, Konark—A service discovery and delivery protocol for ad-hoc networks, in *IEEE WCNC 2003*, Vol. 3. 16–20 March 2003, pp. 2107–2113.
58. C. Fu, R. Glitho, and F. Khendek, A cross-layer architecture for signaling in multihop cellular networks, *IEEE Commun. Mag.*, Vol. 46, No. 9, pp. 174–182, September 2008.
59. M. Chatterjee, S. K. Das, and D. Turgut, An on-demand weighted clustering algorithm (WCA) for ad hoc networks, in *IEEE GLOBECOM'00*, Vol. 3, 27 November–1 December 2000, pp. 1697–1701.
60. Srivastava and M. Motani, Cross-layer design: a survey and the road ahead, *IEEE Commun. Mag.*, December 2005.
61. E. Setton et al., Cross-layer design of ad hoc networks for real-time video streaming, *IEEE Wireless Commun.*, Vol. 12, issue 4, pp. 59–65, August 2005.
62. Q. Wang and M. A. Abu-Rgheff, Cross-layer signaling for next-generation wireless systems, in *Proceding IEEE WCNC*, March 2003.
63. M. Conti, G. Maselli, G. Turi, and S. Giordano, Cross-layering in mobile ad hoc network design, *IEEE Computer*, Vol. 37, No. 2, pp. 48–51, February 2004.
64. R. Winter, J. H. Schiller, N. Nikaein, and C. Bonnet, CrossTalk: Cross-layer decision support based on global knowledge, *IEE Commun. Mag.*, pp. 93–99, January 2006.
65. R. T. Raisinghani and S. Lyer, Cross-layer feedback architecture for mobile device protocol stacks, *IEEE Commun. Mag.*, pp. 85–92, January 2006.
66. H. Choi and D. Cho, Fast and reliable route discovery protocol considering mobility in multihop cellular networks, in *IEEE Vehicle Technology Conference*, Vol. 2, 2006, pp. 886–890.
67. G. Kannan, S. N. Merchant, and U. B. Desai, *Cross layer routing for multihop cellular networks*, in *Advanced Information Networking and Applications Workshops*, Vol. 2, May 2007, pp. 165–170.

13

SELF-COEXISTENCE AND SECURITY IN COGNITIVE RADIO NETWORKS

Shamik Sengupta, Santhanakrishnan Anand, and Rajarathnam Chandramouli

13.1 INTRODUCTION

Recent experimental studies have demonstrated that wireless spectrum suffers from overutilization in some bands and underutilization in others over different points in time and space [1]. This results in a great amount of *white space* (unused bands) being available dynamically that can potentially be used for both licensed and unlicensed services. It is then intuitive that static spectrum allocation may not be the optimal solution for efficient spectrum sharing and usage. In static spectrum allocation, a large number of the radio bands are allocated to the television, government, private, and public safety systems. However, the utilization of these bands is significantly low. Often, the usage of spectrum in certain networks is lower than anticipated, while other bands suffer from crisis because the demands of their users exceed the network capacity. Though it might be argued that the implementation and administration of static allocation policy is very easy, the fact remains that the current allocation policy is ineffective and the penalty trickles down as an increased cost to the end users.

Static spectrum allocation often also faces issues due to the modification in existing technologies. For example, in case of VHF and UHF bands reserved for

Convergence of Mobile and Stationary Next-Generation Networks, edited by Krzysztof Iniewski

television broadcast in the United States, allocation of 6MHz per TV channel was based on the old analog NTSC system even though better-quality video can be now broadcast with almost 50% less spectrum per channel [2]. Given the pervasive penetration of cable TV, this spectrum, though allocated and owned, remains unused in most geographical locations, thus leading to underutilization and wastage. This observation underscores the suboptimality of the fixed spectrum assignment policies and has led to the recent spectrum policy reforms by the Federal Communication Commission (FCC) in the sub-900-MHz bands [3]. This new policy would allow unused, licensed spectrum bands (*white spaces*) to be used by unlicensed users (*secondary users*) under the provision that they would vacate upon the return of the licensed user (*primary user*).[1] The success of this policy depends on the ability of secondary users to dynamically identify and access unused spectrum bands, detect the return of primary users, and switch to a different band promptly upon sensing the primary user.

The newly proposed cognitive radio (CR) paradigm/networks are anticipated to make dynamic spectrum access (DSA) a reality [4]. Unlike conventional radios, cognitive radios can intelligently adjust their transmission/reception parameters based on interaction with the environment and find the best available spectrum bands to use [5]. The core components of a cognitive radio network are the base stations (BSs) and the consumer premise equipments (CPEs) [6, 7]. A BS typically manages its own cell by controlling on-air activity within the cell, including access to the medium by CPEs, allocations to achieve quality of service (QoS), and admission to the network based on network security mechanisms. The operations of BS/CPEs can be divided into two major categories: sensing and transmitting/receiving data. Sensing and avoiding incumbent transmission is the most prioritized task of all CR-enabled secondary devices. If any of the channels used by CR node/network is accessed by the licensed incumbents, the primary task of CR devices is to vacate the channels within the channel move time (e.g., for IEEE 802.22, the channel move time is 2s) and switch to some other channel. To get the knowledge of the presence of licensed incumbents and their usage of channels, BS and CPEs periodically perform channel sensing.

To date, a great deal of research work has been done in spectrum sensing [8–12], most of which falls into three categories: matched filter detection, cyclostationary feature detection, and energy detection. Unfortunately, none of these sensing mechanisms can ensure accuracy in the detection outcomes due to the inherent unreliable nature of the wireless medium and varying physical separation between the primary and secondary users. Such uncertainties in the licensed user detection make the spectrum sensing vulnerable to *denial-of-service* (DoS) threats in the hostile network environment. In this chapter, we discuss a specific class of DoS attacks in cognitive radio networks known as the primary user emulation (PUE) attack. In this type of attack, one or multiple attacking nodes (*malicious users*) transmit in forbidden timeslots and effectively emulate the

[1]Throughout the chapter, we use the terms "user" and "node" interchangeably.

primary user to make the protocol compliant secondary users erroneously conclude that the primary user is present. If the attack is successful, the protocol-compliant secondary users will leave the spectrum band, potentially causing a disruption in their network quality of service (QoS).

Another major challenge in the newly proposed CR standard is ensuring quality of service (QoS) among multiple good (rational) CR networks themselves—in other words, maintaining self–coexistence. In areas with significant high primary incumbents (licensed services), open channels will be a commodity of demand. Therefore, dynamic channel access among CR networks will be of utmost importance so that the interference among CR networks themselves can be minimized; else the throughput and Quality of Service (QoS) will be compromised. Thus in this chapter, we also focus on the issue of self–coexistence among multiple good overlapping and competing CR networks in a geographical region. We use the tools from noncooperative game theory and model the competitive environment as a distributed game. We consider the system of multiple overlapping CR networks operated by multiple wireless service providers that compete for the resources and try to seek a spectrum band void of interference from other coexisting CR networks. If interfered by other CR networks at any stage of the game, the networks face a binary choice of whether to stick to the band (assuming the interferers might move away) or move to another band itself.

Unlike other standards where self-coexistence and security issues are only considered after the specification essentially is finalized, it is required for CR networks to take the proactive approach due to the open nature of dynamic spectrum access. However, the current spectrum etiquette policies only emphasize on the primary user–secondary user interaction (primary avoidance policy) and hardly focus on the policy issues regarding the interaction among multiple secondary networks with good or malicious intentions. In light of these new perspectives, this chapter discusses the approaches to investigate the novel self-coexistence and primary user emulation attack issues for enhanced MAC as revision of the initial cognitive radio standard conception and definition [4].

13.2 COGNITIVE RADIO NETWORK SYSTEM OVERVIEW

Before proceeding further, let us briefly discuss the features of the newly proposed cognitive radio networks [4–7].

13.2.1 System Architecture

A simple architecture of a cognitive radio network consisting of BS and CPEs is shown in Figure 13.1. The BS transmits control, management, and data packets in the downstream direction to the various CPEs, which in turn respond back to the BS in the upstream direction. Based on the feedback received from the CPEs, if any, the BS decides its next actions. The CR network can be used in point-to-point (P2P) or point-to-multipoint (P2MP) mode, using omnidirectional or

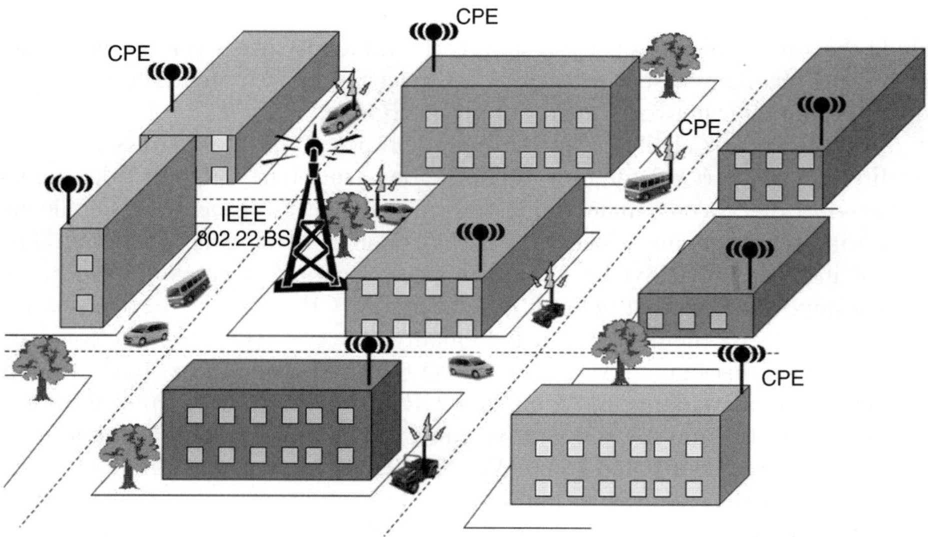

Figure 13.1. An example of CR network system with BS and CPEs.

directional antennas. BS/CPEs typically use omnidirectional antennas for incumbent sensing and receiving data while sectored/directional antennas for transmission purpose. In the United States, TV bands are spread from 54 to 806 MHz and each TV channel is usually 6 MHz. In the initial standard version, the CR-enabled devices can operate/transmit using the 6-MHz channels. The spectral efficiency ranges from 0.5 bit/s/Hz to 5 bit/s/Hz, thus resulting in an average data rate of 18 Mbit/s and maximum up to 30 Mbit/s in a single 6-MHz TV band.

13.2.2 MAC Layer Overview

The existing MAC of proposed cognitive radio devices has most of the features similar to the MAC of 802.11 and 802.16. However, few distinguishing features make the newly proposed MAC worth mentioning.

13.2.2.1 Initial Connection Establishment. Initial connection establishment in CR network differs from that of the previous IEEE 802 standards such as 802.11 or 802.16. Though connection establishment in a true centralized network should be simple, it is not so for the CR paradigm because there is no predefined channel for the CPEs to establish connection with BS as these networks share the spectrum band with licensed devices. Thus there is no way for a CPE to know what channel to use to establish the initial connection with a BS.

In CR network, when a CPE is switched on, it follows the mechanism of *listen before talk* by scanning all the channels in the licensed TV band to determine the presence of any incumbent in the interfering zone and builds a spectrum usage report of vacant and occupied channels. The BS, on the other hand, also follows the same mechanism of sensing spectrum and periodically broadcasts using an unused frequency channel. If a CPE can locate the broadcast sent from the BS, it then tunes to that frequency and then transmits back in the uplink direction with an unique identifier; BS thus becomes aware of the existence of the CPE. Authentication and connection registration is then done gradually. The spectrum usage report is then sent back to the BS from the CPE in the form of feedback. Upon acceptance of the feedback the BS makes a decision on spectrum usage. When more than one CPE tries to establish an initial connection, then a contention-based connection setup similar to that of the IEEE 802.11 takes place after all the CPEs tune to the broadcasted channel.

13.2.2.2 Incumbent Detection. Much of the standard of cognitive radio paradigm is dependent on incumbent sensing and detection. At any point of time, a number of incumbents (TV broadcasting, wireless microphones etc.) may be operating in the same region as that of the CR network. To coexist with the incumbents, it is mandatory that incumbent sensing be done by both the BS and CPEs. CPEs send their spectrum usage reports to the BS in the form of feedbacks. Depending on the incumbent detection algorithms proposed and their efficiencies, the general spectrum sensing process in divided into two categories: *fast sensing* and *fine sensing* [4]. Fast sensing is done typically before fine sensing and uses a quick and simple detection algorithm such as energy detection. It is carried out primarily over in-band channels, and the outcome of these measurements will determine the need and the duration of the upcoming fine sensing. Fine sensing, on the other hand, is of longer duration (on the order of milliseconds for each single-frequency channel, e.g., 24 ms in the case of field-sync detection for ATSC [4]) as more detailed sensing is performed on the target channels. In other words, the fine sensing could be over three orders of magnitude larger than the fast sensing but provides more accuracy.

13.3 CHALLENGES AND RELATED WORK

Since a secondary CR network shares the spectrum bands dynamically with licensed devices and other secondary CR networks, the devices cannot know a priori what frequency bands other devices would be operating on. This gives birth to two very important challenges: (i) efficient dynamic spectrum access among multiple good CR networks (self-coexistence) and (ii) risk of primary user emulation attack from malicious secondary users/networks.

The advancements in software-defined radio (SDR) technology have led to the development of novel algorithms, architectures, and protocols for cognitive radio-based dynamic spectrum access. As far as dynamic spectrum sensing and access are concerned, there is a recent emerging body of works that deal with different decision-making aspects, issues, and challenges in a cognitive radio network setting. A proactive spectrum access approach is used in reference 13, where secondary users utilize past observations to (a) build predictive models on spectrum availability and (b) minimize disruptions to primary users. Energy detection and interference temperature measurements have been largely used in references 14–16 to monitor primary spectrum usage activity. Spectral correlation-based signal detection for primary spectrum sensing in IEEE 802.22 WRAN systems is presented in reference 17. Signature-based spectrum sensing algorithms are presented in reference 18 to investigate the presence of Advanced Television Systems Committee (ATSC) DTV signals. In a similar effort, sequential pilot sensing of Advanced Television Systems Committee (ATSC) DTV signals is carried out in reference 19 to sense the primary usage in IEEE 802.22 cognitive radio networks. In reference 20, a novel frequency-sensing method is proposed known as dynamic frequency hopping (DFH). In DFH, neighboring WRAN cells form cooperating communities that coordinate their DFH operations where WRAN data transmission is performed in parallel with spectrum sensing without interruptions. The aim here is to minimize interrupts due to quiet sensing and increase QoS. Most of the above-mentioned works focus on primary spectrum usage sensing, but the issue of self-coexistence, sharing, or coordinated dynamic spectrum access to minimize interference and maximize self-coexistence is not considered.

In references 2 and 21–23, novel approaches of dynamic spectrum allocation among secondary users (cognitive radio) through a spectrum broker are investigated where the spectrum broker has the knowledge of dynamic availability of spectrum. With the introduction of auctions and pricing, the spectrum broker determines allocations among the secondary users. In reference 24, "economic behavior" of the network nodes are studied using game theory underlimited and finite channel capacity with pricing and purchasing strategies of the access point, wireless relaying nodes, and clients in wireless mesh networks. However, dynamic spectrum access is not considered in this work. Though the pricing models have potential of generating revenue through commercialized secondary spectrum usage, there are a number of challenges in terms of implementations for the pricing in dynamic spectrum allocation—that is, payment transaction method, best-effort-service nature of opportunistic spectrum access, trustworthiness, authentication, and many more.

In a system where unlicensed devices are sharing the spectrum under the presence of licensed incumbents, the issue of self-coexistence among multiple CR operators in an overlapping region is very significant. In areas with analog/digital TV transmissions and wireless microphone services, unused channels are already commodities of demand. The challenge of self-coexistence becomes even tougher because the networks do not have information about which bands other secondary CR networks will choose. In such a scenario (e.g., Figure 13.2), when multiple CR networks operated by multiple operators (or service providers) overlap, it is highly probable that the operators will try to act greedy and use the entire available bandwidth. Because all the operators will act in the same way, this may result in interference among CR networks themselves. Thus an efficient spectrum access method needs to be used such that the interference is minimized.

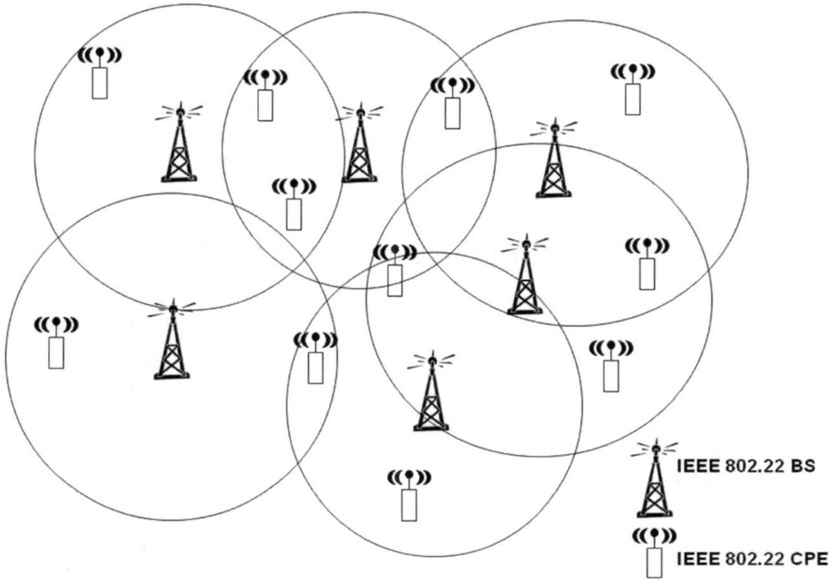

Figure 13.2. Multiple geographically co-located CR networks competing.

Although other aspects of cognitive radio networks have been studied in detail, there is not much literature available on the security [25–29]. In the particular case of DSA networks, it can be argued that in order to stage a denial-of-service (DoS) attack at the sensing level, it is necessary to affect the decision on primary activity during the sensing phase. This can be done in one of the following ways: (a) primary user emulation attacks (PUEA), where some malicious nodes emit signals that emulate the activity of the primary [26, 29–31]; (b) a set of nodes can lie about the the spectrum data (Byzantine attack) [27]; (c) by making use of the weaknesses of existing protocols for evacuation [25]; or (d) by modifying messages passed between the sensing nodes and the centralized decision maker [32].

A class of DoS attacks on the secondaries called primary user emulation attack (PUEA) is studied here. In such attacks, a set of "malicious" secondary users could spoof the essential characteristics of the primary signal transmission to make other "good" secondary users believe that the primary user is present. The secondary users following normal spectrum evacuation process (the good users) will vacate the spectrum unnecessarily, resulting in what are known as the primary user emulation attacks (PUEA).

Chen and Park [30] propose two mechanisms to detect PUEA: distance ratio test and distance difference test based on the correlation between the length of wireless link and the received signal strength. In reference 26, Chen et al. discuss defense against PUEA by localization of the malicious transmission using an underlying sensor network and comparing it with the known location of the primary transmitter. A mitigation technique for DoS attacks arising from fraudulent reporting of sensing results by malicious nodes is studied in reference 27. The PUEA methods described thus far do not take into account the fading characteristics of the wireless environment and require estimation of the location of the malicious users either via a dedicated sensor network or via significant enhancement of the secondary nodes themselves.

13.4 SELF-COEXISTENCE AMONG MULTIPLE CR NETWORKS

In this section, we formulate the self–coexistence problem as a dynamic channel[2] switching game. We assume that N CR networks (players) operated by N separate wireless service providers in a region are competing for one of M separate orthogonal spectrum bands not used by primary incumbents. The CR networks can have partially or completely overlapped geographical overage area with each other. If one network is in the interference range of another, they cannot use the same spectrum band because the QoS of both the networks will suffer. In this scenario, we model the dynamic channel switching as a noncooperative game where the aim of each network is to capture a spectrum band free of interference. We assume that the only control information needed for participating successfully in the game is the number of overlapping competitors in the region, which can be known from the broadcasting beacons by each of the CR networks in the Foreign Beacon Period (FBP) [4].

13.4.1 Decision Problem of CR Network

We assume that each CR network can dynamically choose one of the M allowable spectrum bands for its operations. If two or more overlapped networks operate using the same spectrum band, then interference will occur and their transmissions will fail. Thus the networks will have to make new decisions for channel switching in the next stage of the game. The game ends when all the networks are successful in capturing a clear spectrum band. The optimization problem is to find the mechanism of achieving minimum number of failed transmission stages from the networks. As far as the decision strategy in this game is concerned, if interfered at any stage of the game, network i has the binary strategy set of switching to another band (expecting to find a free spectrum band) or staying on the current band (assuming the interferers will move away).

When an interfered network i chooses either "switch" or "stay," it faces one of two possible costs in terms of time units. If the network i chooses to switch, it faces a cost of finding a clear spectrum band in the game. Note that in a game of N networks competing over M spectrum bands, the network i might find the clear channel just after 1 switching, or it might take more than 1 switching as multiple networks might choose the same band chosen by network i resulting in

[2]Throughout this chapter, we use the words "channel," "band," and "chunk" interchangeably unless explicitly mentioned otherwise.

a subgame. Moreover, note that with varying N and M, the average cost of finding a clear band will also vary. We define the expected cost of finding a clear channel, if the network chooses the strategy of switching, as

$$E[C_i(s_i, \mathbf{s_{-i}})] = c^{f(N,M)} \tag{13.1}$$

over all possible resulting subgames, where s_i and s_{-i} denote the strategies chosen by network i and rest of the networks, respectively. c is the cost of single switching and $f(\cdot)$ is a function that depicts the varying behavior of the cost with N and M. We assume a simple closed form of $f(N, M) = NM/(M - N)$. The intuitive reason behind proposing such a function is that expected cost to find a clear band increases with increasing N but fixed M, while the cost decreases with increasing M but fixed N; however, with both N and M increasing, the cost varies simultaneously with the ratio of M:N and the difference between them. Note that we could choose any other form for $f(N, M)$ as long as the above conditions are satisfied.

At the beginning of the stage, if the network i chooses the strategy of "stay," it might fall in one of three different scenarios: (i) All the other networks that were attempting to operate using the same band as network i might move away, thus creating a clear band for network i. (ii) All the other networks that were attempting to operate using the same band as network i might also "stay," thus wasting the stage under consideration and repeating the original game G, which started at the beginning of the stage. (iii) Some of the networks move ("switch") while some networks end up being in the same band ("stay"), thus wasting the stage under consideration and creating a subgame G' of the original game G. More detailed explanations for subgame G' will be presented later. We define the cost functions as

$$C_i(s_i, \mathbf{s_{-i}}) = \begin{cases} 0 & \text{Case i} \\ 1 + C_i(G) & \text{Case ii} \\ 1 + C_i(G') & \text{Case iii} \end{cases} \tag{13.2}$$

13.4.2 Self-Coexistence Game Analysis

With the strategy set and cost functions defined, the optimization problem in this game is to find a mechanism of switching or staying such that the cost incurred can be minimized and an equilibrium can be achieved. We typically assume that all the players are rational and pick their strategy, keeping only individual cost minimization policy in mind at every stage of the game. We intend to find if there is a set of strategies with the property that no network can benefit by changing its strategy unilaterally while the other networks keep their strategies unchanged (Nash equilibrium) [33].

For this purpose, we study the game with mixed strategy by assigning probabilities to each of the strategies in the binary strategy space. We define the mixed strategy space of network i as $S_i^{\text{mixed}} = \{(\text{switch} = p), (\text{stay} = (1-p))\}$, where network i chooses the strategy "switch" with probability p and chooses the strategy "stay" with probability $(1-p)$. Since all networks are assumed to behave identically, we assume similar mixed strategy space for all the networks. The question now is, What values of $(p, 1-p)$ tuple will help us achieve the optimal solution—in other words, if there exists any finite nonzero probability of "switch" and "stay"?

We start the game with all $(N-1)$ other networks coexisting with network i on one band and choose a strategy from mixed strategy space. Then regardless of the strategy chosen by network i, the resulting subgame will obtain one of the following possible outcomes: All $N-1$ networks choose "switch," or $N-2$ networks choose "switch," · · · , or 0 networks choose "switch." To find the Nash equilibrium, we then determine the expected cost if network i under consideration chooses to "switch" or "stay." Following the switching cost for finding a nonoccupied band as indicated previously in Eq (13.1), the expected cost over all possible resulting subgames for network i, if it chooses to switch, is

$$E\left[C_i^{\text{switch}}\right] = \sum_{j=0}^{N-1} Q_j \times c^{f(N,M)} \tag{13.3}$$

where j denotes the number of other networks choosing to "switch" and Q_j denotes the probability of j networks switching out of other $N-1$ networks and is given by $Q_j = \binom{N-1}{j} p^j (1-p)^{(N-1-j)}$. On the other hand, the expected cost for network i, if it chooses "stay," can then be given as

$$E\left[C_i^{\text{stay}}\right] = \sum_{j=0}^{N-2} Q_j (1 + E[C_i(G'_{(N-j)})]) + Q_{(N-1)} \times 0 \tag{13.4}$$

where $E[C_i(G'_{(N-j)})]$ denotes the expected cost incurred in subgame $G'_{(N-j)}$. Note that if the expected cost of switching is less than the expected cost of staying, network i will always choose the strategy "switch," thus going back to the pure strategy and as a result cannot achieve the Nash equilibrium [34]. Again, if the expected cost of staying is less than the expected cost of switching, similar reasoning can be applied for the strategy "stay" and Nash equilibrium can not be achieved. Thus for the existence of mixed strategy Nash equilibrium, network i must be indifferent between "switch" or "stay" regardless of strategies taken by other networks. In other words, the probability tuple $(p, 1-p)$ helps in choosing the strategy such that network i is never dominated by response from any other networks and thus will not deviate from the mixed strategy space $(p, 1-p)$ unilaterally to obtain lower cost. To find the optimal values for mixed strategy space, we equate Eqs. (13.3) and (13.4) as

$$\sum_{j=0}^{N-2} Q_j (1 + E[C_i(G'_{(N-j)})]) = c^{f(N,M)} \tag{13.5}$$

Note that the expected cost of the game at Nash equilibrium is actually not dependent on j as evident from first part of Eq. (13.5), that is, how many networks are actually switching; rather, the cost varies with N, the number of networks, and M, the number of bands. Thus the expected cost for network i in the subgame $G'_{(N-j)}$ can be deduced to be the same as that in the original game. Using binomial expansion and detailed mathematical derivations, we obtain the closed form for p as

$$p = \left(\frac{1}{1 + c^{f(N,M)}} \right)^{\frac{1}{N-1}} \tag{13.6}$$

For any values of N and M, p has a nonzero finite value, thus proving the existence of a mixed strategy Nash equilibrium point. In other words, the mixed strategy tuple, $(p, 1 - p)$, presented in Eq. (13.6) constitutes the dominant best response strategy in this game.

13.5 RISK OF PRIMARY USER EMULATION ATTACK

In this section, we discuss in detail the threat due to PUEA in DSA networks. We consider a cognitive radio network with a primary transmitter as well as the secondary and malicious nodes as shown in Figure 13.3. We assume that all secondary and malicious users are distributed in a circular grid of radius R as shown in Figure 13.3. A primary user is located at a distance of at least d_p from all other users. We consider energy-based mechanisms to detect the presence of the primary. Typical energy-based detection methods assume that the primary is

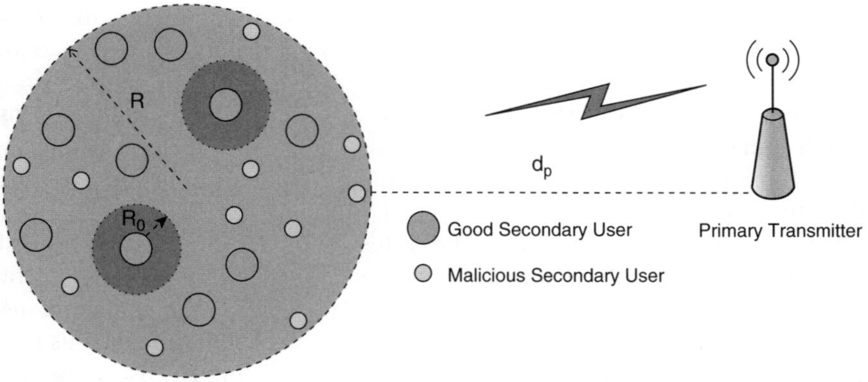

Figure 13.3. A typical cognitive radio network in a circular grid of radius R consisting of good secondary users and malicious secondary users. No malicious users are present within a radius R_0 about each good secondary user. A primary transmitter is located at a distance of at least d_p from all other users.

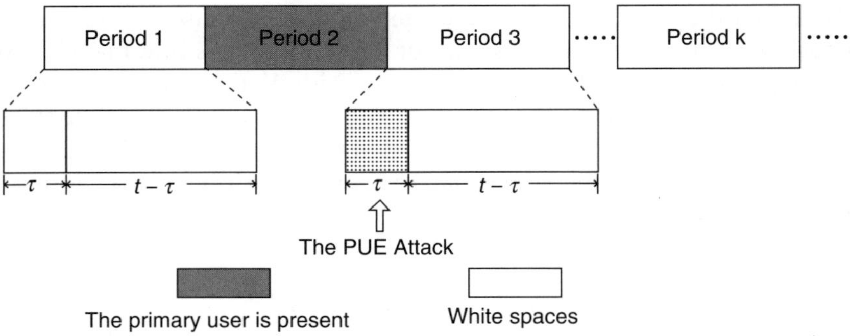

Figure 13.4. Discrete sensing/transmission periods. In each period, τ and $t-\tau$ represent the sensing slot and transmission slot, respectively. Note that the shaded period indicates that the primary user is using the spectrum band and the grid filled sensing slot indicates that the attack has been launched.

present if the received signal strength is –93 dBm [35]. We consider the time epochs in this network model to be divided into discrete periods where each period consists of one sensing slot, τ, and one transmission slot, $t-\tau$. Figure 13.4 shows instances of the presence of the primary, the absence of primary, and an instance of a PUE attack.

In each period, the secondary user measures the received signal power during the sensing slot. Also, we assume that secondary users are not naive, that is, they are aware of the existence of malicious users around the network. However, they know neither the locations of malicious nodes nor the slots when the attacks will be launched. Hence, identifying the presence of the primary user is a challenging task for the secondary user because the received energy might be from the primary user or the malicious user or both.

In order to mitigate this threat, we devise two hypothesis-based testing mechanisms to decide if the primary is transmitting or if an attack is in progress. We consider a system where there is no cooperation between the secondary users. Thus, the probability of PUEA on any user is the same as that on any other user. Hence, without loss of generality, we analyze the probability density function (pdf) of the received signal at one secondary user located at the origin (0, 0). All malicious nodes are then uniformly distributed in the annular region with radii R_0 and R.

The first step to obtain a hypothesis test is to determine the probability density function (pdf) of the received signal at the secondary user due to transmission by the primary and the malicious users. Consider M malicious users located at coordinates (r_j, θ_j) $1 \le j \le M$. The position of the jth malicious user is uniformly distributed in the annular region between R_0 and R. Also, r_j and θ_j are statistically independent $\forall\, j$. The received power at the secondary user from the primary transmitter, $P_r^{(p)}$, is given by

$$P_r^{(p)} = P_t d_p^{-2} G_p^2 \tag{13.7}$$

where P_t is the primary transmit power, $G_p^2 = 10^{\xi_p/10}$, and $\xi_p \sim \mathcal{N}(0, \sigma_p^2)$. Since P_t and d_p are fixed, the pdf of $P_r^{(p)}$, $p^{(pr)}(\gamma)$, follows a log-normal distribution and can be written as

$$p^{(Pr)}(\gamma) = \frac{1}{A\sigma_p\sqrt{2\pi}\gamma}\exp\left\{-\frac{(10\log_{10}\gamma - \mu_p)^2}{2\sigma_p^2}\right\} \tag{13.8}$$

where $A = \ln 10/10$ and

$$\mu_p = 10\log_{10} P_t - 20\log_{10} d_p \tag{13.9}$$

The total received power at the secondary node from all M malicious users is given by

$$P_r^{(m)} = \sum_{j=1}^{M} P_m d_j^{-4} G_j^2 \tag{13.10}$$

where P_m is the transmit power of each malicious user, d_j is the distance between the jth malicious user and the secondary user, and G_j^2 is the shadowing between the jth malicious user and the secondary user. $G_j^2 = 10^{\xi_j/10}$, where $\xi_j \sim \mathcal{N}(0, \sigma_m^2)$. Conditioned on the positions of all the malicious users, each term in the summation on the right-handside of Eq. (13.10) is a log-normally distributed random variable of the form $10^{\omega_j/10}$, where $\omega_j \sim \mathcal{N}(\mu_j, \sigma_m^2)$, where

$$\mu_j = 10\log_{10} P_m - 40\log_{10} d_j \tag{13.11}$$

Conditioned on the positions of all the malicious users, $P_r^{(m)}$ can be approximated as a log-normally distributed random variable whose mean and variance can be obtained by using Fenton's method. Applying Fenton's approximation for the weighted sum, the expression for the pdf $p^{(m)}(\chi)$ can be approximated as a log-normal distribution with parameters μ_χ and σ_χ^2 of the form

$$p^{(m)}(\chi) = \frac{1}{A\sigma_\chi\sqrt{2\pi}\chi}\exp\left\{-\frac{(10\log_{10}\chi - \mu_\chi)^2}{2\sigma_\chi^2}\right\}. \tag{13.12}$$

The details of the derivation of σ_χ^2 and μ_χ can be found in reference 31. The NPCHT can be used to distinguish between two hypotheses, namely,

$$\begin{array}{ll} H_1: & \text{Primary transmission in progress} \\ H_2: & \text{Emulation attack in progress} \end{array} \tag{13.13}$$

The observation space is the sample space of received power measured at the secondary user. It is observed that there are two kinds of risks incurred by a secondary user in this hypothesis test.

- *False Alarm:* When the actual transmission is made by malicious users but the secondary decides that the transmission is due to the primary. In our case, this is also the probability of a successful PUEA.
- *Miss:* When the actual transmission is made by the primary transmitter but the secondary decides that the transmission is due to the malicious users. This is a serious concern if the good secondary does not wish to violate the spectrum etiquette.

The Neyman–Pearson criterion allows the secondary to minimize the probability of successful PUEA while fixing the probability of missing the primary user at a desired threshold, α. The decision variable, Λ, is given by

$$\Lambda = \frac{p^{(m)}(x)}{p^{(\mathrm{Pr})}(x)} \tag{13.14}$$

where x is the measured power of the received signal. In the above, $p^{(\mathrm{Pr})}(x)$ and $p^{(m)}(x)$ are given by Eqs. (13.8) and (13.12), respectively. The decision is then made based on the following criterion:

$$\begin{array}{lll} \Lambda \leq \lambda & D_1: & \text{Primary transmission} \\ \Lambda \geq \lambda & D_2: & \text{PUEA in progress,} \end{array} \tag{13.15}$$

where λ satisfies the constraint that miss probability, $\Pr\{D_2|H_1\}$, is fixed at α.

The WSPRT allows us to specify desired thresholds (α_1 and α_2, respectively) for both the false alarm and the miss probabilities. The decision variable after n sequential tests, Λ_n, is given by

$$\Lambda_n = \prod_{i=1}^{n} \frac{p^{(m)}(x_i)}{p^{(Pr)}(x_i)} \tag{13.16}$$

where x_i is the measured power at the ith stage. In the above equation, $p^{(\mathrm{Pr})}(x_i)$ and $p^{(m)}(x_i)$ are given by Eqs. (13.8) and (13.12), respectively. The decision is then made based on the following criterion:

$$\begin{array}{ll} \Lambda_n \leq T_1 = \dfrac{\alpha_1}{1-\alpha_2} & D_1\text{: Primary transmission} \\ \Lambda_n \geq T_2 = \dfrac{1-\alpha_1}{\alpha_2} & D_2\text{: PUEA in progress} \\ \text{Otherwise} & D_3\text{: Take another observation} \end{array} \tag{13.17}$$

The average number of observations required to arrive at a decision is given by

$$E[n|H_k]=\begin{cases}\dfrac{(1-\alpha_2)\ln T_1+\alpha_2\ln T_2}{E[f(x_1)|H_1]} & k=1\\ \dfrac{\alpha_1\ln T_1+(1-\alpha_1)\ln T_2}{E[f(x_1)|H_2]} & k=2\end{cases} \tag{13.18}$$

where the function $f(x_1) = \ln \Lambda_1$.

13.6 SIMULATION MODEL AND RESULTS

We conducted simulation experiments to evaluate the improvements achieved by the proposed strategies. Source code for the experiment has been written in C under Linux environment.

13.6.1 Self-Coexistence Strategy Evaluation

We assumed that N secondary networks, operated by N separate wireless service providers, compete for one of M available spectrum bands. Each of the networks is associated with a mixed strategy space of "switch" and "stay." The system converges when all the networks capture a spectrum band free of interference from other CR networks. N and M are given as inputs to the experiment.

In Figure 13.5, we present the average system convergence cost with 20 competing cognitive radio (CR) networks. Switching strategy (probability) is

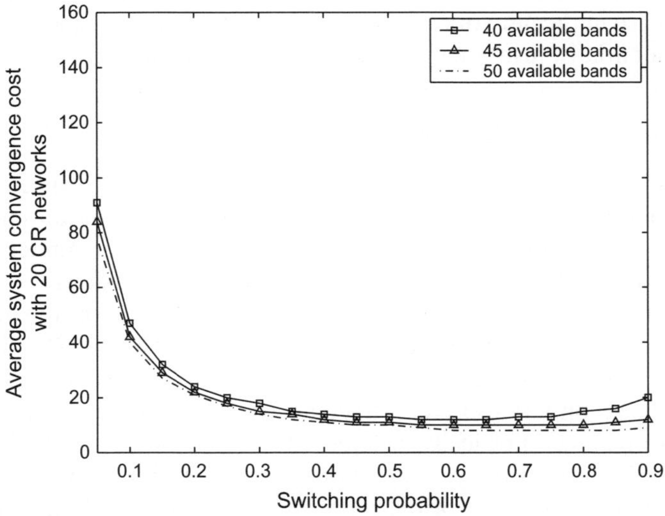

Figure 13.5. Average system convergence cost with 20 CR networks and varying number of bands.

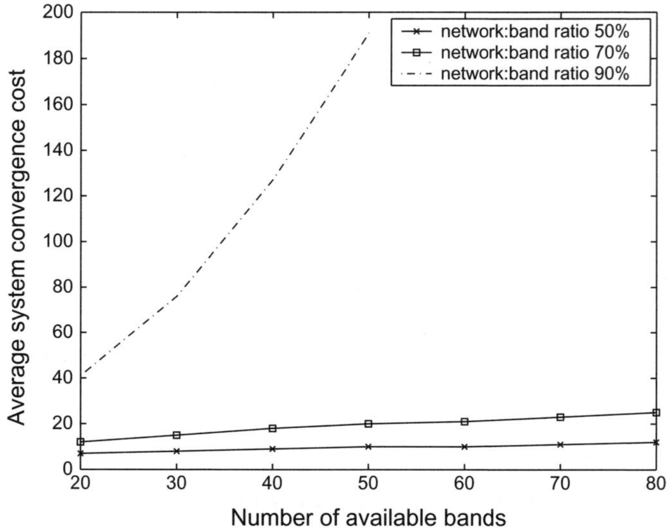

Figure 13.6. Average system convergence cost with varying network : band ratio.

varied for this simulation experiment. With increase in number of available bands, the convergence cost decreases. The convex nature of the curves in Figure 13.5, proves that a point of minima exists for each of the curve. This minima corresponds to the Nash equilibrium strategy (p).

We also present different system convergence costs following mixed strategy space for a varying network : band ratio (50–90%) in Figure 13.6. We find that with an increase in the network : band ratio the system convergence cost increases almost exponentially, justifying the proposed cost function.

13.6.2 Primary User Emulation Attack

The risk of PUEA is measured in terms of the probability of a successful attack. We run numerical simulations in Linux to perform the computations. We consider the following values of the system parameters for our numerical simulations. The variances for the primary and malicious transmissions are assumed to be $\sigma_p = 8$ and $\sigma_m = 5.5$. A primary transmitter (a TV tower), located at a distance of $d_p = 100\,\text{km}$ to the secondary user, has a transmit power of $P_t = 100\,\text{kW}$. The transmit power of the malicious users, P_m, is taken to be 4W. The exclusive distance from the secondary user, R_0, is fixed at 30m. The number of malicious users is assumed to be a geometrically distributed random variable with $E[M] = 25$.

Figure 13.7 presents the probability of successful PUEA (Figure 13.7a) and the probability of missing the primary transmitter (Figure 13.7b) for the NPCHT with the theoretical probability of missing the primary user set to $\alpha = 0.3$. It is observed from Figure 13.7a that the probability of false alarm rises and then falls down with increasing value of R. This is because, for a given R_0, if R is small (i.e.,

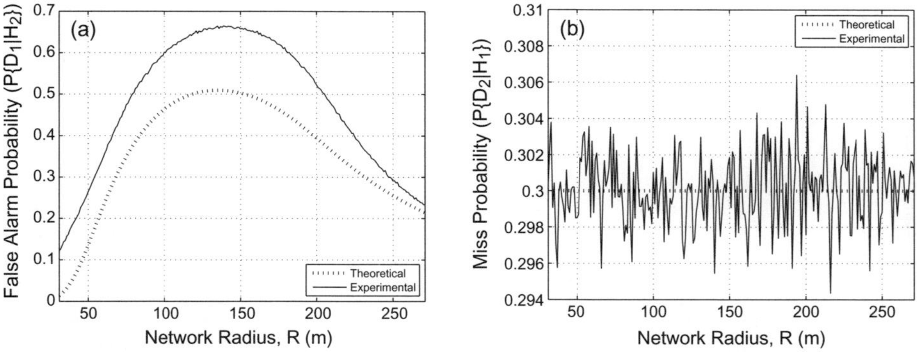

Figure 13.7. NPCHT with theoretical probability of missing primary user $\alpha = 0.3$. **(a)** Probability of successful PUEA using the NPCHT. The average number of malicious users is fixed at 25. **(b)** Probability of missing primary user using the NPCHT.

malicious users are closer to the secondary user), the total received power from all malicious users is likely to be larger than that received from the primary transmitter, thus decreasing the probability of successful PUEA. Similarly, for large R, the total received power from the malicious users may not be enough to successfully launch a PUEA. Figure 13.7b shows that the experimental probability of missing the primary user is always close to the required value (within $\pm$ 0.04 of the desired value).

The performance of WSPRT is shown in Figure 13.8. The thresholds for the probability of successful PUEA and the probability of missing primary user are set to 0.3 each. Although the experimental curve in Figure 13.8a goes above the theoretical one, we achieve much lower probabilities of successful PUEA compared to Figure 13.7a. In fact, the maximum probability of successful PUEA in the NP test can go as high as 0.7, whereas in the Wald's test we can limit this to 0.4. The lower probabilities of successful PUEA are achieved at the cost of more observations as shown in Figure 13.8c and Figure 13.8d. It is observed that number of observation behaves similar to the probability curves. This is because more observations are always taken if a decision cannot be made easily, where decision error probabilities also tend to be relatively high. Note that it is almost always possible to make sure that the probability of missing primary user stays strictly below the required threshold. This is particularly important in CRN to ensure that the secondaries still obey the spectrum-sharing etiquette.

13.7 CONCLUSIONS

In this research, we investigated the spectrum etiquettes from secondary networks' interaction perspectives. We focused on the issues of self-coexistence and primary user emulation attacks in such networks. We discussed how multiple overlapped CR networks controlled by different service providers can operate on the available spectrum and coexist. We used a noncooperative game to model

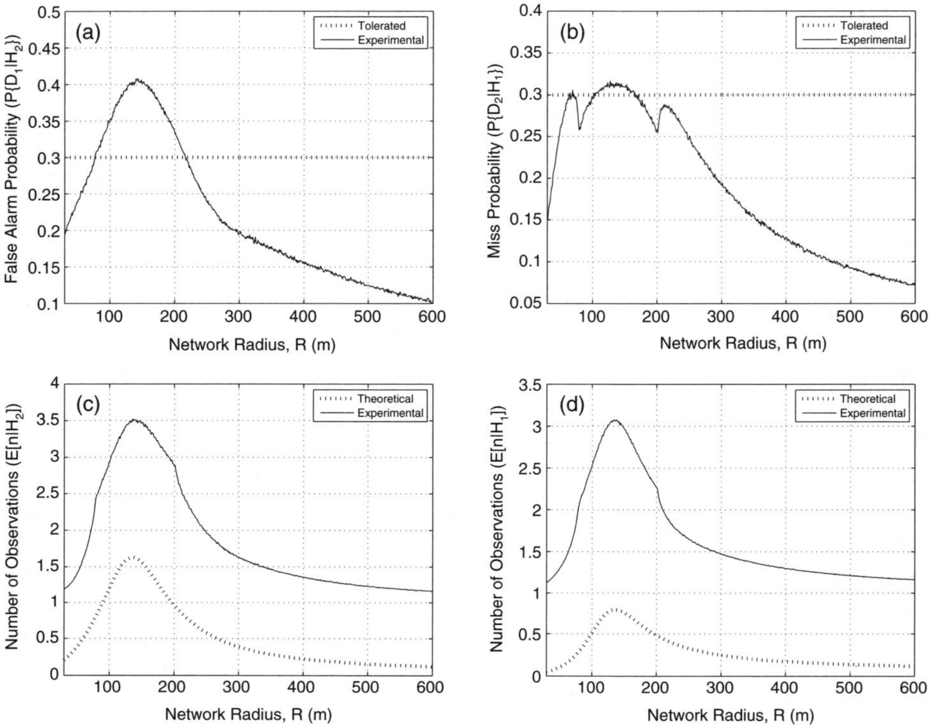

Figure 13.8. WSPRT with theoretical probability of successful PUEA $\alpha_1 = 0.3$ and theoretical probability of missing primary user $\alpha_2 = 0.3$. **(a)** Probability of successful PUEA. **(b)** Probability of missing primary user. **(c)** Average number of observations when malicious users are transmitting. **(d)** Average number of observations when primary user is transmitting.

this problem and presented mixed strategy for the CR networks. In order to mitigate PUEA, we proposed a Neyman–Pearson composite hypothesis test (NPCHT) and a Wald's sequential probability ratio test (WSPRT). Simulation experiments showed that both WSPRT and NPCHT resulted in a range of radii in which PUEA were most successful. For a desired threshold on the probability of missing the primary, WSPRT was found to achieve about 50% reduction in the probability of successful PUEA compared to NPCHT. The extension of our analysis to include power control at the malicious users is a topic for further investigation.

REFERENCES

1. [Online]. Available: http://www.sharedspectrum.com/inc/content/measurements/nsf/NY_Creport.pdf
2. M. Buddhikot, P. Kolodzy, S. Miller, K. Ryan, and J. Evans, DIMSUMnet: New directions in wireless networking using coordinated dynamic spectrum access, in *IEEE*

International Symposium on a World of Wireless, Mobile and Multimedia Networks, 2005, pp. 78–85.

3. Federal Communications Commission (fcc), Notice of proposed rule making, ET Docket no. 04-113, May 2004.
4. IEEE P802.22/D0.1, Draft Standard for Wireless Regional Area Networks Part 22: Cognitive Wireless RAN Medium Access Control (MAC) and Physical Layer (PHY) Specifications: Policies and Procedures for Operation in the TV Bands, 2006. [Online]. Available: http://www.ieee802.org/22/
5. I. Akyildiz, W. Lee, M. Vuran, and S. Mohanty, Next generation/dynamic spectrum access/cognitive radio wireless networks: a survey, *Computer Networks*, Vol. 50, No. 13, pp. 2127–2159, 2006.
6. C. Cordeiro, K. Challapali, D. Birru, and S. Shankar, IEEE 802.22: The first worldwide wireless standard based on cognitive radios, *IEEE International Symposium on Dynamic Spectrum Access Networks*, November 2005, pp. 328–337.
7. S. Sengupta, S. Brahma, M. Chatterjee, and N. Sai Shankar, Enhancements to cognitive radio based ieee 802.22 air-interface, in *Proceedings of IEEE International Conference on Communications (ICC)*, June 2007, pp. 5155–5160.
8. G. Ganesan and Y. Li, Cooperative spectrum sensing in cognitive radio networks, part i: Two user networks and part ii: Multiuser networks, *IEEE Trans. Wireless Commun.* Vol. 6, No. 6, June 2007.
9. Z. Quan, S. Cui, A. H. Sayed, and H. V. Poor, Optimal multiband joint detection for spectrum sensing in cognitive radio networks, *IEEE Trans. Signal Processing*, Vol. 57, No. 3, pp. 1128–1140, March 2009.
10. A. Ghasemi and E. Sousa, Interference aggregation in spectrum-sensing cognitive wireless networks, *IEEE J. Sel. Top. Signal Processing*, Vol. 2, No. 1, pp. 41–56, February 2008.
11. Z. Ye, G. Memik, and J. Grosspietsch, Energy detection using estimated noise variance for spectrum sensing in cognitive radio networks, in *IEEE Wireless Communications and Networking Conference (WCNC)*, April 2008, pp. 711–716.
12. K. Kim, I. Akbar, K. Bae, J. sun Urn, C. Spooner, and J. Reed, Cyclostationary approaches to signal detection and classification in cognitive radio, in *2nd IEEE International Symposium on New Frontiers in Dynamic Spectrum Access Networks (DySPAN)*, April 2007, pp. 212–215.
13. P. Acharya, S. Singh, and H. Zheng, Reliable open spectrum communications through proactive spectrum access, *ACM Int. Conf. Proc. Ser.*, Vol. 222, 2006.
14. G. Ganesan and Y. Li, Cooperative spectrum sensing in cognitive radio networks, in *Proceeding of IEEE DySPAN*, 2005.
15. A. Ghasemi and E. Sousa, Collaborative spectrum sensing for opportunistic access in fading environments, in *First IEEE International Symposium on New Frontiers in Dynamic Spectrum Access Networks (DySPAN)*, November 2005, pp. 131–136.
16. A. Sahai, N. Hoven, and R. Tandra, Some fundamental limits on cognitive radio, *in Forty-Second Allerton Conference on Communication, Control and Computing*, 2004.
17. N. Han, S. Shon, J. Chung, and J. Kim, Spectral correlation based signal detection method for spectrum sensing in IEEE 802.22 WRAN systems, in *Proceedings of 8th*

International Conference on Advanced Communication Technology (ICACT), Vol. 3, February 2006.

18. H. Chen, W. Gao, and D. Daut, Signature based spectrum sensing algorithms for IEEE 802.22 WRAN, in *Proceedings of IEEE International Conference on Communications (ICC)*, June.
19. N. Kundargi and A. Tewfik, Sequential pilot sensing of atsc signals in ieee 802.22 cognitive radio networks, in *IEEE International Conference on Acoustics, Speech and Signal Processing (ICASSP)*, 2008, pp. 2789–2792.
20. W. Hu, D. Willkomm, M. Abusubaih, J. Gross, G. Vlantis, M. Gerla, and A. Wolisz, Cognitive radios for dynamic spectrum access—Dynamic frequency hopping communities for efficient IEEE 802.22 operation, *IEEE Commun. Mag.*, Vol. 45, No. 5, pp. 80–87, May 2007.
21. L. Cao and H. Zheng, Spectrum allocation in ad hoc networks via local bargaining, in *Proceeding of SECON*, 2005.
22. O. Ileri, D. Samardzija, and N. Mandayam, Demand responsive pricing and competitive spectrum allocation via spectrum server, in *Proceedings of IEEE DySPAN*, November 2005.
23. S. Sengupta, M. Chatterjee, and S. Ganguly, An economic framework for spectrum allocation and service pricing with competitive wireless service providers, in *IEEE International Symposium on Dynamic Spectrum Access Networks (DySPAN)*, 2007, pp. 89–98.
24. R. Lam, D. Chiu, and J. Lui, On the access pricing and network scaling issues of wireless mesh networks, *IEEE Trans. Computers*, Vol. 56, No. 11, pp. 1456–1469, November 2007.
25. G. Jakimoski and K. Subbalakshmi, Denial-of-service attacks on dynamic spectrum access networks, in *IEEE CogNets Workshop, IEEE International Conference on Communications (ICC)*, May 2008, pp. 524–528.
26. R. Chen, J. M. Park, and J. H. Reed, Defense against primary user emulation attacks in cognitive radio networks, *IEEE J. Sel. Areas Commun.: Special. Issue on Cognitive Radio Theory and Applications*, Vol. 26, No. 1, pp. 25–37, 2008.
27. R. Chen, J. M. Park, and K. Bian, Robust distributed spectrum sensing in cognitive radio networks, in *Proceedings of IEEE Conference on Computing and Communication (INFOCOM'2008)*, April 2008, pp. 1876–1884.
28. A. Sethi and T. X. Brown, Hammer model threat assessment of cognitive radio denial of service attacks, in *Proceedings of IEEE Symposium of New Frontiers in Dynamic Spectrum Access Networks (DySPAN'2008)*, October 2008.
29. Z. Jin, S. Anand, and K. P. Subbalakshmi, Detecting primary user emulation attacks in dynamic spectrum access networks, in *Proceedings of IEEE Int. Conference on Communication (ICC'2009)*, June 2009.
30. R. Chen and J. M. Park, Ensuring trustworthy spectrum sensing in cognitive radio networks, in *Proceedings IEEE Workshop on Networking Technol. for Software Defined Radio Networks (SDR'2006)*, September 2006, pp. 110–119.
31. Z. Jin, S. Anand, and K. P. Subbalakshmi, Mitigating primary user emulation attacks in dynamic spectrum access networks using hypothesis testing, *ACM Mobile Computing Commun. Rev. (MC2R): Special Issue on Cognitive Radio Networks*, 2009.

32. G. Jakimoski and K. P. Subbalakshmi, Towards secure spectrum decision, in *Proceedings IEEE Intl. Conference on Communication. (ICC'2009)*, June 2009.
33. J. Nash, Equilibrium points in *N*-person games, *Proc. Natl. Acad. Sci.*, Vol. 36, pp. 48–49, 1950.
34. S. Sengupta, R. Chandramouli, S. Brahma, and M. Chatterjee, A game theoretic framework for distributed self-coexistence among ieee 802.22 networks, in *Proceedings of IEEE Global Communications Conference (GLOBECOM)*, December 2008, pp. 1–6.
35. IEEE Standards for Information Technology, Telecommunications and Information Exchange Between Systems; Wireless Regional Area Networks—Specific Requirements, Part 22; Cognitive Wireless RAN Medium Access Control (MAC) and Physical Layer (PHY) Specifications: Policies and Procedures for Operation in the TV Bands, June 2006.

14

MOBILE WiMAX

Aryan Saèd

14.1 INTRODUCTION

Today, most consumers in urban centers are quite familiar with high-speed Internet access. Wired high-speed Internet access is provided to homes and small businesses generally by two means. It can be over a regular twisted-pair phone line, using DSL (Digital Subscriber Lines) and ISDN (Integrated Services Digital Network) technology, or over coaxial cables for cable TV, using Cable Modems. Increasingly, as a third means, FTTH (Fiber to the Home) is becoming available as all-optical Active or Passive Optical Network (AON or PON) architectures.

Fixed and Mobile WiMAX are technologies that provide high-speed wireless Internet access to homes and businesses, as well as cellular data and voice services for phones, laptops, and personal digital assistants.

14.1.1 IEEE 802.16 and the WiMAX Forum

IEEE 802.16 is a technology standard for Wireless Metropolitan Access Networks (WMANs). The WiMAX Forum is tasked with issuing interoperability

Convergence of Mobile and Stationary Next-Generation Networks, edited by Krzysztof Iniewski

profiles and tests for the standard. Profiles are a testable subset of all features, modes, and options in the 802.16 standard, and the forum also issues Radio Conformance Tests for 802.16 equipment. The name WiMAX means *Worldwide Interoperability for Microwave Access*, and it has become synonymous with the subset of 802.16 technology that is defined by the Forum's profiles and conformance tests.

In terms of data rates, WiMAX specifies a broadband rate of at least 1.5 Mbit/s and a channel bandwidth of at least 1.0 MHz. The term broadband has been defined in the Recommendation I.113 of the ITU Standardization Sector, and it refers to a "transmission capacity that is faster than the primary rate Integrated Services Digital Network (ISDN) at 1.5 or 2.0 megabits per second." Some data communication standards consider a 5× improvement over dial-up a speed evolution, others 10×.

High-speed Internet access is more concisely called broadband access and refers (informally) to a minimum down-link data rate of 256 kbit/s. This performance level is based on a 5× improvement over the fastest dial-up analog modems.

Wireless broadband refers to wireless internet access. Earlier versions include MMDS (Multichannel Multipoint Distribution Service), which operates in the 2.5-GHz RF band, and LMDS (Local Multipoint Distribution Systems), which operates in the 24-GHz and 39-GHz RF bands. MMDS is a service that offers broadcast video as a competition to Cable TV, and LMDS was to offer businesses an improved alternative to DSL.

The RF band of a service has a major impact on the technology that enables it. For one, the size of the antenna depends on the RF band. Also, urban environments require lower bands, under 10 GHz. While higher frequencies are cheap and available, the wireless connection between a base station and a subscriber station must be "line-of-sight." For instance, both LMDS and MMDS involve costly installations of roof-top antennae.

14.1.2 Mobile Broadband Wireless Access and 3G Cellular

Mobile Broadband Wireless Access (MBWA) refers to the ability of wireless mobile stations to connect to the internet at broadband rates through cellular base stations. The connection rate is 100 kbit/s up to perhaps 1 Mbit/s. This is the current level of performance of 3G cellular standards, such as UMTS by 3GPP, which is based on GSM, and CDMA-2000 EVDO by 3GPP2, which is based on IS-95.

These 3G standards are based on technologies driven by telecommunications operators. They are rooted in cellular voice communications with significant enhancements to offer data and video. The business model is centered around an operator that is licensed to operate exclusively in a regulatory band and attracts subscribers in its geography by offering voice and data services with subsidized handsets.

Mobile WiMAX, on the other hand, is a technology that is driven by computer or data communication equipment manufacturers, with concepts borrowed from LMDS technology. Significant technological departures from its roots allow it to offer cellular services for voice and data to mobile users. The business model is centered around the sourcing of handsets or wireless computer dongles by independent device manufacturers. The consumer purchases a device at a computer store, and he/she subscribes to services by national or niche operators competing for business in his/her city. The operator may operate in a licensed or even in an unlicensed band.

Both 3G and WiMAX are technology drives to offer wireless internet access at broadband rates. One is a data-rate evolution for cellular systems, the other is a technology migration from wired systems to cellular wireless systems. Of course, ultimately 3G could also migrate to a business model centered around computer retailers, and WiMAX may quite well be the technology of choice for a cellular operator.

With the advent of license-exempt systems, it is also possible for small and independent amateurs or quasi-professionals to build a business as a Wireless Internet Service Provider (WISP), using WiMAX to offer wireless Internet access in a neighborhood.

Mobile WiMAX is based on amendment "e" to the 802.16-2004 Fixed WiMAX standard. The 802.16-2004 standard is sometimes incorrectly referred to as "the 16d standard," to emphasize its pre-mobile capabilities. The latest revision, 802.16REV2, has been published as 802.16-2009 and combines the "e" amendment and the 2004 standard together with several other amendments.

14.1.3 The IEEE 802 Standards Committee

802.16 is the IEEE Working Group on Broadband Wireless Access Standards. It is a Working Group of the IEEE 802 LAN/MAN Standards Committee (IEEE 802).

IEEE-802 has also other active Working Groups, which produce other widely used standards. This includes Wireless LAN (802.11), which is well known as WiFi; Wireless PAN (802.15), well known as Bluetooth, and also ZigBee and UWB. The Ethernet Working Group (802.3) produces the well-known standards for wired Ethernet: 10BASE, 100BASE, and 1000BASE.

The overall LAN/MAN architecture is standardized in 802.1.

Wireless LAN (WiFi) offers wireless connectivity through hot spots in homes and businesses. It reaches up to 54 Mbit/s in 802.11a, and it goes beyond 100 Mbit/s in 802.11n. WiFi plays a different connectivity role than does WiMAX. WiMAX offers a wireless connection from a Base Station to a subscriber unit in a home or business, and WiFi can be used to connect a user station (a laptop, and even a phone) to the subscriber unit.

There are also further alternatives, such as HomeRF (now obsolete) at 1.6 Mbit/s, and various wireless local, metropolitan, and regional networks.

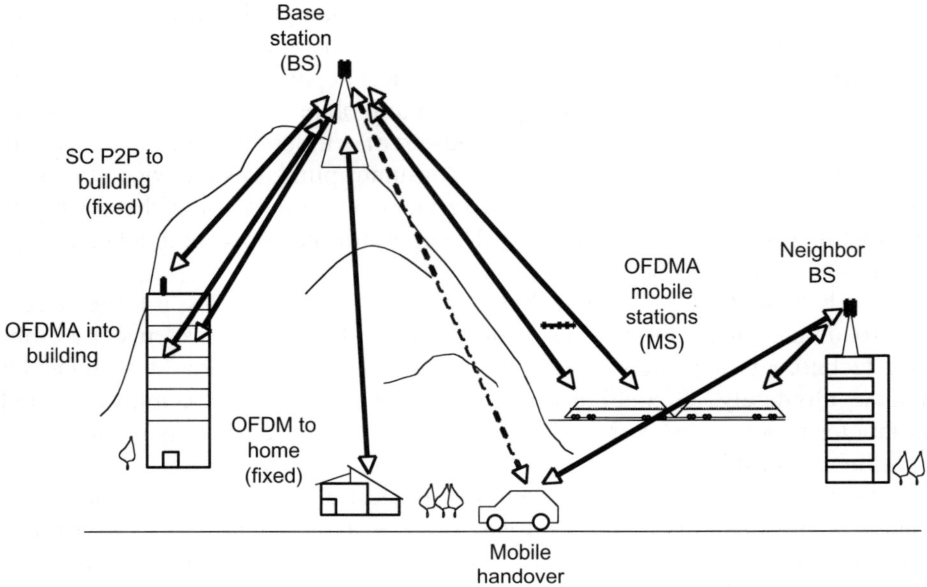

Figure 14.1. Mobile WiMAX in OFDMA mode, and fixed WiMAX in OFDM or single carrier modes.

IEEE 802.16: Metropolitan Broadband Wireless Access. The technical provisions in the standard support networks that are the size of a city. This is also called *a metropolitan area network*. WiMAX is also easily deployed in rural areas. The standard offers many modes and options to optimize for distance, user density, and typical urban or rural RF wave propagation conditions.

An illustration of the application of mobile and fixed WiMAX is provided in Figure 14.1. Mobile WiMAX is designed for users at vehicular speeds in urban environments. Provisions for mobile use particularly deal with handovers as the user moves from one cell to another, and they also deal with fluctuating throughput as channel conditions vary due to blockage and reflections.

Figure 14.1 also illustrates other variants of the WiMAX standard that use single-carrier (SC) modulation for last-mile Internet connections and use OFDM for rural Internet connections.

The 802.16 standard splits the RF bands in two. The lower RF band ranges from 2 GHz to 11 GHz and the upper band ranges from 10 GHz to 66 GHz, with an overlap around 10.5 GHz.

This split is based on the availability of RF spectrum for broadband deployments in the United States, and it takes also other regulatory regions of the world into consideration. The split also considers that toward 10 GHz the benefits of OFDM diminish when compared to a much simpler SC system.

Table 14.1 provides a general overview of the RF spectrum and its suitability for WiMAX.

TABLE 14.1. Suitability of Radio Spectrum for WiMAX

RF Frequency	Band/Wavelengths	Typical Use	Suitability for WiMAX
30 MHz	VHF band 30–300 MHz, 10-m to 1-m wavelength, 2.5-m to 25-cm antenna	Analog TV, future digital TV, toys.	Frequency too low for narrowband channelization. Antennae too large for handsets.
300 MHz	UHF band 300 MHz to 3 GHz, 1-m to 10-cm wavelength, 25-cm to 2.5-cm antenna "Beachfront spectrum"	Digital and analog TV.	
700 MHz		Digital and analog TV.	700-MHz WiMAX Forum proposed profiles.
900 MHz to 1800 Hz	900 MHz is the original ISM band.	General use spectrum (toys, cordless phones, garage door openers etc). Licensed spectrum. GPS at 1575 MHz. 1800–1900 MHz currently used for HSPA and EVDO.	Not available.
2–3 GHz	ISM band	WiFi IEEE 802.11b/g/n at 2.4 GHz , WCS and MMDS at 2.3, 2.5, and 2.7 GHz.	WiMAX Forum profiles for licensed and unlicensed bands at 2.5 GHz and 3.5 GHz. Heavy multipath requires OFDM, enables MIMO.
5 GHz 6 GHz 10 GHz	SHF band 3–30 GHz, 10-cm to 1-cm wavelength, 2.5-cm strip antenna or small dish antenna.	UNII bands (e.g., WiFi). 802.11a/n at 5GHz.	Available. Easier to manage cell–cell interference but less desirable for WiMAX due to difficulty penetrating walls within or into buildings.
30 GHz		Above 5.8 GHz for radar/military use.	Not desirable for WiMAX or other cellular technologies due to wall penetration loss.
		Nothing in United States available to 18 GHz. LMDS	Easy to focus RF waves into beams. Strictly for line-of-sight connections. Original target for WirelessMAN-SC(a). Affected by rain and light foliage.
60–100 GHz		Short-range UWB at 60 GHz. 802.11ad WiFi.	Significant water absorption and noticeable oxygen absorption.

Notes: VHF, very high frequency; UHF, ultra high frequency; SHF, super high frequency; ISM, Industrial, scientific and medical; UNII, Unlicensed National Information Infrastructure; WCS, Wireless Communications Service; MMDS, Multichannel Multipoint Distribution System; LMDS, Local Multipoint Distribution Service; UWB, ultra wideband; GPS, Global Positioning System.

14.1.4 PHY and MAC Components of a Broadband Mobile Air-Interface Standard

The 802.16 standard is an air-interface standard. This means that it describes protocols and methods by which compliant devices can communicate over an air channel. The protocols are grouped in two layers: the MAC layer and the PHY layer.

The PHY layer describes how data bits are transferred to and from radio-frequency (RF) waveforms. This involves the coding and modulation operations, error correction, the use of the RF channel, definition of the burst frame with preambles and pilots, and the use of schemes for multiple antennae. The PHY layers also includes digital signal processing (DSP) for filtering and equalization, as well as RF up- and down-conversion and analog filtering, but their design and specifications are not standardized and instead left to the vendor.

The MAC layer describes the type of connections available to a client of an 802.16 device, and it also describes how the client data are transformed to and from framed data for transmission and reception by the PHY. This involves establishing and maintaining connections between a base station (BS) and a mobile station (MS), assigning transmission slots to supply the desired data rate and Quality of Sevice (QoS), and dealing with temporary and permanent signal drops, encryption and security, and BS-to-BS hand-offs.

In the *layer stack*, the network communicates with the MAC through the link layer at the MAC service access point (SAP), and the MAC communicates with the PHY at the PHY-SAP. In some exceptions where BSs communicate directly with each other, management and control data can be shared over the backbone network, without traversing through the PHY.

The control over the settings in the PHY and the MAC is at the discretion of the operator. The operator has the task to balance the user's QoS requirements against cost and revenue. Capital expenditures (CAPEX) involve the cost of deploying the BSs, and operating expenditures (OPEX) involve the cost of maintaining and servicing the network and the customers. Outages, cell coverage, and even power consumption of mobile devices play a role since they affect the user experience. The operator and device manufacturer must also be compliant with regulatory requirements regarding the use of licensed spectrum, sharing unlicensed or lightly licensed spectrum, and meeting spurious RF transmit emission requirements.

The standard supplies the options and protocols to establish and maintain RF connections between compliant devices. A device contains a vast amount of discretionary algorithms to set system parameters on a connection-by-connection basis. This includes PHY and MAC algorithms for choosing when to change modulation and coding settings, when to perform a hand-off, when to wake-up or put a device to sleep, and how to schedule data for user connections.

14.1.5 History of IEEE 802.16

IEEE 802.16 was originally designed for nonmobile, enterprise-class deployments. The development of the standard started officially in 1999, and it was completed in 2001 with a technical specification for the delivery of high-speed wireless data connections to businesses that would not have access to optical fiber connections. The RF ranged from 10 GHz to 66 GHz, and the system required outdoor antennae with line-of-sight (LOS) connections to a BS. The modulation was based on Single-Carrier Quadrature Amplitude Modulation (QAM).

In 2003 amendment 802.16a was completed, which included modulation with Orthogonal Frequency Division Multiplexing (OFDM) based on a fixed FFT size. This targeted the license exempt RF frequency bands in the 2- to 11-GHz range. These lower frequencies made the use of indoor antennae possible, allowing consumers to subscribe to 802.16-based data services. Indoor reception is heavily impaired by *multi-path* reflections from other buildings, and it causes frequency-selective fading. OFDM was applied to mitigate this impairment.

At sub-11-GHz frequencies, 802.16e (December 2005) provided for mobile services through the addition of mobile handover. A user device such as a cellphone or portable data assistant (PDA) can establish and maintain a service connection across cell boundaries, even at high speeds. An overview of the standards and amendments is provided in Table 14.2.

14.1.6 Mobile Versus Fixed WiMAX

The essence of WiMAX is captured in the definition of its Medium Access Control (MAC) layer.

In the original standard, its task was to supply users with a several levels of QoS for carrier-quality, enterprise-based telecommunications services. The 802.16 BS offers classes of QoS to support services such as T1/E1 guaranteed rates, high-throughput low-latency video conferencing, low-throughput low-latency Voice over IP (VoIP), and a best-effort Internet connection service.

The core of the MAC comprises self-correcting request and grant protocols, and multiple connections per user terminal. The MAC provides an efficient protocol for bursty data that can easily handle high peak data rates in a fading medium.

In the standard there is a distinction between Nomadic use and Mobile use.

In *fixed* use the operator configures a user for one specified cell or cell sector only. This is usually sufficient for Point-to-Point (P2P) broadband services to a residence or business, but there are no provisions in the protocol for a user to dynamically associate with just any of the operator's BSs and negotiate a desired data rate.

Nomadic use implies that the user can and may connect to a different BS or a different sector of a same BS and expect to be recognized and accepted by the operator automatically and promptly. Standard 802.16-2004

TABLE 14.2. Overview of Select IEEE 802.16 Standards and Amendments

802.16.1	LMDS	Became 802.16-2001 Wirel essMAN-SC.
802.16-2001 (1999–2001)	WirelessMAN-SC (Single-carrier TDMA)	Line-of-Sight, fixed outdoor antenna, RF frequency above 10 GHz, >100 Mbit/s fiber extension.
	WirelessMAN-SCa WirelessMAN-OFDM (256 subcarriers) WirelessMAN-OFDMA (128, 512, 1024 and 2048 subcarriers)	Three PHY alternatives for urban wireless DSL service. Below 11-GHz non-line-of-sight (NLOS) deployments use OFDM (256 sub-carriers) and OFDMA (2048 sub-carriers). Line of sight (LOS) uses single carrier "SCa".
802.16.2-2001 (1999–2001)	Recommended Practice for WirelessMAN-SC	Recommendations for operators in licensed bands to deal with co- and adjacent channel inteference, above 10 GHz.
802.16c (2002)	Profiles for 802.16	These are the original profiles, developed with help from the WiMAX Forum. The amendment has now been superseded by activities in the WiMAX forum.
802.16a (2000–2003)	WirelessMAN-SCa WirelessMAN-OFDM (256 subcarriers) WirelessMAN-OFDMA (2048 subcarriers)	Split PHY, OFDM(A) under 11 GHz for indoor and nomadic use, and SC-TDMA above 10 GHz. Single MAC. Licensed and unlicensed bands.
802.16.2-2004 (2001–2004)	Recommended Practice for WirelessMAN-OFDM and WirelessMAN-OFDMA	Recommendations for operators in licensed bands to deal with co- and adjacent channel inteference, below 11 GHz.
802.16d (2004)	Originally: Profiles for 802.16a, under 11 GHz. Later abandoned in favor of a full revision	"Fixed WiMAX"
802.16-2004	Full revision, merging 16a, 16c, 16-2001	
802.16e (2002–2005)	WirelessMAN-OFDMA (128, 512, 1024 subcarriers)	Combined fixed/mobile. "Mobile WiMAX" includes uplink MIMO, scalable OFDMA, and hand-off
802.16h (2004–2009)	License exempt (LE)	Standardized schemes for improving the use of radio resources (RF channels) in license exempt bands, considering other users in the same channel.
802.16j (2004–2009)	Mobile multihop relay (MMR)	Additional capabilities to form a network comprising a single multihhop relay base station (MR-BS), one or more relay stations (RS), and a multitude of mobile stations (MSs)

TABLE 14.2. *Continued*

802.16f-2005	Management information base	MIB and MPPS are used to manage the devices in the network.
802.16g-2007	Management plane procedures and services	
802.16-2009 (2007–2009)	Second revision (REV2)	Obsoletes 802.16-2004 and 802.16e-2005 and several other corrections and amendments. Started as errata fixes, but now covers all amendments except "h" and "j".
802.16m (2006–2010)	Advanced interface	Originally a candidate for IMT-Advanced (4G), competing with LTE. Offers improved spectral efficiency, reduced latency, increased user density, and enhanced localization techniques for emergency services.

specifies how a connection between a BS and an MS is requested by the MS and accepted and managed by the BS. The standard also provides management messaging and access schemes that allow the BS to manage a variable load of MSs in its cell or any of its cell sectors. This is a Point-to-Multipoint (P2M) architecture.

Quality of Service factors such as data rate, latency, and availability are usually not guaranteed during nomadic movement, and the connection may have to be reestablished from scratch. Moreover, the quality of the connection may be impacted significantly during motion, even if the user remains within the cell or sector of a single BS. This is of course not acceptable for cellular voice applications.

Mobile use brings a much tougher requirement: to uphold the connection and data transfer as the user moves, even if the user transitions from one cell or cell sector to another. This involves cell-to-cell or sector-to sector hand-off schemes with sophisticated interactions between the MS and multiple BS, in order to uphold the QoS.

Modulation and coding schemes are optimized for mobility, and they minimize the error rate during motion within a sector or cell. This covers Doppler frequency shifting effects and temporary fading effects at pedestrian and vehicular speeds.

A further change in the mobile version of the standard is the introduction of Scalable OFDMA (S-OFDMA). The "e" amendment provides several options for the FFT size, and this allows the operator to configure the FFT based on channel width. The subcarrier spacing and the symbol duration can be optimized for the RF channel propagation conditions.

Notable features of the "e" amendment for mobility are as follows:

1. Additional OFDMA FFT sizes of 128, 512, 1024, and 2048. This allows the OFDM bandwidth to scale with the channel bandwidth while keeping the subcarrier spacing and the symbol duration independent from the channel bandwidth.
2. Adaptive Modulation and Coding (AMC) in subchannels, to benefit from sections of the channel with notably good SNR performance.
3. Hybrid-Automatic Repeat Request (ARQ) for more efficient use of the Forward Error Control (FEC) schemes during error bursts in mobile fades.
4. Multiple-Input and Multiple-Output (MIMO) diversity for better uplink (UL) cellular throughput.
5. Reduced latency for mobile hand-offs.
6. Sleep modes to extend battery life.

The WiMAX burst-type modulation scheme in the "e" amendment significantly improves data downloads (e.g., web browsing) when compared to cellular standards rooted in voice applications.

14.1.7 WiMAX Forum

The objective of an open standard is to enable independent manufacturers to bring interoperable devices to market. The IEEE standard describes all the details of the technical aspects of interoperability. This includes all types of overhead messaging, frame formats, signal properties, and modes of operation.

The WiMAX Forum is a nonprofit consortium comprising (a) system vendors and (b) component and device suppliers and operators. It provides a certification process for conformance and interoperability. Conformance tests are performed by specialized and certified third-party conformance labs, which test systems against the Radio Conformance Tests (RCT) issued by the Forum. Interoperability tests are performed at so-called wireless "plug-fests." To pass an *interop* test, a vendor must succeed with at least two others during BS and MS connection tests. For any vendor, the goal of these tests is to provide confidence in operators and consumers that its equipment can be mixed and matched with equipment from other vendors.

Table 14.3 specified the channel width, duplexing scheme and the FFT size for various RF bands per the WiMAX Forum System Profiles. These parameters are explained in more detail in later sections.

14.2 MAC OVERVIEW

The MAC manages the traffic load for all user applications, over the physical medium. The PHY is responsible for transmitting and receiving information bits across the air-link, and it has no knowledge of the specific performance requirements for different types of application data.

TABLE 14.3. WiMAX Forum System Profile Specifications

Band	Channel	Duplexing
WCS 2.3 GHz[a]	2 × 3.5 MHz, 2 × 5 MHz, or 2 × 10 MHz[b,c]	FDD[d]
2.3–2.4 GHz (global spectrum)	1 × 8.75 MHz, 1 × 5 MHz, or 1 × 10 MHz[e]	TDD
2.3 GHz (global spectrum)	1 × 3.5 MHz, 1 × 5 MHz, or 1 × 10 MHz	
2.5 GHz	2 × 5 MHz or 2 × 10 MHz	FDD
2.5 GHz	1 × 5 MHz or 1 × 10 MHz	TDD
3.3–3.4 GHz	5 MHz	TDD
3.4–3.6 GHz	2 × 5 MHz, 2 × 7 MHz or 2 × 10 MHz	FDD
AWS 1.7 GHz UL and 2.1 GHz DL[a]	2 × 5 MHz or 2 × 10 MHz	FDD
700 MHz	2 × 5 MHz or 2 × 10 MHz	FDD
700 MHz	1 × 5 MHz, 1 × 7 MHz or 1 × 10 MHz	TDD

[a] AWS is Advanced Wireless Services, and WCS is Wireless Communications Service (both North America/FCC).
[b] 2× refers to uplink (UL) plus downlink (DL) pairing.
[c] 512-pt FFT for 5-MHz channels, 1024-pt FFT for 7, 8.75, and 10 MHz.
[d] For FDD duplex channels, the BS must support FDD, and the MS must support H-FDD. FDD support for MS is not required.
[e] 8.75 MHz is for WiBRO. This is Mobile WiMAX at 2.3 GHz, with 8.75-MHz channelization used in Korea.

In 802.16, an MS establishes multiple independent connections to and from a BS to transfer data. To exchange data units, a connection identifier (CID) is used to address data and management traffic between the BS and its MSs.

The MAC manages the network entry of a station, and it establishes connections to transport data. The MAC also implements the *convergence sublayer*, which maps higher layer addresses such as IP addresses of its service data units to the individual stations served. An MS communicates with a BS through multiple concurrent connections, covering MAC management, initial ranging, user data, bandwidth requests, idle payload padding, and broadcast information.

14.2.1 MAC and PHY Protocol/Service Data Units (PDU/SDU)

The data at the input of a layer is called a service data unit (SDU), and the data at the output of a layer is called the protocol data unit (PDU).

The MAC encapsulates its input SDU (the MAC-SDU, or MSDU) with all necessary framing headers so that the peer MAC at the receiver can process the MAC payload data. The processing by the MAC includes data encapsulation, aggregation, and fragmentation and managing the PHY.

Its output is the MAC-PDU, or MPDU, which is then passed to the PHY as the PHY-SDU (PSDU). The PHY adds any headers and overhead necessary for synchronization and signal decoding, and it transmits the PHY-PDU (PPDU) over the air to the PHY at the receiver end. There, the process is inversed, and ultimately the data are presented to the link layer as a received MSDU. An MPDU consists of a MAC header, optional payload data, and an optional CRC.

The link from the BS to the MS is called the downlink (DL), and the reverse link is called the uplink (UL).

The standard contains hundreds of pages describing the MAC schemes, and the reader is referred to the text for details.

14.2.2 Scheduling Versus Collision Avoidance

The MAC schedules its users based on their traffic load requirements, their QoS requirements, and the conditions of the air link. The BS probes each MS for its capabilities in terms of coding and modulation options, MIMO options (discussed later), and other MAC and PHY options, and it schedules the MS based on the reported capabilities and limitations.

Once the scheduling is completed and communicated to the MS, there are no further air-link resources wasted in arbitration or collision recovery. There is of course some scheduling overhead, and it is the object of the standard to minimize it. To accomplish this, the MAC can reduce header overhead and aggregate short MAC SDUs (e.g., short 40-byte TCP acknowledgment packets). It can also maximize frame utilization by allowing the fragmentation of large MSDUs (e.g., 1-kbyte TCP packets) in order to top-up even small unused parts of the frame.

In contrast, Carrier Sensed Multiple Access Collision Avoidance systems (CSMA-CA) generally do not schedule their users. Stations are required to monitor the channel and avoid collisions with existing transmissions. Congestion challenges arise when multiple transmitters sense that the channel is carrier-free and start their transmissions simultaneously. Despite mandatory sensing, these systems still have to deal with collision rates as high as 30–40% in even lightly loaded systems.

Nevertheless, its simplicity and its lack of a central scheduling entity make CSMA-CA attractive for data-centric applications such as the Wireless Local Area Network, popularly known as WiFi (IEEE 802.11). In WiFi, access points (APs) form a network with their stations, but APs generally have to share the channel with other APs. The network usually tolerates the high collision rates because the channel mostly offers abundant capacity for retransmissions. Collision schemes are used where there is no central entity such as a base station. Service is often without commitment "as-is," "where-is," and "when-is."

In contrast, mobile voice applications with high traffic volumes, high number of simultaneously connected stations, and high costs of RF band licensing make this not a viable candidate for a subscription-based cellular standard. Centralized scheduling is a necessity.

14.2.3 Quality of Service (QoS)

One task of the MAC is to make sure that user applications receive their subscribed Quality of Service. The Quality of Service (QoS) refers to guarantees for a minimum throughput and maximum latency for application traffic in a network. Offerings are priced for different levels of QoS.

Different applications have different demands. For instance, *voice traffic* has tight demands on latency. Excessive delays of the transferred signal between two ends of a call would literally result in irritating echoes. Moreover, variations in the delays, also called jitter, would cause distracting audible voice echoes due to limitations in the delay-tracking ability of the echo cancelers in voice systems. On the upside, voice is quite tolerant to packet losses and high bit error rates. This is different for *real-time video*, where data rates are high, latency is also low, and the tolerance to packet loss and jitter is moderate. Traffic such as Internet *file transfer* has practically no requirements for the rate or latency at which it is transferred, as long as bit error rates are not too high for the application layer to handle.

In wireless systems, QoS must be delivered by the MAC under fluctuating levels of capacity of the channel at the PHY. The task of the scheduler is to allocate user slots in data frames. Under mobile wireless conditions the channel fluctuates dramatically and often unpredictably, and the scheduler relies on many support mechanisms in the PHY to offer the MAC as much throughput as possible. In contrast, QoS in wired access systems is much simpler to implement at the MAC bacause it is based on a fixed-capacity PHY channel.

Quality of Service (QoS) is native to 802.16, and it is modeled after QoS in ATM (Asynchronous Transfer Mode) with some modifications based on DOCSIS. The Data Over Cable Service Interface Specification (DOCSIS) included QoS in its 1999 version, and it is designed for high-speed data transfer on Cable TV systems.

Traffic offered to the MAC is classified with service flow identifiers (SFIDs) for QoS and is then mapped to connection identifiers (CIDs) for scheduling, modulation, and coding.

QoS in 802.16 covers over-the-air service levels in terms of—among other things—minimum and maximum sustained rates, reserved rates, tolerable minimum rates, jitter, and latency. There are four (plus one) *service flow* classes:

1. UGS (unsolicited grant service) for constant bit rate (CBR) requirements as used by legacy Public-Switched Telephony Network (PSTN) systems based on Time Domain Multiplexing (TDM), e.g., DS0 and T1/E1 TDM. In these systems, the data rate is constant even during silence on the line.
2. rtPS (real-time polling service) for real-time variable bit rate (rtVBR) requirements, where multiplexing is done statistically based on increasing, decreasing, and bursty demands for data rate in real-time applications such as Voice over IP (VoIP) and streaming video. The compression and codec algorithms in these applications (such as MPEG) will demand higher data rates or relax to lower rates, depending on the underlying

video or voice signals. In the UL, the BS schedules UL bursts explicitly on demand, based on the subscriber' s burst requests (BRs). A variant, extended rtPS, provides regular UL scheduling with less BR overhead, as in UGS, but with dynamic allocations as in rtPS.

Extended rtPS (ertPS) has been added because the longer frame durations of OFDM and OFDMA versus SC created the need for a scheduling mechanism between UGS and rtPS to accommodate VoIP with reasonable jitter.

3. nrtPS (non-real-time polling service) for non-real-time variable bit rate (nrtVBR) requirements, where multiplexing is done statistically with a minimum guarantee of rate, but where there is no real-time delay or jitter specification.
4. BE (best effort), for applications with no minimum throughput guarantees over some specified short-term time span.

The Service Level Agreement (SLA) promises data rates in a statistical sense, depending on the class required for each application. In setting the SLA data rates, the operator takes into consideration the location and type of subscribers. By considering their device capability (cost and complexity) and location of use (distance and obstructions), subscribers that can communicate at high rates can be promised higher levels of service.

14.2.4 Network Entry

A further task of the MAC is to manage the network entry of subscribers.

When an MS intends to join the network, the BS has no knowledge of its service needs, and it has of course no scheduled slots for its UL transmissions. To obtain entry, a number of unscheduled exchanges with the BS must first be completed, followed by some scheduled exchanges.

14.2.4.1 Scanning, Synchronization, and Authentication. When an MS powers up, it scans RF channels for a suitable BS to establish connections. To this end, an MS is shipped with a list of channel frequencies to scan. This list resides in the driver SW or in a SIM card supplied by the operator.

Scanning is not without challenge. It is possible that the MS simultaneously receives strong DL signals from multiple BS. This can easily happen in a single-frequency deployment or at the cell edge in a multifrequency deployment. Thanks to a pseudo-noise sequence in the downlink preamble transmitted by the BS, the MS can distinguish between multiple overlapping BS cells by correlating the received preamble sequence with a set of locally stored reference sequences. Although reception is heavily interfered, it can still select the strongest signal and establish a connection.

After scanning, the MS receiver synchronizes with the DL frames. This involves RF center frequency adjustments, as well as time alignment of the base band decoder.

After frame synchronization, the MS decodes the broadcasted Uplink Channel Descriptor (UCD) message and uses the supplied information about the frame to start initial ranging. The MS coarsely synchronizes its UL transmission with the UL frame, and it selects a transmit power level based on the power level received from the BS and any additional power information in the UCD. The MS also determines the initial ranging transmission slots from the UCD, and it starts its first transmission with a ranging request to the BS.

This transmission occurs in a special contention-based ranging channel, using Code Division Multiple Access (CDMA). The MS transmits MAC messages without allocation by the BS, but this flexibility comes at the expense of efficiency. CDMA reception quality degrades only slowly and gracefully as the number of overlapping transmissions increases, without coordination or scheduling with other transmitters. In contrast, OFDMA is more bandwidth-efficient but does not tolerate overlaps (collisions) at all.

As part of the ranging response, the BS responds with any further required power adjustments, as well as frequency and frame alignment adjustments to be made by the MS. These fine tunings are directed by the BS to enable the MS to proceed with scheduled communications without interfering with other MS served by the same BS.

14.2.4.2 Authentication. Authentication proves the identity of the MS to the BS. This matters for user-specific parameters related to service agreements and billing. Since user data are shared over the air, which is a notoriously non-secure medium, encryption is used to warrant privacy and protect identity. The BS recognizes the MS by its 48-bit MAC address, and authentication follows through Privacy Key Management (PKM) messages. With PKM, the MS communicates its X.509 certificate during the Security Association (SA). X.509 (1988) is an ITU-T cryptography standard for a public key infrastructure, based on a strict hierarchical system of Certificate Authorities (CAs). In 802.16 the certificate belongs to the manufacturer of the MS.

Data are encrypted using private traffic encryption keys, which are communicated between MS and BS using DES3, AES, or RSA public encryption schemes.

14.2.4.3 Periodic Ranging. Due to fluctuating conditions of the air link, which is typical in mobile conditions, the BS will periodically instruct the MS to adjust its power level, RF carrier synchronization, and frame alignment. These adjustments are performed as part of *maintenance ranging*, also called periodic ranging.

The BS uses periodic ranging to minimize the interference from the MSs it serves. Incident signals from simultaneously transmitting MSd must have receive power levels that are as close together as possible across the subcarriers. The MS transmit level is adjusted based on the power level received by the BS, and thus it automatically accounts for distance and obstructions.

Fine frequency adjustment messages correct for any clock and carrier mistunings by the MS, as well as any offsets caused by the Doppler velocity between different MSs. Frame adjustment also corrects for differences in propagation delays of signals from MSs, which are due to their different distances from the BS.

Maintenance ranging is also used to adapt to changing properties of the air channel. For instance, at 802.16 frequencies above 11 GHz, used in rural or suburban areas, weather and wind conditions play a role (foliage, rain, snow), and at 802.16e frequencies under 6 GHz it is mobility of the SS or even the mobility of obstructions (moving vehicles, bridges, tunnels) that play a role.

The BS performs *unsolicited* periodic ranging if there are no data to communicate to or from the MS. By keeping the power level and synchronization current with a dormant MS, link disruptions are avoided. This allows for a renewed demand for data exchange to ramp-up quickly without reestablishing the connection.

To this end the BS allocates bandwidth for the MS, even though the MS has no demand for it. This is called an unsolicited grant (UG). The MS responds with idle data in the frame pad bits, and the BS evaluates the received signal to perform periodic ranging.

Periodic ranging is also used to maximize battery life in mobile subscribers. The BS can instruct an MS to reduce its transmit power if the volume of data by the MS does not require high modulation rates. With SNR to spare at lower modulation, the RF transmit power is reduced and the battery life is extended. Moreover, transmit power levels of stations at the cell edge can be adjusted through ranging to reduce inter-cell or inter-sector interference.

14.2.4.4 Sleep Mode. Another method to reduce battery power draw by an MS involves sleep mode. The MS negotiates periods of absence from the BS during which the BS will not send any requests or any data to the MS. The MS powers down its RF and DSP subcircuits for transmission and reception, and they only operate a minimal state machine plus timer. Once the scheduled sleep period (or *sleep window*) is over, the MS will decode the following frame and its service flows will be available without any re-negotiation.

14.2.4.5 Idle Mode. When an MS has no traffic to transmit but is available for DL traffic, it can switch to idle mode. During idle mode the station is not registered with any specific BS, which means that there is no need to manage hand-offs. In the event of DL traffic, for instance for a pending VoIP call or text message, the BS will use *paging* to reach the MS.

14.2.4.6 Bandwidth Request. To start a transmission, for instance to initiate a call or an internet data request, a MS issues a bandwidth request (BR) and receives grants from the BS. The MS also uses BR to increase or decrease bandwidth, depending on its application demands.

The BS allocates symbol and subchannels to a MS, and this is broadcast to all MSs in the UL-MAP. The MAP also defines the modulation and coding rates for the MS transmit burst profile.

The norm is to allow MSs to use the CDMA ranging channel for BR, or to allow an MS to piggyback a BW request subheader when transmissions are already ongoing.

The BS can also poll its subscribers for BRs. Polling is generally done one-on-one (unicast), but this may be inefficient if there are many inactive MSs. These inactive stations are better polled through a multicast to a group or through a broadcast to all, or they are left to use ranging as needed.

The BS can also schedule multiple subscribers to receive a common signal for common data. This is called a *multicast connection*. It improves the frame efficiency in terms of the number of connections, but the connection throughput must be lowered to meet the highest rate that all stations in the multicast group can receive.

14.2.4.7 Basic Capability Negotiations. The BS considers the capability limitations and other operational constraints of each MS. These limitations are communicated to the BS during basic capability negotiations.

Cost and size restrictions of devices limit certain capabilities. For instance, modulation rate is often limited by RF distortion specifications of a device. The supported coding techniques are limited by the DSP capabilities. The transmit power is limited by the supplied power amplifier. And the MIMO options are limited by the number of antennae of the device.

To maximize the cell throughput, stations with common capabilities are grouped together in a particular section of the frame, called a *zone*. Zones are also used to manage interference in the same cell and in neighboring cells.

14.2.5 Mobility Management: Handover

A significant new feature in mobile WiMAX (over the fixed variant) is mobility management. Hand-off refers to the transition of a user from one serving BS to another while maintaining connectivity and QoS. Hand-off delays are kept below 50 ms.

The BS advertises the network topology to its MSs by broadcasting the UL and DL channel descriptors (UCD and DCD) of neighboring BSs. This means that the MS does not have to interrupt the connection and leave the BS to scan and decode possible alternate channels.

The MS determines the SNR and RSSI for signals from neighboring BSs during a scanning interval assigned by the serving BS. The MS may also use this interval to associate with a selected target BS before leaving the current serving BS. Two BSs can even communicate over the backbone network to expedite ranging of the MS with the target BS.

There are three handover variants. In hard handover (HHO) the MS maintains its connections exclusively with the serving BS throughout the handover.

After establishing context with the target BS of choice, the MS simply terminates its context with the previous serving BS.

A second variant is macro diversity handover (MDHO), which allows a MS to maintain a list of preferred BSs, called the *active set*. The active set is chosen by the MS, and it is based on the signal quality from neighboring BSs as in HHO.

All data to and from the MS is transmitted to and from all the BS in the active set simultaneously. This comes at the cost of frame inefficiency, but it is temporary and it provides spatial *diversity*. One BS, in the set, the *anchor BS*, provides the necessary scheduling and coordination. At the cell edge the MS can easily maintain its connection to the network as signal conditions with any BS improve and deteriorate. Once the air link is stable and in favor of one particular BS, the multiple contexts are reduced back to a single context in favor of freeing resources in the frame.

A third variant is Fast Base Station Switching (FBSS). The MS maintains connections with multiple BSs, as in MDHO, but only one BS transmits or receives at a time. There is no spatial diversity, but the MS can rapidly switch between BSs of an active set depending on changing signal conditions.

14.2.6 Fragmentation and Packing

In a cellular communication link, the RF link quality will vary over time, and even with sophisticated rate adjustments and resource scheduling, it is inevitable that packets of data will be in error. The target packet error rate is in the range of 0.1% to 1%, which often corresponds to a bit error rate (BER) of 1e-4 to 1e-5. Compare this to a wired communication link, where a BER of 1e-6 to 1e-10 and even lower is desired.

The MAC delivers a BER to the link layer that is at least 100-fold greater than the BER at the PHY. To this end, there are provisions for retransmissions. This includes a data integrity acknowledgment between the two MACs at either end of the connection, as well as means to buffer and possibly retransmit errored data.

The MAC can fragment and pack link-layer packets. Packets larger than 1500 bytes (e.g., large IP data packets) are often fragmented into smaller pieces, and packets as small as 40 bytes (e.g., IP acknowledgment packets) are often packed into a larger MPDU. This offers better efficiency in error recovery, better QoS management and it helps maximize cell throughput.

Fragmentation subheaders (FSH) and packing subheaders (PSH) supply the necessary overhead to reassemble the received data unit.

14.3 PHY OVERVIEW

The physical layer of the standard covers the technical details to modulate signals for communication through the Over-the-Air (OTA) channel. The PHY covers OFDM modulation, coding, MIMO and provisions for synchronization.

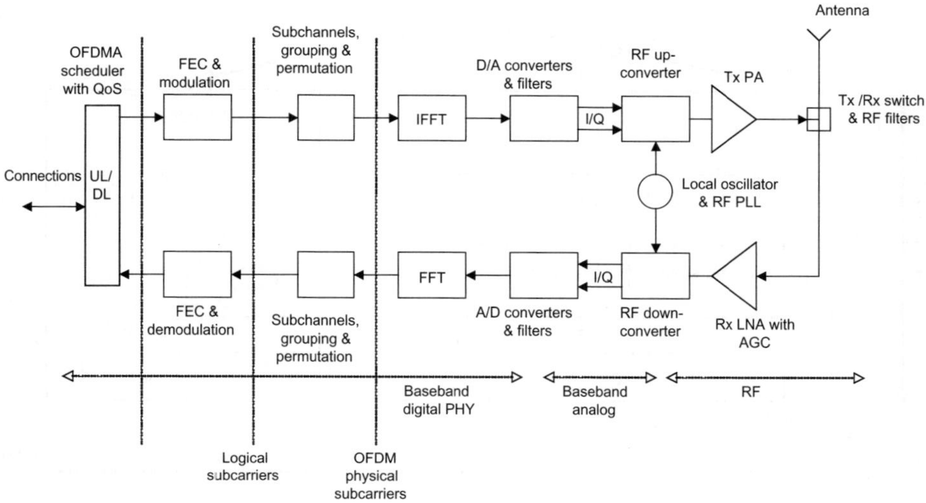

Figure 14.2. Generalized block diagram of a WiMAX modern device, covering the major RF and digital/analog baseband circuit groups.

Figure 14.2 shows a generalized block diagram of a WiMAX modem device, covering the major RF and digital/analog baseband circuit groups. The standard does not specify how to design the circuits or how to partitioning the required functionality. Instead, it specifies the required behavior and performance of the ultimate transmit and receive systems. The vendor chooses between several RF, analog and digital architectures and partitions depending on specific market needs.

14.3.1 Uplink/Downlink Duplexing

The duplexing scheme defines how the downlink (DL) transmissions are separated from the uplink (UL) transmissions.

WiMAX has three duplexing methods:

1. In time division duplexing (TDD), the UL and DL are time multiplexed, which allows the use of a single channel for both directions. To this end the OFDMA frame is split between a DL subframe and a UL subframe. The typical DL:UL split is 26:21 symbols in a 5-ms frame. Frame duration and split are generally not varied during operation.
2. In frequency division duplexing (FDD), the UL and DL occur in two different channels. The BS transmits in one FDD channel while the MS simultaneously transmits in the other.

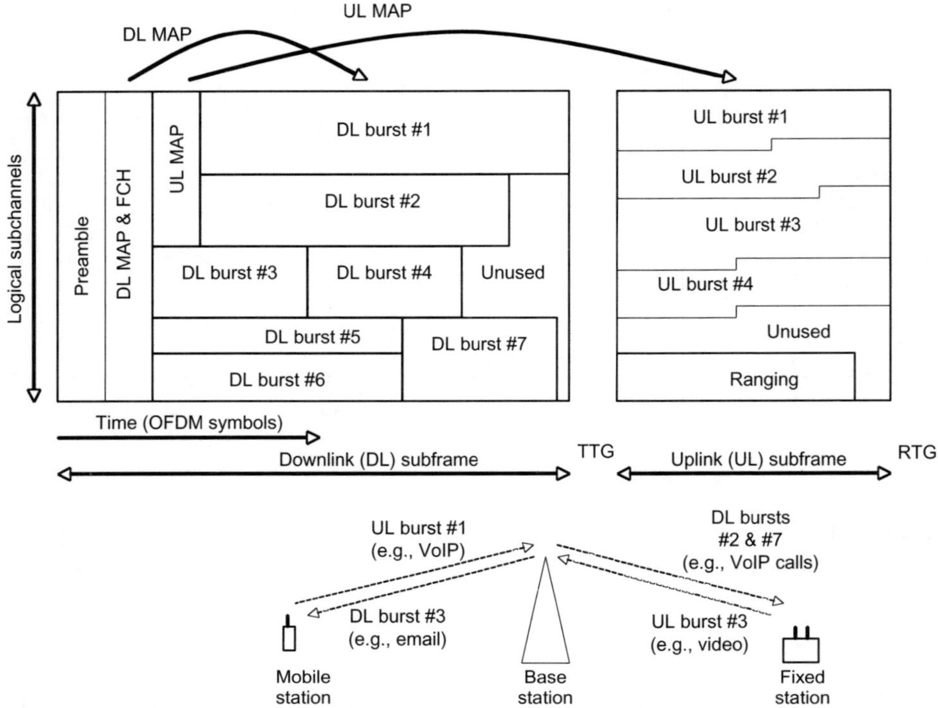

Figure 14.3. OFDMA frame structure for TDD systems.

3. In hybrid FDD (also called half-FDD), a BS can service a mix of FDD and non-FDD stations. The BS is full-FDD, while some MSs are FDD and some are non-FDD. The non-FDD stations do not transmit and receive simultaneously. They operate as in TDD, but with UL and DL in different RF channels.

Figure 14.3 shows the OFDMA frame structure for TDD systems, and Figure 14.4 shows it for hybrid FDD systems. In TDD systems the station at either end must switch from reception to transmission within specified times, called the transmit turnaround gap (TTG) and the receive turnaround gap (RTG).

14.3.1.1 TDD Systems. In contrast to voice traffic, data traffic has significantly more DL traffic than UL traffic. In TDD systems, this asymmetry in demand is easily managed by flexibility in the DL:UL split. In older deployments where traffic is dominated by voice communications, a permanent ratio around 1 would be a good fit, because UL voice has the same data rate requirements as DL voice.

A variable split also allows more flexibility when mixing low-power transmitters with high-power transmitters. Compared to a low-power MS, a high-power

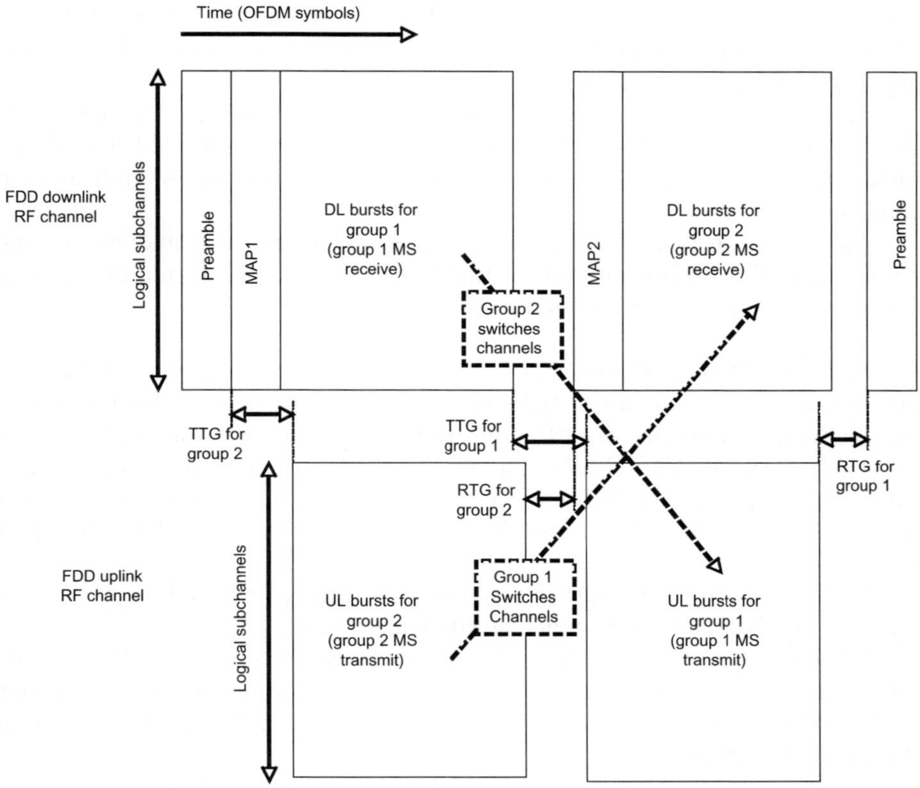

Figure 14.4. OFDMA frame structure for hybrid FDD systems.

MS requires less time to transmit a same amount of data, since it can operate at a higher rate thanks to the higher SNR it delivers at the BS. Thus the optimal split may depend on the mix as well as on the traffic.

Cellular deployments require careful management of interference between cells. This is particularly important in single-frequency deployments, in which an operator occupies only one channel across multiple cells. In TDD the UL and DL subframes between neighboring cells must be synchronized. When an MS at the cell edge receives a DL signal, a nearby MS connected to a neighboring BS should not be transmitting. By agreeing on a split, different operators can synchronize their frames and minimize interference.

The TDD ratio may be adapted depending on the SNR conditions and the bandwidth demands. This technique is called ATDD (adaptive TDD), but if any adjustments are needed, they must be slow varying to best serve the network as a whole.

TDD devices are simpler than FDD devices in terms of RF circuitry, but they require more DSP complexity. The TDD device has simpler RF filters and only one RF oscillator. The DSP and RF however must manage rapid turn-arounds and re-synchronizations. With falling silicon process costs, this disadvantage is becoming insignificant.

TDD is required in unlicensed bands to ensure coexistence with other IEEE devices. A TDD receives in the same channel as it transmits, and thus it can Listen Before Talk (LBT) and avoid interference caused by transmission collisions.

As an aside, in TDD the user can still speak and listen simultaneously during a voice call. The TDD frame rate is rapid enough that DL/UL multiplexing of fragments of voice data remains transparent to the user.

14.3.1.2 FDD Systems. FDD is required in some licensed bands, as these bands were originally specified for the first cellular voice standards. The existing voice bands are an attractive replacement market for WiMAX. The frequency allocations for FDD systems are symmetric, meaning that there is equal bandwidth available for both UL and DL. The DL:UL ratio is thus fixed because the channel bandwidth is fixed, and this offers less flexibility than a TDD system.

Duplex spacing varies significantly for the different bands. In some it is as small as 60 MHz (PCS) or as much as 400 MHz (AWS).

FDD requires stations to transmit and receive at the same time. In comparison to previous voice-based FDD systems that have an unframed "continuous PHY," WiMAX FDD is framed, which provides regular scheduling information at predictable times.

Regardless of the standard, an FDD device must ensure that reception (say at –80 dBm) is not interfered by spurious emissions from its own transmissions (say at +15 dBm) in the alternate duplex channel. This requires the use of fairly large and lossy duplex filters. The filters must be placed after the power amplifier, and their *insertion losses* can result in significant degradation to battery lifetime. It is not unusual for half of the power delivered from a power amplifier to be dissipated as heat before reaching the antenna.

In contrast, the TDD transmitter and receiver are not on at the same time. The filter is replaced by a simple switch, which connects the transmitter and receiver to the antenna. This reduced the component count and insertion losses.

14.3.2 OFDM and OFDMA

The signal modulation scheme in WiMAX is based on Orthogonal Frequency Division Multiplexing (OFDM). Whereas a traditional single-carrier (SC) modulation scheme occupies the complete physical RF channel with a single high-rate stream of modulated bits, in OFDM the channel is first subdivided into multiple subcarriers, and each subcarrier is individually modulated at a lower rate.

14.3.2.1 OFDM: Modulation. Already in 1966 it was shown that OFDM could solve signal impairments caused by multipath impairments, and in 1993 it was adopted for high-speed Digital Subscriber Line (DSL) modems that operate over regular twisted-pair phone lines. In 1999 the IEEE LAN amendment 802.11 (WiFi) adopted OFDM for the 5-GHz 802.11a amendment, and later in 2003 it is also adopted it for 2.4 GHz in the 802.11g amendment. OFDM is also adopted by 802.15 for ultra-wide-band (UWB) systems at high rates and short distances.

An SC-modulated signal can theoretically supply a given symbol rate in a channel of about equal width. The number of bits that are carried by a symbol depends on the modulation order.

An OFDM-modulated signal will yield about the same bit rate but at a much lower symbol rate. To simplify the OFDM processing at the transmitter and receiver, an FFT is used to modulate each subcarrier independently. For instance, a 20-MHz channel is divided into 2048 subcarriers, each with a width of about 10 KHz. The symbol duration is thus 100 μs. This is orders of magnitude longer than that in the SC case.

OFDM is the preferred modulation when the channel has significant multipath interference, since it can combine very low symbol rates with very high data rates. With reflections from other buildings and inside walls easily reaching 10-μs delays, the symbol duration must be long enough to absorb most of the resulting intersymbol interference (ISI).

Single-carrier modulation under these conditions would require impossibly complex equalization to overcome the ISI. In OFDM, however, the ISI is canceled simply by removing a small and designated fraction of the symbol that is affected by it. This fraction is called the guard interval (GI) or cyclic prefix (CP). The remainder of the symbol is practically void of ISI, and it merely requires a simpler form of equalization to help the decoder.

The OFDM symbol is illustrated in Figure 14.5. Figure 14.6 shows how an OFDM signal occupies a designated 5-MHz WiMAX RF channel.

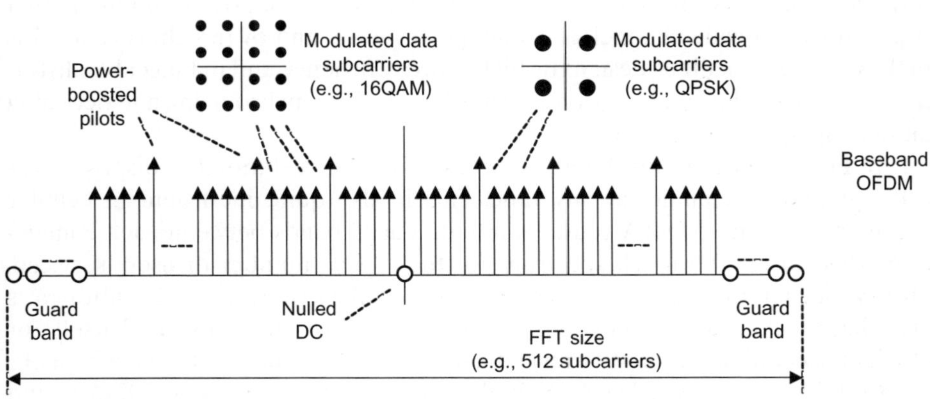

Figure 14.5. OFDMA symbol.

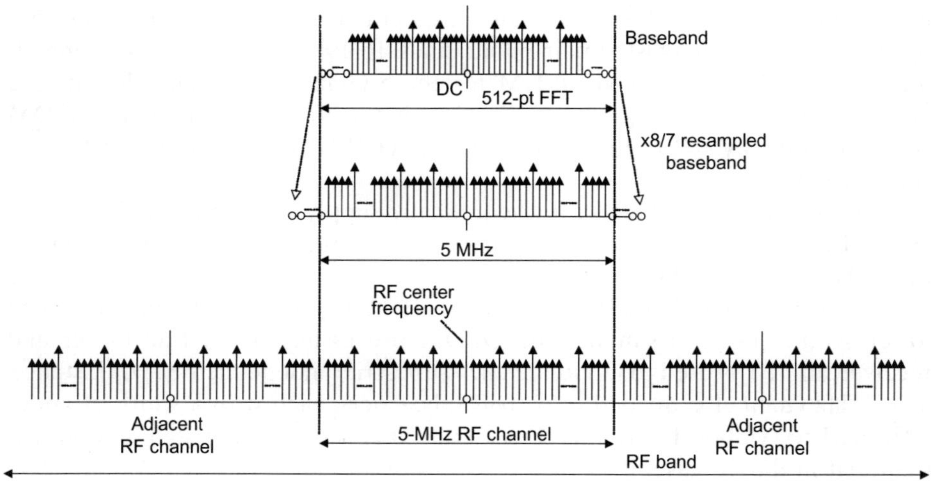

Figure 14.6. OFDMA signal occupying a designed 5-MHz WiMAX RF channel.

14.3.2.2 OFDMA: Access Multiplexing. It is the responsibility of the BS to multiplex its users and provide them access at their required data rate and QoS.

The access scheme has two parts: a physical part at the PHY layer and a management part at the MAC layer. The OFDMA scheme refers to the PHY layer, and it defines how distinct connections share the physical air medium while communicating with a BS.

In Orthogonal Frequency Division Multiple Access (OFDMA), stations share the medium by accessing the medium only in designated short slots of time and narrow slices of the channel.

By contrast, in Time Division Multiple Access (TDMA), a station has disposal over the entire channel during a designated timeslot. For typical user data rates it is very inefficient to allot an entire 5-MHz band to one user, no matter how short the burst of time. Short transmit bursts require power amplifiers that transmit over a wide channel, at a high power level, and over a short time. This makes it very difficult to design for high power efficiency and manageable distortion. Moreover, the receiver is unduly burdened with synchronizing and decoding an unforgiving short data burst.

In Frequency Division Multiple Access (FDMA) a station has disposal over a designated subchannel at any time. This is also quite inefficient in cellular deployments, since FDMA requires minute guard bands between each connection. This wastes too much spectrum for the typical number of users serviced. OFDM does not require guard bands between subcarriers, since the subcarriers are phase-locked and *orthogonal*. Moreover, FDMA does not lend itself for efficient scheduling of bursty user data, no matter how narrow the subchannel.

OFDMA in Mobile WiMAX builds on concepts of TDMA, FDMA, and OFDM.

Before the advent of OFDMA, WiMAX used OFDM to efficiently use the available bandwidth, and connections were scheduled in a TDMA fashion. OFDM with TDMA overcomes some of the drawbacks of TDMA and FDMA.

OFDMA provides a further flexibility for sharing the channel, to more efficiently support a high number of stations with mixed data rate requirements.

14.3.3 Subchannels

The OFDM PHY in 16a describes fixed broadband deployments using OFDM combined with subchannelization. This is a precursor to OFDMA because it can be viewed as a coarse level of OFDM access. Subchannelization is still present in OFDMA and a brief overview is warranted.

In 16a the RF channel is split in groups of 12 subcarriers, which amounts to 1/16 of the total number of usable subcarriers in a 256-pt FFT (edge subcarriers are not usable). Given a limited amount of transmit power, there is a 16× (12 dB) SNR gain when transmitting in only one subchannel rather than in the whole channel. However, the channel is ever only occupied for one user (in TDMA fashion), and so subchannelization reduces the data rate to the subscriber, and ultimately throughput within the BSs cell. Nevertheless, the SNR boost is cautiously used to overcome temporary "rain fades" during adverse weather.

Subchannels are also used to boost range and increase the service area, where low connection numbers allow the allocation of more TDMA to a particular user to offset the cut in channel width.

In the UL, OFDMA overcomes the throughput limitations of TDMA by allowing multiple MSs to transmit at the same time. Interference is avoided by scheduling different subcarriers for different MS connections. Subchannelization is implemented by restricting the scheduling to a subset of subcarriers.

Transmissions over a fraction of the channel, but over a longer period of time, are preferred for the MS. This improves the SNR at the BS, for a given low amount of transmit power radiated by a battery-powered MS. These transmissions are sometimes loosely called "long and thin," named after their occupation of the OFDMA frame.

For power efficiency in the MS receiver it is better to schedule the MS over the shortest possible DL time. This requires the use of the widest possible channel, but it minimizes the receiver on time. It is sometimes loosely called "short and fat." The power amplifiers of a BS transmits at perhaps 40 dBm, which is much higher than a battery-powered MS at perhaps 20 dBm. In the DL, the SNR received at the MS is thus already higher, and "short and fat" is quite feasible.

14.3.4 Scalable OFDMA (SOFDMA)

Scalable OFDMA refers to the adjustment of the FFT size of a device depending on the width of the channel in which the device is deployed.

The intent of the adjustment is to tightly control the subcarrier spacing for mobile use. The spacing affects several core device specifications (at RF and for the DSP), and it has direct influence on the achievable throughput under mobile conditions. The optimum subcarrier spacing is determined by considering several properties of the mobile channel.

14.3.4.1 Typical Mobile Channel Parameters. In the RF bands below 6 GHz, the Doppler shift at 125-km/h mobility is on average 400 Hz, and the worst case is 700 Hz at the upper end of the 5-GHz band. Doppler shift causes inter channel interference (ICI) between subcarriers. To limit ICI, subcarrier spacing must be at least 10 kHz. ICI is then below (27 dB on average) across the sub-6-GHz band.

Another factor is coherence time. It is a measure of how long a specific channel condition remains relatively constant. At 125-km/h mobility, it amounts to about 1 ms. The OFDM symbol duration must be less than that.

The coherence bandwidth is also a factor. It is a measure of how spectrally flat the channel is, despite reflections. For suburban channel conditions it is more than 10 kHz.

Thus at a subcarrier spacing of 10 kHz, it can be assumed that the channel is flat within a subcarrier and it is constant during a symbol. The spacing thus allows the use of OHFDM with simple frequency-domain equalization and channel estimation on a subcarrier basis.

A further factor is the effect of the intersymbol interference caused by multipath reflections. A guard interval of at least 10 μs is needed to cover most of this kind of interference in urban environments. To keep the overhead low, at 10%, this implies a symbol duration of 100 μs.

14.3.4.2 Resulting OFDMA Parameters. The above considerations of the mobile urban channel conditions show that 100 μs is a good choice for the symbol duration and that 10 kHz is a good choice for subcarrier spacing.

Different RF bands across the globe offer different channel widths. Since the OFDMA parameters numbers do not depend on the channel bandwidth, the number of FFT subcarriers has to scale with the width of the channel. Thus, to get the desired symbol duration and subcarrier spacing, a 10-MHz channel requires a 1024-pt FFT, and a 20 MHz requires a 2048-pt FFT.

Table 14.4 provides an overview of the OFDM system parameters for a number of profiles defined by the WiMAX Forum's Mobile Task Group (MTG).

A sampling factor is applied to adjust the channel utilization, depending on the precise channel bandwidth (for instance, 8.75 MHz versus 10 MHz), without changing the number of subcarriers. This keeps the slot scheduling, subcarrier permutation, and bit interleaving parameters constant, regardless minor differences in channel width.

14.3.4.3 WiMAX at 70 Mbit/s. Marketing material for WiMAX often sports a data rate of 70 Mbit/s. A discussion of this number will provide some valuable insights.

To start, the stated rate is based on a 20-MHz channel, which implies a 2048-pt FFT.

TABLE 14.4. Scalability of OFDMA Frame for Different Regulatory Bandwidths

Bandwidth (MHz):	5.00	7.00	8.75	10.00	14.00	20.00
Sampling Factor:	28/25	8/7	8/7	28/25	8/7	28/25
Sampling Frequency (Msps):	5.60	8.00	10.00	11.20	16.00	22.40
FFT Size:	512	1024	1024	1024	1024	2048
Subcarrier Spacing (kHz):	10.9375	7.8125	9.7656	10.9375	15.625	10.9375
Symbol Time, including GI (μs):	102.86	144	115.2	102.86	72	102.86
Guard Interval (μs):	11.429	16	12.8	11.429	8	11.429
Number of Used Subcarriers:	433	865	865	865	865	1703
Channel Utilization (MHz):	4.74	6.76	8.45	9.46	13.52	18.63

After resampling at $\frac{8}{7}$, and insertion of a $\frac{1}{8}$-guard interval, the symbol time becomes 129 μs. Out of the 2048 subcarriers, the standard uses 1536 for data, leaving the rest unused as guard bands.

The pilot overhead varies, depending on the mode of operation. At the low end, it is about 11% to 15% for DL and UL, and at the high end it can reach to 33% for UL.

Then there is a small amount of overhead due to the preamble, some regularly scheduled MAC messages, and ranging. There are also minimal RTG and TTG silence gaps between the UL and DL subframes. All this overhead can be neglected for simplicity.

The highest data rate is provided by 64-QAM rate modulation with rate $\frac{5}{6}$ coding. At this rate, each symbol and each data carrier contains 5 bits of uncoded data.

Putting all this together, the total data rate over a 20-MHz channel then becomes 1536 subcarriers * 6bits * 1/(100 μs*(1 + $\frac{1}{8}$)) *$\frac{8}{7}$ * 89% = 69.3503 (Mbps), or about 70 Mbit/s.

This number scales proportionately with the channel bandwidth. Thus a 10-MHz channel can yield 35 Mbit/s and a 5-MHz channel can yield 17.5 Mbit/s.

The bandwidth efficiency is thus 3.5 bits per second per hertz of channel bandwidth.

It should be noted that this is an approximation of the *maximum* supported rate by the modulation scheme. Under normal deployment conditions, only a fraction of the stations can be addressed at the highest modulation and coding rate. As the distance between subscriber and BS increases, the connection rate will drop since the scheduler will switch to more robust modulation schemes (at fewer data bits per subcrarrier) to keep the BER below limits.

As an aside, the theoretical data rate or bandwidth efficiency could be further increased by the use of even higher modulation rates, such as 256QAM. One could also increase the symbol rate or narrow the subcarrier spacing with a higher-order FFT. But given the mobile channel conditions, these would add complexity that would be rarely, if ever, used,. They are also quite challenging to implement.

The standard offers other schemes at the PHY level that can be used to increase data rate in real deployments. MIMO and subchannel utilization schemes are available to increase spectral efficiency and throughput depending on channel conditions.

14.3.5 Radio Resource Management and Subchannel Utilization

Radio Resource Management (RRM) in cellular networks involves the optimization of transmit power, user scheduling, and the occupation of frequency channels to maximize the cellular throughput per hertz of bandwidth of the radio spectrum. RRM also involves providing coverage over the entire geographic target area, maximizing cell throughput, and meeting broadband service plan commitments. Other factors that come into play are minimizing the overhead from handover and idle stations, consideration of the link budget in the farthest parts of the cell, and planning for terrain and urban obstructions.

14.3.5.1 Adaptive Modulation and Coding (AMC). The term "Adaptive Modulation and Coding" (AMC) refers to the adaptation of the modulation and the coding rate, depending on the channel conditions. The specific term AMC was first introduced in 3G cellular technology, under the revision 1xEVDO-Rev 0. The High-Speed Downlink Packet Access (HSDPA) extension of WCDMA includes the capability to adjust the modulation rate from QPSK to 8PSK and 16QAM as the signal-to-noise and interference ratio of the link improves.

In earlier versions of cellular communications, the modulation and the rate were fixed. Typically, it was BPSK and/or QPSK with rate ½ coding. With advent of higher-speed processing and demand for higher spectral efficiencies, the use of higher-order modulations and coding rates became necessary. EDGE (enhanced GSM) provisions Gaussian Minimum Shift Keying (GMSK) based on 8-PSK. HSDPA (2005) also provisions AMC with QPSK and 16QAM modulation combined with code rates of ¼, ½, ⅝, and ¾.

IEEE 802.16e includes AMC, and it is also used in other wireless technologies. In 802.11 (WiFi), modulation and coding are adjusted as part of the Modulation and Coding Scheme (MCS) algorithms.

AMC operates in supplement to *power control*. The intent of power control is twofold. It minimizes the transmit power in order to minimize the interference within the cell and from cell to cell. In addition, power control reduces the power draw from a mobile's battery. An MS near a BS will simply be controlled to transmit at a lower power level.

AMC is then applied on top of power control, to maximize the data rate at the desired transmit power level. Moreover, AMC provides a fine granularity of packet sizes within a fixed frame, thus adding the ability to minimize unused parts of the frame (the *stuff bits*).

To further improve the robustness for stations at a far distance and to further improve spectral efficiencies for stations near their serving BS, AMC is supplemented by MIMO. Depending on the channel conditions for each MS served, the BS can schedule these stations in groups, addressing some with MIMO modes tailored for robustness [e.g., space time codes (STC)] and others for modes that increase efficiency [e.g., spatial multiplexing (SM)]. This is called adaptive MIMO systems (AMS).

14.3.5.2 CINR and RSSI Channel Measurement and Feedback. For bands below 11 GHz the BS has the option to request channel measurements by the MS. This includes two metrics for the quality of the RF air channel: carrier to interference and noise ratio (CINR) and the receive signal strength indicator (RSSI). The BS uses these to rapidly adapt and optimize the schedules, to map subscribers to subchannels that are best for their reception, and to avoid interference with other cells.

It is also used to minimize interference to and from other IEEE systems (e.g., WiFi) or non-IEEE systems (e.g., radar) in the geographical vicinity. This is particularly addressed in the 802.16h amendment for coexistence.

For MIMO operation (see below) there are additional feedback mechanisms, to allow the transmitter to calculate its MIMO coefficients based on the channel. This includes a channel quality indicator (CQI) and other feedback by the MS, such as a choice of preferred number of BS-activated transmit antennae and a preferred burst profile. Channel coherence time can also be fed back, which matters if the BS is calculating MIMO pre-coding coefficients for a later transmission to the same MS.

14.3.5.3 Subchannel Utilization Modes. Frequency planning is another aspect of RRM. The standard allows for frequency planning at a subchannel level. There are several modes that differ in how subcarriers are allotted to share the channel among users of neighboring cells.

DL FUSC. Downlink full utilization of subchannels (FUSC) involves transmissions using the full breadth of the channel. This is applied where there is no inter-cell or inter-sector interference and where rapidly changing channel conditions make it impractical to optimize the burst profile for any specific MS.

Pilot carriers occur one out of every seven subcarriers, and they are spread evenly across the channel. Data carriers are assigned to the remaining subcarriers, whereby each connection uses a sparse subset of subcarriers from across the entire channel. The MS receiver estimates and tracks the channel based on all pilots across the entire channel, and it applies interpolation to equalize the specific data subcarriers assigned only to it.

A pseudo-randomization scheme permutes the subcarrier assignments from symbol to symbol, which improves the gains from frequency diversity.

DL PUSC. Partial utilization of subchannels (PUSC) is applied in the DL to provide fractional frequency reuse (FFR) with neighbor cells. In PUSC, a BS schedules only part of the channel (often ⅓) for receivers near the cell edge. Such a PUSC *segment* is formed by logically grouping a selection of subcarriers.

Segments do not section the channel into physical sub-bands. Instead they pseudo-randomly map logical subcarriers to a subset of physical subcarriers across the entire channel. Interference is avoided because neighboring cells or sectors are assigned different segments, and thus the subcarriers spectrally interleave without colliding.

Pilots are evenly distributed within subchannels, and data subcarriers are permuted evenly across the subchannel.

UL PUSC. In the uplink, each transmission from an MS requires its own pilots, since each channel from a MS to the BS is different. The BS cannot use pilots from one MS to equalize the data subcarriers from another.

Therefore pilot and data subcarriers are combined in a time–frequency *tile*, and tiles are permuted across the designated PUSC subchannel. The pilots reside on the corner of the tile, and the BS uses them to equalize the transmission from a given MS. There are eight data and four pilot subcarriers that form a tile of three symbols by four subcarriers.

There has been no need for a full-channel UL FUSC, since one MS would rarely ever need to occupy the entire channel. Therefore, tiles are only assigned to a segment.

Optional UL PUSC. There is an option to reduce overhead from pilots where channel conditions permit. PUSC can also operate with eight data and one pilot subcarrier to form a tile of three symbols by three subcarriers. The pilot is in the center of the tile.

TUSC 1 & 2. Tiled utilization of subchannels (TUSC) is the same as PUSC but for the DL. This allows a TDD BS to schedule UL and DL for a specific MS using the same physical part of the channel for both directions. The BS can then infer the transmit channel from the received signal to calculate AAS pre-coding coefficients. The two TUSC modes respectively correspond to a 3×4 and a 3×3 tile for reduced overhead.

Band AMC. In band adaptive modulation and coding (band AMC), the BS scheduler has access to adjacent physical subcarriers of the channel. This is also called the *adjacent subcarrier mode*. It operates with eight data and one pilot subcarrier grouped into a *bin*, which is mapped to an FFT sub-band within the channel.

The BS can operate at a higher burst rate for a selected MS by specifically scheduling it in portions (sub-bands) of the channel where the SNR is high. If the subscriber is not moving, Band AMC can provide higher throughput than the frequency diverse FUSC or PUSC modes.

In FUSC and PUSC, subcarriers are assigned pseudo-randomly and loaded with AMC based on the average SNR across the assigned subcarriers. The maximum achievable rate to an MS with the best SNR is lower than with band AMC, but the BS does not need to rapidly update the burst profile as the channel fluctuates during mobility.

14.3.5.4 Zones. To combine different subchannel utilization modes in a single DL or UL subframe, the subframe is split into zones. The operator synchronizes the zone boundaries in all its BSs across its network, in order for FUSC and PUSC to be effective in interference mitigation.

Zones are also used to schedule MSs with similar requirements for noise and interference robustness together in time. Thus an MS scheduled for a particular zone does not have to attempt to track pilots over the entire frame, but rather can wait to detect and adapt until its designated zone (with suitable SNR) is received.

Although an MS receiver only needs to process the pilots in symbols for which it needs to demodulate data subcarriers, it is advantageous to start estimation of the channel earlier in the frame, even though data are scheduled for other stations. However, if interference levels are significant, the MS can wait until the start of its zone before processing the pilots.

To this end, *zone switches* are broadcast by the BS.

An example of zones and subchannelization is provided in Figure 14.7.

14.3.5.5 Fractional Frequency Reuse (FFR). Frequency reuse refers to the reuse of a channel or a fraction of it so that it can be shared with a neighboring cell.

Operators with access to three channels use *Frequency Reuse 3* cell planning to manage interference. In these cases the interference between cells is reduced at the expense of small inefficiencies in terms of channel spectrum utilization. MSs in neighboring cells operate at different RF frequencies, and so their transmissions do not interfere. In parts of the BS coverage area, such as at the cell edge, this is highly needed, but in other parts, such as close to the BS, this leaves much of the spectrum unused. The net effect is nevertheless a gain in spectral efficiency (bits/s/Hz per BS).

Sectorization is based on the same principles of frequency reuse, and it offers more options to reduce interference and improve efficiency. It comes at the cost of more equipment at the base station, because each sector requires separate high-power RF modules and antennae.

The standard offers PUSC for fractional frequency reuse (FFR). Figure 14.8 shows two different configurations of a cellular plan. The object is to minimize inter-cell and inter-sector interference caused by MS and BS transmissions in neighboring cells and sectors.

At cell edges the interference level from a neighbor BS is often as strong as the intended signal from the serving BS. The standard provides very robust repetition codes for this scenario, but the resulting frame inefficiency is of course undesirable.

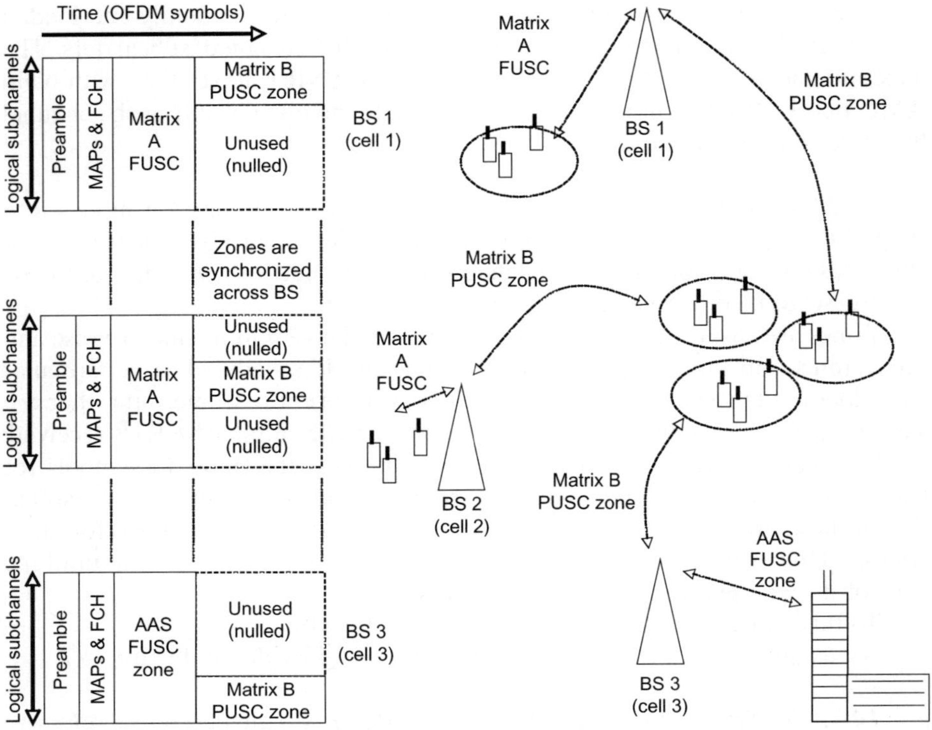

Figure 14.7. Example of zones and subchannelization.

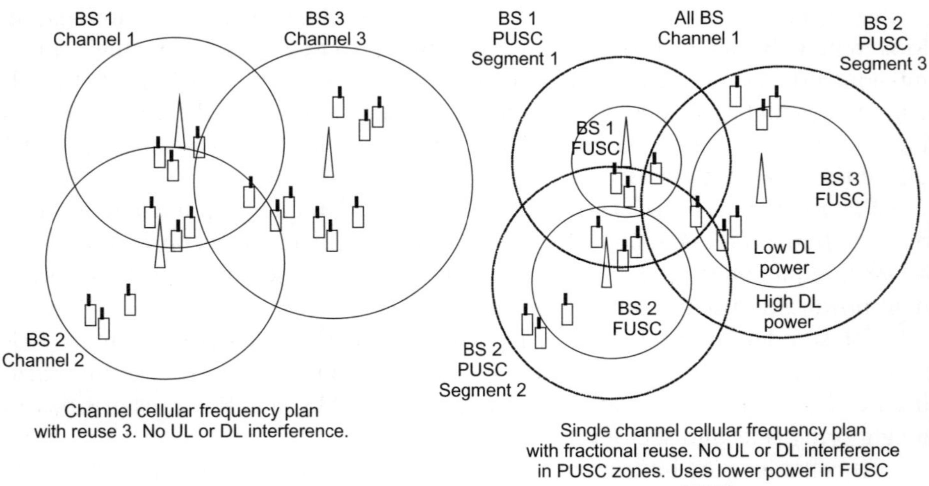

Figure 14.8. Frequency reuse 3 (**left**) and single-channel FRF (**right**).

The BS schedules cell-edge MSs in a PUSC zone, which isolates them from interference. This alleviates the frame inefficiency at the expense of some spectral inefficiency in the cell edge.

The inefficiency does not apply to the entire cell. The BS schedules nearby MSs, which do not experience the interference, in a FUSC zone. The operator synchronizes the FUSC zones among BSs, so that neighbor BSs can do the same to their nearby MSs. Thus the spectrum is fully reused where interference allows.

To further combat the effects of FUSC interference at the receiver's decoder, interfering stations are assigned a different permutation base. The distinct bases ensure that subcarrier "collisions" are rare and random so that the interference is not persistently high for any individual MS.

14.3.6 Error Control

Error control involves two aspects. Forward Error Correction (FEC) is an efficient method to reduce the error rate (the bit error rate and resultant loss of a burst) over-the-air using DSP. Automatic Repeat Request (ARQ) is a method to recover lost bursts. FEC minimizes the need for ARQ, and ARQ minimizes the exposure of errors to the network.

14.3.6.1 Forward Error Correction (FEC) and Interleaving. The *burst profile* defines the precise modulation and coding combination of a scheduled burst between stations. It covers the choice of modulation (QPSK, 16QAM, or 64QAM), the choice of the FEC scheme (CC, CTC, LDPC, ZCC, BTC), and the coding rate FEC parameter (rate ½, ⅔, ¾, ⅚).

The Convolutional Code (CC) produces an output bit sequence out of an input sequence by passing it through a binary feedback shift register. This operation convolves the input sequence with a reference encoding sequence called the code polynomial. The length (also called "depth") of the shift register corresponds to the order of the polynomial, and it is called the constraint length. For CC it is $K = 7$.

The CC codes are based on two polynomials, and for each input bit two output bits are produced. The base rate (or native rate) is thus ½, and for a burst at rate ½, both coded bits are transmitted for each data bit.

A code rate of ⅔ can be attained with the same code. In a process called *puncturing*, the transmitter alternates by sending both coded bits, then just one (dropping the other), then both again, and so on. Puncturing is also used to attain rate ¾.

A base code has more redundancy than its punctured code, but it performs better for noisy receptions. 802.16e supports code rates ½, ⅔, and ¾.

There are two variants to the CC, one with tail biting and one with flush bits. Flush bits add some overhead but provide for a simpler decoder. This is also called a Zero Terminating Convolutional Code, or ZCC.

The standard also provides for a Convolutional Turbo Code (CTC) and a Block Turbo Code (BTC). The CTC is a duo-binary code which means it has two encoding polynomials. The output of the first shift register is interleaved and convolved with a second polynomial. The native rate is ⅓, and puncturing provides the rates ½, ⅔, ¾, and ⅚.

The decoder for CTC is more complex than for CC. It is iterative, and the decoding time depends on the amount of noise in the signal.

The OFDM mode of the standard also includes a Reed–Solomon FEC with codeword length 255 containing 16 check bytes, but this is not required by any of the profiles.

The coding gain would rapidly decline if the decoder were presented with strings of adjacent bit errors rather than a same amount of isolated bit errors. This is a drawback of the type of FEC used, but it is easily avoided. To remedy this, an interleaving step after encoding at the transmitter enables the placement of a de-interleaver before decoding at the receiver.

A de-interleaver merely re-orders coded bits, and it does not directly improve the SNR of the received constellations. Rather, it reduces the probability that errored code bits occur in clusters at the input of the FEC decoder. This improves the probability of error correction, which in turn reduces the BER at the output of the FEC.

In OFDM(A), the adjacency of data bits must avoided by two steps.

In a first step, neighboring bits in the data stream are spread over non-neighboring subcarriers. Often a reduced SNR occurs in several neighboring subcarriers, for instance due to narrowband interference and/or fading (notches) in the channel. The de-interleaver will then cause the good and poor subcarriers to alternate at the decoder input.

The second step applies to higher-order constellations (16-QAM and 64-QAM). Random errors caused by Gaussian noise usually only affect the least significant bit of the constellation, because a small disturbance affects perhaps one bit of a multi-bit constellation point. The interleaver ensures that neighboring bits alternate as most and least significant bits, thus alternating their strength of protection against noise.

14.3.6.2 Automatic Repeat Request (ARQ) and Hybrid-ARQ (HARQ). Automatic Repeat Request (ARQ) is a MAC level operation to attempt recovery from bit errors. It operates independently from the PHY. An MPDU is constructed with a number of blocks of data from one or more SDUs, and the transmitter can resend (automatically repeat) blocks of the MPDU, or even complete MPDUs, that have not been acknowledged by the receiver. ARQ is used to overcome brief air-link interruptions due to temporal fades. While FEC is designed to overcome random and sporadic bit errors within bursts, ARQ is designed to overcome significant frame losses.

The standard supports a few options for sending acknowledgments as a stand-alone management message or as a payload piggyback during a data response.

A 32-bit Cyclic Redundancy Check (CRC) is used to determine whether burst data have been received and decoded without error. The CRC is a data-dependent signature appended to the data unit. A CRC can confirm with sufficiently high probability (but theoretically not with absolute certainty) that the decoded data is error-free.

Hybrid ARQ (HARQ) is an alternative offered by the PHY by tightly coupling retransmissions with the FEC. The MPDU is processed by the PHY, where it is FEC-encoded to produce up to four coded and punctured versions of the same data. In a first transmission of the data, the PHY only transmits one version. If the CRC passes at the decoder, then no further versions are needed, and a new MPDU can be transmitted. However, if the CRC fails, then a different version of the data is sent. This is called a *stop-and-wait* protocol, because the receiver waits for the repeat data before proceeding with the rest of the data.

To reduce latency, the transmitter sends a further ARQ block of data even before it has received an acknowledgment for a previous block. Thus, acknowledgments are lagged and for most of the time, while there are no block errors, latency is shortened.

For rate ¾ Incremental Redundancy HARQ (IR-HARQ), the data are coded at rate ½ and then punctured to ¾. The puncturing sequence is altered for the retransmission. By keeping the retransmission rate at ¾, scheduling is simplified since both transmissions require the exact same number of coded bits. A simple receiver can opt to discard the first transmission and to decode the retransmission at rate ¾ without needing the first transmission. A more complex receiver can merge the two sequences to yield a code that is slightly stronger than rate ½.

Another repetition scheme by the PHY is *Chase HARQ*. In this case, the precise same encoder output bits (with same puncturing) are simply retransmitted. This allows the receiver to sum the received signals before decoding, which averages out some of the noise, yielding a 3-dB SNR improvement.

Repetition codes operate in the same way, except that with Chase Combining the repetition is on-demand, in the event of a CRC failure. In order to get a precise duplicate of the previous transmission, the burst profile and data must be the same. This is different from Hybrid ARQ, where the coded sequence is different. It is also different from ARQ, where the same Data Unit is retransmitted by the MAC, but where the PHY may apply a different modulation and coding rate to it.

14.3.7 PHY MIMO Techniques

Multiple-Input Multiple-Output (MIMO) techniques are used in several cellular standards and other communication protocols to enhance the system capacity. MIMO techniques can improve a poor SNR at the receiver to enable higher modulation and coding rates. And if the SNR is already high, MIMO can be used to further raise the burst rate. Improvement of SNR is achieved with Space Time

Codes (STC), its variants, and beam forming. Rates are increased with Spatial Multiplexing.

MIMO techniques can be split into open-loop and closed-loop techniques. An open-loop transmitter operates without knowledge of the RF channel, and a closed-loop transmitter operates with channel knowledge.

When the transmit channel is inferred from the receive signal, feedback is said to be *implicit*. In TDD systems the MIMO transmitter can infer its channel from the channel conditions during previously received bursts, since both directions use the same RF channel.

In contrast, an FDD system employs different channels for UL and DL. The MS must send a designated message to the BS, containing information about the DL channel. When channel information is messaged back from the target receiver to the MIMO transmitter, feedback is said to be *explicit*.

MIMO systems gain over single-antenna systems through *array gain* and *diversity gain*.

Array gain is the improvement of the signal strength attributed to the reception of a larger proportion of the radiated signal power. Quite simply put, an array of two antennae together in one receiver captures twice the amount of RF power compared to a single antenna. Array gain can equivalently be attributed to the transmitter, when two transmitters of an array radiate twice the RF power of one transmitter. The challenge in realizing array gain arises when processing the array signals. Combined signals must be synchronized and equalized in order to sum constructively.

Diversity gain is the improvement of a decoded signal due to the reception or transmission of diverse versions of a same signal. *Temporal diversity* occurs when the two versions are delivered at distinct instances in time. *Frequency diversity* occurs when the two are delivered in distinct subcarriers.

In MIMO, *spatial diversity* occurs when the versions are received and/or transmitted by distinct multiple antennae. Separation at the MS is usually half the wavelength. At the BS it is often several times the wavelength.

14.3.7.1 Antennae and Antenna Arrays. The antenna *beam width* quantifies the directionality of an antenna or antenna array. It is a measure related to the antennae *radiation pattern*. The pattern usually features a dominant *beam*, and the width of the beam is called the *spatial angle*.

The antenna array is designed such that its pattern matches the coverage requirements for the base station. In some cases the pattern is omnidirectional, and the entire cell around the base station is serviced as one sector. This would typically be the case for smaller in-building *pico cells* or *femto cells* that service under 100 calls. In *macro* BSs, where the radio head resides on top of an outside tower, the antenna array is more complex, often comprising of four antennae per sector, with three sectors per cell. The beam of a sector covers a 120-degree division. The directionality of the antenna array offers further range within the sector, along with less cell-to-cell interference.

14.3.7.2 Spatial Multiplexing and Virtual (Collaborative) MIMO. At frequencies below 11 GHz, in particular at those envisioned for 16e, namely 700 MHz to 5.8 GHz, there is plenty of spatial diversity to allow the use of MIMO techniques.

Transmit spatial diversity refers to the use of multiple antennae for transmission, and receive diversity refers to the use multiple antennae for reception. Usually the same antennae used for transmission are also used for reception. To reduce the cost of an MS, transmission drives one antenna, but reception uses two. Such an MS would be called a 1×2 MIMO device.

Spatial Mutliplexing (SM) involves the transmission of multiple streams of data simultaneously, in the same subcarriers and at the same time. Each stream is transmitted by a separate antenna. To decode these streams the receiver must have an antenna count at least equal to the number of streams. For instance, a 1×2 MS can receive two spatially multiplexed signals. Each antenna requires its own RF and DSP processing, plus additional MIMO decoding across both received signals to separate the multiple streams.

In Collaborative MIMO, pairs of MS are scheduled so that both transmit simultaneously and the two signals blend in a SM fashion. This is also called Space Division Multiple Access (SDMA) or *Virtual MIMO*. Thus the cell throughput can theoretically be doubled during the UL, using MS that have just a single antenna. This is illustrated in Figure 14.9. The requirement to operate pairs of stations poses a challenge on the scheduling algorithms in the BS. Both

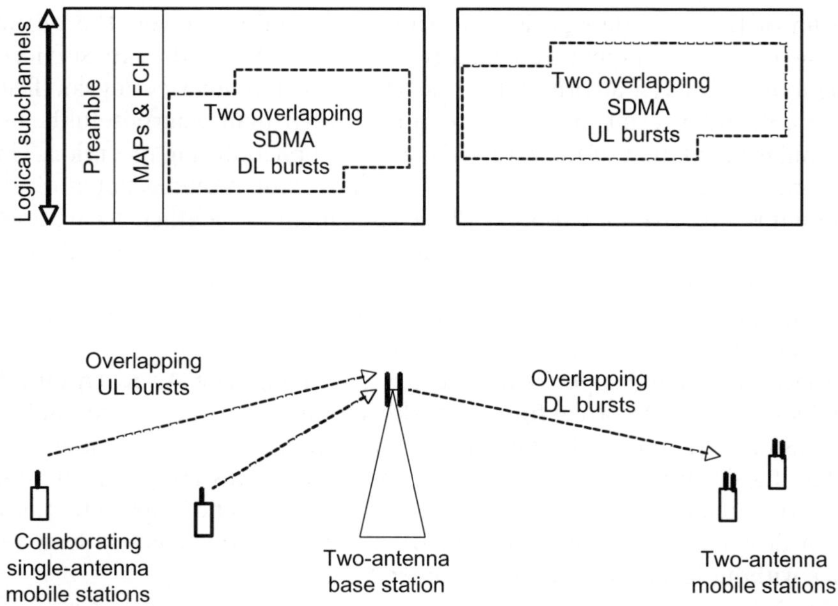

Figure 14.9. Collaborative MIMO and SDMA.

stations must be capable of Virtual MIMO operation, and they must both be positioned to deliver signals to the BS with close SNR and close reception levels.

On the DL, two MSs can be scheduled to receive overlapping signals. Each MS requires two antennae to separate the SDMA signal intended for it and discard the other.

14.3.7.3 Adaptive Antenna Systems (AAS), Smart Antennae, and Beam Steering. Adaptive Antenna Systems (AAS) refer to the adaptation of the transmit signal by *precoding* the signals from each of several antennae. Precoding consists of optimal phase and amplitude adjustments on a per-subcarrier basis, so that the sum of the multiple signals coherently add-up at the receiver. Of course, the adjustment depends on the channel from transmitter to receiver. Feedback is used to determine the channel and make the adjustments, and the transmitter is thus said to be "smart" about the channel.

Precoding at the transmitter is often also called *beam forming* or *beam steering*. This name finds its origins in the directed radiation pattern that forms from the antenna array. Such a pattern appears when there is line-of-sight (LOS) radiation, void of any reflections on the path to the receiver. This is typical in fixed outdoor-to-outdoor type transmissions. In outdoor-to-indoor applications, however, there are substantial reflections and as a result the channel phase and amplitude depend greatly on the subcarrier. Thus there can be a completely different beam per subcarrier, but these non-line-of-sight (NLOS) conditions do not necessarily limit the benefits.

Precoding can also be used for *null steering* rather than beam steering so that the sum of the multiple signals coherently vanish at a receiver. The object of steering a null is to minimize the energy to a cluster of MSs that are serviced by a neighboring BS. The algorithm for calculating the antenna steering coefficients is different, and it results in a purposeful null rather than a purposeful beam.

Similar adaptation of the array can also be applied during reception. This is sometimes called *receive beam forming*. It does not require standardization, because it is a receive-only process. Phase and amplitude coefficients are applied per subcarrier, so that the sum of the multiple received signals coherently add-up at the decoder. Maximum Ratio Combining (MRC) is an example of such a technique. The coefficients can also be calculated so that the signal from a nearby interferer vanishes at the decoder input.

Special zones and superframe structures accommodate AAS. An OFDMA superframe is a set of normal frames, of which some are regular frames and some are AAS-only frames. Under this kind of superframing, a great number of MSs can be efficiently served with AAS. The MSs simply do not even attempt to receive the preamble or decode any channel descriptors and maps in frames that are not designated to them. Thus the BS can beam-form select MSs in select frames without losing connections to other MSs.

An AAS-Zone provides a similar mechanism, yet at a smaller scale. This zone has its own preamble, channel descriptors, and maps for DL and UL scheduling. The entire zone is beam-formed to select MS.

By using a zone or superframe, distant stations can decode preambles and scheduling parameters, which extends the reach of the BS. Without such a zone or superframe, AAS would only be able to boost the rates of stations that are already within reach.

14.3.7.4 MIMO with Two Transmit Antennae. The standard supports various MIMO techniques for transmitters equipped with two antennae.

2 × 1 Space–Time Coding (STC). STC is also known in the standard as *Matrix A*, and it provides spatial transmit diversity and array gain. The code spreads data over two antennae, and its salient feature is that the receiver requires only a single receive antenna. It operates independently on each subcarrier. Encoded constellation points (e.g., a two-bit pair in a QPSK symbol) are grabbed pairwise. One symbol point is transmitted on one antenna, the other on the second antenna. In a subsequent OFDM symbol the same point is transmitted, but the first point is conjugated, and transmitted on the second antenna, and the other is negated, conjugated, and transmitted on the first antenna. It offers the highest spatial diversity for the given antenna configuration, but no rate increase. It is therefore a rate 1 code.

This code is also referred to as an Alamouti code, named after its inventor.

STC improves the link budget by transmitting spatially diverse signals to the receiver. The implementation cost resides primarily at the transmitter, since the receiver requires only one antenna and RF circuit. This technique is classified as an *open-loop* MIMO technique because the transmitter requires no knowledge of the RF channel. It is quite suitable for mobile conditions where channel information is inconsistent from frame to frame. Moreover, it can be used for broadcasting to stations across completely different channels. STC is applied to all subcarriers in the symbol, and STC bursts are joined in a designated zone with special pilots.

Frequency Hopping Diversity Coding (FHDC). FHDC is a Space Frequency Code (SFC) that requires OFDM and is equivalent to STC. The conjugate complex retransmission occurs in a different subchannel, rather than in a different symbol. SFC provides the same spatial gains as STC, but offers additional frequency diversity when the channel is heavily affected by multipath reflections.

Precoding. The standard covers several options for antenna selection and beam forming with various levels of complexity. An MS can advise its serving BS which antenna to best use for the next burst, based on signal evaluations from previous bursts. An MS can calculate precoding coefficients for the BS and can communicate them to the BS for a following transmission. Alternatively, the MS communicates channel information back to the BS, and the BS performs the calculation of the precoding coefficients itself.

Precoding techniques provide transmit array gain and greater diversity gain compared to STC. They work well with stationary channels.

Cyclic Delay Diversity (CDD). CDD operates at the transmitter and generally requires no special processing at the receiver. An OFDM symbol is transmitted from one antenna, and a replica of it is transmitted from the second antenna. Before transmission, the waveform of the replica is cyclically rotated in time. This avoids unintentional constructive signal summation at the output of the transmitter.

CDD provides mainly array gain with a small amount of diversity gain. The cyclic rotation of the time domain waveform can equivalently be applied as a frequency domain operation at the input of the FFT. CDD changes the end-to-end channel transfer function as perceived by the receiver, and therefore CDD must be applied to the pilots (UL and DL) as well as to the preamble (DL). The delay must be restricted to a small amount, in order to maintain the integrity of the receiver's synchronization algorithms.

Spatial Multiplexing (SM). SM is also known in the standard as *Matrix B*, and it is used to boost the data rate where SNR permits. This is a rate 2 code, and it requires at least two transmit and receive antennae. Each antenna simultaneously transmits constellation points pertaining to different data. In *Horizontal Encoding*, the data are provided by two distinct FEC encoders, each with independent coding and modulation rates. The data streams can be scheduled independently and the MCS can be optimized separately. In a simpler version called *Vertical Encoding*, the output of a single FEC encoder is multiplexed over two antennae. The benefit is that SNR differences between the streams are averaged out at the decoder, which provides simpler scheduling.

SM increases the data rate by a factor proportionate to the number of antennae, and in effect it multiplies the spectral efficiency. SM decoding introduces self-interference, which somewhat degrades the effective Signal to Interference and Noise Ratio (SINR) at the FEC decoder.

Receive Diversity. To further increase the diversity gains from these techniques, the receiver can optionally be equipped with additional antennae. This provides further array gain as well as diversity gain. Common receiver-only techniques are based on antenna selection (Switched Diversity), RF signal combining [Equal Gain Combining (EGC)], and DSP-signal combining [Maximum Ratio Combining (MRC)]. These techniques are at the discretion of the device manufacturer and do not require standardization.

Figure 14.10 illustrates MIMO processing on a subcarrier basis. This applies to all OFDM spatial diversity techniques, such as spatial multiplexing, beam forming, and space time/frequency codes. Spatial multiplexing requires multiple receive antennae to decode the multiple streams of data. Space time/frequency codes require MIMO decoding to realize the transmit diversity gain. Beam

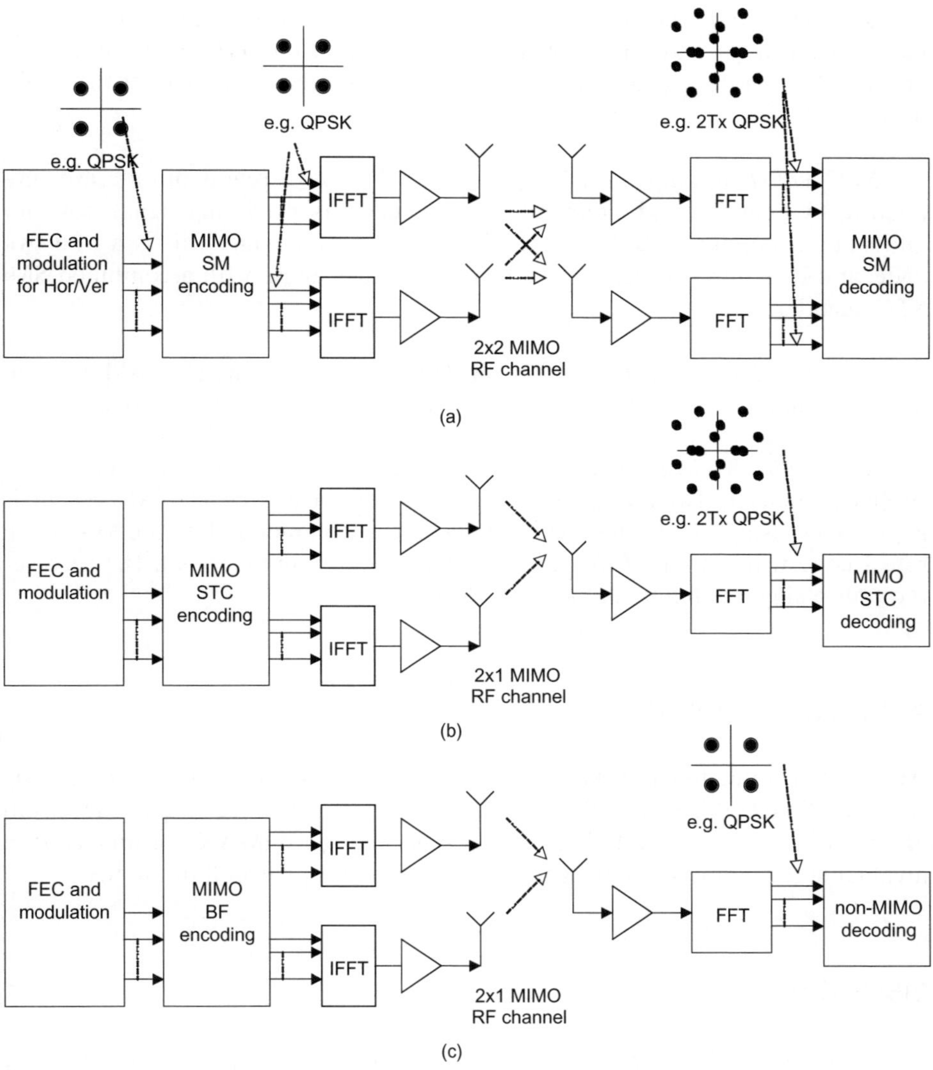

Figure 14.10. **(a)** Spatial Multiplexing, **(b)** Space Time/frequency Codes, and **(c)** Beam Forming.

forming requires little extra at the receiver, but this comes at the cost of coefficient calculations at the transmitter.

14.3.7.5 MIMO with Three or Four Transmit Antennae. The standard also defines techniques for three and four transmit antennae based on the MIMO techniques for two transmit antennae. As with two antenna techniques, separate zones can be configured for different MIMO techniques in order to dramatically improve reach and spectral efficiency.

STC with Four Antennae, Using "Matrix A". This scheme place four antennae in two groups, and for every symbol it alternates the STC between the two groups. It is a rate 1 code and requires a single-antenna receiver with a STC decoder.

STC with Four Antennae, Using "Matrix B". Two streams of data are supplied to two parallel and independent STC encoders, providing signals for four antennae. It combines 2× STC with 2× spatial multiplexing. This is a rate two code, and it requires a receiver with two antennae, along with a combined SM/STC decoder.

STC with Four Antennae, Using "Matrix C". This scheme is 4× SM, without STC. It requires a four antenna receiver. This option is used for fixed stations.

STC with Two Antennae, Using Directivity for Four Antennae. One STC supplies signals for two antennae. In addition, a duplicate of each of these signals is precoded using MIMO feedback coefficients recommended by the MS. It is a rate 1 code with a total of four signals transmitted simultaneously. The receiver requires a single antenna and an STC decoder.

ACKNOWLEDGMENTS

Aryan Saèd would like to acknowledge the detailed chapter reviews provided by Kenneth Stanwood, Darcy Poulin, and Peter Stewart. Their experience from direct participation in the IEEE 802 meetings and the WiMAX Forum has been invaluable for many of the insights and backgrounds provided in the text.

BIBLIOGRAPHY

J. G. Andrews, A. Ghosh, and R. Muhamed, *Fundamentals of WiMAX—Understanding Broadband Wireless Networking*, Prentice-Hall, Upper Saddle River, NJ, 2007.

T. Cooklev, *Wireless Communication Standards—A study of 802.11, 802.15 and 802.16*, IEEE Press, New York, 2004.

Draft Amendment to IEEE Standard for Local and Metropolitan Area Networks, Part 16: Air Interface for Fixed and Mobile Broadband Wireless Access Systems Improved Coexistence, Mechanisms for License-Exempt Operation, P802.16h/D8, 2008-11-22.

Draft Amendment to IEEE Standard for Local and Metropolitan Area Networks Part 16: Air Interface for Fixed and Mobile Broadband Wireless Access Systems, Multihop Relay Specification, P802.16j/D9, 2009-02-04.

C. Erklund, R. B. Marks, S. Ponnuswamy, K. L. Stanwood, and N. J. M. Van Waes, *WirelessMAN—Inside the IEEE 802.16 Standard for Wireless Metropolitan Networks*, IEEE Press, New York, 2006.

V. Genc, S. Murphy, Y. Yu, and J. Murphy, *IEEE802.16j relay-based wireless access networks: An overview, IEEE Wireless Communi.*, Vol. 15, No. 5, pp. 56–63, October 2008.

Intel Technol. J., Vol. 8, No. 3, pp. 173–258, August 2004.

Mobile WiMAX—Part I: A Technical Overview and Performance Evaluation, WiMAX Forum, 2006.

Mobile WiMAX—Part II: A Comparative Analysis, WiMAX Forum, 2006.

A. Molisch, *Wireless Communications*, John Wiley & Sons, Hoboken, NJ, 2005.

R. van Nee and R. Prasad, *OFDM for Wireless Multimedia Communications*, Artech House, Norwood, MA, 2000.

S.W. Peters and R.W. Heath Jr., *The future of WiMAX: Multihop relaying with IEEE802.16j, IEEE Communi. Maga*, Vol. 47, No. 1, pp. 104–111, January 2009.

Standard for Local and Metropolitan Area Networks, Part 16: Air Interface for Broadband Wireless Access Systems, 802.16-2009, 29 May 2009.

D. Sweeney, *WiMAX Operator's Manual—Building 802.16 Wireless Networks*, Apress 2006.

J. Sydor, *Messaging and Spectrum Sharing Between Ad-Hoc Cognitive Radio Networks*, ISCAS, Island of Kos, Greece, 2006.

WiMAX Forum Mobile Certification Profile Release 1.0 Approved Specification (Revision 1.1.0), 2008-12.

WiMAX Forum Mobile Certification Profile, Release 1.5 Approved Specification (Revision 1.0.0), 2008-12.

WiMAX Forum Mobile Protocol Implementation Conformance Statement (PICS) Proforma, Release 1.0 Approved Specification(Revision 1.5.0), 2008-09.

WiMAX Forum Mobile Radio Conformance Tests (MRCT) Release 1.0 Approved Specification (Revision 2.2.1), 2008-10.

15

ULTRA-WIDEBAND PERSONAL AREA NETWORKS: MIMO EXTENSIONS

Cheran Vithanage, Magnus Sandell, Justin P. Coon, and Yue Wang

The focus of this chapter is on short-range, wireless communication using so-called ultra-wideband (UWB) technology.* The purpose of the chapter is to illustrate the advantages that can be gained from using multiple transmit and (possibly) receive antennas in such systems when appropriate precoding, or beamforming, techniques are employed at the transmitter. Several precoder designs are considered, all of which are based on the optimization of some objective, such as minimizing biterror rate or maximizing the received signal-to-noise ratio or the mutual information between the transmitted and received signals. Importantly, these precoder designs adhere to the strict system and regulatory constraints that relate to UWB transmissions. In fact, these constraints cause the UWB precoder designs to be significantly different from precoders that are used in many narrowband scenarios. Despite the aforementioned restrictions, it will be shown that multi-antenna precoding is a promising practical method of achieving robust, high-rate communication in ultra-wideband networks.

*This chapter is based on "Precoding in OFDM-based multi-antenna ultra-wideband systems," by C. Vithanage, M. Sandell, J. Coon, and Y. Wang, which appeared in the IEEE Communications Magazine, January 2009. © 2009 IEEE.

Convergence of Mobile and Stationary Next-Generation Networks, edited by Krzysztof Iniewski

15.1 INTRODUCTION

State-of-the-art ultra-wideband systems transmit data using bandwidths of hundreds of megahertz in frequency bands that are shared with many existing licensed narrowband devices. Transmission of UWB signals over licensed frequency bands will cause interference to the existing licensed narrowband systems. To help mitigate this problem, various regulatory agencies have placed a low transmit power limit on UWB transmissions. For example, the Federal Communications Commission (FCC) in the United States has set an extremely low transmit power spectral density limit of –41.3 dBm/MHz [1, 2]. Despite these severe power restrictions, it has been shown through both theory and practical demonstrations that UWB systems are capable of communicating over short distances at data rates of hundreds or even thousands of megabits per second.

Although these systems have been shown to perform well, it is likely that, with the advent of high-definition video streaming in the home, UWB devices will be required to transmit *more robustly*—that is, less susceptible to channel fading in frequency—at these high data rates in the near future. One approach to achieving this goal is to employ multiple antennas at the transmitter and (optionally) the receiver, with an aim to exploit the spatial diversity in the channel. Such an approach has been shown to give good results in wireless local area networks (WLANs), such as those based on the IEEE 802.11n specification, and has recently been proposed for use in next-generation UWB systems based on the WiMedia specification [3]. In the case of the IEEE 802.11n standard, provisions have been made to allow systems to obtain channel state information at the transmitter (CSIT) [4]. This information can be exploited to direct the energy of the transmitted signal along the best spatial paths such that information is conveyed robustly and at the highest possible rate. Unfortunately, these *beamforming*, or *precoding*[1], solutions are not all directly transferable to UWB networks for one reason in particular: The equivalent isotropic radiated power (EIRP) of UWB transmissions must not exceed –41.3 dBm for each megahertz of bandwidth employed (in the United States). Consequently, conventional optimal beamforming techniques based on transmission over the principal eigenmodes of the channel cannot be employed without a power back-off since they create *spatial directivity*, thus leading to a violation of the aforementioned EIRP restrictions. Moreover, using conventional techniques *with* a power back-off leads to poor performance [5].

This result triggers several immediate questions, such as What type of precoding scheme achieves capacity in multi-antenna systems with stringent EIRP restrictions? Moreover, can such a scheme be implemented efficiently in practice? Do

[1]Traditionally, the term *beamforming* has been related to the manipulation of radiation patterns in the spatial sense, while *precoding* has been related to a baseband processing of data taking into account the channel state information. In the modern era of digital communication, precoding can be used to achieve beamforming; thus, these two terms are used interchangeably throughout this chapter.

suboptimal approaches exist, which facilitate implementation while achieving near-optimal performance? Answers to these questions are provided in this chapter.

In the next section, an overview of the concept of UWB is given, which is followed by a brief description of its different guises (i.e., impulse radio and carrier-based). The focus of the chapter, however, is on multicarrier, multi-antenna UWB systems and the optimal precoding schemes that can be employed with such technology; such systems can be thought of as extensions to the single-antenna systems specified by ECMA for UWB communication [6]. Three approaches to designing precoding schemes are described. The first approach is to design the precoder to maximize the mutual information of the transmitted and received messages. It will be shown that the practicality of using precoders that satisfy this criteria is highly dependent upon the operating received signal-to-noise ratio (SNR) and the number of antennas that are used at the transmitter. The second approach to precoder design that is discussed in this chapter is a simple antenna selection scheme whereby information is conveyed from only one antenna on any given frequency bin, noting that the *selected* antennas for two different frequency bins may not be the same. This precoding strategy actually arises as a solution to the problem of mutual information maximization in some practical scenarios. Finally, the third approach is to design the precoder to maximize the received SNR. This approach has several practical benefits compared to the first approach mentioned above.

Note that throughout this chapter, it is assumed that the amount of channel state information (CSI) that is required to implement a given precoding scheme is available at the transmitter; the practicalities of CSI feedback are not considered, although this can be achieved through a dedicated feedback channel or, when time division duplex (TDD) communication is employed, through channel reciprocity.

15.2 ULTRA-WIDEBAND COMMUNICATIONS

UWB technology is defined primarily according to the bandwidth that is utilized. In particular, the FCC in the United States specifies that a UWB signal must occupy at least 500 MHz or a fractional bandwidth of more than 20% [1, 2]. The typical maximum operating bandwidth of UWB systems is over 7 GHz. This amount of available spectral resource can theoretically facilitate enormous information transfer rates in power-constrained systems. Figure 15.1 provides a simple illustration of these gains by depicting the Shannon capacity[2] of an AWGN channel as a function of bandwidth for a fixed total transmit power [i.e., variable

[2]This chapter makes frequent references to the *capacity* of a channel. Following the pioneering work of Shannon [37], this refers to the maximum error-free data transmission rate supported by the channel. In a nutshell, the *mutual information* between *some* transmit signal and the corresponding signal received through the channel indicates the largest error-free data rate supported for that particular transmit signal. Optimization over all possible transmit signals then leads to the maximization of such mutual information, which is the channel capacity. p. 566

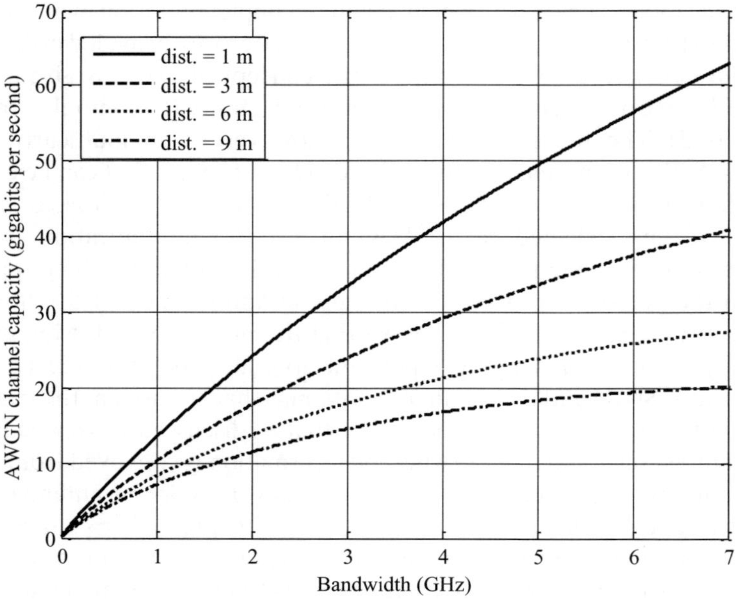

Figure 15.1. AWGN capacity versus bandwidth for various transmitter/receiver separations. A line-of-sight, free space transmission model has been adopted, and the total transmit power is fixed at 1 mW, which is divided equally across the band.

power spectral density (PSD)] and various transmitter/receiver separations. Although this illustration is not, perhaps, a practical representation of the capacity of a UWB channel, it gives a clear and simple view of the benefits of utilizing a large amount of bandwidth over short distances in power-limited systems.

It is also interesting to compare the theoretical limits of communication over a UWB channel to that which can be achieved over other channels. In Figure 15.2, the Shannon capacity of an AWGN channel is plotted as a function of the distance between the transmitter and the receiver for two bands: the UWB band ranging from 3.1 to 10.6 GHz and the ISM band ranging from 2.4 to 2.483 GHz. The standard Friis transmission model was used to generate this graph, where a distance-related path loss exponent of 3.3 was applied. This is a pessimistic assumption for the path loss exponent at short range; nevertheless, the advantage of UWB transmission at close range is clear from this illustration.

With regard to modulating data, clearly, many different approaches that meet the FCC's criteria can be employed. Most practitioners divide the admissible approaches into two categories: impulse radio (IR) and multiband systems. Both systems have advantages, disadvantages, and target applications. A brief outline of these two categories is provided below, where the focus is mainly on multicarrier techniques. Following this outline, a short discussion on the coexistence of UWB systems and other networks is given.

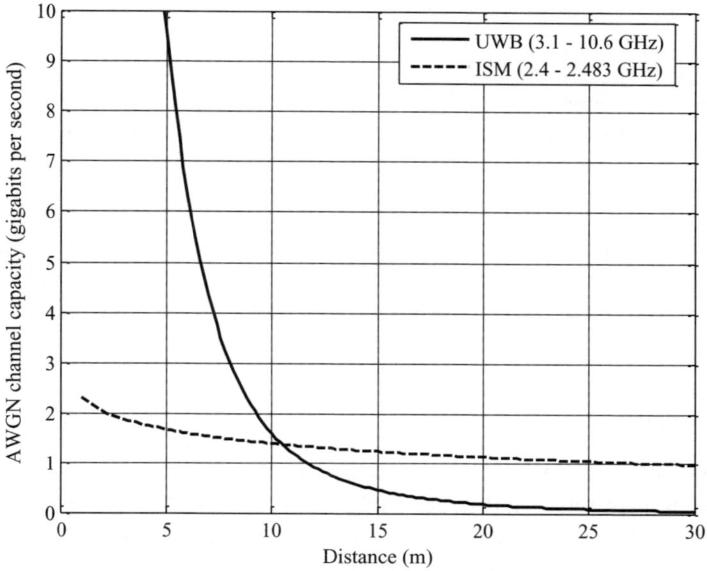

Figure 15.2. AWGN capacity versus distance for UWB and ISM bands.

15.2.1 Impulse Radio

IR for UWB communication is based on the transmission of very short, discontinuous-pulse waveforms (e.g., Gaussian or Hermitian pulses) that occupy the required bandwidth stated above. The duty cycle of pulse transmission in IR-UWB systems can be very low. One advantage of a low duty cycle is that the problem of intersymbol interference (ISI) is significantly reduced or even eliminated [7]. Many different modulation schemes can be used with IR systems, including both orthogonal and antipodal signaling. However, in order to ensure a low level of interference on other systems operating in the same band, IR schemes require some sort of randomization of the pulse train prior to transmission [7]. The primary application of IR-UWB is for robust communication and accurate positioning. In fact, the IEEE 802.15.4a standard has adopted IR-UWB for these reasons [8].

15.2.2 Multiband UWB

The main alternative to IR-UWB is multiband UWB. In such systems, the total available bandwidth is divided into smaller bands, each of which satisfies the regulatory definition of a UWB signal. Several signaling schemes have been proposed for multiband operation, including direct sequence code-division multiple access (DS-CDMA), multicarrier CDMA, and multiband orthogonal frequency-division multiplexing (MB-OFDM) [7]. The latter of these techniques is the focus of this section. In essence, MB-OFDM is akin to the classic

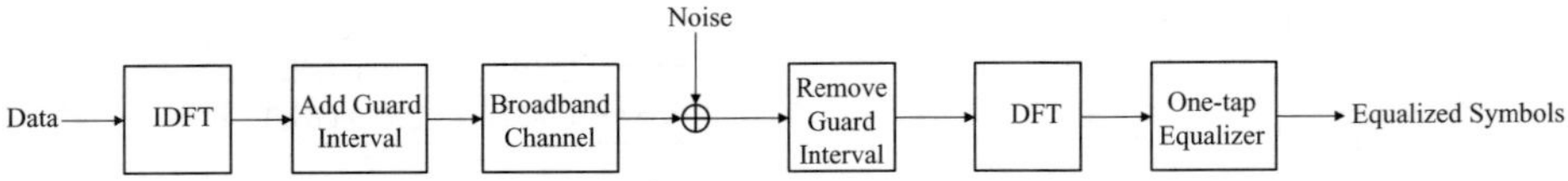

Figure 15.3. Baseband block diagram of an OFDM system.

frequency-hop version of spread spectrum, but where hopping is performed on a macroscopic level. In the next section, a brief review of OFDM is given. Following this review, a description of the MB-OFDM approach adopted by much of the UWB industry[3] is given.

15.2.3 OFDM

In its simplest form, OFDM is a so-called multicarrier transmission technique whereby data symbols—that is, constellation symbols such as BPSK or QPSK—are transmitted over separate, narrowband subchannels. These subchannels, known as subcarriers, are chosen to be orthogonal, which ensures that—assuming that the receiver is appropriately synchronized and all carrier frequency/phase offsets are accounted for—data symbols transmitted on different subcarriers do not interfere with each other.

The concept of OFDM was developed in the late 1960s [9]. In modern communications, OFDM systems are realized in the following manner. First, data symbols are arranged into blocks, each of which is processed with an inverse discrete Fourier transform (IDFT) prior to the addition of a guard interval and subsequent transmission. The guard interval—which can take several forms, such as padded zeros or inserted redundant symbols—has two primary purposes: (1) It separates adjacent transmitted blocks in such a way as to prevent interblock interference, and (2) it establishes an orthogonality condition among subcarriers, which allows the signal received on a given subcarrier to be detected independently of other subcarriers. At the receiver of an OFDM system, the guard interval for each block is removed or otherwise processed, and a DFT is applied to each resulting block. These steps effectively convert the broadband channel to multiple narrowband subchannels. Therefore, each original data symbol is simply transmitted through its corresponding subchannel and is, ideally, not affected by intersymbol interference. The received message can be equalized by filtering each received symbol by the reciprocal of the appropriate channel transfer function, which, if the system is designed correctly, can be implemented by a single scalar multiplication for each subcarrier. A baseband block diagram of an OFDM system is illustrated in Figure 15.3.

[3]MB-OFDM was, perhaps, most famously adopted by a consortium of companies known as the WiMedia Alliance, which focused on building upon the ECMA 368 standard. However, following the financial troubles that plagued the international community at the end of 2008, the WiMedia Alliance disbanded, and the work of that group was absorbed into the Bluetooth SIG [36].

15.2.4 Multiband OFDM

The MB-OFDM technique specified by ECMA 368 is similar to standard OFDM, but with the added feature of allowing different OFDM symbols to be transmitted on different frequency bands. ECMA 368 specifies an MB-OFDM technique whereby the total available bandwidth (3.1–10.6 GHz) is partitioned into six *band groups*, each of which is further divided into a number of smaller bands [6]. There are a total of 14 such bands, each with a bandwidth of 528 MHz. In particular, as described by Figure 15.4, the first four band groups contain the first 12 bands, the fifth band group contains bands 13 and 14, and the sixth band group contains bands 9–11.

In the specified OFDM transmission, the information bits are encoded with an error-correcting code, which are then spread using a time-frequency code (TFC). This spreading operation enhances the diversity of the transmission, thus making it more robust to fading and interference, both from other UWB networks and from third-party technology (e.g., WLANs). The current ECMA standard specifies three types of TFCs: three-band time-frequency interleaving (TFI), two-band TFI, and fixed-frequency interleaving (FFI), representing cases where the coded data are interleaved over three bands and two bands and simply transmitted over a single band, respectively. The application of a particular TFC in a given band groups determines the bands that are used for transmission. For example, TFC 1 for band group 1 indicates that OFDM symbols hop over the first three bands, with each hop to a new band occurring with each new OFDM symbol, following a hopping pattern illustrated in Figure 15.5.

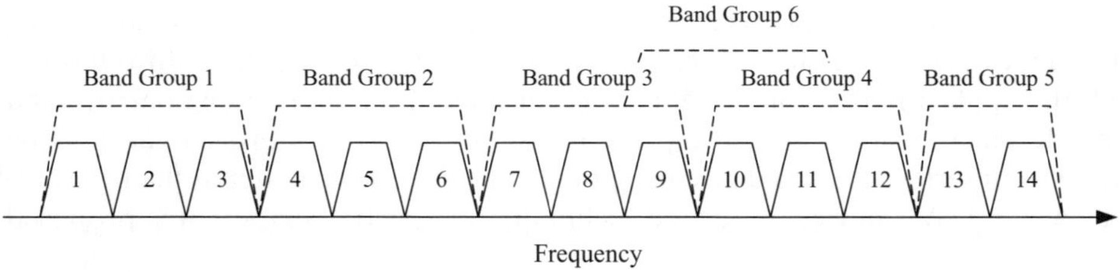

Figure 15.4. Diagram of band group allocation for ECMA 368 multiband OFDM standard.

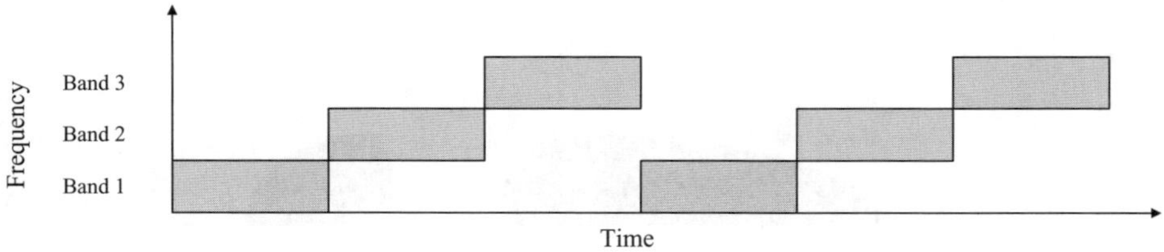

Figure 15.5. Illustration of band hopping in an MB-OFDM system based on the ECMA 368 specification using TFC 1 and band group 1.

All other aspects of the MB-OFDM system specified by ECMA 368 are similar to those found in typical OFDM-based WLAN systems, including modulation/coding schemes, packet construction, and pilot subcarrier utilization. The reader is referred to references 6 and 10 for more information on the subject of OFDM and MB-OFDM UWB.

15.2.5 Interference Detection and Avoidance in UWB

As previously mentioned, many UWB systems occupy a large portion of the 3.1- to 10.6-GHz band. Clearly, this bandwidth can also be used by other *licensed* narrowband systems. Consequently, problems with the coexistence of UWB systems and licensed third-party systems may arise in practice. In the worst case, a UWB transmission may significantly interfere with a third-party transmission. In order to minimize the impact of this interference, in addition to regulating the UWB transmission power spectrum density (PSD) to an extremely low level of –41.3 dBm/MHz, it has also been proposed that all UWB devices should come equipped with an interference detection and avoidance (DAA) subsystem. By first detecting the signal transmitted by a licensed user and then transmitting at a very low power (or not at all) in that particular frequency band, UWB devices implementing DAA would effectively transmit in an "opportunistic" manner—that is, only when the licensed bandwidth is free for use. Figure 15.6 illustrates the concept of narrowband interference avoidance. In most UWB systems, *detection* of a third-party transmission can be performed by measuring power levels in certain parts of the band. However, IR and MB-OFDM UWB systems typically implement *avoidance* techniques in a manner that best suits their respective transmission schemes.

In IR-UWB systems, interference avoidance can be achieved by modifying the underlying pulse shape according to a desired spectral profile. In reference 11, the application of a notch filter was considered for suppressing narrowband interference arising from third-party transmissions. This approach can also be used for interference avoidance by applying the same notch filter at the transmitter of the UWB device. A related technique for spectral shaping was proposed

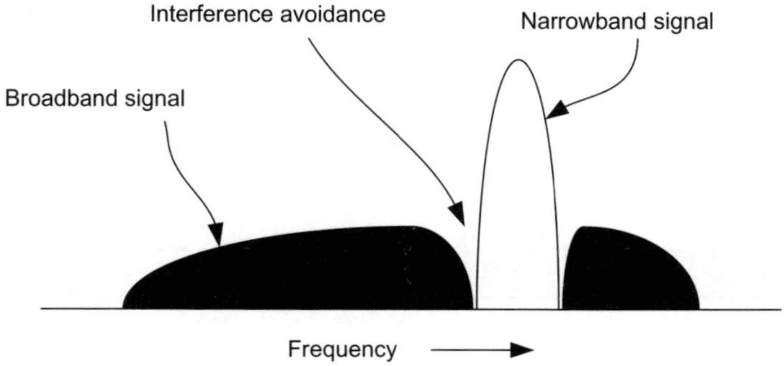

Figure 15.6. Illustration of narrowband interference avoidance.

in [12], whereby a sequence of Gaussian monocycle pulses are weighted, delayed, and summed to create a composite pulse, which can be modulated using any number of IR-UWB techniques. The weights and delays are designed to fulfil requirements on the spectral shape.

In OFDM-based UWB systems, one method of performing interference avoidance is to null one or more OFDM subcarriers to reduce the transmission power within the interference band. In general, however, this method is not capable of providing a sufficiently deep notch due to residual signal power that remains in the so-called *interference band*, which originates from data transmitted on subcarriers that are adjacent to the band. Advanced signal processing methods have been developed to reduce this residual signal power beyond that which simplistic subcarrier nulling can achieve. In one such method, known as active interference cancellation (AIC), one or more subcarriers located on either side of the interference band are loaded according to an optimization procedure that ensures the residual signal power that remains in the interference band after subcarrier nulling is minimized [13]. Although AIC has been shown to generate notches of 30 dB or more in depth, this approach can also lead to a power amplification at the edges of the interference band [13], which in turn violates the UWB EIRP limit unless a power back-off is applied. A number of approaches have been proposed to mitigate this problem. These include a regularized AIC algorithm that takes the unwanted power amplification into account when designing the signals to be transmitted on the subcarriers adjacent to the interference band [13] and, for multi-antenna systems, a joint optimization procedure that minimizes the signal power in the interference band subject to a constraint on the power transmitted on subcarriers adjacent to this band for all transmit antennas [14].

15.3 MULTIPLE ANTENNAS AND PRECODING FOR UWB SYSTEMS

The focus so far has been on OFDM-based UWB communications with single transmit and receive antennas. This section seeks extensions to the use of multiple transmit and/or receive antennas. The corresponding channels between the transmitter and receiver are popularly known as MIMO (multiple-input multiple-output) channels. Please see Figure 15.7 for an illustration.

15.3.1 Use of Multiple Transmit-Receive Antennas

During the 1990s, theoretical works of Foschini and Gans [15] and Telatar [16] established that the use of multiple transmit and receive antennas can greatly enhance the channel capacities in frequency flat fading channels. One of their key results is that the high SNR capacities of such MIMO channels are a factor $\min(M_{tx}, M_{rx})$ times that of single antenna channels, where M_{tx} is the number of transmit antennas and M_{rx} is the number of receive antennas. In practice, the increased dimensions due to the multiple antennas are used to either increase

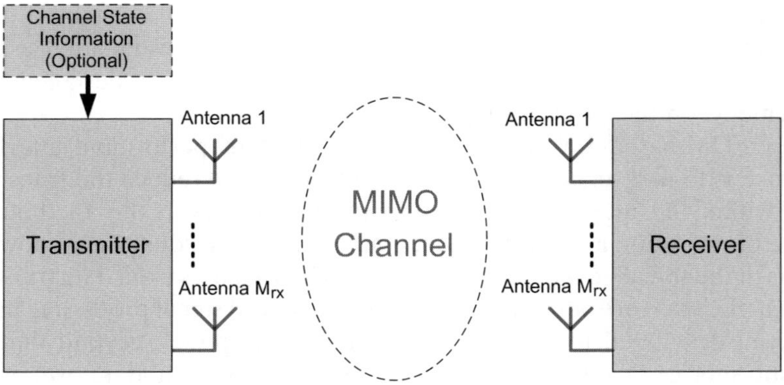

Figure 15.7. MIMO communications. Transmissions can be with or without knowledge of the intermediate channels. Channel knowledge enables better conditioning of the transmit signals.

the throughput (i.e., data rate) or reduce error rates at reception. Such attempts at utilizing the spatial dimension, in addition to the time domain, are also known as space-time coding. An interesting aspect is that even without the transmitter knowing the forward channels to the receiver; space–time coding enables significant performance improvements compared to single antenna systems. Also note that the MIMO channel induces a mixing of the signals transmitted from different antennas, complicating the symbol detection at the receiver, in general. However, space-time coding with proper design can lead to simple receiver architectures that are optimal for the detection of the transmitted signals. Alamouti's space–time block code [17] is such an example of a space–time code with an associated simple detection process.

These results naturally extend to frequency-selective channels where OFDM systems dominate. Since the OFDM communication system breaks the frequency selective channel into a set of non interfering flat fading channels, it is possible to consider such systems with N subcarriers and multiple transmit–receive antennas as a set of N MIMO channels that do not interfere with each other [18].

15.3.2 Precoding at the Transmitter

When the transmitters also have access to channel state information (i.e., some knowledge about the channels to the receiver); as might be expected, it is possible to transmit signals that are more suited to these channels and achieve performance better than that given by simple space–time coding. Such *precoded* transmissions have been widely investigated in the literature. Conventionally, transmit precoding is investigated along with a total transmit power constraint. This is rightly so, because in many systems such as WLANs and cellular systems, total transmit power is the practical constraint. Optimal precoding strategies for total power restricted systems have been derived, for example, in reference 16.

Optimality obviously depends on the criterion that needs to be improved. When achieving capacity is the criterion, it is shown that optimal precoding for flat fading MIMO channels reduces to transmitting on the right singular vectors of the matrix composed of fading coefficients of channels between each transmit and receive antennas. Such transmissions, when coupled with receiver processing that utilizes the left singular vectors of the above-mentioned matrix, essentially transforms the MIMO channel into a set of parallel noninterfering channels. These channels are also called *eigenmodes*, since the received powers of these parallel channels are related to the eigenvalues of the MIMO channel matrix. The total transmit power needs to be properly allocated among these eigenmodes to realize the capacity of the MIMO channels. Thus for total transmit power restricted systems, capacity achieving transmit–receive processing are known in simple and closed form. As will be discussed in the next section, the transmit constraints are different when UWB transmissions are concerned and capacity-achieving transmission schemes are known only for some special cases. Finally, note that as the MIMO channel is decomposed into a set of noninterfering channels by signal transmission on the eigenmodes, the associated symbol detection at the receiver also takes a simple form.

15.3.3 Capacity Optimal Transmitter Precoding for UWB

UWB devices in conception are defined by their ability to coexist with other licensed communication systems, although certain regulatory domains further impose the requirement of completely avoiding the spectral bands where a licensed device is active, as mentioned previously. As a consequence, UWB devices are subject to stringent peak restrictions on their emitted spatial radiation (e.g., FCC part 15.503 [1]). In other words, these systems are equivalent isotropic radiated power constrained. This is to ensure that an unfortunate relative spatial location of a licensed device would not result in a noticeable interference from some UWB device. In fact, the transmit EIRP of UWB devices are restricted on each megahertz of their transmission, which is also necessary due to the wide bandwidths considered.

Due to such EIRP restrictions applicable throughout the bandwidth, OFDM-based UWB systems conforming to the ECMA368 specification [6] that specifies subcarrier bandwidths of 4.125 MHz are subject to EIRP restrictions over each of the subcarriers. Thus for a transmitter equipped with CSI, precoding is ideally applied on each of the subcarriers of the OFDM system, *independently*, as illustrated pictorially in Figure 15.8. From here onwards, most of our focus is on the signal processing applied on some generic subcarrier.

The problem considered is transmit precoding such that the regulatory EIRP restrictions are not violated. Let us see how the EIRP can be accounted for in the design of transmit precoding methods. For single transmit antenna systems, EIRP restrictions can be accounted for by scaling the transmit power by the peak gain of the antenna radiation pattern. When a transmit antenna array is employed, it produces a further beamforming effect, which needs to be accounted for. When

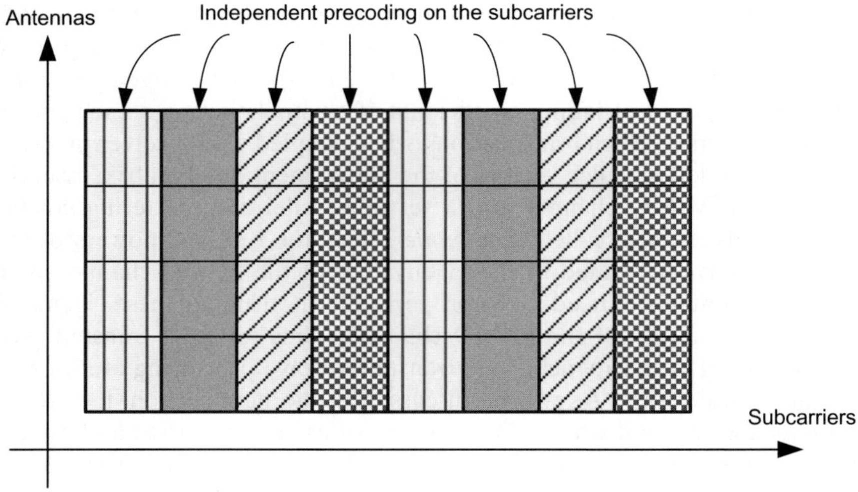

Figure 15.8. Pictorial Illustration of a MIMO OFDM system with eight subcarriers and four transmit antennas. EIRP restrictions on each subcarrier promotes transmit precoding on a per-subcarrier basis.

the component transmit antennas are identical, this beamforming effect separates from the individual element patterns and is quantified by the so-called "array gain" of the antenna array [19]. For such transmitters, a two-step approach can be taken to design transmission schemes that satisfy the required EIRP restrictions. First, as in the case of single-transmit antennas, the spatial directivity of the elemental radiation is taken into account. Second, the spatial directivity induced by the use of an array of antennas is considered. A graphical illustration of examples of such spatially directive transmissions are given later on in Figure 15.15. In this chapter we will restrict attention to the control of this second component, the array gain, in utilizing CSI at the transmitter. Essentially, the overall radiation pattern is considered, assuming that the individual component antennas produce isotropic radiation.

An immediate question that arises is whether it is optimal to have the array gain itself be isotropic when one is interested in maximizing some objective function, such as the mutual information between the transmitted and received signals (i.e., achieve capacity) or the received SNR. It was shown in reference 20 that isotropic radiation is indeed optimal in terms of maximizing the mutual information when the transmitter consists of two antennas spaced apart by at least half a wavelength. However, this result is not true for general transmitter configurations.

In general, constraining a transmission to radiate *isotropically* subject to some EIRP constraint is obviously much more restrictive than simply constraining its *EIRP*. In other words, the set of transmission schemes that radiate isotropically is a subset of the EIRP-constrained transmissions, as illustrated pictorially

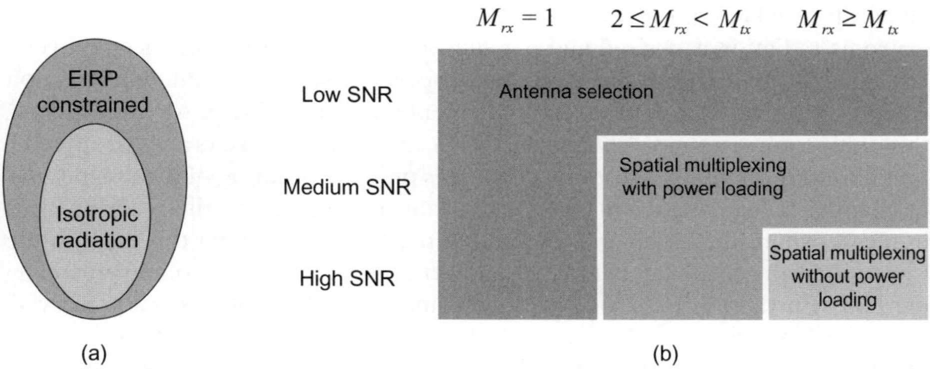

Figure 15.9. **(a)** Pictorial representation of the sets of solutions that have isotropic radiation constraints and EIRP constraints. **(b)** diagram summarizing optimal isotropic transmission schemes for various SNR regimes and numbers of antennas.

in Figure 15.9a. Nevertheless, it is convenient to ensure that the array gain is isotropic for practical reasons. Isotropism results when the signals transmitted from the multiple antennas are uncorrelated, which is easily achieved by simple transmission schemes such as spatial multiplexing (i.e., the transmission of independent data streams on the transmit antennas) with arbitrary power allocations on the antennas or other well–known space–time coding schemes such as Alamouti's space–time block code [17]. For the case of mutual information optimization subject to an isotropic array gain, the spatial multiplexing scheme with power loading applied across the antennas, *not the eigenmodes*, is optimal [20]. This is an interesting contrast to the case of more conventional precoding approaches designed for total transmit power restricted systems. The power loading represents the freedom for exploiting the available CSIT. For some particular system configurations, power loading across the antennas leads to simpler transmission schemes as summarised by the following points.

- For systems with a single receive antenna, transmit antenna selection is optimal for all SNR.
- For arbitrary transmit and receive antenna numbers, transmit antenna selection based on the column norms of the channel matrix is optimal at low SNR.
- When the number of receive antennas is greater than or equal to the number of transmit antennas, standard power balanced spatial multiplexing is optimal at high SNR.

These results are summarized in Figure 15.9. Again, the points outlined above refer to precoding schemes applied on a per-subcarrier basis, and it should be highlighted that an isotropic array gain is not optimal for all EIRP-constrained systems as discussed above and in references 5, 20, and 21. This point will be elaborated on later; however, the next section focuses on the practical case where isotropic radiation is desired and per-subcarrier antenna selection is employed.

Apart from its optimality described above, per-subcarrier antenna selection is also attractive from a practical implementation perspective. At transmission, this only requires a simple switching (ON or OFF) of subcarriers at each antenna. The information required for this purpose is only the ordering of the magnitudes of channel fading coefficients from each transmit antenna to the receiver. Thus in a case where the receiver is estimating the CSI and feeding this back to the transmitter, the feedback overhead is lower compared to most other precoding methods. Furthermore, as long as the channel magnitude orders are preserved, an optimal antenna-to-subcarrier selection can be made. Perfect knowledge of the fading coefficients is not necessary. Thus, this scheme is also tolerant to errors in CSIT. In practice, CSIT is never perfectly obtained due to effects such as channel estimation errors and the inherent fluctuations of the channels with time, which causes their estimates to be outdated. Another attraction of per-subcarrier antenna selection is that the symbol detection at the receiver is (or can be made to be—c.f. Section 15.4) the same as that of a single transmit antenna system. Receiver complexity increases with most other multiple antenna transmission schemes. However, there are some practical issues to consider in applying per-subcarrier transmit antenna selection. These are discussed next.

15.4 TRANSMIT ANTENNA SELECTION

It is apparent from the above that antenna selection, although a simple transmission scheme, is a powerful method of exploiting CSIT for UWB systems. In this section, several practical aspects related to transmit antenna selection are investigated. A comparison is made between per-subcarrier and bulk selection and we explore the issue of transmit antenna selection in a manner that is compatible with legacy receivers. We also consider the issue of power amplifier efficiency optimization as well as combining peak-to-average power ratio (PAPR) reduction with antenna selection.

15.4.1 Per-Subcarrier Versus Bulk Selection

Antenna selection in OFDM systems can broadly be divided into two types: per-subcarrier and bulk selection. With per-subcarrier selection, also called per-tone selection, we choose to transmit from the best antenna on each subcarrier individually, whereas in bulk selection we choose the best antenna to transmit all subcarriers on. The advantage of per-subcarrier selection is that greater selection gains can be obtained since the frequency selectivity of the channel can be exploited. The channel on two subcarriers spaced sufficiently apart (greater than the coherence bandwidth) will be essentially independent and hence it is likely that the best antenna on those subcarriers are different. On the other hand, bulk selection means that only one antenna can be used for the whole band and for some subcarriers this means that transmission must be made on a suboptimal choice of antenna. This exploitation of frequency selectivity obviously depends on the channel characteristics. For the extreme case of a flat fading channel (no time dispersion), the two methods would have the same antenna selection. However, for larger time dispersion of the channel (more frequency selectivity), per-subcarrier antenna selection can be expected to outperform bulk selection. A performance comparison between per-subcarrier and bulk selection was done in reference 22, where it was shown that although they have the same diversity order, per-subcarrier selection has a superior coding gain.

A practical aspect of antenna selection in OFDM systems is the number of radio-frequency (RF) chains required. If per-subcarrier antenna selection is used, transmission is done on all antennas and hence the same number of RF chains is required. On the other hand, with bulk selection, only one RF chain is required since only one antenna is used for transmission. This can make it an attractive solution from a practical aspect, although there are also disadvantages. Some switching losses are associated with antenna selection of this type and there is also the issue of channel estimation. Since only one antenna is accessible at a time, some protocol must be set up to regularly switch antennas to estimate changes; this may incur delays, and outdated channel estimates could be used which degrade performance. An investigation of the implementation aspects of antenna selection in OFDM systems is done in reference 23. In the next section we will address one problem with per-subcarrier antenna selection, which is the power imbalance between the antennas that might occur if most subcarriers are chosen to be on one particular antenna.

15.4.2 Antenna Selection with Power Amplifier Efficiency Optimization

If antenna selection is done independently on each subcarrier, one antenna may have many more active subcarriers than other antennas. This is illustrated in Figure 15.10a, where the top antenna has six subcarriers allocated to it while the bottom one only has two. If the system has EIRP constraints, a maximum transmit power limit is applied to each subcarrier, which excludes power loading; hence the transmit power per antenna is proportional to the number of allocated subcarriers to transmit on. This may cause problems with the PAs because their working range must be extended to cope with cases of very large per-antenna transmit powers. An obvious solution is to reduce the power on *some* antennas to balance the distribution of power across *all* antennas. However, this would result in performance degradation, which is obviously not desirable.

Another solution is to require the per-subcarrier antenna selection to have the same number of active subcarriers per antenna, which would result in a power balance across antennas. This is shown in Figure 15.10b. In this case, performance can be optimized with respect to some cost function subject to equal power on all antennas. If $x_{n,m}$ denotes the selection variable—that is, $x_{n,m} = 1$ if antenna m is selected on subcarrier n, otherwise it is 0—then the *constrained antenna*

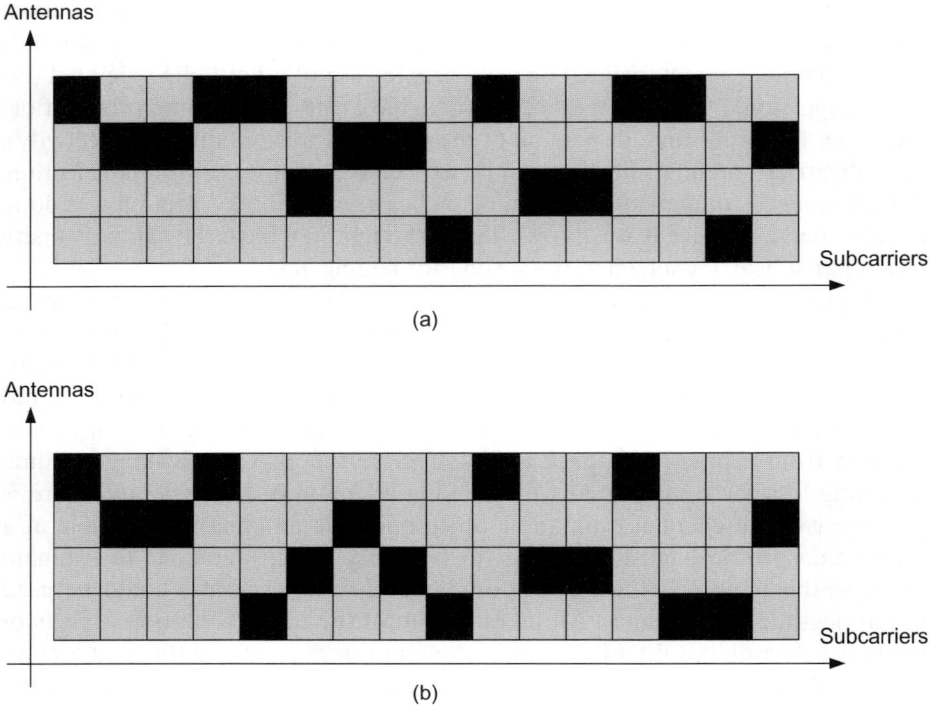

Figure 15.10. Examples of per-subcarrier antenna selection, where a black box indicates that transmission will take place: **(a)** Unbalanced selection, where the antennas *do not* have the same number of subcarriers allocated to them. **(b)** Balanced selection, where the antennas have the same number of subcarriers allocated to them.

selection problem can be formulated as an integer programming problem [24] in an OFDM system with N subcarriers and M_{tx} transmit antennas (it is assumed that M_{tx} divides N, although the same idea can be applied in the general case):

$$\min_{x\in\{0,1\}^{NM_{tx}}} \sum_{n,m} x_{n,m} P_{n,m}$$

$$s.t. \sum_{m} x_{n,m} = 1$$

$$\sum_{n} x_{n,m} = \frac{N}{M_{tx}}$$

The cost $P_{n,m}$ can be the BER resulting from transmitting on antenna m on subcarrier n, although other costs such as received SNR or mutual information can also be considered (which would then be maximized instead of minimized). The first constraint requires that only one antenna is transmitting on each subcarrier, while the second constraint ensures that all antennas have the same number of active subcarriers and hence the same transmit power.

This type of integer programming problem can be solved by techniques such as branch and bound or cutting planes; however, in practice the number of subcarriers is quite large, which means that the computational complexity would be high. To overcome this problem, it can be shown that the constrained antenna selection problem can be linearly relaxed without any loss of performance [25]; that is, the integer constraints on $x_{n,m}$ can be dropped and we may consider it as a continuous variable between zero and one. This means that the problem can be solved by much simpler linear programming algorithms, such as the simplex and interior point methods [26].

It is worth noting a few properties of the constrained antenna selection. Firstly, the BER is a better objective function than the SNR, which is proved in reference 25. The constraints force a trade-off between a small total cost and power balance, but the cost penalties are different when measured in BER or SNR. A fixed SNR difference does not mean a fixed BER difference; the relative differences are dependent upon the operating SNR region. This is an interesting observation because the choice of cost function is irrelevant in the case of unconstrained antenna selection; the selection is done independently on each subcarrier and the antenna that optimizes SNR also optimizes BER and mutual information.

15.4.3 Antenna Selection with Phase Precoding for Legacy Compatibility

As discussed above, antenna selection applied on a per-subcarrier basis exploits the frequency selectivity of the channel to obtain diversity gains. This is illustrated in Figure 15.11a, where the magnitude of the frequency-domain channels are shown before and after antenna selection. Note that the channel after antenna selection will assume the value of one of the individual channels, hence the lines are overlapping. In TDD systems, channel knowledge at the transmitter can be obtained by the reverse channels due to the reciprocity of the wireless medium. Thus it is possible to envisage proprietary multiple antenna devices employing transmit antenna selection, which can seamlessly interact with legacy single antenna devices that do not provide the CSIT to the other end. Note that channel reciprocity is not applicable per se, since communication devices employ different radio paths for certain stages in transmission and reception. This can be addressed by calibrating transmit–receive sections of the multiple antenna device, again without any support from the legacy device at the other end.

In OFDM, an initial channel estimate is often obtained from a training sequence, typically in the form of a preamble (a whole OFDM symbol containing only known symbols). In this case, the receiver can be oblivious to the precoding done by the transmitter because it only estimates the compound channel (precoding plus actual propagation channel); this facilitates proprietary precoding which is compliant with the standard or is for system upgrades that must still support legacy devices. However, some care must be taken not to cause performance degradation when the receiver is unaware of the precoding.

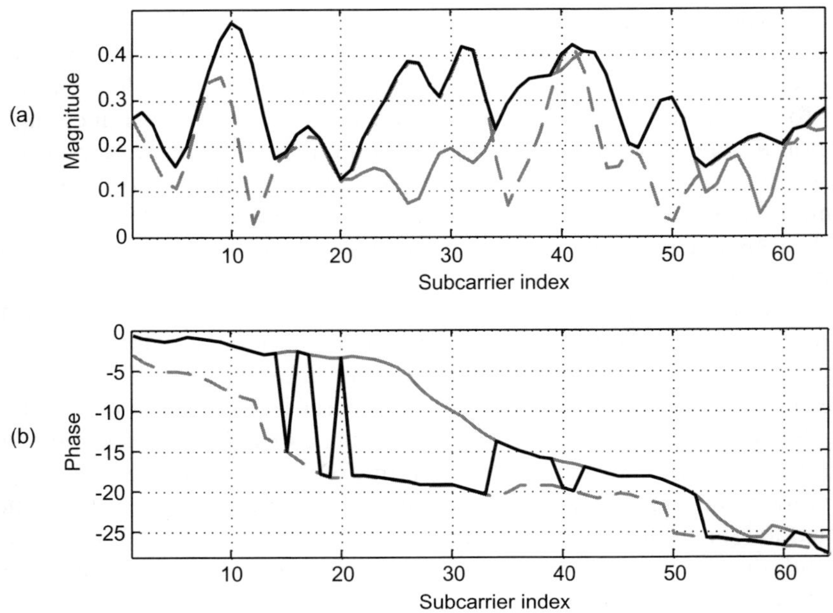

Figure 15.11. Per-subcarrier antenna selection illustrating the channels before and after antenna selection: **(a)** Frequency-domain magnitude. **(b)** Frequency-domain phase. The two individual channel responses are shown in gray, while the channel response after antenna selection is shown in black.

One side effect of the per-subcarrier antenna selection is that while the magnitude of the frequency response remains reasonably smooth, the phase experiences sudden changes (see Figure 15.11b). This is natural because the antennas are selected according to their magnitude; the phase is not considered at all. This lack of smoothness might cause a problem at the receiver if some advanced channel estimation is used which exploits the frequency correlation of the channel (see, e.g., reference 27). Conventionally, the received signal on each subcarrier is divided by the known symbol in the preamble to obtain the least-squares (LS) estimate. This initial estimate can then be enhanced by exploiting the nature of the channel as explained in the following.

The OFDM system is designed to have a guard interval larger than the time dispersion of the channel. This can be exploited by transforming the LS estimate of the channel to the time domain with an inverse discrete Fourier transform. In the noiseless case, this impulse response should now be inside the length of the guard interval. Hence we can apply, for example, a windowing filter that removes everything outside this window and only keeps the signal inside. Then the time-domain estimate can be transformed back to the frequency domain with a discrete Fourier transform where the final estimate is obtained. Details about this type of time-domain (or more generally, subspace) channel estimation can be found in reference 27. Alternatively, the filtering can be done directly in the

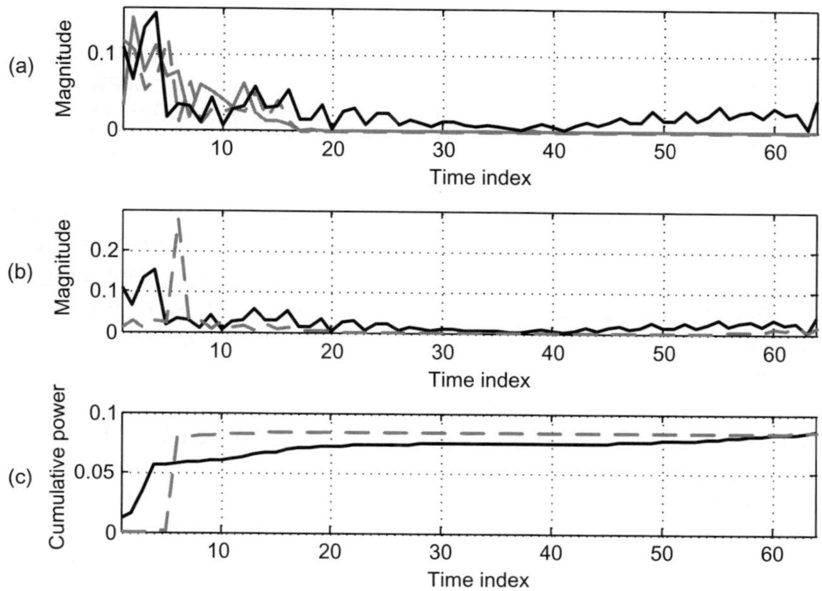

Figure 15.12. Examples of per-subcarrier antenna selection with and without phase precoding: **(a)** Impulse responses of the individual channels (gray) and after antenna selection (black) without phase precoding. **(b)** Impulse responses after antenna selection with (gray) and without (black) phase precoding. **(c)** Cumulative power of the impulse response with (gray) and without (black) phase precoding.

frequency domain. Since the impulse response of the channel is shorter than the length of the guard interval, which in turn is much shorter than the length of the OFDM symbol, the subcarriers will be significantly correlated. This large frequency correlation can be used to remove some of the noise by applying, for example, a Wiener filter that optimally reduces the noise. Examples of this type of filtering are given in reference 28.

However, with per-subcarrier antenna selection there will be a problem with applying these advanced channel estimation techniques. Because the phase of the channel after antenna selection will experience sudden changes, frequency-domain filtering may actually degrade the performance. This can also be viewed in the time domain: In Figure 15.12a, the magnitude of the impulse response of the channel after antenna selection is shown. It is clear that the time dispersion has been significantly increased due to the antenna selection and the impulse response is no longer contained within a small window. Hence, if windowing is applied, a large part of the channel will be removed and although the noise will be reduced, the mean-squared error (MSE) of the channel estimate will be large.

This problem can be combated with *phase precoding* [29]. The antenna to transmit on is chosen based on the magnitudes; the phase is not considered and can be viewed as a parameter that can be used to "smoothen" the channel response. Since the channel will be estimated at the receiver using a preamble, we can modify the phase as much as we want; it will not degrade the performance of antenna selection. Hence we select the phase on the transmitting subcarriers such that the compound channel is as smooth as possible. This means combining the phase of the actual channel with the phase of the precoding; together they should form a desirable value. One possible approach is to consider the magnitude of the compound channel as given (by the antenna selection) and then choose the phase to produce a minimum-phase signal [30]. This guarantees that the energy of the channel is maximally concentrated near the origin and hence as much energy as possible is within the guard interval window. Another approach is to use phase linearization, which aims at making the phase of the compound channel linear; the slope can be chosen by using ideas from channel shortening in equalization [31] to optimize performance. In Figure 15.12b, this technique has been used for phase precoding, where it produces a channel impulse response with reduced delay spread. This can be seen even more clearly in Figure 15.12c, where the cumulative power of the impulse response after antenna selection is shown. It is clear that phase precoding can significantly concentrate the power to a short interval, which results in better channel estimation. More crucially, phase precoding allows the benefits of per-subcarrier antenna selection to be obtained when interacting with legacy receivers.

The benefits of antenna selection with phase precoding can be illustrated with an example. In Figure 15.13 the packet error rate (PER) is shown as a function of SNR for an antenna selection-based UWB system with four transmit and one receive antenna and advanced channel estimation at the receiver (for full simulation details, see reference 29). As a reference, the performance of a system with only one transmit antenna is also shown. If no phase precoding is applied, the advanced channel estimation will actually perform worse at medium-to-high SNR since the channel estimation is very poor due to the discontinuities of the frequency response of the compound channel. However, if phase precoding is applied, the performance is significantly improved and a gain of about 6 dB is observed. Linear phase is slightly better than the minimum phase precoding, indicating that the performance may be further improved by optimizing the method of phase precoding. Furthermore, it is shown in reference 29 that this proposed scheme is also robust to calibration errors in transmit–receive chains of the multiple antenna device. Thus, use of per-subcarrier antenna selection with phase precoding is suitable for link enhancement in interacting with existing TDD systems [6], which currently do not provide an explicit mechanism to provide the CSIT to the other end.

As a final point, it should be noted that the constrained antenna selection described earlier may be combined with the phase precoding if advanced channel estimation is desirable at the receiver.

15.4.4 PAPR Reduction in Antenna Selection Systems

All OFDM transmissions suffer from a high PAPR, a problem that can lead to power amplifier (PA) inefficiencies and distortion of the transmitted signal. This

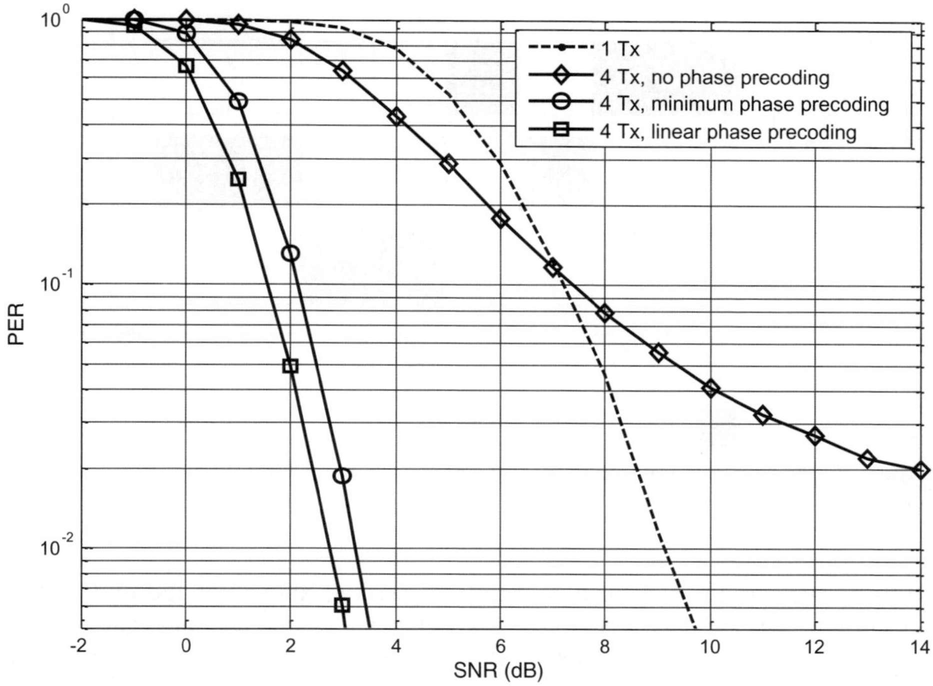

Figure 15.13. Performance with and without phase precoding in antenna selection.

problem results from the linear combination of modulated data signals prior to transmission, which is performed by the IDFT. A significant amount of research has been devoted to reducing the PAPR of OFDM signals in the past decade. The most straightforward method consists of applying a back-off to the PA output, which is clearly undesirable from an efficiency point of view. Other techniques for mitigating the PAPR problem in OFDM transmissions include active constellation extension (ACE), tone injection (TI), and tone reservation (TR), to name a few [32]. In particular, TR is a promising technique that is based on reserving a subset of subcarriers from data transmission, instead opting to excite these subcarriers, which are termed peak reduction carriers, in such a way as to reduce the PAPR of the transmitted signal. The peak reduction signals can be designed in a number of ways, such as by executing an exhaustive search over a finite set of possible signals to find the signal that minimizes the PAPR or by implementing a gradient descent algorithm to find the (nearly) optimal signal.

Peak reduction signals do not carry information, and thus the implementation of conventional TR algorithms leads to a reduction in throughput. However, for multi-antenna systems employing per-subcarrier antenna selection, an opportunistic TR strategy has been developed that makes use of "bad" subcarriers (as seen by a particular antenna) to transmit peak reduction signals [33]. In this scheme, antenna selection is first performed according to some optimization

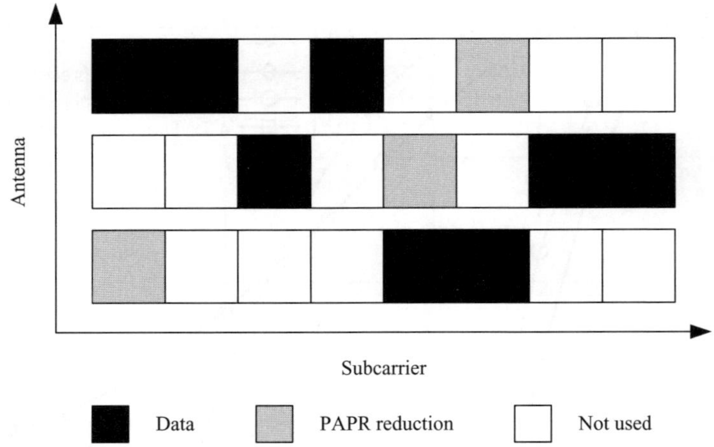

Figure 15.14. Illustration of opportunistic PAPR reduction in OFDM systems employing per-subcarrier antenna selection.

strategy, such as the balanced technique described above. At the end of the selection process, a single antenna will have been allocated for data transmission on each subcarrier. In the opportunistic TR scheme, peak reduction signals are designed and transmitted from a subset of the remaining "inactive" antennas. Note that since data are still transmitted on each subcarrier, this TR strategy is rate lossless. Figure 15.14 illustrates the allocation of peak reduction carriers to inactive antennas across an OFDM symbol with eight subcarriers.

It is evident from the description above, as well as from Figure 15.14, that the peak reduction signals will interfere with the data signals at the receiver, which will lead to degradation in performance. Crucially, the peak reduction signals are transmitted on "bad" channels—that is, channels with a large attenuation, a condition that results from the initial antenna selection process. When these channels exhibit severe fading characteristics, the peak reduction signals are heavily attenuated and thus cause minimal interference to the received data signal. Of course, the severity of the interference depends on a number of factors, including the fading statistics of the channel, the number of peak reduction signals, and the power that is allocated to each of these signals. Interestingly, it has been shown that the performance degradation due to interference can be made arbitrarily small by limiting the peak reduction signal power, even if a large number of subcarriers are used for peak reduction [33]. Moreover, the reduction in PAPR that can be achieved with this approach is on the order of several decibels.

15.5 TRANSMIT BEAMFORMING IN UWB SYSTEMS

In this section, we consider the conditioning of transmit signals to improve the received SNR in UWB communication systems.

15.5.1 Conventional Beamforming Approaches

Conventionally, the term "beamforming" is associated with the adjustment of transmit signals in order to effect the transmitted spatial radiation such that some measure related to the received SNR is optimized. Such spatial beamforming of transmit signals is a widely investigated topic and has found application, for example, in signal transmission from cellular base stations to mobile devices [34]. The idea is to first identify the spatial directions of individual mobile users and then properly direct the transmissions from the base stations such that the received signals of the users are optimized.

Alternatively, it is also possible to *beamform* purely in the domain of baseband signal processing. Here, the channels are viewed based on the fading coefficients as seen in the baseband, and the transmit signals are suitably optimized. Identification of the corresponding optimal beamforming vectors simplifies for systems with a total transmit power constraint. Essentially, it is optimal to transmit on the channel eigenmode corresponding to the largest eigenvalue of the MIMO channel matrix. Such transmission schemes are known as *eigenbeamforming*. While such eigenbeamformed transmissions do not explicitly take into account the spatial direction of the receiver, they do result in spatially directed transmissions that are matched to the baseband channel.

15.5.2 Transmit Beamforming with EIRP Constraints

Let us now focus on the case where transmit beamforming is used to improve the received SNR in UWB systems. Again, the difference to conventional beamforming is the existence of EIRP restrictions rather than total transmit power constraints. Similar to the case of capacity optimal transmissions, it turns out that received SNR optimal transmission schemes for UWB systems *do not* lead to isotropic radiation in general. Had it been so, transmit antenna selection would have been the EIRP optimal beamforming method. Interestingly, even eigenbeamforming is not optimal for these EIRP-constrained systems. Due to the high directivity of transmissions from eigenbeamforming, the transmit power must be backed off (i.e., scaled) to satisfy the EIRP restrictions, resulting in a substantial performance degradation.

It can be shown that the optimal beamforming scheme for systems with EIRP restrictions can be formulated as the solution to a convex optimization problem [21]. However, numerical solutions necessary for this optimal method suffer from high computational complexity since an optimization over a complicated region in a multidimensional space is required. Conventional suboptimal methods, such as transmit antenna selection and use of eigenbeamforming scaled to satisfy the EIRP restrictions, require lower computational complexity at the expense of a degradation in performance [5, 21].

Figure 15.15 illustrates the radiation patterns due to various transmit beamforming methods for a particular channel realization. Transmit antenna selection

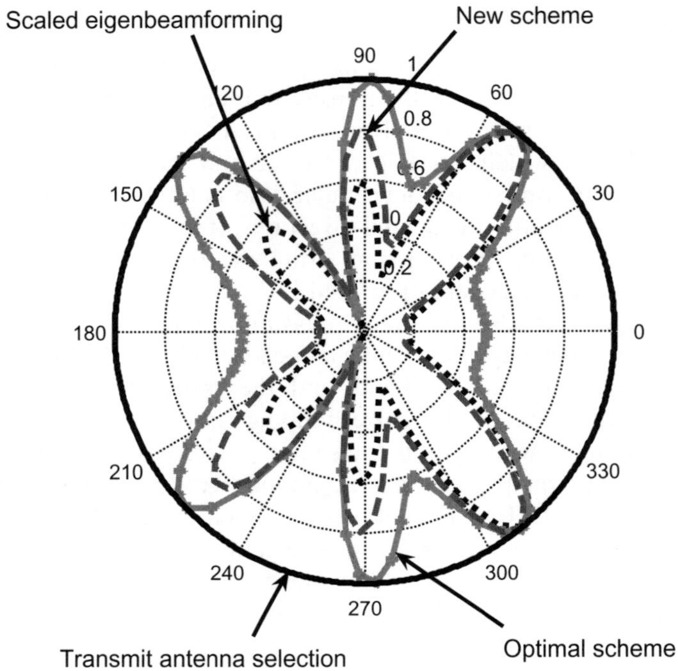

Figure 15.15. Radiation patterns due to some transmit beamforming methods.

leads to an isotropic array effect. Generally, the following properties of the radiation due to the optimal scheme can be observed:

- It resembles the radiation due to scaled eigenbeamforming.
- Compared to scaled eigenbeamforming, it has a lower spatial PAPR, while the directions of peak radiation are approximately preserved.
- For particular channel realizations, it reduces to transmit antenna selection.

Essentially, it appears that the scaled eigenbeamforming solution does not fully exploit the fact that it is the EIRP that is restricted rather than the total transmit power. The optimal scheme attempts to rectify this. These observations form the basis of a transmit beamforming methodology for EIRP-constrained systems, as explained with more detail in reference 35. Following the observations given above, it is possible to develop a beamforming scheme with the objective of designing beamforming vectors by perturbing the scaled eigenbeamforming vector such that (a) the PAPR of its spatial radiation is reduced and (b) its direction of peak radiation is approximately preserved.

When the transmitter consists of a linear array of antennas, samples of the spatial radiation are given by the IDFT of the beamforming vector. The problem of reducing the spatial PAPR of the radiation therefore becomes similar to PAPR reduction of the time-domain OFDM signals (which was discussed earlier), where many algorithms have been proposed [32]. However, these algorithms do not necessarily preserve the direction of the peak radiation in their attempts at PAPR reduction, and therefore they will yield inferior performance when directly applied to beamforming vector design. In reference 35, this issue was addressed by introducing phase adjustments, which ensure that the IDFT operations sample the direction of peak radiation, thereby facilitating its preservation. For the task of spatial PAPR reduction itself, many of the algorithms found in the OFDM literature, such as the iterative soft-clipping and filtering method, can be adopted.

The PER performance of this new scheme, when implemented with the iterative soft-clipping and filtering method for PAPR reduction, is illustrated in Figure 15.16, where a system with four transmit antennas and one receive antenna is considered. The simulated channels correspond to the IEEE 802.15.3a channel model CM3. For further details of the simulation setup, please refer to reference 35. It can be observed that at a PER of 10^{-3}, antenna selection is more than 1.5 dB suboptimal, while the performance of the proposed beamforming method is

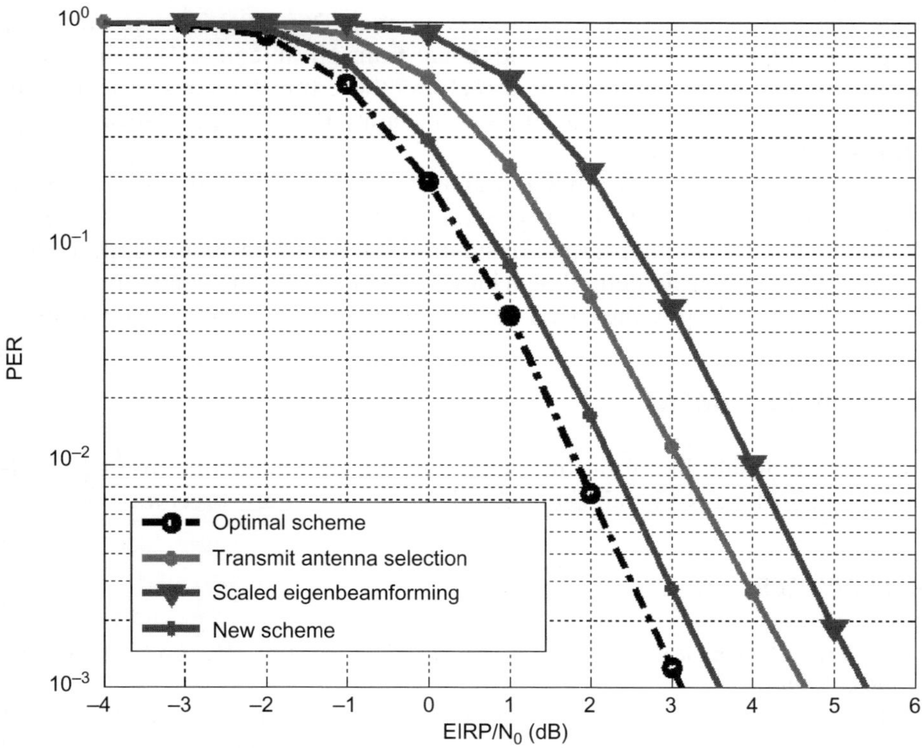

Figure 15.16. PER performance of some EIRP-constrained beamforming schemes.

only 0.5 dB inferior compared to the optimal scheme. The performance using the scaled eigenbeamforming vector, however, is about 2.5 dB suboptimal, due to the power back-off required to satisfy the EIRP restrictions.

15.6 SUMMARY

Exploitation of CSIT in multiple transmit antenna equipped UWB systems was considered in this chapter. An investigation of capacity-optimal precoding reveals that with only two transmit antennas, it is optimal to employ a spatial multiplexing scheme with a power allocation across the antennas, on each of the subcarriers. The power allocations represent the freedom to exploit the CSIT. For the cases of low SNR or when there is only a single receive antenna, this power allocation reduces to antenna selection, performed on each of the subcarriers.

Per-subcarrier antenna selection is also attractive as a general precoding method for EIRP-constrained systems due to its practicality. It was shown that antenna selection coupled with phase precoding can be used to design legacy ECMA368 compatible transmission schemes that can give significant performance improvements over conventional systems. Also, antenna selection can be performed to improve link performance while ensuring that the transmit power is evenly distributed across all transmit antennas. This optimizes the efficiency of power amplifiers used on the individual transmit antennas. The absence of signals in some of the subcarriers also allows these transmissions to send additional signals purely for the purpose of alleviating PAPR issues associated with OFDM transmit signals.

Finally, noting that per-subcarrier antenna selection is not the received SNR optimal beamforming scheme in general, existences of better transmit beamforming schemes were illustrated. In summary, one can conclude that precoding is a promising practical method of achieving robust, high-rate communication in UWB networks, despite the strict regulatory limitations on the transmissions.

ACKNOWLEDGMENT

The authors would like to acknowledge the fruitful discussions with their colleagues at Toshiba Research Europe and the support of its directors.

REFERENCES

1. Federal Communications Commission, Title 47, Section 15, *Code of Federal Regulations*.
2. Federal Communications Commission, First report and order, revision of part 15 of the commission's rules regarding ultra-wideband transmission systems, *ET Docket 98-153*, February 2002.

3. C. Razzell, J. Yang, and D. Birru, Approaches and considerations for evolution of OFDM-based UWB PHY solutions beyond 1 Gbps, in *IEEE International Conference on Ultra-Wideband (ICUWB)*, Singapore, 2007.
4. T. K. Paul and T. Ogunfunmi, Wireless LAN comes of age: Understanding the IEEE 802.11n amendment, *IEEE Circuits Systems Mag.*, Vol. 8, No. 1, pp. 28–54, 2008.
5. A. M. Kuzminsky, Downlink beamforming subject to the equivalent isotropic radiated power constraint in WLAN OFDM systems, *Signal Processing*, Vol. 87, No. 5, pp. 991–1002, May 2007.
6. ECMA, ECMA368: High Rate Ultra Wideband PHY and MAC Standard, December 3, 2008.
7. I. Oppermann, M. Hämäläinen, and J. Iinatti (editors), *UWB: Theory and Applications*, John Wiley & Sons, Chichester, 2004.
8. J. Zhang et al., UWB systems for wireless sensor networks, *Proc. IEEE*, Vol. 97, No. 2, pp. 313–331, February 2009.
9. R. W. Chang, Synthesis of band-limited orthogonal signals for multi-channel data transmission, *Bell System Technical J.*, No. 46, pp. 1775–1796, 1966.
10. R. van Nee and R. Prasad, *OFDM for Wireless Multimedia Communications*, Artech House, Norwood, MA, 2000.
11. J. Wang and W. T. Tung, Narrowband interference suppression in time-hopping impulse radio ultra-wideband communications, *IEEE Trans. Commun.*, Vol. 54, No. 6, pp. 1057–1067, June 2006.
12. Y. Wang, X. Dong, and I. J. Fair, Spectrum shaping and NBI suppression in UWB communications, *IEEE Trans. Wireless Commun.*, Vol. 6, No. 5, pp. 1944–1952, May 2007.
13. J. Balakrishna and H. Yamaguchi, Ultra wideband interference cancellation for orthogonal frequency division multiplex transmitters by protection-edge tones, 2006. US Patent US 2006/0008016 A1.
14. Y. Wang and J. Coon, Active interference cancellation for systems with antenna selection, in *IEEE International Conference on Communications*, Beijing, 2008, pp. 3785–3789.
15. G. J. Foschini and M. Gans, On limits of wireless communications in a fading environment when using multiple antennas, *Wireless Personal Commun.*, Vol. 6, No. 3, pp. 311–355, March 1998.
16. E. Telatar, Capacity of multi-antenna Gaussian channels, *Eur. Trans. Telecommun.*, Vol. 10, No. 6, pp. 585–596, 1999.
17. S. M. Alamouti, A simple transmit diversity technique for wireless communications, *IEEE J. Select Areas Commun.*, Vol. 16, No. 8, pp. 1451–1458, October 1998.
18. G. L. Stuber et al., Broadband MIMO-OFDM wireless communications, *Proc. IEEE*, Vol. 92, No. 2, pp. 271–294, February 2004.
19. C. A. Balanis, *Antenna Theory*, John Wiley & Sons, Toronto, 1997.
20. C. M. Vithanage, J. P. Coon, and S. C. J. Parker, On capacity-optimal precoding for multiple antenna systems subject to EIRP restrictions, *IEEE Trans. Wireless Commun.*, Vol. 7, No. 12, part 2, pp. 5182–5187, December 2008.
21. P. Zetterberg et al., Performance of multiple-receive multiple-transmit beamforming in WLAN-type systems under power or EIRP constraints with delayed channel estimates, in *IEEE Vehicular Technology Conference (VTC Spring)*, Vol. 4, Birmingham, 2002, pp. 1906–1910.

22. H. Zhang and R. Nabar, Transmit antenna selection in MIMO-OFDM systems: Bulk versus per-tone selection, in *International Conference on Communications (ICC)*, IEEE, Beijing, 2008, pp. 4371–4375.
23. A. Molisch et al., Implementation aspects of antenna selection for MIMO systems, in *International Conference on Communications and Networking in China (ChinaCom)*, ICST, Beijing, 2006, pp. 1–7.
24. L. Wolsey, *Integer Programming*, John Wiley & Sons, New York, 1998.
25. M. Sandell and J. Coon, Per-subcarrier antenna selection with power constraints in OFDM systems, *IEEE Trans. Wireless Commun.*, Vol. 8, No. 2, pp. 673–677, February 2009.
26. S. Boyd and L. Vandenbergh, *Convex Optimization*, Cambridge University Press, Cambridge, 2004.
27. O. Edfors et al., OFDM channel estimation by singular value decomposition, *IEEE Trans. Commun.*, Vol. 46, No. 7, pp. 931–939, July 1998.
28. T. Onizawa et al., A simple adaptive channel estimation scheme for OFDM systems, in *Vehicular Technology Conference (VTC Fall)*, Vol. 1, IEEE, Amsterdam, 1999, pp. 279–283.
29. C. M. Vithanage, S. C. J. Parker, and M. Sandell, Antenna selection with phase precoding for high performance UWB communication with legacy WiMedia multiband OFDM devices, *International Conference on Communications (ICC)*, IEEE, Beijing, 2008, pp. 3938–3942.
30. A. W. Oppenheim and R. W. Schafer, *Discrete-Time Signal Processing*, Prentice-Hall, Englewood Cliffs, NJ, 1989.
31. N. Al-Dhahir, FIR channel shortening equalizers for MIMO ISI channels, *IEEE Trans. Commun.*, Vol. 49, No. 2, pp. 213–218, February 2001.
32. S. H. Han and J. H. Lee, An overview of peak-to-average power ratio reduction techniques for multicarrier transmission, *IEEE Wireless Commun.*, Vol. 12, No. 2, pp. 56–65, April 2005.
33. J. P. Coon, PAPR reduction in OFDM systems with per-subcarrier antenna selection, in *IEEE Wireless Communication Networking Conference*, 2009.
34. L. C. Godara, Application of antenna arrays to mobile communications, Part II: Beam forming and direction-of-arrival considerations, *Proc. IEEE*, Vol. 85, No. 8, pp. 1195–1245, August 1997.
35. C. M. Vithanage, Y. Wang, and J. P. Coon, Spatial PAPR reduction based beamforming scheme for EIRP constrained systems, in *Global Telecommunications Conference (Globecom)*, IEEE, New Orleans, 2008, pp. 1–5.
36. WiMedia Alliance, *Multiband OFDM Physical Layer Specification*, WiMedia Alliance, 2005.
37. C. E. Shannon, A mathematical theory of communication, *Bell System Tech. J.*, Vol. 27, pp. 379–423, 623–656, July, October 1948.

PART IV

METROPOLITAN, CORE, AND STORAGE AREA NETWORKS

16

NEXT-GENERATION INTEGRATED METROPOLITAN-ACCESS NETWORK: TECHNOLOGY INTEGRATION AND WIRELESS CONVERGENCE

Shing-Wa Wong, Divanilson R. Campelo, and Leonid G. Kazovsky

Since the past decade, broadband access networks have undergone significant changes. Fiber optics is reaching to homes, and wireless data access networks are becoming ubiquitous. As emerging multimedia applications continue to demand for larger bandwidth, the evolution of broadband access networks is expected to remain. As a result, next-generation metropolitan area networks must be able to support the architectural changes and growing traffic demands of emerging broadband access infrastructures. The integration between metropolitan and access networks is becoming a promising solution for future networks to accommodate these challenges. This chapter exploits future integrated metropolitan-access network architectures that can seamlessly support high-bandwidth, pervasive applications to come.

In this chapter, we first review important recent developments in metropolitan and access networks. With this foundation, the integration of metropolitan and optical access networks is discussed in some detail, with focus on the trade-offs between integrated and nonintegrated metro-access solutions. We then examine the convergence of optical and wireless access networks as a solution to overcome the bandwidth and coverage limitations of current broadband access networks. A number of examples are included in the chapter to demonstrate how

Convergence of Mobile and Stationary Next-Generation Networks, edited by Krzysztof Iniewski

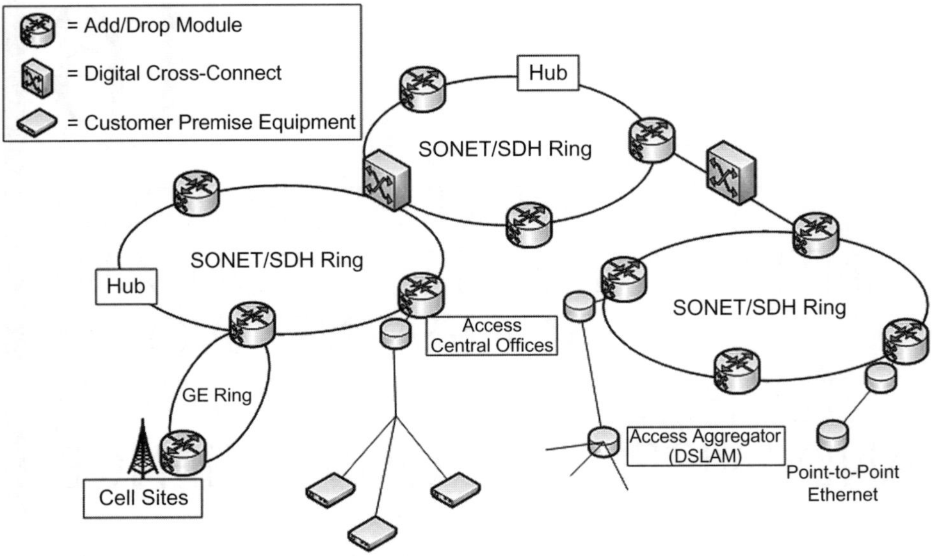

Figure 16.1. Metropolitan and access network (MAN) architecture.

next-generation networks can meet the emerging traffic challenges and gracefully evolve from existing networks. A future direction of integrated metro-access networks concludes the chapter.

16.1 RECENT DEVELOPMENTS IN METROPOLITAN AND ACCESS NETWORKS

16.1.1 Metropolitan Area Networks

16.1.1.1 SONET/SDH: Metropolitan Ring Network and Flexible Adaptation of Data Traffic. Initially, metropolitan area networks (MANs) have predominantly relied on interconnected rings (as shown in Figure 16.1) based on the Synchronous Optical Network (SONET) hierarchy, which was standardized in the United States in the late 1980s. SONET is closely related to Synchronous Digital Hierarchy (SDH), the standard proposed almost simultaneously in Europe. SONET/SDH networks are typically deployed in rings, because this topology allows survivability of connections upon the occurrence of a single failure, such as a fiber cut, at the expense of some efficiency loss. SONET/SDH ring nodes are composed of add/drop multiplexers (ADMs), whose primary function is to electronically aggregate traffic into the fiber and drop tributary lower-speed traffic destined for the node. SONET/SDH networks still represent a significant share of the current service providers' infrastructure [1].

The SONET/SDH standards emerged mainly as a transport mechanism for carrying a large number of plesiochronous digital hierarchy (PDH) payloads, which are comprised of several lower-speed digitized-voice signals. Because the voice traffic represented the major service SONET/SDH networks initially were designed to support, the SONET/SDH standards were built on a time-division multiplexing (TDM) circuit-switched hierarchy with a synchronous frame structure for the transport of digital-voice traffic. The key attribute of circuit-switched networks is the existence of a fixed, guaranteed bandwidth circuit in all links between the source and destination nodes before they may communicate.

Recently, SONET/SDH networks incorporated mechanisms for mapping generic data transport protocols into the synchronous frames. This was motivated by the fact that whereas the growth of the voice traffic was smooth since the introduction of the first SONET/SDH metro optical networks, the Internet traffic was increasing very rapidly, close to doubling every year since 1997 [2]. Although SONET/SDH networks can be combined with the wavelength division multiplexing (WDM) technology to scale up their transmission capacity due to the aggregation of access traffic, the variety of services and types of traffic presented in today's metro networks require a more efficient usage of the bandwidth than that provided by TDM circuits. This is because the capacity of the payload was rigidly defined to accommodate PDH streams. The limited number of protocols that could be mapped into SONET/SDH frames did not reflect the emergence of new technologies for transporting data in the metro segment, especially 100 Mbits Ethernet and Gigabit Ethernet. In order to define an efficient and interoperable mapping of generic protocols in SONET/SDH networks, the International Telecommunication Union (ITU) defined the Generic Framing Procedure (GFP) [3] in 2001. To provide SONET/SDH systems with high efficiency in the accommodation of a variety of protocols, GFP must be combined with two other technologies [4]:

(a) *Virtual Concatenation (VCAT).* VCAT defines a mechanism for carrying payloads with flexible bandwidth, which is a desirable feature for transporting data service; the payload is comprised of a virtual concatenation of several smaller payloads that are separately carried from the source to destination terminals and eventually combined to reconstruct the original payload at the destination.

(b) *Link Capacity Adjustment Scheme (LCAS).* LCAS dynamically uses VCAT to enable efficient transport of generic formats by allowing the number of concatenated payloads to be changed dynamically. The SONET/SDH networks that incorporate data traffic adaptation features have been referred to as next-generation SONET/SDH systems.

16.1.1.2 Resilient Packet Ring. Resilient Packet Ring (RPR) is a ring-topology network architecture designed to provide packet-aware service provisioning in optical fiber ring networks such as SONET/SDH [5]. RPR aims at combining SONET/SDH's resilience and reliability functionalities with Ethernet's simplicity, low-cost, and efficient bandwidth utilization. The choice of a ring topology allows for fast protection switching (within 50 ms) upon the occurrence of a single link or node failure, whereas the definition of a medium access control (MAC) protocol with a client interface similar to Ethernet's improves the bandwidth utilization of the optical fiber ring. RPR works on a dual-ring configuration in which each ring is called a ringlet. The ringlets are set up by the creation of RPR stations at the network nodes. RPR enables spatial reuse within the ring, since the receiving RPR station completely removes its received packet from the ring rather than replicating its content and sending it back to the sender [5]. This feature, often referred to as destination stripping, allows the released bandwidth to be used for carrying additional traffic if needed. Moreover, in contrast to traditional SONET/SDH rings in which the backup ring is only utilized when the primary ring fails, RPR makes full use of the bandwidth of both counterdirectional fiber rings under no-fail operation.

The RPR architecture provides class of service support and can bridge to Ethernet with fairness. RPR gives priority to transit traffic over the traffic a station is ready to add, and guarantees that no in-transit traffic is lost. To maintain this lossless property, RPR relies on a fairness control that reduces the amount of admissible transmit traffic during high transit flows. The objective of the fairness control is to distribute a fair traffic insertion rate among congested RPR stations. However, it is known that the RPR fairness control may not always reach a steady state under many realistic traffic scenarios. This problem significantly underutilizes the useable part of RPR bandwidth during fairness rate oscillation [6]. Current RPR standard assumes single-channel networks and does not provide yet a clear guideline for a WDM upgrade. Emerging integrated metro-access architectures should preserve the beneficial properties of RPR, including protection, spatial reuse, QoS support, fairness, and introduce more stable bandwidth utilization and scalability features through a multi channel WDM layer.

16.1.1.3 G.709 Optical Transport Network. The Optical Transport Network (OTN) is a recently proposed standard to meet metro-optical network convergence of wide-ranging services in a common platform [7]. OTN consists of a multiservice transport architecture that transparently supports packet-based data transport as well as SONET/SDH circuits over a dense wavelength division multiplexing (DWDM) layer. The main functionality of OTN is its "digital wrapper" technology, which provides a mechanism for encapsulating a number of existing frames into an entity that can be successfully transported with a small amount of overhead and forward error correction (FEC) bytes. Essentially, client traffic of any protocol can be wrapped into a frame that carries information about both the client and the optical wavelength it uses as transport medium. Moreover, this protocol-agnostic, digital wrapper technology provides an upper layer that allows end-to-end monitoring of connections, even if they traverse several networks from different service providers with different SLAs. In summary, the digital wrapping mechanism provides intelligence and OAM capabilities to optical wavelengths, leading to a scalable platform capable of carrying a number of protocols with almost all benefits of the SONET/SDH performance management. Initial clients of OTN were predominantly SONET/SDH signals, but data clients like Ethernet and IP are increasingly being transported over this converging platform.

The key elements of the DWDM layer are the fixed optical add/drop multiplexers (OADMs), whose function is to statically add and drop traffic in the wavelength granularity and bypass wavelengths in the optical domain [7, 8]. Most of the OADMs are built with fixed optical filters, which are inserted within the optical path along the ring to add and/or drop pre-determined wavelengths. OADM nodes based on fixed optical filters may be considered a very cost-effective mechanism to access the optical spectrum within a WDM signal, as these nodes are easily designed and deployed. However, the main disadvantage of a solution based on fixed optical filters is the requirement of significant information about the expected traffic growth in the network. Eventual traffic reallocations require the insertion of additional optical filters within the signal path, a procedure that cannot be performed without traffic interruption and the presence of trained personnel in the node site. Moreover, in situations where an unforeseen traffic demand surpasses the traffic expectations in some network location, the lack of flexibility of fixed OADMs prevents the reallocation of unused optical spectrum to support this demand. The novel integrated metropolitan-access architectures presented in this chapter address these problems by making use of a reconfigurable optical layer, which is discussed in the next subsection.

16.1.1.4 ROADM and Reconfigurability. The reconfigurable OADM (ROADM), whose advent resulted from recent developments of optical technologies [9], is an alternative solution to overcome the absence of flexibility of fixed OADMs. In essence, ROADMs are network elements that enable an automated optical layer by allowing dynamic wavelength add/drop functionality within a WDM signal. ROADMs are capable of offering fast wavelength service provisioning and are the key elements for the emerging dynamic WDM networks in metropolitan environments [8, 10]. Such dynamic WDM MANs must be characterized by fast optical circuit provisioning (namely, fast optical circuit switching, OCS), in contrast to sporadic connection requests that can be successfully accomplished by SONET/SDH, OTN, or point-to-point optical Ethernet transmission systems. As a result, the rapid establishment of wavelength services will be severely dependent on the ROADM switching times, which are on the order of milliseconds with current optical component technology. Moreover, driven by the emergence of new optical switching techniques for future metropolitan environments such as optical burst switching (OBS), ROADM switching times will have to be as low as a few microseconds or even nanoseconds to successfully accomplish the switching of an optical burst or packet [11].

In this chapter, new integrated metro-access architectures based on alternative solutions to ROADMs are presented. Such alternative solutions employ fast tunable transceivers to enable reconfigurability. Tunable transceivers place reconfigurability at the source and/or destination terminals, allowing transmission from the source to destination without intermediate switching. Moreover, this type of architecture allows smooth WDM upgrade without requiring changes at intermediate nodes. Current tunable transceivers can typically switch under a microsecond, and some recent demonstrations have reported switching times within the nanosecond range [12, 13]. As a result, networks that employ such fast tunable transceivers can usually achieve high flexibility.

16.1.2 Broadband Access Networks

16.1.2.1 Passive Optical Network. A passive optical network (PON) is a low cost fiber-optic network designed to provide broadband fiber access to end users. In general, PON systems employ a point-to-multi point architecture that consists of an optical line terminal (OLT), a remote node (RN), and several optical network units (ONUs). A PON employs a passive RN in its optical distribution network (ODN) to lower capital and operational expenditure (CAPEX/OPEX) requirements. Figure 16.2 illustrates the generic PON architecture. The terminology FTT*x* refers to where the optical fiber termination is located. Some common examples are FTTH (fiber-to-the-home), FTTN (fiber-to-the-neighborhood), and FTTB (fiber-to-the-business).

Currently deployed PONs are based on three dominating standards, all of them TDM-based: Broadband PON (BPON), Gigabit-capable PON (GPON), and Ethernet PON (EPON). A TDM-PON typically supports between 16 and 64 users, and its physical-layer (PHY) rates can support up to 2.488-Gbps downstream and 1.244-Gbits upstream traffic. The key characteristics and differences between these three current-generation optical access networks are summarized in Table 16.1. Overall, each standard differs primarily in their

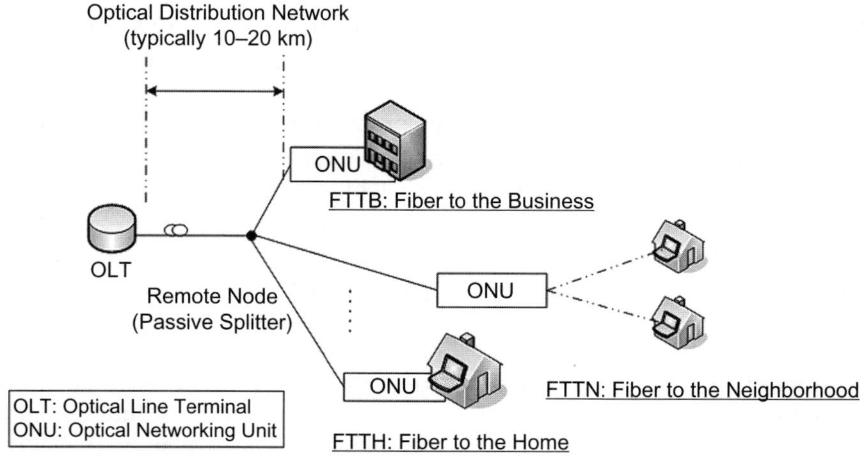

Figure 16.2. Passive optical network (PON) architecture.

TABLE 16.1. TDM-PON Comparison

	EPON	BPON	GPON
Standard	IEEE 802.3 ah	ITU G.983	ITU G.984
Framing	Ethernet	ATM	GEM/ATM
Bandwidth (down/up)	1.25/1.25 Gbit/s	622/155 Mbit/s	2.488/1.244 Gbit/s
Typical split ratio[a]	16	32	64
Typical span[a]	10 km	20 km	20 km
Estimated costs	Lowest	Low	Medium

[a]Split-ratio and span combination depends on the supported optical budget.

TABLE 16.2. Next-Generation PON (NG-PON) Comparison

	10GE-PON	WDM G-PON	LR G-PON
Enhancement type	Line Rate	WDM	Physical reach
Standard work group	IEEE 802.3 av	ITU G.984.5	ITU G.984.6
Downstream bandwidth	10 Gbit/s	2.488 or 10 Gbit/s per λ	2.488 or 10 Gbit/s per λ
Upstream bandwidth	1.25 or 10 Gbit/s	1.244 Gbit/s per λ	1.244 Gbit/s per λ
Optical budget	20/24/29 dB	29 dB	27.5 dB both span[b]
Coexistence support	ONU retrofit	No change[a]	New RN
Downstream	WDM separation	WDM separation/ blocking	Open
Upstream	10G/1G dual rate	TDM	Open
Estimated cost	Low–medium	Medium	High

[a]ITU G.984.5 recommends addition of wavelength blocking filter at ONU during initial deployment.

[b]An active mid-span extender would allow the optical line to be extended to a total of 60 km.

protocol adaptations and offered bitrates. In essence, they share similar operating principles. Without loss of generality, this chapter focuses on EPON technology.

Anticipating the growth of demand for bandwidth, both the ITU-T and IEEE standardization bodies are studying options to expand the capacity of current TDM-PONs. Table 16.2 summarizes three standardized upgrading approaches: line rate enhancement, WDM enhancement, and physical reach extension. These standard activities are also summarized in Section 16.4. The key challenge in upgrading a TDM-PON is the ability to make incremental upgrades over an existing ODN—that is by using a pay-as-you-grow strategy. Such an evolution strategy reduces the occurrence of service disruption and delays capital expenses to only when they become necessary. Section 16.2 presents a detailed discussion about integrated metro-access networks as an evolutionary framework to enable smooth and flexible expansion towards next-generation access networks.

16.1.2.2 Wireless Broadband Access Network. This chapter focuses on wireless broadband Internet access technologies. In particular, wireless broadband networks such as Worldwide Interoperability for Microwave Access (WiMAX) and Wireless Fidelity (WiFi) access systems are considered.

In general, WiMAX is a wireless broadband access technology based on a subset of profiles recommended by the IEEE 802.16 standards. WiMAX systems can operate in several modes, and this chapter considers the 802.16 time division duplex (TDD) mode without multiple hop extension. WiFi is a system based on IEEE 802.11-standards and is mostly used in wireless local area networks (WLANs). For the purpose of this chapter, we restrict our analysis to the IEEE 802.11s-based wireless mesh networks (WMNs) due to their suitability for large-scale deployment.

16.1.2.3 Convergence of Optical and Wireless Access Network. As wireless technology has become more popular with ever-increasing quality attributes, telecom operators are interested in supporting services with QoS requirements in their core mobile segment. Smooth cooperation between the wireline and wireless networks is necessary to facilitate the support of wireline-like services over wireless networks. For this reason, the Fixed-Mobile Convergence (FMC) Alliance [14] was formed in 2004 to address the convergence between wireline and wireless service network. FMC was initially established to study the convergence between PSTN (Public Switched Telephone Network) and PLMN (Public Land Mobile Network) networks. Recently, FMC has been also expanding and considering the integration with broadband wireline networks such as FTT*x*. However, it focuses on the application layer and incorporates a session initiation protocol (SIP) to provide high-capacity and seamless connection across fixed and mobile networks. Section 16.4 summarizes the FMC's efforts to integrate fiber and wireless network using an application layer convergence.

In Section 16.3, a new trend called optical-wireless integration (OWI) is presented. OWI is expected to provide FMC service by exploring complementary characteristics between optical and wireless networks. Whereas optical networks are robust and offer high bandwidth, wireless networks support mobility and ubiquitous coverage. In particular, new designs of metropolitan area networks are presented to provide a backbone for large-scale broadband wireless access networks. Moreover, readers are introduced to a new integrated control framework to enhance existing wireless network performance.

16.2 METROPOLITAN AREA AND BROADBAND ACCESS NETWORK INTEGRATION

Future broadband access networks are expected to support ever-increasing demands from end users, and metropolitan area networks must employ architectures that can gracefully scale up with the growing volume of access traffic. In this context, efficient traffic aggregation and transparent optical transmission are desirable features for future MANs, since they alleviate the scalability problem in backbone networks. For this purpose, two promising architectures for future optical MANs are presented in this section. The first architecture is optical burst transport (OBT), which provides an effective mechanism to groom bursty traffic at transmit nodes and employs optical bypassing to efficiently manage aggregated traffic without opto-electro-optical (OEO) conversion at intermediate nodes. OBT is a ring-based adaptation of optical burst switching (OBS), but it has its own characteristics, as presented in Section 16.2.1. The second architecture is metropolitan-access ring integrated network (MARIN). MARIN is an integrated architecture that can seamlessly combine the metropolitan network with the access network. Unlike in OBT, in MARIN only the source and destination nodes are involved in the transmission, and there is no switching in intermediate nodes. This is an important feature for access integration because intermediate nodes in optical access networks are desired to remain passive.

16.2.1 Optical Burst Transport Technology

The basis of OBT is to accommodate bursty data traffic using burst mode transmission and fast optical switches. Like in OBS, the transmission in OBT is initiated after a control header is sent to the destination node to set up a lightpath that remains active until the transmission ceases. Control signals in the OBT network use a dedicated control channel, and they are processed at every node. Once the transmission is initiated, there is no switching at intermediate nodes, and data bursts see a single optical hop between the source and destination nodes within a given data channel. Each data channel is specified by a wavelength. Figure 16.3a illustrates the architecture of the OBT network proposed by [15].

16.2.1.1 OBT Protocol. OBT requires an appropriate medium access control (MAC) in order to avoid collision among the various data paths that can be established from source to destination nodes. To this end, OBT uses tokens to allocate transmission bandwidth among the nodes. Another approach for evenly sharing the medium could be a timeslotted WDM ring that uses an optical supervisory channel (OSC) to control the use of time slots. Such a timeslotted WDM ring has been shown to outperform the token WDM ring when the network data rate increases [16]. However, OBT does not employ an OSC because its use requires strict synchronization between the control and data channels. Since synchronization is a challenge in a WDM network due to group velocity dispersion (GVD), each data channel would have a time offset with respect to the OSC. Using a token control, OBT sacrifices some bandwidth utilization because it does not allow immediate channel access, but it avoids the GVD problem. Token access further enables flexibility to adapt variable size Ethernet packet and provides fairness using limitation on holding times.

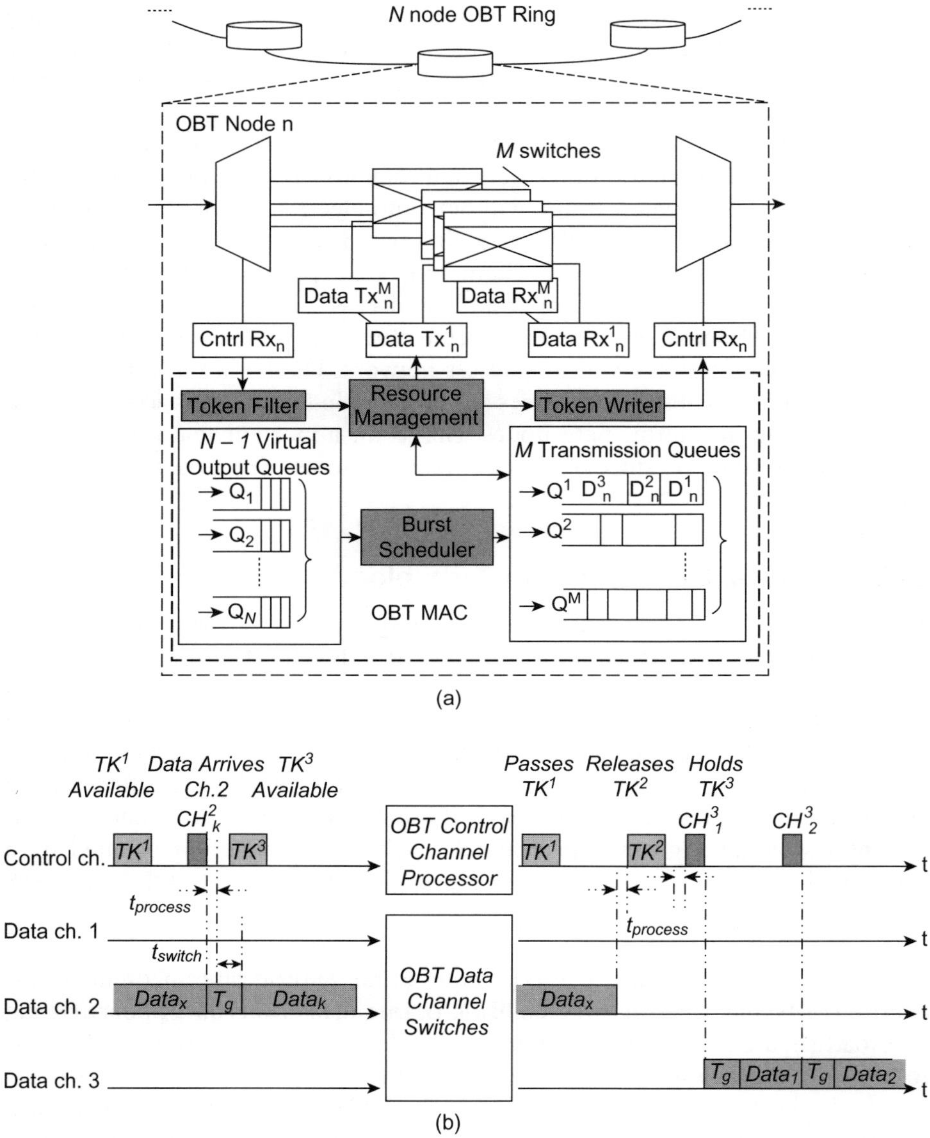

Figure 16.3. **(a)** Optical burst transport (OBT) network architecture. **(b)** Example of OBT wavelength token and burst control protocol.

Figure 16.3b illustrates the OBT protocol through an example. In this example, the node passes the first data token—that is, TK^1 (for channel 1)—because it has not accumulated enough data at the time it sees the token. At the arrival of the second data token, TK^3, it accumulates enough data and holds onto the third wavelength channel. Traffic is groomed in virtual output queues (VOQ) according to its destination node. The transmit node can send a burst with several sub-bursts, each one addressed to a different destination. A constant offset time is inserted between control header and the first sub-burst. The length of the offset time is determined by the time to configure the lightpath and process the control packets. Under the OBT architecture, the amount of time required to configure a lightpath is around 260 ns, which is made possible by an ultra fast switch such as the (Pb,La)(Zr,Ti)O_3-based switch. The 1×2 PLZT switch has reported less than 2.5-ns switching times [17].

16.2.1.2 Spatial Reuse. OBT employs spatial reuse to enhance its bandwidth utilization. This is possible because OBT drops receiving bursts at the destination, leaving the data path from the destination to the source unused. This path could be reused for secondary transmission when the primary transmission takes place between source and destination. Collision is not a concern since wavelength token is held by the source. Thus, it is possible for a node to initiate secondary transmission upon receiving a control channel indicating it as the destination. The length of this secondary transmission is determined by the duration of the sub-burst and the node can utilize the data path up to this duration, minus the guard time and processing time. The performance of spatial reuse in OBT is shown in [18] and the results showed nearly 100% throughput enhancement over OBT without spatial reuse.

16.2.1.3 Traffic Grooming. OBT provides robust sub-lambda granularity by aggregating and grooming individual data into singular wavelength bursts. Because the burst size can be adjusted dynamically, OBT can adapt to either circuit-oriented or data-oriented services. Figure 16.4 shows the comparative performance between OBT and RPR under single channel operation using simulation. Figure 16.4a shows that under balanced traffic load, OBT and RPR have similar throughput performance. OBT outperforms RPR during high network load because RPR transmit traffic has strictly lower priority than its transit traffic [5, 19], [5]. Figure 16.4b shows that RPR outperforms in terms of delay during low load because OBT nodes cannot obtain immediate channel access. Under unbalanced traffic load (Figures 16.4c and 16.4d), OBT is shown to outperform RPR because token control enables more graceful adaptation to asymmetrical traffic. RPR fairness control invokes undesired oscillating response to highly asymmetrical traffic because it aggressively shuts down transmit traffic rate when congestion is observed in one part of the ring.

16.2.2 Metropolitan-Access Ring Integrated Network

The MARIN metropolitan transport architecture differs from OBT in two main features: (a) The transmission does not involve intermediate switching, and (b) its WDM scalability does not require additional switch for each newly added

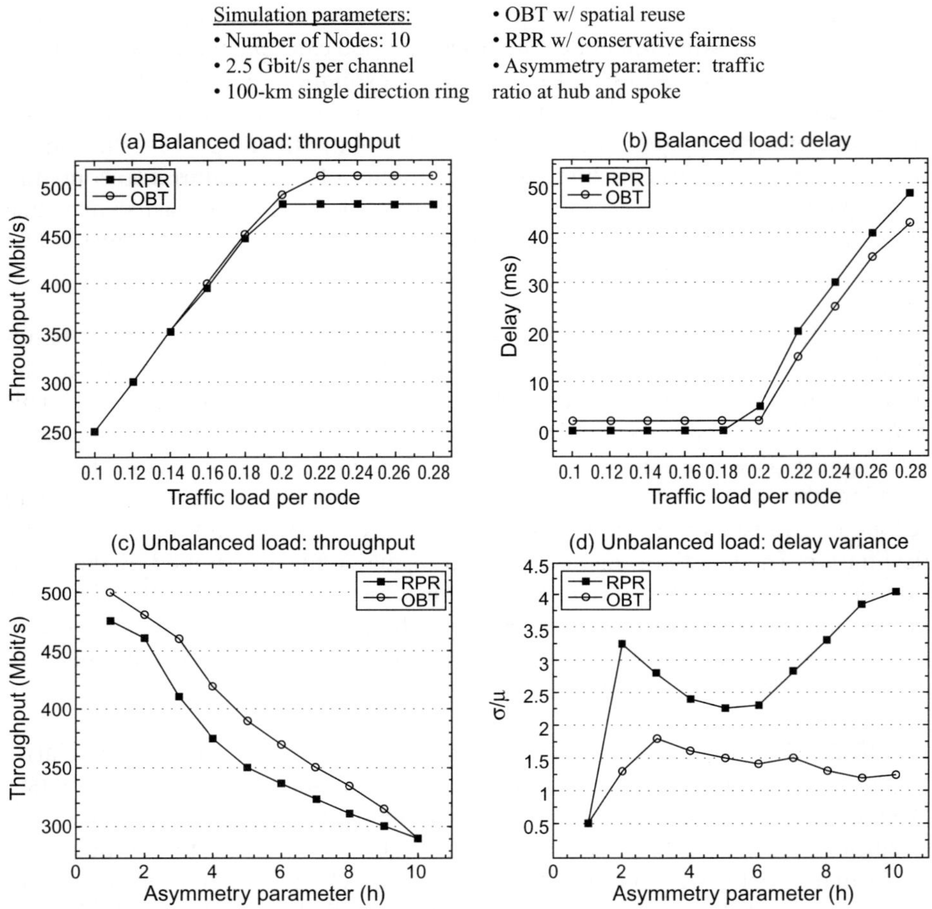

Figure 16.4. Performance comparison of OBT against RPR network.

wavelength. MARIN backhauls aggregated access traffic using passive a wavelength router and fast tunable optics. Moreover, MARIN combines metropolitan and access networks by sharing light sources.

Figure 16.5 illustrates the MARIN architecture proposed by [20]. In the MARIN network, metropolitan and access traffic are transmitted/received via the same physical sources. Receiving metro and access nodes employ fixed wavelength receiver(s), and traffic is routed by the passive a wavelength router at the source. The integrated transmission scheduler manages wavelength and laser resources and performs their allocation to either metropolitan or access traffic, depending on traffic conditions.

16.2.2.1 Access Traffic Aggregation. The uplink portion of MARIN access traffic is groomed and added to the metro network without requiring

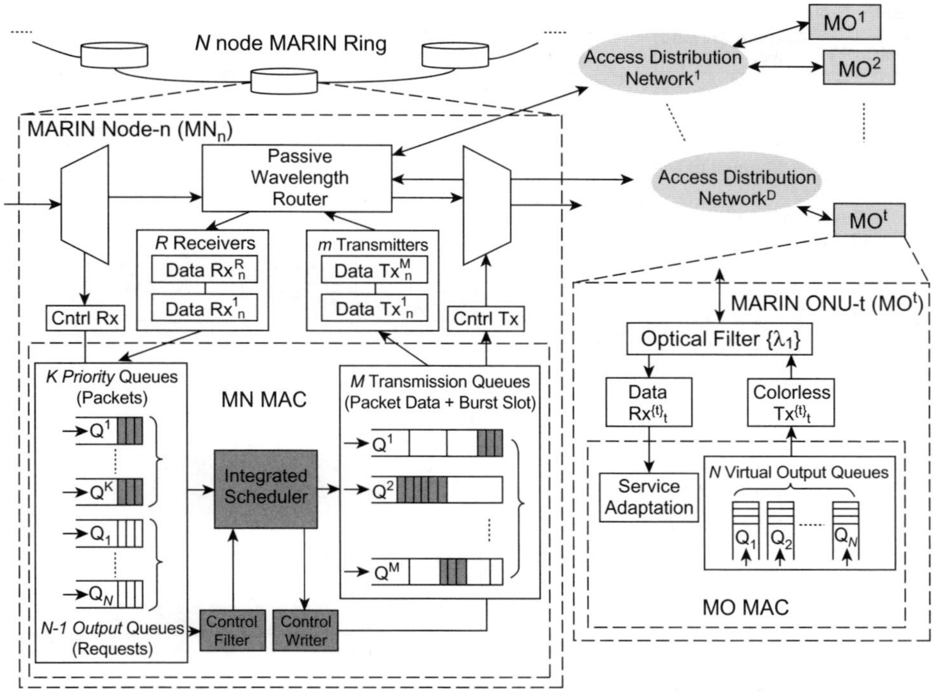

Figure 16.5. Metropolitan access ring integrated network (MARIN) architecture.

intermediate queuing. In the access traffic aggregation process, the MARIN node (MN) grooms access traffic from MARIN ONUs (MOs) and aggregates it into data bursts. Each MO patches its traffic into a VOQ according to its destination in the metropolitan ring, and classes of traffic within each VOQ are preserved. The access traffic aggregation process in the MARIN architecture is agnostic to the technology of the access network. This is possible as long as the MOs can report their traffic information to the MN and allow the MN to groom their uplink traffic. In the following, EPON based optical access network is used to illustrate the operation of the access traffic aggregation process.

Due to their point-to-multi-point topology, EPON networks employ multipoint control protocol (MPCP) to perform bandwidth assignment and polling [21]. EPON MPCP protocol relies on GATE and REPORT messages for bandwidth grant and request. MARIN utilizes the same MPCP control in the access segment to perform upstream traffic allocations. Figure 16.6 illustrates an example of the MARIN MPCP protocol window for access traffic aggregation. The integrated transmission scheduler at source MN_0 sends downlink $GATE_1^1$ message to MO^1 when there is enough data aggregated for the destination MN_1 on the ring. MO^1 selects data from VOQ_1 and send its uplink data following $GATE_1^1$ message. Following the data transmission, MO^1 sends a REPORT message containing the updated VOQ_1 lengths to the MN_1. In the transmission from MN_0 to MN_2, the

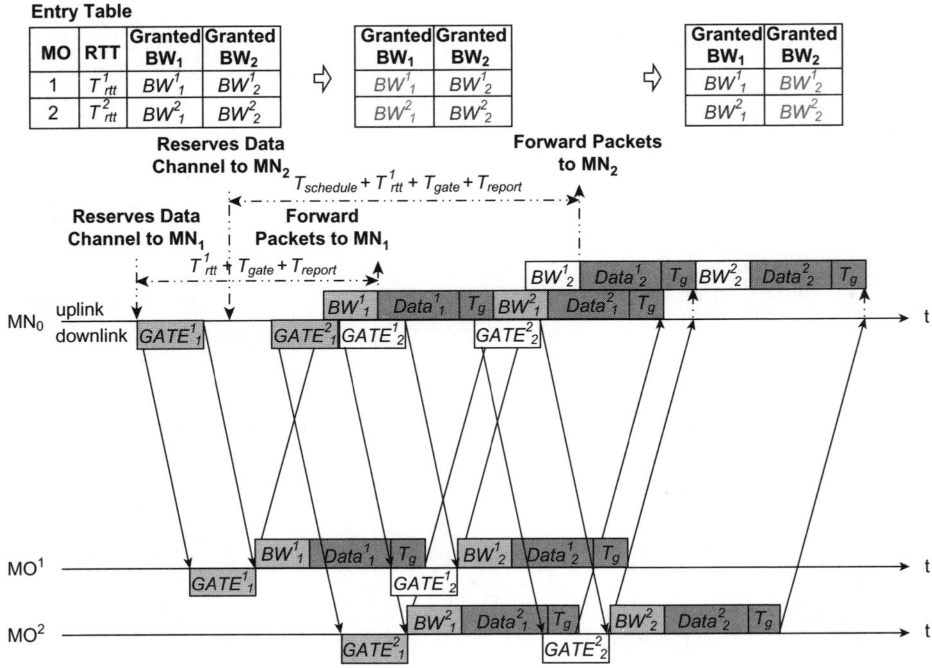

Figure 16.6. Example of MARIN multi-threaded traffic aggregation protocol.

bursts suffer from additional scheduling delay. In the example shown in Figure 16.6, this additional delay is illustrated by the fact that the new sub-burst should wait for the completion of the previous sub-burst. In the MARIN architecture, MOs employ colorless transmitter, which allows them to flexibly transmit the allocated wavelength.

After upstream traffic is groomed into a burst using MPCP control, the aggregated burst is received and reassigned to a wavelength in the metro wavelength set $\{\lambda_m\}$. Because the metropolitan area resource is acquired prior to access traffic grooming is scheduled, source MN_o can immediately add the burst directly to destination MN_i without undergoing further queuing. As a result, the transmission queue in MN does not have to queue uplink access packets. Whereas the integrated scheduler is more complex than individual metro or access scheduler, MARIN reduces hardware complexity in the MN and delay associated with intermediate queuing.

16.2.2.2 Metropolitan Data Transport Protocol. MN receiver relies on a coarse WDM (CWDM) filter to strip off consecutive bands of dense WDM wavelengths from the ring and a cyclic AWG-based passive router to forward added traffic. The allocation of the wavelength resource is determined by the dynamic switching and wavelength allocation protocol (DSAWP). Figure 16.7 illustrates the wavelength routing table used by the DSWAP [22] method. When

MARIN Wavelength Allocation Table																
input, i	1	1	1	1	2	2	2	2	3	3	3	3	4	4	4	4
output, j	1	2	3	4	1	2	3	4	1	2	3	4	1	2	3	4
FSR(1)	1	2	3	4	4	1	2	3	3	4	1	2	2	3	4	1
FSR(2)	5	6	7	8	8	5	6	7	7	8	5	6	6	7	8	5
FSR(3)	9	10	11	12	12	9	10	11	11	12	9	10	10	11	12	9
FSR(4)	13	14	15	16	16	13	14	15	15	16	13	14	14	15	16	13

Figure 16.7. Example of MARIN wavelength routing table (4 × 4 AWG, FSR = 4, λ = 16). In the example, the target user employs filter FSR(1) and is connected to input port 1.

a laser connected to input port *i* wants to transmit to output port *k*, it can transmit on a set of wavelength λ^J, where $J \in \{j = (k - i + 1) + n \cdot B | 0 < j < max\}$. The parameter *B* represents the free spectral range (FSR) and wavelengths separated by *B* channels apart could be reused at the same fiber output. A typical AWG has FSR between 4 and 32 DWDM channels and can support up to 128 DWDM wavelengths. The index *n* is selected such that the wavelength index *j* is between 1 and 128. The set *J* represents all admissible wavelengths that can be sent from input laser *i* to output access network *k*. The objective of the DSWAP is to allocate a laser-wavelength (i, j) pairing when access or metro traffic demands bandwidth resource. All wavelengths are further partitioned into the sets $\{\lambda_a\}$ and $\{\lambda_m\}$, which represent the wavelengths for access and metro transport. The passive router utilizes a WDM filter to separate wavelengths $\{\lambda_m\}$ and inserts them into the metropolitan ring. The remaining wavelengths $\{\lambda_a\}$ are unaffected and continue onto the access network.

In the MARIN access segment, each MO is connected to the MN through the access distribution network via one of its cyclic AWG output ports. Similar to the MN receiver, each MO receiver also employs a CWDM filter to strip off consecutive bands of DWDM wavelengths. When a downlink access packet is received from the ring, DSAWP first identifies a laser source *i* for the packet based on backlog in laser transmission queue. Once the laser source *i* with the shortest backlog is identified, the scheduler finds an appropriate AWG output port *k* to connect to the destination MO. After input laser *i* and output access network *k* are defined, a row is selected from the allocation table. Each column in the $\{\lambda_a\}$ set of the allocation table corresponds to the stripping waveband of a MO. The final (i, j) pairing of the allocation is the intersection of the identified row and column in the table. To summarize, the allocation of (i, j) pairing for access traffic depends on the input port of the earliest available laser and the output port that connects to the MO.

In the MARIN metro network segment, the metropolitan ring is connected to all outputs of the cyclic AWG. This allows every laser to connect the ring with all the available wavelengths in the metro wavelength set $\{\lambda_m\}$. Each MN receiver is assigned to a fixed wavelength and employs a DWDM filter to strip a single wavelength for every receiver. Every MN could have more than one receiver. The MN schedules a burst to node *N* when it accumulates enough traffic requests for node *N* from the MOs. Once the burst threshold is reached, the MN schedules a downlink GATE message to the MOs and grooms uplink data into a single burst from the MOs. When the uplink burst arrives, the MN forwards the received burst to laser *i* and directly sends them over the allocated wavelength *j*.

The MARIN metro transport adapts token control to support lossless burst transport. Unlike in OBT, each wavelength token represents a free receiver in MARIN. To transmit a burst from the source MN to the destination MN without intermediate switching, the source node acquires a wavelength token λ^j from the ring and transmits the aggregated burst using the laser *i* with shortest transmission backlog. Metro traffic is given priority over the access traffic and all active or scheduled access packets for laser *i* are suspended during the transmission of the metropolitan traffic.

16.2.2.2 Metropolitan and Access Resource Sharing. MARIN reuses the same light source for downlink access and metropolitan traffic. The integrated transmission scheduler arbitrates the usage of the light sources, allowing better resource utilization through the use of statistical sharing of them. To evaluate its performance against the nonintegrated architecture, simulations have been performed. In the simulation, the MARIN architecture adapts a variant of the token control protocol. The token control is adapted because it was shown to have robust performance compared with an RPR network. In MARIN, each waveband token represents a free receiver whereas in OBT a wavelength token only refers to free wavelength. This is because MARIN does not utilize receiver switching and each receiver can receive a unique set of wavelengths.

Figure 16.8 shows the performance evaluation of MARIN network. First, MARIN is compared against a metropolitan area network with fixed transceiv-

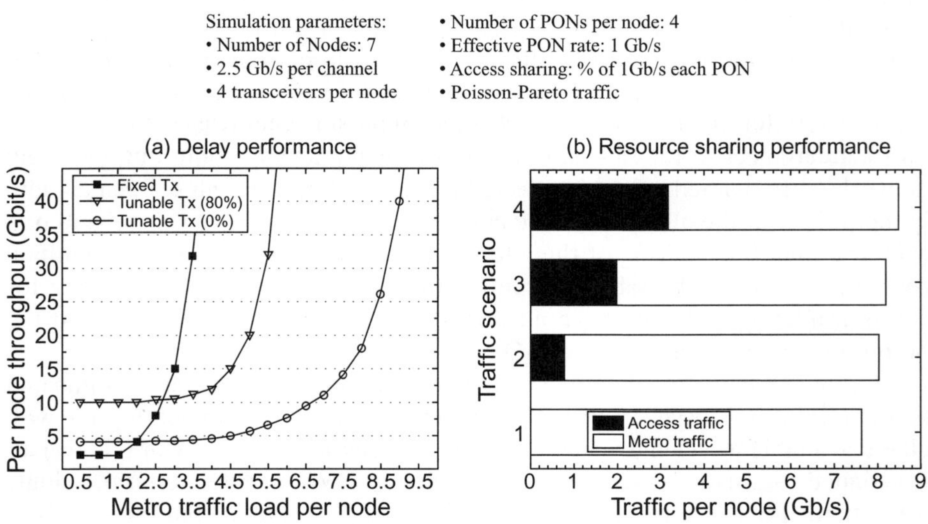

Figure 16.8. MARIN metro-access resource sharing performance.

ers, such as OBT and RPR. The simulation scenario simulates a metropolitan area ring with seven nodes. Both fixed- and tunable-transmitter-based architectures employ four 2.5-Gbit/s transceivers. Each metro node further supports four access networks at 1.25-Gbit/s downlink rate (about 1 Gbit/s useful data after discounting the overhead). RPR is selected as the fixed transmitter based architecture in the simulation. The simulation results in Figure 16.8a are used to show that the tunable transmitter-based architecture is far more scalable than fixed transmitter architecture. This is because while there are only four common data channels available for the entire Fixed-Tx network, there is no limit to the number of data channels that can be used in the Tunable-Tx network. In this example, the Tunable-Tx architecture can use a total of 28 (4×7) wavelength channels and is able to reach any receiver using any four dedicated wavelengths without contending with another receiver node.

In Figure 16.8a, the transmitters in MARIN node also have to support downlink access traffic. For example, the differences in metropolitan traffic throughput between two MARIN scenarios are caused by increasing access traffic loads. To demonstrate the performance of this transmitter resource sharing property, Figure 16.8b compares the total transmitter utilization under four different downlink access traffic loads. They are 0%, 20%, 50%, and 80% loads and correspond to series 1, 2, 3, and 4, respectively. The results show the maximum metropolitan traffic that the MARIN node can support when simultaneously supporting the required access demands. Results show slight total transmitter utilization improvement. This is because MARIN enables statistical multiplexing in the access network by making all four of its transmitters available to the receivers at any given time.

16.2.2.3 WDM Scalability and RPR Upgrade. In MARIN, if there can be a maximum of 64 wavelengths for metropolitan traffic, there can be a maximum of n nodes in the network, where $n = 64/m$ and m represents the number of receivers per node. For the access network, since each access distribution network connects only to one output port of the AWG, each access network can support $n = 64/4 = 16$ terminals with unique waveband passing characteristics.

Standard RPR currently does not provide a clear guideline to upgrade from its single-channel platform into a multiple-channel one. The MARIN architecture provides a clear method to scale to multichannel platform using WDM technology. Figure 16.9a illustrates the method to upgrade MARIN from single-channel to a multiple-channel platform. The figure shows a modified MN to integrate a legacy RPR node. VOQs are added in the MN to convert excess RPR traffic into MARIN traffic. Figure 16.9b shows the performance of the network when it upgrades from a single-channel RPR into an integrated RPR-MARIN platform. The results show that the integrated RPR-MARIN network can gracefully transit excess RPR traffic without scarifying bandwidth utilization. During the transition phase, where the metropolitan traffic load first exceeds the maximum bandwidth that RPR can support, integrated MARIN-RPR traffic suffers from additional delay because MARIN cannot provide immediate channel

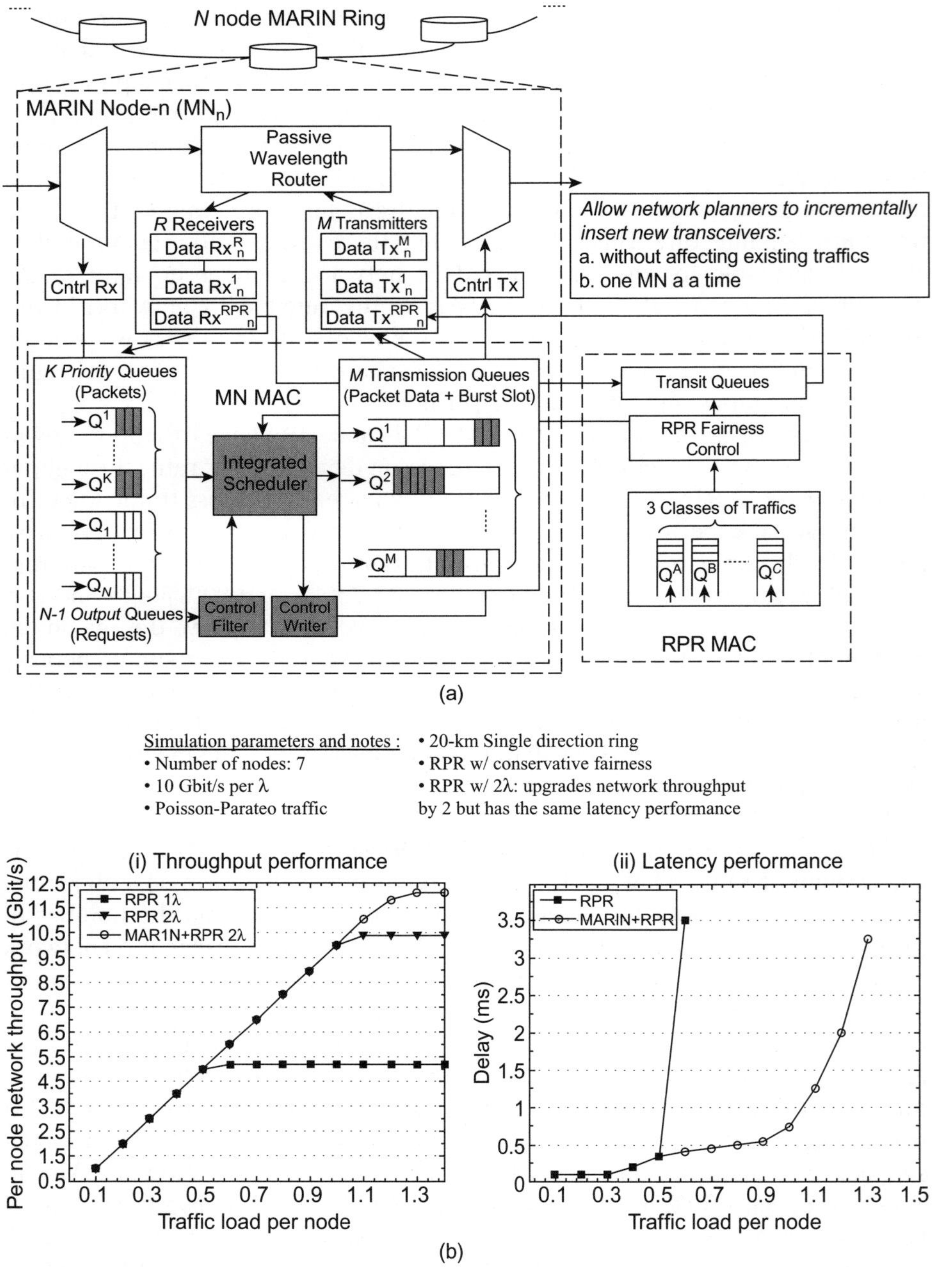

Figure 16.9. **(a)** MARIN-RPR integration. **(b)** MARIN-RPR integration performance.

TABLE 16.3. Comparison of Next-Generation Metropolitan Area Network (MAN)

	Next-Generation SONET	RPR	OBT	MARIN
Standard	ITU G.704.1	IEEE 802.17	N/A	N/A
Bandwidth granularity	SONET hierarchy	SONET/ Ethernet	Flexible	Flexible
Bandwidth provisioning time	>1 s	<1 ms	<1 ms	<1 ms
Access resource sharing	None	None	None	Yes

TABLE 16.4. EPON and WiMAX Comparison

	Standard	Framing	Data Rates	Max Reach	Bandwidth Reservation	QoS
EPON	IEEE 802.3 ah	Ethernet	60 Mbit/s per user (average)[a]	20 km	Request and grant	IEEE 802.Q (up to 7 traffic classes)
WiMAX	IEEE 802.16	Ethernet	1.5 ~ 144 Mb/s[b]	15 km	Request, grant, and confirm	Connection-oriented

[a]Bandwidth depends on the number of users, and the number listed here is typical values.
[b]Achievable data rate depends on channel conditions and is shared by multiple users.

access. However, the delay performance does not further degrade up to the full capacity of the second light source. Figure 16.9b also shows that the utilization of a second RPR would result in a lower throughput compared to the case where a MARIN architecture is used.

Table 16.3 presents a comparison among MARIN, OBT, RPR, and Next-Generation SONET/SDH in terms of bandwidth granularity, bandwidth provisioning time, and access resource sharing [15]. Regarding bandwidth granularity, next-generation SONET utilizes SONET hierarchy; that is, the granularity of bandwidth is a multiple of SONET virtual tributary groups, whereas RPR supports SONET and Ethernet frames. In turn, both OBT and MARIN can support flexible traffic due to their fast tunable or transmitting capability. Regarding bandwidth provisioning times, next-generation SONET is based on conventional management systems, whose setup times of connections take from seconds to some minutes, whereas RPR, OBT, and MARIN are all in the submillisecond range of connection provisioning times. The major benefit of MARIN over the other three technologies is its ability to share access resources due to its integrated architecture.

16.3 CONVERGENCE OF OPTICAL AND WIRELESS ACCESS NETWORK

The increasing convergence between optical and wireless access networks is also bringing significant changes to the design of metropolitan-access networks. This section examines the design of new metropolitan-access architectures based on optical-wireless access convergence. Specifically, a metro-access optical infrastructure for municipal WiFi mesh network is presented. The architecture employs a metropolitan backbone that relies on the aforementioned MARIN architecture. The example demonstrates an evolutionary architecture that is capable of physically scaling to support future demands and technologies. Moreover, the concept of integrated control framework is introduced to demonstrate potential performance enhancements in the municipal WiFi mesh network and as well as in other standard-based converged access networks.

16.3.1 Municipal WiFi

WiFi network is one of the most popularly used wireless technologies. Moreover, comparing to infrastructure-based technologies like WiMAX or Long-Term Evolution (LTE), WiFi requires a smaller upfront investment on installation and equipment. Because of these reasons, new municipal access networks are considering to leverage WiFi networks. In past years, a number of companies pursued aggressive plans to deploy municipal WiFi (Muni-WiFi) networks in American

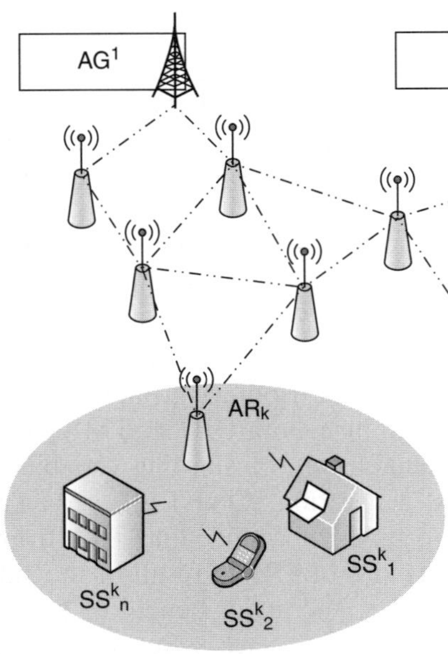

Figure 16.10. Municipal WiFi mesh architecture.

cities. However, until now, Muni-WMN networks have limited performance due to their inability to deliver broadband-like experience to end users across a large area.

Figure 16.10 illustrates the architecture of a common Muni-WiFi employing a wireless mesh network (WMN). In this architecture, access gateways (AGs) connect to the fiber backbone and provide the point of ingress and egress for aggregated downstream and upstream traffic, respectively. The access routers (ARs) serve as the intermediate node and relay transit traffic before they arrive to the destination. Traffic is routed from the AG to the AR or vice versa typically via a k-nearest-neighbor routing strategy. This section will first present a standard compatible metropolitan optical backbone for Muni-WiFi based on EPON technology.

16.3.1.1 Challenges in Large-Scale WiFi Mesh Network. WiFi-based WMN is viewed as a cost-effective solution to provide network coverage over a large area. The ARs could be deployed throughout the city without requiring significant operational or capital expenses. The operator can conveniently extend the coverage of the network by incrementally placing ARs at desirable locations. Traffic can be routed to the backbone network through the access gateways, and the network can be managed in a distributed fashion without requiring centralized planning. Whereas this approach enables rapid network deployment, the resulting network usually suffers from significant relay traffic and low spectral efficiency, leading to a reduction of its effective bandwidth. Since broadband experience would require significantly higher end-user throughput rate than currently provided rates, the throughput bottleneck in current WMN based Muni-WiFi must be addressed.

The first approach is to enhance individual wireless link capacity through the use of high-throughput wireless PHY technologies. High-throughput (HT) technology such as P802.11n multiple-input multiple-output (MIMO) draft standard can increase the maximum PHY bit rate from 54 Mbit/s in current 802.11 a/g standard to more than 300 Mbps. Currently, the IEEE P802.11ac very high throughout (VHT) task group studies advanced technologies such as multiple channels and/or spatial division multiple access (SDMA) to further increase link capacity. The VHT specification aggressively predicts more than 1-Gbit/s MAC throughput rate when a combination of multiple channels and SDMA technologies is used.

As promising as these physical layer enhancements are, they cannot sufficiently alleviate the throughput bottleneck in mesh networks for one glaring reason. An alternative approach to enhance the effective network throughput is to reduce the amount of intermediate transit traffic in the WMN. The effective throughput rate of a WMN increases when the physical hop count reduces in the network. This approach, also referred to as cell splitting in WMN, requires increasing number of AGs connecting to backbone. Figure 16.11a illustrates the cell splitting gain using a simulation study of WMN capacity with a high-fidelity simulator [23]. The simulation study emulates a square torus universe with 12 km

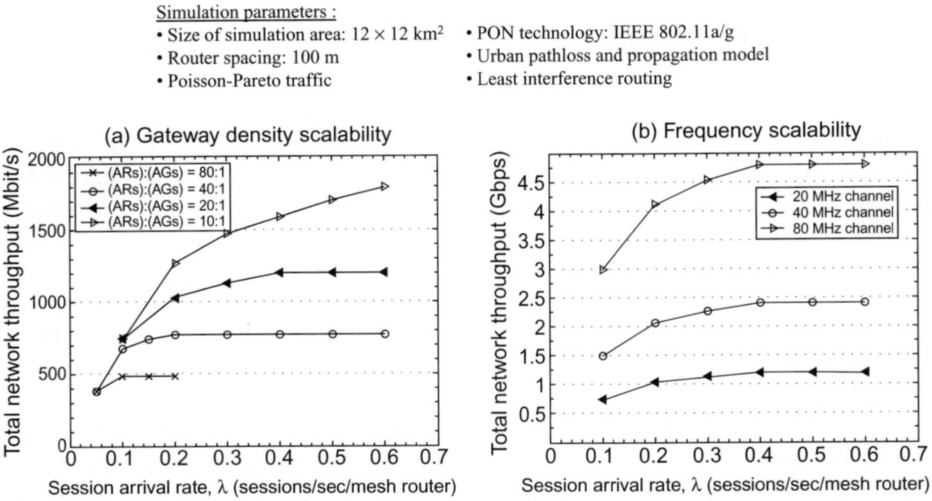

Figure 16.11. Effective network throughput gain using cell splitting.

in each side that wraps interference around the edge of the universe. ARs are placed 100 m apart and uniformly distributed in this universe. AGs are placed at the intersections of the streets to allow maximum neighbor visibility. The simulator considers effects of propagation characteristics, shadowing, and co-channel interferences to the calculation of the signal-to-interference-plus-noise ratio (SINR). The PHY layer employs 802.11 a/g technology and the physical communication rate depends on the instantaneous SINR.

Simulation results in Figure 16.11a shows that when a single 20-MHz channel is used, an aggregate throughput of 773 Mbit/s and 1.2 Gbit/s is supported under AR:AG ratio of 40 and 20, respectively. The overall network throughput experiences a diminishing return with respect to the number of connected gateway routers. This is because the co-channel interference effect is quite high when gateways are located at nearer distances. To alleviate the bottleneck, multiple frequency bands are desired and Figure 16.11b shows the expected throughput performance using wider frequency band when limit the AR:AG ratio to 20. Thus, to optimally support the traffic demand in future metropolitan area access networks, a combination of multiple channel and spectrally efficient techniques, as well as careful placement of the ONU to increase the AR:AG density with minimum infrastructure cost, is desirable.

To support the increase of AG density, a scalable and physically extensible infrastructure must be in place. Current wireless backbone solutions primary rely on T1 or T3 circuits [24], since these lines provide bit rates of Mbit/s and are widely common. However, these fixed circuits are very costly and their infrastructure cannot be flexibly expanded due to the point-to-point characteristic of these lines. An alternative to these two problems is to employ a TDM-PON as the backbone to provide a logical point-to-point connection over a cost-effective and flexible point-to-multipoint topology. Nevertheless, a large-scale wireless network exhibits widely varying traffic demands within its subscription groups and a TDM-PON has limited reach and bandwidth flexibility beyond a fixed group of terminals. Furthermore, a TDM-PON is not expected to completely satisfy the bandwidths especially when the WMN interfaces start to evolve from current 802.11 a/g into HT or VHT standards. To address these problems, two architectures are proposed to support the backbone network of a large-scale WMN.

16.3.1.2 MARIN: Metropolitan WDM Ring Backbone. In reference 25, an integrated metropolitan-access optical network is proposed to provide the necessary backbone network to large-scale WiFi-based WMN. This architecture, shown in Figure 16.12, employs an integrated optical backbone that relies on a WDM metropolitan ring and is considered a variant of the MARIN network. In the MARIN hybrid architecture, the EPON distribution network is mostly unchanged and remains passive. The backbone fiber ring replaces current central offices (Cos) with a passive splitter/filter to connect these distribution networks. The ring integrates existing EPON distribution networks and consolidates their OLTs into a unified CO. Wavelength resources and traffic controls are all centrally managed by the new CO. The MARIN hybrid architecture leverages existing standards and equipments to allow flexible placement of AGs at the edge of the distribution network.

The architecture makes selected changes to existing EPON at the physical media dependent (PMD) sublayer. MARIN moves each EPON network onto a different pair of operating wavelengths instead of reusing the same pair of wavelengths. This allows MARIN to support multiple EPON networks via a single metropolitan ring backbone and a single consolidated CO. Moreover, these changes occur only in the PHY layer of the terminals and are not visible to the system. In the new CO, each OLT employs a fixed CWDM transceiver and is assigned a unique wavelength pair to connect to the subscribed AGs.

The EPON-WiFi further integrated AG employs a tunable receiver to allow flexible subscription to multiple OLTs. This allows load balancing in the backbone network when aggregated traffic overutilizes part of the wavelengths. In large local area networks, this may occur when the network load increases due to varying geographical traffic patterns. Utilizing existing EPON MPCP and auto-discovery process, the CO is able to move the AG from an overutilized wavelength to an underutilized one. An illustration of this wavelength load balance process is shown in Figure 16.13a and discussed below.

The backbone wavelength load balancing process starts with OLT^1 sending a wavelength reallocation GATE message to the identified AG_j^1. Upon receiving the reallocation command, the AG_j^1 stops its operation and tunes its wavelength from λ^1 to λ^2. After a fixed waiting time to wait for the wavelength tuning operation to complete in AG, OLT^2 starts the auto-discovery process by sending a registration $GATE^2$ message and expects for $REGISTER\_REQ^2$ from the

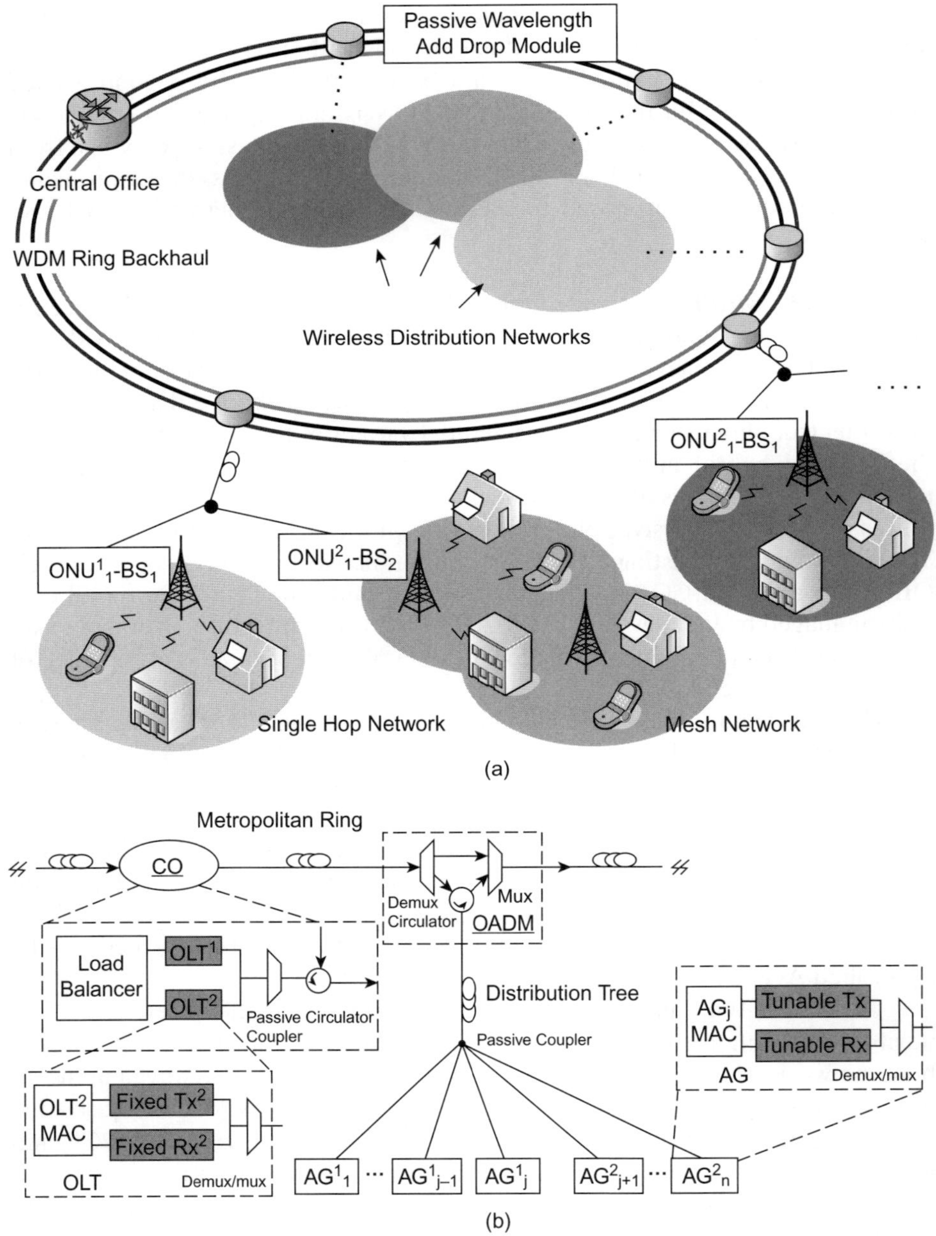

Figure 16.12. **(a)** MARIN hybrid wireless metro-access network. **(b)** MARIN backhaul architecture.

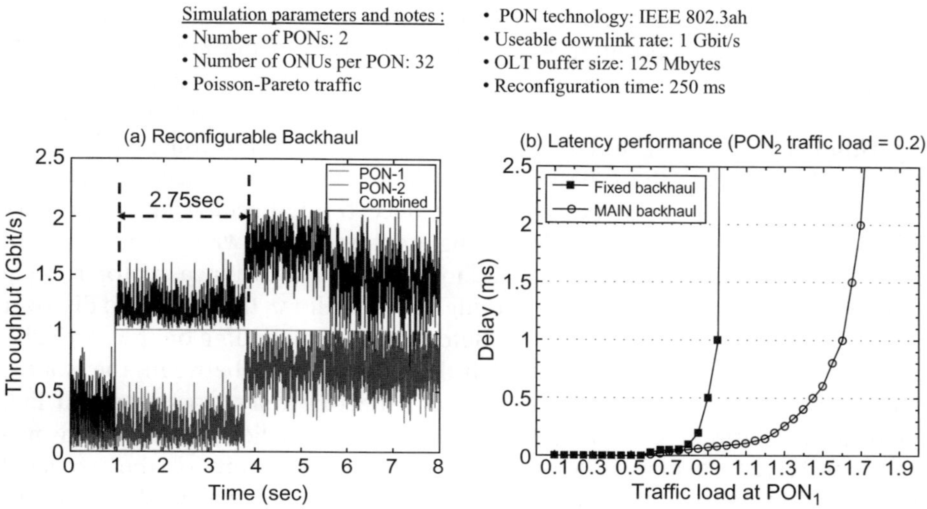

Figure 16.13. **(a)** Effective network throughput. **(b)** Backhaul wavelength load balancing performance.

reallocated AG_j^x during the discovery window. Upon successfully receiving $REGISTER\_REQ^2$ message from the AG_j^x, OLT^2 sends a $REGISTER^2$ message containing the unique link layer identification (LLID) number for the AG. The load balance operation is successfully completed when the AG_j^2 responds with the $REGSTER\_ACK^2$ message to the OLT^2. If the operation fails or OLT^2 cannot support the new AG, the AG moves back to the original home wavelength for subscription and subscribes to OLT^1 through its auto-discovery process.

The backbone load balance operation and its protocol are demonstrated in reference 25. In the paper, the AG employs a tunable receiver based on MEMS tunable filter that reports a 33.6-μs response time. Figure 16.13a shows the simulation results based on this protocol and set the guard time to the tuning time plus the typical uplink guard time between frames. The expected reconfiguration time is found to be 250 ms on average, and the reconfiguration time is fast enough such that the OLT can hold onto the packets for the reallocated AG during its reconfiguration. In the example here, the OLT can buffer up to 125 Mbytes of data for each ONU during reconfiguration state. In the simulation, the access traffic loading for PON_1 increases from 0.2 to 1.2 at $t = 1.0$ s. It takes 2.75 s for PON_1 to notify the overloading problem (i.e., with traffic reaching buffer limit), and it takes another 250 ms to complete the reconfiguration. Figure 16.13b shows the network performance enhancement of the reconfigurable backbone architecture versus a fixed architecture, where the traffic load for one of the EPONs exceeds the maximum load for a single wavelength and the load for the other EPON remains fixed at 0.2.

16.3.1.3 Integrated WMN Control: Downlink Load Balanced Routing. Besides dynamically allocating wavelength resources in the network, MARIN hybrid architecture introduces an integrated control plane to support dynamic wireless resource allocation. In reference 26, an integrated load balance routing algorithm is proposed with the assistance from an integrated control plane. The control plane utilizes the *k*-best routing scheme and routes excess traffic load via the secondary (second best) AG to reduce traffic backlog in the primary (best) AG. Current 802.11s WMN routing employs hybrid wireless mesh protocol (HWMP), which is a hybrid of adhoc on-demand distance vector routing (AODV) and spanning-tree based routing. HWMP allows the routers to discover the most advantageous path in a distributed fashion, depending on the particular metric it employs. To achieve a high-throughput path between the source and destination, a link quality metric that reflects wireless loss rate and link bandwidth is desired. A common link quality metric is called expected transmission time (ETT), which is a bandwidth adjusted measure of the expected number of transmission (ETX) [27]. ETX estimates the expected number of transmission and re-transmission(s) and estimates the impact of interference to link quality at the MAC layer. The ETT enhances the ETX by reflecting the different transmission modes for individual links. Using an adaptation of the ETT metric, HWMP can provide good throughput and distributed routing strategy within one mesh network. However, HWMP cannot support load balancing through multiple AGs. MARIN allows such load balanced routing enhancement by dynamically assigning the packets with the same destination address (DA) to different AGs.

In the MARIN WMN, most of the traffic is infrastructure-based (i.e., routed to and from the AG), and there is little peer-to-peer traffic. When the downlink part of the WMN is overloaded, the AG will reflect the congestion through extended backlog. The MARIN integrated control can monitor the downlink congestion by examining the backlog condition reported by the AGs. MARIN modifies the REPORT message from current EPON MPCP to include the statistics for downlink queues. MARIN initially assigns the packets to the primary AG based on shortest path first (SPF) routing. Note that once the packets are routed from the OLT to AG, the AG and AR will determine the best routing path using HWMP, which is very different from SPF. However, SPF routing allows the OLT to choose a good enough approximation of the best AG for most traffic. When congestion is observed, the OLT allocates the traffic to a less loaded secondary AG according to SPF by relabeling the LLID address on the header and relieves the excess traffic load at the primary AG.

Figure 16.14 shows the simulated performance of the load-balanced WMN-EPON network versus WMN. In the simulation scenario, 196 ARs are placed over a 1400-m by 1400-m square. Four of these routers are connected to the fiber backbone. The simulation creates a traffic hotspot to one of the AG, and the performance is compared with an integrated load-balanced network against a nonintegrated network. The integrated control plane adjusts to the traffic hotspot

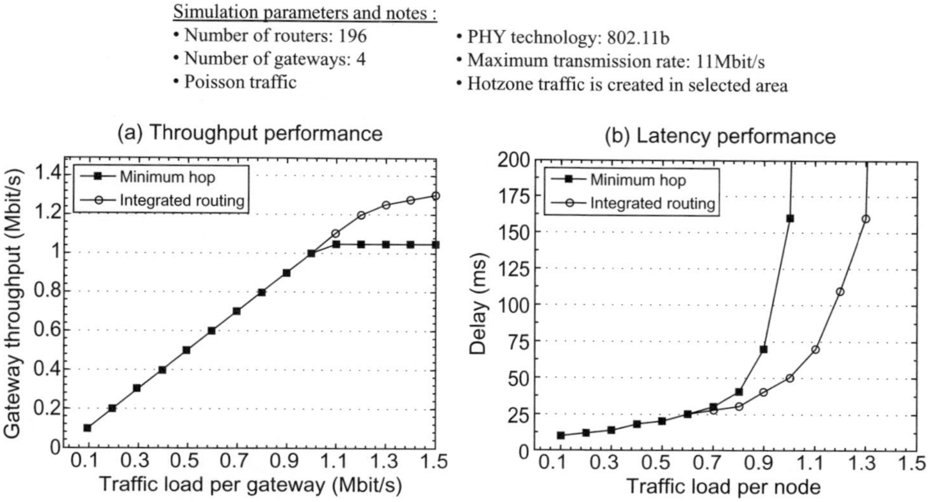

Figure 16.14. Performance of MARIN integrated routing.

by allocating part of the traffic to neighbor AG and can achieve 20% throughput enhancement versus the nonintegrated WMN.

So far, the performance of the integrated network focuses on backbone capacity and wireless throughput enhancements. References 26 and 28 present pioneering works in this area by first integrating WiFi with an EPON network. An integrated control plane could provide many additional benefits for the OWI network. One benefit is that an integrated control plane can provide end-to-end QoS and better management. Integrated OWI control framework, in fact, is an active research area. The majority of the published works in this area studies WiMAX and EPON systems due to their similarity in terms of bandwidth and bandwidth grant/reservation mechanism. Section 16.3.2 presents the OWI integrated framework and its benefits by using WiMAX and EPON standards as the example.

16.3.2 Convergence Between EPON and WiMAX

EPON and WiMAX share a number of important properties. Most importantly, 802.16 base stations (BS) are equipped with an Ethernet interface that can be easily connected to the EPON ONU. Both networks employ similar bandwidth request/allocation and QoS supporting mechanisms. They also support similar physical bandwidth and employ the same point-to-multi-point architecture. Table 16.4 summarizes key similarities between these two networks. The convergence would further facilitate the combination of complementary characteristics of EPON and WiMAX networks.

Figure 16.15a illustrates the integrated EPON and WiMAX architecture. Under this integrated architecture, WiMAX extends the coverage area and

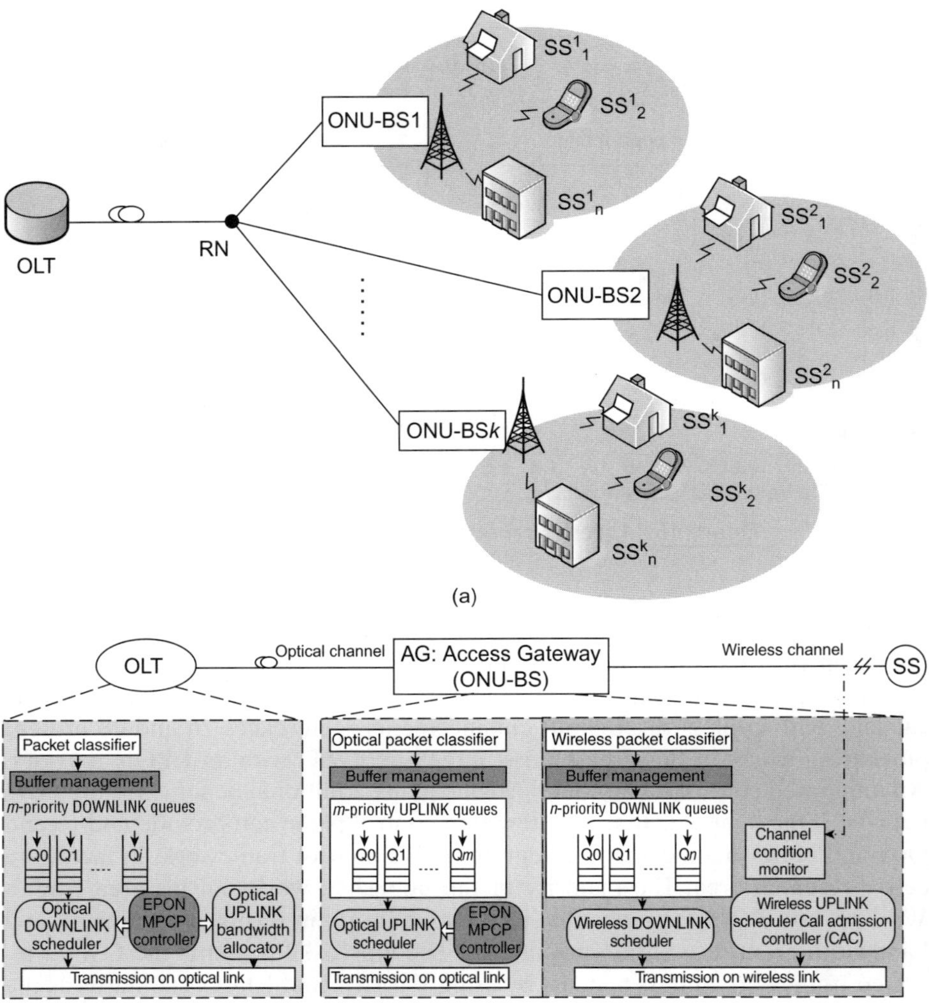

Figure 16.15. **(a)** EPON-WiMAX integration architecture. **(b)** Control module details of EPON-WiMAX integrated network.

supports mobility to subscriber stations (SS). EPON supplies a low-cost and dynamic backhaul infrastructure to the WiMAX network. This architecture saves significant CAPEX associated with the infrastructure comparing to conventional point-to-point backhauls, such as T1 leased lines.

Despite the fact that they share many similarities, EPON and WiMAX use different operational protocols in their bandwidth control and QoS mechanisms. A converged interface with integrated control could further enhance the performance of the networks. The following analysis summarizes the state-of-the-art research in this area. Figure 16.15b shows the module details of the integrated system. In this illustration, the ONU and the BS are combined into a single hybrid (AG). The two devices could be independent, in which case the interworking relies on an external software bridge to exchange control and data. The highlight boxes represent key enhancement functions and they are subjects of the following discussions. The discussion will be separated into uplink and downlink enhancements.

16.3.2.1 Uplink Enhancement: End-to-end QoS, and Call Admission Control. EPON MPCP supports dynamic bandwidth allocation (DBA) using methods such as interleaved polling with adaptive cycle time (IPACT) [29]. IPACT schedules uplink traffic in a just-in-time architecture based on reported traffic requests. IPACT and other EPON DBA are strictly queue based; that is, they make the allocation decision based on reported uplink queue lengths. Unlike EPON, the physical layer of WiMAX supports adaptive modulation and coding (AMC) to adapt to channel conditions. Thus, its bandwidth allocation process is based on the channel rate and employs variants of latency-rate (LR) schedulers [30]. In networks employing LR schedulers, end-to-end delay and buffer requirement are determined by latency and allocated rate, instead of queue length. A solution to consolidate their differences is to employ a hierarchical scheduler. Hierarchical scheduler is also better suited to in an integrated network rather than a direct scheduler because the latter cannot scale with a large aggregate number of queues.

Moreover, in order to support better end-to-end differentiated service (DiffServ), the converged interface at the AG needs to convert 802.16 differentiated traffic to appropriate EPON queues and vice versa. The IEEE 802.16 standard defines four types of scheduling services—that is, unsolicited grant service (UGS), real-time polling service (rtPS), extended real-time polling service (ertPS), non-real-time polling service (nrtPS), and best effort (BE) [31, 32]. The EPON standard supports up to eight different priority queues in the ONU and uses the IEEE 802.Q guideline to distinguish the following traffic classes: network control, voice, video, controlled load, excellent effort, best effort, and background. Yang et al. [33] demonstrate a simplified mapping using three priorities at EPON: best effort (BE), assured forwarding (AF), and expedited forwarding (EF). In the proposed scheduling algorithm, UGS traffic is mapped to EF and rtPS/ertPS traffic is mapped to AF. The scheduler employs a three-level scheduling hierarchy where local scheduling takes place inside the SS, BS, and OLT. To provide basic fairness among subscriber stations, the algorithm splits the allocated bandwidth evenly among SS that have made the corresponding class of connection requests. Under this simple scheme, the converged network is able to support coarse end-to-end DiffServ from the SS to the OLT.

Performance enhancement could also be achieved by employing call admission control (CAC) at the AG. The CAC algorithm considers end-to-end delay across WiMAX and EPON network for each connection request to guarantee QoS. Because WiMAX is connection-oriented, bandwidth requests are made for every connection. CAC helps sift through connection requests and rejects

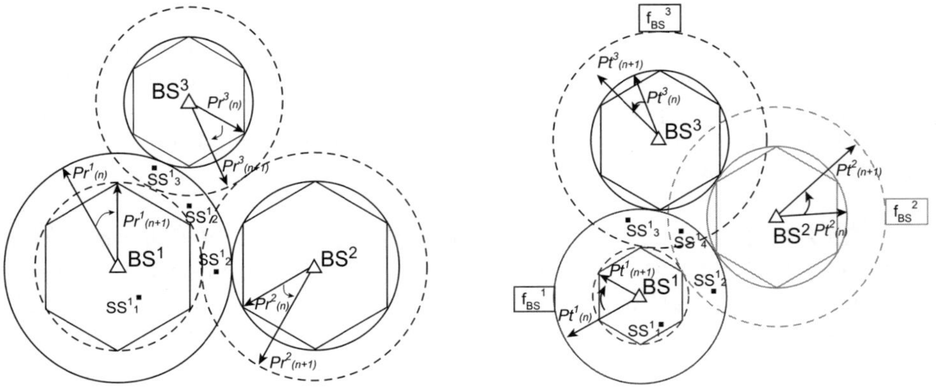

Figure 16.16. Cell breathing concept.

unserviceable requests, preventing them from competing network resources with serviceable packets. When a new SS connection request arrives, the CAC algorithm would accept the request based on existing queue size and expected service time, T_{exp}. T_{exp} depends on three parameters: queuing time, polling delay, and wireless transmission and propagation delay. The queuing time and polling delay depends on the AG queue occupancy and polling cycle time T_{cycle}, respectively. The second parameter could be known or estimated based on the particular MPCP control that is used. Furthermore, the wireless transmission delay is known based on the negotiated transmission mode between the SS and the AG. If the channel conditions vary and the transmission fails, the CAC will discard the transmission request and reevaluate based on new admission criterion when the re-transmission request arrives.

16.3.2.2 Downlink Scheduler: Power Control and Cell Breathing. In broadband wireless access networks, the base station power is typically controlled to minimize mutual interference among adjacent cells. By lowering transmit power, a base station could further reduce coverage dynamically during heavy network load. Reducing coverage could reduce the number of subscribed users and relieve itself from excess traffic burden. Similarly, a neighboring base station could share the excess traffic load and expand its coverage. This operation is called cell breathing, and such a dynamically balanced network could provide better QoS within a single cell or statistically support more mobile users. Figure 16.16 illustrates the dynamic cell breathing concept in WiMAX network. In each of the two examples, BS^1 reduces its transmission power and leaves SS_{2-4} outside of its coverage. SS_4 is not covered by any of BB^{1-3} during cell breathing if frequency reuse is not allowed.

For downstream traffic, cell breathing averts extended backlog at the AG. Backlog at the AG could occur due to either overutilization or poor channel condition. To determine whether cell breathing is necessary, the OLT calculates

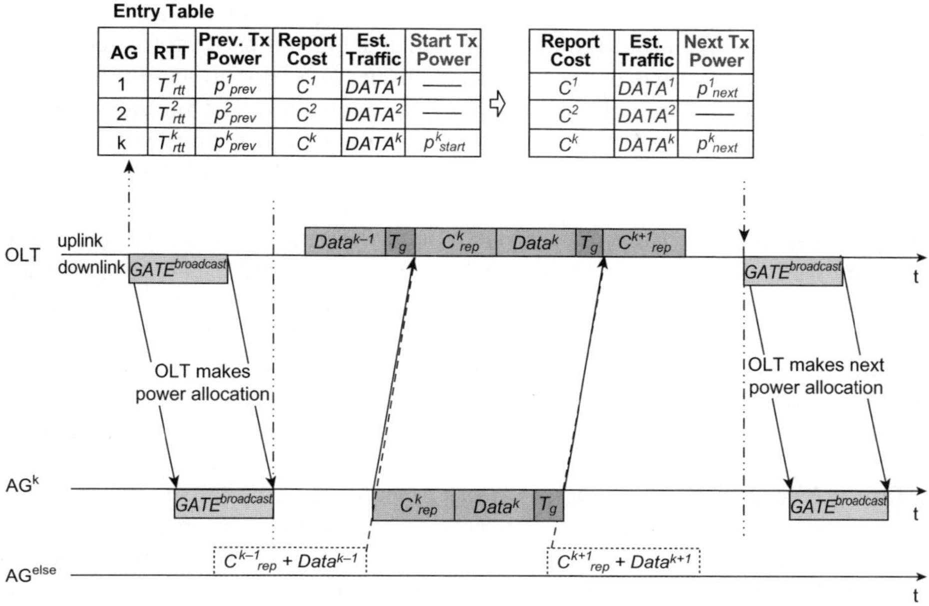

Figure 16.17. MPCP control window for cell breathing.

the estimated total service time and equalizes them across all AGs. Figure 16.17 illustrates the protocol window and corresponding entry at the OLT. The computation at OLT relies on the residual backlog and control costs C^i feedback from $AG^i$and knowing the current transmit power level P^i. In the example, C^k is determined by AG^k locally based on connection queue sizes and corresponding connecting rates as well as perceived channel qualities. AG^k sends C^k in an extended $REPORT^k$ message. Upon receiving the renew C^k message, the OLT determines if a further action is necessary to load-balance the network. In the example, the OLT utilizes a broadcast GATE message immediately afterward to issue power re-allocation to AG^1 and AG^k.

During cell breathing, an AG could lose connections to some subscribers and handover might occur for these subscribers. The handover problems could be effectively mitigated by using the broadcast nature of PON and queue duplicate version of the traffic at nearby handoff stations to reduce handover costs.

16.4 FUTURE OUTLOOK FOR METRO-ACCESS NETWORKS

16.4.1 100-Gbit/s Ethernet

Currently, the standardization of a next-generation 100 Gbit/s Ethernet data transport is under development in the IEEE P802.3ba 40 Gbit/s and 100 Gbit/s

Ethernet Task Force [34, 35]. The transport of data traffic in metropolitan area distances through Ethernet-based equipments has been mostly motivated by the fact that the Ethernet has been very successful in LANs due to its simplicity, low cost, standardization, and compelling market penetration. The migration of the Ethernet protocol from LAN to MAN environments has been possible mainly because of the separation of the medium access control (MAC) layer from the physical (PHY) layer since the very former times of this protocol. As a result, the progress to newer Ethernet standards over different physical layer technologies and transmission media has preserved almost the same frame format of the previous standards, which have allowed the leverage of existing Ethernet installations. The choice of the 100-Gbit/s rate came naturally from the conventional evolution of the Ethernet over its previous generation—that is, a 10-fold increase in the speed over 10 Gbit/s.

Ethernet as a metro-service architecture must be able to support a large number of terminals and high-capacity links in a scalable way, as well as offer OAM capabilities and network resilience guarantees for service providers. In other words, the Ethernet as a carrier-grade platform should be able to provide almost all the benefits that the SONET/SDH networks have, but using a packet-switched platform. It is not clear yet, however, if the inclusion of all these acclaimed benefits in the Ethernet will keep its cost-effectiveness. Most of the standards activities to accomplish these goals are developed by the IEEE 802 LAN/MAN Standard Committee (LMSC) [36], who is responsible for the evolution of Ethernet in LANs and MANs. The Metro Ethernet Forum (MEF) [37], an industry consortium that aims at fostering the adoption of the Ethernet as a carrier-class platform, has been also defining Ethernet service types, management, and service level agreements (SLA) for metro and wide area environments. Parallel standardization efforts led by the ITU have mostly focused on the definition of the transport of Ethernet over TDM circuits, Ethernet protection switching, and OAM functionalities.

The 100-Gb Ethernet transmission over metropolitan networks will require very stringent component tolerances with respect to the chromatic dispersion (CD) and polarization mode dispersion (PMD) effects, since the impact of these impairments in the optical transmission increases with the square of the data rate [38, 39]. Moreover, whereas interchannel effects strongly affect mostly dispersion-compensated optical transmission systems at per-channel rate of 10 Gbaud and below, intrachannel nonlinearities (e.g. intrachannel cross-phase modulation (iXPM) and intrachannel four-wave mixing (iFWM)) severely impact such systems at per-channel rates of 10 Gbaud and above [40]. Advanced, spectrally efficient modulation formats will play a central role in the design of 100-Gbit Ethernet optical transmission systems, since an appropriate choice of the data modulation scheme can alleviate the impact of such impairments on the transmission and make a more efficient utilization of the channel spectrum at such high-speed data rates [41].

16.4.2 Future Mobile Backhaul and Fiber-Wireless Convergence

Currently, the standardization of 3G partnership project (3GPP) specifies next generation of cellular wireless networks to deliver broadband-like data rates. Besides MIMO and advanced antenna configurations, next generation cellular standards look into aggressive use of radio spectrum and advanced inter-cell interference (ICI) mitigation techniques to increase network capacity. In the former case, cellular operators have adopted the use of universal frequency reuse, that is, no frequency reuse among neighboring cells (and fractional frequency reuse at cell edge), and are considering the use of additional radio spectrum if necessary. However, with such an aggressive use of spectrum, ICI is fast becoming the limiting factor not only to the throughput performances of cell edge users but also cell throughput. ICI mitigation techniques are proposed to reduce or even avoid interferences altogether by coordinating decisions among multiple cellular sites. Therefore, they promise more efficient use of the scarce radio spectrum resource.

The performance of ICI mitigation hinges on how well the coordination is performed in the backhaul. In general, cell throughput can increase with higher degree of coordination and with greater number of coordinating cells. Recent studies have suggested the use of fiber backhaul with very high bandwidth capacity to support ICI mitigation [42, 43]. These studies employ Common Public Radio Interface (CPRI) standard, or CPRI-like digital formats, to transport radio signals over simple fiber backhaul. While these backhauls can effectively support ICI mitigation for a very small number of cells, cellular operators have very strong interests to increase both the degree and number of coordinating cells. This is because reduction of signal processing performed at the base station level can reduce the energy footprint and complexity on these remote sites. Moreover, consolidation on the number of centralized control sites can further lower the total cost of ownership (TCO) for operators.

Very recently, the Cloud Infrastructure Radio Access (C-RAN) architecture has been proposed to centralize base station processing to a super central site [44]. In particular, C-RAN proposes a very large-scale and centralized architecture to encourage centralized coordination and management. In C-RAN or other large-scale next-generation cellular architectures, advanced fiber backhaul architecture like the ones presented in this chapter are desirable over conventional ones. Reconfigurable and WDM fiber architectures such as MARIN and GROW-Net would allow the network to flexibly allocate backhaul resources to meet the dynamic needs of the wireless networks. Moreover, they enable the operators to scale their wireless network infrastructure in a graceful way. In addition, in anticipation of greater wireless broadband capacity demands, radio frequency over fiber (RoF) technologies are the focus of many active researches due in large part to their ability to substantially reduce the complexity of the remote base stations. Therefore, the combination of advanced fiber backhaul architectures and RoF technologies are expected to play a key role in the development of next generation mobile backhauls.

ACKNOWLEDGMENTS

This work has been supported in part by the U.S. National Science Foundation (NSF) under grant no. 0627085 and by the Brazilian National Council for Scientific and Technological Development (CNPq).

REFERENCES

1. R. Doverspike and P. Magill, Commercial optical networks, overlay networks and services, in *Optical Fiber Telecommunications V B: Systems and Networks*, I. Kaminow, T. Li, and A. E. Willner, editors, Academic Press, New York, 2008, pp. 511–560.
2. A. M. Odlyzko, Internet traffic growth: Sources and implications, in *Proceeding SPIE—Optical Transmission Systems and Equipment for WDM Networking II*, B. Dingel, W. Weiershausen, A. K. Dutta, and K.-I. Sato, editors, Vol. 5247, 2003, pp. 1–15.
3. ITU-T Rec. G.7041/Y.1303, Generic Framing Procedure (GFP), 2001.
4. D. Cavendish, K. Murakami, S.-H. Yun, O. Matsuda, and M. Nishihara, New transport services for next-generation SONET/SDH systems, *IEEE Commun. Mag.*, Vol. 40, No. 5, pp. 80–87, May 2002.
5. F. Davik, M. Yilmaz, S. Gjessing, and N. Uzun, IEEE 802.17 resilient packet ring tutorial, *IEEE Commun. Mag.*, Vol. 42, No. 3, pp. 112–118, March 2004.
6. V. Gambiroza, P. Yuan, L. Balzano, Y. Liu, S. Sheafor, and E. Knightly, Design, analysis, and implementation of DVSR: A fair high-performance protocol for packet rings, *IEEE/ACM Trans. Networking*, Vol. 12, No. 1, pp. 85–102, February 2004.
7. ITU-T Rec. G.709, Interfaces for the Optical Transport Network (OTN), 2001.
8. T. E. Stern, G. Ellinas, and K. Bala, *Multiwavelength Optical Networks*, Cambridge University Press, New York, 2008.
9. E. B. Basch, R. Egorov, S. Gringeri, and S. Elby, Architectural tradeoffs for reconfigurable dense wavelength-division multiplexing systems, *J. Select. Quant. Electron.*, Vol. 12, No. 4, pp. 615–626, July-August 2006.
10. C. R. Giles and M. Spector, The wavelength add/drop multiplexer for lightwave communication networks, *Bell Labs Tech. J.*, Vol. 4, No. 1, pp. 207–229, January–March 1999.
11. S. J. Ben Yoo, Optical packet and burst switching technologies for the future photonic Internet, *J. Lightwave. Technol.*, Vol. 24, No. 12, pp. 4468–4492, December 2006.
12. K. Shrikhande, I. M. White, M. S. Rogge, F.-T. An, A. Srivatsa, E. S. Hu, S. S-H. Yam, and L. G. Kazovsky, Performance demonstration of a fast-tunable transmitter and burst-mode packet receiver for HORNET, in *IEEE Optical Fiber Communication Conference (OFC 2001)*, Vol. 4, ThG2, San Francisco, CA, March 2001.
13. R. Chen, H. Chin, D. A. B. Miller, K. Ma, J. S. Harris, Jr, MSM-based integrated CMOS wavelength-tunable optical receiver, *IEEE Photonics Technol. Let*, Vol. 17, No. 6, pp. 1271–1273, June 2005.
14. M. Vrdoljak, S. I. Vrdoljak, and G. Skugor, Fixed-mobile convergence strategy: Technologies and market opportunities, *IEEE Commun. Mag.*, Vol. 38, No. 2, pp. 116–121, February 2000.

15. J. Kim, J. Cho, S. Das, D. Gutierrez, M. Jain, C.-F. Su, R. Rabbat, T. Hamada, and L. G. Kazovsky, Optical burst transport: A technology for the WDM metro ring network, *IEEE J. of Lightwave Technol.*, Vol. 25, No. 1, pp. 93–102, January 2007.

16. C. Elger and J. Elmirghani, A slotted MAC protocol for efficient bandwidth utilization in WDM metropolitan access ring networks, *IEEE J. Selected Areas Commun.*, Vol. 21, No. 8, pp. 1295–1305, October 2003.

17. K. Nashimoto, PLZT waveguide devices for high speed switching and filtering, in *IEEE Optical Fiber Communication Conference (OFC 2008)*, OThE4, San Diego, CA, March 2008.

18. S. Das , J. Kim, D. Gutierrez, L. Kazovsky, R. Rabbat, C.-F. Su, and T. Hamada, Protection and spatial reuse in Optical Burst Transport (OBT) networks, in *3rd International Conference on Broadband Communications, Networks and Systems*, pp. 1–10, San Jose, CA, October 2006.

19. J. Kim, M. Maier, T. Hamada, and L. G. Kazovsky, OBT: Optical burst transport in metro area networks, *IEEE Commun. Mag.*, Vol. 45, No. 11, pp. 44–51, November 2007.

20. S.-W. Wong, W.-T. Shaw, K. Balasubramanian, N. Cheng, and L. Kazovsky, MARIN: Demonstration of a flexible and dynamic metro-access integrated architecture, in *Proceedings, IEEE GLOBECOM'07*, pp. 2173–2177, Washington, DC, November 2007.

21. G. Kramer, B. Mukherjee, and G. Pesavento, IPACT: A dynamic protocol for an Ethernet PON (EPON), *IEEE Commun. Mag.*, Vol. 40, No. 2, pp. 74–80, February 2002.

22. S.-W. Wong, W.-T. Shaw, N. Cheng, C. Qiao, L. Kazovsky, Dynamic wavelength allocation in a converged and scalable interface for metro-access ring integrated networks, in *IEEE Optical Fiber Communication (OFC 2008)*, OTuI6, San Diego, CA, March 2008.

23. H. Lee, V. Manshadi, and D. C. Cox, High-fidelity and time-driven simulation of large wireless networks with parallel processing, *IEEE Commun. Mag.*, Vol. 47, No. 3, pp. 158–165, March 2009.

24. P. Magill, Optical/wireless access architecture and field trials, *IEEE Optical Fiber Communication Conference (OFC 2007)*, NThB2, Anaheim, CA, March 2007.

25. W.-T. Shaw, S.-W. Wong, N. Cheng, and L. G. Kazovsky, MARIN hybrid optical-wireless access network, in *IEEE Optical Fiber Communication Conference (OFC 2007)*, OThM2, Anaheim, CA, March 2007.

26. W.-T. Shaw, S.-W. Wong, N. Cheng, K. Balasubramanian, X. Zhu, M. Maier, and L. G. Kazovsky, Hybrid architecture and integrated routing in a scalable optical-wireless access network, *IEEE J. Lightwave Technol.*, Vol. 25, No. 11, pp. 3443–3451, November 2007.

27. R. Draves, J. Padhye, and B. Zill, Routing in multi-radio, multi-hop wireless mesh networks, in *Proceedings, ACM MobiCom'04*, pp. 114–128, Philadelphia, PA, September 2004.

28. S. Sarkar, H.-H. Yen, S. Dixit, and B. Mukherjee, Hybrid wireless-optical broadband access network (WOBAN): network planning and setup, *IEEE J. Selected Areas Commun.*, Vol. 26, No. 6, pp. 12–21, August 2008.

29. G. Kramer, B. Mukherjee, and G. Pesavento, IPACT, a dynamic protocol for an Ethernet PON (EPON), *IEEE Commun. Mag.*, Vol. 40, No. 2, pp. 74–80, February 2002.

30. D. Stiliadis and A. Varma, Latency-rate servers: A general model for analysis of traffic scheduling algorithms, *IEEE/ACM Trans. Networking*, Vol. 6, No. 5, pp. 611–624, October 1998.

31. C. Cicconetti, L. Lenzini, E. Mingozzi, and C. Eklund, Quality of service support in IEEE 802.16 network, *IEEE Network*, Vol. 20, No. 2, pp. 50–55, March–April 2005.

32. G. Shen, R. S. Tucker, and C.-J. Chae, Fixed mobile convergence architecture for broadband access: integration of EPON and WiMAX, *IEEE Commun. Mag.*, Vol. 45, No. 8, pp. 44–50, August 2007.

33. K. Yang, S. Ou, K. Guild, and H.-H. Chen, Convergence of Ethernet PON and IEEE 802.16 broadband access networks and its QoS-aware dynamic bandwidth allocation scheme, *IEEE J. Selected Areas Commun.*, Vol. 27, No. 2, pp. 101–116, February 2009.

34. IEEE P802.3ba 40 Gb/s and 100 Gb/s Ethernet Task Force: http://www.ieee802.org/3/ba/

35. J. McDonough, Moving standards to 100 GbE and beyond, *IEEE Commun. Mag.*, Vol. 45, No. 11, pp. 6–9, November 2007.

36. IEEE 802 LAN/MAN Standards Committee: http://www.ieee802.org/index.html

37. The Metro Ethernet Forum (MEF): http://www.metroethernetforum.org/

38. C. F. Lam and W. I. Way, Optical Ethernet: Protocols, management, and 1–100 G technologies, in *Optical Fiber Telecommunications V B: Systems and Networks*, I. Kaminow, T. Li, and A. E. Willner, Academic Press, 2008, pp. 345–400.

39. M. Duelk and M. Zirngibl, 100 Gigabit Ethernet—Applications, features, challenges, in *Proceedings, INFOCOM'06*, Barcelona, Spain, pp. 1–5.

40. R.-J. Essiambre, G. Raybon, and B. Mikkelsen, Pseudo-linear transmission of high-speed TDM signals: 40 and 160 Gb/s, in *Optical Fiber Telecommunications IV*, I. Kaminow and T. Li, editors, Academic Press, New York, 2002, pp. 232–304.

41. P. J. Winzer and R.-J. Essiambre, Advanced optical modulation formats, *Proc. IEEE*, Vol. 94, No. 5, pp. 952–985, May 2006.

42. V. Jungnickel, L. Thiele, T. Wirth, T. Haustein, S. Schiffermüller, A. Forck, S. Wahls, S. Jaeckel, S. Schubert, C. Juchems, F. Luhn, R. Zavrtak, H. Droste, G. Kadel, W. Kreher, J. Mueller, W. Stoermer, and G. Wannemacher, Coordinated multipoint trials in the downlink, in *GLOBECOM Workshop*, pp. 1–9, Honolulu, HI, November 2009.

43. C. Hoymann, L. Falconetti, and R. Gupta, Distributed uplink signal processing of cooperating base stations based on IQ sample exchange, in *Proceedings, IEEE ICC'09*, pp. 1–5, Dresden, Germany, June 2009.

44. China Mobile Research Institute, C-RAN: The Road Toward Green RAN. White paper, April 2010.

17

RESILIENT BURST RING: A NOVEL TECHNOLOGY FOR NEXT-GENERATION METROPOLITAN AREA NETWORKS

Yuefeng Ji and Xin Liu

17.1 INTRODUCTION

In existence for over a decade now, the metropolitan area network (MAN) has traditionally been designed for voice transmission based on the time division multiplexing (TDM) technology. At that time, voice traffic was significantly more important than data traffic. As a result, synchronous optical network/synchronous digital hierarchy (SONET/SDH) became the dominant standard on these networks. Those technologies have been able to meet the initial requirements of MANs quite adequately, and today most of them are based on SONET/SDH technology.

With the explosion in the demand for bandwidth for data transmission, it became quite clear that SONET/SDH networks needed to be reengineered to handle data traffic in a more efficient way. The new MAN technology should connect all the various access networks and provide everything from real-time services to traditional data-transfer services. It should also provide Quality of Service (QoS) and handle any kind of traffic, from constant bit-rate traffic to packet- or cell-based traffic. While such a multiservice network would minimize

Convergence of Mobile and Stationary Next-Generation Networks, edited by Krzysztof Iniewski

overall operating costs, the existing SONET/SDH infrastructure is unfortunately unable to meet these requirements. The worst-case scenario is that the operators must maintain their legacy networks for legacy services while the new services need overlapping network infrastructure.

In recent years, SONET/SDH-based transport networks have come to be considered as too inflexible, inefficient, and overly complex for the purposes of data communication. As the importance of data communication has increased, research has begun for the replacement for the traditional SONET/SDH networks. That led to the emergence of the next-generation SONET/SDH [1, 2] and the resilient packet ring (RPR) [3, 4], and it also led to an increasing number of networks based on the dense wavelength division multiplexing (DWDM), such as optical burst switching (OBS) [5, 6], optical burst transport (OBT) [7, 8], and so on.

17.1.1 Next-Generation SONET/SDH Network

Next-generation SONET/SDH network offers a major step in combining new multiservice metro-access/transport equipment into legacy traditional SONET/SDH networks. It enables new types of services with more efficient network usage to be easily implemented by utilizing existing infrastructure.

Next-generation SONET/SDH extends the utility of the existing SONET/SDH network by leveraging existing layer 1 networking and including technologies such as generic framing procedure (GFP), virtual concatenation (VCAT), and the link capacity adjustment scheme (LCAS), and it integrates the customer interface into the network elements. GFP is defined in ITU-T G.7041 and ANSI T1-105.02 standards. It is an all-purpose protocol for an encapsulating packet over SONET (POS), ATM, and other Layer 2 traffic on the SONET/SDH networks. VCAT technology is the solution to the inefficiencies inherent in fixed payloads and contiguous concatenation. LCAS, a VCAT control mechanism, allows for the addition or reduction in the payload capacity of a virtual concatenation group (VCG) to meet the bandwidth needs of the application. Moreover, LCAS also offers other benefits, such as dynamically replacing failed member links within a VCG without removing the entire VCG. Unidirectional control of a VCG allows for asymmetric bandwidth, allows interworking of an LCAS transmitter with a non-LCAS receiver and vice versa, and aids in the creation of customer-based, on-demand services, and so on.

The next-generation SONET/SDH networks are core and metro networks with edge transport and traffic aggregation equipment that represents a significant generational improvement in functionality, size, and capacity compared to traditional SONET/SDH. New equipment is being designed that retains the traditional SONET/SDH functionality but in a package size five to ten times smaller. These systems are also inherently designed for multi-wavelength applications. Their SONET/SDH structure might also be "data aware," enabling variable packet sizes to be carried and therefore making more efficient use of the available bandwidth.

Traditionally, the different services are provided through technology-specific transport pipes. However, the next-generation SONET/SDH enables the simultaneous transport of heterogeneous services over one wavelength, thereby saving network-building and maintenance costs. Moreover, using the next-generation SONET/SDH, which maps 8B/10B-coded data into 64B/65B-coded sequences, the required bandwidth is substantially decreased.

The most important merit of this technology is that the existing SDH network would not be replaced. Next-generation SONET/SDH has a transport mechanism that enables the concurrent existence of legacy and new services over the same network, without disturbing each other. However, the demerit of over-provisioning and LCAS latency will appear when it comes to supporting highly dynamic data traffic [9].

17.1.2 Resilient Packet Ring

Resilient packet ring, or RPR as it is commonly known, is the IEEE 802.17 standard designed for the optimized transport of data traffic over fiber rings. Its design is to provide the resilience found in SONET/SDH networks (50-ms protection) but instead of setting up circuit-oriented connections, providing a packet-based transmission. This is to increase the efficiency of Ethernet and IP services.

RPR works on a concept of dual counter-rotating rings called ringlets. These ringlets are set up by creating RPR stations at nodes where traffic is supposed to drop, per flow (a flow is the ingress and egress of data traffic). RPR uses MAC (Media Access Control protocol) messages to direct the traffic, which traverses both directions around the ringlet. The nodes also negotiate for bandwidth among themselves using fairness algorithms, avoiding congestion and failed spans. The avoidance of failed spans is accomplished by using one of two techniques known as "steering" and "wrapping." In steering, if a node or span is broken, all nodes are notified of a topology change and they reroute their traffic. In wrapping, the traffic is looped back at the last node prior to the break and routed to the destination station.

All traffic on the ring is assigned a Class of Service (CoS), and the standard specifies three classes. Class A (or high) traffic is a pure CIR (committed information rate) and is designed to support applications requiring low latency and jitter, such as voice and video. Class B (or medium) traffic is a mix of both a CIR and an EIR (excess information rate, which is subject to fairness queuing). Class C (or low) is best effort traffic, utilizing whatever bandwidth is available. This is primarily used to support internet access traffic.

Another concept within RPR is what is known as "spatial reuse." Because RPR "strips" the signal once it reaches the destination (unlike SONET/SDH, which consumes the bandwidth around the entire ring), it can reuse the free space to carry additional traffic. The RPR standard also supports the use of learning bridges (IEEE 802.1D) to further enhance efficiency in point to multipoint applications and VLAN tagging (IEEE 802.1Q, Virtual Local Area Network).

RPR is a new data transport technology designed to meet the requirements of a packet-base MAN. It is optimized for robust and efficient packet networking over a fiber ring topology and combines the best features of legacy SONET/SDH and Ethernet into one layer to maximize profitability while delivering carrier-grade service. This technology incorporates extensive performance monitoring, proactive network restoration, and flexible deployment capabilities. RPR answers for the future Quality of Service (QoS) requirements with three state QoS classes, and it provides less than 50 ms of protection with two different methods, steering and wrapping. The key features of RPR are differentiated services handling, fast resilience, and ring-wide bandwidth management. However, RPR implements a MAC protocol for access to the shared ring communication medium that has a client interface similar to that of the Ethernet. All the packets will be processed on Layer 2 at each node on the ring. Therefore, it will make the processing delay of RPR much longer than that of the SONET/SDH technology, which only needs the O/E/O conversation and some add/drop or OXC (optical cross connect) processing. Moreover, due to the process complexity, the max data rate that RPR can support now is only 10 Gbit/s. That is another restriction to RPR.

17.1.3 Optical Burst Switching

Optical burst switching (OBS) is a switching concept which lies between optical circuit switching and optical packet switching. It operates at the subwavelength level and is designed to better improve the utilization of wavelengths by rapid setup and teardown of the wavelength/lightpath for incoming bursts. In the OBS network, incoming traffic from clients at the edge of the network are aggregated at the ingress of the network according to a particular parameter [commonly destination, Type of Service (ToS), Class of Service (CoS), and Quality of Service (QoS)]. At the OBS edge router, different queues represent the various destinations or class of services. Based on the assembly/aggregation algorithm, packets are assembled into bursts by using either a time-based or threshold-based aggregation algorithm. In some implementations, aggregation is based on a hybrid of timer and threshold. From the aggregation of packets, a burst is created and this is the granularity that is handled in the OBS network.

In the OBS network, the burst header generated at the edge of the network is sent on a separate control channel which could be a designated out-of-band control wavelength. At each OBS node, the control channel is converted to the electrical domain for the electrical processing of the header information. The header information precedes the burst by a set amount known as an offset time, in order to give enough time for the switch resources to be made available prior to the arrival of the burst. Different reservation protocols have been proposed, and their efficacy has been studied and published in numerous research publications. Obviously, the signaling and reservation protocols depend on the network architecture, node capability, network topology, and level of network connectivity. The reservation process has implications on the performance of the OBS network due to the buffering requirements at the edge. The one-way signaling paradigm obviously introduces a higher level of blocking in the network because connections are not usually guaranteed prior to burst release. Again numerous proposals have sought to improve these issues.

An all-optical path for the data burst makes the transmission delay in the OBS network much shorter than that in the next-generation SONET/SDH and the RPR network. However, the efficiency of the OBS network can be reduced by resource contention of bursts directed to the same transmission links. Moreover, with the unidirectional resources reservation scheme, combining with the effect of buffering incurred by fiber delay lines (FDL) and the impact caused by the multiple paths transmission with the deflection routing scheme, the delay variation problem in the OBS network will be much worse than in the other two networks; in some cases, the packet will arrive in disorder, which will make the performance of the OBS network much worse, especially for the time-based services and applications. Finally, due to the signal impairment in the high-speed transmission systems and the lack of the ultra-high-speed large-scale optical switch matrix, OBS technologies do not seem to be practical in the near future.

17.1.4 Optical Burst Transport

As advances in tunable lasers and filters have made rapidly reconfigurable lightpaths a reality, to better support bursty and rapidly changing data traffic, a variation of OBS, which is called optical burst transport (OBT), was proposed by Kim et al. [7]. It can perform the no-blocking data transmissions and guarantee the QoS of the bursts.

The OBT is based on burst-mode transmission between senders and receivers on a WDM-enabled ring topology and does not require complex electronic processing. The burst transmission is made possible by swift reconfiguration of short-term lightpaths. The network architecture exhibits the following fundamental capabilities:

1. WDM-based technology with minimal electronic processing at the transit nodes.
2. Simple coupler-based wavelength add/drop. No MUX/DEMUX and active components on the rings.
3. Intelligent packet aggregation at OBT nodes.
4. Burst transmission via tunable lasers or tunable filters, or combinations of both.
5. Token-based wavelength access control.

In the OBT network, burst collisions can be easily avoided by using a token mechanism, which can also guarantee fairness among nodes. The traffic scheduler of a node manages the available capacity at multiple wavelengths for optimal transmission to destination nodes. In addition, control messages are exchanged between nodes, making the overall network reliable.

However, in the OBT network, the increase of the girth of the ring network or the number of stations on the ring will lead to the small value of the network efficiency and the long transmission latency [10]. In other words, to achieve a better data transmission performance, the scale of the OBT network needs to be limited.

In this chapter, we present a new efficient network technology for the next-generation 100-gigabit and terabit carrier-grade MAN, which is called resilient burst ring (RBR) [11]. It combines the concept of RPR with the burst-mode all-optical data transmission scheme to implement the ultra-high-speed data transmission with scalability and reliability. On the one hand, it inherits all the best features of the RPR networks and improves its data transmission rate to 100 Gbit/s or higher with the burst-mode all-optical data transmission scheme. On the other hand, it reduces the conflict probability of the data burst with the sample ring architecture and solves the packet loss problem with the two different buffering schemes. Moreover, it can provide different QoS classes to transmit different applications and services, which can cover all the applications and services in the existing networks, even including the TDM services.

The chapter is organized as follows. An overview of the RBR network is presented in Section 17.2. Then, node structure, control scheme, and QoS strategy of the RBR network are discussed in the following sections, respectively. After that, the performance analyses with simulation by computer are presented in Section 17.6, and we conclude the chapter in Section 17.7.

17.2 OVERVIEW OF THE RESILIENT BURST RING

RBR answers for the next-generation ultra-high-speed carrier-grade Optical Ethernet. It can not only provide the ultra-high-speed data transmission over the WDM ring network, but also provide the carrier-grade services, including scalability, reliability, QoS, service management, and TDM (time division multiplexing) support [12]. RBR has a two-sublayer architecture and adopts a bidirectional ring topology. It transmits data packets by the burst-mode all-optical data transmission scheme. The objective of the burst-mode all-optical data transmission is to assemble large bursts (i.e., optical burst packets, OBP) of electronic data packets (EDP) and transmit them optically by looking at the burst header (i.e., electronic control packets, ECP) through the downstream nodes. ECP is an out-of-band control signal. It can be processed electronically at each downstream node and sent ahead of the OBP in order to allow enough time for the downstream nodes to reconfigure. RBR adopts a two-layer buffering scheme to resolve the bursts' contentions, including the optical buffering scheme and the electronic buffering scheme. With this kind of buffering scheme, RBR can provide three different data transmission mode. In the remainder of this section, we will discuss those features of RBR, respectively.

17.2.1 Ring Topology with Burst-Mode Data Transmission

As shown in Figure 17.1, the same as RPR network, RBR adopts a bidirectional ring topology. That enables RBR to inherit all the best features of RPR. RBR adopts burst-mode data transmission scheme on the WDM ring network to implement the ultra-high-speed data transmission. Each data packet transmitted by RBR will be queued at the source node first and will then be assembled into an OBP before it is injected into the ring. After the burst-mode data transmission from the source node to the destination node, those packets will be disassembled at the destination node and then leave the RBR network.

In more detail, when data packets arrive at an RBR node, they will be scheduled and then queue at the source node, according to the information carried by each packet, such as the destination node, the QoS class, and others. When the whole length of those packets to the same destination node is larger than the minimal burst length, or there are some packets that have waited for a maximum assembly time (i.e., the min-burst-length and max-assembly-period assembling algorithm [13]), they will be assembled into an OBP by the source node. After that, the source node will select a proper ringlet and a free wavelength for the generated OBP and create the corresponding ECP that carries the information of the OBP. The ECP will be sent ahead of the OBP in order to allow enough time for the downstream nodes to reconfigure. Then, a proper offset time later, the OBP will be sent out on the selected data wavelength. The ECP will be processed electronically at each downstream node to reserve the resources for the corresponding OBP going through successfully. When the OBP arrives at the destination node, it will be dissembled and the electronic data packets in it will be scheduled and then exported.

Actually, the assembling, disassembling, and the resource reservation approaches of the above burst-mode data transmission are the same as those in the OBS technology. Since burst-mode all-optical transmission avoids the by-hop electronic data processing at the intermedial nodes, RBR can perform more

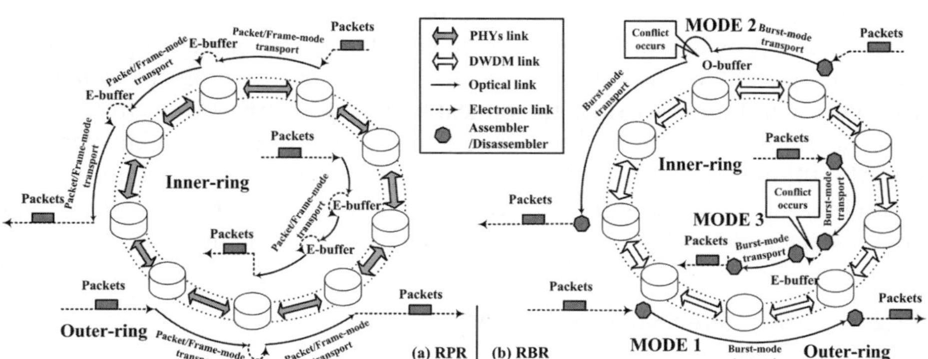

Figure 17.1. Differences between RPR and RBR network. E, electronic; O, optical.

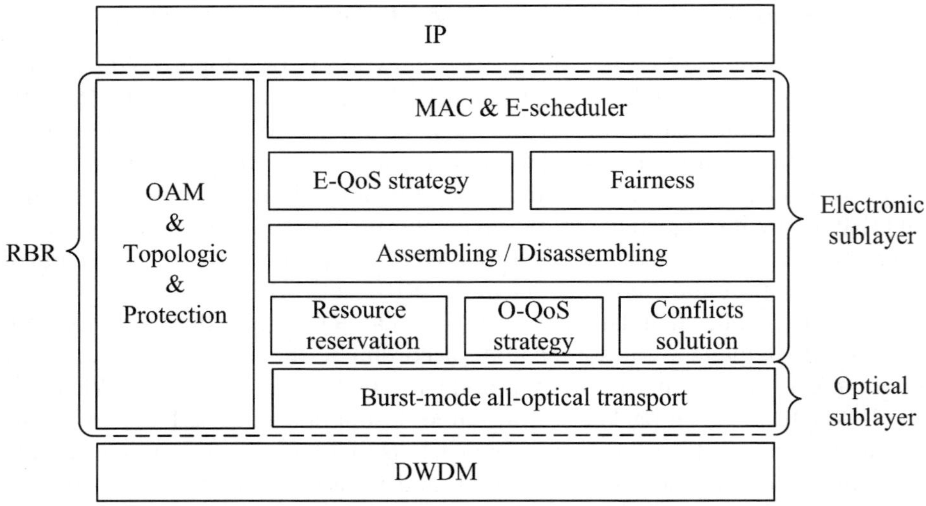

Figure 17.2. Hierarchical model of RBR. OAM stands for Operation, Administration, and Maintenance. E, electronic; O, optical.

efficiently than RPR. Moreover, because RBR does not use tokens to resolve the bursts' contentions, its network performance is also better than the OBT technology.

17.2.2 Hierarchical Model of RBR

The hierarchical model of RBR is shown in Figure 17.2. It can be divided into two sublayers, the electronic sublayer (ESL) and the optical sublayer (OSL).

The ESL implements medium access control (MAC), inherited QoS strategy (E-QoS strategy), and fairness scheme from RPR. Besides, it can realize the function of the assembling and disassembling between the EDPs and the OBPs. Moreover, resource reservation, QoS strategy (O-QoS strategy), and conflicts resolution for OBPs in the OSL are also performed in the ESL.

The OSL provides the interface to the WDM layer and implements the burst-mode all-optical data transmission. It can perform the OBP-based adding and dropping operations, as well as the optical buffering schemes.

In addition, there is an exclusive Operation, Administration, and Maintenance (OAM) [14] module in RBR. It is used not only to implement the control of the two sublayers and harmonize their works, but also to perform the topology and fault management and the network performance management.

17.2.3 Two-Layer Buffering Scheme

The two-layer buffering scheme is used to resolve the conflicts of the OBP in the OSL. In accordance with the name of the two sublayers of RBR, the two-layer

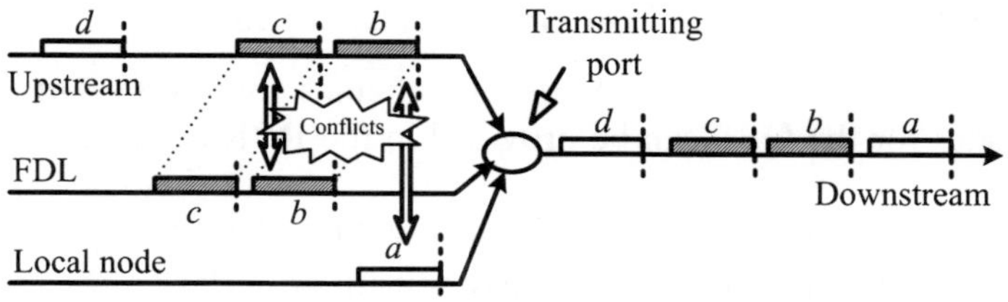

Figure 17.3. Burst contentions in the RBR network.

buffers are the optical buffer (O-buffer) for the OBPs in the OSL and the electronic buffer (E-buffer) for the electronic packets in the ESL. The O-buffer is implemented by the fiber delay lines (FDL) and the E-buffer is realized by the electronic buffers, such as the assembling/disassembling buffers at each RBR node.

When the conflict between an OBP to be forwarded and an OBP being inserted on the ring occurs at an RBR node, the node will buffer the OBP from the upstream with the O-buffer first and then export it directly to the downstream nodes. As shown in Figure 17.3, if OBP *b* reaches the local node at the time that OBP *a* is just being transmitted, conflict between OBP *a* and OBP *b* will occur. With the FDL, OBP *b* will be delayed. Then, if the arrival time of the following OBP from the upstream is not later than the time that the OBP in the O-buffer has been transmitted, the following OBP will be also delayed by the FDL, as shown in Figure 17.3. In other words, when O-buffer is used, if there is not enough available time between the two adjacent OBP from the upstream on the wavelength or the latter OBP is not received by the local node, all the following OBP will have to enter the FDL to avoid the conflict at the transmitting port.

When E-buffer is used, the OBP can be received by the intermedial RBR node, though it is not the destination of the OBP. After the O/E conversion and being disassembled, the OBP will be divided into data packets. Then, those data packets will be rescheduled and reassembled into a new OBP and retransmitted to the destination.

Obviously, the E-buffer scheme is an aggressive way to end the O-buffer data transmission mode. However, more importantly, it is the key scheme to make RBR inherit all the best features of RPR, including sharing the E-buffer of all the nodes on the ring, realizing the fairness of the traffic, recovering the network from incidents, and reducing the lost caused by link or node faults. In addition, it can make RBR support the TDM services, which has a quite strict restriction to the performance of the data transmission.

Since RBR adopts the burst-mode all-optical transmission scheme, a large electronic buffer should be required. That is the same as the electronic buffer required in the OBS edge node. For details on the parameters and performance of this electronic buffer, we refer the interested reader to reference 15. Finally, since O-buffer is much faster than E-buffer, the two-layer buffering scheme can make RBR provide flexible QoS strategy. Detailed discussions will be provided in Section 17.5.

17.2.4 Three-Mode Data Transmission Scheme

With the two sublayers and their buffering schemes, three different data transmission modes for the OBP can be provided in the RBR network.

- MODE 1: All-optical data transmission with no buffer.
- MODE 2: All-optical data transmission with O-buffer.
- MODE 3: Hybrid data transmission with E-buffer.

With the resource reservation by the ECP, if there is no conflict occurring, the OBP will be transmitted directly to the destination node through an all-optical path without any buffering. This is the all-optical data transmission with no buffer, as MODE 1 in Figure 17.1b shows.

However, resource contentions cannot be avoided in the burst-mode all-optical transmission. When resource contention occurs at an intermedial RBR node, the node will buffer the OBP from the upstream with the O-buffer first and then export it. That is the all-optical data transmission with O-buffer, as MODE 2 in Figure 17.1b shows.

Alternatively, when the conflict occurs, the OBP can also be buffered by the intermedial RBR node with the E-buffer. That is, the OBP can be received by the intermedial RBR node though it is not the destination of the OBP. After the O/E conversion and being disassembled, the OBP will be divided into data packets. Then, another burst-mode data transmission process from this intermedial RBR node to the destination will be performed. That is the hybrid data transmission with E-buffer, as MODE 3 in Figure 17.1b shows.

In the hybrid data transmission with E-buffer, packets in the OBP will be disassembled and rescheduled at the intermedial node after O/E conversion. It should be pointed out here that when those packets are being rescheduled by the intermedial RBR nodes, both Cut-through and Store-and-Forward algorithms [16] of RPR can be adopted in the RBR network. In more detail, when the Cut-through algorithm is adopted, the OBP received by the intermedial RBR node will be reassembled as soon as it is converted into electronic signals, and a separate transmitting buffer will be used to buffer those electronic packets. On the contrary, when the Store-and-Forward algorithm is adopted, no separate transmitting buffer will be used and those packets will queue with the packets of the intermedial RBR node. Obviously, the same as that in the RPR network, data packets transmitted by the Cut-through algorithm in the RBR network will have a better performance than by the Store-and-Forward algorithm.

17.3 NODE STRUCTURE OF THE RBR NETWORK

To implement the two-layer buffering scheme and the three-mode data transmission scheme, an appropriate RBR node structure is illustrated in Figure 17.4.

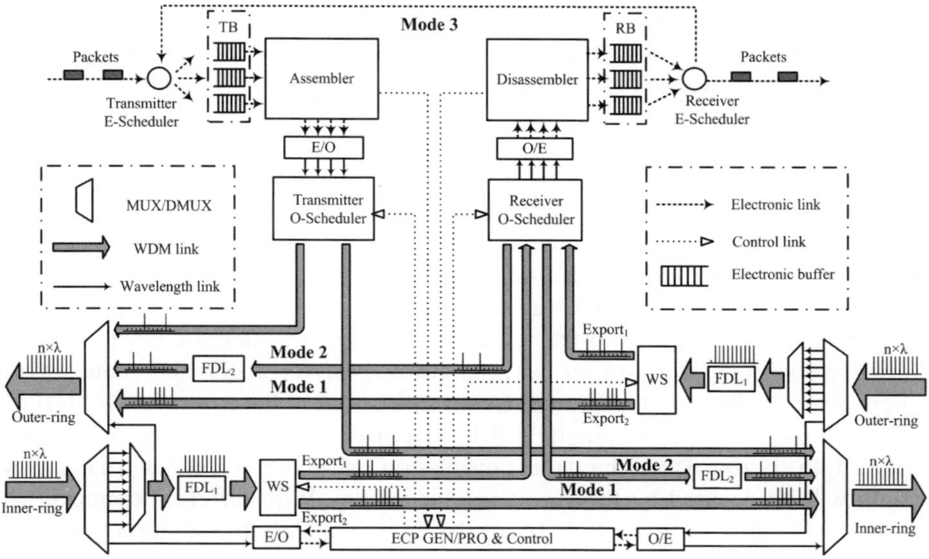

Figure 17.4. Node structure of the RBR network. ECP GEN/PRO, generator and processor of electronic control packets; WS, wavelength splitter; FDL, fiber delay lines; E, electronic; O, optical, TB, transmitting buffer; RB, receiving buffer; Mode 1, all-optical data transmission with no buffer; Mode 2, all-optical data transmission with O-buffer; Mode 3, hybrid data transmission with E-buffer.

There are two electronic queuing buffers in the RBR node. One is the transmitting buffer, and the other is the receiving buffer. Both of them are used to buffer the electronic data packets. The function of the assembler and the disassembler are used to perform the conversions between the EDPs and the OBPs.

The O-scheduler is used to implement the selective connections and constructed by several 1×2 optical switches. For the transmitter O-scheduler, it is used to implement the alternative connection of the assembler to the two ringlets. For the receiver O-scheduler, it is used to implement the alternative connection of the wavelength splitter to the disassembler or the O-buffer.

The core component of the RBR node is the wavelength splitter (WS), which is used to implement the program-controlled wavelength splitting. When there are several wavelengths input into the wavelength splitter, with the different control signals, the wavelength splitter can divide those wavelengths into two different groups. As shown in more detail in Figure 17.4, different wavelengths will be exported from the $Export_1$ of the wavelength splitter with the different control signals, while the rest will be exported from its $Export_2$. This wavelength splitter can be realized by many optical devices, such as a tunable flitter like the acousto-optical tunable filter (AOTF), a certain number of 1×2 optical switches, and so on. The fast optical switches of the test-bed in reference 17 can be considered as a material form of the wavelength splitter. Since the wavelength splitter is only used to implement the wavelengths splitting, its cost will be much lower than that of the complex optical switch matrix in the OBS node.

There are two types of the FDL (FDL_1 and FDL_2) in the RBR node. FDL_1 is used to compensate the processing delay of the corresponding ECP at the intermedial node. FDL_2 acts as the O-buffer buffering the OBP exported from the wavelength splitter when conflict occurs.

The three-mode data transmission scheme is performed by the wavelength splitter and the O-schedulers. MODE 1 is the all-optical data transmission with no buffer. To achieve this, the wavelength splitter will let the wavelength that carrying an OBP from the upstream node directly go through the local node. As shown in Figure 17.4, when the wavelength is exported from FDL_1, the wavelength splitter will export it through $Export_2$ directly. On the contrary, for MODE 2 and MODE 3, the wavelength will be exported through the $Export_1$ of the wavelength splitter. Then, the data transmission mode of the OBP on that wavelength will be decided by the receiver O-scheduler. With different connections of the wavelength splitter with FDL_2 or the disassembler, different data transmission modes will be performed. In more detail, when the receiver O-scheduler connects the wavelength splitter with FDL_2, the OBP exported from the wavelength splitter will enter FDL_2 and then be exported to the downstream node. That is MODE 2, the all-optical data transmission with O-buffer. Otherwise, when the receiver O-scheduler connects the wavelength splitter with the disassembler, the OBP exported from the wavelength splitter will be disassembled by the local node. Then, the packets disassembled from the original OBP will be rescheduled and reassembled into a new OBP and will then be transmitted to the downstream node. That is MODE 3, the hybrid data transmission with E-buffer.

17.4 CONTROL SCHEME OF THE RBR NETWORK

The three-mode data transmission scheme in the RBR network is realized by a novel control protocol, which is called the Priority Only Destination Delay (PODD) protocol. This protocol is a variation of the Only Destination Delay (ODD) control protocol that was proposed in reference 18. In the ODD protocol, as the processing delay of the intermedial nodes is compensated by the FDL, the offset time between the bursts and their control packets is only related to the processing delay of the destination node. The difference between the PODD and ODD protocol is that the PODD protocol has an extra offset time that is used to guarantee the QoS of the OBP in terms of the lower blocking probability for the higher-priority OBP.

Besides, another key point of the PODD protocol is that the offset time between each OBP and its ECP is invariable during the whole transmission approach. This is used to keep the firm scheduling relationship between the OBP and its ECP and avoid the conflicts caused by the variable offset time. As shown in detail in Figure 17.5a, when the conflict between OBP *a* and OBP *b* occurs, the OBP *a* will be buffered by the FDL. When OBP *a* exports from the FDL, a new conflict between OBP *a* and OBP *c* will occur. All the controls and conflict solutions of the OBP are implemented by their ECP. If the offset time can be variable and the OBP *c* has already been delayed by the upstream node, its offset time will be changed from "Offset time" to "Offset time + T_{FDL2}." Since ECP *c* has already transmitted to the downstream node when ECP *a* arrives and OBP *a* cannot be buffered by the FDL again, there is no way to resolve the conflict and the packet loss will occur. On the contrary, if the offset time between each OBP and its ECP is invariable, since ECP *c* will arrive at the local node after ECP *a*, the conflict above can be resolved by buffering the OBP *c* at the local node again, as Figure 17.5b shows.

Detailed resource reservation schemes of the PODD control protocol in the three different data transmission modes are shown in Figure 17.5c–e. When there is no conflict, the OBP will be transmitted through an all-optical path. Each time it goes through an intermedial node, it will be delayed the length of FDL_1, which is used to counteract the processing delay of the corresponding ECP, as shown in Figure 17.5c.

For the O-buffer mode, as shown in Figure 17.5d, when conflict occurs at node 2, the node will use FDL_2 to resolve the resource contention. As a result, the corresponding OBP will be delayed the length of FDL_2 more. Due to the reason discussed above, the ECP will be also delayed a period that is equal to the length of FDL_2 by node 2 in order to keep the offset time between the OBP and its ECP invariable.

For the E-buffer mode, as shown in Figure 17.5e, the resource reservation can be considered as two independent all-optical data transmission processes. If node 2 decides to receive the OBP whose destination is node D, the corresponding ECP will be destroyed by node 2 first. When the OBP have been received by node 2, the packets will be rescheduled and reassembled there. Finally, a new OBP (OBP′) and its corresponding ECP (ECP′) will be created by node 2 and transmitted to Node D, respectively.

Finally, it should be pointed out here that in the PODD control protocol, the OBP with a high priority will have a long extra offset time and can adopt MODE 2 to resolve contentions, while the OBP with a low priority will have a short or even no extra offset time and can adopt MODE 3 to resolve contentions.

17.5 QOS STRATEGY OF THE RBR NETWORK

The two-sublayer architecture enables RBR to have more flexible QoS strategy. On the one hand, in the OSL, with the PODD control protocol, the QoS strategy for different class OBPs can be provided by the different extra offset time. That is, the OBP with a high priority will have a long extra offset time and the OBP with a low priority will have a short or even no extra offset time. On the other

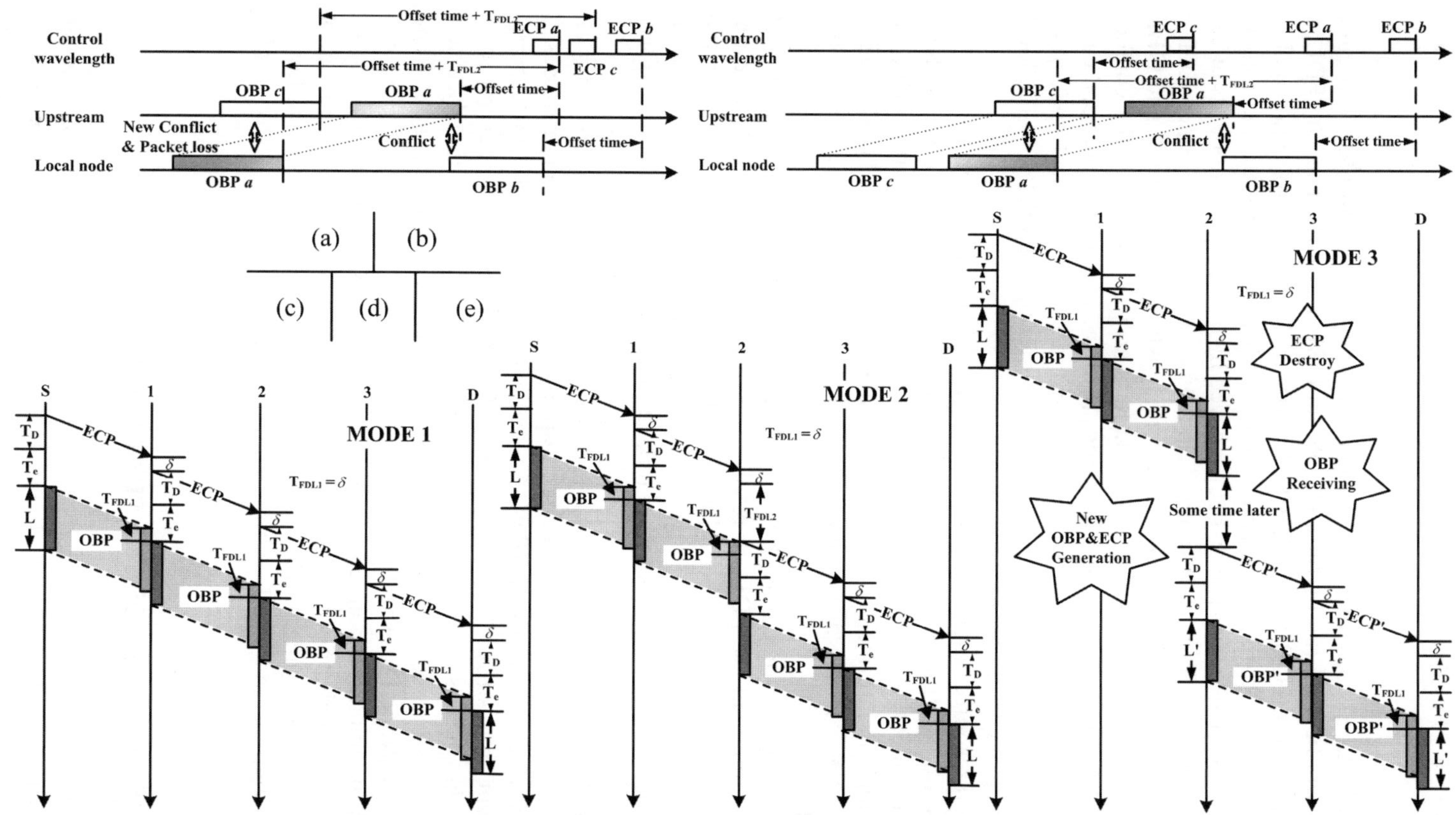

Figure 17.5. Control scheme of the RBR network. T_{FDL1} and T_{FDL2} are the length of FDL_1 and FDL_2 in Figure 17.3, respectively. S, source node; D, destination node; 1 to 3, intermedial nodes; T_D, the processing time for the destination node; T_c, the extra offset time for higher priority OBP; δ, the ECP processing time for the intermedial node; L, the length of the OBP; MODE 1, all-optical data transmission with no buffer; MODE 2, all-optical data transmission with O-buffer; MODE 3, hybrid data transmission with E-buffer.

hand, in the ESL, with the QoS strategy and fairness algorithm inherited from RPR, QoS strategy of RBR can be provided by the Cut-Through and the Store-and-Forward data transmission schemes. Besides, in the RBR network, resource contentions caused by the OBPs with the high priority can be resolved by the optical buffering mode, while resource contentions caused by the OBPs with the low priority can be resolved by the electronic buffering mode.

Detailed QoS strategy of the RBR network is presented in Table 17.1. It can satisfy the different performance requirements of the different applications and services in the existing networks, such as audio, video, data applications, TDM services, and so on. In addition, it should be pointed out here that in order to implement the proposed QoS strategy in the RBR network, packets with different QoS class and different destination will be placed into different queues and then assembled into different OBPs at the source RBR node. Concerning the generated OBP, according to their different QoS classes, different transmission scheme will be performed.

17.5.1 Audio and Video Applications

Audio and video applications here mean the real-time applications—for example, IP phone, video conference, and others. There are two major performance requirements for these kinds of audio and video applications: One is the low latency, and the other is the small jitter.

As shown in Table 17.1, because the all-optical data transmission modes (including the no buffer mode and the O-buffer mode) are adopted in Class A, it is easy to satisfy the requirement on the low latency. Concerning the requirement on jitter, because those applications are real-time and high interactive, jitter less than 1 ms [19] should be satisfied.

To do this, several parameters of the RBR network should be considered, including the maximum assembly time, the length of the O-buffer, and the maximum number of the nodes on the ring. If we define that T_{assemble} is the maximum assembly time, $T_{\text{O-buffer}}$ is the length of the O-buffer, and h represents the hops from the source node to the destination node, the maximum jitter $J_{\max}$ of the packets transmitted by all-optical data transmission modes can be calculated by Eq. (17.1).

$$J_{\max} = T_{\text{assemble}} + h \times T_{\text{O-buffer}} \tag{17.1}$$

From Eq. (17.1), we can see that if we assume that the maximum assembly time of the bursts is 500 μs and the latency of the O-buffer at each RBR node is 50 μs, to satisfy the requirement on jitter that is less than 1 ms, the maximum hops from the source RBR node to the destination RBR node should be not more than 10. In other words, the maximum number of the node on the ring should not exceed 21.

Compared with the other applications, real-time audio and video applications need a higher priority. Therefore, to reduce the conflict probability of those packets, a long extra offset time is adopted in QoS Class A.

TABLE 17.1. QoS Strategy of the RBR network

Class of Services		QoS Requirements			Implementations in RBR			
Class	Examples of Use	Guaranteed Bandwidth	Latency	Jitter	Transmission Mode	Extra Offset Time	Rescheduling Mode in ESL	Fairness Eligible
A	Audio, video, and high Interactive data transfer	Yes	Low	<1 ms [19]	MODE 1 or MODE 2	Long	—	No
B	Traditional real-time data transfer (HTML, Transaction services)	Yes	Middle	—	MODE 1 or MODE 3	Short	Cut-Through	No
C	Traditional best-effort applications of default IP networks	No	—	—	MODE 1 or MODE 3	0	Store-and-Forward	Yes
D	TDM services	Yes	Middle	<<1 ms [20]	MODE 3(by-hop)	0	Cut-Through	No

Note: MODE 1, All-optical data transmission with no buffer; MODE 2, all-optical data transmission with O-buffer; MODE 3, hybrid data transmission with E-buffer.

17.5.2 Data Applications

Data applications can be divided into three types: the high interactive data applications, the traditional real-time data applications, and the non-real-time data applications with the scheme of best effort. Different data applications have different performance requirements [19]. According to their different performance requirements, different QoS classes are provided in the RBR network.

First, for the high interactive data transfer, such as interactive games, telnet, command/control, and so on, because those applications have the same performance requirement on latency as the real-time audio and video applications [19], the data transmission modes for those applications provided by the RBR network are the same as those for the real-time audio and video applications, as shown in Table 17.1.

Then, as the traditional real-time data applications, including Web browsing, E-mail (server access), high-priority transaction services (E-commerce), bulk data, still image, fax ("real-time"), and so on, can abide a certain bound of latency [19] and have a lower priority compared with the applications in QoS class A, RBR adopts the e-buffer data transmission mode with a short extra offset time to transmit those applications, as QoS Class B in Table 17.1 shows. In addition, when the E-buffer is used during their transmission, to ensure their performance requirement on the latency, the Cut-though data transmission algorithm is adopted.

Finally, QoS Class C of the RBR network is applied to all the non-real-time data applications with the scheme of best effort. The non-real-time data applications include E-mail (server to server transfer), low-priority transaction services, usenet, fax (store and forward), and so on. The same as QoS Class B, E-buffer data transmission mode can be adopted in QoS class C. Moreover, because those non-real-time data applications have no special performance requirements and have the lowest priority among all the applications, when E-buffer is used during their transmission, the Store-and-Forward data transmission algorithm and the fairness algorithm can be adopted in the RBR network.

17.5.3 TDM Services

Although Internet traffic has been explosively grown in the existing networks, TDM service still plays an important role in data transmission due to its high quality service. TDM support is an important feature of the next-generation Metropolitan Area Network. However, to realize the transmission of the TDM traffic with the jitter much less than 1 ms [20] in the RBR network, it means that all those OBP that contain TDM frames should have the same processing time at each intermedial RBR node. Therefore, in order to support the TDM services, RBR provides a by-hop electronic data transmission mode. It can set up a steady data transmission pipeline for the TDM services and realize the transmission with the jitter much less than 1 ms.

The assembling and disassembling schemes for the TDM frames are the same as those for the other data packets. When the TDM frames arrive at an RBR node, they will be buffered in the transmitting buffer first. TDM traffic is characterized as length-fixed and interarrival-time-fixed. Therefore, regardless of whether the assembling scheme is time-based or length-based, the generated OBP traffic that contains TDM frames will be also characterized as length-fixed and interarrival-time-fixed. Because all the other data transmission modes will bring the delay variation when conflicts between OBP occur, the by-hop electronic processing is the best choice, and concerning the frames rescheduling algorithm at each node, Cut-Through algorithm, is the optimal option compared with Store-and-Forward algorithm, which will make the packets have different queuing delay.

17.6 PERFORMANCE EVALUATION OF THE RBR NETWORK

Different parameters setting will provide different data transmission performance in the RBR network. This enables the RBR network to satisfy the performance requirements of all the applications and services in the existing networks.

By computer simulations, a 17-node RBR network is used to investigate the data transmission performance. There are 10 data wavelengths in each ringlet, and the data transmission rate of each wavelength is 10 Gbit/s. The distance between each two adjacent RBR nodes is 10 km, and the processing delay of ECP at each RBR node is 50 ns.

Figure 17.6a–c plot the transmission latency and jitter of traffic with different QoS classes between two given RBR nodes under different network load conditions. Latency is measured from the time that a packet or TDM frame arrives at an RBR node to the time that it leaves its destination node. The average latency in Figure 17.6 is defined in each 10 ms of the simulation time.

From Figure 17.6, we can see that with the different data transmission modes, the transmission latency and jitter of the different QoS classes are different. For class A, with the all-optical data transmission mode, they have a very low latency and a small jitter. However, for class B and C, with the adoption of the E-buffer mode, they have a high latency and a large jitter. Concerning class D, with the by-hop E-buffer mode and Cut-Through algorithm, its transmission latency is almost a straight line.

Moreover, it should also be observed that the impact on the traffic with different QoS classes caused by the variation of the network load is different. Traffic with QoS class C is the most sensitive to the change of the network load. The transmission latency and jitter of the traffic with QoS class C under the high network load condition are much higher than that under the low network load condition. That is because they carry the traditional best-effort applications of default IP networks and have the lowest priority in the RBR network. Concerning traffic with QoS class A and D, as shown in Figure 17.6, they are insensitive to the change of the network load. That is because they have the highest priority in the RBR network and adopt the by-hop E-buffer transmission mode, respectively.

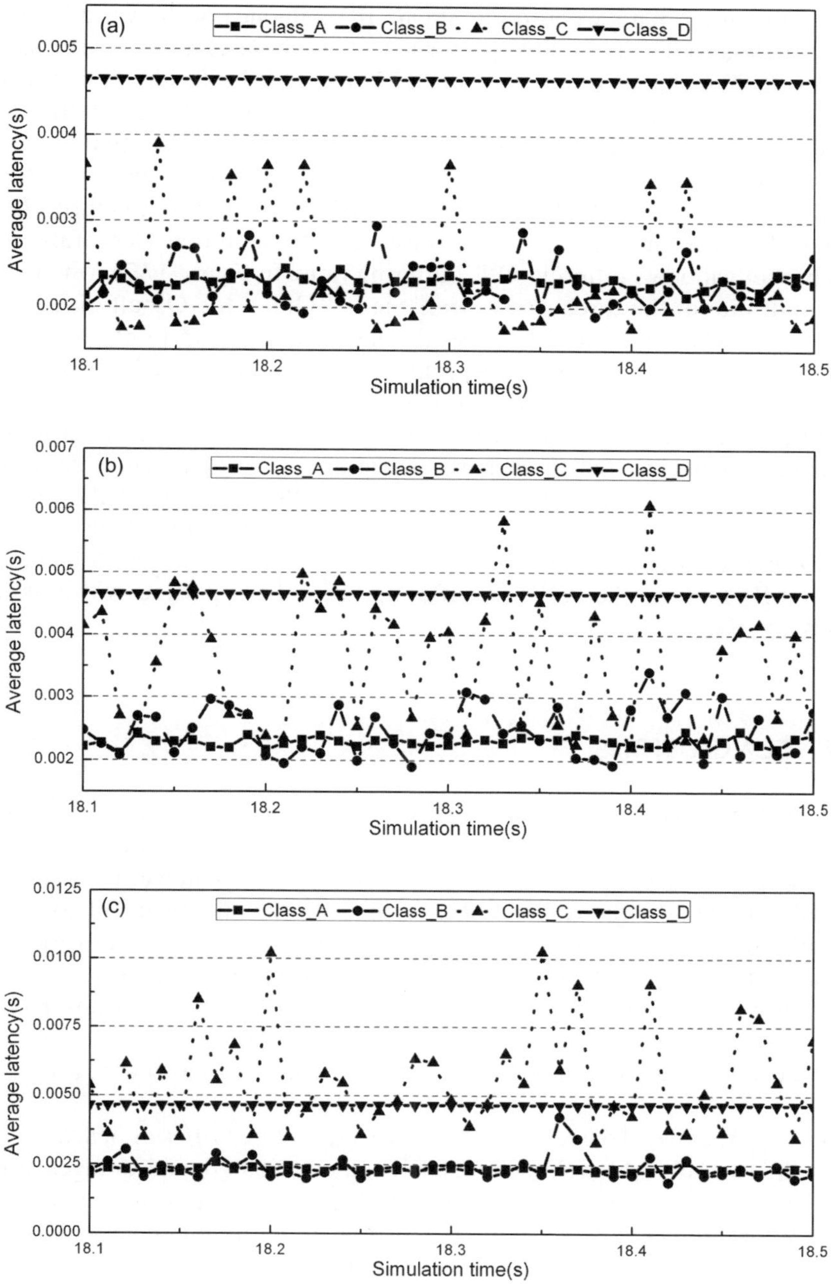

Figure 17.6. Transmission latency for different QoS class traffic in the RBR network with different network load. **(a)** Low network load. **(b)** Middle network load. **(c)** High network load.

On the other hand, we also investigate the influence of the variation of the different class traffic on the QoS performance of the RBR network. The simulation results have been shown in Figure 17.7. The O/E buffering ratio is defined as the ratio of the packets; that have been transmitted through the O-buffer mode or the E-buffer mode to all the packets with the same QoS class that have been transmitted. This parameter shows the conflict probability of the packets; and the higher it is, the longer the average packets' latency is. From Figure 17.7, we can see that with the different extra offset time for different class traffic, each class packet will have different O/E buffering ratio. In Figure 17.7a, when all the data input rate of the three classes traffic are 30 Gbit/s, the O-buffering ratio of packets with class A is only 3.7%, while the E-buffering ratio of packets with class B and C are 29% and 58%, respectively. This is because the long extra offset time makes the packets have a high reservation priority. The longer the extra offset time, the lower the conflict probability and the lower the buffering ratio. Moreover, comparing the three parts of Figure 17.7, we can see that another function of the extra offset time is that it make the variation of the low class traffic have no influence on the transmission performance of the high-lass traffic. As shown in Figures 17.7b and 17.7c, no matter how much the input data rate of the class B and C traffic is, the O-buffering ratio of the packets with class A is always invariable.

17.7 CONCLUSIONS

The world of transport is on the verge of significant change. The days of T1/E1s being the driving force are fading as packet services and high-capacity access networks grow. To meet the ever-increasing demand for bandwidth and network flexibility, integrating the resilient packet ring with the intelligent WDM, a novel network technology for the next-generation Metropolitan Area Network, which is called a resilient burst ring (RBR), is presented in this chapter. It can implement the WDM-supported ultra-high-speed carrier-grade data transmission with scalability and reliability.

We can now summarize the RBR network with the following features:

1. *Two-Sublayer Architecture and the Corresponding Buffering Schemes.* With the two-sublayer architecture, RBR can inherit all the advantages of RPR and implement the burst-mode ultra-high-speed data transmission in the WDM ring networks. With the two-layer buffering scheme, RBR can resolve the conflicts in the optical domain completely.
2. *Three-Mode Data Transmission Scheme and the Corresponding Control Algorithms.* With the proposed Priority Only Destination Delay (PODD) protocol in the OSL and the Cut-Through and Store-and-Forward algorithms inherited from RPR in the ESL, both the ultra-high-speed data transmission and the carrier-grade services are realized.

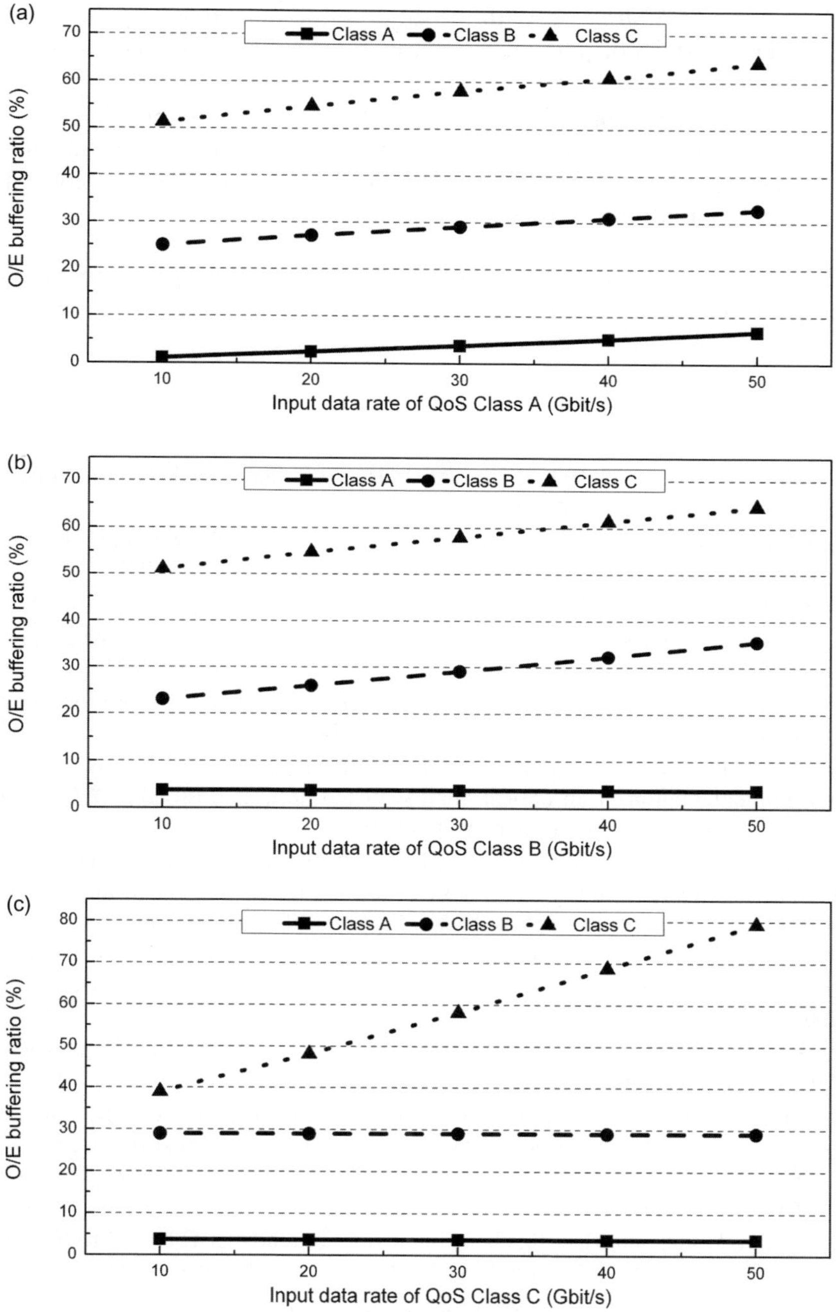

Figure 17.7. Influence of the variation of the different class traffic on the QoS performance of the RBR network. **(a)** The data input rates of QoS class B and C are 30 Gbit/s. **(b)** The data input rates of QoS class A and C are 30 Gbit/s. **(c)** The data input rates of QoS class A and B are 30 Gbit/s.

3. *Flexible QoS Strategy.* With the different parameter setting and different data transmission modes, RBR can satisfy the requirements of all the applications and services in the existing networks.
4. *TDM Services Support.* The steady data transmission pipeline set up by the by-hop electronic processing with Cut-Through algorithm at each intermedial node can realize the synchronous data transmission on the packet-based optical transport network.

On all accounts, with the four features above, RBR will be a very competitive solution for the next-generation ultra-high-speed Metropolitan Area Networks.

REFERENCES

1. T. Hills, Next-Gen SONET, Lightreading Rep. [Online]. Available on http://www.lightreading.com/document.asp?doc_id=14781, 2002.
2. L. Choy, Virtual concatenation tutorial: Enhancing SONET/SDH networks for data transport, *J. Opt. Networking*, Vol. 1, No. 1, pp. 18–29, December 2001.
3. Resilient packet ring (RPR) IEEE 802.17.
4. F. Davik, M. Yilmaz, S. Gjessing, and N. Uzun, IEEE 802.17, Resilient packet ring tutorial, *IEEE Commun. Mag.*, Vol. 42, No. 3, pp. 112–118, March 2004.
5. C. Qiao and M. Yoo, Optical burst switching (OBS)—A new paradigm for an optical internet, *J. High Speed Networks*, Vol. 8, pp. 69–84, 1999.
6. Y. Sun, T. Hashiguchi, V. Q. Minh, X. Wang, H. Morikawa, and T. Aoyama, Design and implementation of an optical burst-switched network testbed, *IEEE Commun. Mag.*, Vol. 43, No. 11, pp. 48–55, November 2005.
7. J. Kim, J. Cho, S. Das, D. Gutierrez, M. Jain, C.-F. Su, R. Rabbat, T. Hamada, and L. G. Kazovsky, Optical burst transport: A technology for the WDM metro ring networks, *J. Lightwave Technol.*, Vol. 25, No. 1, pp. 93–102, January 2007.
8. J. Kim, M. Maier, T. Hamada, and L. G. Kazovsky, OBT: Optical burst transport in metro area networks, *IEEE Commun. Mag.*, Vol. 45, No. 11, pp. 44–51, November 2007.
9. L. Xu, H. G. Perros, and G. N. Rouskas, Access protocols for optical burst-switched ring networks. *Information Sci.*, Vol. 149, 75–81, 2003.
10. X. Liu, G. Wen, H. Wang, and Y. Ji, Analyses, simulations, and experiments on the performance of the token-based optical burst transport ring networks, in *Proceedings of SPIE—The International Society for Optical Engineering, Optical Transmission, Switching, and Subsystems*, 2007.
11. X. Liu, H. Wang, and Y. Ji, Resilient burst ring: Extend IEEE 802.17 to WDM networks, *IEEE Commun. Mag.*, Vol. 46, No. 11, pp. 74–81, November 2008.
12. Metro Ethernet Forum. http://www.metroethernetforum.org/
13. X. Cao, J. Li, Y. Chen, and C. Qiao, Assembling TCP/IP packets in optical burst switched networks, in *IEEE Globecom 2002*, Taipei, Taiwan, 2002, 11.
14. D. O'Connor, Ethernet service OAM—Overview, applications, deployment, and issues," in *OFC2006*, NWF2, 5–10 March 2006.

15. H. Li and I. Thng, Edge node buffer usage in optical burst switching networks, *Photonic Network Commun.*, Vol. 13, No. 1, pp. 31–51, January 2007.
16. D. A. Schupke and A. Riedl, Packet transfer delay comparison of a store-and-forward and a cut-through resilient packet ring, *Proc. Broadband Communi.*, pp. 12-1–12-5, 2002.
17. H. Yoo, J. P. Park, S. Han, J. S. Cho, Y. H. Won, M. S. Lee, M. H. Kang, Y.-K. Seo, K. J. Park, C. J. Youn, H. C. Kim, J.-K. Rhee, and S. Y. Park, Microsecond optical burst add–drop multiplexing for WDM ring networks, in *OFC2006*, OWP6, March 5–10 2006.
18. L. Xu, H. G. Perros, and G. N. Rouskas, A simulation study of optical burst switching access protocols for WDM ring networks, in *Networking 2002*, LNCS 2345, pp. 863–874.
19. ITU-T Recommendation G.1010 (11/2001), End-User Multimedia QoS Categories.
20. ITU-T Recommendation G.825 (03/2000), The Control of Jitter and Wander within Digital Networks which Are Based on the Synchronous Digital Hierarchy (SDH).

18

MULTIPROTOCOL LABEL SWITCHING

Mario Baldi

18.1 WHY MPLS?

Since its inception, Multiprotocol Label Switching (MPLS) has received a lot of attention in the research community, has been in the headlines in the technical news, and has achieved importance in the marketing strategies of vendors. Noteworthily, MPLS is one of the few examples of networking technology that found the consensus of virtually all vendors of core IP routers, Internet Service Providers, and telecommunications operators, that is, it is the one technology that finally achieves the long sought convergence between the information technology and the telecommunications worlds.

MPLS importance stems from its potential to make IP networks in general and the Internet in particular what have been traditionally called *public networks*—that is, networks on which operators provide services to be sold. Public networks, domain of the telecommunications world, have been typically providing services with known quality so that they can be the object of contracts between providers and their customers and can be sold for a significant amount of money.

Convergence of Mobile and Stationary Next-Generation Networks, edited by Krzysztof Iniewski

18.1.1 The Onion That Makes Telcos Cry

Mainly due to the success of the Internet with its steady growth and ubiquitous reach, IP has for a long time shown its predominance as the preferred protocol and has more lately become the one protocol on which to base each application. However, network operators have been for many years compelled to deploy several other technologies to implement the services they wanted to sell. This led to the "networking onion" shown in Figure 18.1 where various technologies are deployed in layers to finally support the provision of IP connectivity, as shown by the outer layer. A wavelength division multiplexing (WDM) core is used to (i) maximize the transport capability of each fiber by transmitting over multiple optical channels on the same fiber and (ii) create long haul optical channels across multiple fibers among devices that are not directly connected. This is achieved through the deployment of (i) optical multiplexers and (ii) optical switches, a.k.a. lambda (λ) switches. Although more recently optical channels are being used to connect IP routers directly (as represented in Figure 18.1 by the pipes extending from the WDM core to the IP outer layer) or even customer equipment, most often a circuit-switched network is built on them by means of SDH/SONET multiplexers and switches in order to enable the provision of lower-capacity channels. While the circuit-switching network can be used to sell services—specifically telephony and leased lines—directly to the customers, it does not feature the flexibility required for handling what is traditionally called *data traffic*. ATM (Asynchronous Transfer Mode) [1] was standardized by ITU-T [2] with the specific goal of realizing a Broadband Integrated Services Digital Network (ISDN)—that is, an infrastructure capable of providing the determin-

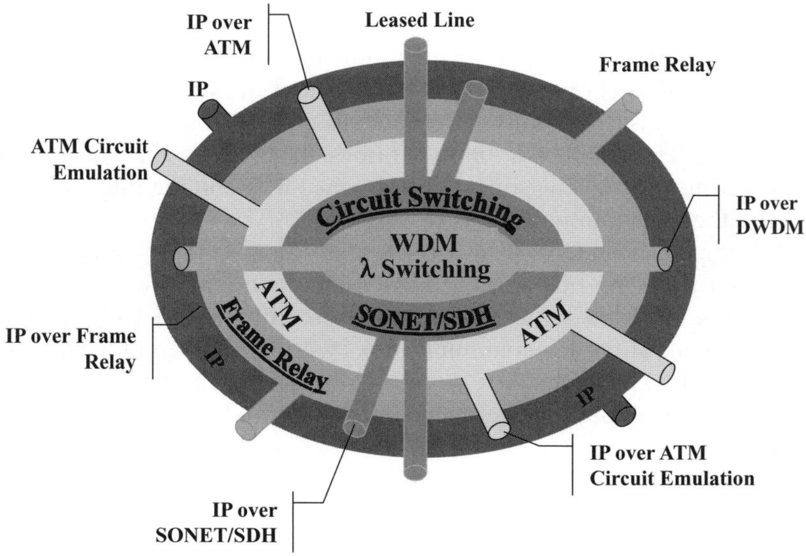

Figure 18.1. The "networking onion."

istic quality of service required by traditional telecom applications (such as telephony and videoconferencing) while efficiently supporting data traffic. Packet switching rather than circuit switching was chosen to achieve this goal, as highlighted by the term "asynchronous" that is juxtaposed to the synchronous nature of circuit switching.

ATM failed to achieve the anticipated success and widespread reach because end users preferred the simplicity and low cost of technologies such as Ethernet and IP to the more sophisticated services of the more costly and complicated ATM. As a consequence, ATM has been confined to the operators' networks with few customers requiring ATM services (as represented by the pipes extending through the outer layers in Figure 18.1), possibly to interconnect their IP routers, or taking advantage of appealing commercial offers for circuit emulation over the ATM network to interconnect their PBXs (private branch exchanges). Instead, operators resorted to their ATM infrastructure to interconnect their IP routers and provide IP connectivity services.

Although the deployment of an ATM backbone to interconnect IP routers provides some advantages in controlling the traffic and hence the service provided (as discussed later), it is problematic due to the high cost of ATM interfaces on IP routers, which is due to the high complexity of ATM. Consequently, equipment manufacturers designed access devices with Frame Relay [3] interfaces mainly used for the interconnection of IP routers (both the operators' and the customers' ones). Frame Relay is a convenient solution for the interconnection of IP routers because, being originally proposed by a consortium of IP router manufactures named the Frame Relay Forum,[1] interfaces are available on virtually all high end routers at basically no cost. However, this required operators to add yet another layer to their "networking onion."

Such a multitude of technologies represents a significant cost for operators that need personnel trained on each of them, gear from several vendors, and spare parts for a large number of devices. Especially as the popularity and ubiquity of IP, further reinforced by the increasing deployment of IP telephony, results in most of the customers requiring IP connectivity services, operators would certainly benefit from getting rid of the "networking onion" in favor of an all IP network—that is, a mesh of IP routers interconnected by optical fibers.

However, controlling the quality of the services offered on such a mesh interconnection of IP routers was until recently very complex, unless the network was engineered so that actual traffic would be a small fraction of the capacity. The sophisticated traffic management support offered by ATM enabled the realization of services with known quality through the ATM backbone while efficiently utilizing its resources. Consequently, *commercial* IP services have been provided by interconnecting peripheral IP routers through a partial mesh of virtual connections across an ATM backbone.

[1]The MPLS forum later merged with into the MPLS/FR Alliance, which in 2008 joined the DSL Forum into the Broadband Forum [4].

18.1.2 Enters MPLS

MPLS introduces the opportunity of controlling traffic within a mesh of interconnected IP routers basically by enhancing IP with a powerful *traffic engineering* capability. The reason why this is important, its consequences, and the technological solutions through which this is realized are discussed in the rest of this chapter. In a nutshell,

MPLS enables IP to work as a connection-oriented protocol while still retaining the valuable properties of a connectionless one.

This statement, which might sound cryptic and oxymoronic at first, should become clear at the end of the chapter. One key consequence of connection oriented operation is the possibility of facilitating *traffic engineering*, which is instrumental for an operator to control the quality of the services provided by its network. The way in which MPLS enables traffic engineering and the differences with respect to performing traffic engineering with the connectionless operation of IP are discussed in detail in Section 18.2.5.

By replacing ATM switches with MPLS capable IP routers, the ATM layer can be removed from the "networking onion" in Figure 18.1, and with it the Frame Relay layer can be peeled off as well. After a closer look at the way MPLS uses label switching to forward packets through the network, one might argue that in its basic principles it is not different from ATM. This is true, but the key for MPLS's popularity as a backbone technology for carrying IP traffic versus ATM's failure in the same role lies in the former using the same control plane as IP routers, with the latter requiring to operate two different, hard-to-integrate control planes, one for ATM switches and one for IP routers. This concept will be elaborated further in Sections 18.2.4 and 18.4.5.

The distinctive features of MPLS and the widespread adoption of IP to carry voice and telephony enable operators to peel the SONET/SDH layer off the "networking onion" in Figure 18.1. In fact, thanks to the traffic engineering and fast fault restoration capabilities of MPLS operators are able to multiplex and demultiplex IP packet flows on high capacity (2.4 Gbit/s, 10 Gbit/s, 40 Gb/s) optical channels with quality levels that, although lower than SONET/SDH, are suitable to both voice/telephony over IP and IP connectivity services as required by their customers.

As a result of MPLS deployment, Telcos' IP-based networks can realize the network architecture graphically represented in Figure 18.2 where a WDM core based on optical switches is deployed to provide connectivity to MPLS-capable IP routers.[2] MPLS as a packet switching solution is not expected to replace the WDM core because of the scalability and potential inexpensiveness of (all) optical switching where large volumes of traffic are to be switched along the same

[2]Note that although the architecture is here described as "future," it is already being used by a number of operators.

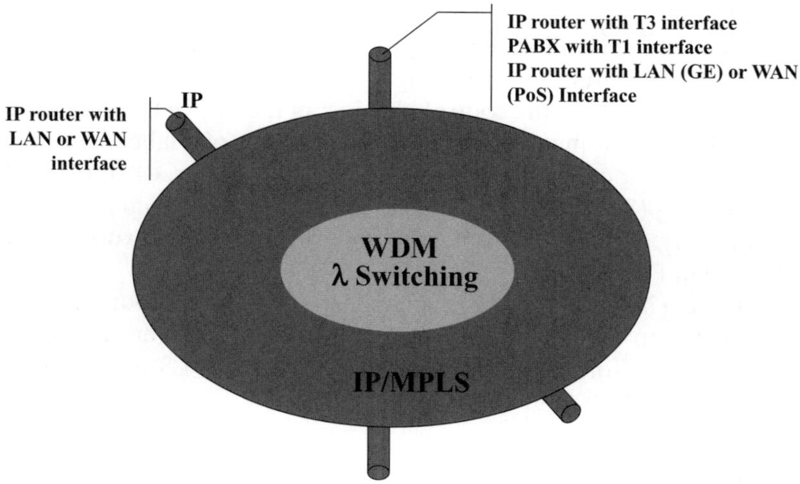

Figure 18.2. The future protocol architecture.

route—that is, where the switching of whole optical channels is effective, such as in the very core of the network. However, as briefly discussed in Sections 18.2.4 and 18.5.1, MPLS plays a major role also in the WDM core and its integration with the outer part of the network because the protocols of the MPLS control plane can be adopted as the control plane for the optical network.

Moreover, when introduced, MPLS had an important role in creating new market opportunities for, and fostering competition between, telecommunications equipment vendors and networking equipment vendors. In fact, the former had label switching hardware (from ATM) and know-how in connection oriented technologies and manufacturing of carrier grade equipment (acquired with ATM, Frame Relay, and SONET/SDH), while the latter have competence in running software for the routing and signaling protocols deployed in the MPLS control plane. Consequently, MPLS was seen as an opportunity and endorsed by both classes of vendors, which fostered its importance and sped up its deployment.

18.2 BASIC UNDERLYING IDEA AND OPERATING PRINCIPLES

18.2.1 Label

The basic idea underlying MPLS is to add a label to packets to be used for routing instead of the destination address carried in the IP header. The label was originally added as a means to enable a lookup operation faster than the longest prefix matching required on addresses when routing IP packets (see the origins of MPLS in Section 18.3). As integration technology advanced and router manufactures refined their ASIC (Application Specific Integrated Circuits) design and fabrication skills, IP addresses longest prefix matching became possible at wire speed, thus eliminating the need for a different type of lookup. Consequently, the original label purpose was meaningless, but the importance of another capability enabled by the label became predominant: facilitating traffic engineering. The key MPLS feature in facilitating traffic engineering is the separation of the routing functionality in the control plane from the one in the data plane (see Section 18.2.4), which is made possible by the presence of a label in each packet. This enables network operators to deploy on (MPLS-powered) IP networks the traffic engineering (Section 18.2.5) and fast fault recovery (Section 18.5.2) solutions traditionally deployed in connection-oriented public networks from X.25 to ATM through Frame Relay.

Multiprotocol in the MPLS acronym indicates that the solution was originally targeted not just to IP, but to any routed protocol. However, although some initial developments for AppleTalk took place, today MPLS is being deployed only in conjunction with IP.

18.2.2 Deployment Architecture

Learning from the failure of ATM (which, although expressly engineered to replace all existing networking technologies, has not been widely adopted largely because of its inability to "reach" the desktop dominated by IP), IETF (the Internet Engineering Task Force; see Section 18.3.1) has targeted MPLS to backbone deployment. This way no change is required to end-systems that generate plain IP packets and do not need to be aware of MPLS being used somewhere in the networks on the path of their packets. Labels are used in a limited area, called an *MPLS network* or *MPLS cloud*, where *label switch routers* (*LSRs*), shown in Figure 18.3, route packets based on their MPLS label. A label is prepended to a packet entering an MPLS cloud by an *Ingress LSR* and is stripped

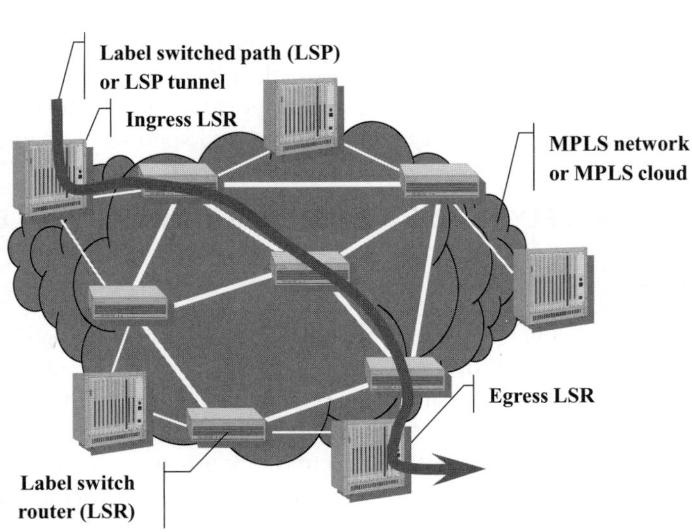

Figure 18.3. MPLS deployment architecture.

from the packet upon exiting the MPLS cloud through an *Egress LSR*. (Due to their position in the network, Ingress and Egress LSRs are also generally called *Label Edge Routers*.)

Packets travel across an MPLS cloud over a *label switched path* (LSP) or *LSP tunnel* that must have been set up before the forwarding of the packet toward its egress point can begin. From this point of view the LSP resembles a virtual circuit in a connection-oriented network. However, a unique feature of MPLS is that LSP setup does not necessarily require an *explicit end-to-end* signaling operation, as explained in Section 18.4.3, which enables the MPLS cloud to retain the valuable properties of a connectionless IP network.

18.2.3 Label Switching Operation

Packets are forwarded through the network according to the well-known *label switching* paradigm that has been deployed for a long time in the history of networking protocols, especially those designed for public networks, such as X.25, Frame Relay, and ATM. Figure 18.4 shows how the label switching paradigm is applied in MPLS networks to realize an LSP. Upon creation of the LSP, an entry is added to the *forwarding table* of each switch on the path from the ingress LSR to the Egress LSR containing a *mapping* between the label attached to packets traveling on the LSP when entering the node, the label attached when exiting, and the port through which the packet with the new label shall be forwarded. An LSR, upon receiving an MPLS packet searches the table for the entry whose input label matches the label of the packet, retrieves the new label that replaces the input label in the packet (operation often called label swapping), switches the packet to the output port identified in the relevant

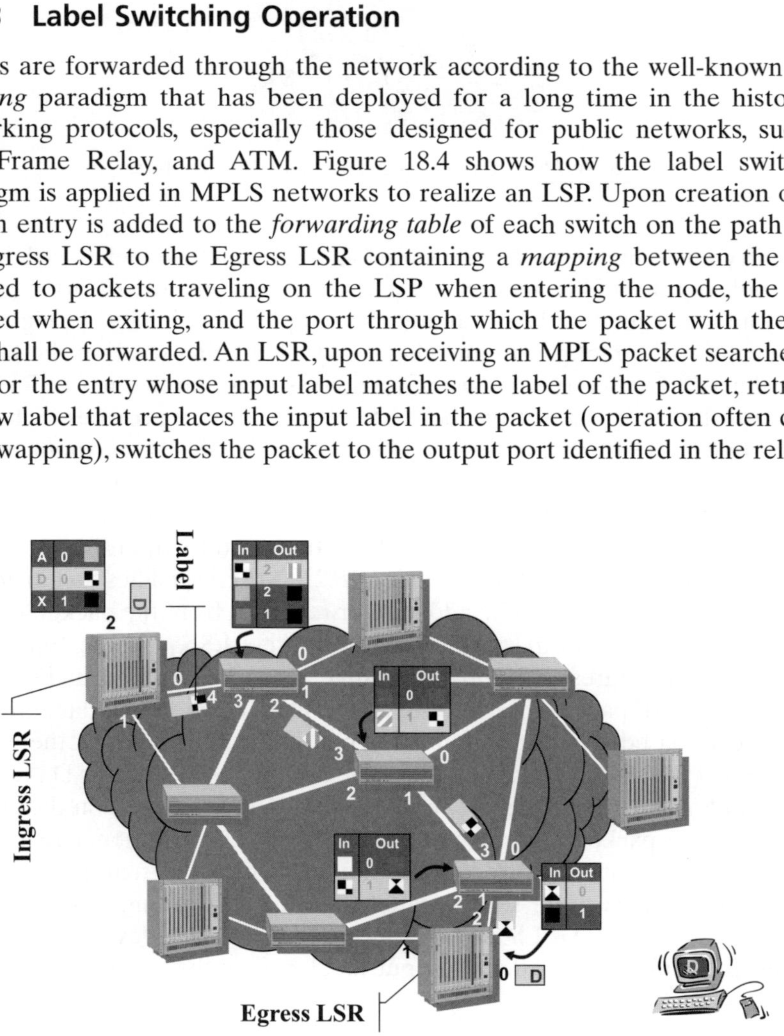

Figure 18.4. LSP working principle: label switching.

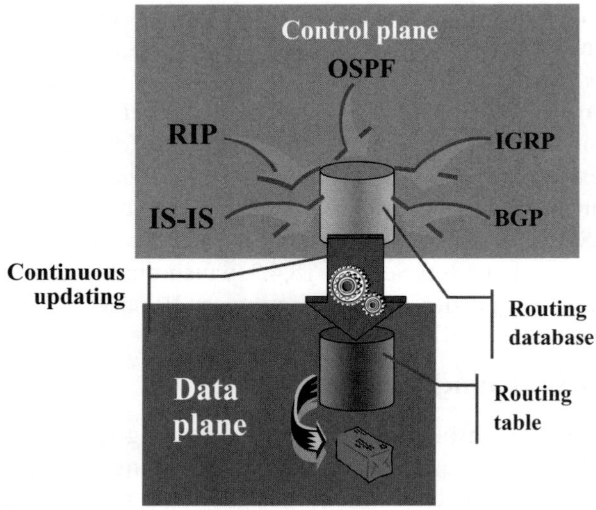

Figure 18.5. Control and data plane in IP routers.

table entry, and schedules it for transmission. Changing the label at each hop has the complexity of modifying a packet each time it is forwarded, but it provides the advantage of limiting label validity to each pair of switches.[3] In Figure 18.4, label locality is exemplified by the checkered label being used on the first link within the MPSL cloud (i.e., between the Ingress LSR and the first LSR) and on the next-to-last link, but not on the others. Label locality provides higher scalability—that is, the number of bits encoding the label limits the maximum number of LSPs across a link, not across the entire network—and makes label choice when creating an LSP a simple, local matter rather than requiring a distributed agreement on a label unused on all the links traversed by the LSP.

As discussed in more detail in Section 18.2.1, MPLS specifies that in addition to being swapped, a label can be added to or removed from the packet; in Figure 18.4 these operations are performed by the Ingress LSR and the Egress LSR, respectively. The Ingress LSR in the sample scenario depicted in Figure 18.4 receives a plain IP packet (i.e., without a label) and decides what label to attach to the packet based on information contained in the IP header. In the example in Figure 18.4 the label is chosen based on the destination IP address D. According to the MPLS specification, LSRs should be able to route (based on the IP destination address) packets that do not have a label (i.e., traditional IP packets); however, many commercial products do not implement this feature.

A data structure more sophisticated than the forwarding tables shown in Figure 18.5 is used in MPLS switches in order to create a level of indirection between the input label and the output label, rather than a simple table. The

[3]Actually, in MPLS the scope of a label might include several input links to a switch, as explained in more detail in Section 18.2.2. However, the underlying concept and advantages are analogous.

purpose is to have multiple input labels mapped to the same output port and label pair, as well as one input label mapped to multiple output port and label pairs. This provides maximum flexibility in the construction of LSPs (e.g., point-to-point, point-to-multipoint, and multipoint-to-point) and their dynamic modification (e.g., for load sharing, overflow, and fast fault protection purposes). The actual data structure supporting the routing operation in MPLS is described in Section 18.4.2.

Looking up a 20-bit label has lower complexity than performing longest prefix matching—that is, finding the entry in the routing table whose destination address has the largest number of bits in common with the destination address in the IP header of the packet being routed. Although initially this was a key advantage of MPLS and the motivation behind its initial design, currently the main benefit stemming from the label is the possibility of controlling the path taken by packets independently of their destination address and the route to it as calculated by dynamic routing protocols, which stems from the network operating according to a connection-oriented paradigm. This enables traffic engineering and fast fault protection, as will be explained later.

However, label-based routing also comes with the disadvantages typical of the connection-oriented paradigm: having to create an LSP and set up forwarding table entries in all the nodes along its path, represents a significant overhead and delay for short, datagram-type communications. Moreover, because packets are routed along a path previously chosen (when creating the connection), connection-oriented protocols lack the adaptability to network faults and conditions that originally inspired the design of the Internet Protocol (IP) and have long been at the core of its success. However, MPLS includes an operation mode (resulting from the combination of independent LSP control and topology-based label binding, presented in Section 18.4.2 and 18.4.3, respectively) in which LSPs are created without the need of explicit end-to-end signaling and as an automatic consequence of route discovery; that is, MPLS has the capability to behave as a connectionless protocol. In other words, an MPLS network can operate according to both the connection-oriented and connectionless paradigms, thus offering the network administrator the flexibility of choosing the one that fits best the needs and requirements of the supported applications and services.

18.2.4 Separation of Control Plane and Data Plane

The distinctive features, whether advantageous or disadvantageous, of the connection-oriented paradigm versus. the connectionless one stem from the different relationship between data plane and control plane each one implies.

Figure 18.5 shows the relationship between control and data plane in a typical IP router. In the control plane, routing protocols continuously update a routing database, taking into account changes in the topology of the network; a routing table, mapping each destination prefix with its corresponding next hop, is used in the data plane for routing incoming packets and is updated whenever the routing database changes. In other words, a topological change possibly

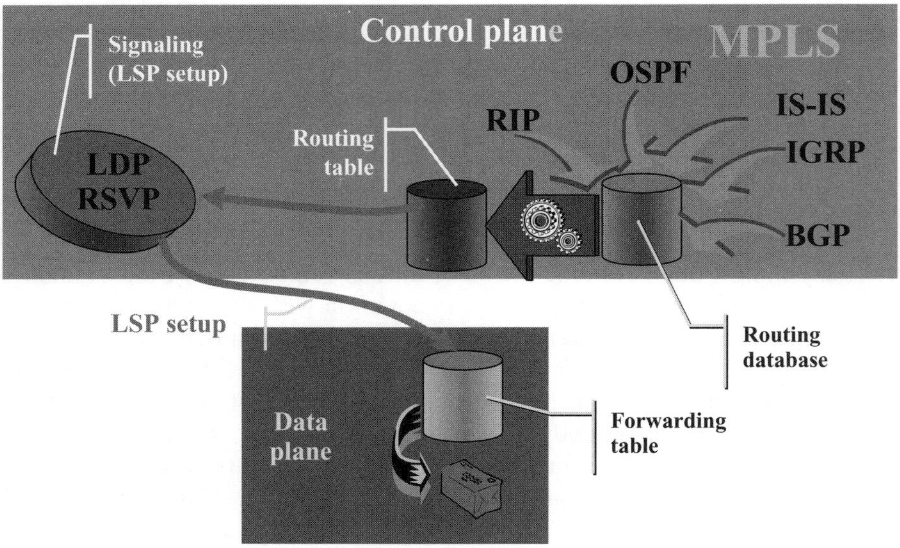

Figure 18.6. Control and data plane in MPLS switches.

results in an update in the routing table so that whenever a new packet will arrive to one of the input ports of the routes, it will be routed according to the updated information in routing table, thus possibly following a different path to the destination with respect to previous packets if a more convenient route has been found by the continuous update process based on the information gathered by routing protocols. This enables automatic adjustments to topological changes and specifically link and node failures.

On the other hand, Figure 18.6 shows the functions in the control and data plane of a typical MPLS switch and the relationship among them. A routing table, automatically and dynamically calculated from the routing database continuously updated by routing protocols, is deployed within the control plane by signaling protocols. Whenever an LSP is to be set up, messages are exchanged according to the signaling protocols described in Section 18.4.3 by network nodes along a path determined according to the most recent routing information. Specifically, a node receiving a message requiring the creation of an LSP toward a given destination will use the routing table to establish to which of its neighbors it should send a corresponding LSP setup message. Moreover, as shown by the arrow between the control and data plane in Figure 18.6, the LSP setup procedure results in the installation of a new entry in the forwarding table within the data plane that provides the mapping between a label tagging incoming packets and the next hop to which the packet shall be forwarded (and a new label to be substituted when forwarding the packet, as will be further detailed in Section 18.4). Consequently, the path followed by packets traveling through a given LSP is determined by the corresponding forwarding table entry, which is in turn determined by the dynamic routing information gathered in the routing database at the time of LSP setup. Although such routing database is continuously updated to reflect topology changes, such changes do not impact the forwarding table, hence the path followed by packets of existing LSPs. Consequently, even though the control plane of MPLS switches might be aware of topology changes due to routers or link failures, routing of packets is not adapted accordingly and packets traveling on LSPs through failed network elements are no longer able to reach their destination.

As much as such separation between the control and data plane might be a problem, it is a valuable asset to MPLS because it enables the implementation of traffic engineering as discussed in the following section.

18.2.5 Traffic Engineering with and without MPLS

Traffic engineering refers to the action and/or capability of controlling the distribution of traffic across the network with the goal of reaching a given objective, such as (a) keeping link utilization below a certain threshold value or (b) making link utilization uniform. Traffic engineering is key to a network operator to be able to control the quality of the services so that it can abide by the contract with its customers. Moreover, spreading the traffic across the network in a way that no links are underutilized allows the network operator to exploit all the resources (in terms of routers and links) it has invested in to carry customer traffic, thus maximizing the number of customers besides their satisfaction.

Traditional IP routing does not necessarily pursue the above goals by itself. In fact, traffic routed is based on the destination, that is, packets coming from various sources and going to the same destination tend to be aggregated toward the destination, as graphically exemplified in Figure 18.7. If traffic from each source was uniformly distributed among destinations, link utilization would be uniform. Instead, current Internet traffic is mostly directed toward a relatively small subset of the connected hosts that are servers and major portals offering services to clients around the world. Consequently, traffic tends to *aggregate* in the vicinity of such preferred destinations and on some links interconnecting autonomous systems containing highly popular sites. In such a situation, graphically represented in Figure 18.7, some of the links turn out to be overloaded, which possibly results in loss, high delay, and ultimately poor performance perceived by customers, while others are underutilized, which implies a waste of resources invested by the network operator. Through a traffic engineering operation, the network administrator could spread the traffic for a given destination (or a group of destinations located in the same topological area) across the network, for example, as exemplified by Figure 18.8.

Achieving such a traffic distribution in a network of traditional IP routers is not straightforward. One possible way is by using some sort of static policy routing within routers, that is, make them route packets based on some criteria other than the destination address (alone). For example, the traffic distribution exemplified in Figure 18.8 could be achieved by configuring in each router a static route based on the pair source-destination address. However, in general cases

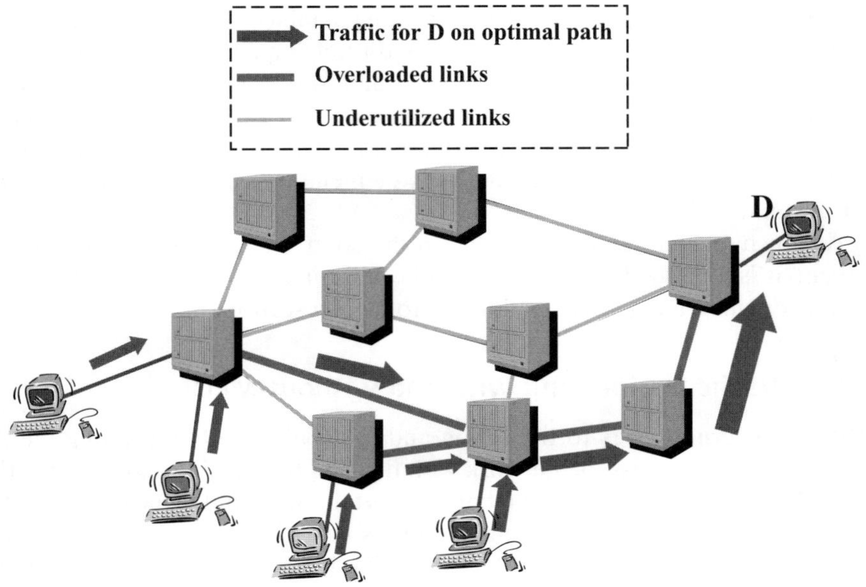

Figure 18.7. Natural traffic aggregation with destination-based IP routing.

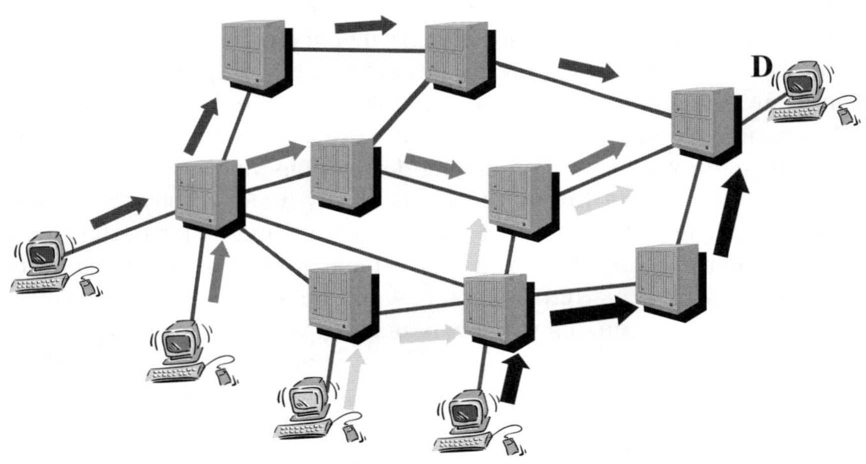

Figure 18.8. Traffic distribution Across the network.

such a solution requires a large amount of configuration work, is prone to errors, and does not react to topological changes in the network, that is, like any other static routing approach manual, intervention is required to overcome node and link failures. Contrary to what it might appear after superficial consideration, traffic aggregation cannot be achieved automatically by using link load as a routing metric. When routers chose routes that involve least-loaded links, traffic begins flowing through such links (for example, some of the links in the top part of the network shown in Figure 18.7), thereby increasing their utilization. As a result, other links become less loaded, routing tables are re-computed, and traffic is re-routed through the newly least loaded link. In the sample network in Figure 18.7, routing will become unstable and keep oscillating between transferring packet through the bottom part of the network and the top part of the network. As a result, the network will be, on average, as congested as in the case of topology-based (i.e., shortest path) routing, but the performance might be lower because routers will spend (a significant) part of their processing power re-computing routing tables over and over, rather than processing incoming packets.

The instability problem stems from the close relation between the control plane and the data plane in IP routers. As updated link utilization information is gathered by the routing protocols in the control plane, routing table entries in the data plane are updated (see Figure 18.5), which affects routing of all subsequently incoming packets. In MPLS, link utilization-based routing can be effectively used thanks to the separation between control and data plane. Routing table updates due to link utilization information changes in the routing database do not affect routing of packets in the data plane because it is based on the forwarding table entries (see Figure 18.6). However, whenever a new LSP will be created, it will be routed according to current information in the routing table and a new entry will be created in the forwarding table of switches along a path involving least-utilized links. Consequently, in the sample network considered in this section, subsequently opened LSPs will be spread across the network, as shown in Figure 18.9, where the number of the LSPs indicates the order in which they have been created and routing chooses the least-loaded path.

Prior to the availability of MPLS network, operators that wanted to perform traffic engineering on their backbone, would choose a different technology to implement it. For many years the solution of choice has been a backbone based

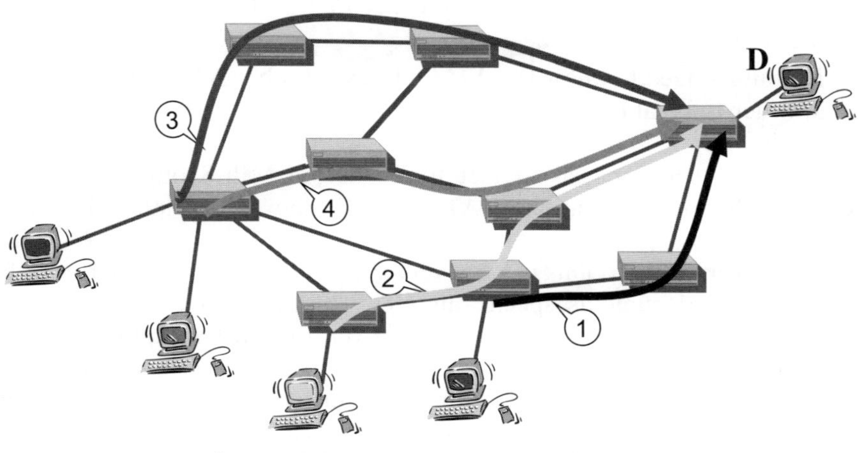

Figure 18.9. Traffic engineering with MPLS.

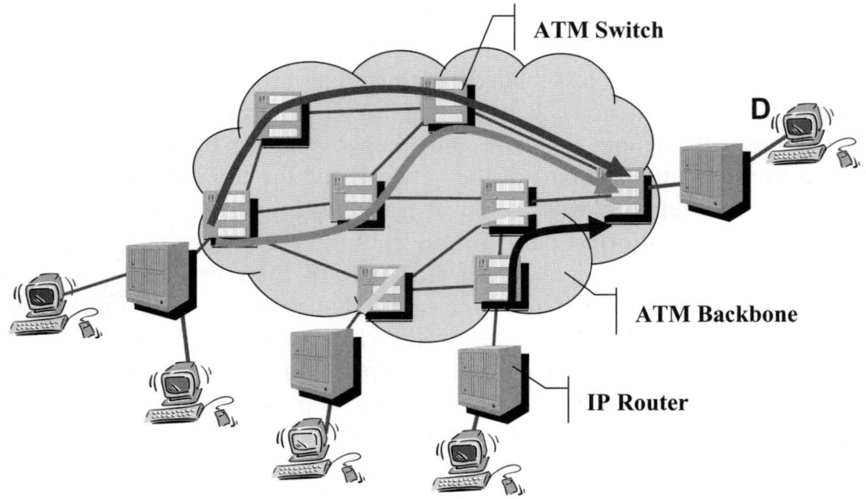

Figure 18.10. Traffic engineering with ATM.

on ATM (Asynchronous Transfer Mode) [1] interconnecting peripheral IP routers, as shown in Figure 18.10, thus reinforcing the onion architecture discussed in Section 18.1.1 (see Figure 18.1) and the issues stemming from it. Moreover, the network architecture depicted in Figure 18.10 raises many problems that were extensively addressed for many years by the vast research and development in the context of internetworking between IP and ATM. The main issues can be summarized as follows:

- IP routers connected to the "ATM cloud" have a large number of adjacencies, that is, all routers connected to the cloud are one-hop neighbors, which affects the number of exchanged message and the amount of computation required by routing protocols, that is, it limits the scalability of the solution;
- IP routers need to know the ATM address of other ATM attached routers in order to be able to forward packets to them.

No satisfactory solution was in fact found before MPLS offered itself as a valid option for the integration of ATM and IP network, as later discussed in Section 18.3.

ATM is from many points of view very similar to MPLS; as a matter of fact, MPLS standard documents specify an ATM "packet format" to be used in MPLS, as presented in Section 18.4.1. Moreover, the MPLS deployment architecture presented in Section 18.2.2 and shown in Figure 18.3 is basically the same as the one depicted in Figure 18.10 for ATM. So, why is MPLS well-integrated with IP—actually, in many cases even presented as a variant of or enhancement to IP, as in Section 18.2.1—while ATM cannot be easily and effectively integrated with IP? The answer—as well as the origin of the two issues listed previously in the internetworking between IP and ATM—lies in ATM switches not being aware of IP routers and vice versa, which, in turn, stems from ATM switches and IP routers not deploying the same control plane. *The strength of MPLS in terms of being effectively deployable in IP networks is the adoption of the traditional IP control plane, that is, the same addressing scheme and routing protocols deployed in IP networks.* This has certainly played a pivotal role in its technological and market success.

18.3 MPLS HISTORY

The major market and technological success encountered by MPLS stems from it having been multiple times the right solution to a significant problem of the moment, as the short history below highlights.

MPLS originated by *Tag Switching* a solution originally designed at the end of the 1990s by Cisco Systems [5] to speed up packet forwarding in their routers as part of an effort to build more powerful and scalable devices. Tag Switching was presented in the context of IETF [6], where it gained popularity especially as a solution for the deployment of IP in ATM network. As previously mentioned, there has been a lot of activity around the so-called *IP over ATM* problem; and among all the various proposals, each with its advantages, but especially disadvantages, Tag Switching stuck out as particularly effective. In fact the basic paradigm underlying Tag Switching was the same as ATM's—that is, routing based on a label in each packet that is changed at each hop on the way. However, having been conceived as an extension to IP, Tag Switching uses the IP control plane—that is, the same addressing architecture and routing protocols as IP routers, which provided the key to seamless integration between Tag Switches and IP routers. The same basic idea can be applied to solve the IP over ATM problem by including the IP control plane within ATM switches. Given that ATM is a layer 2 technology (with reference to the ISO-OSI layered protocol architecture), while IP is a layer 3 technology, the application of Tag Switching in the context of IP over ATM is categorized as a multilayer switching solution as the data plane works at layer 2 by switching ATM cells, while the control plane works at layer 3 by using IP addresses to identify destination and IP routing protocols to find a path in the network of ATM switches toward them. As was adopted as an IETF solution, the technology was renamed MPLS in order to differentiate from the vendor-specific Tag Switching; notwithstanding the different terminology, the two solutions and their basic components and protocols are the same. Although MPLS popularity began with ATM, it was not over once the latter had failed. On the contrary, once more MPLS was a handy solution to a widespread problem: the death of ATM. In fact, all the players in the ICT domain—that is, telecommunications and computer network equipment manufacturers as well as telecom operators and Internet Service Providers—had invested heavily in ATM throughout most of the 1990s. However, before all these players could get their due return on their investments—the former in the design of ATM equipment, the latter in the deployment of ATM infrastructures—it had become clear that future customers wanted IP connectivity and that the deployment of ATM networks to interconnect IP routers, as exemplified in Figure 18.10, was not effective and led to multilayered infrastructures (see Figure 18.1). MPLS offered a way to convert ATM equipment into IP equipment, or at least IP-aware MPLS equipment, by "replacing" software implementing the ATM control plane with the MPLS one. This could be done in unsold equipment to the benefits of manufacturers and vendors and, at least in principle, on installed one, to the benefit of operators. As a matter of fact, the first MPLS switches were converted ATM switches, which not only enabled manufacturers to recover the investment spent in designing them, but also helped to reduce the time to market for MPLS.

The success and popularity of MPLS continued because once again it was a convenient solution for each of the players of the ICT domain. Operators had a chance to slash their costs by eliminating the onion architecture discussed in Section 18.1.1. IP equipment manufacturers had a chance to provide carrier grade devices suitable to provide reliable and controllable services (once exclusive domain of telecom devices) by equipping their IP routers with MPLS capability, and telecom vendors that had pursued the doomed ATM technology had a chance to enter the growing IP business by evolving their ATM switches into MPLS ones.

The next important milestone in MPLS history was in the early 2000s when optical technologies gained popularity as the means to deliver plentiful bandwidth to all to provide the necessary support to any existing application and open the way to the deployment of many new ones. Notwithstanding the confidence in the advantages of optical networking, partly shaken in the following years, it was by then clear to anyone that applications were going to be based on IP for the time being. Moreover, even though not everyone might have agreed, optical networking was for a while going to be a backbone technology. In other words, optical networks were, at least for some nonnegligible time, to be deployed according to the network architecture shown in Figure 18.10 where the central cloud was a lambda-switched network composed of arbitrarily interconnected optical switches. Such architecture had proven fatal to ATM due to the lack of integration with the external IP routers. Hence, when designing the control plane for the ASON (Automatically Switched Optical Network) [7], it was apparent that being IP-aware was key to success and acceptance. Consequently, ITU-T easily accepted the idea of deploying the MPLS control plane in optical networks, which matured within IETF as MPλS (Multi-Protocol Lambda Switching) [8]. The basic idea is that although data are not switched in the form of IP packets, but as whole optical channels (a.k.a. lambda from the Greek letter commonly used to indicate wavelength), typical IP routing protocols (i.e., OSPF, IS-IS, and BGP), are deployed to find routes to destinations that are identified by IP addresses and MPLS signaling protocols are deployed to create optical path, similarly to the way they are deployed to set up LSPs. While an LSP is uniquely associated with the label deployed to identify and route packets belonging to it, lambdas are uniquely associated to optical channels and deployed to route them into optical switches.

The parallel between MPLS switches routing packets based on a label and optical switches routing bits or optical signals based on a wavelength (lambda)—more specifically the parallel between labels and lambdas as well as between LSPs and optical channels—could be carried even further to any switching device capable of routing data units based on a connection-oriented paradigm. Consequently, as much as MPLS switches use a set of protocols to automatically set up connections (LSPs) along properly computed routes, the same protocols could be used to set up any other type of connections if:

- End systems were identified by means of IP addresses;
- The format of the control messages (i.e., routing and signaling protocols) enabled the system to carry the piece of information corresponding to the MPLS label.

This led to the definition of GMPLS (Generalized MPLS) [9] that, through a generalization of the label concept and a flexible syntax in the control messages, provides a control plane suitable for any switching technology such as packet switching (i.e., MPLS or ATM), circuit switching (e.g., SONET and SDH), optical switching.

This represents the latest reason for the success of MPLS—that is, as a unifying control plane for various technologies whose highly needed integration would otherwise be difficult and inefficient.

18.3.1 Standardization Effort

Various entities produced documents describing MPLS, additional features, and protocols required to implement them. The main actor in the definition of MPLS is undoubtedly the *IETF* (Internet Engineering Task Force) [10] and more specifically its *MPLS Working Group* [9] that produced most MPLS-related documents in the form of RFC (request for comments). Although the MPLS Working Group has been chaired by a Cisco Systems' engineer and people from the manufacturer contributed significantly to various MPLS documents produced, the working group has consisted of individuals working for virtually all major IP and telecom equipment manufacturers.

Another major player in the production of MPLS "standards" has been the *MPLS Forum*, a consortium of vendors aiming at fostering the deployment of MPLS. Given that the organizations involved in the MPLS Forum were basically the same involved in the MPLS Working Group, in order to avoid duplication of work, the former focused on aspects outside the scope of IETF. For example, given that the IETF considers MPLS as a technology strictly aimed at the network core, the IETF Forum addressed issues related to the deployment of MPLS at the periphery, such as Voice over MPLS (VoMPLS) and MPLS-based ADSL.

At some point the MPLS Forum merged with the Frame Relay Forum to form the MPLS/FR Alliance because the same organizations were involved in the two fora and they were tackling similar issues in the same domain; more recently the MPLS/FR Alliance merged with the DSL Forum into the Broadband Forum [4] as one major application for MPLS networks, as it has been for ATM and Frame Relay, is the collection and delivery of traffic generated by DSL modems. Moreover, as discussed in Section 18.4.1, standard documents specify a Frame Relay "packet format" to be used in MPLS, which officially makes frame relay switches a type of MPLS switch.

Because of its suitability to solve enable IP routers to effectively interconnect through an ATM network, MPLS-related documents have been produced also by the *ATM Forum*, a consortium of companies interested in fostering the adoption of ATM (merged with the MPLS/Frame relay alliance in 2004, which later joined the DSL forum into the Broadband Forum [4]). Given the great importance and wide diffusion of IP, proper interworking between ATM and IP was a very crucial issue to be addressed.

All of the above are not official standardization bodies, but one, the ITU-T (International Telecommunication Union—Telecommunication Standardization Sector) [2], got involved when the MPLS control plane was chosen for the ASON.

Because a significant amount of work on MPLS and related documents was not performed by official standardization bodies, most MPLS specification documents are not standards; however, they are commonly referred as such. If nothing else, MPLS is certainly a *defacto standard*.

18.4 PROTOCOLS AND FUNCTIONS

MPLS operation requires the specification of three basic protocol elements: (i) a label format and how it can be attached to packets, which led to the specification of an MPLS header format, (ii) mechanisms and protocols required by nodes to handle labels (i.e., signaling protocols), and (iii) routing protocols to determine the route for LSPs, possibly taking into account constraints on the route selection. These three protocol elements together with a few key mechanisms and concepts are the object of this section.

18.4.1 Header Format

MPLS standard documents define a format for the label and how it can be attached to any packet by defining a *shim header* to be inserted between a generic layer 2 header and a generic layer 3 header [11]. In fact, although MPLS has come to be seen as strictly related to IP—possibly even a variant to IP—originally it was meant to be protocol—independent (from which the "Multi-Protocol" in its name), where the packets carried within the MPLS shim header might be any type of packets, not even necessarily belonging to a layer 3 protocol. For example, in certain current deployments, Ethernet frames are carried on LSPs—that is,

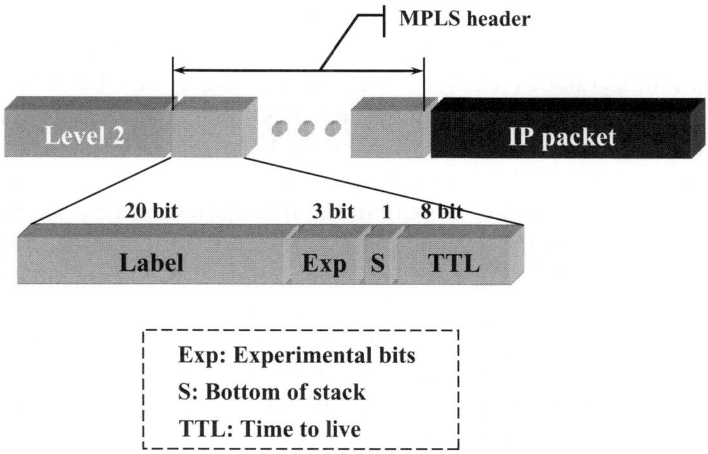

Figure 18.11. Label and shim header format.

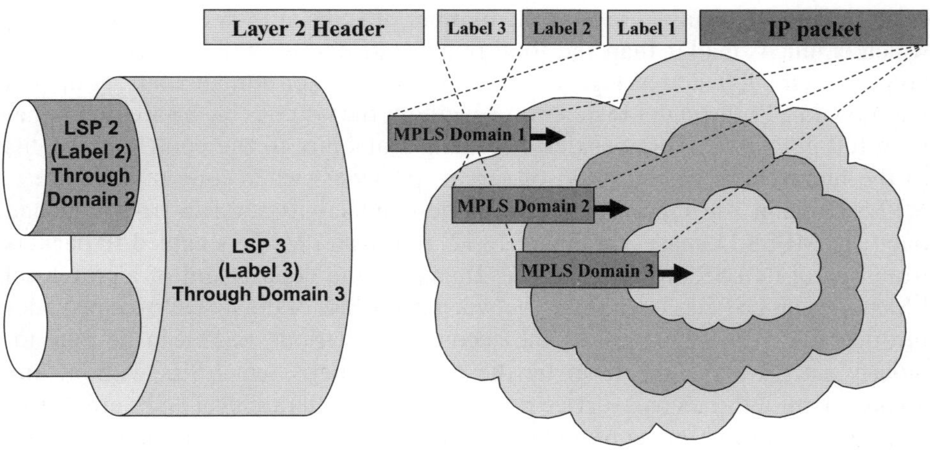

Figure 18.12. Hierarchical LSPs; label stack and nested MPLS domains.

within MPLS headers. The remainder of the chapter, when not otherwise relevant, refers to IP packets as carried within MPLS headers.

As shown in Figure 18.11, the shim header consists of one or more modules, each containing a label and other information. The modules are said to realize a *label stack* as modules can be added (pushed) or removed (popped) only at the beginning (in terms of transmission order) of the sequence. A node receiving a packet processes only the top module—that is, routes the packet based on the label at the top of the stack. This enables the realization of hierarchical LSPs, graphically represented in the left part of Figure 18.12, wherein an LSP identified by a specific value as outermost label includes several LSPs identified by different values for the second label in the stack. Each level of the LSP hierarchy possibly corresponds to a nested MPLS domain, as exemplified in the right part of Figure 18.12. When a packet enters MPLS Domain 1, an ingress LSR inserts a shim header containing Label 1 between the Layer 2 header and the IP packet that will consequently travel on LSP 1 until it reaches a label edge router at the egress of MPLS Domain 1. When on its way the packet reaches the border of MPLS Domain 2, an ingress LSR adds to the shim header a module containing Label 2 so that the packet continues its way to the destination on LSP 2. The Domain 2 LSRs are not aware of the various LSPs within LSP 2 and handle the same way all packets belonging to LSP 2. However, once a packet gets to a Domain 2 egress LSR, the Label 2 module is removed and the packet is handled as belonging to LSP 1—that is, differently from other packets exiting LSP 2 but having a different value in Label 1 (thus traveling on other LSPs through Domain 1). Given the possibility to extend the shim header with further modules, several levels of LSP hierarchy and domain nesting can be realized.

The capability to deploy such a hierarchy of LSPs and MPLS domains has a twofold advantage. Firstly, it enables high *scalability* because the forwarding tables of the LSRs in the inner domains (possibly part of a high-capacity backbone) include as many entries as the number of LSPs at their hierarchical level, which is much smaller than the total number of end-to-end LSPs (the smaller "pipes" on the left side of Figure 18.12). Secondly, outer domains can set up and tear down LSPs through the inner ones, without the latter being aware of it—that is, in full *autonomy*. This, besides extending scalability to the control plane, fits to the hierarchical organization of service providers where one provider (e.g., MPLS Domain 1 in Figure 18.12) buys interconnectivity services from a second one (e.g., MPLS Domain 2 in Figure 18.12) to transfer MPLS-enabled IP packets from one of its LSRs to another one. Thanks to the deployment of hierarchical LSPs, once an LSP has been set up between two LSRs of the first service provider through the MPLS Domain of the second one, various LSPs can be autonomously set up and torn down by the first provider's two LSRs without any involvement of the second service provider's LSRs. Hierarchical LSPs are instrumental also in other areas of application including virtual private networks and fast fault recovery.[4]

In addition to the label, which is 20 bit long, each module of the shim header contains the following fields:

- The `Exp` or *experimental* field contains three bits whose purpose is not strictly specified in the shim header format document [11], but has been later considered for the realization of Differentiated Services (DiffServ) [12] in MPLS networks [13].
- The `TTL` or *time to live* field has the same size (8 bits) and purpose as the homonymous filed in the IP header, namely preventing (temporary) routing

[4]Although the role of MPLS in both applications is presented in Section 18.5, the deployment of hierarchical LSPs is not discussed because it would require a level of understanding of the solution that is beyond the scope of this chapter.

loops from bringing (part of) the network to a halt by overloading routers with copies of packets traveling endlessly on circular routes. Each time an MPLS switch routes a packet, is causes decrements in the value in the TTL field of the outermost label module. If the field reaches value 0, the packet is not further forwarded; that is, it is discarded.

- The `S` or *bottom of stack* field indicates whether a label module is the one at the bottom of the stack, in which case a switch popping the label knows that it should expect an IP packet, rather than another label module, when parsing the remaining part of the packet.

The encapsulation of the shim header in Ethernet and PPP (Point-to-Point Protocol) frames—that is, the two most deployed layer 2 protocols—is defined in [11] by specifying the value to be used for the `ethertype` and `protocol` fields, respectively. Moreover, for proper operation with PPP, an MPLS Control Protocol is defined for MPLS parameter negotiation when a PPP link is initialized.

In addition, the label and MPLS header format specification [11] defines that when the layer 2 protocol in use makes itself use of labels (possibly under a different name, such as virtual circuit identifier or data-link channel identifier), the outermost label shall be carried within the layer 2 header. Specifically,

- When IP packets are carried within ATM cells, the VCI/VPI (virtual circuit identifier/virtual path identifier) fields of the ATM header is used [14];
- When IP packets are carried within Frame Relay frames, the DLCI (data-link connection identifier) in the Frame Relay header is used [15].

Consequently, an MPLS switch upon receiving an IP packet encapsulated in an ATM cell (Frame Relay frame) will route it based on the content of the VPI/VCI (DLCI) field, which is in fact what an ATM switch (Frame Relay switch) does. In other words, the above specifications fully entitle ATM and Frame Relay switches to be considered MPLS switches. The inclusion of the above label encoding options was a very smart decision in the MPLS standardization process because it avoided the fact that ATM or Frame Relay manufacturers would oppose or delay the adoption of technology in order to protect their investment in the legacy technologies. Instead, it provided them with a smooth migration path to MPLS. Moreover, it reduced the time to market for MPLS because networking and telecommunication equipment manufacturers did not need to start the design of MPLS switches from scratch.

18.4.2 Key Concepts

The *Forwarding Equivalence Class* (*FEC*) is an important component of the MPLS architecture [16] defined as the set of packets that are forwarded the same way. While in traditional IP packets that receive the same forwarding treatment are those with the same destination address (or prefix thereof), the MPLS FEC can be defined in a more flexible way. In other words a FEC might be identified by any filter on header fields and other information not encoded in the packet (e.g., the port through which a packet is received). Packets belonging to a given FEC are associated with the same MPLS label by an Ingress LSR; and MPLS switches inside the MPLS cloud do not need to reclassify packets to figure out which FEC they belong to, but simply look at the attached label to decide what to do with the packet.

The correspondence between a label and a FEC is called *label binding* and has to be agreed upon by the two nodes at the ends of the link on which the label binding is used. MPLS specifies that LSRs deploy *downstream binding*, which implies that a label binding is chosen by the switch at the downstream end of a link—that is, the node that will receive packets with the given label. Bindings can be performed in either an *unsolicited* way or *on-demand*—that is, upon an explicit request from the upstream LSR. In MPLS the scope of a label is a single node, i.e., the same label cannot be used for two LSPs traversing the same node. Thanks to the deployment of downstream label binding, the same binding can be used on the links connected to multiple input interfaces, which enables the construction of per-destination delivery trees, which is equivalent to the way routing is performed in traditional IP networks. As a consequence, when LSRs are performing unsolicited topology-base label binding (see Section 18.4.3) with independent LSP control (see below), an MPLS cloud features the same operation and properties of a traditional IP network. Moreover, downstream label binding enables the deployment of LSRs that can operate only with a subset of the possible labels because they just have to bind only labels in the range they can handle.

In any event, once the downstream node has decided a binding between a label and a FEC, it notifies the upstream node, which is referred to as *label distribution*. MPLS encompasses both ordered and independent LSP control. According to *independent LSP control*, each LSP independently decides when to make a label binding for a FEC (e.g., upon discovering a given destination) and distribute it. When using *ordered LSP control*, an LSR performs a label binding for a FEC and distributes it only upon receiving a label binding for the FEC from the next hop to the destination corresponding to the FEC.

A set of protocol has been defined to support label distribution as well as to solicit label binding and are discussed in Section 18.4.4.

Once a node has decided a binding for a given FEC (i.e., for an incoming label) and has received through label distribution by one of its downstream neighbors its label binding (i.e., an outgoing label) for the same FEC, it performs a *label mapping* by creating in its forwarding table an entry that relates the incoming label with the outgoing label and the port through which the binding had been received. Label mapping results in the actual creation of an LSP, or more precisely a section of it around the LSR performing the mapping.

Actually, MPLS documents [16] define a set of tables to handle label binding and label mapping as shown in Figure 18.13. The *Next Hop Label Forwarding Entry* (*NHLFE*) table contains information on outgoing labels, specifically:

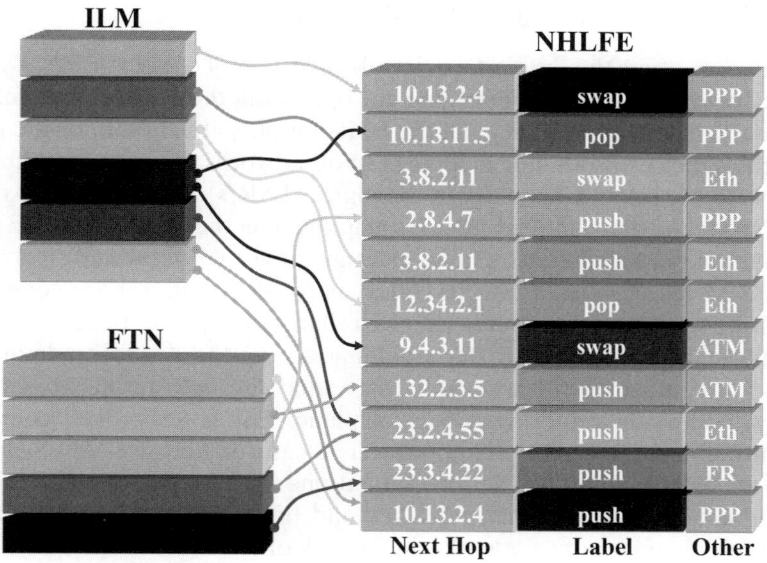

Figure 18.13. Label mapping: ILM, STM, and NHLFE tables.

- Labels to be used for packets that are being forwarded and whether the label should be swapped, pushed on the stack, or the top label module should be popped.
- The next hop to which a packet should be forwarded.
- Other information that might be useful to handle the packet transmission, such as, for example, the link layer protocol deployed.

The FEC-To-NHLFE (FTN) table is used to handle label binding in an Ingress LSR: The table contains an entry for each FEC for which the LSR acts as an ingress LSR and each FTN table entry is mapped to one or more NHLFEs. The NHLFE(s) to which an FTN entry is mapped provide the label to be attached to packet (i.e., pushed on the empty label stack) before forwarding the packet.

The *Incoming Label Map* (*ILM*) table contains an entry for each label bound by the LSR to a FEC and is used to handle label mapping: Each ILM entry is mapped to one or more NHLFEs, providing the next hop to which incoming packets with the label in the ILM should be forwarded and their outgoing label. When an NHLFE instructs the (egress) LSR to pop the label at the bottom of the stack, the packet being forwarded is a plain IP packet on which the next hop must act as a traditional IP router.

The possibility to map one ILM or FTN table entry to multiple NHLFEs can be used for various purposes, such as multicasting, load sharing, or fast fault protection; the specific use is not specified as part of the basic MPLS architecture [16].

18.4.3 Static and Dynamic Label Binding

Static binding offers the simplest, yet less flexible, way of creating LSPs: By means of a device configuration or management operation, the network administrator associates a label with a FEC in all the LSRs on the path chosen for the packets and creates the corresponding mappings. In other words, the network administrator must create (a) an FTN entry in the ingress LSR(s) and (b) an ILM entry in each LSR on the path, and he/she must map them on the proper NHLFE entries. This solution is equivalent to creating permanent virtual circuits (PVCs) in ATM networks or leased lines in SONET/SDH networks or, in a more general way, configuring static routes in routers; hence it shares their properties and shortcomings. The main advantages of static binding are (i) the network administrator having full control on the way traffic is routed through the network and (ii) network nodes not requiring a control plane. The latter proved convenient, especially in speeding up the commercial release of the first MPLS products. However, static binding has various limitations such as (i) poor scalability, (ii) error proneness due to human mistakes, and (iii) unsuitability to set up LSPs across multiple operators' networks due to lack of interoperability among different management systems and access rights to network devices.

Dynamic binding offers more flexibility and scalability because LSRs can cooperate to create LSPs in response to a number of different events.

- In *data-driven* label binding, a new binding is created upon arrival of a packet for which the binding is needed—for example, a packet that does not belong to any of the FECs for which a binding already exists.
- In *control-driven* label binding, the triggering event pertains to the control plane:
  - According to the *topology-based* variant, a label binding is created in response to routing information, such as the discovery of a new route by the routing protocols (e.g., OSPF, IS-IS, and BGP).
  - According to the *explicit event* variant, a label binding is created in response to an explicit event, such as a signaling procedure started, for example, by a label edge router or by a network control center. Such an event could be in response to a traffic engineering condition or policy or the detection of a fault in the network.

All of the above dynamic label binding options require protocols for label distribution and possibly event notification, which is the subject of Section 18.4.4. While some of the protocols are suitable to all the dynamic label binding options, others can work only with some of them. Moreover, the choice of which LSRs participate in the label distribution for a given LSP represents a routing decision that determines the path of the LSP and, ultimately, routing of packets belonging to it. Such a decision can be based on routing information either statically configured (i.e., in the case of *static routing*) or dynamically gathered by *routing protocols*, which is the subject of Section 18.4.5.

18.4.4 Label Distribution Protocols

Two alternative approaches have been used to support label distribution: (i) adoption of an existing protocol, possibly modified to include new information and functionalities, and (ii) definition of an ad hoc label distribution protocol. The protocols described in Sections 18.4.4.1 and 4.4.2 represent the first approach, while the second one has been followed when defining the Label Distribution Protocol (LDP) discussed in Section 18.4.4.3.

18.4.4.1 Border Gateway Protocol. The Border Gateway Protocol (BGP) [17] has been designed for interdomain routing in IP networks—that is, to support the exchange of routing information between routers at the border of an autonomous system. Such border routers (or "gateways," according to the obsolete terminology initially deployed in the Internet) are interconnected to at least one router belonging to a different (neighboring) autonomous system from which they receive quite articulate information about routes to possibly each destination connected through the Internet. Border routers use such information to choose through which of the neighboring autonomous system to forward packets for each destination.

Due to the nature of its application, BGP has two valuable features:

- Packet format has been designed to be customizable and extensible so that new types of information can be easily encoded and exchanged among routers.
- BGP implementations have flexible mechanisms to filter routing information received, transmitted, and deployed in route calculation.

For this reason, BGP is being used also for intradomain routing and in particular is being proposed for routing in MPLS networks. In this case the BGP message format can be extended to carry for each destination, together with the routing information, also the label to be used when sending MPLS packets to the destination. The solution is thus particularly suitable to support topology-based label binding (see Section 18.4.3): an LSR advertising a route to a destination includes in the BGP message also its label binding—that is, the label its neighbors shall use to create an LSP to such a destination. Dually, an LSR receiving a BGP message advertising a route to a destination shall find in the message the label binding used by the neighbor; hence, once the LSR has decided its own label binding and chosen the advertised route to reach the destination, it can create a label mapping for an LSP to such destination.

18.4.4.2 Resource Reservation Protocol. The Resource Reservation Protocol (RSVP) [18] was introduced in the context of the Integrated Services (IntServ) [19] architecture to provide services with deterministic quality over the Internet. In this context, RSVP is used by hosts to signal to the network the type of service required and its quality (QoS, quality of service) parameters; while relaying RSVP messages between source and destination network nodes can use them to negotiate the amount of resources (e.g., buffer space, output link capacity, processor capacity) needed in each node in order to ensure the required QoS (while messages propagate from source to destination) and to reserve them at each hop on the path (while messages propagate back from destination to source).

Although MPLS at its inception is a routing and transport solution not aiming at addressing QoS issues, it is reasonable to assume that sooner or later it will be a key requirement that packets traveling on an LSP are not only effectively switched to their destination, but also guaranteed a deterministic service. In such a scenario, creating an LSP encompasses, besides finding a route through the network, distributing labels, and installing a label mapping in each node, also requesting a specific service and reserving resources on an LSP. Given that provisioning of services with previously negotiated quality over the Internet is based on the IntServ architecture of which RSVP is the centerpiece, the protocol becomes an attractive solution to enable label distribution and QoS parameter signaling and resource reservation all at once. Consequently, RSVP messages have been extended to include additional information to support both label distribution and traffic engineering (as further detailed in Section 18.4.5.3) resulting in the specification of RSVP-TE (RSVP for Traffic Engineering) [20].

RSVP delators criticize the fact that having been designed for a different purpose, the protocol is not optimized for label distribution and, more in general, signaling in MPLS networks. For example, RSVP has been designed to be a softstate protocol; that is, all information network nodes obtained from RSVP are considered to have limited temporal validity and expire; consequently, in order to be kept valid, it needs to be refreshed by periodic exchanges of RSVP messages. This results in a significant burden on the network, in terms of both (a) bandwidth required to transfer the messages and (b) processing power needed to process them. Moreover, as a consequence of the softstate design choice, RSVP has not been equipped to provide reliable message exchange, which is a problem when the protocol is to be used, in the context of MPLS signaling, to notify network nodes (e.g., in fault recovery to notify the ingress LSR of a failure in an LSP, as discussed in Section 18.5.2).

18.4.4.3 Label Distribution Protocol. The Label Distribution Protocol (LDP) [21] was explicitly conceived to provide support to label distribution by deriving its design from a protocol deployed for the same purpose in the context of the Cisco System's tag switching architecture [5]. LDP, efficient and optimized for deployment in the context of MPLS, supports all label binding flavors: on-demand as well as unsolicited (presented in Section 18.4.3), topology-driven as well as triggered by explicit events (introduced in Section 18.4.3). However, its delators criticized the fact that it is yet another protocol that LSRs need to implement and execute.

18.4.4.4 Remarks. Although the three label distribution alternatives presented in the previous section are incompatible (an LSP cannot be set up across subnetworks of LSRs deploying different label distribution protocols), the MPLS Working Group [22] ratified all three for deployment in MPLS networks. In fact, all of them met the requirements expressed by the IETF motto "we believe in rough consensus and running code," and the Working Group agreed to let the market and the users decide which alternative was preferable. Some vendors supported and implemented one solution, others the other, and all further developments and additions involved all the solutions. Finally, in 2003 the MPLS Working Group decreed that LDP, although a standard label distribution solution was not to be developed any further.

Consequently, the two solutions for LSP setup supported by current and future MPLS equipment are RSVP-TE and BGP; the former is mainly used when MPLS is deployed for traffic engineering purposes, and the latter is mainly used for topology-based LSP creation and for the realization of Virtual Private Network (VPN) services (see Section 18.5.3 for more details).

18.4.5 Routing and Routing Protocols

Usually two different levels of routing are involved in packet networks:

- *On-the-Fly Routing.* This deals with determining the output interface through which a packet received by an intermediate network nodes is to be forwarded toward its destination.
- *Proactive Routing.* This deals with determining the output interface of an intermediate network node through which a destination should be reached (i.e., the route to the destination), *independently* of whether packets for that destination are actually received by the node.

On-the-fly routing is usually the performed within the data plane, while proactive routing is usually the responsibility of the control plane, but the two operations are tightly interrelated because the former is more or less directly based on the outcome of the latter.

In the following, the basic principles of routing in traditional IP networks are reviewed together with the most widely deployed routing protocols. The traditional IP routing paradigm is then compared with packet routing in MPLS networks (Section 18.4.5.2) discussing how notwithstanding the differences in on-the-fly routing, the same paradigm and protocols can be used for proactive routing. However, when aiming at performing traffic engineering, a new routing paradigm, supported by the MPLS architecture for on-the-fly routing, must be deployed in proactive routing, which requires an extension to existing routing and label distribution protocols (Section 18.4.5.3).

18.4.5.1 Routing in IP Networks. A traditional IP network node bases on-the-fly routing on information contained in its *routing table* that stores the *next hop* to each destination—that is, the neighboring node to which packets to such destination are to be forwarded. The routing table is built during proactive routing either by configuration (static routing) or by exchanging routing information with neighboring nodes by means of a routing protocol. Although the (proactive) routing decisions of the various nodes on a path from source to destination have to be coherent to ensure consequent routing (i.e., to avoid routing loops that lead to packets being randomly forwarded through the network without possibly reaching their destination), the decision of each node is independent. Consequently, the node has no impact on the route that packets will take all the way to their destination, but only on the one-hop portion to the next hop. Traditional IP routing is thus based on a *hop-by-hop routing* paradigm.

Various routing protocols are being used in IP networks to realize dynamic proactive routing, with their specific properties making them suitable for different situations. The exchange of information between routers belonging to different autonomous systems (i.e., interdomain routing), virtually uniquely relies on the Border Gateway Protocol (BGP) [17], already briefly described in Section 18.4.4.1. Instead, Open Shortest Path First (OSPF) [23] and Intermediate System to Intermediate System (IS-IS) [24] are the most widely adopted solutions for intradomain routing—that is, in the exchange of information between routers belonging to the same autonomous system—and their deployment is envisioned also in MPLS networks. Both protocols are based on the link state routing algorithm that can find the best (or shortest) route also in networks with complex topologies ensuring not to generate routing loops as long as all routers have consistent information on the network topology. Moreover, both protocols enable routers to be organized *hierarchically* to achieve high scalability—that is, to limit the amount of information handled by each router independently of the overall size of the network. This is achieved by dividing the network in areas and having routers inside the area having detailed information on the internal topology and limited information of the exterior. A subset of the routers—called area border routers in OSPF and level 2 routers in IS-IS—"summarize" information about areas before distributing it to the other routers throughout the network. While OSPF has been designed specifically for deployment in IP networks, IS-IS initially represented an interim solution for scalable routing before commercial OSPF implementations were finally stable realized by extending a protocol originally implemented for OSI networks. Since several Internet service providers have not been willing to migrate their large networks deploying IS-IS to OSPF even after stable implementations have become available, IS-IS is still widely used.

18.4.5.2 The Role of Routing in MPLS Networks. In the context of MPLS, on-the-fly routing is based on the label mapping installed in LSRs—that is, the information in the ILM and NHLFE tables (Figure 18.13). Each LSR performing dynamic proactive routing also maintains a routing table updated by means of routing protocols; the routing protocols deployed in IP networks—specifically OSPF, IS-IS, and BGP—are used also in this context. However, while in IP networks the routing table is used in the data plane to directly drive on-the-fly routing (see Figure 18.5), in MPLS it is deployed within the control plane (see Figure 18.6) to drive label mapping. An LSR wanting to create a label mapping for a certain FEC (e.g., in response to a signaling event) maps a new entry in the ILM table containing its local label binding onto a new NHLFE containing the label distributed by one of its neighbors and the information required to forward packets to it. If hop-by-hop routing is performed, such a neighbor is the one listed in the routing table as the next hop to the destination. Consequently, routing protocols and the routing table they build indirectly determine the path of LSPs and consequently the route taken by packets through the network.

18.4.5.3 Constraint Based Routing and Protocol Extensions. As discussed in Section 18.1 when presenting the motivations for the success of MPLS, one of the key values of MPLS is the capability of enabling traffic engineering. Such a capability requires that, in addition to topological information, (proactive) routing decisions take into account also other types of information, such as load levels of links and nodes, delay experienced by packets across links and nodes, dependencies[5] among links and nodes. This is called *constraint-based routing* since the above additional, more articulated, information is usually deployed as a *constraint* in the choice of a route to a destination. Since additional information must be exchanged by routers, constraint-based routing must be explicitly supported by routing protocols whose messages and data structures need to be designed to hold *constraint data*. Specifically, the OSPF-TE (OSPF for Traffic Engineering) [25] and IS-IS-TE (IS-IS for Traffic Engineering) [26] extensions have been specified, where the reference to traffic engineering stems from it being the main application for constraint-based routing.

Examples of the application of constraint-based routing include automatic load balancing and fault protection. In order to achieve load balancing, LSRs when creating a new LSP choose the shortest route to a destination whose links have available bandwidth above a given threshold. In order to implement fault protection on an LSP, LSRs set up a backup LSP on the shortest route to the destination whose links and nodes do not have any dependencies with the ones of the protected LSP (see Section 18.5.2).

Hop-by-hop routing, traditionally deployed on IP networks, is not suitable for constraint-based routing. Hop-by-hop routing is a fully distributed solution since each network node takes its (proactive) routing decisions "independently." However, in order to avoid routing loops, routing decisions taken by nodes across the network must be coherent. This is ensured by the routing algorithms most commonly deployed (i.e., distance vector and link state), as long as (i) routing information is properly distributed to all network nodes and (ii) a criterion to

[5]An example of dependency is two nodes sharing the same power supply, or two links whose cables cross the same duct. Dependency among network elements is fundamental when choosing the path for a LSP to be used as a backup to another one in fast fault recovery, as discussed in Section 18.5.2.

uniquely choose one route among all those available to the same destination can be defined (e.g., the shortest route in terms of number of hops, or the route with the minimum average end-to-end delay).

While the two above conditions can be satisfied when the network is expected to provide a best-effort service (as the Internet has long done), problems arise when a more sophisticated service is to be provided—for example, when a certain service quality in terms of maximum loss probability, delay, or down time is to be offered—and constraint-based routing is to be offered. Specifically, given the time required by routing messages to propagate throughout the network, the change frequency of constraint date (such as the load level on a link) limits the fraction of time during which all routers have updated and coherent information, as required by condition (i) above. Moreover, the existence of multiple routes to a destination satisfying the required constraints violates condition (ii) above.

Consequently, hop-by-hop routing cannot be deployed by LSRs when performing label mapping to create an LSP. Instead, *explicit routing* is to be used, where one LSR chooses the path for an LSP and imposes it to the other nodes on it. In this case, there is no risk of creating routing loops even if the LSR chooses the route arbitrarily (i.e., not following a unique criterion as required by condition (ii) above) and based on information not shared with other nodes (i.e., violating condition (i) above). At worst the route might not be optimal if computed using outdated information. However, explicit routing requires a communication means for the LSR choosing a route for an LSP to provide it to the other LSRs on the path; the most natural solution is to deploy label distribution protocols that need to be modified for the purpose. Consequently, RSVP-TE (RSVP for traffic engineered tunnels) [20] CR-LDP (LDP with constraint-based routing) [27] have been specified to support explicit routing; that is, label distribution messages explicitly include the route along which they are to be forwarded and label mapping is to be performed.

18.5 MPLS AND OTHER BUZZWORDS

Having addressed the most important concepts and mechanisms related to MPLS applications and operation, the reminder of the chapter tackles the various areas in which MPLS is mentioned and deployed today.

18.5.1 Generalized MPLS

Although MPLS has been originally conceived as a solution to speed up the (on-the-fly) packet routing and enable more scalable network devices (i.e., for its role in the data plane of routers), its current relevance is mostly related to its role as a control plane solution enabling key features not supported by traditional IP, such as traffic engineering and fast fault protection. The control plane functionalities and protocols developed for the MPLS control plane, namely routing and signaling protocols to support the setup of LSPs, are mostly independent of the specificities of the data plane. In particular, the requirements to be able to utilize the MPLS control plane are to (i) identify network nodes by means of IP addresses and (ii) route data units based on some sort of information (the label, in the case of MPLS) associated with a flow of data units, possibly changed at each hop. Following this considerations, in the years in which optical networking was receiving utmost attention and promising to become the underlying technology for all future networks, the MPLS control plane was considered as a candidate to route and set up lightpaths in a solution named Multiprotocol Lambda Switching (MPλS) [8]. In optical switching, the wavelength of an optical channel (a.k.a. lambda from the Greek letter commonly used to indicate wavelengths) fulfills the same purpose as the label of packet-based MPLS: While MPLS packet switches use the label of an incoming packet to determine the output interface on which it shall be switched, lambda switches route all the bits received on an incoming optical channel (or the whole electromagnetic wave encoding them in case of all-optical switches) based on the wavelength of the channel signal. Like an MPLS packet switch might change the label associated to an outgoing packet, in an optical switch performing wavelength conversion the wavelength of the outgoing optical channel might be different from the wavelength of the incoming one. In MPλS the sequence of wavelengths to be used to create a lightpath through the network and the path to be followed—that is, the output interface to which each node must switch incoming optical channels—is established by means of a label distribution protocol (such as RSVP-TE and LDP) and a routing protocol (such as OSPF, IS-IS, and BGP), respectively.

The idea can be further generalized to (a) other types of labels, such as ATM virtual circuit identifier/virtual path identifier (VCI/VPI) and frame relay data-link channel identifier (DLCI), and (b) SONET and SDH slot numbers and data units, such as ATM cells, frames, and slots. This led to the definition of generalized MPLS (GMPLS) [9] as a set of extensions to the protocols of the MPLS control plane—for exapmle, support for encoding various generalized labels in RSVP-TE messages and constraint data relevant to different types of networks in OSPF and IS-IS messages.

18.5.2 Fast Fault Protection

One very important application of MPLS is in fast fault protection. Although IP has been designed since its very inception to be fault-tolerant,[6] the timing in the reaction to malfunctions was not among the initial concerns. Routing protocols are at the basis of fault tolerance in IP networks: When a link or a node fails, neighboring nodes detect it and modify their routing messages to reflect the resulting topological change. Routers, by computing their routing tables using the updated information, change the routes followed by packets on the network

[6] IP originated from a research funded by the Defense Advanced Research Projects Agency (DARPA) to design a network capable of adapting to topological changes due to failures of links and nodes, as it could be in wartime or in a battlefield.

to include only operational nodes and links—that is, avoiding the failed element. The time elapsed between a failure and the update of the routing tables in all network nodes can be on the order of several tens of seconds to few seconds if adhoc mechanisms are implemented to reduce routing convergence time on specific topologies. While such timescale is sufficient for most applications traditionally deployed on IP networks, it is too large for some new applications, such as voice, telephony (including interconnection of central offices over IP networks), and some forms of electronic commerce, including stock trading and financial transactions.

Due to its connection-oriented nature, MPLS provides a solution to the issue by enabling restoration of service following a fault within a fraction of a second, possibly limiting the loss of connectivity to less than the 50 ms, the maximum tolerated by central offices in telephone networks before dropping all the phone calls traversing a trunk link interconnecting them. This allows using an LSP for the trunk interconnections on circuit-switched networks, thus finally enabling the long-sought convergence of the telecommunications (circuit) and information technology (packet) worlds onto a single infrastructure of which MPLS is the centerpiece.

Fast recovery from a fault, commonly called *fast fault protection*, can be achieved by means of *backup* LSPs that are proactively created and kept ready for use in case of fault. When a fault is detected (usually by the LSRs directly connected to the failing network element), a light and fast mechanism is deployed to notify the LSR at the ingress of a backup LSP so that it switches traffic from the affected *primary* LSP(s) to the backup one. Since the switching operation is very fast (it requires changing an ILM mapping from NHLFE of the primary LSP to the NHLFE of the backup LSP), the downtime due to the failure basically depends on the time required for the notification. If the signaling protocol is lightweight, such time mostly depends on the propagation delay on the links; that is, if the network is not too large, a 50-ms recovery time can be achieved.

Figure 18.14 exemplifies two common protection strategies. In *edge-to-edge re-routing*, a backup LSP is created between the ingress and egress LSRs of the primary LSP (A and E in Figure 18.14); the backup LSP must be routed on a path

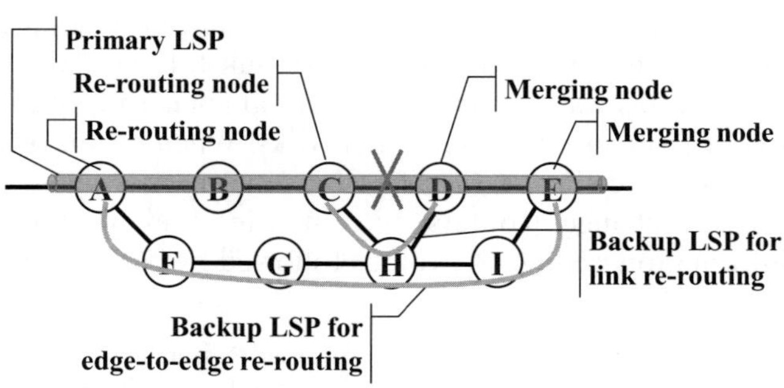

Figure 18.14. Fast fault protection through link rerouting and edge-to-edge rerouting.

(F-G-H-I in Figure 18.14) that does not have any dependency with the path of the primary LSP[7] to avoid that a single failure affects both LSPs. If a link fails, compromising the operation of the primary LSP (link CD in Figure 18.14), the LSRs at its edges (C and D in Figure 18.14) detect the failure and the upstream one notifies the node terminating the backup LSP (A in Figure 18.14) that switches the traffic from the primary LSP to the backup LSP by changing the ILM or FTN entry corresponding to the LSP. In the context of fast fault protection, such an LSR is called a *re-routing node*. Note that no action is required on the side of the other end of the backup LSR (i.e., the so-called *merging node*), since the ILM of the backup LSP is already mapped on the same NHLFE as the primary LSP; that is, packets entering the node through the primary LSP and through the backup LSP are handled the same way and specifically routed to the same output port and assigned the same output label (or forwarded as regular IP packets).

In *link re-routing*, a backup LSP is created between the two LSRs at the edges of a link that is to be protected from failure. Link re-routing has the advantage that setting up a single backup LSP provides protection of all LSPs routed through the corresponding link. However, when assurance on the service quality is required, determining the amount of resources to be reserved to the backup LSP is not straightforward because they are going to be shared by several LSPs in case of failure of the link.

18.5.3 MPLS-Based VPNs

A *virtual private network* (VPN) aims at providing connectivity to private networks through a *shared* infrastructure while enabling the enforcement of policies as if interconnected through a dedicated infrastructure. At least in principle, the nature of such policies ranges from security, to addressing, to quality of service; the typical application for VPNs is the interconnection of remote corporate sites allowing the same private addressing scheme to be used across the various sites and ensuring isolation of corporate traffic from other traffic traveling on the shared communication infrastructure. The shared infrastructure might be the backbone of a service provider or even the public Internet. Although various solutions are available and being implemented, MPLS enables a scalable, manageable solution for VPNs where the shared infrastructure is the MPLS network of a service provider [28].

Figure 18.15 shows the network architecture for providing VPN services over an MPLS network [28]. A customer edge (CE) router at the edge of a private (corporate) network interconnects to a provider edge (PE) router that is a label edge router in charge of adding an MPLS header with a proper label to packets received from the private network and strips the header of packets to be forwarded to the CE router. Because an MPLS backbone [where (on-the-fly) routing

[7]Besides the primary and backup path not sharing any LSR and link, their elements must be independent. For example, LSRs of the backup path should not share power supply with any LSR of the primary path and no link on the backup path should not be contained in a duct containing also a link of the primary path.

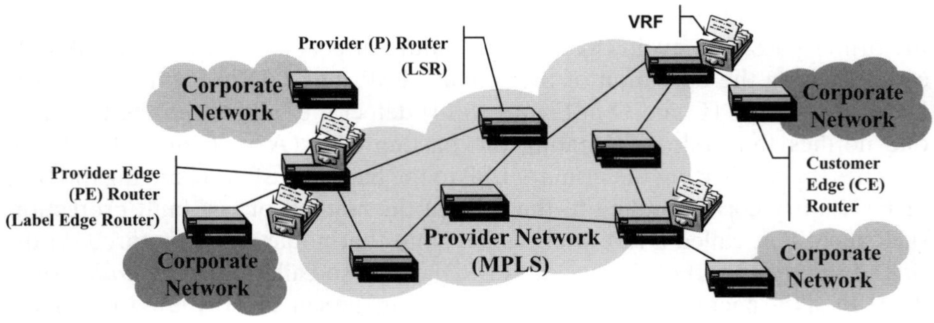

Figure 18.15. MPLS-based VPN network architecture.

is based on labels (rather than addresses in the IP packet headers)] is used for interconnecting the private networks, deployment of private addresses, possibly duplicated on different corporate networks and in the backbone, does not affect proper packet forwarding. In order to allow the same addresses to be used by more than one corporate network it is connected to, each PE router builds multiple forwarding tables (each one with its FTNs and NHLFEs) called Virtual Routing and Forwarding (VRF), one for each corporate network it is connected to (see Figure 18.15). Consequently, provider edge routers attach to each packet entering the backbone a label looked up in the VRF corresponding to corporate network the packet is coming from, which ensures that the packet is delivered to the appropriate site of the same corporate network, thus guaranteeing separation of traffic among different corporate networks.

A similar separation must be implemented also in the control plane; that is, when exchanging routing information about the reachability of the various private destinations, advertisements related to different corporate networks must be kept and processed separately. Each PE router gathers information about the corporate networks it is connected to by exchanging routing information, usually through an intradomain protocol such as OSPF, IS-IS, or RIP, with the CE routers it is connected to. Such information is then redistributed across the MPLS backbone reaching all other PE routers connected to CE routers of the same corporate network. To each of the CE routers to which it is connected, PE router must redistribute to only the advertisements concerning destinations of the corporate network to which the CE router belongs. Although different variants for the implementation of MPLS-based VPN services exist, one widely accepted relies on the (preexisting) multiprotocol extensions of the border gateway protocol (BGP) [18] and the definition of a *VPN-IPv4 address family* to segregate route advertisements pertaining different corporate networks. Addresses of the VPN-IPv4 family are obtained by prepending a 8-byte *route distinguisher* to each 4-byte IP address. Since each VPN active on the MPLS backbone is assigned a unique route distinguisher, even if different corporate networks deploy the same addresses, their VPN-IPv4 addresses are different. The multiprotocol extensions enable BGP to advertise routes to VPN-IPv4 addresses; and routers can filter advertisements and compute their forwarding table, taking into account the route distinguishers. For example, a PE router will use in the computation of each of its VRFs only the advertisements related to VPN-IPv4 addresses corresponding to one of the corporate networks that the VRF refers to.

Another concern when realizing a large-scale VPN solution is the scalability of routing because, in order to enable dynamic routing, all the destinations of the corporate networks must be advertised through the MPLS backbone in addition to the information on the backbone topology. BGP features help also in this respect. Since BGP messages can be exchanged by routers that are not directly connected, internal routers do not have to participate into the exchange of information about corporate networks that is carried out directly by PE routers. Consequently, internal routers run an intradomain routing protocol, such as OSPF or IS-IS, to gather information exclusively on the MPLS backbone, which limits the size of their data structure and the amount of routing traffic generated. On the other hand, although PE routers participate in the exchange of information about private destinations, their burden is somewhat limited by the fact that they can use the filtering functionalities of BGP to avoid storing and processing locally all the advertisements.

In summary, the BGP/MPLS VPN solution leverages on the flexibility and versatility of the specification and the implementations of BGP to ensure separation of routing information and scalability. Note that BGP can also be used to distribute labels through the MPLS backbone, so LSPs can be properly set up through it to carry traffic among the corporate sites.

Compared to other solutions for the realization of VPN services over IP networks at both the third ISO/OSI level, mostly relying on tunnels based on the encapsulation of IP packets into IP packets and the second ISO/OSI level, mostly relying on setting up a mesh of permanent circuits on a connection-oriented backbone, the BPG/MPLS solution features a high level of automatism stemming from the deployment of the sophisticated routing solution outlined above. As a result, adding new corporate networks requires limited administrative action: Once a new CE router is connected to a PE router, the network administrator needs only to configure the proper VPN identifier to be used as a route distinguisher. The routers, through the exchange of routing information, will automatically discover the destinations on the new corporate site, advertise them to the other sites of the same corporate network, and create the needed label bindings in the PE routers and in the internal routers to provide connectivity. Delators of the BGP/MPLS VPN solution criticize that, contrary to other solutions, data traveling through the backbone unencrypted represents a possibly threat to confidentiality. As in many solutions deployed for interconnection of remote networks, including virtual channels on frame relay networks, leased lines on circuit switched networks, virtual circuits on ATM networks, data confidentiality relies on the trust the customer grants to its VPN service provider. For customers or applications that require a higher security level, specific encryption solutions can be implemented on the corporate network—that is, beyond the scope and the support of VPN provisioning.

18.5.4 MPLS and Quality of Service Support

Independently of the specifics of a solution, providing deterministic quality entails an initial negotiation or *signaling* phase in which the "network user," possibly an application, specifies the quality parameters required when transferring data through the network. During this phase the network verifies whether it has sufficient resources to satisfy the request and possibly reserves such resources to the traffic generated by the application. This operational procedure does not fit with the traditional IP paradigm according to which (i) a host wanting to transfer packets through the network simply transmits them without prior negotiation and (ii) each packet represents a stand-alone unit that travels through the network on a route not necessarily related to the one followed by previous and following packets generated by the same host or application. While statement (i) clearly contrasts with the principle of negotiating quality of service (QoS) parameters and reserving resources before transmitting, statement (ii) potentially makes useless the resource reservations made through the network because (some of the) packets might take a different route than the one along which resources have been reserved. These issues have worked around in the Integrated Services (IntServ) [19] architecture by introducing the Resource Reservation Protocol (RSVP) [18] to perform the initial negotiation phase and a complex mechanism called *route pinning* to ensure that all packets are actually routed through the network nodes that have reserved resources for their flow.

The basic principles underlying MPLS are instead more in line with the basic operational requirements of QoS support, even through MPLS has not been explicitly designed for this purpose since the aim of the MPLS Working Group design effort was to produce a routing, not a QoS, solution. Specifically, although in the topology-based label binding mode no explicit and coordinated signaling is used, MPLS includes the concept of creating an LSP before transferring data through the network and the architecture encompasses mechanisms and protocols to do so. Moreover, all packets belonging to the same FEC are handled the same way and specifically are routed through the same route.

In summary, although MPLS was not designed as a QoS solution and is not used as such, its basic underlying principles are compatible with it and potentially facilitate QoS support.

18.5.5 MPLS and Differentiated Services

Although QoS support, in the sense of providing a deterministic service with resources dedicated to a specific application or data flow as in an integrated services network [19], is not implemented today over the Internet and IP/MPLS networks, the provision of services more sophisticated than the traditional best-effort one is of great importance. Service differentiation, based on the Differentiated Services (DiffServ) architecture [12], is at the basis of this and in traditional IP networks it relies on a field in the IP header whose value identifies the traffic class to which a packet belongs. Each router might handle the packet in a specific way based on the content of this field, called *Type of Service* (*ToS*) in the original IP specification and renamed *DS Field* in the context of the DiffServ architecture. Although this does not ensure that packets generated by a given application receive a specific service with known quality parameters, as in the IntServ approach, packets belonging to applications that need a different service might be handled differently from the others by routers. For example, when a router is congested and needs to discard packets, it tries to discard only packets belonging to classes for which loss is acceptable. Although this does not guarantee that packets belonging to such classes are not lost, it ensures a lower probability of loss with respect to other classes. Moreover, if it is ensured that the amount of traffic *belonging to that class* is *small* compared to the overall capacity of the network, packets belonging to such class in fact experience a small or null loss probability, *independently* of the amount *of other types of traffic* and loss probability experienced by their packets.

Due to the low complexity of its implementation and deployment, the DiffServ solution is widely used today to provide different levels of service to customers willing to pay a different service fee and, by limiting the amount of traffic of a given service class, to guarantee a certain service (e.g., delay below a given threshold and loss probability below a given threshold). In traditional IP networks, traffic is limited through *network engineering*, namely by limiting the number of users (generating traffic of a specific service class) and the amount of traffic each user can transmit. This has very low efficiency because being based on (almost) worst-case scenarios, some parts of the network might turn out to be heavily underutilized.

MPLS with its traffic *traffic engineering* capability enables efficiency to be improved; for example, when traffic of a given class has reached the maximum acceptable level on a certain link, route, or area of the network, part of it can be diverted on a different route. In other words, the combination of MPLS and DiffServ enables *per-class traffic engineering* through which higher efficiency can be achieved in the utilization of the network when provisioning services with assured quality.

REFERENCES

1. The ATM Forum, ATM User–Network Interface Specification-Version 3.1, September 1994.
2. International Telecommunication Union–Telecommunication Standardization Sector (ITU-T), http://www.itu.int/ITU-T/.
3. International Telecommunications Union, ISDN Data Link Layer Specification for Frame Mode Bearer Services, ITU-T Recommendation Q.922, 1992.
4. Broadband Forum, http://www.broadband-forum.org.
5. Cisco Systems, Tag switching, in *Internetworking Technology Handbook*, available on-line at http://www.cisco.com/en/US/docs/internetworking/technology/handbook/Tag-Switching.html.

6. D. Katz et al., Tag switching architecture—Overview, Internet Drafts, IETF, 1997.
7. ITU-T, Architecture for the Automatically Switched Optical Network (ASON), Recommendation G.8080/ Y.1304, November 2001 (and Revision, January 2003).
8. D. Awduche and Y. Rekhter, Multiprotocol lambda switching: Combining MPLS traffic engineering control with optical crossconnects, *IEEE Communi. Maga.*, Vol. 39, No. 3, pp. 111–116, March 2001.
9. G. Swallow, J. Drake, H. Ishimatsu, and Y. Rekhter, Generalized Multiprotocol Label Switching (GMPLS) User–Network Interface (UNI): Resource ReserVation Protocol-Traffic Engineering (RSVP-TE) Support for the Overlay Model, RFC 4208, Standards Track, October 2005.
10. IETF, The Internet Engineering Task Force (IETF), http://www.ietf.org.
11. E. Rosen, D. Tappan, G. Fedorkow, Y. Rekhter, D. Farinacci, T. Li, and A. Conta, MPLS Label Stack Encoding, RFC 3032, Standards Track, January 2001.
12. S. Blake, D. Black, M. Carlson, E. Davies, Z. Wang, and W. Weiss, An Architecture for Differentiated Services, RFC 2475, Standard Tracks, December 1998.
13. F. Le Faucheur et al., Multi-Protocol Label Switching (MPLS) Support of Differentiated Services, RFC 3270, Standards Track, May 2002.
14. B. Davie, J. Lawrence, K. McCloghrie, Y. Rekhter, E. Rosen, and G. Swallow, MPLS Using LDP and ATM VC Switching, RFC 3035, Standards Track, January 2001.
15. A. Conta, P. Doolan, and A. Malis, Use of Label Switching on Frame Relay Networks Specification, RFC 3034, Standards Track, January 2001.
16. E. Rosen, A. Viswanathan, and R. Callon, Multiprotocol Label Switching Architecture, RFC 3031, Standards Track, January 2001.
17. Y. Rekhter, T. Li, and S. Hares (eds.), A Border Gateway Protocol 4 (BGP-4), Standards Track, RFC 4271, January 2006.
18. R. Braden (ed.), L. Zhang, S. Berson, S. Herzog, and S. Jamin, Resource ReSerVation Protocol (RSVP)—Version 1 Functional Specification, RFC 2205, Standards Track, September 1997.
19. R. Braden, D. Clark, and S. Shenker, Integrated Services in the Internet Architecture: An Overview, RFC 1633, Informational, June 1994.
20. D. Awduche, L. Berger, D. Gan, T. Li, V. Srinivasan, and G. Swallow, RSVP-TE: Extensions to RSVP for LSP Tunnels, RFC 3209, Standards Track, December 2001.
21. L. Andersson, P. Doolan, N. Feldman, A. Fredette, and B. Thomas, LDP Specification, RFC 3036, January 2001.
22. IETF MPLS Working Group, Multiprotocol Label Switching, http://www.ietf.org/html.charters/mpls-charter.html.
23. J. Moy, OSPF Version 2, RFC 2328—STD 54, April 1998.
24. R. W. Callon, Use of OSI IS-IS for Routing in TCP/IP and Dual Environments, RFC 1195, Standards Track, December 1990.
25. K. Ishiguro, V. Manral, A. Davey, and A. Lindem (ed.), Traffic Engineering Extensions to OSPF Version 3, RFC 5329, Standards Track, September 2008.
26. T. Li and H. Smit, IS-IS Extensions for Traffic Engineering, RFC 5305, Standards Track, October 2008.

27. B. Jamoussi, (ed.), L. Andersson, R. Callon, R. Dantu, L. Wu, P. Doolan, T. Worster, N. Feldman, A. Fredette, M. Girish, E. Gray, J. Heinanen, T. Kilty, and A. Malis, Constraint-Based LSP Setup Using LDP, RFC 3212, Standards Track, January 2002.
28. E. Rosen and Y. Rekhter, BGP/MPLS IP Virtual Private Networks (VPNs), RFC 4364, Standards Track, February 2006.

19

OVERVIEW OF STORAGE NETWORKING AND STORAGE NETWORKS

Eugene Ortenberg and Christian van den Branden

19.1 INTRODUCTION

This chapter presents an overview of storage networks and highlights the synergies and parallels between storage technologies and networking. While lesser known to computer networks specialists, storage networking technology is a field rapidly growing in importance. Additionally, recent protocol developments are merging common data and emerging storage networking technologies in the data centers.

Section 19.2 covers the history of storage networks and describes how they came to existence in data centers. The key concepts in storage networks are then introduced in Section 19.3, which goes through the storage networking stack from raw physical disks to file systems. This section also introduces the difference between storage area networks (SAN) and network attached storage (NAS).

Section 19.4 is devoted to the various business and operational applications either realized or greatly facilitated by storage networks. Section 19.5 then dives into the various protocols specific to storage area networks, starting with the SCSI protocol and covering recent developments that start mixing pure storage technologies with IP-based networks. File serving protocols are discussed in

Convergence of Mobile and Stationary Next-Generation Networks, edited by Krzysztof Iniewski

Section 19.6. Finally, Section 19.7 is dedicated to virtualization, which is an emerging technology destined to significantly improve efficiency, scalability, and cost effectiveness of the data center architecture.

19.2 INTRODUCTION TO STORAGE NETWORKS

19.2.1 Overview of Computer Storage

For a long time, storage capabilities were just implemented within a computer subsystem. This was the case for data center servers and mainframe as well as for individual workstations and personal computers. In fact, due to limited bandwidth and high cost of input/output (I/O) channels and also because of severe disk capacity limitations, the most common storage attachment paradigm was to directly connect multiple storage subsystems to a single computer. Additionally, in the early mainframe days, each manufacturer had a proprietary I/O stack and protocol. Standardization of I/O protocols and disk interfaces came much later.

This remained the case throughout the evolution of the data centers from early mainframes to general-purpose servers until the early 1980s when enterprise computing started shifting to client–server computing. By then, enterprises were heavily relying on mainframes for compute-intensive data processing applications while minicomputers were also being commonly deployed and connected through local area networks (LANs).

Managing these assets became increasingly difficult for the enterprises. While the cost of the infrastructure was rapidly decreasing, the management overhead became very significant and its complexity increased dramatically:

- Various operating systems and their revisions had to be deployed across a large number of computers.
- A variety of software applications had to be maintained across that collection of computers, each having its set of versions, patches, and updates.
- Most importantly, crucial company digital assets—that is, the data being manipulated by these computers—were scattered over the collection of computers, making it difficult and expensive to back up and protect. Backup was typically done to tape, which required dedicated hardware and incurred long recovery time window when needed.
- Ensuring high availability of applications required an enterprise to purchase custom-made fault-tolerant servers and disk subsystems, both of which were very expensive.

Client–server computing leveraged technological advances, which greatly simplified that picture. The paradigm of enterprise computing is depicted in Figure 19.1. In this paradigm the general-purpose client computers are separated

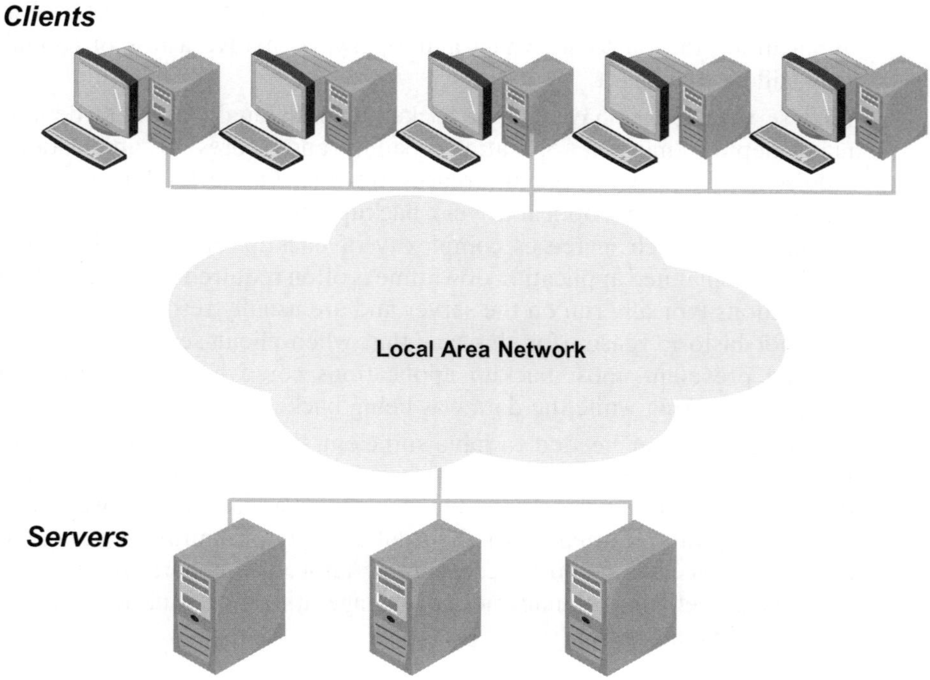

Figure 19.1. Early data center composed of servers connected through clients over a local area network.

from the high-end/high-performance servers. The servers are placed in data centers and are responsible for running most enterprise applications as well as locally maintaining the data for these applications. The client computers connect to the servers over a local area network. Most compute cycles and all the data for mission critical applications impact the servers but not the host computers.

The management overhead and reliability of the information technology (IT) infrastructure of the enterprise are now much improved. The IT staff maintains the servers, backs them up, protects them against intrusion and viruses, and maintains their application software and operating system versions. Client computers, operated by the users only, run a minimal set of applications.

Even though this picture offers an improved solution, it still suffers from a major shortcoming: Each server still maintains its own little island of storage. The major drawbacks of this problem are as follows:

- Each server (or some of its subcomponents, including the storage subsystem) becomes a single point of failure. Should a server, its networking infrastructure, or one of its applications fail, the enterprise may be unable to provide a crucial service until the failure is repaired. Such outages cannot be tolerated by mission critical applications, because they may lead to catastrophic outcomes including potential loss of life. For less critical applications, the outcome, which almost always involve a loss of capital, may still be significant.

- Each server needs to have its data backed up individually as the IT department deploys multiple operating systems and scores of applications. Additionally, because of the differences in OS backup technology involved (such as drums, tapes, optical drives), backup applications themselves may be different, which increases complexity of backup. Finally, maintaining windows of planned application downtime is often required because backup applications typically run on the server and are usually resource-intensive. Another historic reason for this was that when client–server computing became prevalent, most backup applications could not deal with data changes occurring while the data was being backed up.
- Finally, each server needed to have sufficient storage capacity. The addition/increase of storage (i.e., storage provisioning) was cumbersome and often required server downtime. On the other hand, if storage was over-provisioning (and it often was!), the total capacity of unutilized storage in a given data center could be exceedingly large. The general problem of effective and efficient management of storage utilization is difficult at best and infeasible at worst.

19.2.1.1 External Storage Systems. The next step in data center evolution happened in the 1980s when the storage subsystems started migrating out of the server enclosure. The basic data center architecture changed from the one depicted in Figure 19.1 to the one depicted in Figure 19.2. The storage subsystem (or array) is now a separate entity, which is connected to servers over a network. This architecture is typically referred to as a *Storage Area Network* (SAN). Several technologies have been used for networking the storage array to the servers or computers. Mainframe systems relied on proprietary technologies such as ESCON.[1] Non mainframe systems (often referred to in the storage industry as open systems) relied on Fibre Channel. While we will not be discussing mainframe topology in detail in this chapter, we will dive into Fibre Channel and other SAN infrastructures in Section 19.5.

The technological advances that enabled this new SAN architecture are:

- Increase in disk capacity.
- Increase in disk drive performance: drives have been able to achieve a large number of I/O operations per second and have achieved lower seek delays.
- Increase in network bandwidth.
- Decrease in network latency.

[1]ESCON (Enterprise Systems Connection) is an IBM data connection, used to connect mainframes to peripheral devices. ESCON is an optical fiber, half-duplex, serial interface.

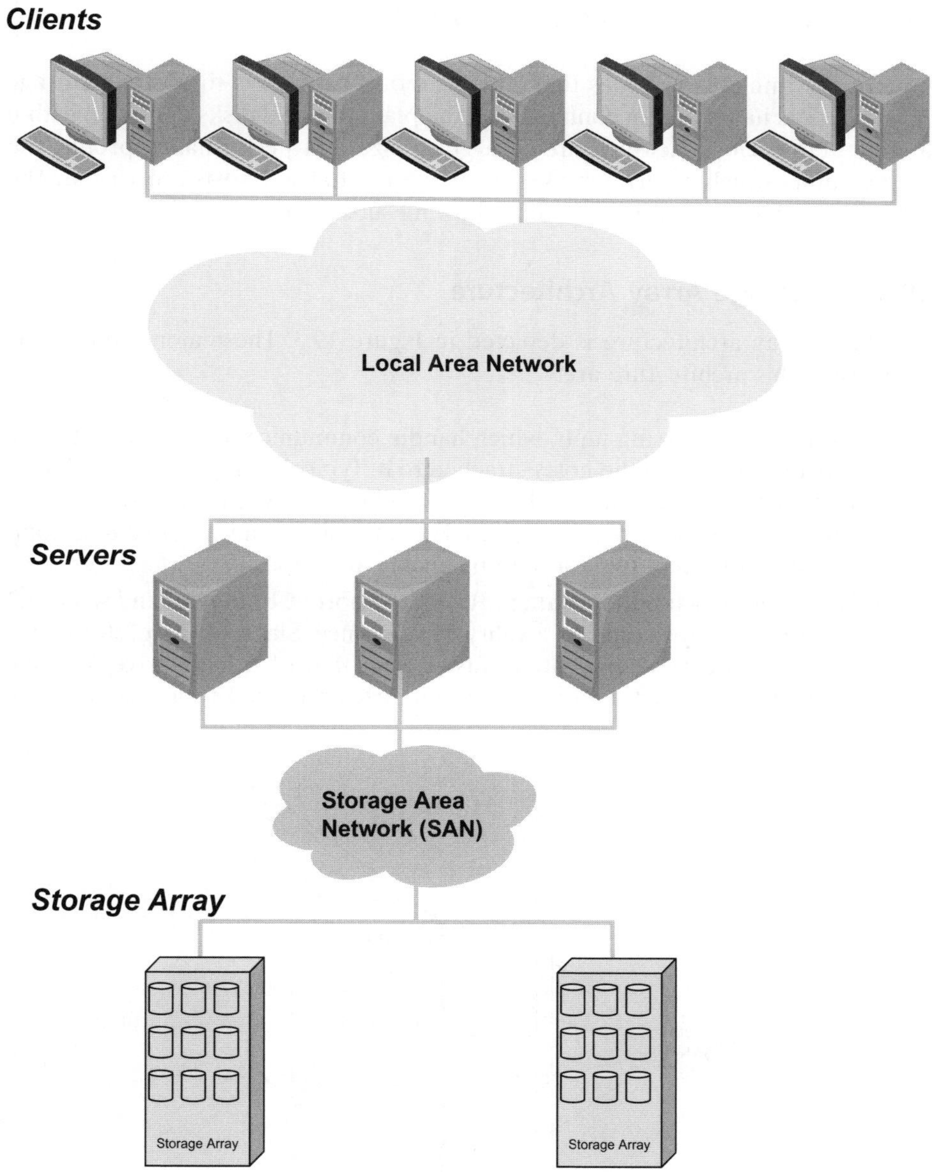

Figure 19.2. Data center with a consolidated storage area network.

Note that storage arrays provide an immediate consolidation benefit: By aggregating a large number of disks in a single enclosure, they are able to provide improved throughput, reliability, and flexibility. This is achieved by adding an abstraction layer above the level of the physical drives as described in Section 19.3.2. A more detailed discussion of this topic as well as many associated technologies can be found in references 1 and 2.

19.3 STORAGE NETWORKING CONCEPTS

One of the immediate values that storage arrays provide is to abstract storage from physical hardware. In contrast with the plain physical disks deployed within a computer enclosure, which are of a given size and performance profile, the storage units which an array makes available to attached hosts are virtual. The major concepts are described in the following subsections.

19.3.1 Storage Array Architecture

A typical array architecture is depicted in Figure 19.3. The major components comprising this architecture are:

- Front-end processing units, which handle communication with the storage network fabric and the hosts attached to it. Typical storage arrays will have at least two such units for high availability and will have a transparent failover mechanism between them. Higher-end systems are able to accommodate more than two for load balancing and capacity.
- The cache, which buffers data in RAM, to improve both read and write I/O performance by exploring locality of reference. Since RAM performance is several orders of magnitude faster than that of a hard drive, keeping frequently accessed data in cache minimizes the need to access the slow

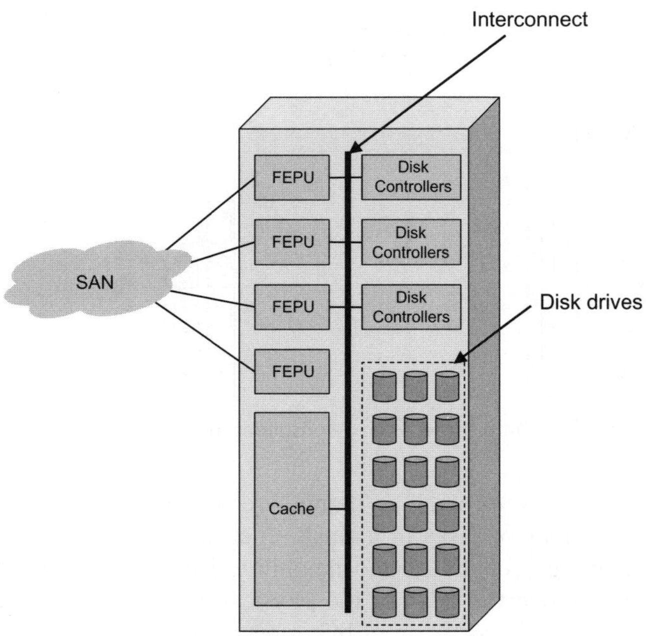

Figure 19.3. Typical storage array architecture.

drives and significantly decreases the processing latency. Also, allowing the write I/Os to complete before the data have been de-staged to the hard drives dramatically improves write performance. Storage arrays typically rely on standby battery power supply in order to write the content of the cache (held in volatile RAM) to disk in the event of a loss of power to the array. This allows arrays to guarantee the integrity of data in the presence of power failures. This concept is called *cache vaulting*.

- Interconnect that handles communication between the front-end processing unit and the disk controllers.
- The disk controllers that handle I/O to physical disks.
- The physical disks themselves.

19.3.2 Physical Layers Abstraction

The physical disks in a storage array are never directly exposed to a host connected to it. The array typically provides some level of protection through a mirroring mechanism (as described in Section 19.3.2.1). The useable storage capacity (the total raw storage capacity minus the mirroring overhead) is then pooled in volumes. A volume is a virtual disk that consists of logical slices from one or more physical disks. This operation, called striping, allows distributing the load of I/O's across multiple physical disks when the host writes to a given volume as its operating system sees it. A volume is often referred to as a logical unit, or LU (see Section 19.5.1). The layering of abstraction layer from raw physical disks to the host is depicted in Figure 19.4.

19.3.2.1 RAID Protection and Mirroring. RAID stands for *Redundant Array of Independent Disks* and is a set of technologies that provide mirroring, protection, and performance enhancements through the use of several disks to achieve higher level of performance, reliability, or capacity. The common RAID modes are:

- RAID 0, also referred to as striping, distributes I/O across several physical disks in order to improve performance.
- RAID 1 is a basic mirroring capability in which disks are paired in groups of 2 or more and each I/O is replicated to all disks in the group. As long as at least one disk in the group survives, this mode provides protection against disk hardware failure. While RAID 1 offers strong protection, its overhead is quite heavy. A RAID 1 scheme on an array where each disk is paired with another has a useable capacity that is half the total raw capacity of the array.
- RAID 3 offers a higher usage of capacity. This scheme combines striped disks and parity. Disks are combined in groups of three or more. One disk in the group is dedicated to parity calculation, which protects against the loss of any single disk.

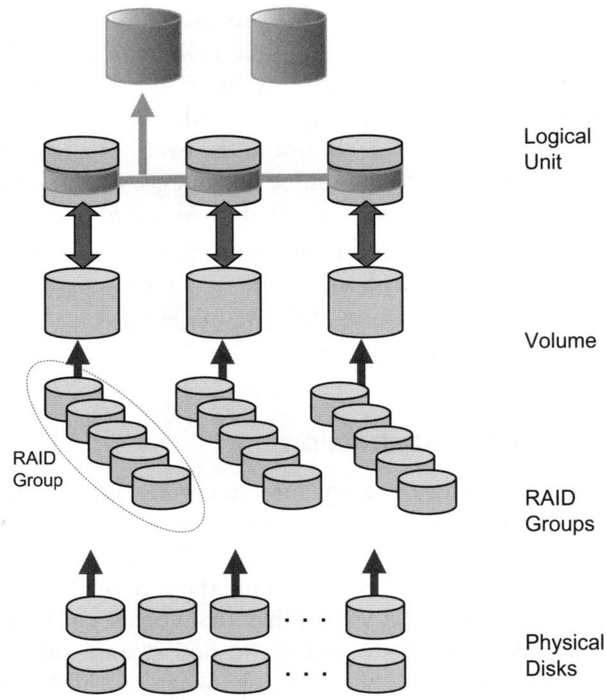

Figure 19.4. Layers of abstraction in a storage array.

- RAID 4 is identical to RAID 3, the difference being that striping is done at the block level instead of at byte level.
- RAID 5 is similar to RAID 3 but has a rotated parity stripe instead of a dedicated parity disk. This scheme offers an advantage over RAID 3 and RAID 4 by removing the bottleneck created by the parity disk (as each I/O requires a write operation to the parity disk)
- RAID 6 is very similar to RAID 5 but has dual parity and protects data against the loss of two disks in a RAID group.
- RAID 10, also termed RAID 1 + 0 combines striping and mirroring (RAID 1), where each disk in the RAID 0 group is mirrored. A variation of this mode, termed RAID 0 + 1, performs striping first across a set of disks, and then it mirrors that set in a RAID 1 scheme. RAID 0 + 1 is not as common as RAID 1 + 0 because RAID 1 + 0 performs better in the case of a drive failure.

Figure 19.5 depicts the various RAID schemes. In this figure, a stream of data blocks (*D0*, *D1*, ...) is being written to a RAID group. In practical deployments, most systems use RAID 5 or RAID 6. RAID 10 is also used for high-performance high-reliability situations, though far less commonly than RAID 5 or 6 due to the significant overhead it incurs.

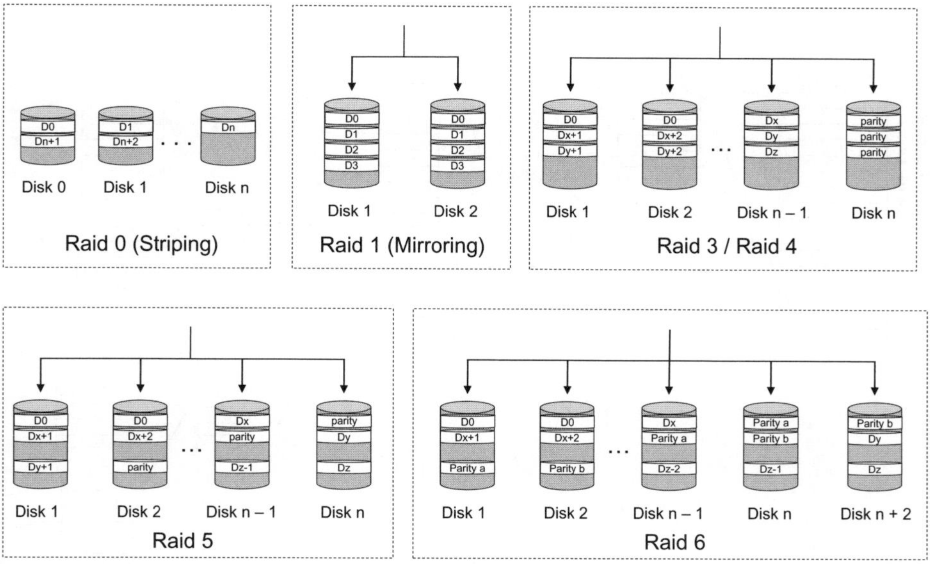

Figure 19.5. A representation of various RAID schemes. A stream of data blocks (D0, D1, ...) are being written to RAID groups.

19.3.3 File Systems

To an operating system, an LU is a raw disk. The OS, however, needs to be able to organize data for applications and users though files and directories. The file system layer is the layer that maps files and directories to the collection of blocks on disk.

Most file systems today provide convenient features such as:

- Ability to create and modify files, including changing file size.
- Allowing the user to name files using variable length strings.
- Organization of files in directories or folders.
- Providing a hierarchical directory structure.
- Maintaining access control lists, allowing files to be owned and be visible by single users or shared, managing editing privileges amongst users for shared files.
- Maintaining access rights (read, write, execute) for shared files.
- Maintaining meta-data about files, such as creation date and modified date.

Because files are fairly dynamic entities—as they are frequently modified and moved—the file system must provide an efficient and flexible mechanism for space allocation and management. Simply allocating fixed amount of contiguous space to a given file is not sufficient. Typically, a volume is broken down into blocks of a given fixed size. The block size is a file system level parameter with

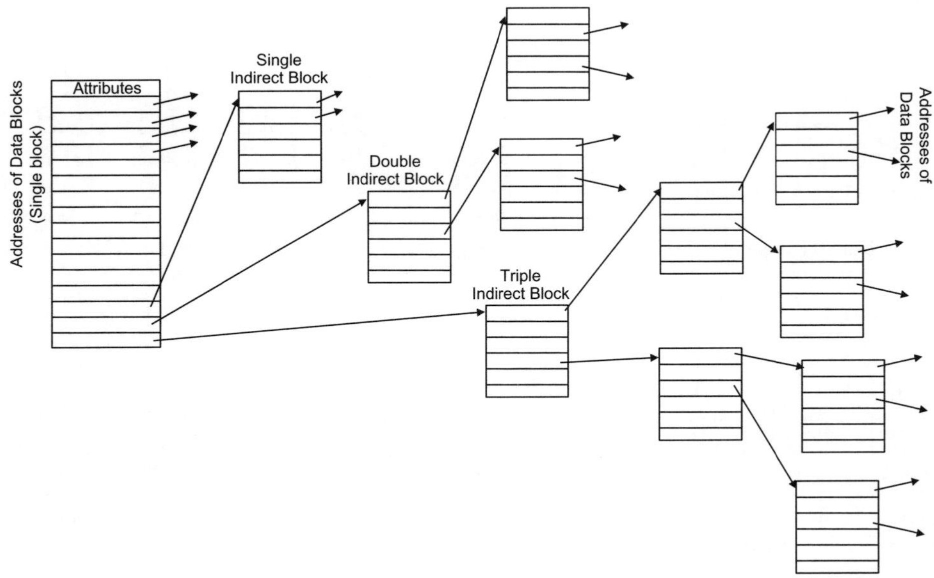

Figure 19.6. Example of an i-node structure.

values commonly ranging between 0.5 Kbytes and 16 Kbytes. When the file is created, its data are recorded in several possibly noncontiguous sets of blocks on a specified LU. The file system also maintains meta-data, which allows it to keep track of the blocks allocated for a particular file as well as their order. A detailed overview of various related file system techniques and their historic development can be found in reference 3. One of the most common file system implementations in UNIX relies on the use of *i-nodes* (or index nodes). An *i-node* is essentially a fixed-size data structure, which identifies the attributes of a given file as well as the data blocks allocated to it. Each data block referenced by the *i-node* is a contiguous block of data (called a direct block), which is used for storing the file content. The fixed-size nature of *i-nodes* makes them ideal for working with small files since a single *i-node* can track an entire file. In order to manage files larger than a single *i-node* can track, the *i-node* entries may also point to other blocks of disk addresses (called indirect blocks). This hierarchical structure enabled by this technique is depicted in Figure 19.6.

In addition to files, the file system must also manage directories or folders. Typically, a directory itself is implemented as a file whose content tracks the files stored within the directory as well as their names and their respective *i-nodes*. Thus, if an OS had to find a file *abc.xyz* stored in *directory_a*, which itself is stored within *directory_b* on disk *X* (in Unix the location of this file is */x/directory_b/directory_a/abc.xyz*), the following actions would take place:

- The file system locates the root directory of the system or that of the volume (top level directories are located in a specific place).

- The file system then parses the content of the root directory to locate *directory_b* and its corresponding *i-node*.
- The file system accesses the content of *directory_b* (via its *i-node*) and locates the entry corresponding to *directory_a* (and its respective *i-node*).
- Then, the file system accesses the content of *directory_a* (via its *i-node*) and looks for the *i-node* of file *abc.xyz*.
- Once the *i-node* for the sought file is known, the file system can open it in order to read or write its content.

19.3.3.1 Block Size. The file system block size is a key performance parameter that may be highly dependent on application characteristics and its I/O profile. Larger block size will help optimize throughput for streaming applications, which require the data to be accessed sequentially. On the other hand, smaller block size results in improved I/O latency and better disk utilization (because there is less wasted space) and is beneficial in environments where large numbers of small files need to be processed.

Studies identified that a standard Unix environment's average file size is only one kilobyte (see reference 4). On the opposite end of the spectrum, a server dedicated to serving video content will be optimized at block size as high as 64 Kbytes because the files it manages are typically read-only and sequentially organized.

19.3.3.2 Quotas. Multi-user concurrent operating systems typically have the ability to limit storage space allocated to individual users. The file system maintains quotas for every known user. It continuously tracks the space currently used up by its users, and it also tracks the list of open files. Furthermore, the file system monitors the file size changes to see whether it affects the specified quotas. This method allows the file system to manage quotas for shared files, which may be modified by someone other than the file owner.

19.3.3.3 File System Reliability. As can be inferred from the previous discussion, a file system is a critical layer in maintaining data integrity. While many topics discussed in this chapter are aimed at protecting the content of disks or volumes, none of these methods are of any worth if a file system itself becomes corrupted.

File system corruption may occur as a result of either (a) disk subsystem failure or (b) file system consistency issues. Disk failures arise when specific blocks in a given file become corrupted due to use. While this issue is very relevant to personal computers which often have single disks (not protected by RAID technology), this threat is of far lesser concern for users of storage arrays where volume level integrity is often guaranteed. File system consistency is a bigger threat to data integrity in these environments.

Consistency in file systems can be compromised due to the fact that, for performance reasons, many operating systems will cache their I/Os before they write them to disk. While many I/Os are held in cache, waiting to be de-staged (these I/Os may include updates to the file content itself as well as the updates to the corresponding *i-nodes*), a system failure can prevent these I/Os from being committed to permanent storage. Such failures can result in:

- Incorrect map of blocks that might be available to the OS.
- Corrupted files if I/Os are time-dependent from the application standpoint but written asynchronously from the OS standpoint.

Many operating systems provide utilities (or services) that can check a file system validity and integrity and attempt to repair it. Such utilities typically build two lists: one consisting of free blocks and one consisting of blocks allocated to files. Then they attempt to reconcile the list content and identify inconsistencies. Please note that this description provides only a high-level overview of the file system reliability issues. A more in-depth discussion of this topic can be found in reference 3.

19.3.3.4 Security. File systems are also a key component of maintaining security of the data and the system. They implement file level or directory level security and provide a way to configure and enforce user access privileges, often leveraging a central user database and an authentication mechanism.

Some file system implementations also allow for versioning and/or encrypting of their core operations.

19.3.4 Networked File Systems

There is an obvious benefit in sharing files managed by an operating system across users and servers. This can be done by a single server that offers concurrent access. However, a typical environment will have multiple servers and the same desire to share files and data across machines exist. A better alternative is provided by a network file systems employing file sharing protocols. These protocols essentially allow the same file level operations and access as in a centralized file system, but across a network. Hence a file system owned by a given server could be mounted on another server and appear as if it is a local file system.

In the Unix world, the most commonly known and most widely used file serving protocol is NFS (which stands for Network File System) and was originally developed by Sun Microsystems. In the Windows world, the file serving protocol which is proprietary to Microsoft Corp. is the CIFS (Common Interchange File System). NFS is covered in Section 19.6.1 while CIFS is briefly discussed in Section 19.6.2.

19.3.5 SAN, NAS and Unified Storage

As we discussed earlier in this chapter, storage area networks (or SANs) consolidate the storage pool of a datacenter. Application servers can be attached to the SAN, and each server is deployed with a set of volumes allocated to it. The

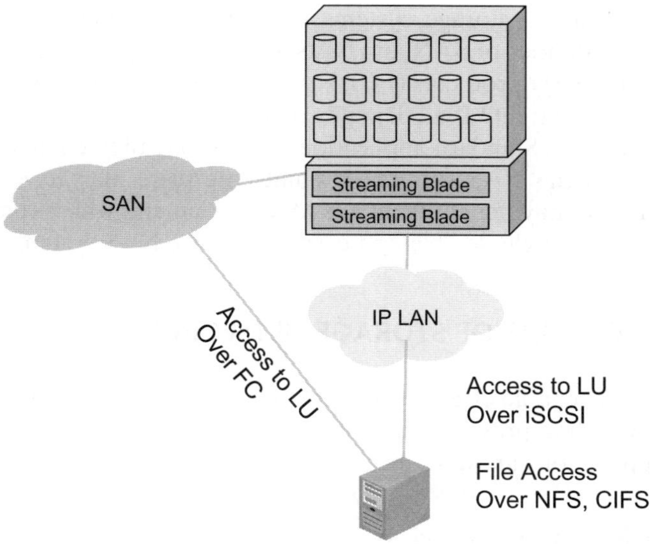

Figure 19.7. A unified storage system allowing data access through Fibre Chanel, iSCSI and file access protocols.

servers then typically format provided volumes for the file system compatible with OS they are running.

Following the development of SANs, storage equipment vendors developed the concept of *Network Attached Storage* (NAS). A NAS system is a file server that uses a dedicated storage array or a portion of a SAN and provides file level services over an IP network. The difference between a SAN environment and a NAS environment is depicted in Figure 19.7. While a SAN will export volumes to servers at a block level, a NAS system creates the file system layer and allows to share file systems across a network. Typical features provided by NAS systems are:

- High availability and failover, through multiple file serving engines.
- Sharing of files by all NAS clients.
- Multi-protocol access to the data: File serving can be done over multiple protocols (such as CIFS or NFS) for the files managed by the NAS system. In those cases the NAS system also handles the reconciliation of access rights and user credentials across multiple operating systems (e.g., managing the fact that the user John Doe may have a Unix account *jdoe* but also a Windows account *WORKDOMAIN\john_doe* and that both logins have similar access rights to files managed by the NAS system).
- Centralized management of all the file systems owed by the NAS systems. These can be protected (replicated), backed up, and snapped from a central location (snapping a file system refers to the operation of creating a point in time snapshot as described in Section 19.4.2).

More recently, unified storage systems have become popular. These systems combine the capabilities of SAN and NAS into a single system and provide data access over pure SAN protocols (FC), NAS (NFS, CIFS), or IP SAN (iSCSI).

Which system is right? It all depends on the need. A large enterprise usually deploys a large SAN. Standard application servers may then connect to both the SAN system and the NAS system. A small enterprise may have much more modest needs and may benefit from a NAS environment, which consolidates all of their needs in a single flexible and easy-to-manage NAS system.

19.4 APPLICATIONS OF STORAGE NETWORKS

Storage networks enable effective implementation of a large number of applications that can both improve the operations of data centers and lower their cost. Some key applications are described below.

19.4.1 Backup and Archiving

Backup and archiving are key needs of data centers. Backup allows the IT personnel to take copies of volumes or file systems so as to be able to restore them in case of need. Archiving allows migration of the data that are no longer active to a secondary form of storage such as cheaper disk or tape.

19.4.1.1 Centralized Backups. As storage networks centralize the data in storage arrays or file servers, backups can now be taken from that central location. In addition to removing the complexity of backing up every system in the enterprise, it enables the backup to be done on the SAN itself, at higher speed than on the local area network connecting all the hosts in the enterprise. This also enables a higher utilization and sharing of tape library or other backup targets and removes the need to have a backup application run on servers. This architecture is depicted in Figure 19.8, showing a backup server connected to the SAN for data traffic and to the LAN for meta-data traffic.

In order to implement this, the storage array vendors have integrated their products with major backup application providers. The integration is often done through the use of the NDMP protocol (Network Data Management Protocol) [5]. Such integration provides a multitude of benefits:

- It guarantees interoperability with key backup applications.
- It shortens the backup window thanks to the performance gain as well as the centralized point of backup.
- It shares libraries or other backup targets across the storage network.
- It makes backup automation simpler.

The backup automation benefit comes from the fact that a data center usually needs to take several snapshots of the data at various time intervals based on the

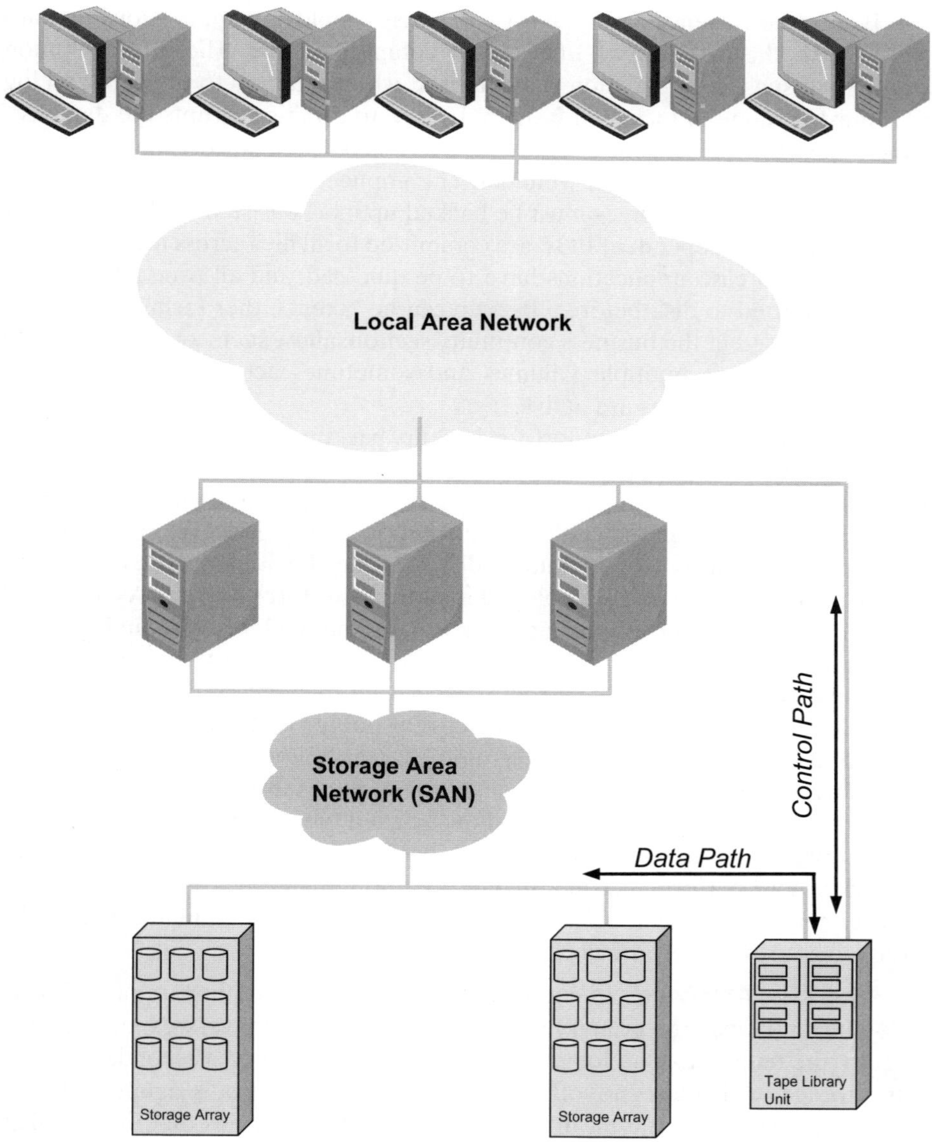

Figure 19.8. Centralized backup across a SAN.

application specifics as well as sensitivity of the data. Sensitive data require frequent backups to protect not only against hardware or software failures but also against data corruption and/or erroneous data entry. Backup administrators typically schedule a combination of full backups scheduled at coarse grain intervals (from once per day to once a month, depending on application sensitivity) as well as incremental backups taken very frequently. We will come back to some of these concepts in Section 19.4.2, dedicated to business continuity.

Backup consistency is critical for enterprises deploying applications that are sensitive to the ordering of I/Os across volumes or files. The most common example of such an application is a database, where changes to the data and log files must be done in lock step fashion. Failure to do this corrupts the database, which may prevent it from being restarted. Databases typically maintain their log and data files on separate volumes. The implication to backup is that these separate but related volumes must be backed up in a consistent fashion—that is, at a time when all dependent I/Os[2] are committed to all files across these volumes. In the simplest case, applications have to be quiesced, and all transactions have to be committed to disk before a backup can be taken. Other techniques, which will be discussed in the business continuity section, allow such consistent backup to be taken across multiple volumes and sometimes across multiple storage arrays while applications are active.

In terms of hardware support for backup, backup targets have traditionally been tape libraries. They consist of robotic arms that manage a set of tapes and a tape drives. The use of tape is motivated by its long-term reliability as well as low price point compared with disk drives. Recently, however, part of that market has shifted toward backup to disk. This has been enabled by the availability of cheaper disk drives deployed in enterprise storage arrays. As discussed in Section 19.5, the core storage network is based on the Fibre Channel protocol as its enabling technology. Disk drives used by storage arrays are usually Fibre Channel drives as well. ATA[3] and SAS[4] drive offer higher capacity at lower price point than Fibre Channel drives, at the cost of poorer performance.

Storage array vendors have integrated adapters allowing the use of SAS and ATA drives in their arrays (effectively employing a Fibre Channel to ATA or SAS protocol converter) in order to allow a mixed type of drives to be used. This has motivated the use of such drives as backup targets, allowing much faster backup/restore performance and higher reliability than tapes. Backup to disk is done either on the SAN itself (sometimes using tape library emulators) or on the local LAN, using a NAS device as a backup target.

19.4.1.2 Archiving. In contrast to backup, the main purpose of archiving is to efficiently manage primary storage (which is typically very expensive) and migrate the data that is no longer needed on a daily basis to a secondary or tertiary storage (which does not offer as high a performance but is significantly less expensive). Archiving is typically done on the basis of policies that identify files to be archived based on age, creation date, content, and its use pattern.

Advanced archival functionality provided by NAS devices maintains a link between the archived file and its primary location. The file logically appears to

[2]Dependent I/Os are I/Os that have to be committed in a specific timely sequence to ensure data consistency for an application.

[3]ATA and SATA (serial-ATA) is a storage interface commonly used for connecting hard disks to personal computers. It is an evolution of the older IDE protocol. ATA is a technology with a lower price point than fiber channel.

[4]SAS drives are serial attached SCSI drives: drives connected in a serial point-to-point fashion but using the SCSI protocol.

be still present in the file system it was originally a part of, even though it has been physically moved to a different medium. The archival policy engine and the NAS device then manage recalling the file if it is accessed by a user or application. A sophisticated policy engine can manage the storage pool, ensuring that it never exceeds a given capacity utilization by moving files to secondary storage based on their aging or any other criteria.

19.4.2 Business Continuity

Business continuity capabilities are meant to facilitate various applications by extending the concept of mirroring discussed earlier. Considering the SAN-based capability first, business continuity allows the storage administrator to create one or more real-time copies of a volume or set of volumes. In this scenario, while an application issues I/Os to the array, the write operations are replicated across all the copies of the volumes [these are called business continuity volumes (BCVs)]. The array then provides an interface enabling the user (or application) to programmatically suspend or resume mirroring. Interrupting the mirroring operation is called splitting a mirror. When a mirror is split, the data on disk remains as it was at the moment of the split operations. In the meantime, the array keeps track of all blocks that are updated on the primary volume (i.e., the active volume exposed to the application) and when mirroring is resumed, the array automatically updates only those blocks that have changed on the primary volume.

This capability can be provided on a set of related volumes and can be implemented in such a way that the mirrored volume set is updated consistently as seen from the application (i.e., I/Os to various volumes in the set are committed in the same order as issued by the application). These sets of volume are often called *consistency groups*. When the mirroring of a consistency group is split, it creates a consistent copy of the set.

There are many benefits of this technology:

- Backups can now be performed on active data. To do so, array automation capability can be used to:
  - Establish the mirrors prior to the backup.
  - Split one mirror at the time when backup is to take place.
  - Perform the backup on the BCVs (that is, a consistent image of the volumes in a consistency group can be backed up while the application remains active and continues issuing I/Os to the primary volumes).
  - When the backup operations completes, mirroring is automatically restored and the BCVs get synchronized with the primary volumes, preparing for the next backup operation.

  It is also relatively easy to integrate the automation of such operations with the backup/snapshot logic of major application (such as a database) so that this can be performed automatically and triggered by the application itself.

- One or multiple point-in-time copies of a set of volumes can be maintained. This allows users to inspect previous states of a volume set or allows rolling back in time to a previous snapshot of data if required. This can be very useful if data can be corrupted either by process or by user error. When corruption is identified, the storage administrator can roll back to the last known good copy of the data to resume operations.

This mirroring technique is obviously expensive because the amount of additional storage required to implement it is equal to the number of copies times the size of the data set. More advanced implementation will allow a scheme that is essentially similar to that of full and incremental backup, by tracking the changed blocks between the primary volumes and the BCVs, therefore reducing the overall storage requirements of BCVs.

The same concepts of mirroring can be applied to file servers (i.e., NAS devices), where copies of a given file system can be created. A regular application of this is to create multiple snapshots of the file system and give users the ability to access any previous copies should a file become corrupted. These are typically implemented as point-in-time checkpoints at the file system level. These checkpoints then offer the functionality of simple file restore or a full file system restore from a consistent and stable backup source.

19.4.3 Disaster Recovery

A key application of storage networks is to centralize, consolidate, and optimize a business's disaster recovery plan. The basic concept behind disaster recovery is to extend the capabilities of mirroring across networks and sites.

A basic disaster recovery strategy is to create a mirror image of a data center miles away from the primary location. The two data centers are connected through a wide-area network (WAN), and software running on the storage arrays replicates their content from the primary site to the secondary site. Should a catastrophic event happen to the primary site, the secondary site is activated and takes over operation in lieu of the primary. Furthermore, if the data center has been designed to have all application servers boot from the SAN, there is no need to replicate the applications or settings from these servers because these will be automatically discovered when servers in the secondary site boot from the their local SAN. Figure 19.9 depicts this basic setup.

Similarly to business continuity applications, a storage array provides automation capabilities allowing a user to programmatically establish a mirror, stop replication, and fail over from a primary to a secondary site. This capability can be tied to larger business process tools and serve as part of network management applications, which manage the underlying storage network replication.

Based on the application and business need, several types of disaster recovery technologies (each with its own set of tradeoffs) can be deployed.

19.4.3.1 Synchronous Replication. In synchronous replication mode, the primary and secondary sites are kept in perfectly synchronized state at all times.

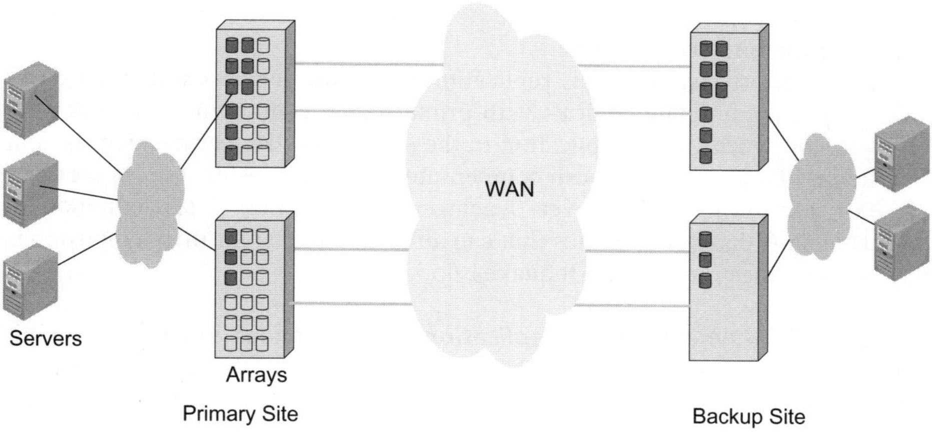

Figure 19.9. A basic disaster recover setup.

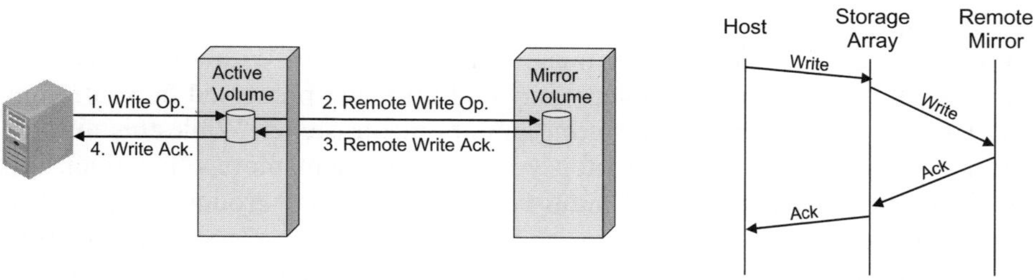

Figure 19.10. Synchronous replication.

This is achieved as follows: When a host issues a write I/O to the array in a primary site, the array writes the data to the local disk and simultaneously forwards this I/O to the remote mirror replica at a secondary site. When both the local disk and the remote replica confirm the completion of the write I/O, it is acknowledged back to the host. This is depicted in Figure 19.10. Similarly to business continuity, consistency of I/O can be enforced not just for a single volume but also for a set of related volumes (a consistency group). The replication is completely transparent to the host and its application, except for the fact that the write I/O processing time increases due to the additional latency of transmitting the I/O and its confirmation over the WAN, therefore potentially reducing the performance of the application.

While synchronous replication offers the best guarantee and virtually eliminates any data loss in the presence of disasters, it comes at the price of either significant performance degradation or major infrastructure cost increase. For this reason, a synchronous replication scheme can only be used at the metropolitan area scale, over distances of 60 miles or less. Alternatively, to guarantee low latency over long-distance transmission links, synchronous replication requires dedicated high-performance networking infrastructure, whose cost is typically prohibitively expensive.

Still, at short distances, this replication technique remains widely used. As an example, many financial firms with primary data centers in New York City often deploy synchronous replication to the secondary data centers deployed in New Jersey. This architecture can compensate for local disasters (i.e., September 11, 2001, for example). However, synchronous replication is less common to protect against disaster scenarios that can affect an area larger than a metropolis (e.g., geographies prone to earthquakes, floods, etc.).

19.4.3.2 Asynchronous Replication. Asynchronous replication eliminates WAN performance impact on the host application. While a synchronous replication scheme requires the secondary site to acknowledge write I/O completion over the WAN link, an asynchronous operation allows the host to acknowledge the write I/O as soon as the data is stored in the cache of the primary array. The I/O is then destaged to local disk and then independently forwarded to the remote mirror replica at the secondary site. This is depicted in Figure 19.11.

A major challenge of asynchronous replication technique is to guarantee consistency of the remote mirror image with the primary. Without special care, I/Os issued to the secondary image may end up being processed in a different order than at a primary one, resulting in data inconsistency. While this may be fine for some applications, consistency is a requirement more often than not. This is particularly important if multi-volume consistency groups are required. A common approach is to use a journal where all writes are recorded in FIFO order. Portions of the journal (deltas) may then be periodically transmitted over a WAN link, maintaining the FIFO order. The reader is referred to reference 6 for a list of other mechanisms to guarantee write order fidelity in asynchronous replication systems.

19.4.3.3 Hybrid Solution. A simple solution mitigating the network latency issues and the lack of consistency for asynchronous replication can be obtained by combining data replication with business continuity. In this scheme, the primary site is configured to have BCVs running on the production volumes to be replicated. At regular intervals the BCV is split, then replicated asynchro-

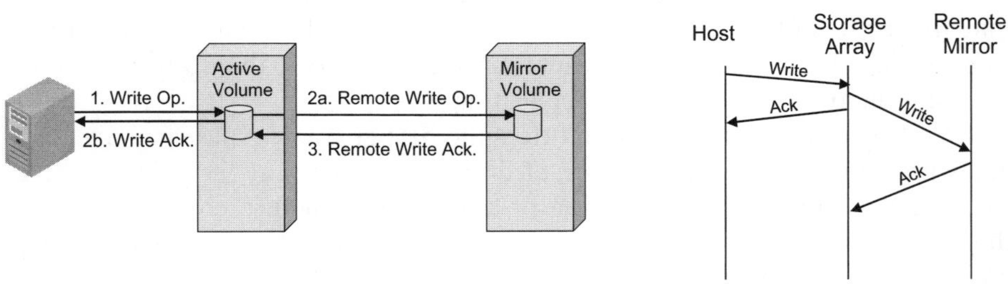

Figure 19.11. Asynchronous replication.

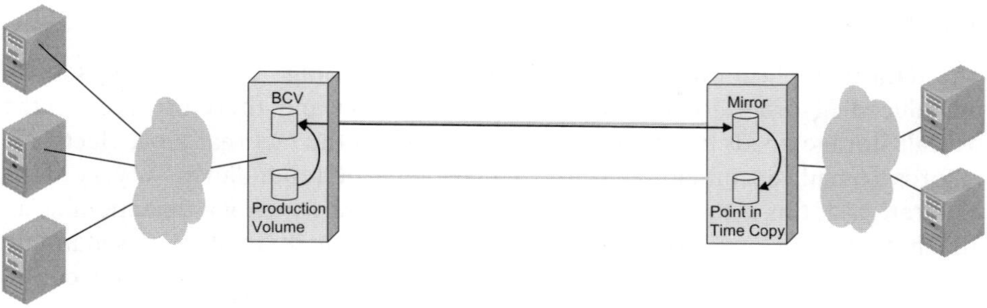

Figure 19.12. Hybrid replication.

nously over the WAN. When asynchronous replication is complete, the mirror is re-established until the next cycle starts. A similar setup on the remote site is used, where the replication target also has a BCV. When the replication function is updating the secondary site, its mirror is split so that a known valid copy of the data is maintained to protect against a disaster taking place in the middle of this refresh cycle. This is shown in Figure 19.12.

While this solution suffers from two major drawbacks—namely, a latency problem (the volumes on the remote site are typically two cycles old) and a cost overhead (four copies of the data are needed to perform this operation, which is often cost prohibitive)—this deployment tends to be used more frequently than expected as it does allow consistent replication across long distances.

19.4.4 File Retention and Compliance

File-level retention is a capability, which allows file servers to set file-level attributes to have the file system limit file write access for a specified retention period. This enables an administrator to prevent specific files from being deleted or modified based on business level policies or processes.

File-level retention can also be implemented in order to implement compliance requirements from market authority organizations (such as the SEC in the United States) that impose regulations to guarantee the availability and integrity of core business documents. In the United States, the SEC has issued a rule, SEC Rule 17A-4 (similarly to the SEC's counterparts in many other countries), that imposes requirements on storage systems. The SEC rules 17A-4 requirements on storage systems can be summarized as follows:

- Provide a non-rewriteable, nonerasable format for storing records.
- Prevent records from being overwritten or erased.
- Automatically verify the quality and accuracy of the storage media recording process,
- Serialize the original and, if applicable, duplicate units of storage media, and time–date for the required period of retention,
- Provide indexing capabilities.

19.4.5 Data Deduplication

The data deduplication capability is a feature of storage arrays or file servers introduced in order to optimize storage utilization and therefore reduce the overall storage needs of an enterprise. It recognizes the fact that most electronic records (email, documents) exhibit a high level of redundancy. A very basic illustration of this is the classical example of someone sending a document as an attachment to an email to several colleagues. Traditionally, mail servers maintain a copy of this email in each recipient's mailbox, thereby creating as many copies of the attached file and the email itself as there were recipients.

Data deduplication reduces storage needs by eliminating redundant data. To do so, a deduplication solution relies on a range of sophisticated algorithms that search for redundant data and replace copies of the data with a reference to its master copy. Several implementations of deduplication exist, and they can operate at different levels: file, block, and even bit level. File-level deduplication eliminates duplicate files. Block- and bit-level deduplication parse file contents and identify replicas of blocks or bit patterns.

A typical implementation of a data deduplication algorithm leverages a hash to build an index referencing pieces of data in a given volume or a collection of volumes. The index is then used then to identify and potentially eliminate redundant data. Some implementations also apply data compression algorithms to further reduce storage space utilization.

Available statistical data on the relative strength of deduplication algorithms (based on a real usage sample) are as follows:

- File-level deduplication offers moderate gains on the order of 10% savings in storage space but does not require significant computing cycles or memory usage.
- Fixed-block deduplication can provide on the order of 20% savings at the cost of higher computational power and memory usage.
- Variable-block deduplication provides slightly better performance than fixed block algorithm (25–30% savings) and is more efficient for data sets where data are not aligned to the media block structure such as backup data, or virtual tape library environment.

On the other hand, compression (whether it is paired with deduplication or not) can provide much more significant savings, in some cases up to 50%.

19.5 STORAGE NETWORK PROTOCOLS

19.5.1 SCSI

Small Computer System Interface (SCSI) was introduced as a standard in 1986 [7], and it defined both the physical interface and the protocol to enable attachment and communication of hosts and peripheral devices. Since 1986, the physical parallel bus interface defined by SCSI was replaced by faster and more flexible technologies such as USB, IEEE 1394 (FireWire), SAS, and so on. The SCSI protocol, however, evolved to become the most dominant application-level protocol in storage networks. Both Fibre Channel and iSCSI (described in Sections 19.2 and 19.4, respectively) ultimately serve as transports for SCSI traffic. Currently, SCSI is maintained in a large body of standards defined by ANSI and INCITS T10 committee [8]. The following list identifies some of the most relevant of these standards:

- *SCSI Architecture Model (SAM).* Provides an overview of SCSI communication model functionality.
- *SCSI Primary Commands (SPC).* Specifies generic commands used for communication with all SCSI compliant devices.
- *SCSI Block Commands (SBC).* Specifies commands used for communication with block-oriented storage devices such as disk drives.
- *SCSI Stream Commands (SSC).* Specifies commands used for communication with stream-oriented storage devices such as tape drives.
- *Multimedia Commands (MMC).* Specifies commands used for communication with multi-media devices such as CD-ROMs, CD-RW, and DVD players.
- *SCSI Controller Commands (SCC).* Specifies commands used for communication with RAID controllers.
- *SCSI Enclosure Services (SES).* Specifies commands used for controlling device enclosure parameters such as cooling and power.

The remainder of this section provides a brief overview of the SCSI Architecture Model based on the SAM-3 specification [9]. The reader is encouraged to refer to specifications identified above if more detail is required.

19.5.1.1 SCSI Architectural Model. At its core, the SCSI protocol is based on a client–server architecture, which enables client devices (called Initiators) to request services and exchange information with server devices (called Targets). A SCSI Initiator is typically implemented by an I/O subsystem in the client's OS. An Initiator may actively originate requests to Targets, which in a context of storage networking may be implemented by a variety of devices such as storage arrays, RAID controllers, optical storage, JBODs, tape libraries, jukeboxes, and so on. Targets respond to incoming requests but never originate their own.

Figure 19.13 depicts an example deployment of a SCSI Initiator and Target devices. In the context of a SCSI Architectural Model (SAM), these devices are connected via a service delivery subsystem, which provides error-free transport for transmission of I/O requests, responses, and data. The properties of the service delivery subsystem are not further defined by SAM. Instead, additional specifications document the mapping of the service delivery subsystem onto

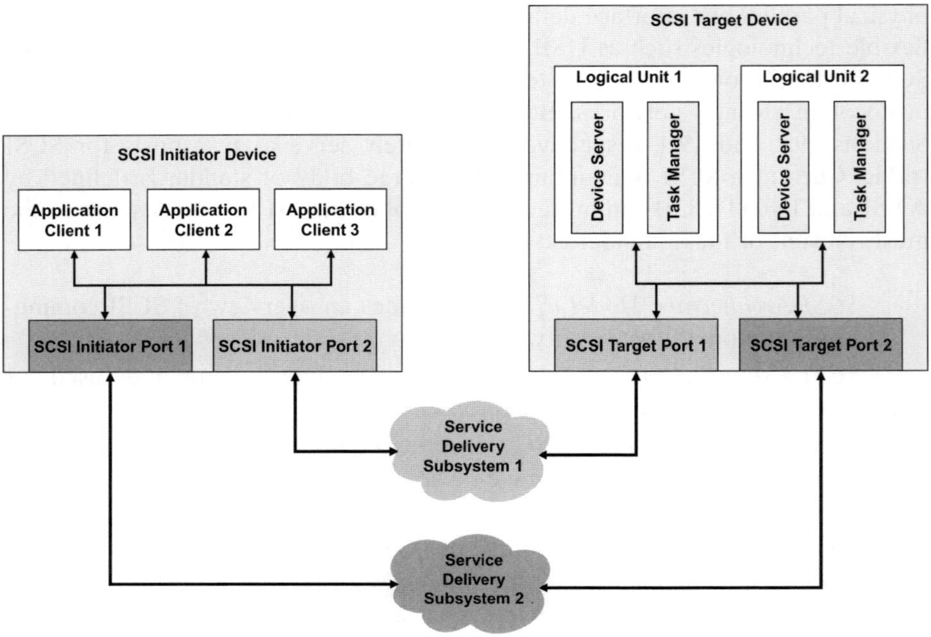

Figure 19.13. SCSI architectural deployment example.

existing transports including SCSI Parallel Interface, Fibre Channel (FCP), TCP/IP (iSCSI), Serial Attached SCSI (SAS), IEEE 1394 (SBP), and so on.

Both Initiator and Target devices are not elemental; they can aggregate other entities. A slightly simplified version of the SCSI architectural model is depicted in Figure 19.14. It defines an Initiator device as aggregating one or more Initiator ports and one or more application clients. An Initiator port provides connectivity with Target devices via the service delivery subsystem. An application client, which may, for example, correspond to a volume manager, a file system, or a database, leverages Initiator ports in order to exchange information with remote Targets. A Target device consists of one or more Target ports and Logical Units (LUs). An LU is an addressable device within a target, which is capable of processing application client's requests. LUs in turn consist of the following logical entities:

- One or more task sets, each of which contains one or more tasks (each task is a work item corresponding to command(s) issued by Initiator's application clients).
- A device server, which processes tasks corresponding to Initiator's I/O requests.
- A task manager, which handles task management requests issued by the application client and coordinates the processing of tasks by the device server.

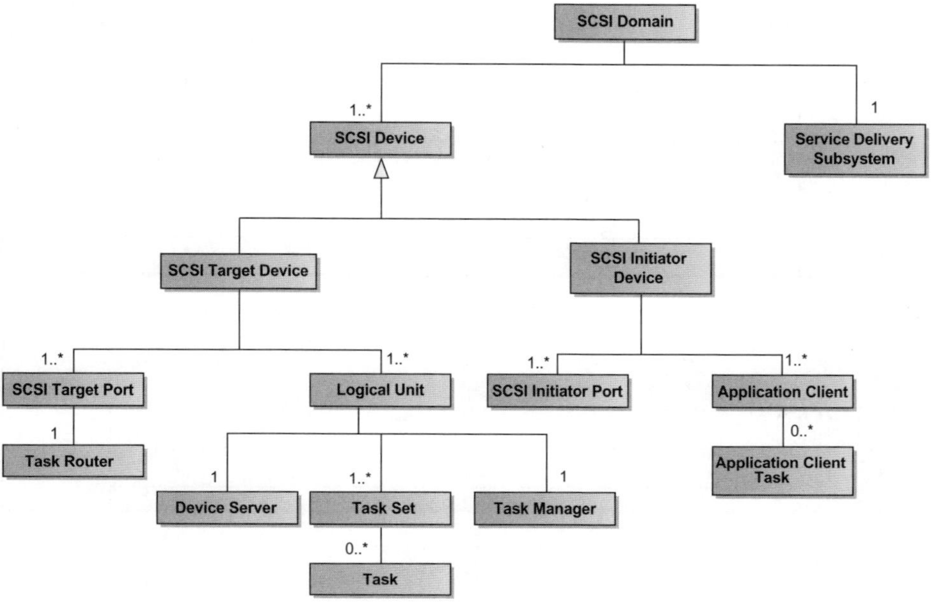

Figure 19.14. SCSI domain model.

19.5.1.2 Communication Model. SAM does not define how Initiators discover and establish connectivity with Targets because this information is transport-specific (please, refer to Sections 19.5.2 and 19.5.4 for a description of how this is done in Fibre Channel and iSCSI respectively). Instead, SAM identifies three generic nexus objects, which model the following relationships between SCSI devices:

- *I_T nexus.* Defines a relationship between a specific Initiator (I) and specific Target (T) SCSI ports; this relationship is equivalent to a logical connection/login session between the two ports.
- *I_T_L nexus.* Same as above but limited to a specific LU (L).
- *I_T_L_Q nexus.* Same as above but limited to a specific Task (Q is a task tag, which is unique for a given I_T or I_T_L nexus).

Once an I_T nexus is established, an Initiator may initiate I/O requests to the corresponding Target. Each I/O operation consists of multiple phases as can be seen in Figure 19.15, which depicts an example READ command processing. The first phase is a mandatory Command Phase, during which the Initiator sends a Command to the Target's device server. Each request is identified by a Command Descriptor Block (CDB), which specifies the operation to be performed as well as its parameters. Table 19.1 enumerates a small subset of commonly used CDBs and describes their high-level function. The last phase is a mandatory Status

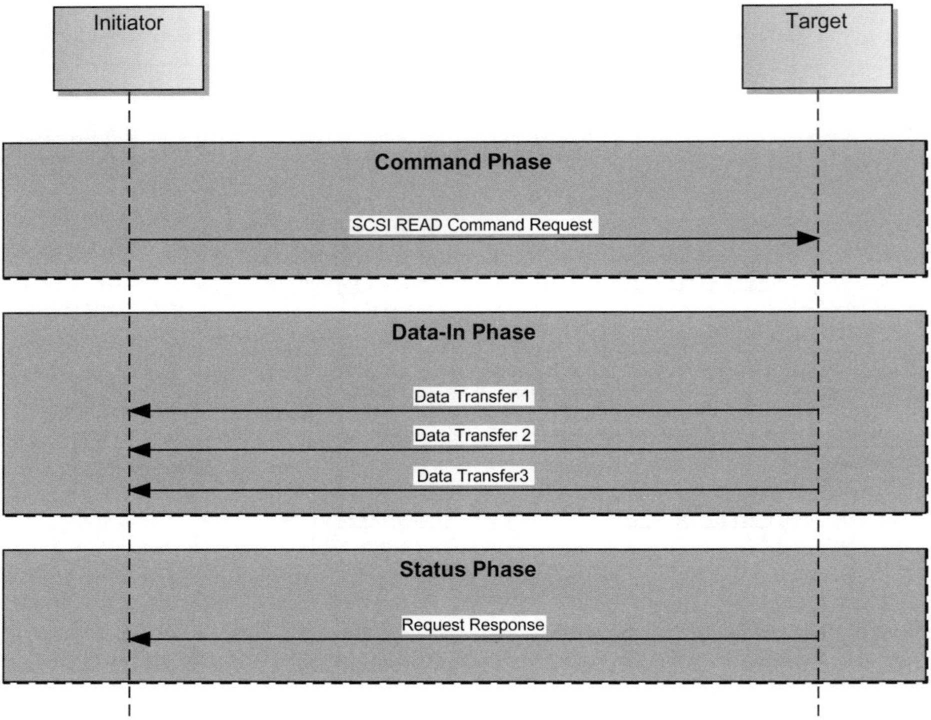

Figure 19.15. SCSI command sequence.

Phase, which allows the Target's device server to send a status response back to the Initiator to indicate whether command execution completed successfully or not. In between these mandatory phases there may be an optional data phase, which enables the transfer of data from application client to device server or vice versa. Some commands (such as TEST UNIT READY) do not need to transfer data and therefore omit the data phase. Other commands (such as READ) need to transfer data from the Target to the Initiator (this is known as Data-In phase), while commands (such as WRITE) transfer data in the opposite direction (known as Data-Out phase). There are also commands, which may require bidirectional transfer.

Compared with Initiators, Targets have limited amount of memory available for data transfer. As a result, a Target may decide to complete a Data Phase in several small transfers in order to accommodate its buffering limitations. It is important to note that the transfers within a data phase are always controlled by the Target.

19.5.1.3 Task Management. An application client may direct task management requests to a Logical Unit's task manager in order to affect task execution by the corresponding device server. A task manager executes the

TABLE 19.1. Example SCSI CDBs

CDB	Specified in	Description
TEST UNIT READY	SPC	Checks the readiness of the specified Logical Unit.
REPORT LUNS	SPC	Queries the inventory of Logical Units accessible by the I_T nexus.
INQUIRY	SPC	Queries detailed information about the specified Logical Unit and its Target device.
REQUEST SENSE	SPC	Requests data describing error condition encountered by the device server.
MODE SENSE	SPC	Queries parameters associated with the corresponding device server.
MODE SELECT	SPC	Changes parameters associated with the corresponding device server.
READ CAPACITY	SBC	Queries capacity and medium format of the Logical Unit.
READ	SBC, SSC, MMC	Reads and transfers specified data from the Logical Unit.
WRITE	SBC, SSC, MMC	Transfers and writes specified data to the Logical Unit.
FORMAT UNIT	SBC, MMC	Formats the medium into logical blocks that can be accessed by the application client.

request (typically by interacting with the device server) and replies to the application client by sending it a task management response indicating whether the processing completed successfully or not. Table 19.2 provides some examples of task management functions defined by SAM.

19.5.1.4 Error Handling. As discussed in Section 19.5.1.2, whenever a device server encounters a problem while processing a command, it returns an error status code in its response to the application client during the status phase. SAM documents several applicable error status codes and their causes. One of the most important status codes, which may be used in such a scenario, is CHECK CONDITION. This status code alerts the application client that additional error handling may be required. More detailed information describing the error condition may be provided in the sense data supplied by the device server during the status phase. Alternatively, the application client may request the sense data by issuing a REQUEST SENSE command to the logical unit on the affected I_T nexus.

In certain situations, based on application client's direction, a device server may establish an Auto Contingent Allegiance (ACA) condition on an I_T nexus (known as a faulted I_T nexus) when it returns a CHECK CONDITION status. This condition enforces strict error handling rules and enables the device server

TABLE 19.2. Example Task Management Functions

Task Management Function	Description
ABORT TASK	Requests that the execution of a specified task (I_T_L_Q nexus) be aborted by the device server.
ABORT TASK SET	Requests that all tasks received on a specific I_T nexus (as identified by the I_T nexus) be aborted by the device server.
CLEAR TASK SET	Requests that all appropriate tasks be aborted by the device server regardless of the I_T nexus, on which they were received.
CLEAR ACA	Clears Auto Contingent Allegiance (ACA) condition on the faulted I_T nexus (refer to Section 19.5.1.4 for more detail).
LOGICAL UNIT RESET	Requests that a specified LU should abort all tasks in its task sets and reset its internal condition.

to either abort or block processing of all commands on the faulted I_T nexus, except for commands responsible for resolving or clearing the ACA condition. An ACA condition may be cleared when one of the following takes place:

- A power-on condition resulting in a hard reset of the corresponding logical unit
- A receipt of a LOGICAL UNIT RESET task management function for the corresponding LU
- A loss of a faulted I_T nexus (based on a failure of the service delivery subsystem between the corresponding Initiator and Target devices)
- A receipt of a CLEAR ACA task management function for the faulted I_T nexus

Normal command processing resumes once the ACA condition is cleared.

19.5.2 Fibre Channel

The development of Fibre Channel in the early 1990s enabled large-scale deployment of Storage Area Networks. The Fibre Channel protocol suite has been specifically designed to provide:

- A flexible, cost-effective, highly scalable, and reliable communications environment
- High performance (both in terms of high bandwidth[5] and low latency)

[5]At the time of this writing, 1-, 2-, 4-, 8-, and 10-Gbit/FC devices are widely deployed. FC protocol revisions to support even higher bandwidth are under development.

- Reliable transmission with very low error rates
- Long-distance connectivity (up to 10s of kilometers),
- Flow control and congestion avoidance.

The large family of Fibre Channel standards is defined and maintained by ANSI and the INCITS T11 committee [10]. Fibre Channel remains by far the most dominant SAN technology available today, and it is likely to maintain this status quo for many years to come.

Fibre Channel architecture forms a stack of five functional levels (FC-0 to FC-4). FC-0, the bottommost level, identifies the physical interface. FC-4, the topmost level, defines mappings between Fibre Channel transport and application-level protocols called Upper-Level Protocols (ULPs). SCSI is the most commonly deployed ULP, but mappings to other ULPs such as IP, ESCON, FICON, and so on, have also been developed.

Table 19.3 captures the most fundamental services of each functional level in the Fibre Channel stack and also identifies the relevant standard specifications.

TABLE 19.3. Fibre Channel Protocol Stack

Fibre Channel Level	Relevant T11 Specification	Description of Services
FC-0	FC-PI, FC-10GFC, FC-FS	Defines physical interfaces including transmission media, connectors, cables, etc. Supports both copper (electrical) and optical technologies with a variety of data rates and distance limitations.
FC-1	FC-FS or FC-PH	Defines a transmission protocol, which utilizes an 8b/10b information encoding/decoding method in order to enhance error detection and recovery. Also defines a primitive link control protocol.
FC-2	FC-FS or FC-PH	Defines the FC frame structure, framing and sequencing protocols, as well as additional services, which allow FC nodes to connect and communicate with one another. Also defines the classes of service (refer to Section 19.5.2.3) and their corresponding buffer-credit based flow-control mechanisms.
FC-3	Currently undefined	Currently, FC-3 is a placeholder for common services that were supposed to help handle multiport applications on a single FC node.
FC-4	FCP (Fibre Channel Protocol)	Defines how SCSI commands, task management requests, and data are to be transported and processed by Fibre Channel networks.
	RFC 4338 (IPFC)[a]	Defines how IPv4, IPv6, and ARP packets are to be transported and processed by Fibre Channel networks.

[a]IPFC is defined by IETF (http://www.ietf.org).

The remainder of this section provides a high-level overview of the fundamental concepts of the FC architecture. For detailed overview the reader is referred to Kembel [11].

19.5.2.1 Fibre Channel Topologies and Ports. Fibre Channel networks provide connectivity to host and storage devices, known as nodes. The standards define the following three network topologies for attachment of FC nodes:

- **Peer-to-peer or direct connect topology** *(defined by FC-FS, the framing and signaling specification):* This simplest topology is used for direct attachment of two FC nodes. This topology was designed to enable high-bandwidth full-duplex connectivity between a server and a backup storage device over distances much larger than allowed by SCSI.
- **Arbitrated loop topology** *(defined by FC-AL specification):* This topology daisy-chains FC nodes over a shared unidirectional ring. Access to the loop is arbitrated by protocol, which ensures access fairness for connected devices. An arbitrated loop can connect up to 126 devices. One of the typical use cases of this topology is attachment of disks within a FC RAID controller. This topology can be implemented either by physical daisy-chaining of devices together or by leveraging an FC hub. The latter provides a more efficient and economical way to deploy an arbitrated loop topology.
- **Switched fabric topology** *(defined by FC-SW specification):* This is the most flexible and scalable FC topology, which supports simultaneous communication between devices. This topology relies on deployment of multiport FC switches, which attach to many FC nodes and also to one another (via Inter-Switch Links or ISLs) to form a fabric. FC switches are high-performance devices, whose primary responsibility is to route FC traffic through the fabric. Theoretically, a single switched fabric can connect up to 2^{24} (or 16,777,216) devices. To provide access control between servers and storage devices, a physical fabric can be logically partitioned into separate zones. Access and visibility between devices are only provided within a zone but not between zones.

Each FC node aggregates one or more physical FC ports, which enable attachment to the FC network. Both the node itself and each of the ports are identified by 64-bit-long globally unique identifiers, called World-Wide Names (or WWNs). To provide connectivity, FC client devices (such as hosts) are often outfitted with a Host Bus Adapter (or HBA). An HBA combines one or more FC ports and serves a function similar to that of a NIC in a LAN environment. Different types of FC ports offer connectivity to different topologies. Table 19.4 provides information about most commonly used port types, and Figure 19.16 depicts the way these ports are connected in the corresponding topologies.

A switched fabric topology also provides several generic services, as documented by FC-GS specification. These services are implemented by FC switches. The following list enumerates some of the most important generic services:

TABLE 19.4. Commonly Used FC Port Types

FC Port Type	Where Is It Deployed?	Description
N_Port	Node	**Node Port:** A port that can either attach to another N_port in a point-to-point topology or attach to an F_Port in a fabric topology.
NL_Port	Node	**Node Loop Port:** A port, that attaches to other NL_Ports to form an arbitrated loop or to an F_Port to connect the loop to a fabric.
F_Port	Switch	**Fabric Port:** A port that enables one N_Port to connect to the fabric.
FL_Port	Switch	**Fabric Loop Port:** A port that connects to one NL_Ports to enable loop attachment to the fabric.
E_Port	Switch	**Expansion Port:** A port used for fabric expansion. E_Ports enable connectivity between FC swithes.
G_Port	Switch	**Generic Port:** A port that can auto-sense whether to operate as an F_Port or an E_Port.

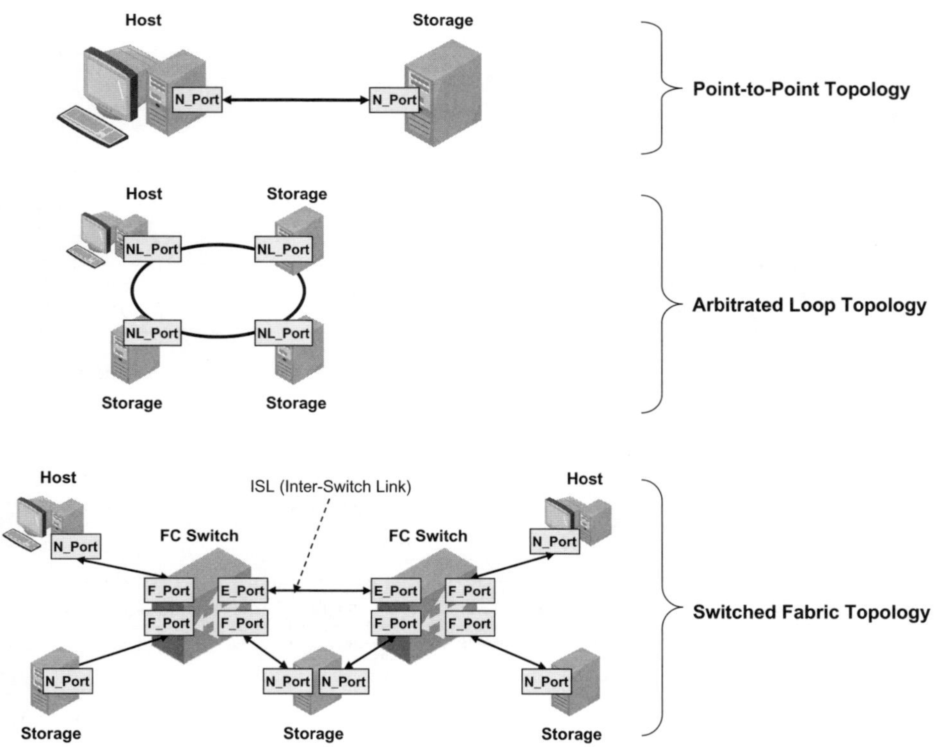

Figure 19.16. FC ports and topologies.

- *Fabric Controller:* Provides state change notifications to N_Ports, which register for this service. This allows participating ports to learn about fabric changes as other N_Ports log into or log out of the fabric.
- *Directory/Name Server:* Enables N_Ports to register pertinent information about themselves and their nodes. N_Ports can also query information stored by the Name Server to discover devices and other ports present in the fabric.
- Time, Management, and Secure Key Distribution Services are also among the provided generic services.

19.5.2.2 Information Transmission. All FC-2 level data are transmitted by frames. The structure of an FC frame is depicted in Figure 19.17. Every frame begins with a start-of-frame delimiter, which is followed by the frame header and up to 2112 bytes of data payload. The payload contains data transmitted by a higher-level protocol and is completely transparent to FC-2. Every frame ends with a CRC (used for detection of transmission errors not detected by the lower protocol levels) and an end-of-frame delimiter. Among other parameters, the frame header contains a 24-bit-long source and destination addresses called S_ID and D_ID. The D_ID plays the crucial role in frame routing, which in the switched fabric topology is performed by the Fabric Shortest Path First (FSPF) routing algorithm.

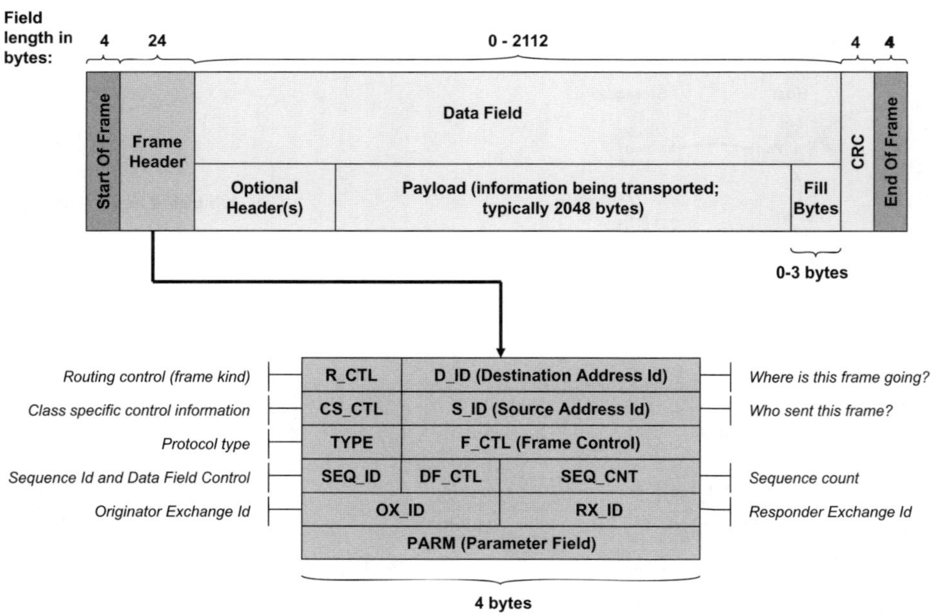

Figure 19.17. FC frame structure.

A collection of one or more related frames sent by one N_Port (or NL_Port) to another is called a sequence. Each sequence corresponds to one Information Unit (or a message) transmitted through FC by ULP. If an Information Unit is small enough to fit in a 2112-byte frame, the sequence consists of a single frame. Otherwise, the sending port fragments the Information Unit into a sequence of multiple frames during transmission, and the receiving port reassembles the sequence back into the Information Unit. The SEQ_ID and SEQ_CNT fields in the frame header identify, respectively, the sequence to which the frame belongs as well as its order in the sequence. Sequences play an important role for higher-level protocols because error recovery often applies to entire sequences as opposed to individual frames within them.

Sequences are further aggregated into an exchange—a transaction-like request–response conversation between two communicating N_Ports. FC-4 protocols define the rules for formation and management of exchanges. Figure 19.18 illustrates an example processing of a SCSI READ command as it is issued by the SCSI ULP on the initiator side and traverses FCP and FC-2 on its way to the target. The entire flow corresponds to a single exchange, which consists of three sequences carrying the Command, Data, and Response Information Units.

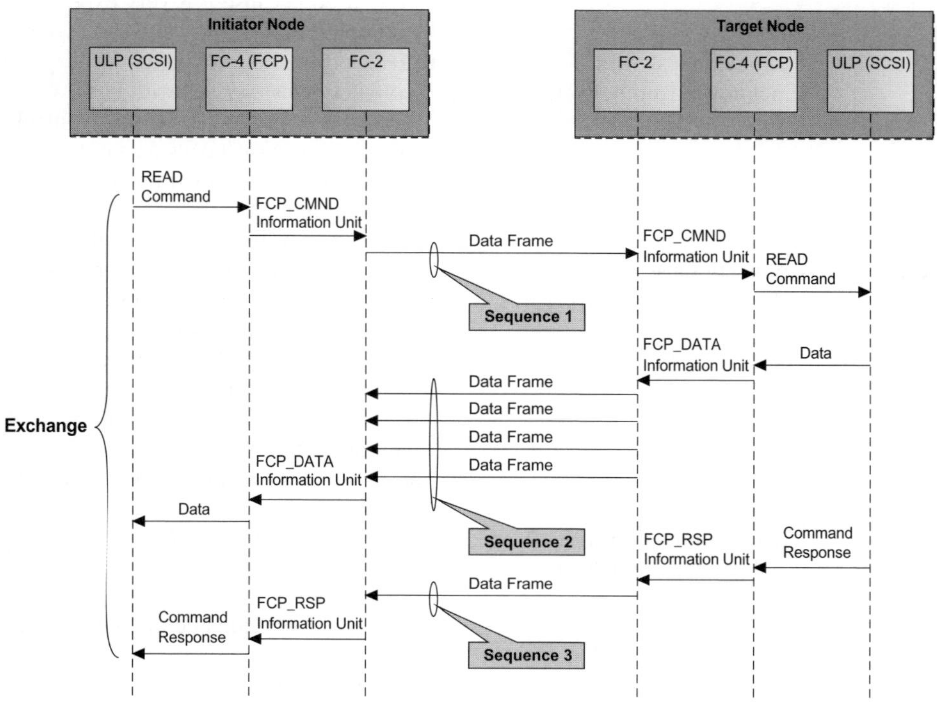

Figure 19.18. FC command processing flow.

19.5.2.3 Classes of Services. Fibre Channel Framing and Signaling Specification (FC-FS) defines six classes of service, which determine link and frame delivery properties for every pair of communicating FC ports. These classes of service, listed in Table 19.5, are independent of the underlying network topology and are established by negotiation during a port login process.

Storage systems typically use Classes 2 and 3. Because Class 3 is similar to the Internet UDP protocol, it does not provide services for frame retransmission in the presence of errors. If Class 3 is used, the error handling policy and frame retransmission must be implemented at a higher stack level (either in FC-4 or in ULP).

TABLE 19.5. FC Classes of Service

Name	Brief Description	Service Description
Class 1	Dedicated physical connection with acknowledged delivery	Provides fully dedicated physical connection between two FC ports with fully reserved bandwidth, guaranteed ordered delivery, and end-to-end flow control. Receiver acknowledges frame delivery or omission. Rarely used because it is very expensive to implement.
Class 2	Connectionless with acknowledged delivery	Provides frame multiplexed connectionless service with both end-to-end and link-level flow control. Receiver acknowledges frame delivery or omission. In-order frame delivery is not guaranteed. May be used in Inter-Switch Links (ISL).
Class 3	Connectionless without acknowledged delivery	Similar to Class 2, this service provides a frame multiplexed connectionless service but without frame acknowledgment. Only link-level flow control mechanism is provided. In-order frame delivery is not guaranteed. This is the most commonly used class of service.
Class 4	Virtual connection with acknowledged delivery	This service is similar to Class 1, except that it uses a virtual connection instead of a physical one. This class of service is being deprecated.
Class 5	Currently undefined	
Class 6	Reliable multicast (derivative of Class 1)	This service is functionally similar to Class 1, except that it is designed for one-to-many multicast communication.

19.5.3 SAN Bridging: IP to the Rescue

As described in Section 19.5.2, Fibre Channel defines a flexible transport with high reliability and performance characteristics, capable of meeting the ever-growing demands of storage applications. However, because of the high cost associated with supporting FC traffic over long distances, native FC SAN deployments are typically limited to individual data centers. It should be noted that even though FC-0 supports single-mode fiber technology, which can sustain high transmission quality for distances close to 80 km, the cost associated with this type of deployment is often prohibitive. While FC-based long-haul infrastructure is fairly uncommon, dedicated networking infrastructure (either leased or privately built) for transmission of IP-based IT traffic over LAN and MAN is more commonplace. Both FCIP and iFCP protocols aim at leveraging this infrastructure in order to enable interoperability between SAN islands in a distributed enterprise. The following two sections describe these protocols in more detail. A comprehensive overview of IP technologies and their relevance to SANs is provided in Clark [12].

19.5.3.1 FCIP. FCIP is a tunneling protocol, which relies on TCP/IP connections to merge geographically distributed FC fabrics into a single virtual fabric. To achieve this, one or more FCIP gateways (or FCIP-capable switches) need to be deployed in each fabric. Figure 19.19 depicts an example deployment where two FC SANs leverage FCIP for interconnection over an IP network.

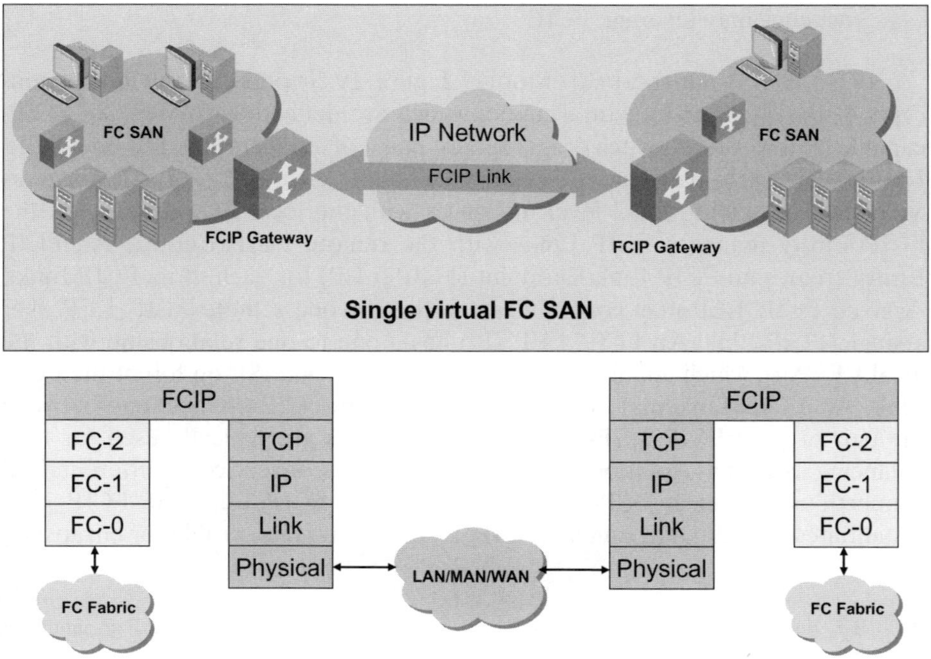

Figure 19.19. FCIP deployment model.

TCP/IP connections may then be established between these gateways over existing LAN, MAN, or WAN infrastructure. An FCIP gateway encapsulates FC frames destined to the remote fabric into payload, adds FCIP encapsulation header, and sends it over a TCP/IP connection. The recipient gateway strips the FCIP encapsulation header, de-encapsulates the FC frame, and transmits it to the appropriate destination in its local FC fabric. FCIP does not perform any FC address translation and fully relies on existing FC and IP routing mechanisms. As a result, an FCIP link between two FC gateways (or switches) essentially functions as a virtual ISL link. FCIP relies on security mechanisms defined by FC-SP to authenticate FC gateways with one another, and it also relies on IPsec and IKE to provide authentication, confidentiality, and key management for communication over the IP network. Other than the gateways themselves, no FC device in the attached FC fabrics has any direct knowledge of FCIP being used, and no device on the IP network is aware of the FC-specifics being transmitted over IP between the gateways.

FCIP protocol is defined by the following standards:

- IETF RFC 3821 documents the protocol, its architectural model, and control functions.
- IETF RFC 3643 documents the rules for encapsulation of FC frames into TCP/IP stream.
- IETF RFC 3822 documents FCIP discovery mechanism based on SLPv2.
- INCITS T11 FC-BB-5 provides additional information describing how FC switches may leverage FCIP.

19.5.3.1.1 Architectural Model. Figure 19.20 presents the major concepts of the FCIP architectural model, which are defined as follows. An FCIP-capable gateway (or switch)[6] aggregates pairs of cooperating FC and FCIP Entities. In each pair, the FC Entity contains Virtual E_Ports (known as VE_Ports) providing FC-2 level interfaces with the local FC fabric, while the FCIP Entity manages FCIP Links with the remote FCIP Entities. An FCIP Entity creates an FCIP Link Endpoint (FCIP_LEP) for each of its FCIP Links. A given FCIP_LEP may communicate with only one remote FCIP_LEP over a single FCIP Link. An FCIP_LEP also has a one-to-one relationship with the local VE_Port, which enables establishment of a Virtual ISL on top of the FCIP Link. An FCIP Link may aggregate one or more TCP connections between the pair of peer FCIP_LEPs. Multiple connections are typically used for load balancing as well as isolation of different FC classes of service to different TCP connections. For each TCP/IP connection in the FCIP Link, the FCIP_LEP instantiates a FCIP Data Engine (FCIP_DE), which is responsible for encapsula-

[6]FC-BB-5 documents both the switch-based and bridge-based deployments. For simplicity, the remainder of this section focuses on the switch-based deployment.

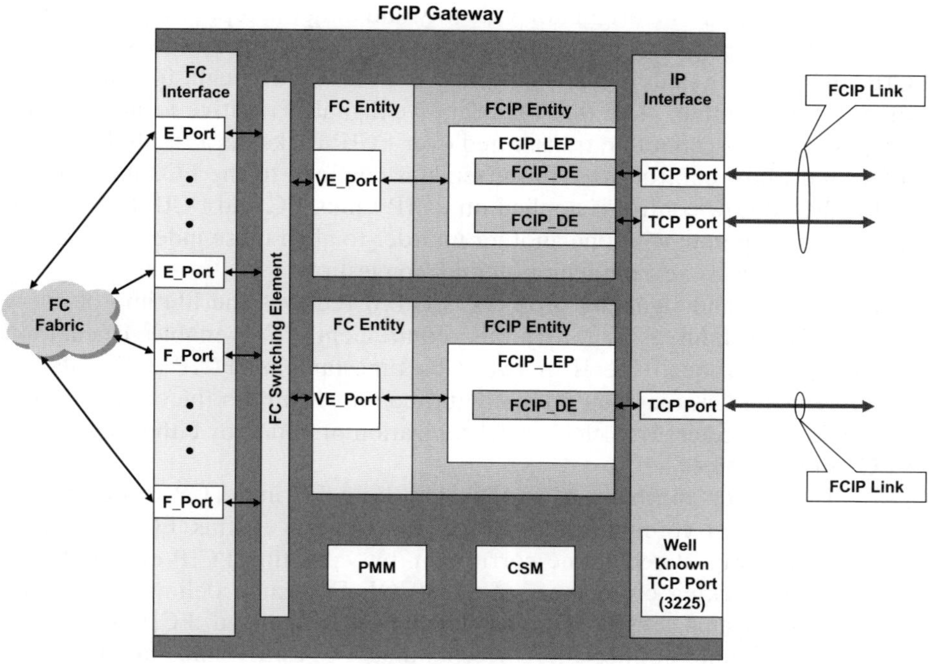

Figure 19.20. FCIP architectural model.

tion and de-encapsulation of FC frames as well as their forwarding between the corresponding VE_Port and TCP/IP connection. FCIP relies on TCP/IP to guarantee FIFO-ordered delivery for each connection within a FCIP Link. No ordering guarantees across connections are maintained; but once a frame corresponding to an FC flow between a given pair of FC ports has been assigned an FCIP_DE for transmission, all subsequent frames in the same flow are transmitted by the same FCIP_DE.

Finally, an FCIP-capable switch also contains a Control and Services Module (CSM) to manage establishment and closure of TCP/IP connections and the Platform Management Module (PMM) to handle clock synchronization, discovery, and security aspects of the FCIP protocol.

19.5.3.1.2 Protocol Operation. An FCIP Entity must know an IP address and a TCP port number of all other FCIP Entities it is expected to connect to. This information can be either statically configured or dynamically discovered via SLPv2.[7] Once the TCP connection is established, the originator and the acceptor FCIP Entities exchange an FCIP Special Frame (FSF) as the very first frame to be sent in either direction. This exchange enables the acceptor FCIP

[7]RFC 3822 documents the use of SLPv2 specifically for this purpose.

entity to learn about the originator's identity and either accept or reject the connection. If the exchange completes successfully and connection is accepted, FCIP_DEs on both ends of the connection are allowed to begin forwarding FC frames to one another. Note that FC Primitive Signals, Primitive Sequences, and Class 1 FC frames cannot be transmitted over FCIP Links. As a result, FC-based buffer credit flow control mechanism remains confined to the FC fabrics, while window-based flow control is applied on FCIP Links. FC and FCIP Entities are required to cooperate with one another in order to align these independent flow control mechanisms and mitigate potential congestion.

FC framing and signaling protocol (FC-FS) requires the lifetime of every FC frame to be limited. To satisfy this requirement, FCIP enables calculation of transit time through the IP network by timestamping FCIP encapsulation headers. This feature may be optionally turned on when both the originator and the recipient have access to clock synchronization provided by either SNTP [13] or FC timing services.

FCIP relies on mechanisms available in both FC and TCP to detect and recover from most communication errors. In addition to this, FCIP_DE also helps detect encapsulated frame corruption, by validating FCIP encapsulation headers and checking correctness of SOF/EOF FC frame delimiters and the corresponding frame length. If frame corruption is detected, FCIP_DE drops the damaged frame, notifies the corresponding FC Entity, and initiates error recovery procedures. As part of error recovery, FCIP_DE may close its TCP connection (which would later need to be reestablished), or it may attempt to skip ahead in the TCP stream to identify the next valid FC frame to reestablish frame synchronization. If successful, FCIP_DE keeps the connection open and resumes frame delivery. In this case, affected FC Entities are expected to recover from the loss of discarded FC frames.

19.5.3.2 iFCP. iFCP is a gateway-to-gateway protocol, which provides IP-based connectivity between distributed FC fabrics. Similar to FCIP, iFCP utilizes TCP/IP as its transport. Unlike FCIP, iFCP enables individual FC Nodes to communicate with one another via TCP/IP. iFCP protocol is defined by RFC 4172. It is complemented by RFC 3643, which defines FC frame encapsulation rules (the same RFC is used by FCIP), and by RFC 4171, which documents iSNS (see Section 19.5.5). From security perspective, iFCP relies on both IPSec and IKE for confidentiality, authentication, and key management. RFC 3723 provides additional information about security in the context of iFCP.

19.5.3.2.1 Architectural Model. As shown in Figure 19.21, an iFCP gateway device aggregates F_Ports, which provide connectivity for the local FC fabric (known in iFCP as a gateway region) as well as one or more Network Interfaces (known as iFCP portals), which provide IP connectivity with remote iFCP gateways over a LAN, MAN, or WAN infrastructure. iFCP functionality is transparent to the FC devices within a gateway region; as a result, such devices see its local gateway as a regular FC switch. Communication between FC devices

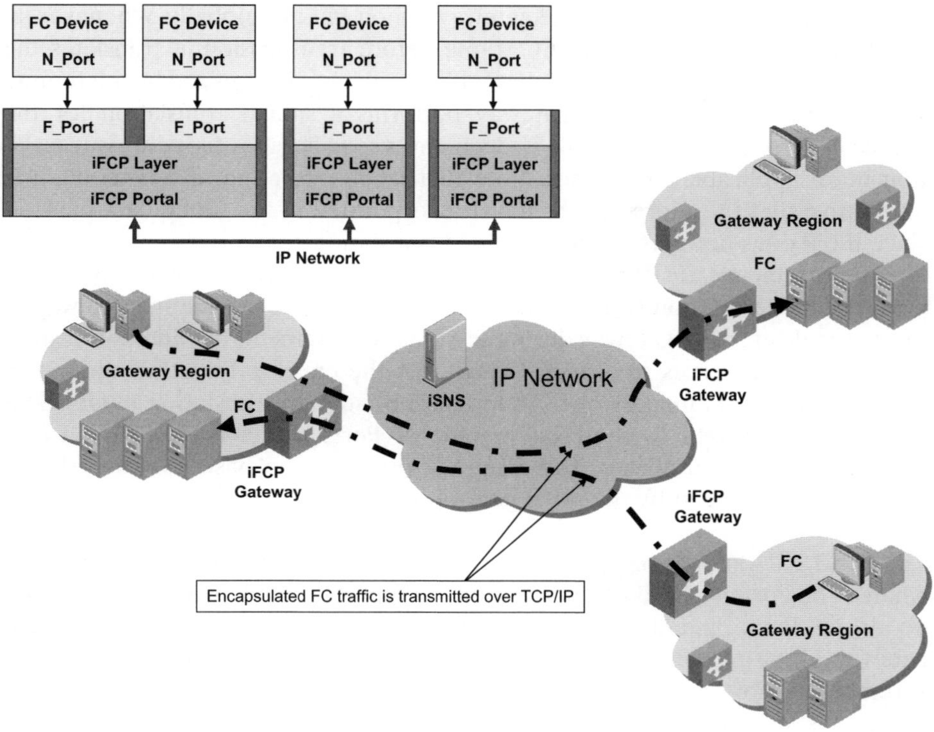

Figure 19.21. iFCP deployment model.

in the same gateway region does not involve iFCP. However, when an N_Port in a local gateway region needs to communicate with an N_Port in a remote gateway region, the local and remote iFCP gateways establish an iFCP session over TCP/IP and then forward encapsulated FC frames within this session. An iFCP session is an association between a pair of N_Ports in different iFCP regions, communicating with each other over a single TCP/IP connection. Only a single session may be established between a given pair of N_Ports at any point in time. However, to reduce session establishment time, iFCP allows gateways to maintain a pool of TCP/IP connections, which are not actively used by a session. These connections are called unbound. During session establishment, if an appropriate unbound connection exists, it becomes bound and is removed from the pool.

iFCP supports two operation modes: address transparent mode and address translation mode. An iFCP fabric may be configured to operate in either of the following two modes but not both. In address transparent mode, an FC address assigned to each N_Port is globally unique across all iFCP gateway regions. While this mode is simpler of the two, it limits the scale of the iFCP fabric because FC only allows 239 FC switches (including iFCP gateways) in it. The scalability issue is avoided when iFCP operates in address translation mode. In this mode, FC addresses are unique within iFCP gateway region but not globally. Whenever an iFCP gateway encapsulates an FC frame before transmission, it translates the destination N_Port address to an alias address, which is recognized by the remote iFCP gateway. The receiving gateway performs a similar translation for the source N_Port addresses. RFC 4171 documents specific rules for address translation in all applicable FC frames. Note that when operating in this mode, it is imperative that FC addresses are never incorporated in the frame data payload since iFCP does not perform any payload translation.

In both modes, iFCP relies on iSNS to identify the remote gateway's iFCP portal given destination N_Port's WWN. This enables the originating gateway to locate the destination gateway during session establishment. In address transparent mode, iSNS serves as a name service for the entire fabric, thus ultimately ensuring global uniqueness of FC addresses. In address translation mode, iSNS maps addresses used to represent remote N_Ports within a local region into remotely recognized aliases. This mapping provides the basis for the address translation performed by the iFCP gateways.

19.5.3.2.2 Protocol Operation. iFCP session establishment is triggered when an N_Port in a local gateway region attempts to login into a remote N_Port. To do so, the originating N_Port sends a PLOGI request to its local iFCP gateway. Upon receipt of PLOGI, the local gateway consults iSNS to construct a remote N_Port descriptor if it does not have one already. This descriptor aggregates information identifying the remote N_Port (including the WWN, address, and possibly alias) as well as IP address and TCP port of the remote gateway's iFCP portal. Armed with this information, the local gateway either takes an existing unbound connection to the remote gateway from the pool or creates a new one. To establish the iFCP session, the local gateway sends a CBIND (connection bind) control request to the remote gateway, which is expected to construct a remote descriptor for the initiating N_Port and then reply with a CBIND response. The CBIND request/response allows the initiating and terminating gateways to exchange information about the initiator and target N_Ports as well as negotiate parameters of the iFCP session. After a successful exchange, the iFCP session is considered open for FC traffic while the PLOGI frame, which initiated iFCP session establishment, is forwarded to the remote N_Port via the remote iFCP gateway.

In addition to CBIND request/response messages, iFCP introduces an LTEST (liveness test) message. iFCP gateways may periodically transmit this message to one another to proactively test liveness and performance of their bound connections and also to prevent TCP/IP connection closure due to inactivity. Periodicity of LTEST heartbeat messages is determined during the CBIND exchange. Failure to receive the LTEST heartbeat within a predefined interval results in termination of the iFCP session.

Like FCIP, iFCP enables calculation of IP network transit time by timestamping its encapsulation headers. Unlike FCIP, this feature is mandatory in iFCP and must be implemented by gateway clock synchronization via SNTP. The gateways are required to discard any "stale" frames, whose in-flight time exceeds the configured threshold. In addition to this, iFCP fully relies on TCP to perform flow control, congestion avoidance, and corresponding error detection and recovery.

19.5.3.3 Summary. FCIP and iFCP provide bridging mechanisms for extending FC fabrics across geographical distances by leveraging IP. Both FCIP and iFCP are responsible for transporting FC frames over TCP/IP connections. Both of these protocols have strong affinity with FC and cannot function without it. In the next section, we will analyze iSCSI, which is the first native IP-based storage protocol, which can penetrate both Initiator and Target devices without any dependency on FC.

19.5.4 iSCSI

iSCSI is a transport protocol developed by IETF (refer to reference 15) in order to enable transmission of block-oriented SCSI traffic through an IP network. Unlike FCIP and iFCP, iSCSI provides a native mapping of SAM-2 [14] SCSI Architecture Model on top of TCP/IP without any dependency on Fibre Channel. This makes iSCSI a key enabler for building and deploying IP SANs.

19.5.4.1 Typical Deployments. Initial adoption of FC SANs is often prohibitively expensive for small businesses and enterprises. Costs associated with the new FC equipment are often coupled with a significant investment to either train or hire personnel to administer the new equipment. On the other hand, an IP SAN does not require either of these expenditures since it can be deployed over an existing Ethernet LAN infrastructure and be administered by the existing personnel using familiar LAN management tools. Initially, an IP SAN may be deployed over an existing corporate LAN and share resources with other traffic. Over time to achieve better performance an IP SAN may be transitioned to a dedicated LAN. As the business continues to expand, so does its demand for storage with better availability, performance, and scalability characteristics. This often necessitates investment into high-end FC SAN equipment coupled with physical distribution of the SAN over several geographic locations (as shown in Figure 19.22). At this point, iSCSI storage switches (or gateways) can be introduced to bridge-independent iSCSI and FC SAN islands over MAN or WAN links in order to form a single highly scaled geographically distributed SAN. These storage switches devices are responsible for translating iSCSI traffic to FCP and vice versa, thus providing seamless connectivity and interoperability between the IP and FC storage devices.

19.5.4.2 Architectural Model. iSCSI defines a connection-oriented client–server model to enable communication between Network Entities.

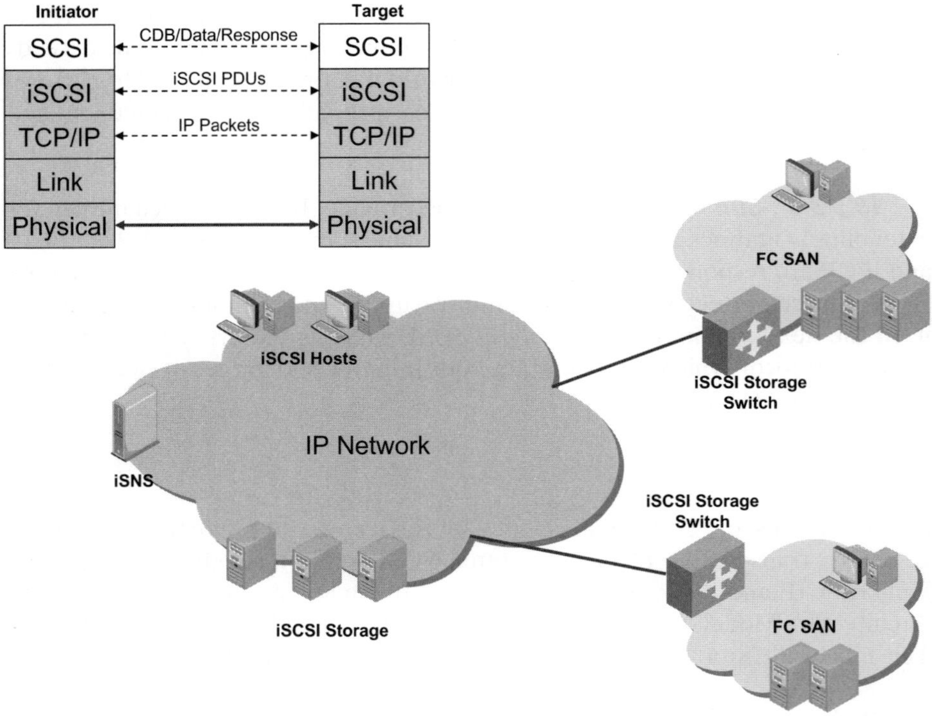

Figure 19.22. iSCSI deployment model.

Figure 19.23 depicts an example architectural configuration. A Network Entity is a single device (or gateway) that aggregates one or more iSCSI Nodes and Network Portals. An iSCSI Node represents a specific Initiator or Target. A Node is identified by a network-independent, globally unique, human-readable Name, typically used for authentication and discovery purposes [16]. A Network Portal represents a TCP/IP endpoint, which may either be used exclusively by a particular Node or be shared between Nodes. A Network Portal is identified by an address; for Initiators, it is just an IP address; for Targets it is an IP address and a TCP port, on which the Target listens for incoming connections.

In order to communicate with a given iSCSI Target, an iSCSI Initiator must establish a Session, which is equivalent to an I_T nexus in SCSI terminology. A Session may aggregate one or more TCP/IP connections between Initiator and Target. Multiple connections (over the same or different Network Portals) may be needed in order to enable traffic load balancing and connection failover. iSCSI introduces the concept of a Portal Group to identify which local Network Portals may be used by multiple connections within the same Session. If a Network Entity contains multiple Portal Groups, their membership is always disjoint.

19.5.4.3 Communication. iSCSI defines a protocol, which enables Initiator and Target to exchange messages, known as iSCSI Protocol Data Units

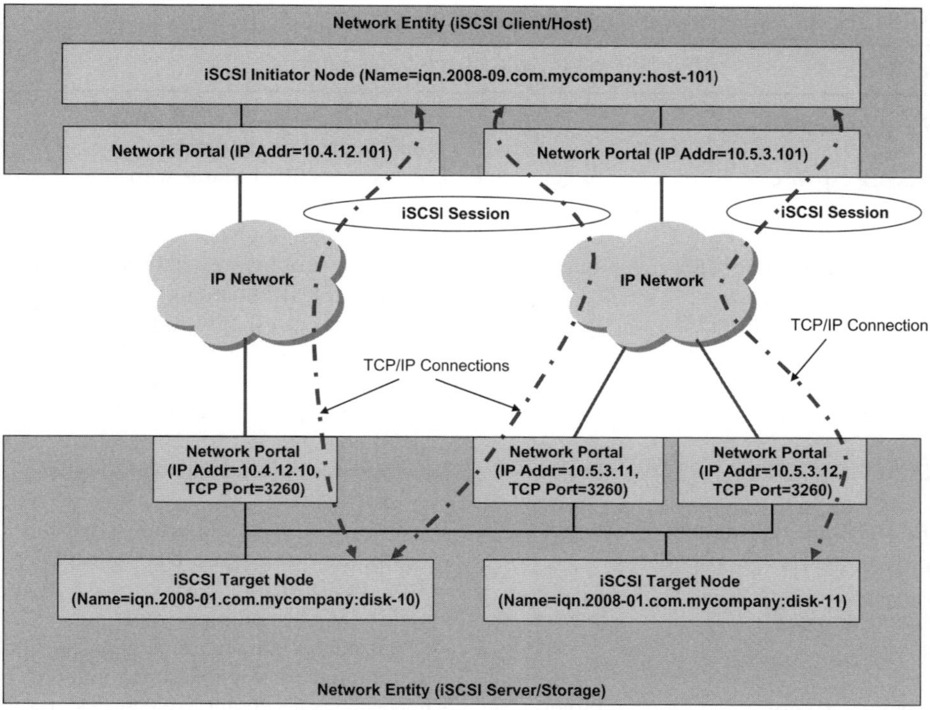

Figure 19.23. iSCSI architectural deployment example.

(PDUs) over TCP connections in a given Session. Table 19.6 identifies PDU types defined by RFC 3720. Each PDU contains a 48-byte-long Basic Header Segment, followed by several optional segments, which include additional headers (with more information), a header digest, a data segment, and finally a data digest. iSCSI requires each Command PDU and all the corresponding data and response PDUs to be sent over the same connection within a Session. This is known as connection allegiance. Different commands may be sent over different connections. Because iSCSI traffic may be sent over multiple TCP connections, iSCSI provides sequence numbers to enable the recipient to restore FIFO order and identify potentially missing PDUs. iSCSI also allows a Target to limit the number of commands an Initiator may send over a given Session before waiting for their responses. This enables flow control between a slow (or overloaded) Target and a fast Initiator.

To establish a Session, an Initiator connects to the Target's Network Portal and sends a Login Request PDU identifying its capabilities. A Login Phase then ensues and multiple Login Requests and Responses may be exchanged as an Initiator and a Target negotiate their security and operational parameters in two separate negotiation phases. Note that at least one of these phases is required by iSCSI; however, both may be optionally performed. The security negotiation phase allows the Initiator and the Target to agree on whether authentication is

TABLE 19.6. iSCSI Protocol Data Units

PDU Kind	Originator	Description
SCSI Command	Initiator	Encapsulates a SCSI CDB (may optionally contain data).
SCSI Response	Target	Encapsulates SCSI Status with corresponding Sense Data (auto-sense is required in iSCSI).
Task Management (TM) Function Request	Initiator	Enables control over execution of SCSI Task or iSCSI Functions.
Task Management (TM) Function Response	Target	Encapsulates a response to the corresponding request including special information for failure responses.
SCSI Data-Out	Initiator	Encapsulates payload for WRITE-type SCSI commands.
SCSI Data-In	Target	Encapsulates payload for READ-type SCSI commands.
R2T (Ready To Transfer)	Target	Encapsulates Target's indication that it is ready to process more Data-Out PDUs from the Initiator.
Login Request	Initiator	Initiates iSCSI login phase and encapsulates connection/Session-specific parameters for capability negotiation.
Login Response	Target	Encapsulates acceptance/rejection status as well as connection/Session-specific parameters for capability negotiation.
Logout Request	Initiator	Initiates orderly connection/session shutdown for recovery or maintenance.
Logout Response	Target	Encapsulates the status for the corresponding request.
NOP-Out	Initiator	Pings the Target to determine its current state.
NOP-In	Target	Pings the Initiator to determine its current state.
Text Request	Initiator	Provides additional information to the Target. May be used for future extension of iSCSI.
Text Response	Target	Provides additional information to the Initiator. May be used for future extension of iSCSI.
SNACK Request	Initiator	Provides a positive or negative acknowledgment of Target PDU(s).
Asynchronous Message	Target	Encapsulates unsolicited SCSI Asynchronous Events or iSCSI asynchronous messages.
Reject	Target	Enables the Target to reject a corrupt PDU or report a protocol error to the Initiator.

to be performed and if so which authentication method should be used. Several authentication methods including CHAP, SRP, SPKM1, SPKM2, or KRB5[8] may be supported for this purpose. Once the authentication method is chosen, the Initiator and the Target authenticate each other by exchanging challenges and responses via additional Login Request/Response PDUs. Operational phase enables the Initiator and the Target to exchange both session-wide and connection-wide parameters, which among others include Initiator and Target Names and Aliases, Target Addresses and Portal Groups, the total number of connections allowed in this session, maximum size of a data segment, flags to determine whether header and data digests are to be used, and so on. Some parameters may only be exchanged on the very first connection in the Session (known as the leading connection) and may not subsequently change. If negotiation completes successfully, the target sends a final Login Response indicating success and the corresponding connection transitions to the Full Feature phase, in which the connection is fully operational and SCSI I/Os may be transported. Otherwise, the Target rejects the Login, sends a Login Response indicating failure, and terminates the connection.

The way that iSCSI encapsulates SCSI I/O in the Full Feature Mode is as follows. An Initiator creates a SCSI Command PDU identifying the SCSI operation, affected LUN, position and length of the transfer, and so on, and sends it to the corresponding Target over an existing Session. For READ-type operations, the Target transfers the data back to the Initiator in one or more Data-In PDUs as the data becomes available and depending on the maximum allowed data segment length. Figure 19.24 depicts an example of a WRITE-type command processing. For WRITE-type commands, the Target is responsible for driving the data transfer from the Initiator by issuing R2T PDUs indicating its readiness. Thus, once the Target receives the Command PDUs, it issues R2T PDUs back to the Initiator, which then provides the requested data in Data-Out PDUs. If no errors have been encountered in this process, once the requested data transfer is completed, the Target sends a SCSI Response PDU indicating successful completion of the operation back to the Initiator. For performance reasons, iSCSI enables "phase collapse": An Initiator may send the first data segment in a Command PDU, while a Target may provide a response in the last Data-In PDU of the exchange.

19.5.4.4 Error Detection and Recovery. iSCSI relies on TCP/IP for most error detection and handling. However, there are two kinds of errors that require special handling by iSCSI Nodes:

- Header/data digest errors leading to PDUs being dropped despite successful delivery from TCP (note that the likelihood of this type of errors is significantly diminished in IPSec is used by the underlying IP stack).
- Failures of one or more TCP connections within a Session.

[8]CHAP is defined by RFC 1994, SRP by RFC 2945, SPKM1 and SPKM2 by RFC 2025, and KRB5 by RFC 1510.

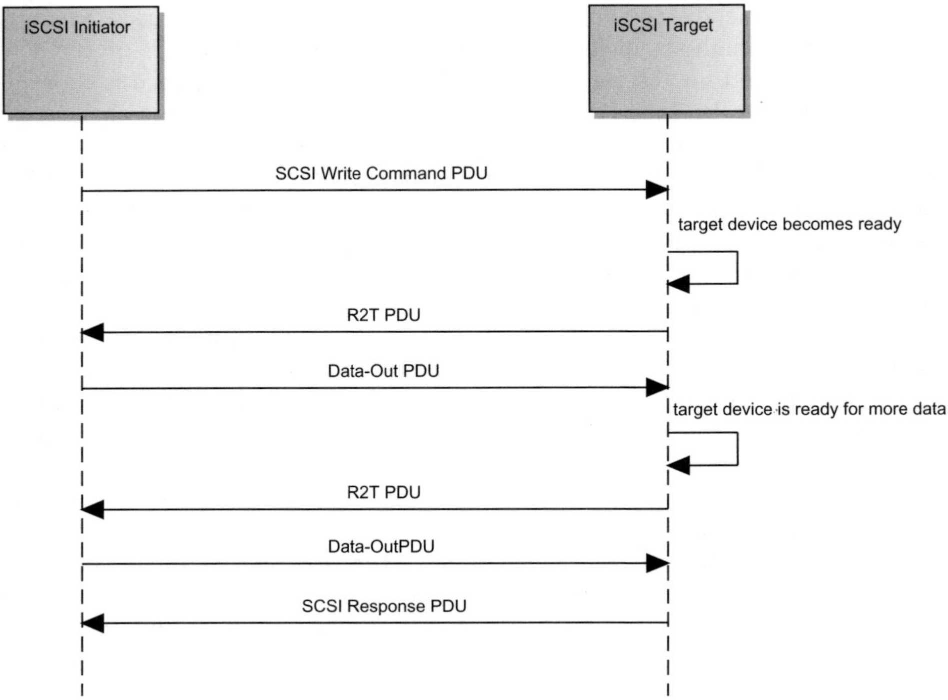

Figure 19.24. iSCSI Write sequence diagram.

To deal with these errors, iSCSI protocol defines a hierarchy of Error Recovery Levels, which provide a framework of error handling policies for the iSCSI-compliant implementation:

- Error Recovery Level 0, known as Session Recovery, requires the Session to be closed and cleaned up and then re-initiated over newly established TCP/IP connections.
- Error Recovery Level 1, known as Digest Failure Recovery, enables a connection to recover from a header/data digest failure by detecting and retransmitting failed PDUs. NOP-Out/NOP-In/SNACK PDUs as well as Command, Data, R2T, and Response retransmissions over the existing TCP connection may be used to facilitate recovery.
- Error Recovery Level 2, known as Connection Recovery, enables failover of tasks outstanding on the failed connection onto a redundant connection in the same Session (this is known as connection allegiance reassignment). Task Reassign TM Request may be used for this purpose. All outstanding SCSI operations are then resumed with the help of mechanisms from Level 1 to recover potentially lost PDUs.

Each higher Level subsumes the semantics of the lower Levels. Support of Level 0 is mandatory; others are optional. Note that Level 2 is useful not only for dealing with failed TCP connections but also for maintenance scenarios requiring nondisruptive hardware replacement, and so on. Error Recovery Level is a Session-wide property negotiated between the Initiator and the Target during operational negotiation login phase of the leading connection.

19.5.4.5 Discovery. An iSCSI Initiator learns about iSCSI Targets to which it has access through the process of discovery. Three main discovery mechanisms are supported [17]. First, an Initiator may statically configure the Name, IP address, and TCP port of the Target it wishes to communicate with. This option is least flexible and is typically viable only for the smallest and simplest setup. Alternatively, an Initiator may initiate a special "discovery" Session with a known Target and send it a "SendTargets" query encapsulated into the Text Request PDU. The Target responds with a list of acceptable Targets in the Text Response PDU. The Initiator may then use received information to establish normal Sessions with specified Targets on as needed basis. Finally, the most flexible mechanism assumes the use of a configuration service such as SLPv2 [18] or iSNS [19]. In this case, the Initiator first establishes a connection with the corresponding service and then queries it for the list of available Targets. Note that use of iSNS has additional benefits such as the ability to restrict Initiator access to Targets through Discovery Domain as well as providing asynchronous state change notification events about changes taking place in the Discovery Domain.

19.5.4.6 Security. Historically, security of Fibre Channel SANs was not a major risk due to their relative isolation from the nontrusted network domains.[9] This is not the case for iSCSI and consequently both RFC 3720 and RFC 3723 [20] provide recommendations for secure deployments of iSCSI. Several mechanisms are suggested:

- IPSec[10] should be used to provide integrity, authentication, and confidentiality at the packet level.
- Authentication methods described in Section 19.5.4.3 to provide mutual authentication of iSCSI Initiators and Targets during connection login.
- SLPv2 and iSNS may be used to store public keys required for authentication.
- Through use of Discovery Domains in iSNS, it is possible to limit Initiator access to only those Targets, which they are authorized to access.
- VPN services must be deployed in the iSCSI gateways to protect traffic traveling over WAN/MAN links.

[9]This claim is no longer true in today's SAN environments. T11 FC-SP defines the standard to implement security in Fibre Channel fabrics.

[10]IPSec is documented by RFC 4301 (Security Architecture for the Internet Protocol), RFC 4302 (IP Authentication Header), RFC 4303 (IP Encapsulating Security Payload (ESP)), RFC 5282 (Using Authenticated Encryption Algorithms with the Encrypted Payload of the Internet Key Exchange version 2 (IKEv2) Protocol), and RFC 4308 (Cryptographic Suites for IPsec).

19.5.4.7 Summary. iSCSI is a flexible, extendible, and cost-effective protocol, which enables creation of IP-based SANs. Because of its cost benefits, iSCSI has become very popular in the storage networks deployed by small businesses and enterprises. However, there are several issues preventing iSCSI from being a viable option for mid-range and high-end enterprises.

CPU overhead associated with TCP/IP retransmissions and flow and congestion control algorithms as well as iSCSI compute-intensive operations is often significant. The need to steer iSCSI traffic to application buffers in the presence of TCP packet loss and reordering in order to avoid unnecessary data copying further adds to the CPU overhead.

This may be prohibitive for high-performance servers and storage devices, which are expected to handle numerous parallel workflows under significant load. This issue may be mitigated by the use of TCP Offload Engines (TOE) as well as iSCSI ASICs, which offload compute intensive operations associated with both protocols to specialized hardware and thus free up the main CPU(s). The negative side of such a solution is that it increases the overall cost of deployment.

As 10-Gbit/s Ethernet becomes cheaper and more readily available, it remains to be seen how iSCSI competes with other technologies such as FCoE.

19.5.5 iSNS

Internet Storage Name Service (iSNS) provides discovery, configuration, and management functionality mimicking that of a FC Name Server and FC Controller (as documented by T11's FC-GS specification) for use in IP networks. iSNS relies on client–server architecture to implement its services. It is actively used by iFCP and iSCSI protocols described above. iSNS and its services are defined by IETF in the following RFCs:

- RFC 4171—defines the iSNS protocol and object model.
- RFC 4939—defines the iSNS Management Information Base (or MIB) and the hierarchy of managed objects for integration with SNMP.

The following list enumerates the most important features of iSNS:

- Name Service for naming and discovery of storage resources and their properties.
- Discovery Domain and Login Control Service for zoning purposes.
- State Change Notification Service to allow devices to learn about network changes.
- Open Mapping between FC and iSCSI devices enable further interoperability between FC and iSCSI.

For detailed description of the iSNS functionality the reader is referred to the corresponding RFCs.

19.5.6 FCoE

Fibre Channel over Ethernet (FCoE) is a newly emerging standard, being defined by the INCITS T11 committee.[11] It enables native transport of FC (and consequently SCSI) traffic over lossless 10-Gigabit Ethernet. Similar to iSCSI, FCoE aims to converge LAN and SAN technologies. Unlike iSCSI, FCoE avoids IP altogether. Instead, it leverages the Fibre Channel stack but abandons physical and signaling levels[12] (FC-0 and FC-1) of the stack and maps the framing level (FC-2) directly onto lossless 10-Gigabit Ethernet.

19.5.6.1 Ethernet Enhancement for Data Center. Traditional Ethernet is highly prone to packet loss and congestion, which makes it unacceptable for the QoS required by FC protocol. The IEEE Data Center Bridging Task Group has defined a number of extensions, which collectively facilitate creation of Converged Enhanced Ethernet (CEE) network based on 10-Gigabit Ethernet. The key objective of CEE is to provide a lossless, high-performance networking infrastructure, which facilitates convergence of SAN and LAN technologies. When this is realized, it will bring about significant savings in deployment, maintenance, and management of the converged network. The following list enumerates the most significant Ethernet enhancements at the core of CEE:

- Support of 2500-byte-long Ethernet mini-jumbo frames allows FCoE avoid fragmentation and reassembly of 2112-byte-long FC frames.
- IEEE 802.1Qbb Priority-Based Flow Control enables creation of independently flow-controlled virtual traffic lanes over the same physical Ethernet link. This allows differentiation of QoS based on need.
- IEEE 802.1Qau Congestion Notification makes it possible to throttle traffic-originating nodes in the presence of congestion and remove the throttling once the traffic conditions improve. Note that at the time of this writing, this extension is not being used by FC-BB-5 devices.
- IEEE 802.1Qaz Enhanced Transmission Selection enables pre-allocation and adjustment of bandwidth based on defined classes of traffic and its actual patterns.
- DCBX Data Center Bridging eXchange protocol allows nodes to discover each other's capabilities and adjust the behavior appropriately.

It is important to note that iSCSI will also benefit from the 10 Gbit/s CEE. However, given the lack of a centralized management application as well as iSCSI's reliance on TCP/IP and the corresponding need for sophisticated ToEs

[11]FC-BB-5 standard (available from http://www.t11.org/fcoe) has been ratified in June of 2009.

[12]Refer to Section 19.5.2 for detailed information about the FC levels.

to reduce server CPU utilization, it is unlikely that iSCSI will be able to compete with FC and FCoE in large-scale environments. iSCSI will definitely continue to be a viable and cost-effective option for small-scale and entry-level data centers.

19.5.6.2 Architectural Model. As shown in Figure 19.25, FCoE follows the FC architectural model in defining three kinds of ports, each of which represents a different kind of an endpoint capable of FCoE communication:

- VN_Port (Virtual N_Port) can originate and terminate FCoE traffic.
- VF_Port (Virtual F_Port) enables attachment of VN_Port to FCoE fabric.
- VE_Port (Virtual E_Port) provides inter-switch connectivity or bridging via another VE_Port.

These ports are designated virtual because they are logical entities; that is, they emulate the behavior of the corresponding physical FC ports. Multiple virtual ports may be associated with the same physical lossless Ethernet port. A virtual port is addressed by its MAC address, which is provided by the FCoE fabric at login and initialization time. Virtual ports may establish connectivity between one another by creating virtual links. FCoE supports virtual links between a VN_Port

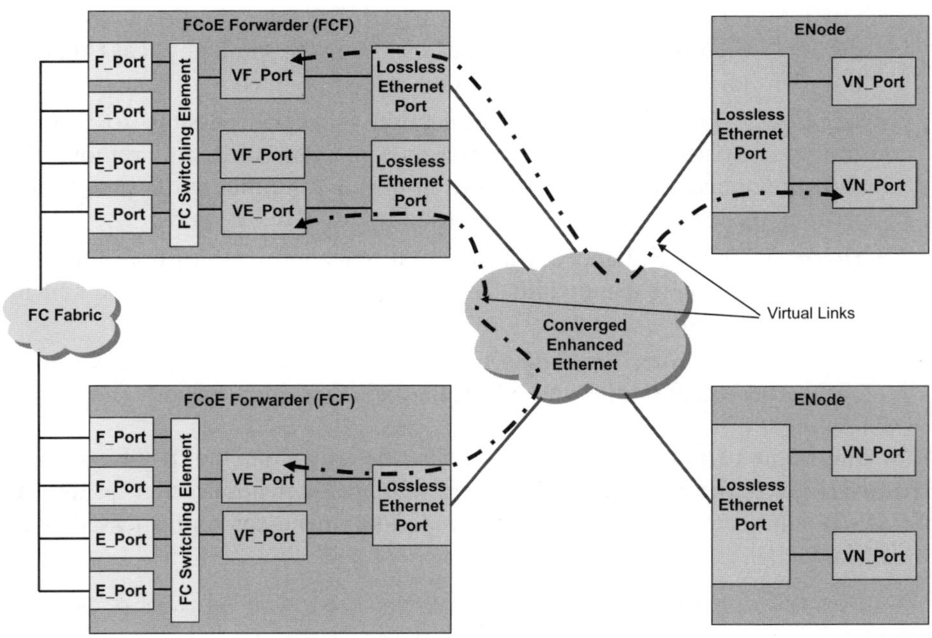

Figure 19.25. FCoE architectural deployment example.

and a VF_Port and between a pair of VE_Ports but not between a pair of VN_Ports (i.e., point-to-point).[13] Typically, virtual links are established over a VLAN to improve security and provide logical isolation of FCoE fabric from unrelated Ethernet traffic. To further improve security, FCoE leverages Access Control Lists (ACLs) in order to specify a set of operations permitted or denied to a given port.

In addition to ports, FCoE defines an ENode as an FC node with one or more lossless Ethernet ports. Since FCoE does not currently support a point-to-point topology, ENodes must be connected to the FCoE fabric via the switching element called FCoE Forwarder (FCF). An FCF typically aggregates multiple lossless Ethernet ports (for connectivity with ENodes and other FCFs) and possibly one or more native FC interfaces. The latter allow an FCF to bridge FCoE and FC fabrics. Whenever an ENode logs into a fabric, a VN_Port and a VF_Port are instantiated on the ENode and FCF, respectively, and a virtual link is established between them. Virtual links between VE_Ports are established in a similar way. FC frames can then be sent over established virtual links. FCF relies on existing FSPF[14] protocol to forward FC frames through the FCoE fabric based on the destination address field (D_ID) in the encapsulated FC frame header. Each time an FCF forwards a frame, it substitutes the source and destination MAC addresses of the corresponding Ethernet packet but leaves the encapsulated FC frame intact.

To enable device discovery, initialization, and maintenance, FCoE defines a subprotocol called FCoE Initialization Protocol or FIP. FIP uses special frames (different from regular FCoE frames) for communication. Its main responsibilities include:

- Discovery of FCFs and their capabilities by ENodes and other FCFs.
- Virtual link formation and teardown.
- Virtual link liveness detection based on keep-alive messages.

19.5.6.3 Deployment. FCoE has not been designed to replace FC. Given the current dominance of FC SANs, FCoE and native FC will coexist for many years to come. As shown in Figure 19.26, FCoE's primary goal today is to enable network convergence and thus explore a cheaper and simpler alternative of connecting servers to FC SANs. It is expected that server and switching devices will lead in adoption of FCoE and the storage devices will follow. An FC HBA and a NIC will be consolidated into an FCoE-capable 10-Gbit/s CEE interface called Converged Network Adapter (CNA). Similarly, Converged Network Switches (CNS) with FCF functionality will be deployed to enable CEE LAN connectivity between servers as well as server connectivity to storage via FC SANs. The crucial point is that every server already connected to the CEE LAN via its CNA

[13]Support of virtual links between a pair of VN_Ports is expected to be added in FC-BB-6.

[14]FSPF (Fabric Shortest Path First) protocol is defined by T11's FC-SW-4 standard.

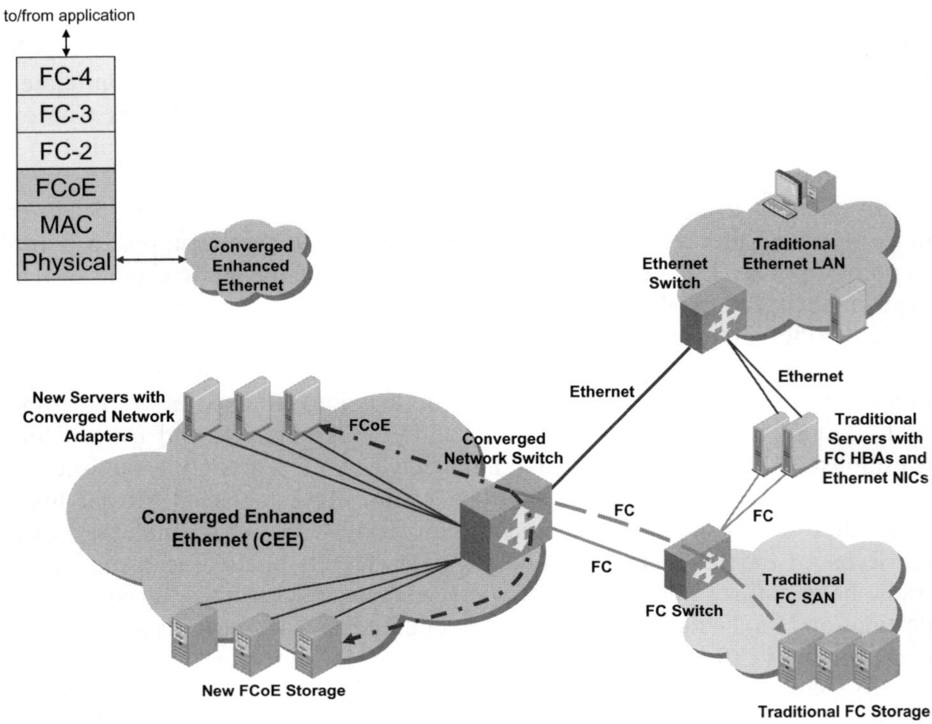

Figure 19.26. FCoE deployment model.

may be connected to a FC SAN without having to install HBAs and provide additional cabling. Over time, as native FCoE storage devices become available, they will be attached to the CNS to provide end-to-end FCoE connectivity between servers and storage.

19.5.6.4 Summary. FCoE enables network convergence of LAN and SAN infrastructures without negatively impacting performance, reliability, and cost associated with existing technologies. New servers deployed with CNAs will have a smaller footprint, reduced power, and cooling needs as well as greatly simplified cabling requirements. CNS will help create a core fabric with differentiated QoS, which is capable of transferring reliable high-performance storage-related traffic as well as unreliable LAN traffic.

From the perspective of storage management and administration, FCoE is identical to native FC. This is a major benefit to storage administrators of large and mid-size data centers, who already deploy hundreds or even thousands of FC ports; and it will enable easier penetration of FCoE into such environments.

19.6 NAS PROTOCOLS

19.6.1 NFS

The NFS protocol was initially defined by SUN Microsystems, and its implementation is based on SUN's Remote Procedure Call (RPC) protocol. RPC (discussed in detail in reference 21) allows a client to perform system calls on a remote server across a network. RPC removes the need for an application to implement a client–server protocol. NFS performs its system calls using RPC as an underlying protocol.

Using NFS, a server exports one or more file systems to its clients (more accurately, the server manages which directory tree appears as a file system to remote clients). Clients mount the file system within their own directory tree. Several clients can mount the same file system and therefore directly exchange or share files through the mounted file system.

NFS is currently in its fourth revision. Versions one to three have been developed by SUN Microsystems, while version four is the first one that has been developed by the Internet Engineering Task Force (IETF). Version three is still largely deployed today; and since version four is a significant architectural departure from version three, they both deserve separate discussion.

19.6.1.1 NFSv3. Because NFS is aimed at being used across a network of heterogeneous operating systems, the protocol is defined to be platform-independent. NFS achieves this by defining two protocols: one for mounting file systems and the other for file serving itself. A client can send a mount request to the server indicating the path name of the directory to be mounted. If that directory exists and is exported, the server returns a file handle, which amongst other things contains the *i-node* of the corresponding directory. The client uses this file handle for all subsequent read/write commands, and so on. The file serving protocol supports all Unix system calls defined for file manipulation files, with the exception of *open* and *close*. This is done intentionally so that the server does not have to track information about files being manipulated by remote clients, making NFSv3 a stateless protocol. Because of this, a server can reboot even though its file systems may be mounted by clients. Clients remain operational and don't lose data during server recovery time (however, any remote calls made by clients during this time either fail or time out). This stateless nature of NFSv3 does not match Unix calls exactly. One of the major consequences of this is manifested in the file locking semantics: NFSv3 does not provide a mechanism for exclusive file access; other mechanisms must be employed for that purpose.

In order to improve network communication performance and reduce overhead, NFS implements the following optimizations, which are also different from regular file handling of standard Unix file systems:

- All data transfers are done in 8K chunks, regardless of the actual amount of data being exchanged.

- For *read* operations, if the client requests less than 8K of data, the server reads ahead and fills in the 8K block with the data following the requested chunk. The NFS layer at the client caches the extra data. If the client continues to read the data, the NFS layer is often capable to satisfy its local client's requests from the cache without having to communicate with the server over the network.
- NFS employs a similar technique for *write* operations, *writes* are buffered until an 8K block is accumulated (or until the file is closed).
- Caching is done on *i-node* changes as well as for file data. This policy can cause consistency issues across the network when multiple clients share a file. Several methods have been put in place to mitigate this issue. For further discussion on this, the reader is referred to Tannenbaum [3].

NFS was originally implemented over the Internet UDP protocol (with the intermediate RPC layer handling timeouts and retransmits). However, along with version 3, the option of using TCP as a base transport layer was formalized. This addition made the implementation of NFS across wide area networks possible.

19.6.1.2 NFSv4. The fourth version of NFS was influenced by concepts from CIFS as well as from the work in distributed computing and distributed network file system of the Andrew project at Carnegie Mellon University, where the Andrew File System (AFS) was developed. NFSv4 integrates a file locking solution, provides several performance improvements over NFSv3, mandates a strong security framework, and implements a stateful protocol. A minor revision (NFS v4.1) provides further support for clustering as well as parallelism of data access through parallel NFS (pNFS). The parallelism is accomplished in the protocol by separating the handling of file system meta-data and data, allowing operations across server clusters. In a pNFS implementation, multiple servers are controlled by a meta-data server. NFS v4.1, while finalized, has not yet received an RFC number at the time of this writing.

19.6.2 CIFS

The CIFS protocol is a proprietary protocol defined and implemented by Microsoft Corporation. It is largely based on the Server Message Block protocol (SMB). SMB was originally developed by IBM on top of NetBIOS, though it could also run over TCP. As Microsoft extended SMB and added support for symbolic links, hard links, and larger file sizes, it also changed the name of its implementation to CIFS.

Besides allowing file system access, SMB and CIFS allow sharing of any resource on a network (such as a printer) and leverage the Windows Domain services such as authentication. The core authentication of an NT domain infrastructure is the Kerberos protocol developed at MIT [22].

19.7 VIRTUALIZATION

19.7.1 Server Virtualization

Server virtualization is a new trend in information technology that has gained tremendous traction over the past several years. The virtualization technology allows running of several operating system instances on a single server. This is realized through the means of a *hypervisor*, which is a software layer that abstracts the hardware and monitors the various *virtual machines* (i.e., virtualized OS instances running in a server). This is illustrated in Figure 19.27. Traditionally, a hypervisor can be a level 1 hypervisor if it operates directly above the hardware, or it can be a level 2 hypervisor if it runs within a guest operating system.

From a storage perspective, server virtualization creates a few more layers of abstractions which are depicted in Figure 19.28. This is required to handle the complexity and differentiation of the various physical storage elements involved across several host OSs and various storage systems that can be leveraged. Typically, each virtual machine will have a SCSI adapter with SCSI disks connected to it. These disks are actually virtual disks that are provisioned by a data store layer, which manages the physical LUs as well as file systems exported by the storage system. Virtual machines and hypervisors typically boot from the storage network, which allows for convenient mobility of the hypervisors across servers.

As the technology became more sophisticated, it enabled virtual machines to fully utilize advanced storage services such as multi-pathing (availability of redundant paths to the storage array). Likewise, it is possible to use replication and

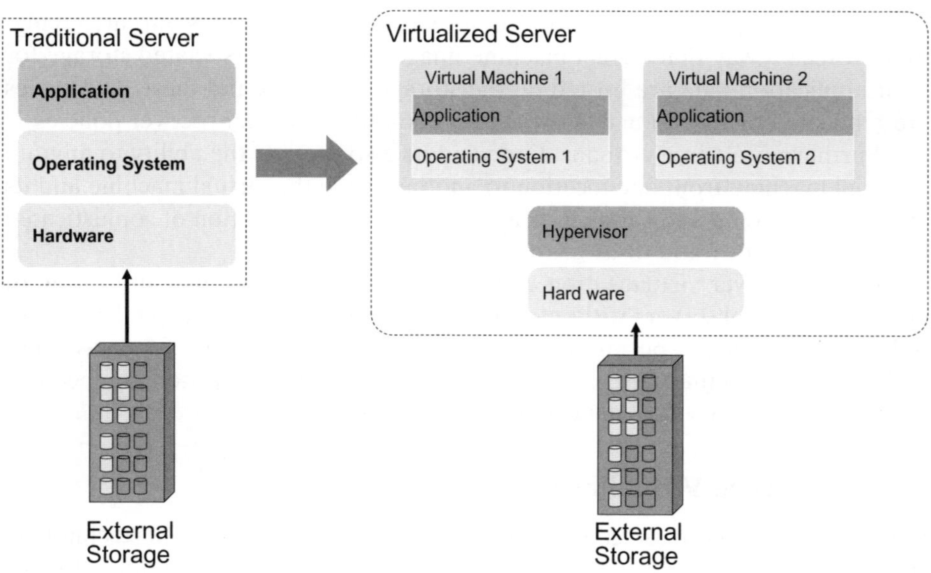

Figure 19.27. A virtualized server.

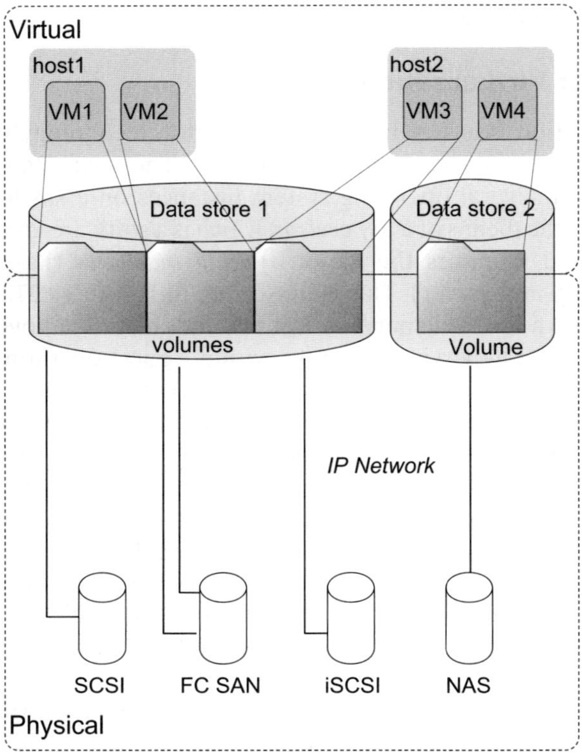

Figure 19.28. Layers of storage abstraction in a virtualized server.

disaster recovery technologies with server virtualization. The systems are configured in such a way that virtual machine images as well as associated storage for their application data are hosted on the storage networks, and these data stores are replicated by the storage networks as required by disaster recover policies.

Furthermore, some virtualization vendors also provide the ability to migrate a virtual machine from one location to another while the virtual machine and its applications are up and running. This allows for implementation of sophisticated business continuity and disaster recovery functionality.

Overall, server virtualization enables enterprises to maximize utilization of deployed physical servers without significant increases in either capital expenses or lab footprint. It also permits implementation of sophisticated disaster recovery and business continuity use cases. As a result, server virtualization has become very popular in corporate data centers.

19.7.2 Storage Virtualization

Storage virtualization also gained a lot of interest in recent years. This technology creates an abstraction layer between servers and the corresponding storage. The benefits that this technology brings are as follows:

- Abstracts storage array specifics, allowing hosts to deal with heterogeneous storage pools created by deploying systems from different suppliers.
- Enables transparent migration of data between heterogeneous and (potentially geographically distributed) storage arrays.
- Provides efficient local or long-distance replication across heterogeneous storage systems.
- Enables cost-effective and performance-efficient storage resource usage policy by automatically and nondisruptively moving storage resources based on application's importance and performance profile.
- Enables virtual provisioning of storage resources, which allows more storage than physically available to be allocated for application use; as application storage demands increase, additional storage capacity can be added over time without incurring any application downtime.

Storage virtualization technology allows advanced services and applications described earlier in this chapter to be moved into the specialized storage virtualization layer, which provides feature-rich functionality and significantly simplifies management. As this is done, care must be taken to ensure that the virtualization layer itself meets the availability requirements of the enterprise and is capable of tolerating and recovering from failure scenarios that can affect both the storage subsystems as well as the storage network.

In a virtualized SAN fabric, there are three architectures, which enable delivery of storage virtualization services:

- An in-band appliance architecture.
- An out-of-band appliance architecture.
- A hybrid architecture called SPAID (Split Path Architecture for Intelligent Devices).

Figure 19.29 depicts the three architectures. Regardless of the architecture, all storage virtualization solutions must do three essential things:

- Maintain a map between virtual and physical storage and also manage additional configuration meta-data.
- Manage configuration and execute storage management tasks.
- Most importantly facilitate transmission of data between hosts and storage.

The three architectures differ in the way they handle these three separate services, which are often referred to as meta-data, control, and data path services. The differences hold implications for performance, scalability, and availability.

With in-band virtualization appliance architecture, an appliance is placed in the storage network between the hosts and the physical storage systems.

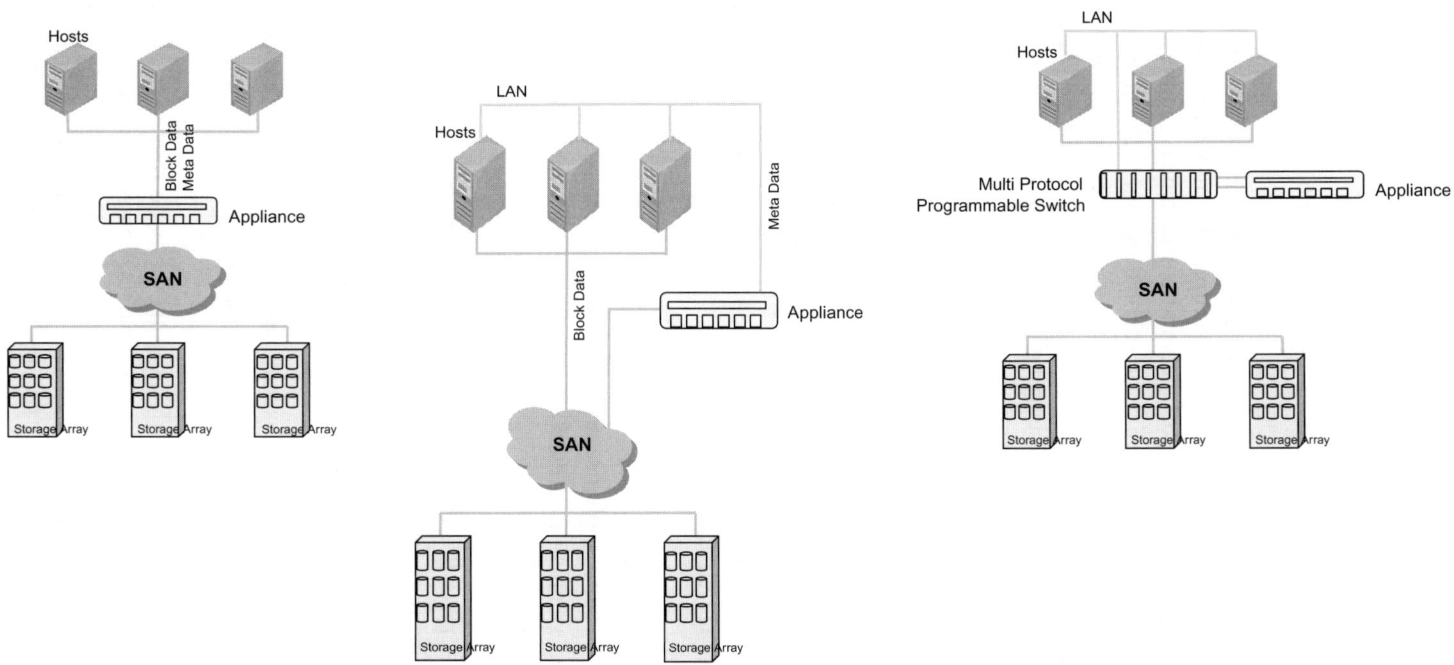

Figure 19.29. The three storage networks virtualization architectures. From left to right: In-band, out-of-band and SPAID.

The appliance handles meta-data, control, and actual data communication. The appliance maintains a map between virtual storage presented to the hosts and available physical storage. Provided with this map, it routes all host I/Os accordingly. It also manages the back-end physical storage systems. In this architecture, performance and availability are the two major concerns. Since the appliance is placed in the data path, there is a risk that it may become a performance bottleneck because any I/O processing overhead may negatively impact application performance. This is often mitigated by adding a significant amount of cache memory to the virtualization appliance. Also, individual appliance failure results in the inability of the application to access data. This is unacceptable for most mission-critical applications. As a result, in-band virtualization appliances are typically deployed in a cluster, which may increase complexity and cost of this solution.

In out-of-band virtualization appliance architecture, the meta-data management and control operations are handled via a separate network (typically a LAN) while the data I/O is handled by the SAN. While performance impact of this architecture may be superior to the in-band appliance, it has the drawback of requiring host agents to separate the data traffic from meta-data and control traffic. This is often prohibitive for large enterprises because maintenance (i.e., configuration, installation, and upgrading) of host agents across hundreds or even thousands of application servers is either fragile or infeasible.

A hybrid architecture is often termed network-based or SPAID (Split Path Architecture for Intelligent Devices). It leverages programmable multi-protocol switches, which offer port-level programmable processing capabilities, which can map incoming I/Os from virtualized to physical devices without requiring host based agents. These switches are highly specialized devices, which implement the corresponding mapping tasks at wire speed without any impact to I/O performance. In this architecture, the meta-data and control traffic is split from data traffic at the switch ports, which direct it to an out-of-band control engine. The role of this control engine is to handle control and meta-data traffic, configure and manage the switches and the back-end physical storage, and instruct the switches in handling of I/O exceptions, which cannot be handled by the switches themselves.

19.7.3 NAS Virtualization

Similar to SAN virtualization, NAS virtualization (also called file virtualization) attempts to abstract some of the complexity of enterprise file servers and move that intelligence into the network. Specifically, NAS virtualization provides a uniform namespace across file servers. Namespace refers to a hierarchy of directories and files, along with their meta-data, which are typically associated with a physical file server. A uniform namespace abstracts the physical limitation of the servers by providing a uniform logical hierarchy across multiple physical enclosures.

This technology allows consolidation of servers as well as nondisruptive migration of file systems. When combined with clustering technology, NAS virtualization enables increased scalability and performance.

NAS virtualization is typically implemented as a virtualization appliance located between NAS devices and their clients. The appliance mounts the file systems exported by the NAS devices and maps their hierarchy into a uniform one. Similar to SAN virtualization appliances, both in-band and out-of-band architectures have been deployed. Just as with SAN virtualization, the main difference between these architectures lies in the ways they handle data versus control and meta-data.

19.8 CONCLUSION

The aim of this chapter was to introduce storage technologies and storage networking to readers with a strong computer networking background. Storage networks evolved independently from computer networks and, as a result, introduced a wealth of their own protocols, largely unknown in the computer networking world. In recent years, the two networking technologies began merging more intimately than ever before as performance and availability of the computer networking caught up with the storage domain while its cost stayed relatively low. This technological advancement provides data centers with an exciting opportunity to introduce new feature-rich functionality, simplified management, and consolidation of networking and storage infrastructures. At the time of this writing, Fibre Channel is the dominant presence in most data centers. We, however, see a very rapid growth of iSCSI together with other newer protocols, such as FCoE. These protocols have a chance to bring even further consolidation and uniformity into the networking world. As various technologies compete for their role in the data center, the concepts introduced in Sections 19.5.3, 19.5.4, and 19.5.6 of this chapter will play a crucial role in shaping this technology roadmap.

REFERENCES

1. R. Barker and P. Massiglia, *Storage Area Network Essentials*, John Wiley, & Sons, Hobohen, 2001.
2. U. Troppens, R. Erkens, and W. Müller, *Storage Networks Explained*, John Wiley, & Sons, Hoboken, NJ, 2004.
3. A. S. Tannenbaum, *Modern Operating Systems*, Prentice-Hall, Upper Saddle River, NJ, 2007.
4. S. J. Mullender and A. S. Tanenbaum, Immediate files, *Software—Practice and Experience*, Vol. 14, pp. 365–368, April 1984.
5. See http://www.ndmp.org.
6. M. Ji, A. Veitch, and J. Wilkes, *Seneca: remote mirroring done write*.

7. ANSI X3.131.
8. http://www.t10.org.
9. ANSI INCITS 402-2005 (ISO/IEC 14776-413:2007).
10. http://www.t11.org.
11. R. W. Kembel, *Fibre Channel: A Comprehensive Introduction*, Northwest Learning Associates, Hingham, MA, 2002.
12. T. Clark, *IP SANs: A Guide to iSCSI, iFCP, and FCIP Protocol for Storage Networks*, Addison-Wesley, Reading, MA, 2002.
13. IETF RFC 4330.
14. SCSI Architecture Model 2, T10 Committee, http://www.t10.org.
15. IETF RFC 3720, http://www.ietf.org.
16. IETF RFC 1737, Functional Requirements for Uniform Resource Names, http://www.ietf.org/.
17. IETF RFC 3721, Internet Small Computer Systems Interface (iSCSI) Naming and Discovery.
18. IETF RFC 2608, Service Location Protocol, Version 2.
19. IETF RFC 4171, Internet Storage Name Service (iSNS).
20. IETF RFC 3723, Securing Block Storage Protocols over IP.
21. W. Richard Stevens, *TCP/IP Illustrated*, Vol. 1: The Protocols, Addison-Wesley, Reading, MA, 1994.
22. B. C. Neuman and T. Ts'o. Kerberos: An authentication service for computer networks, *IEEE Commun.* Vol. 32, No. 9, pp. 33–38, September 1994.

PART V

PHOTONIC AND ELECTRONIC COMPONENT TECHNOLOGY

20

ROADM ARCHITECTURES AND WSS IMPLEMENTATION TECHNOLOGIES

Neo Antoniades, Georgios Ellinas, Jonathan Homa, and Krishna Bala

20.1 INTRODUCTION

20.1.1 Optical Network Evolution

Optical networks are expanding at a dramatic rate to support the explosive growth of bandwidth-rich Internet video applications along with traditional voice and data services. Fiber-optic-based networks are ideal for this task because they can carry information further and in greater density than previous copper-based transmission systems. In particular, using dense wavelength division multiplexing (DWDM), optical fibers can carry upwards of 100 wavelength channels of information simultaneously, with each wavelength channel operating at high speeds (e.g., 2.5, 10, 40 Gbit/s). Fiber-optic communications has evolved from point-to-point transmission where information is transmitted between two nodes in the network, to intelligent fully meshed optical networks, where individual channels are added, dropped, and routed at individual nodes in the network. Thus, optical network architectures as we envision them now and as they are currently being deployed not only provide transmission capacities to higher transport levels, such as inter-router connectivity in an IP-centric infrastructure,

Convergence of Mobile and Stationary Next-Generation Networks, edited by Krzysztof Iniewski

but also provide the intelligence required for efficient routing and fast failure restoration. This is possible due to the emergence of optical network elements that have the intelligence required to efficiently manage such networks.

In the past, adding, dropping, and routing wavelength channels at individual nodes in the network was done by demultiplexing the wavelength channels at each node, manually rearranging the wavelengths using an optical patch panel, and then multiplexing them again into the desired fiber output for transmission to the next node. Needless to say, this was time-consuming and, far worse, very prone to human error. Over the years, manual switchboards were replaced by automated switches across all types of technologies. For example, in the last few decades, digital cross-connects (DCSs) and SONET/SDH add/drop multiplexers (ADMs) have completely replaced the practice of manually rearranging individual circuits using back-to-back channel banks. Similarly, in today's optical networks, optical cross-connects (OXCs) and reconfigurable add/drop multiplexers (ROADMs) are two of the main network elements that were introduced to quickly and efficiently route the optical connections to their desired destinations. These elements have considerably simplified some of these manual operations and in the future they are expected to eliminate them completely.

Over the last decade, the initial focus was on large port count OXCs, and following a wave of timely technologic breakthroughs, in the early 2000s, optical network equipment vendors were announcing a variety of optical switching systems capable of exchanging and redirecting several terabits of information per second. The dimensions of the proposed switches were colossal, ranging from a few tens to a few thousand ports with each single port capable of carrying millions of voice calls, or thousands of video streams. These switches were opaque (with either an electronic switch fabric or an optical switch fabric but with electrical interfaces at the edges of the switch) or transparent, with either a large port-count switch or several smaller port-count switches (one per wavelength).

In the last few years the focus has shifted from these large port count OXCs to ROADMs that are smaller network elements that can efficiently be used to direct wavelength channels in an optical network. Section 20.1.2 that follows argues for the need for ROADMs, while Section 20.1.3 discusses some of the initial OADM architecture designs used. Section 20.1.4 then describes how these ROADMs are utilized in today's networks. In Section 20.2 we describe several promising ROADM architectures and two-degree, multidegree, colorless, and directionless-type architectures. Section 20.3 describes in detail the wavelength-selective switch (WSS) technologies used to implement the ROADM architectures of Section 20.2. Section 20.4 discusses several experimental and commercial ROADM deployments, and the chapter concludes in Section 20.5 with a discussion on future ROADM design and deployment issues.

20.1.2 The Need for ROADMs

Today's optical networks have historically evolved from the long-haul backbone through regional and metro networks, and are now being deployed in the access

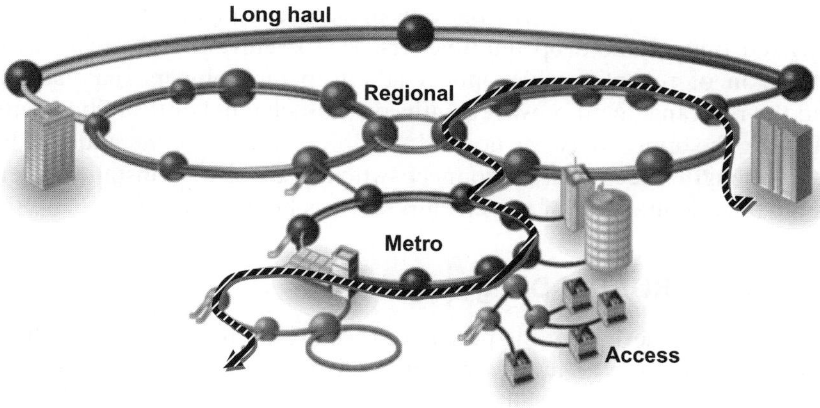

Figure 20.1. Ubiquitous ROADM deployment.

networks (last mile). Reconfigurable optical add/drop multiplexers (ROADMs) are rapidly following this evolutionary path so that they are now being deployed as a basic building block throughout optical networks, for the flexible network provisioning and optical transmission management benefits they provide. Two parallel evolutions taking place are the emergence of mesh networking to interconnect multiple optical network segments, and higher-speed optical signals, with 40 Gbits/channel beginning to occur and 100 Gbits/channel on the very near horizon. To meet these evolutionary trends, ROADMs will need to be able to maintain signal quality at high levels to transport higher-speed optical signals through a large number of cascaded nodes across interconnected networks.

In Figure 20.1, four tiers of a generic ring hierarchical design are presented and the ubiquitous deployment of ROADM network elements is demonstrated. The ROADM has emerged as one of the key building blocks for networks at the metro and regional level. Used for adding and dropping wavelengths from intermediate points along a transmission link or a ring, the role of ROADMs in a typical network is illustrated in Figure 20.1. They offer a simple and effective way to access a general mesh network along links or within ring subnets. Specifically, in tier 1 of the hierarchy, the access network, traffic is electronically aggregated and placed onto the metro rings. The metro network (tier 2) spans a few tens of kilometers and composes of interconnected ROADMs in a ring configuration. Connectivity on these rings is mostly hubbed to a central office (CO) location. In tier 3 of the hierarchy, ROADMs in the regional network interconnect edge rings to the larger regional rings that span possibly tens to hundreds of kilometers. Finally, in tier 4 of the hierarchy, regional rings can also be interconnected using ROADMs forming mesh connectivity optical networks (long-haul) as shown in the figure.

The implementation of the add/drop multiplexers shown in Figure 20.1 can be static (fixed) (denoted as optical add/drop multiplexers [OADMs]) or dynamic (denoted as ROADMs) depending on the presence of switching elements. Until recently, some of the connectivity patterns presented in Figure 20.1 were achieved using fixed optical add/drop multiplexers. This is currently changing with the introduction of reconfigurable optical add/drop multiplexers that have automated the rearrangement of wavelengths on multichannel optical fibers entering and leaving the optical network nodes. In addition, in some of the nodes of Figure 20.1, wavelength selective cross-connect switches can also be installed to enable dynamic reconfigurability of wavelength services between rings.

20.1.3 Initial ROADM Designs

Two fundamental designs of the precursor OADM are the parallel and the serial architectures shown in Figures 20.2a and 20.2b, respectively. Parallel designs are better suited to network architectures where high numbers of add/drop wavelengths are a requirement, mainly because of the fact that the insertion loss involved in the multiplexing/demultiplexing process is encountered only once, compared to the serial case where it is encountered several times. However, in the serial configuration, an added advantage is that individual modules can be incrementally added on a per-needed basis [1].

A traditional reconfigurable OADM architecture called Broadcast and Select (B&S) is shown in Figure 20.3. In this architecture tunable receivers and transmitters are utilized and a reconfigurable wavelength-selective device (e.g., a dynamic spectrum equalizer (DSE) [2]) is used to selectively pass any wavelength channel on the pass-through path [3]. Traffic entering the node is split using a 3-dB splitter. The traffic that is directed to the drop port is again split so as to access several receiving stations. Optical filters are used to dynamically direct different wavelength channels to different stations. DSEs are utilized on the pass-through path to selectively block each channel that has been dropped, while also equalizing all the channels that were allowed to pass through. Channels can then be added on the network by using a combiner to combine traffic on different wavelengths from a number of tunable transmitters onto the output fibers.

A second ROADM architecture based on broadcast and select is shown in Figure 20.4. The DSEs are now replaced by wavelength-selective switches

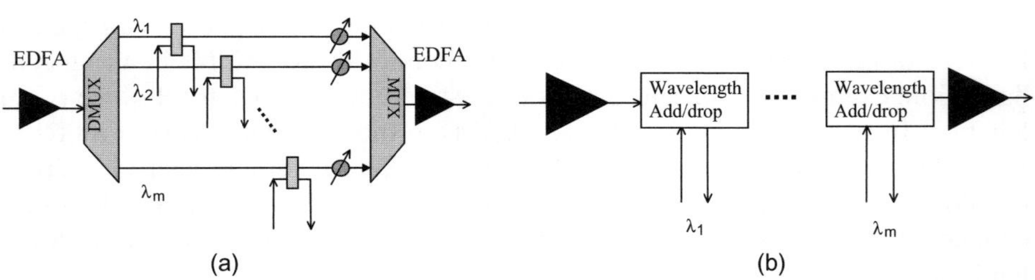

Figure 20.2. **(a)** Parallel and **(b)** serial OADM architectures. (From Stern [8, Figure 4.82]. Copyright 2008 Cambridge University Press. Used by permission of Cambridge University Press.)

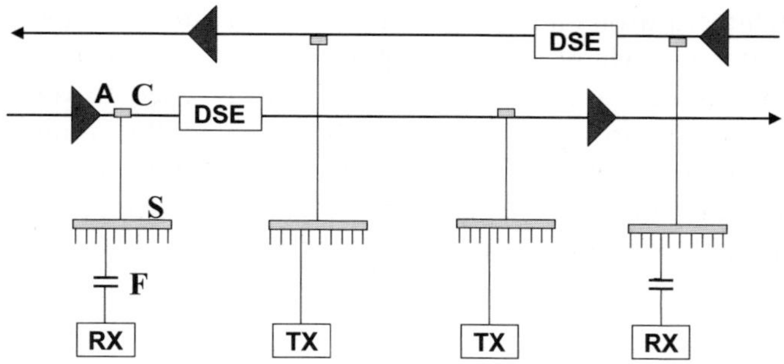

F - Optical filters
S - 1 to N power splitter
A - Optical amplifier
C - 3-dB coupler
DSE - Dynamic Spectrum Equalizer

Figure 20.3. Broadcast and select OADM design. (From Boskovic [3, Figure 2]. Copyright 2002 IEEE. Used by permission of The Institute of Electrical and Electronics Engineers, Inc.)

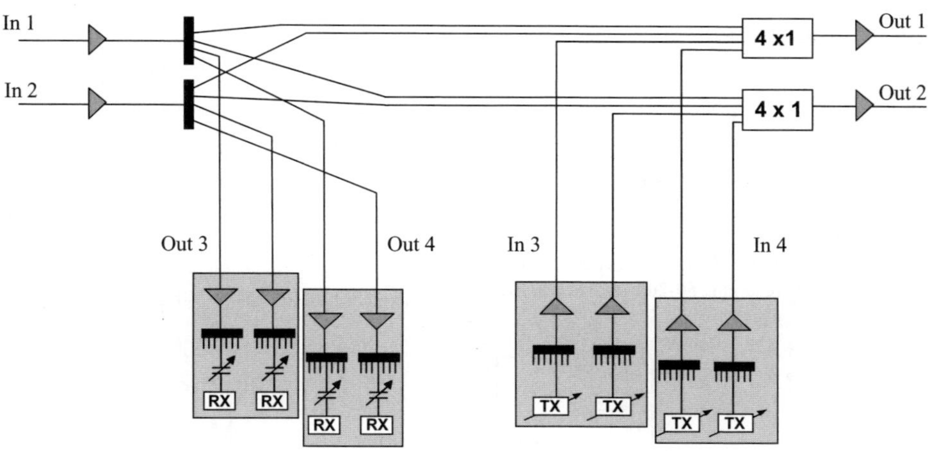

Figure 20.4. Broadcast and Select (B&S) ROADM architecture. (From Antoniades et al. [25, Figure 15]. Copyright 2004 IEEE. Used by permission of The Institute of Electrical and Electronics Engineers, Inc.)

that are used to selectively pass through or add channels on the network. This architecture is more flexible than the previous one because it allows for all paths from an input to an output fiber (because of the wavelength-selective switches used). Signals from the input ports are sent to all output ports using passive splitters. Outputs on the drop side can then receive selected wavelength channels using $1 \times N$ splitters and tunable filters in the same manner described for the architecture of Figure 20.3. This ROADM design mostly resembles a switch architecture where a passive splitter is used to distribute the incoming wavelength channels and controllable devices are used to direct the desired wavelengths (pass-through and added channels) to the appropriate outputs. Clearly, the number of wavelengths that are split/combined on the drop/add sites is dictated by how many channels need to be accessed from each fiber. Typically, it is anticipated that approximately 25% of the traffic will be added/dropped at each network node and the rest will be pass-through traffic.

20.1.4 ROADMs in Networks

Metro architectures traditionally consisted of interconnected fiber SONET/SDH rings as shown in Figure 20.5 [4]. This architecture is similar to the interconnection hierarchy between different networks shown in Figure 20.1. In Figure 20.5 there exist three levels of SONET/SDH ring hierarchy. Add/drop multiplexers (ADMs) on the edge rings are utilized to aggregate traffic from the users, while digital cross-connect switches (DCSs) are used to interconnect edge rings to the larger inter-office (IOF) rings. Finally, IOF rings are also interconnected via DCSs in a ring or mesh fashion.

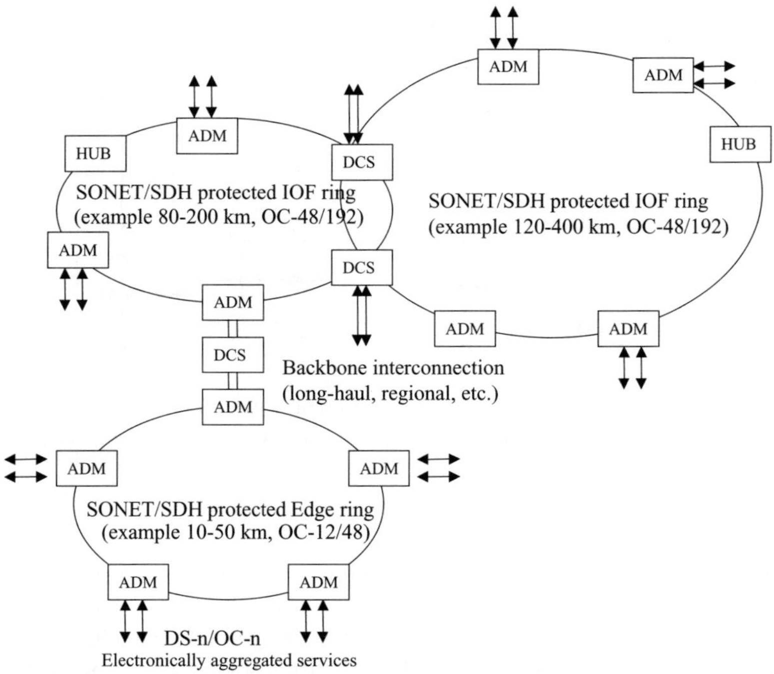

Figure 20.5. Legacy SONET/SDH metro infrastructure. (From Ghani et al. [4, Figure 8.2]. Copyright 2002 Academic Press. Used with permission of Elsevier.)

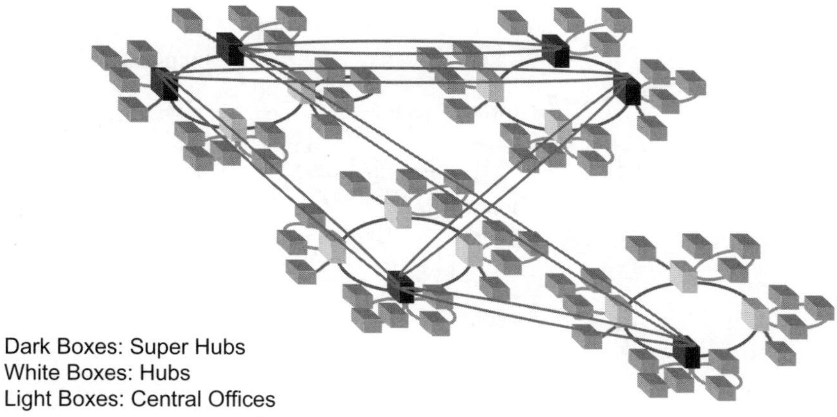

Figure 20.6. ILEC metro ring interconnection architecture. (From Elby [5, Slide 8]. Copyright 2002 Verizon. Used by permission of Verizon.)

These networks have evolved to interconnected optical rings utilizing wavelength division multiplexed (WDM) fiber-optic links as shown in Figure 20.6 for an Incumbent Local Exchange Carrier (ILEC) metro network [5]. The architecture comprises of the feeder sections (IOF sections) that aggregate traffic from distribution sections (that interface with the customers) at the hub or super-hub nodes and send them to the long-haul network.

As demand for single-wavelength services continues, the superhub architecture in a typical metro network of an ILEC evolved from one that uses electronic add/drop multiplexers (ADMs) and digital cross-connect systems (DCS) (Figure 20.7) to ones that utilize ROADM modules as shown in Figure 20.8 [5]. This evolution clearly demonstrates the increased need for flexible and cost-effective ROADMs that will eventually populate all types of networks and architectures.

20.2 NEXT-GENERATION ROADM ARCHITECTURES

20.2.1 Two-Degree Versus Multi-degree ROADMs

In this section we present various ROADM architectures with new configurations. There are two general types of ROADMs, namely two-degree and multi-degree, where the degree refers to the numbers of DWDM fibers entering and exiting the ROADM node. (Note that this refers to only a single direction of traffic entering and exiting the ROADM. In practice, pairs of fibers are generally used, with each set carrying traffic in an alternate direction, so there would be twice as many fibers entering and exiting the ROADM as its degree.) A two-degree ROADM resembles a stretch of highway with off and on ramps to drop off and accept local traffic. Its functions are to terminate an incoming DWDM fiber, drop specified wavelengths and in most cases block these wavelengths from

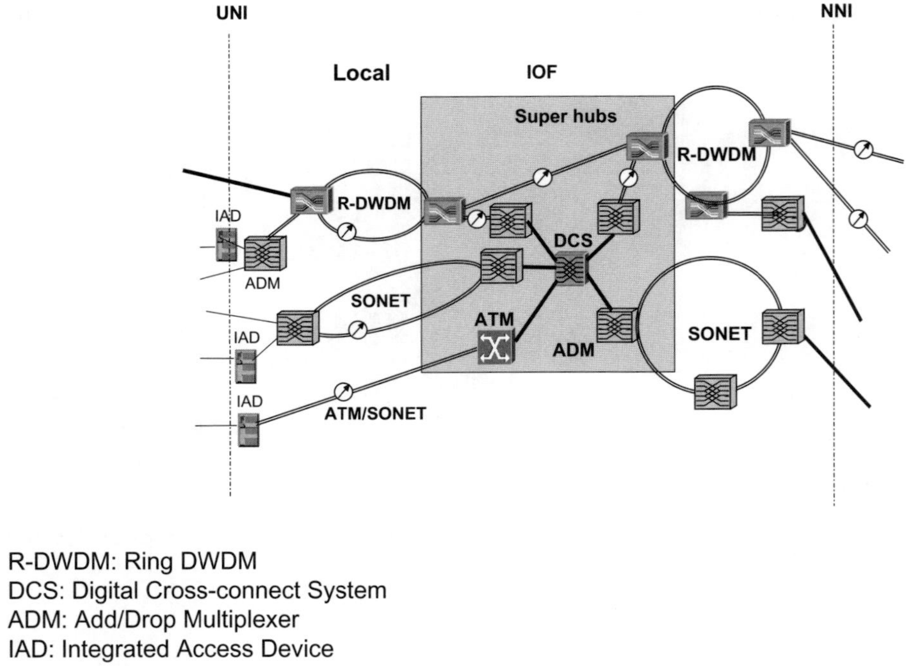

Figure 20.7. Legacy superhub architecture in an ILEC metro network. (From Elby [5, Slide 13]. Copyright 2002 Verizon. Used by permission of Verizon.)

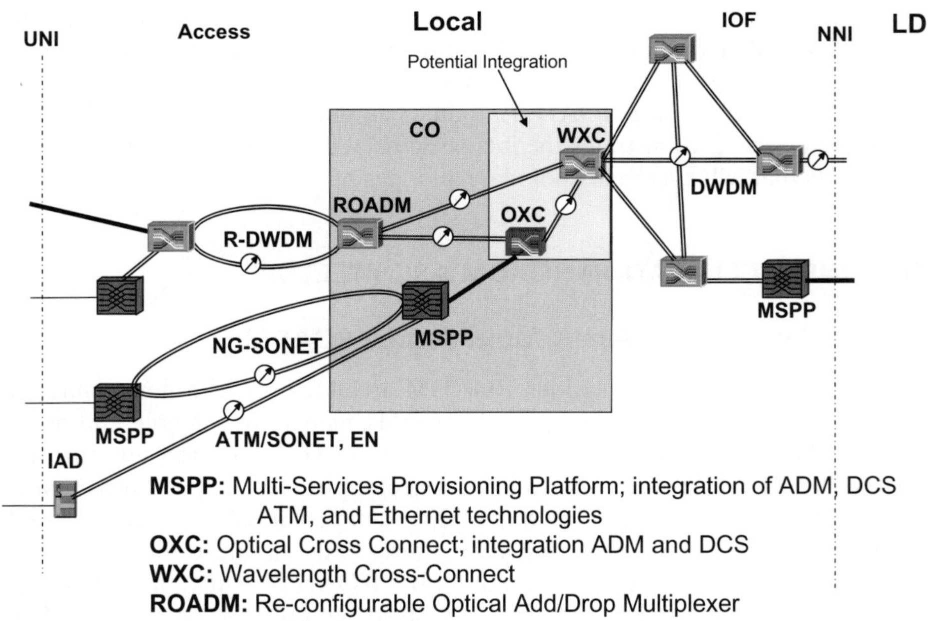

Figure 20.8. Current superhub architecture in an ILEC metro network. (From Elby [5, Slide 14]. Copyright 2002 Verizon. Used by permission of Verizon.)

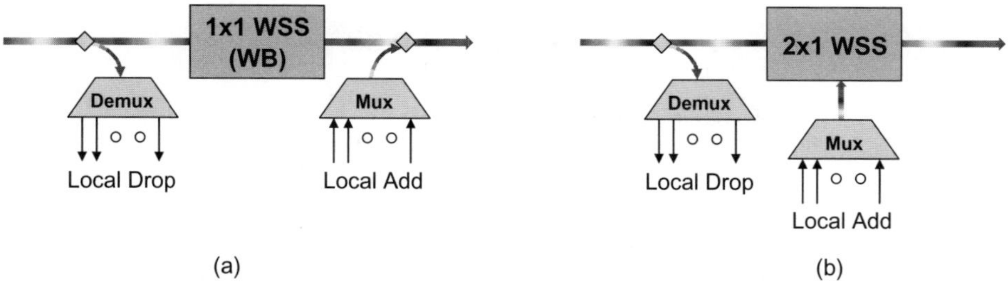

Figure 20.9. Two-degree ROADMs with fixed add/drop. **(a)** Classic broadcast and select architecture using MUX/DMUX on the add/drop sides. **(b)** Modified broadcast and select with the add going through the ports of a 2 × 1 WSS.

propagating further, add local wavelengths, equalize the combined traffic of pass-through wavelengths and added wavelengths, and provide an egress for this traffic toward the next ROADM node. A multidegree ROADM, on the other hand, is like an interchange where highways meet and is used for interconnecting DWDM rings or for mesh networking. Its functions are to accept and rearrange wavelengths from the multiple fibers entering and leaving the multidegree node, as well as adding and dropping local wavelength traffic. The majority of ROADMs expected to be deployed in networks in the coming years will be predominantly two-degree nodes (with a figure of about 75% often cited) that are somewhat less complex than their multidegree counterparts.

As discussed in Section 20.1.2, initial deployments of ROADMs began in the late 1990s with the architecture shown in Figure 20.9a [3]. The multichannel DWDM fiber enters the node and the optical power is immediately split, making all the wavelengths available for a local drop through a demultiplexer. Typically an 80/20 split is used here, resulting in about 1-dB loss on the through path. This through traffic enters a 1×1 WSS (Wavelength Selective Switch) that under automated control either passes through, equalizes, or blocks any or all wavelengths. The WSS or wavelength blocker (WB) architecture has been widely deployed since the turn of the century and is used to block dropped wavelengths so that their slots become available for added wavelengths and also to equalize the optical power of passed-through wavelengths to improve transmission; that is, we use a device that has just one input and one output port so that there is no switching. New wavelengths are added by passive combination after the WSS through a multiplexer, typically using a 60/40 split. The WSS blocks any wavelengths that are the same as the added wavelengths so that there are no two same wavelengths carrying traffic on the same output fiber. Discrete variable optical attenuators (VOAs) are also used to equalize the optical power of the added wavelengths. An optical power monitor (OPM) is further used to provide feedback for the optical power equalization controls of the WSS and the VOAs.

Figure 20.9b shows a variation on this architecture where the locally added wavelengths are still combined at a multiplexer but are now directed to the *Add* port of a 2×1 WSS. In others words, it has two input ports and one common

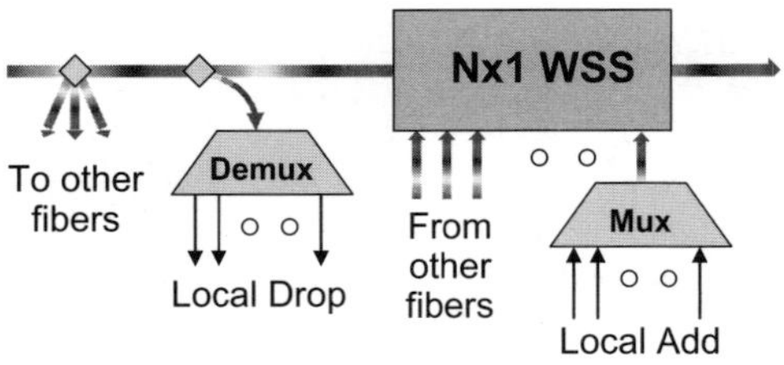

Figure 20.10. Multi-degree (mesh) ROADM with fixed add/drop.

output port. The WSS selects specific wavelengths from either the *In* or the *Add* ports and routes these to the *Out* port for transmission to the next network node. The WSS in this architecture can also individually equalize the optical power of the locally added wavelengths eliminating the need for discrete VOAs as in the design of Figure 20.9a. Both the architectures in Figures 20.9a and 20.9b are termed fixed optical add/drop multiplexers because the dropped and added wavelengths are associated with specific or fixed ports on the multiplexers. While these wavelengths are still connected manually to specific service line cards (e.g., 10-Gigabit Ethernet, SAN protocol, etc.), there is a school of thought that believes that this is not a major concern because this is usually done in conjunction with the manual provisioning of the service line cards themselves. The main advantage of these ROADM architectures is that the multiple wavelengths passing through the node are routed and equalized in an automated fashion.

For multi-degree ROADM nodes, the incremental requirement over two-degree nodes is the ability to send optical traffic to and accept optical traffic from other DWDM fibers. For instance, to interconnect two fiber rings a total of four WSS are required, making this a four-degree node. Figure 20.10 demonstrates a multidegree node with fixed add/drop. The first splitter routes DWDM fibers to the other WSS in the node, and the second splitter drops local traffic as in the case of the two-degree node. An $N \times 1$ WSS is then used to accept traffic from the other DWDM fibers in the node as well as local traffic. For a four-degree node a 4×1 WSS is required, for an eight-degree node an 8×1 WSS, and so forth.

20.2.2 Colorless Capabilities

In the first two architectures the dropped and added wavelengths had fixed physical associations with the demultiplexer and multiplexer ports. Figure 20.11 shows a two-degree ROADM configuration that eliminates this fixed association. This feature is called a "colorless" property because any color (wavelength) can be directed to any *Drop* port and from any *Add* port. In Figure 20.11a, this is achieved using tunable receivers (that nowadays are implemented using a tunable filter feeding a fixed receiver) and tunable transmitters that are passively split from and added to the optical path. Figure 20.11b shows a variation of this architecture using a $1 \times N$ WSS to dynamically select and drop selected wavelengths. For instance, a 1×9 WSS provides an ability to drop any eight wavelengths with

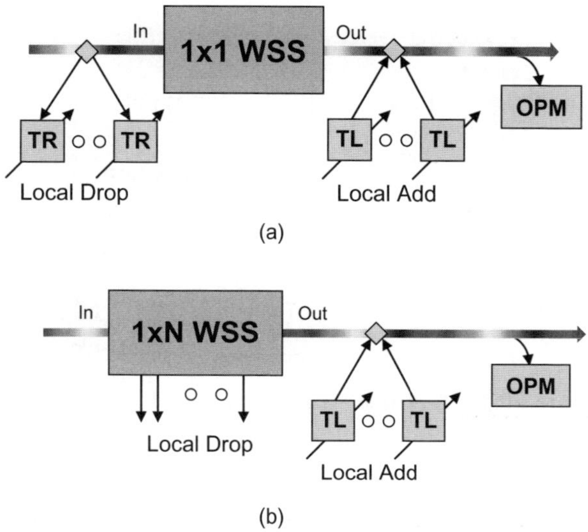

Figure 20.11. Two-degree colorless add/drop ROADM. **(a)** Drop using tunable receivers (TR). **(b)** Drop using ports on a 1 × N WSS. TL, tunable lasers on the transmitter side.

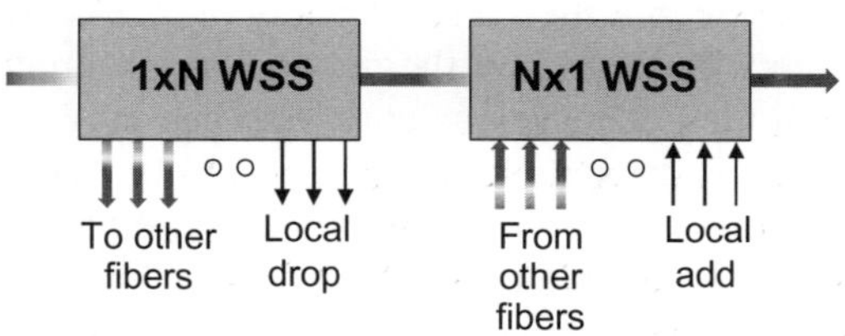

Figure 20.12. Multi-degree ROADM with colorless add/drop.

the ninth port used for the DWDM output fiber. On the add side, a scheme to add wavelengths in a colorless way is shown using tunable lasers that are coupled to the outgoing fiber. Colorless architectures are most efficient if it is anticipated that only a limited number of wavelengths need to be dropped and added at a node due to the optical power losses associated with passive coupling or the fixed port size of the WSS.

Figure 20.12 shows a multidegree colorless add/drop architecture. Here a 1 × N WSS is used to both route the DWDM traffic to the other WSS in the node and also drop individual local wavelengths, and an $N \times 1$ WSS performs the corollary function of accepting traffic from the other DWDM fibers and adding local wavelengths. This architecture has advantages in terms of flexibility and reducing the optical power budget, but of course is much more expensive because it requires two large WSS for each segment of the node; thus in an eight-degree node, for example, a total of 16 WSSs are required.

A variation of the architecture in Figure 20.12 is shown in Figure 20.13 in which a "directionless" functionality is also introduced, where apart from being

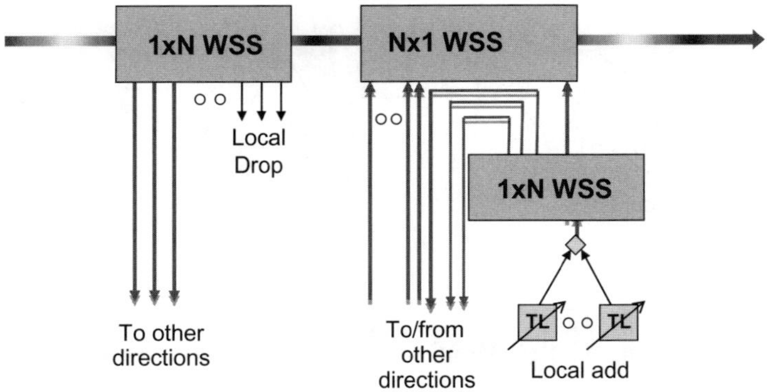

Figure 20.13. Directionless and colorless add/drop ROADM.

colorless, the architecture also allows the insertion of any wavelength in either direction. Clearly, this increased functionality comes at a price, because now WSSs with much larger port count are required and more than two WSSs are needed for each node, thus increasing the cost of the network node.

In summary, it is clear that there are a variety of ROADM architectures and that the usage of each will depend on the network requirements for two-degree versus multi-degree nodes, edge versus core network applications, fixed versus colorless, and fixed versus directionless add/drops. In turn, this is driving a need for a broad range of wavelength selective switching (WSS) engines of different sizes and configurations to enable these ROADM architectures. Different WSS designs and technologies are discussed in detail in Section 20.3.

20.3 WAVELENGTH-SELECTIVE SWITCHING IMPLEMENTATION TECHNOLOGIES

As discussed above, the key enabling technology for current ROADMs is the wavelength-selective switching (WSS) module. WSS is an advanced fiber-optic component that can be used under software control to dynamically select individual wavelengths from multiple DWDM input fibers and switch them to a common output fiber and at the same time perform signal attenuation. A generic WSS module implementation is shown in Figure 20.14 [6] where input/output optics steer the light beam to/out of the device; typically the input/output optics can be fiber and micro-lens arrays. Specialized optics then shape the collimated beam and introduce polarization diversity, typically using prisms, cylinders, birefringent crystals and waveplates, if needed, at the input and output of the device. The above is followed by the dispersion system, which can be a common grating that is polarization-sensitive and thus requires incident light to be of a specific polarization.

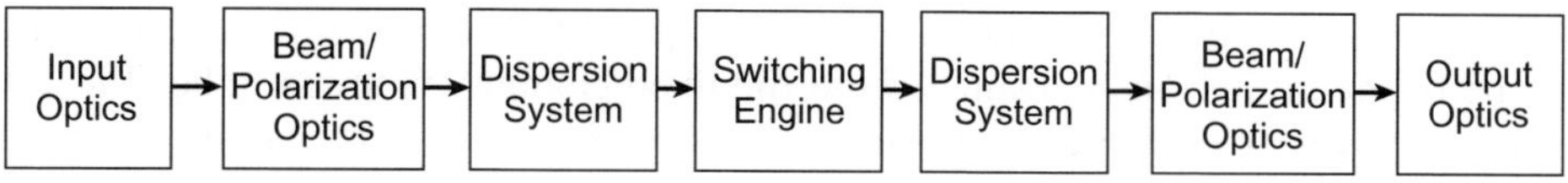

Figure 20.14. Typical wavelength-selective switch (WSS) implementation components.

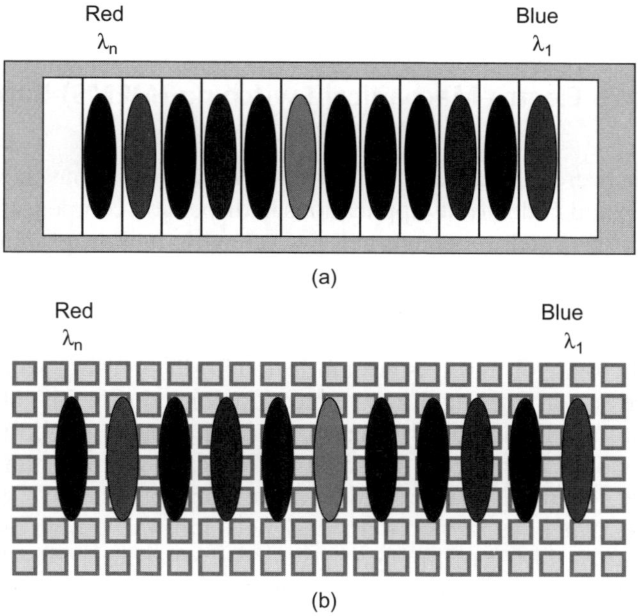

Figure 20.15. **(a)** 1D-switching array configuration. **(b)** 2D-switching array configuration.

At the heart of such a device sits a switching technology that is uniquely engineered to achieve the above required switching functionality. Different products in the marketplace use different switching technologies; however, all available switching engines are able to control wavelength-separated beams using one or more of the following: angle, phase, polarization, and displacement. The effective channel shape of each device is the ratio of the channel spacing (d) in μm and beam spot radius (ω). The optics system focal length and grating dispersion directly affect the channel spacing, whereas the overall optics size affects both the channel spacing and the spot radius. It is clear that the choice of the switching engine will affect the device's main characteristics such as band-pass shape and bandwidth, insertion loss, crosstalk, size, choice of internal optics components, and ultimately the device overall cost.

Depending on the technology, 1D and 2D arrays of switching elements can be constructed to serve as the switching engines in the WSS. Figures 20.15a and 20.15b show the 1D- and 2D-switching array implementations, respectively. In the former, one switching element per wavelength channel is used, whereas in the latter the switching element size is less than the beam spot size. The advantages for the 1D array are simpler control electronics and software, whereas some of the limitations pertain around the fixed channel bandwidth. For the 2D array case, variable channel bandwidth is possible at the expense of more complex electronics and control algorithms.

Below is a breakdown of the most commonly employed switching implementation technologies along with a brief comparison and a set of achievable requirements.

20.3.1 Micro-Electro-Mechanical Switching (MEMS)-Based Approach

These systems were introduced in the early 1990s using various micromachining techniques [7] and had various applications in dispersion compensation, dynamic gain equalization, and reconfigurable wavelength add/drop multiplexing [8, Chapter 4]. However, the predominant application has since been as integrated optical switching fabrics. The main principle of operation in MEMS is that voltage applied to electrodes causes the micromirror to tilt due to electrostatic attraction. Micromirror matrices are fabricated in silicon using wafer scale lithographic processes originating in the semiconductor industry. The above concept is demonstrated in Figure 20.16 where a translation plate is used to "flip" each mirror up or down. Because of the small mirror size, switching times of submilliseconds are achieved. Based on the above concept, the first generation of MEMS-based switching fabrics were designed and are known as 2D MEMS switches or one-axis MEMS. In Figure 20.17a typical crossbar $N \times N$ switch configuration is shown consisting of an integrated array of N^2 moving mirrors.

1D-array switching elements based on MEMS are used in the WSS in the generic configurations shown in Figure 20.18. In both configurations the array switching elements are 2D MEMS switches that have single rotation axes to

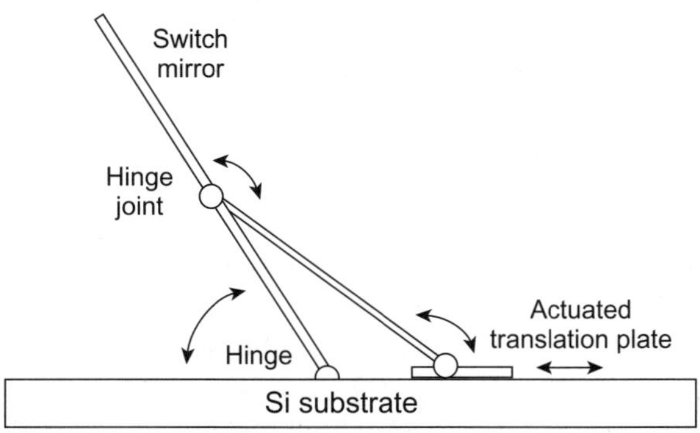

Figure 20.16. Main operation of the micro-machined mirrors. (From Lin [7, Figure 1b]. Copyright 1998 IEEE. Used by permission of The Institute of Electrical and Electronics Engineers, Inc.)

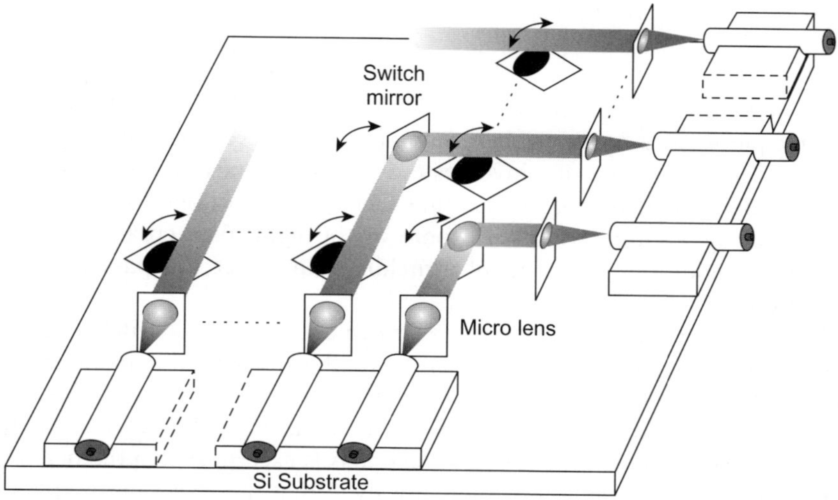

Figure 20.17. Schematic drawing of an $N \times N$ micro-machined free-space matrix switch. (From Lin [7, Figure 1a]. Copyright 1998 IEEE. Used by permission of The Institute of Electrical and Electronics Engineers, Inc.)

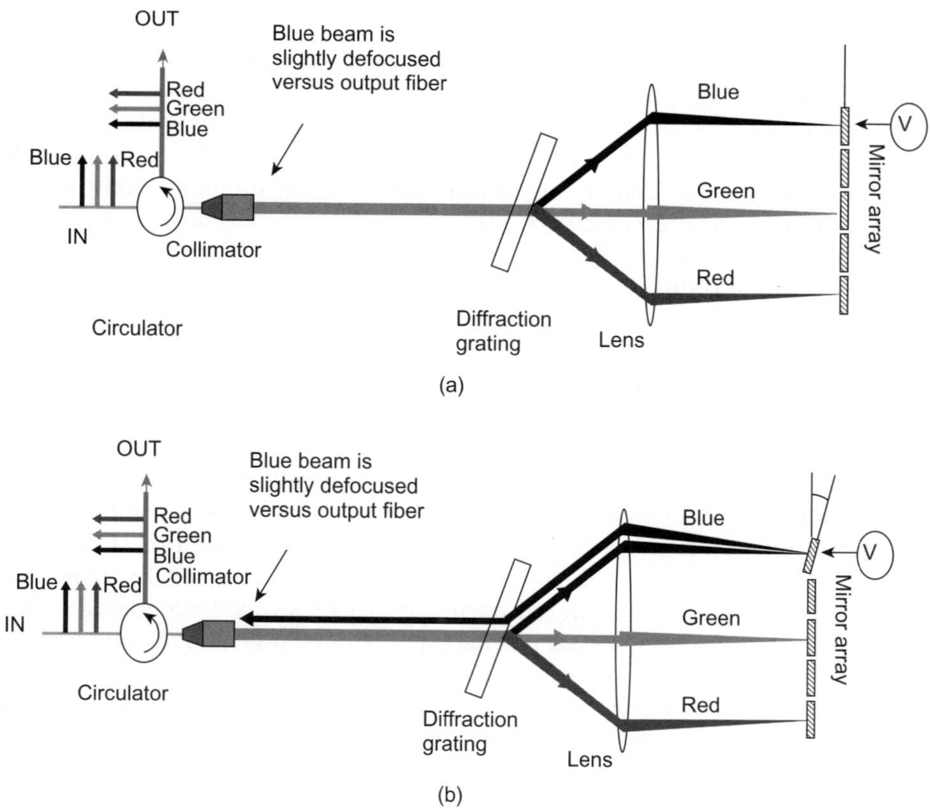

Figure 20.18. **(a)** WSS switching engine optics configuration using 1D-switching array technology. **(b)** Selective attenuation functionality.

slightly deviate the wavelength beam from its nominal path. In Figure 20.18b this principle is used for controlled attenuation; however, the same applies for switching. The main issue that involves the use of one-axis MEMS (2D MEMS) configuration for switching is the fact that each output port that the switch passes through during rotation will "see" some of the light, thus creating unwanted crosstalk. In addition, this kind of switching will affect the passband during attenuation.

The second generation of switches known as 3D MEMS switches (two-axis MEMS) can tilt in two directions [9]. As a result, the effect of dynamic crosstalk during switching mentioned above can be eliminated (allowing for hitless operation) due to the multiple axis rotations. The first axis is used to deviate the beam from output port alignment and the second axis is used to switch the beam from one port to another. It must be noted, however, that the effect on the passband shape during attenuation is still present.

20.3.2 Liquid Crystal (LC) Approach

Liquid crystals (LCs) are materials that have anisotropic refractive properties that can be changed when an electrical field is applied. These materials can thus exhibit special polarization properties when an external voltage is applied. When used along mirrors, polarizers, lenses, and other birefringent (polarization-sensitive) crystals as shown in Figure 20.14, LCs can be used to spatially deflect light signals and modulate their phase and intensity. As a result, these materials can be used as the switching engines in various applications including wavelength-selective cross-connects [2, 10]. One type of LCs in which we are interested for the WSS application is the smectic class of crystals. These exhibit ferroelectric polarization behavior, acting as polarizers whose axis of polarization can be switched between two states by the application of an electric field; they are relatively fast, with switching times on the order of $10\,\mu s$. The principle of operation is shown as a 1×2 switch based on the polarization diversity scheme in Figure 20.19. For each wavelength λ, input light is polarized and sent to an LC slab that can rotate polarization up to $90°$ with the application of voltage. Depending on the routing decision on a per-wavelength basis, the birefringent crystal following the LC will passively route the beam to one of two output ports. A birefringent crystal here is a simple polarization beam splitter (PBS) that separates an incoming beam into the vertically and horizontally polarized components on the outputs. Based on this principle, switching among a larger number of ports can be achieved by enabling the same wavelength, for example, λ_1, from different ports to impinge on the liquid crystal with the wavelength from each port having a different polarization. It must be noted here that a PBS is needed before the LC element in cases where the input light is nonpolarized.

Liquid crystals are most commonly used in spatial light modulators (SLMs) where a slab of LC material sandwiched between other elements is arranged to produce an array of pixels that are controlled through a grid of electrodes to produce prescribed light patterns. An SLM is a device capable of modulating the

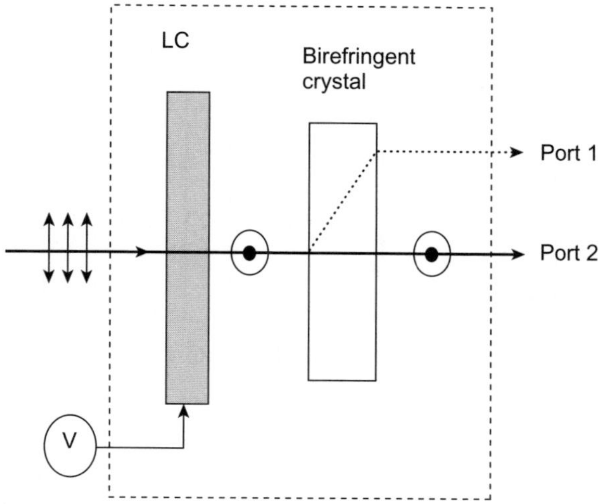

Figure 20.19. The principle of switching using an LC is based on controlling the voltage applied to the LC slab and using a birefringent crystal to route the signal to one of two ports.

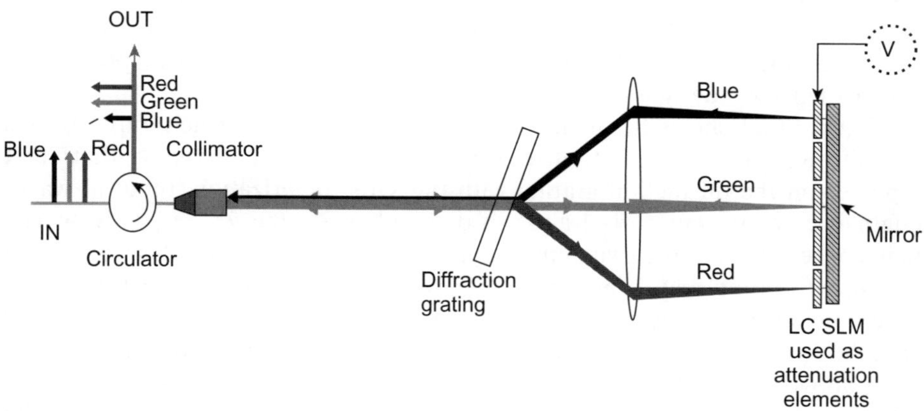

Figure 20.20. An LC SLM used as selective attenuation element in a WSS design.

amplitude, direction, and phase of a beam of light within the active area of the modulator. An implementation of an SLM is the 1D array of switching elements shown in Figure 20.15a using LC elements. In a binary switching configuration N LC cells are needed to be able to select among 2^N possible output ports. The use of a SLM in a WSS is shown in Figure 20.20 as a basic principle where LC elements are used for selective attenuation. In this case the blue wavelength is attenuated by applying a voltage at the top LC pixel. The applied voltage is such that it will partially rotate the polarization to a state in between the vertically and horizontally polarized states effectively achieving controlled attenuation.

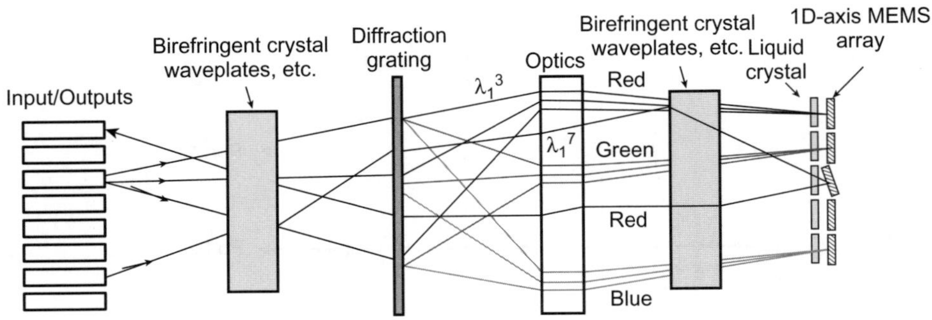

Figure 20.21. An LC+MEMS hybrid approach WSS switching engine.

The clear positive feature of an LC switching engine is the fact that it does not contain any movable parts, which tend to always raise reliability issues.

20.3.3 Liquid Crystal (LC) + MEMS Hybrid Approach

A unique and novel approach in building cost-effective and scalable WSSs is to use a combination of LC and MEMS technologies as discussed in reference 11. That work discusses this hybrid technique in the context of high port WSSs (larger than 1×4 designs).

Figure 20.21 shows the main concept for a 1×8 WSS switch. Input beams pass through a series of birefringent/waveplate combinations so that light impinges on the diffraction grating with the same polarization. Input at wavelength λ_1 coming from ports 1 and 2 is denoted as λ_1^1 and λ_1^2, respectively. The main concept behind the switching is that the 1D-MEMS array will switch λ_1^2 to a different output—say 1, becoming λ_1^1 by simple tilting of the mirror. At the same time during the switching, the LC slab will completely eliminate the power from the switched channel to avoid any issues with crosstalk discussed in Section 20.3.1 above. This is done based on the selective attenuation concepts discussed in Section 20.3.2. Once the switching to the output port has been achieved, the LC slab will condition the output power to the desired level. As a result of the above process, the original signal passband shape is maintained.

20.3.4 Liquid Crystal on Silicon (LCoS) Approach

Liquid crystal on silicon (LCoS) is a display technology that combines liquid crystal and semiconductor technologies to create a solid-state switching engine with very high resolutions. Liquid crystals are applied directly on the silicon, hence the name of the technology: liquid crystal on silicon (LCoS). As shown in Figure 20.22, LC elements are sandwiched between a silicon wafer and the top glass. The actual implementation of the switching engine follows the 2D-switching array described in Figure 20.15b. This 2D matrix of LC elements

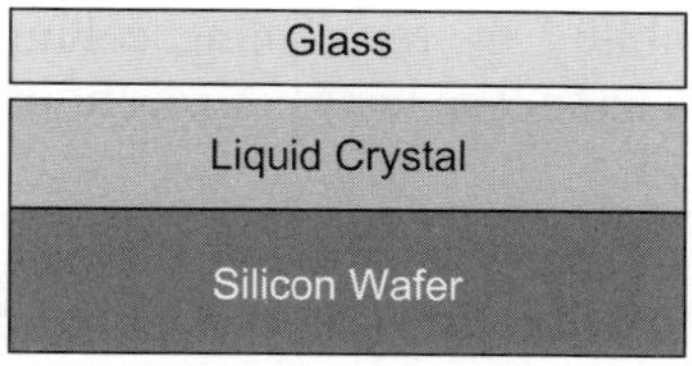

Figure 20.22. An LCoS cross section.

Figure 20.23. Schematic of optical design of LCoS-based WSS. (From Baxter et al. [13, Figure 1]. Copyright 2005 IEEE. Used by permission of The Institute of Electrical and Electronics Engineers, Inc.)

referred to as LCoS controls the phase of the incident light by changing the voltage that controls each pixel of the array. The theory behind these controls was first demonstrated in reference 12. An implementation of a WSS switching engine using the LCoS technology is shown in Figure 20.23 [13]. Light enters the switch and goes through the polarization imaging optics that ensure maximum efficiency polarization when impinging the conventional diffraction grating. The input light from a chosen input fiber is reflected from the imaging mirror and then angularly dispersed by the conventional grating reflecting the light back to the optics which route each wavelength to a different portion of the LCoS 2D array. The path for each wavelength is then retraced upon reflection from the LCoS, with the beam-steering pitch and phase pattern of the LCoS routing the light to a particular port of the fiber array with the proper attenuation. Wavelength channels are separated on the LCoS, and thus the switching of each wavelength is independent of all others and shows no signs of crosstalk.

The main principle of operation in a WSS using LCoS as the optical switching engine is light diffraction through the grating formed by the LC elements and the fact that the phase of light at each pixel is controlled to produce an electrically programmable phase front modification. A change in the phase results in controlling the attenuation applied on each channel, whereas a change in the pitch of the LCoS array (distance between the LC pixels) changes the wavelength selectivity. The pixilated 2D LC array creates a stepped phase response on each wavelength; and just like in a conventional grating the line spacing controls the diffraction properties, in LCoS the pixel pitch is the corresponding parameter. The 2D nature of the switching array technology allows for variable channel capability by reassigning columns of pixels to different channels as well as mixed channel spacings. A recent overview of the technology is presented in reference 14.

20.3.5 Digital Light Processing (DLP) Approach

In Section 20.3.1 the WSS switching engine uses MEMS components whose mirrors have a relatively large surface area on the order of 1- to 2-mm mirror diameter. These are structured in a 1D switching array configuration and are commonly referred to as "analog" mirrors. In Section 20.3.4, LC elements are used in a 2D-switching matrix configuration to form digital pixels for light diffraction. A combination of the above two technologies uses a pixilated SLM comprised of tiny mirrors to achieve the switching. The discrete pixel elements are micrometer-size mirrors that are thus referred to as "digital" mirrors. The above array, which acts as a "light switch" and consists of thousands of individually addressable, tiltable, mirror-pixels, is called a digital micromirror device (DMD) [15]. Typical pixels that make-up a DMD are shown in Figure 20.24. The physics that govern the operation are described in detail in reference 15 and are based on the fact that the DMD is a pixelated reflector that behaves like a diffraction grating with the maximum power reflected (diffracted) in a direction dependent on pixel pitch, wavelength, and angle of incidence. Light modulation is thus achieved, and switching and attenuation are performed with software control of the tiny mirrors and proper optics alignment as described in all the previous sections.

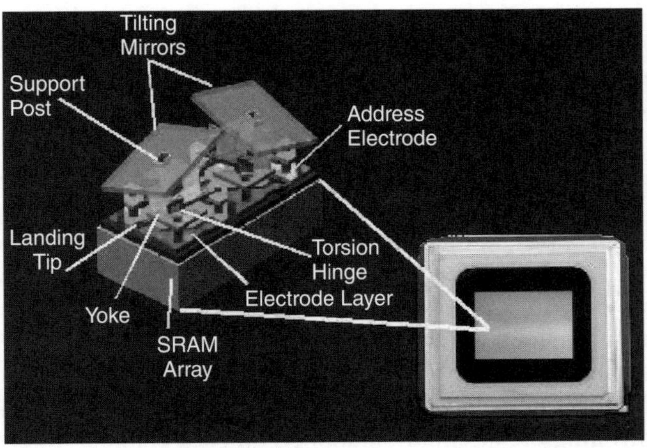

Figure 20.24. Schematic of two DMD mirror pixels next to a typical DMD light modulator consisting of 1024 × 768 individually addressable mirror pixels (From Yoder et al. [15, Figure 1]. Copyright 2005 IEEE. Used by permission of The Institute of Electrical and Electronics Engineers, Inc.)

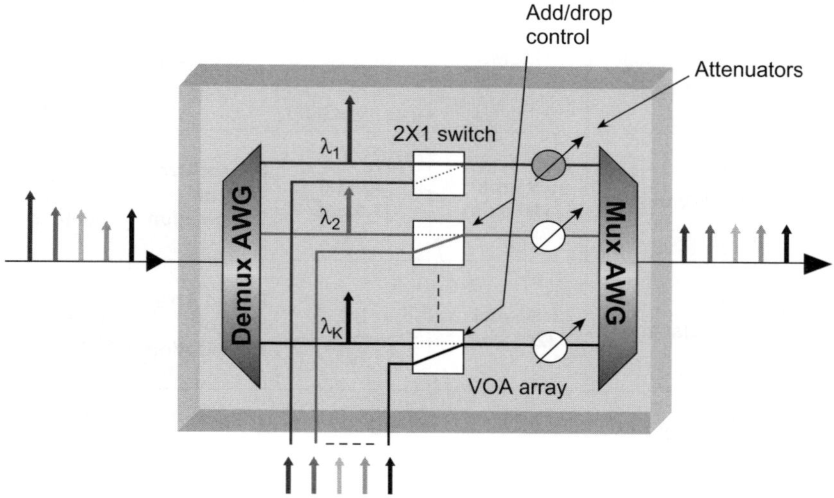

Figure 20.25. Typical functionalities of an IPLC device used in today's optical networks.

20.3.6 Integrated Planar Lightwave Circuit (IPLC)

This is a rather "old" technology that appeared in the late 1990s–early 2000s timeframe that offered exceptional integration level based on silica over silicon (SiO_2/Si) polymers. The technology provides full-wavelength access with local add/drops, switching, equalization, and blocking, all on a single chip as shown in Figure 20.25. IPLCs are now a very mature technology and have been widely deployed especially in ROADM designs in metro and edge serving 100-GHz channel spacing 10-Gbit/s systems. Their compatibility for 40-Gbit/s systems has been their main drawback so far.

20.3.7 Requirements and Technology Comparisons

Figure 20.26 summarizes the operation principle of each WSS switching technology described above in relation to the desired switching and attenuation functionalities. Figure 20.27 provides some typical optical performance specification requirements for these switching technologies. Commercial products are now available for all the above-mentioned technologies, and most of the products target the performance guidelines of Figure 20.27. Clearly, the choice of technology depends on the capability of the device to meet these specifications and also depends on the application space for which the WSS is intended. The "edge" and the "core" of optical networks are two areas which, although tough to accurately define, represent both application spaces for WSSs. In the core, which might represent the main metropolitan optical ring or even mesh optical topology, the switch port requirement (as will be discussed in the next section) is larger. In the

	Liquid Crystal (LC)	2-Axis MEMS (i.e., 3D MEMS)	Hybrid: 1-Axis MEMS plus LC (i.e., 2D MEMS)	Liquid Crystal on Silicon (LCoS)	Planar Lightwave Circuit (PLC)
Switching	**Polarization**	**Spatial (1st axis)**	**Spatial (1 axis)**	**Phase Array or Polarization**	**Waveguide 2×1 switches**
Attenuation	**Polarization**	**Spatial (2nd axis)**	**LC Polarization**	**Phase Array or Polarization**	**Waveguide VOA arrays**

Figure 20.26. Summary of WSS switching technology operation principles.

Performance Parameter	Through Path	Add/Drop Path
Supported bands	C first then L	Same
Channel spacing	100 GHz	Same
Number of channels	>44	Same
Passband @ 0.5 dB	±25 GHz	±20 GHz
Insertion loss	<6 dB	<6 dB
PDL	<0.3 dB	<0.5 dB
CD	±10 ps/nm	Same
PMD	<0.2 ps	Same
Switching speed	<100 ms	Same
Operating temperaure	–5°C to 70°C	Same

Figure 20.27. Switching technology typical optical performance specification requirements.

edge, which can be the typical collector ring of an optical network, interconnection port requirements are smaller.

Figures 20.28 and 20.29 provide a comprehensive comparison and ranking of the above switching technologies for the edge and core application spaces, respectively. For low switch port count requirements (more representative of the edge environment), LC technology is ranked the highest mainly because of control simplicity, cost, and upgrade to higher bit rate considerations. Red color-coded areas represent weaknesses for the corresponding technology, whereas yellow color-coded ones represent borderline performance. For high switch port count requirements (more representative of the core environment), the hybrid

	LC	2D MEMS	1D MEMS plus LC	LCoS	IPLC
Passband	☑	☑	☑	☑	☑
40/100 Gbit/s support	☑	☑	☑	☑	
Insertion loss	☑	☑	☑	☑	☑
PDL	☑	☑	☑	☑	☑
Crosstalk	☑	☑	☑	☑	☑
Control complexity	☑		☑	☑	☑
Volume manufacturing	☑				☑
Cost	☑			☑	☑
Overall rank	**1**	**5**	**4**	**3**	**2**

Figure 20.28. Low-port-count edge WSS technology ranking. Dark-shaded areas represent weaknesses for the corresponding technology, whereas the light-shaded ones represent borderline performance.

	Multi-Layer LC	2D MEMS	1D MEMS plus LC	LCoS	IPLC
Passband		☑	☑	☑	
40/100 Gbit/s support		☑	☑	☑	
Insertion loss	☑	☑	☑	☑	☑
PDL	☑	☑	☑	☑	☑
Crosstalk	☑	☑	☑	☑	☑
Control complexity	☑	☑	☑		☑
Scaling # of ports (20+)		☑	☑		
Volume manufacturing	☑	☑	☑	☑	☑
Overall rank	**4**	**2**	**1**	**3**	**5**

Figure 20.29. High-port-count core WSS technology ranking.

approach of LC and one-axis MEMs ranks the highest mainly because of scaling and upgrading considerations with the other competing technologies.

20.4 ROADMS IN NETWORKS

20.4.1 Case Studies

Interoffice (IOF) rings that traditionally span tens to hundreds of kilometers in the metro to regional networking space as shown in Figure 20.1 are at the heart of new RBOC LATAs (Regional Bell Operating Company Local Access and Transport Areas). Traffic from central offices in these LATAs is hubbed through collector rings at the hubs or superhubs of these networks. A typical view of a current superhub in an Incumbent Local Exchange Carrier (ILEC) network is shown in Figure 20.8. The introduction of ROADM modules and latest wavelength cross-connects (WXCs) at the superhub nodes has led to integration of transmission and switching technologies as described in reference 16. Multiservice Provisioning Platforms (MSPP), Multiservice Transport Platforms (MSTP) and Multiservice Switching Platforms are being integrated to the MSxP term as discussed in reference 17, where a traditional RBOC network is presented in the form of a case study. The LATA network consists of between 50 and 150 central offices, and various case studies are presented in terms of the network interconnections of the hub and superhub sites. In Figure 20.30 a typical network is presented along with the currently implemented network scenario that interconnects the 13 hubs via three- to five-node small SONET rings. The introduction of next-generation ROADMs as discussed in this chapter enables DWDM among

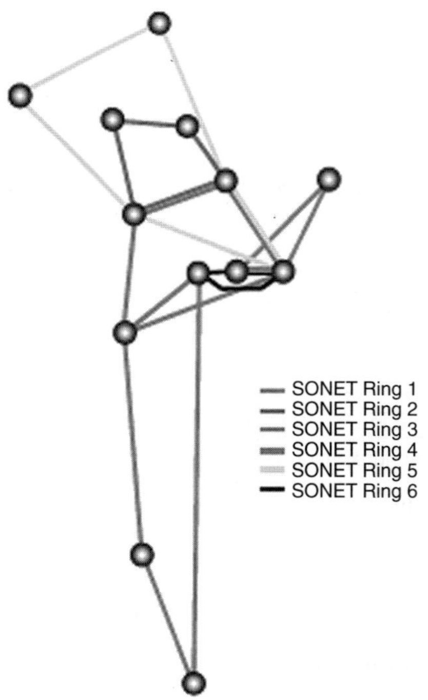

Study Network:
- Typical RBOC LATA
- 150 nodes
- 284 fiber links

PMO:
- 13 hub sites interconnected via 3-5 node SONET rings
- No optical bypass between rings
- SONET and DWDM in separate network elements

Figure 20.30. Currently implemented multiple ring interconnection of hub sites in a LATA case study. PMO, present mode of operation. (From Bonenfant et al. [17, Figure 2]. Copyright 2005 IEEE. Used by permission of The Institute of Electrical and Electronics Engineers, Inc.)

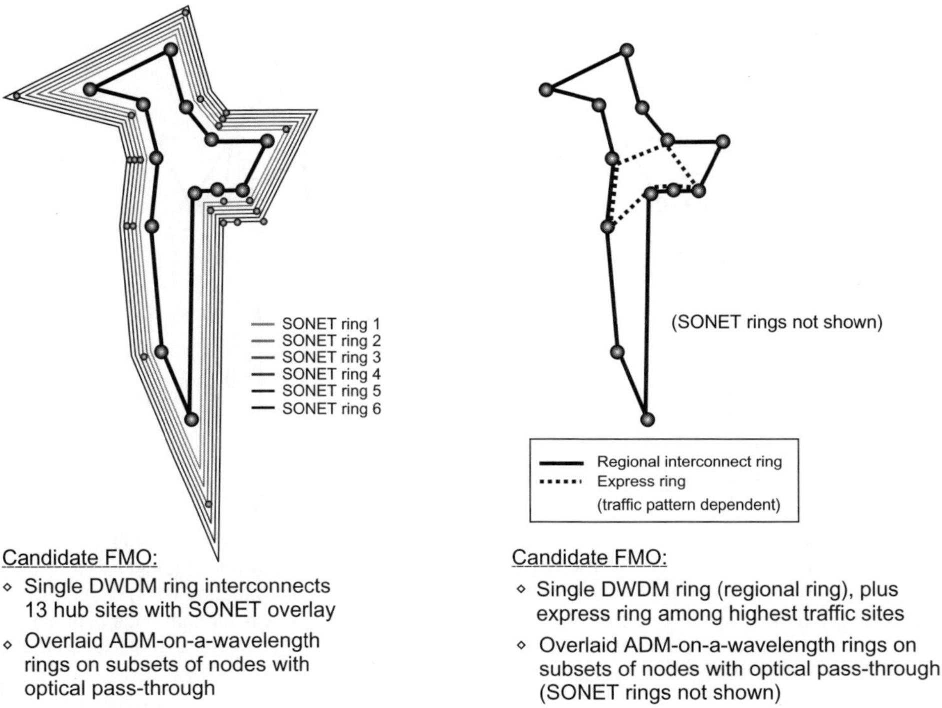

Figure 20.31. Future candidate implementation design showing single regional ring or single regional ring and express ring for interconnection of hub sites in a LATA case study. FMO, future mode of operation. (From Bonenfant et al. [17, Figure 3]. Copyright 2005 IEEE. Used by permission of The Institute of Electrical and Electronics Engineers, Inc.)

rings and more effective architectures. For example, Figure 20.31 presents the case where all hub nodes in the LATA are interconnected using a single DWDM ring. In this case, overlaid SONET rings on subsets of nodes are enabled using optical pass-throughs, and/or express rings among superhubs (highest load nodes) can be established.

Another alternative of the case study is to use the concept of smaller rings as currently used in SONET (Figure 20.30), however, we optically interconnect these rings to leverage the multi-degree capabilities of modern ROADMs as shown in Figure 20.32. In this case, logical SONET rings can be created generating a shift toward a highly interconnected DWDM mesh. On all the above, considerations of cost, flexibility, protection and efficiency are central to the choice of the selected scenario and are discussed in detail in reference 16.

20.4.2 ROADM Commercial Deployments

Over the last five years, long-haul as well as regional-to-metropolitan network transport has seen a tremendous number of both laboratory and commercial

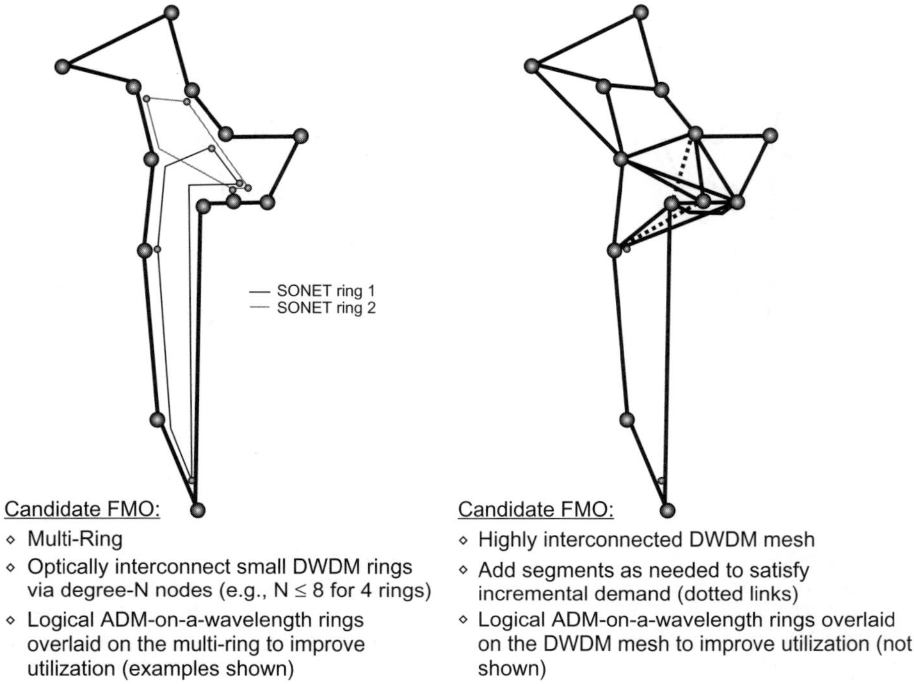

Figure 20.32. Future candidate implementation design showing multi-ring or highly interconnected mesh for interconnection of hub sites in a LATA case study. (From Bonenfant et al. [11, Figure 4]. Copyright 2005 IEEE. Used by permission of The Institute of Electrical and Electronics Engineers, Inc.)

deployments of high-speed optical networks containing ROADMs. In the laboratory, new modulation formats and long-haul propagation dominate the scene. In reference 18 a 50-Gbaud DQPSK has been transmitted over 1200 km including six ROADMs on a 100-GHz grid, whereas in reference 19 a 25-Gbaud Quaternary Phase Shift Keying (QPSK) in combination with coherent detection has been transmitted over 2400 km on a 50-GHz grid. Baud represents the use of advanced modulation formats where multiple bits per symbol can be transmitted; clearly, lower symbol rates result in narrower signal spectra, which can more easily be deployed in existing optically routed networks due to the reduced optical filtering concatenation effect in ROADMs. In reference 17, ROADMs are implemented using a combination of $1 \times N$ WSS devices. One ROADM uses WSS in LCoS technology, whereas another uses a WSS that combines IPLC and MEMS technologies. In reference 20, polarization-division multiplexed Differential QPSK (DQPSK) at 25 Gbaud is used to demonstrate transmission over 1280 km of SSMF and four ROADMs.

An early example of ROADM deployment in a network setting was the DRAGON network, which stands for Dynamic Resource Allocation over Generalized Multiprotocol Label Switching (GMPLS) Optical Network [21].

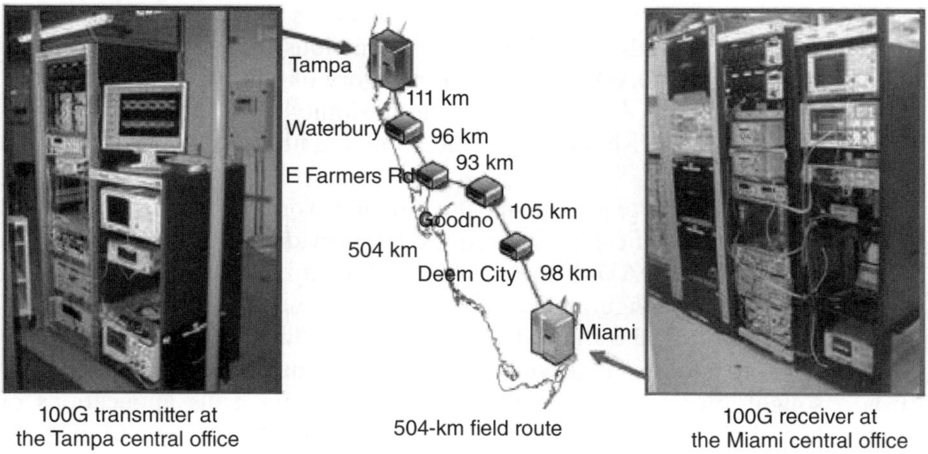

Figure 20.33. The field trial conducted over live Verizon data traffic on a Tampa to Miami Florida 504-km segment. (From Xia et al. [22, Figure 1]. Copyright 2008 IEEE. Used by permission of The Institute of Electrical and Electronics Engineers, Inc.)

DRAGON was essentially a GMPLS grid network with an experimental testbed deployed in the Washington, DC metro area to interconnect various campuses and agencies. The deployed ROADMs were first-generation WSS devices based on the MEMS technology described above. DRAGON was aimed at developing dynamic connection-oriented transport capabilities to support emerging globally distributed e-Science applications, and it materialized as an all-optical GMPLS-controlled metro-area network.

In reference 22 a commercial deployment of ROADMS is demonstrated in the Verizon network through a field trial transmission of 107-Gbit/s DQPSK over a 504 km of the commercial LambaXtreme transport system. The field trial was demonstrated on a live Verizon network segment between Miami and Tampa, Florida with a test 100G DQPSK channel inserted along five 10-Gbit/s On-Off Keying (OOK) channels carrying Verizon live traffic. The ROADMs were based on asymmetric interleavers with alternating pass-bands of approximately 60-GHz and 35-GHz 3-dB bandwidth to support alternating-channels of 40-Gbit/s and 10-Gbit/s binary modulation. Figure 20.33 shows the details of the optical path used for the field trial.

In reference 23, AT&T's photonic network is described and illustrated. The network provides dedicated MPLS services across the globe and is OC-768-capable. Modern ROADM technologies are used throughout the regional side of the network enabling local add/drops and pass-through traffic along each route.

20.5 CONCLUSIONS—FUTURE TRENDS

Although the all-optical vision of the late 1990s has taken a while to fully materialize due to the unprecedented economic downturn of the early 2000s as well as recent shift in carrier's needs, transport networks continue to evolve and equipment with increased optical features and functions are constantly deployed. As presented above, ROADMs fall in this category of equipment whose added functionalities will reduce cost, increase the flexibility and efficiency of carrier networks, speed up provisioning time, and eliminate human error from manual reconfiguration.

Specifically, in order to provide meaningful interworking between the IP and the optical layers, the optical layer is required to provide flexibility, or the ability for reconfigurations. ROADMs are proving to hold a key position in providing this flexibility and reconfigurability. Specifically, in the metro and regional environment where statistical traffic variations are higher than long haul (due to the presence of more nodes and the constant add/drop requirements), reconfigurability is a sought-after feature by the network designers. Thus applications of ROADMs and control planes promise to have a particularly powerful effect in these areas.

A variety of ROADM architectures are emerging to fulfill different requirements for two-degree versus multi-degree nodes, edge versus core network applications, and fixed versus colorless add/drop. In turn, this is driving a need for a broad range of WSS engines to enable these ROADM architectures, including LC-based, MEMS-based, and a mix of liquid crystal and one-axis MEMS core technologies to provide a family of scalable WSS.

Currently, WSS size is limited and does not allow the amount of connectivity that will be required for the large-scale photonic networks predicted in the future. As the technology matures, the degree of the ROADMs will continue to grow and will allow for more connectivity to avoid transponders in the way of service provisioning. A current area of research involving ROADMs is that of dispersion compensation. Current ROADM designs are capable of signal conditioning (equalization and amplification), so the next critical functionality is dispersion compensation. A particularly interesting technique demonstrated by Fujitsu and discussed in Doerr [24, slides 58 and 59] uses a virtually imaged phased array to do the dispersion compensation. The use of 3D mirror technology and optics similar to that of WSS will create a technology that can easily be integrated in current WSS designs. It remains to be seen if dispersion compensation within ROADMs will make it to a commercial product.

Based on the network architecture design, reconfigurable optical add/drop multiplexers with automated rearrangement of wavelengths on multichannel optical fibers entering and leaving the optical network nodes will definitely play a key role in the development of intelligent, flexible, and reconfigurable optical networks with advanced provisioning, routing, and restoration functionalities. Even though wavelength-selective cross-connect switches can also be installed to enable dynamic reconfigurability of wavelength services, it is expected that next-generation ROADMs will be widely deployed in next-generation metro, regional, and access networks.

REFERENCES

1. R. E. Wagner, R. C. Alferness, A. A. M. Saleh, and M. S. Goodman, MONET: Multiwavelength optical networking, *IEEE/OSA J. Lightwave Technol.*, Vol. 14, No. 6, pp. 1349–1355, 1996.
2. A. R. Ranalli, B. A. Scott, and J. P. Kondis, Liquid crystal-based wavelength selectable cross-connect, in *Proceedings of the European Conference on Optical Communications*, Vol. I, Nice, France, September 1999, pp. 68–69.
3. A. Boskovic, M. Sharma, N. Antoniades, M. Lee, Broadcast and select OADM nodes application and performance trade-offs, in *Proceedings of the IEEE/OSA Optical Fiber Communications Conference*, 2002, Anaheim, CA, pp. 158–159.
4. N. Ghani, J.-Y. Pan, and X. Cheng, *Optical Fiber Telecommunications*, Academic Press, Vol. IV-B, San Diego, CA, 2002, Chapter 8.
5. S. Elby, Public network reliability and resiliency, Presentation at Columbia University, http://www2.cvn.columbia.edu/course/elene6901/session/S.ElbyNetworkReliability TalkPT1.ppt, October 30, 2002.
6. P. Wall, P. Colbourne, C. Reimer, and S. McLaughlin, WSS switching engine technologies, in *Proceedings of the IEEE/OSA Optical Fiber Communication Conference*, paper OWC1, San Diego, CA, February 2008.
7. L. Y. Lin, Micro-machined free-space matrix switches with sub-millisecond switching time for large-scale optical cross-connect, in *Proceedings of the IEEE/OSA Optical Fiber Communication Conference*, paper WH5, San Jose, February 1998, pp. 147–148.
8. T. Stern, G. Ellinas, and K. Bala, *Multiwavelength Optical Networks: Architectures, Design, and Control*, 2nd edition, Cambridge University Press, New York, 2008, Chapter 4.
9. D. S. Greywall, P. A. Busch, F. Pardo, D. W. Carr, G. Bogart, and H. T. Soh, Crystalline silicon tilting mirrors for optical cross-connect switches, *J. Microelectromechanical Systems*, Vol. 12, pp. 708–712, 2003.
10. J. S. Patel and Y. Silberberg, Liquid crystal and grating-based multiple-wavelength cross-connect switch, *IEEE Photonics Technol. Lett.*, Vol. 7, No. 5, pp. 514–516, 1995.
11. J. Homa and K. Bala, ROADM architectures and their enabling WSS technology, *IEEE Commun. Mag.*, Vol. 46, No. 7, pp. 150–153, June 2008.
12. K. Johnson, D. J. McKnight, and I. Underwood, Smart spatial light modulators using liquid crystals on silicon, *IEEE J. Quantum Electron.*, Vol. 29, No. 2, pp. 699–714, February 1993.
13. G. Baxter, S. Frisken, D. Abakoumov, H. Zhou, I. Clarke, A. Bartos, and S. Poole, Highly programmable wavelength selective switch based on liquid crystal on silicon switching elements, in *Proceedings of the IEEE/OSA Optical Fiber Communications Conference*, paper OTuF2, Anaheim, CA, March 2006.
14. S. Frisken, Advances in liquid crystal on silicon wavelength selective switching, in *Proceedings of the IEEE/OSA Optical Fiber Communications Conference*, paper OWV4, Anaheim, CA, March 2007.
15. L. Yoder, W. M. Duncan, E. M. Koontz, J. So, T. A. Bartlett, B. L. Lee, B. Sawyers, D. A. Powell, and P. Rancuret, DLP technology: Applications in optical networking, in *Proc. SPIE*, Vol. 4457, pp. 54–61, 2001.

16. S. Elby, An ILEC's view of next generation networking, in *Proceedings of the IEEE/OSA Optical Fiber Communications Conference*, paper ThAA1, pp. 606–607, Atlanta, GA, March 2003.
17. P. Bonenfant, S. Gringeri, and B. Basch, Optical transport network evolution: The MSxP reality check, and opportunities going forward, in *Proceedings of the IEEE/OSA Optical Fiber Communications Conference*, paper NTuG3, Anaheim, CA, March 2005.
18. P. J. Winzer et al., 10 × 107-Gb/s NRZ-DQPSK Transmission at 1.0 b/s/Hz over 12 × 100 km including 6 optical routing nodes, in *Proceedings of the IEEE/OSA Optical Fiber Communications Conference*, paper PDP24, Anaheim, CA, March 2007.
19. C. R. S. Fludger et al., 10 × 111 Gbit/s. 50 GHz spaced, POLMUX-RZ-DQPSK transmission over 2375 km employing coherent equalization, in *Proceedings of the IEEE/OSA Optical Fiber Communications Conference*, PDP22, Anaheim, CA, March 2007.
20. S. Chandrasekhar et al., Hybrid 107-Gb/s polarization-multiplexed DQPSK and 42.7-Gb/s DQPSK transmission at 1.4-bits/s/Hz spectral efficiency over 1280 km of SSMF and 4 bandwidth-managed ROADMs, in *Proceedings of the European Conference in Optical Communications*, PDS 1.9, Dresden, Germany, 2007.
21. http://dragon.maxgigapop.net/twiki/bin/view/DRAGON/WebHome.
22. T. J. Xia, G. Wellbrock, W. Lee, G. Lyons, P. Hoffmann, T. Fisk, B. Basch, W. Kluge, J. Gatewood, P. J. Winzer, G. Raybon, T. Kissel, T. Carenza, A. H. Gnauck, A. Adamiecki, D. A. Fishman, N. M. Denkin, C. R. Doerr, M. Duelk, T. Kawanishi, K. Higuma, Y. Painchaud, and C. Paquet, Transmission of 107-Gb/s DQPSK over Verizon 504-km commercial LambdaXtreme transport system, in *Proceedings of the IEEE/OSA Optical Fiber Communications Conference*, paper NMC2, San Diego, CA, February 2008.
23. K. Tse, AT&T's photonic network, in *Proceedings of the IEEE/OSA Optical Fiber Communication Conference*, paper NMC1, San Diego, CA, February 2008.
24. C. R. Doerr, Optical compensation of system impairments, in *Proceedings of the IEEE/OSA Optical Fiber Communications Conference*, paper OThL1, Anaheim, CA, March 2006.
25. N. Antoniades, I. Roudas, G. Ellinas, and J. Amin, Transport metropolitan optical networking: Evolving trends in the architecture design and computer modeling, *IEEE/OSA J. Lightwave Technol.*, Vol. 22, No. 11, pp. 2653–2670, 2004.

21

INTEGRATED CIRCUITS FOR DISPERSION COMPENSATION IN OPTICAL COMMUNICATION LINKS

Anthony Chan Carusone, Faisal A. Musa, Jonathan Sewter, and George Ng

21.1 MOTIVATION

Demand for highdata rates has motivated researchers to work on optical communication links in many applications. However, the most powerful computing and storage platforms we have today operate on electrical signals. The use of optical signals in a system that is predominantly electrical presents some challenges. Electrical links use copper wires and are advantageous in terms of cost but suffer from channel impairments such as skin effect and dielectric losses, especially at high frequencies. Crosstalk can be a major concern in closely spaced electrical links at high data rates. On the other hand, optical links employ optical fiber and are colloquially labeled an infinite bandwidth medium due to their low loss. Furthermore, relatively little noise is introduced along an optical fiber, and fibers can be bundled tightly together with much less crosstalk than copper wires. However, optical links mandate the use of expensive components to convert the signals between the electrical and optical domains, and their bandwidth is limited by the dispersive property of light. Dispersion occurs because the propagation delay of a light signal through an optical fiber depends on the wavelength of light and also on the light's mode of propagation. Called dispersion, these impairments cause different portions of an optical pulse to propagate with different velocities.

Convergence of Mobile and Stationary Next-Generation Networks, edited by Krzysztof Iniewski

Consequently, closely spaced pulses carried by the optical fiber interfere with each other.

Currently, dispersion compensation in both the optical and electrical domain is a topic of active research. Although some dispersion can be easily compensated for in the optical domain, electronic dispersion compensation (EDC) is generally more cost-effective and flexible. Since dispersion can change dynamically, compensators must be adaptive. Adaptation is not easily achieved in the optical domain because of the relative lack of flexibility in optical components and because of the difficulty in extracting an appropriate error signal to control the adaptation. Moreover, silicon-integrated circuits have reached a pivotal point where they can easily provide flexible, cost-effective, low-power, and ultra-high-speed solutions that were previously impossible.

Several integrated circuit solutions to dispersion compensation in optical communication links are discussed in this chapter. In Section 21.2, different types of dispersion are briefly described. Section 21.3 introduces the use of linear filters for dispersion compensation. As examples of linear filters, finite impulse response (FIR) and infinite impulse response (IIR) filter topologies are described and measurement results are presented in Sections 21.4 and 21.5, respectively. Section 21.6 discusses nonlinear (decision feedback) dispersion compensation techniques. Alternative approaches to dispersion compensation such as maximum likelihood sequence estimation and adaptive optics are discussed in Section 21.7. Finally, Section 21.8 summarizes the findings in this chapter.

21.2 DISPERSION IN OPTICAL FIBERS

Fibers are broadly categorized into two groups. Fibers with core diameters of 50 and 62.5 μm are referred to as multimode fiber (MMF) because they can support many modes of light propagation at the optical wavelengths for which laser sources are readily available. Fibers with a core diameter of 8–10 μm support only one mode of propagation and are, hence, referred to as single-mode fiber (SMF).

The larger core size of MMF makes alignment of the fiber with another fiber or laser chip less critical. Consequently, MMF is more mechanically robust and less expensive, but suffers from more dispersion than SMF. Therefore, MMF is currently employed for links less than 1 km in length and data rates of 14 Gbit/s and lower (such as data communications within buildings via computer interconnects). On the other hand, SMF supports data rates of 40 Gbit/s over hundreds of kilometers. Currently, the rate and reach of both MMF and SMF links are limited by our ability to efficiently compensate for dispersion. The dominant forms of dispersion in commercial optical links today are chromatic dispersion and modal dispersion.

21.2.1 Chromatic Dispersion

Chromatic dispersion (CD) occurs because the refractive index of the material used to produce optical fibers has a wavelength dependence $n(\lambda)$. The velocity of light in the fiber v is related to the speed of light in free space c and the refractive index $n(\lambda)$ as shown by

$$v = \frac{c}{n(\lambda)} \tag{21.1}$$

Any laser source modulated by random data produces spectral components spread over a range of wavelengths. Each wavelength results in a particular refractive index in the fiber and according to Eq. (21.1) propagates through the fiber at a certain speed. Consequently, different spectral components emitted by an optical source have different arrival times at the receiver, thus producing a distorted signal. Both MMF and SMF are effected by this type of dispersion.

Fortunately, chromatic dispersion compensation is not very costly and can be easily performed in the optical domain. For example, if the propagation delay of an optical fiber exhibits a known wavelength-dependency, it can be cascaded with a short section of fiber exhibiting the exact inverse wavelength-dependency. Ideally, the resulting link will be free from chromatic dispersion. Fibers can also be designed with propagation velocities that are nearly constant around the wavelength of a transmitting laser. Hence, chromatic dispersion is not the focus of most current research on integrated circuit dispersion compensation.

Modal dispersion has proven more difficult to combat.

21.2.2 Modal Dispersion

Both MMF and SMF suffer from modal dispersion. The physical mechanisms are different, but both can result in similar time-varying channel responses.

In MMF links, a single transmitted light pulse excites multiple modes of propagation along the fiber, each experiencing a different delay and attenuation. As illustrated in Figure 21.1, each optical pulse transmitted by an optical source results in multiple pulses of light at the receiver with different arrival times and amplitudes. In this manner, the transmitted optical pulse effectively "spreads out." Note that if the optical source were transmitting only one pulse through the fiber, then pulse spreading would not be a problem. But in practice, the optical source is being modulated by random data and a train of optical pulses

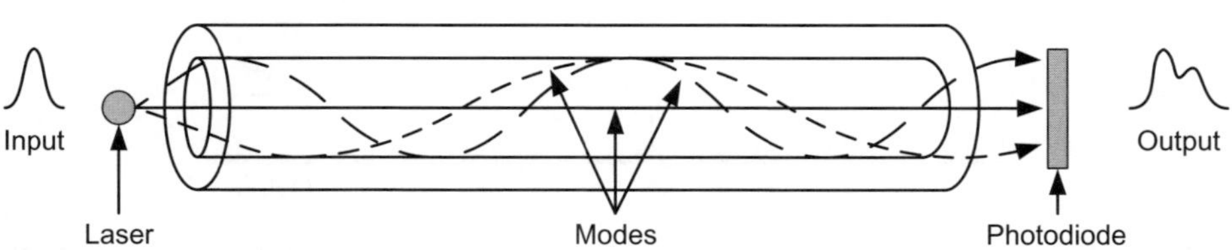

Figure 21.1. Pulse spreading due to modal dispersion in MMF. A light pulse launched at the input of the fiber takes different paths (modes) to the photodetector, resulting in different arrival times at the receiver. This causes intersymbol interference (ISI) at the receiver.

is being generated over the link. Due to pulse spreading, adjacent pulses interfere with each other, rendering them indistinguishable in the receiver. This phenomenon is widely known as intersymbol interference (ISI). Significant ISI results when the difference in delay between the short path (low-order modes) and the long path (high-order modes) is comparable to a bit period. To complicate matters, energy from one propagation mode can spill over into another at discontinuities along the fiber [1]. Hence, an MMF channel response can change dramatically when a fiber or connector is mechanically stressed. A practical trick that mitigates dispersion in MMF is to introduce an intentional offset between the light source and the center of the main fiber by using a short section of fiber called a "patch cord." Doing so excites fewer modes of propagation and can therefore reduce the spread in propagation delay.

In SMF, although only a single mode of light propagation is supported, dispersion still arises due to asymmetries in the fiber cross section. The asymmetries may be due to imperfections during manufacture, or due to mechanical stress after manufacture. Each pulse of light has energy in two orthogonal modes of polarization, which, due to the asymmetries, propagate with different velocities and attenuation. As a result, again, single optical pulses may split into two pulses by the time they reach the receiver. This phenomenon is referred to as polarization mode dispersion (PMD).

PMD is a result of birefringence, which affects all real optical fibers. Birefringence refers to the difference in refractive index experienced by light in the two orthogonal polarization modes of the fiber. It is caused by ellipticity of the fiber cross section due to asymmetric stresses applied to the fiber during or after manufacturing. Birefringence leads to fast and slow modes of propagation and consequently dispersion [2].

In terms of digital communications, PMD results (to a first order) in an input pulse being split into a fast and slow pulse that arrive at the receiver at different times, as shown in Figure 21.2. If the differential delay of the two pulses is significant compared to the bit period, ISI and an increase in bit error rate (BER) will result.

To a first order, the impulse response of an optical fiber with PMD is [3]

$$h_{\mathrm{PMD}}(t) = \gamma\delta(t) + (1-\gamma)\delta(t-\Delta\tau) \tag{21.2}$$

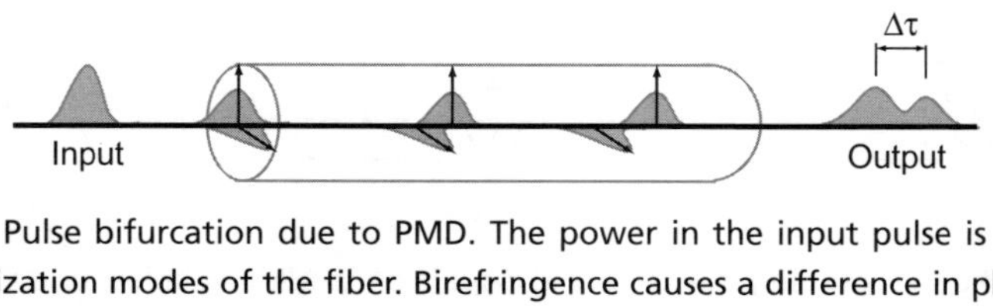

Figure 21.2. Pulse bifurcation due to PMD. The power in the input pulse is split between the two polarization modes of the fiber. Birefringence causes a difference in phase velocities between the two modes, resulting in ISI at the output.

where γ is the proportion of the optical power in the "fast" state of polarization (SOP), $(1 - \gamma)$ is the proportion of power in the "slow" SOP, and $\Delta\tau$ is the differential group delay (DGD) between the fast and slow components.

Both γ and $\Delta\tau$ vary, depending on the particular fiber and its associated stresses. The value of γ can be anywhere from zero to one, with uniform probability throughout this range [4] whereas $\Delta\tau$ varies statistically according to a Maxwellian distribution [5], given by

$$\rho(\Delta\tau) = \sqrt{\frac{2}{\pi}} \frac{\Delta\tau^2}{\sigma^3} e^{-\frac{\Delta\tau^2}{2\sigma^2}} \tag{21.3}$$

The distribution is defined by σ, which is related to the average DGD, $\Delta\tau_{\text{avg}}$, by [6]

$$\sigma = \frac{\sqrt{2\pi}\Delta\tau_{\text{avg}}}{4} \tag{21.4}$$

Therefore, though it can vary to large values, $\Delta\tau$ will for the most part remain close to some average value. Furthermore, $\Delta\tau$ varies with time, and significant variations can be observed on the order of milliseconds [7].

The average DGD per unit length of a fiber is defined as its PMD parameter, which has units of $\text{ps}/\sqrt{\text{km}}$. Typical installed fibers exhibit a PMD of 0.5–$2.0\ \text{ps}/\sqrt{\text{km}}$ [8]. New fibers can be manufactured with a PMD of as low as $0.05\ \text{ps}/\sqrt{\text{km}}$ [9]. Given the PMD parameter, the average DGD of a fiber of length L is given by

$$\Delta\tau_{\text{avg}} = \text{PMD} \times \sqrt{L} \tag{21.5}$$

It has been calculated that to prevent PMD from causing system outages amounting to more than 30 s per year (corresponding to an outage probability of 10^{-6}), the average DGD must be less than approximately 15% of a bit period, T_B [10]:

$$\Delta\tau_{\text{avg}} < 0.15 T_B \tag{21.6}$$

This has severe implications as the data rate of these systems is increased to 10 and 40 Gbit/s. As the data rate is increased on a given fiber, the maximum useful length of the fiber decreases according to the square of the increase. For example, given a fiber with a PMD of $1.0\ \text{ps}/\sqrt{\text{km}}$ and using (21.6), the maximum length of a 2.5-, 10-, and 40-Gbit/s system is 3600, 225, and 14 km, respectively, if PMD is the limiting factor.

Dispersion due to multimode propagation in MMF and polarization mode dispersion in SMF are referred to, together, as modal dispersion. Figure 21.3 illustrates three possible effects of modal dispersion on the pulse response of

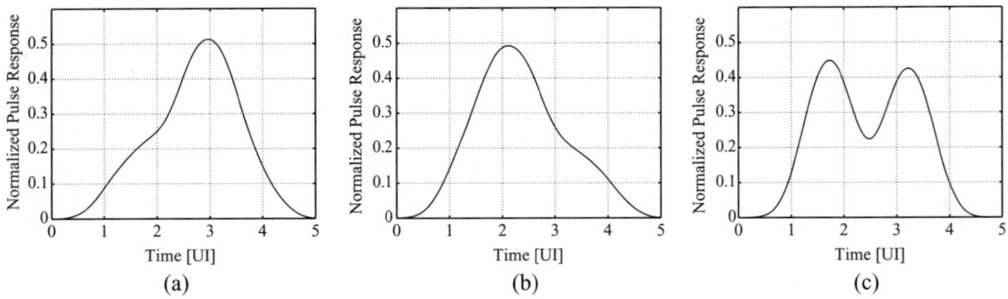

Figure 21.3. Example fiber channel pulse responses. **(a)** Precursor ISI pulse. **(b)** Postcursor ISI pulse. **(c)** Split pulse response. (Copyright IEEE 2008.)

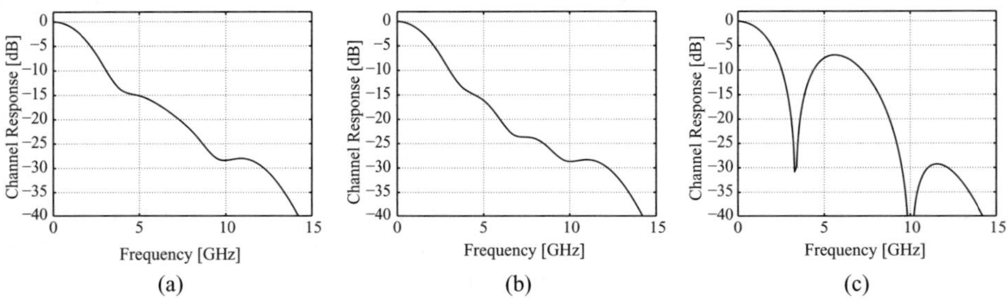

Figure 21.4. Example fiber channel frequency responses. **(a)** Precursor ISI pulse. **(b)** Postcursor ISI pulse. **(c)** Split pulse response. (Copyright IEEE 2008.)

a 10-Gbit/s link over 220 m of MMF with a patch cord. In all three cases, the pulse response is confined to roughly 3 baud unit intervals (UI) or 300 ps at 10 Gbit/s. In Figure 21.3a, a fast mode of propagation is visible preceding the main pulse causing each bit to interfere with two preceding bits. This is called a precursor response because it has mostly precursor intersymbol interference (ISI).The pulse response in Figure 21.3b, on the other hand, has a slow mode of propagation following the main pulse resulting in mostly postcursor ISI. In Figure 21.3c, the channel response has two distinct pulses that are nearly equal in amplitude and clearly separated in time. Any receiver for this channel will face some ambiguity in determining which pulse is the "main" one. We shall see that the resulting confusion necessitates nonlinear receiver architectures.

The normalized frequency responses corresponding to the three test cases in Figure 21.3 have a bandwidth of approximately 2 GHz as shown in Figure 21.4. This is well below the bandwidth required for a 10-Gbit/s optical system. Furthermore, modal dispersion can vary significantly within a few milliseconds [7], demanding an adaptive approach to EDC. A receiver architecture that performs EDC and can be easily configured to adapt to variations in modal dispersion is shown in Figure 21.5. It comprises a transimpedance amplifier (TIA), a feedforward equalizer (FFE), and a decision feedback equalizer (DFE). After necessary amplification by the TIA, the FFE mitigates ISI by linearly filtering

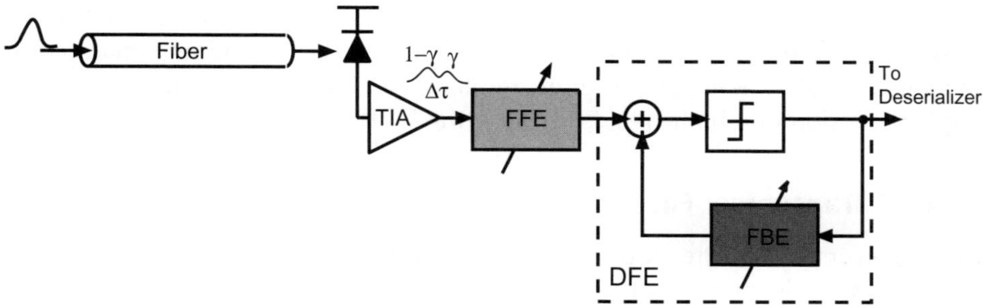

Figure 21.5. Electronic dispersion compensation (EDC) by using linear and nonlinear filters in a fiber-optic receiver.

the received signal, inverting the channel response. Such linear equalization of the channel is considered in Section 21.3–21.5. However, the split pulse response in Figure 21.4c proves to be more problematic. Since the frequency response has deep nulls, a portion of the transmitted spectrum is essentially lost and cannot be recovered by linearly filtering the received signal. For these cases, nonlinear methods for recovering the data using a DFE are discussed in Section 21.6.

21.3 ELECTRONIC DISPERSION COMPENSATION (EDC) USING LINEAR EQUALIZATION

A straightforward and potentially low-power method for mitigating ISI is to employ an adaptive linear filter that equalizes the channel response. Although it is possible to place the equalization filter at the transmitter, modal dispersion is time-varying and it is difficult to communicate information about the channel from the receiver back to a transmit equalizer in real time. Hence, EDC is generally performed at the receiver. The employed linear filters are of two types:

1. Finite Impulse Response (FIR). This is discussed in Section 21.4.
2. Infinite Impulse Response (IIR). This is discussed in Section 21.5.

21.4 FIR FILTERS FOR EDC

Programmable finite impulse response (FIR) filters can accommodate the wide variety of fiber responses attributable to modal dispersion, making them a popular choice for EDC applications. FIR equalizers have long been employed in magnetic storage applications at data rates exceeding 1 Gbit/s. The filters are guaranteed stable, and established techniques exist for the adaptation of their coefficients. However, their low-power implementation remains an open research topic.

FIR filters of two types are considered: transversal (TVF) and traveling-wave (TWF) filter topologies. Section 21.4.1 describes the design and implementation of a high-speed transversal FIR filter, and Section 21.4.2 discusses the design and implementation of a high-speed traveling-wave FIR filter.

21.4.1 Transversal FIR Filter Topology

The transversal FIR filter topology is shown in Figure 21.6. Here a delay line is tapped at specific intervals (τ) to generate N delayed versions of the input signal $x(t)$. These delayed versions are scaled by the tap weights c_k and then combined to form an output $y(t)$ of the form

$$y(t) = \sum_{k=1}^{N} c_k x(t-(k-1)\tau) \tag{21.7}$$

where τ is the tap spacing. While Figure 21.6a describes the TVF topology in its most general form, Figure 21.6b illustrates the TVF topology as it would appear for a high-speed implementation [11]. The delay line is implemented using passive components and terminated to prevent reflections. The tap multipliers are implemented as transconductors, with the summation performed in the current domain.

An FIR filter requires analog delays. A clock can be used to define the delay intervals [12], but its distribution with low skew and low jitter generally burns a lot of power. Alternately, analog continuous-time delays may be synthesized in two ways: using an artificial L-C transmission line or a microstrip transmission line. For a two-tap FIR implementation, a microstrip transmission line may be wrapped in a "U" shape, as shown in Figure 21.7a, making it simple to route both equalizer taps to a common summation node. This long narrow shape may be easier to fit in some receiver layouts. Since there are no inductors, it is easier to

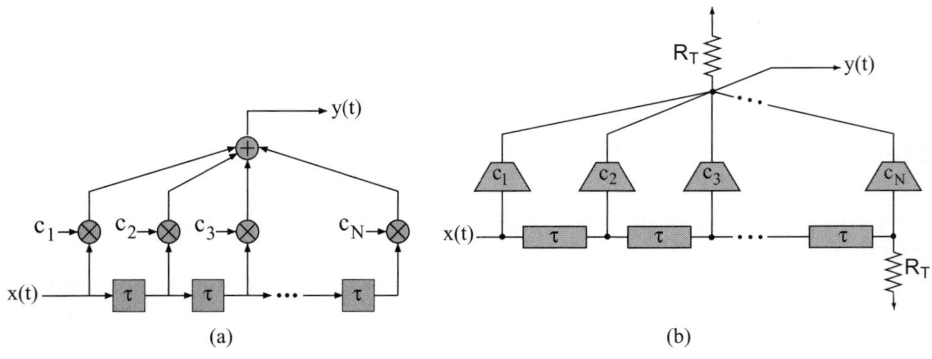

Figure 21.6. Transversal FIR filter topology. **(a)** Conceptual block diagram. **(b)** High-speed implementation.

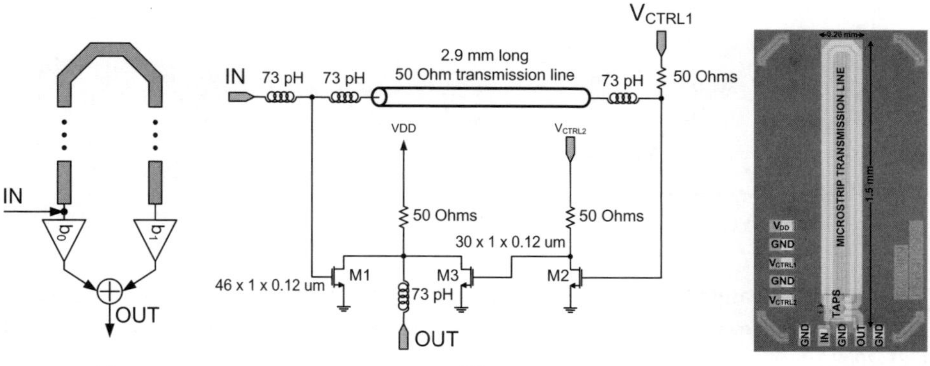

(a) (b) (c)

Figure 21.7. Implementing a two-tap transversal FIR filter. **(a)** Transversal filter with microstrip line shaped as "U". **(b)** Circuits chematic. **(c)** Die photo. (Copyright IEEE 2008.)

accommodate metal fill rules. Microstrip lines also have an inherently higher bandwidth due to their distributed nature. Whereas it is necessary to use special techniques to achieve baud-spacing with sufficient bandwidth due to the lumpiness of artificial transmission lines [13, 14], a single length of microstrip line can simply be used to achieve baud-rate tap spacing. This simplifies both the design and modeling of the interconnects.

The schematic of a single-ended two-tap transversal FIR equalizer is shown in Figure 21.7b [15]. Common-source amplifiers are used as the tap transconductors. Located at the input of the equalizer, transistor M1 forms the first equalizer tap. The cascaded transistors M2 and M3 form the second tap of opposite polarity, a condition needed to create peaking. Signal inversion was performed with an additional common-source stage at the end of the transmission line instead of at the beginning to introduce additional signal delay, thus increasing the effective inter-tap delay and shortening the required transmission-line length. The majority of the inter-tap delay is provided by a 50-Ω microstrip transmission line that is terminated by a matching load resistor. In this design the tap outputs are summed in the current domain on a 50-Ω resistor which also provides output impedance matching, although a higher load resistance could have been used to provide high gain if driving on-chip loads only.

The die photo of the circuit is given in Figure 21.7c. Implemented in 130-nm CMOS, the size of the die excluding pads is 1.5 mm × 0.26 mm. The die includes a 2.9-mm-long, 9-μm-wide microstrip line that provides 18.2 ps of group delay and 50-Ω characteristic impedance and is laid out in a U-shaped fashion. In order to improve the input and output matching of the TVF, long narrow strips of metal each with 73-pH inductance are introduced around the nodes where the transconductors are connected.

Eye diagram measurement results of the two-tap FIR filter are shown in Figure 21.8. Based on the channel frequency response shown in Figure 21.8a, a 38.2-Gbit/s signal experiences 14.3 dB of attenuation at one-half the bit rate

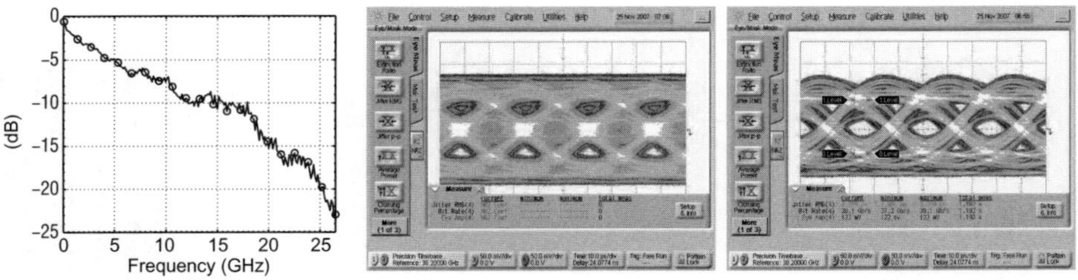

(a) (b) (c)

Figure 21.8. Measured 38.2-Gbit/s eye diagrams curve. **(a)** Frequency response of channel. **(b)** Unequalized eye: channel output. **(c)** Equalized eye: two-tap TVF output. (Copyright IEEE 2008.)

(19.1 GHz). Figure 21.8b shows the closed and unequalized eye through the channel. Setting the tuning voltages V_{CTRL_1} at 0.82 V and manually tuning the peaking control voltage V_{CTRL_2} to 0.88 V, results in the equalized eye shown in Figure 21.8c. The chip consumes a maximum of 30 mW from a 1.2-V supply.

21.4.2 Traveling-Wave FIR Filter Topology

The TVF topology discussed so far has two major disadvantages: Firstly, the outputs of the tap multipliers are tied together at the summation node. As a result, the capacitance at this node is very large and potentially speed-limiting. In addition, if the delay line is implemented using passive elements the inter-tap spacing may be quite large, making it physically difficult to tie these outputs together at a single node without introducing skew and signal degradation.

Secondly and more importantly, because the delay line is equal to the span of the equalizer, reflections that occur at the end of the delay line can be detrimental to the equalizer perform ance. For example, consider a three-tap symbol-spaced equalizer. A reflection at the end of the two-section delay line will traverse back to the first tap after four symbol periods. Thus, if Γ_T is the reflection coefficient at the end of the delay line, the output will not be equal to (21.7), but rather

$$y(t)=c_1[x(t)+\Gamma_T x(t-4\tau)]+c_2[x(t-\tau)+\Gamma_T x(t-3\tau)]+c_3 x(t-2\tau) \quad (21.8)$$

assuming that the higher-order Γ_T terms (Γ_T^2, Γ_T^3, etc.) are negligible. Therefore, the output contains terms involving $x(t-4\tau)$ and $x(t-3\tau)$ that are outside the span of the equalizer and that cannot be eliminated by adjusting the equalizer coefficients. In effect, these terms serve to increase the ISI at the output [11]. An alternative topology is the traveling-wave filter FIR topology that was first suggested in Jutzi [16], and is shown in Figure 21.9. The main difference between this topology and the TVF is that this topology makes use of delay lines at both the input and the output. While the output $y(t)$ is still given by Eq. (21.7), this

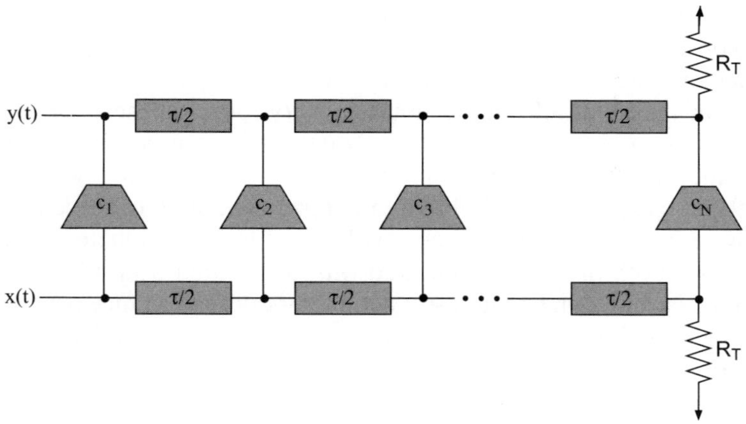

Figure 21.9. High-speed implementation of traveling-wave FIR filter topology.

topology addresses some of the major issues with the TVF topology that arise at high speeds.

First, there is no longer a lumped node at which all of the outputs are tied together. The capacitance at the inputs and outputs of the tap multipliers serve only to increase the capacitance of the input and output delay lines, respectively. In other words, the device capacitances are distributed along the length of the delay lines.

Also, this topology lends itself to an efficient layout because the inter-tap spacing is the same for both the input and output lines. The input and output lines can be laid out parallel to one another, with the tap multipliers interspersed between them.

Finally, this topology offers an improvement over the TVF in terms of robustness in the presence of reflections. Using the example of a three-tap filter, the output in the presence of reflections is

$$y(t) = c_1[x(t) + 2\Gamma_T x(t-2\tau)] + c_2[x(t-\tau) + 2\Gamma_T x(t-2\tau)] + c_3 x(t-2\tau) \quad (21.9)$$

In contrast to Eq. (21.8), Eq.(21.9) does not contain any terms outside the span of the equalizer. Therefore, the extra $x(t - 2\tau)$ terms caused by the reflections can be compensated for by properly adjusting the weighting coefficient for the third equalizer tap [11].

Because of these benefits, many of the reported FFE implementations for EDC make use of the TWF topology [17–20].

TWF Design. The basic TWF design is shown in Figure 21.9. The input and output transmission lines are chosen such that their characteristic impedance is matched to the system impedance and their delays implement the proper tap spacing. The input and output capacitances of the tap multipliers serve to capacitively load the input and output transmission lines, respectively. This loading effectively reduces the characteristic impedance of the transmission lines and must be considered in the design.

To achieve the required delays for a 40-Gbit/s equalizer, the transmission lines used in the input and output must be very long (on the order of several millimeters). As a result, the size of the equalizer IC can become prohibitively large. In addition, long transmission lines introduce significant series loss. For both of these reasons, it is desirable to minimize the length of these lines.

Using artificial transmission lines made up of lumped inductors and capacitors addresses both problems associated with distributed transmission lines. By winding the transmission line into spiral inductors, the inductance per unit length is greatly increased because of the mutual inductance between adjacent windings. Thus, the overall length of transmission line is decreased, reducing the chip area as well as any resistive losses.

The design of a TWF with artificial transmission lines is fairly straightforward, because most of the design variables are fixed by the desired configuration of the equalizer. The delays of the input and output transmission lines are determined by the desired tap spacing. The delay ΔT of a lumped L–C transmission line section is approximately equal to

$$\Delta T = \sqrt{L'C'} \tag{21.10}$$

where L' and C' are the inductance and capacitance of each section of the transmission line, respectively. Also, the characteristic impedance of the input and output transmission lines (Z_0) is determined by the system impedance (e.g., $50\,\Omega$) and is equal to

$$Z_0 = \sqrt{\frac{L'}{C'}} \tag{21.11}$$

Substituting (21.11) into (21.10), we can solve for L' and C':

$$L' = Z_0 \Delta T \tag{21.12}$$

$$C' = \frac{\Delta T}{Z_0} \tag{21.13}$$

As an example, consider a three-tap symbol-spaced equalizer operating at 40 Gbit/s with a system impedance of $50\,\Omega$. The delay per transmission line section should be $T_B/2$, or 12.5 ps. Using Eqs. (21.12) and (21.13), L' is equal to 625 pH and C' is equal to 250 fF. The input and output capacitances of the tap multipliers effectively add to the capacitance of the input and output transmission lines, respectively. Therefore, C' is made up of the sum of the transmission line capacitances and the device capacitances. The termination resistors R_T are set equal to the system impedance. The resulting equalizer design is shown in Figure 21.10a. Note that $L'/2$ inductors have been added to the ends of each transmission line. This

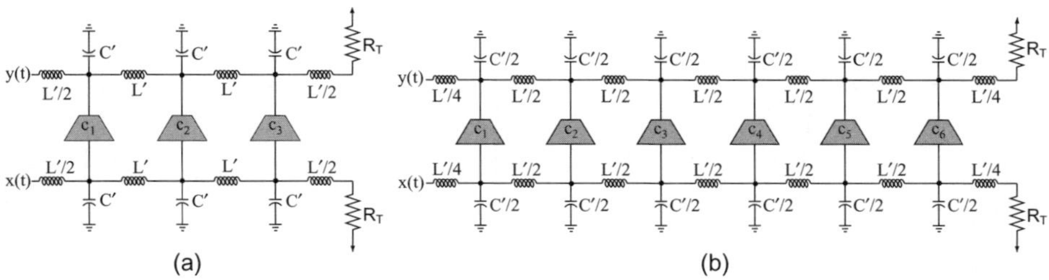

(a) (b)

Figure 21.10. TWF design. **(a)** Three-section input and output transmission lines. **(b)** Six-section input and output transmission lines.

improves the symmetry of the transmission lines and ensures that the total inductance ($3L'$) matches the total capacitance ($3C'$).

The 3-dB bandwidth of an L–C section is equal to

$$f_{3\text{dB}} = \frac{1}{\pi\sqrt{L'C'}} = \frac{1}{\pi\Delta T} \tag{21.14}$$

For a TWF, the delay of each section, ΔT, is equal to one-half the tap spacing, or $T/2$. Thus, there is an inverse relationship between the bandwidth and the tap spacing of a TWF, as given by

$$f_{3\text{dB}} = \frac{2}{\pi\tau} \tag{21.15}$$

For our example above, Eq. (21.15) yields a bandwidth of 25 GHz. This is insufficient for a 40-Gbit/s equalizer. To extend the bandwidth, the L–C stages must be made smaller, or the transmission line must be made less "lumpy." Figure 21.10b shows an equalizer that is half as lumpy as the equalizer shown in Figure 21.10a.

The bandwidth of each transmission line section is doubled to 50 GHz, using Eq. (21.14). Plots of the magnitude response and group delay for three- and six-section lumped transmission lines are given in Figure 21.11. These plots demonstrate the doubling of the bandwidth corresponding to a reduction of the lumpiness by a factor of two. From the group delay plots it is also observed that the group delay is flat only within the bandwidth of the transmission line, another important reason for making the transmission line more distributed.

Note that since the node capacitances C' are made up in part by the device capacitances, when C' is scaled the device sizes must scale accordingly. Therefore, the gain through each stage of a six-section equalizer is only half that through each stage of a three-section equalizer, for the same equalizer span. Splitting the three gain cells into six gain cells effectively halves the maximum possible gain through any particular tap. This limits the performance of the equalizer for operation in the absence of channel impairments, when only one tap needs to be on, for example.

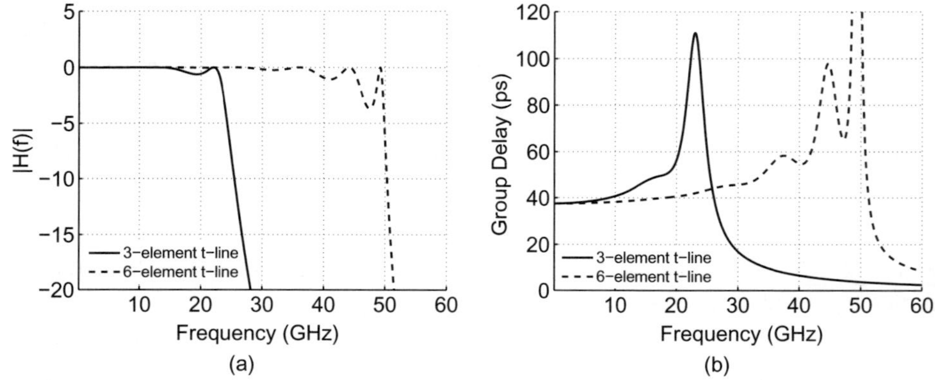

Figure 21.11. Comparison between three- and six-element transmission lines. **(a)** Magnitude response. **(b)** Group delay. (Copyright IEEE 2006.)

Crossover TWF Topology. A slightly modified version of the TWF topology, which will be referred to as the crossover TWF topology, can be used to allow symbol-spaced equalization without increasing the lumpiness of the transmission lines or sacrificing any gain [13]. The crossover TWF topology is illustrated in Figure 21.12a.

In Figure 21.12a, each L–C section is replaced with two sections having component values of $L/2$ and $C/2$ to maintain the same characteristic impedance. Thus each L–C section would have a delay ΔT equal to one-quarter of the tap spacing, or $\tau/4$. Therefore, the bandwidth of the TWF would be given by

$$f_{3\mathrm{dB}} = \frac{4}{\pi\tau} \tag{21.16}$$

Hence, the delay through each section is halved while the bandwidth is increased. At data rates that push technologys limits, the capacitances C are comprised mostly of the transconductors' parasitic input and output capacitances. Therefore, reducing these capacitances to $C/2$ implies scaling the size and of each transconductor by one-half. Fortunately, by crisscrossing outputs of consecutive transconductors, their gains add with equal group delay effectively forming a distributed amplier. Furthermore, each tap delay is now formed by four L–C sections (instead of two) giving the same tap spacing as the conventional TWF topology. Since this topology uses six-section transmission lines, it has a bandwidth given by Eq. (21.16), which is twice the bandwidth of the three-tap symbol-spaced equalizer described by Figure 21.10a.

A block diagram of the entire circuit is given in Figure 21.12b. A lumped preamplifier stage accepts differential inputs and performs a variable gain function to condition the input signal so as to maximize the dynamic range of the equalizer. It also performs a single-ended to differential conversion function

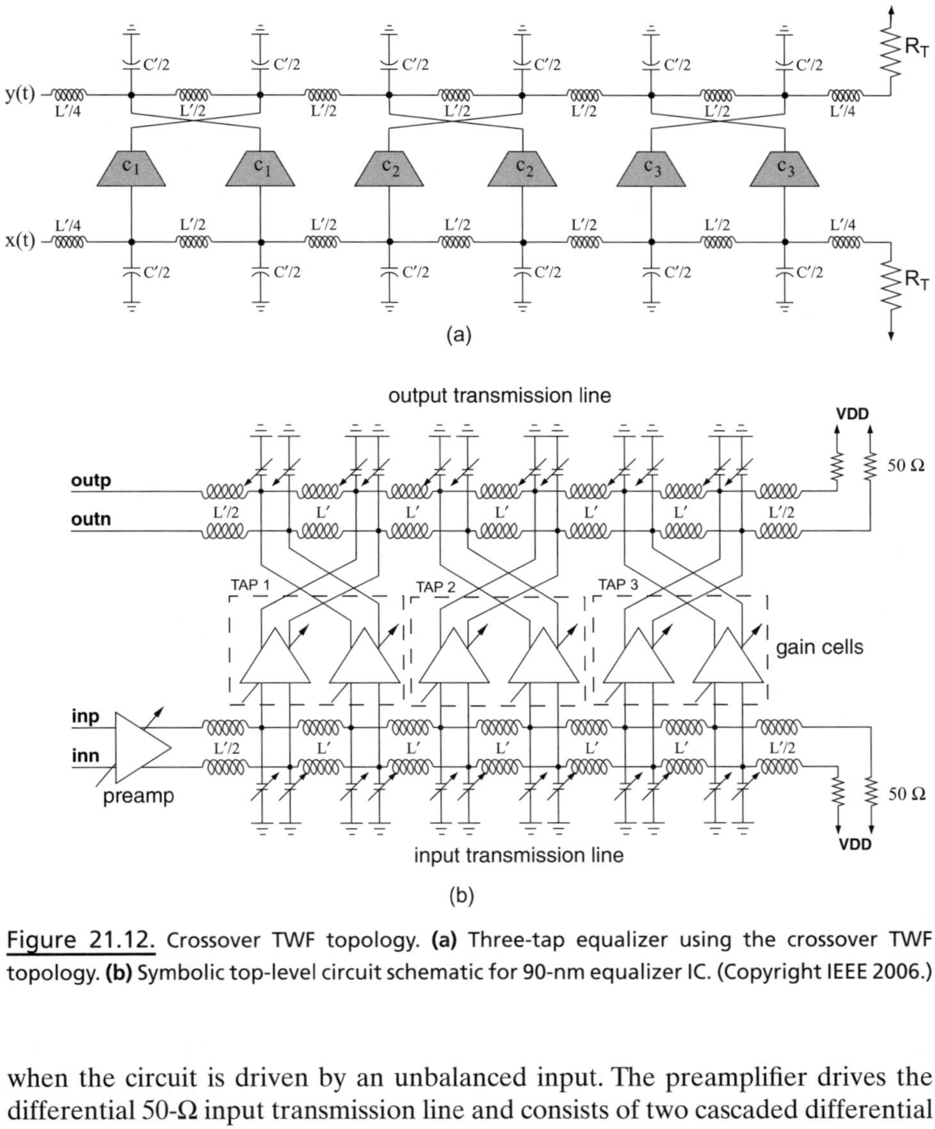

Figure 21.12. Crossover TWF topology. **(a)** Three-tap equalizer using the crossover TWF topology. **(b)** Symbolic top-level circuit schematic for 90-nm equalizer IC. (Copyright IEEE 2006.)

when the circuit is driven by an unbalanced input. The preamplifier drives the differential 50-Ω input transmission line and consists of two cascaded differential pairs. A schematic of the lumped preamplifier stage is given in Figure 21.13a. The gain of the preamplifier stage is digitally-controllable. This gain control is provided both to allow tuning of the preamplifier bias currents for performance optimization and to allow compensation for possible variations in input power. The input and output transmission lines are made up of differential inductors and capacitances, and they generate the delays necessary in an FIR filter. Three gain cells tap this transmission line at intervals such that the difference in delay from one tap to the next is 25 ps (or one symbol period at 40 Gbit/s).

Each of the gain cells is composed of two differential pairs. The two differential pairs are connected with opposite polarity. This allows the filter to

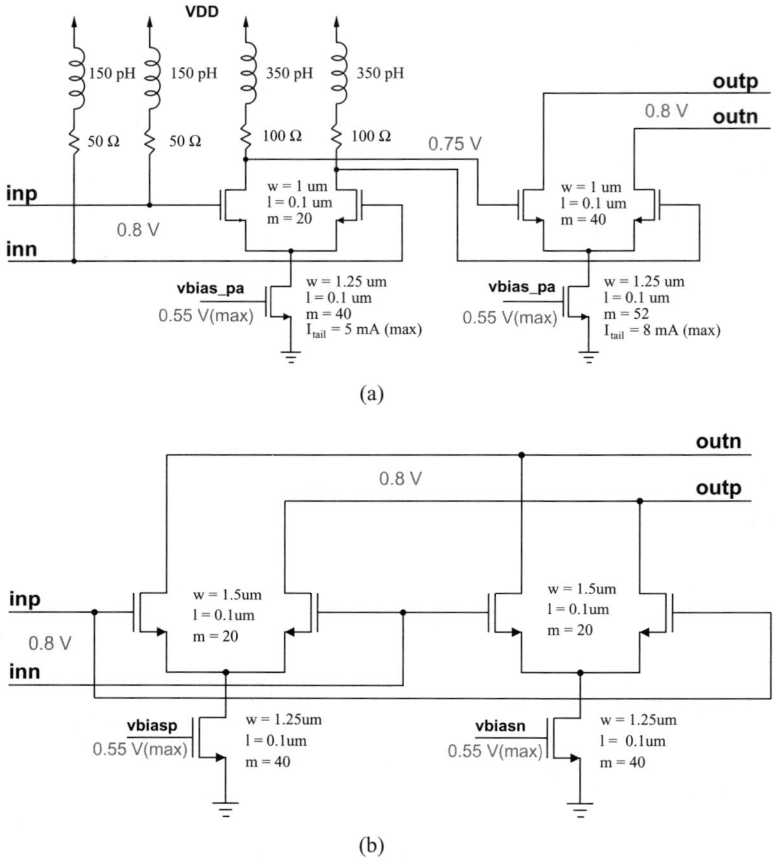

Figure 21.13. Circuit schematics. **(a)** Preamplifier block. **(b)** Variable gain cell. (Copyright IEEE 2006.)

implement both positive and negative tap weights. To implement a positive tap weight, the bias current for the differential pair connected with negative polarity is first zeroed, leaving only the positive path from input to output. The gain in this positive path is controlled by adjusting the bias current of the differential pair connected with positive polarity. A negative tap weight is implemented using the converse of this procedure.

From Eq. (21.12), the inductance per element (L') required for these transmission lines is 312.5 pH per side. From Eq. (21.13), the capacitance per element (C') is 125 fF per side. The transmission lines contain six nodes, one for each gain cell. The capacitance at each of these nodes is C'. Between these six nodes are five inductances of L' each. Inductances of $L'/2$ are added to each end of the transmission line so that the total inductance of the line is $6L'$, which matches the total capacitance of $6C'$.

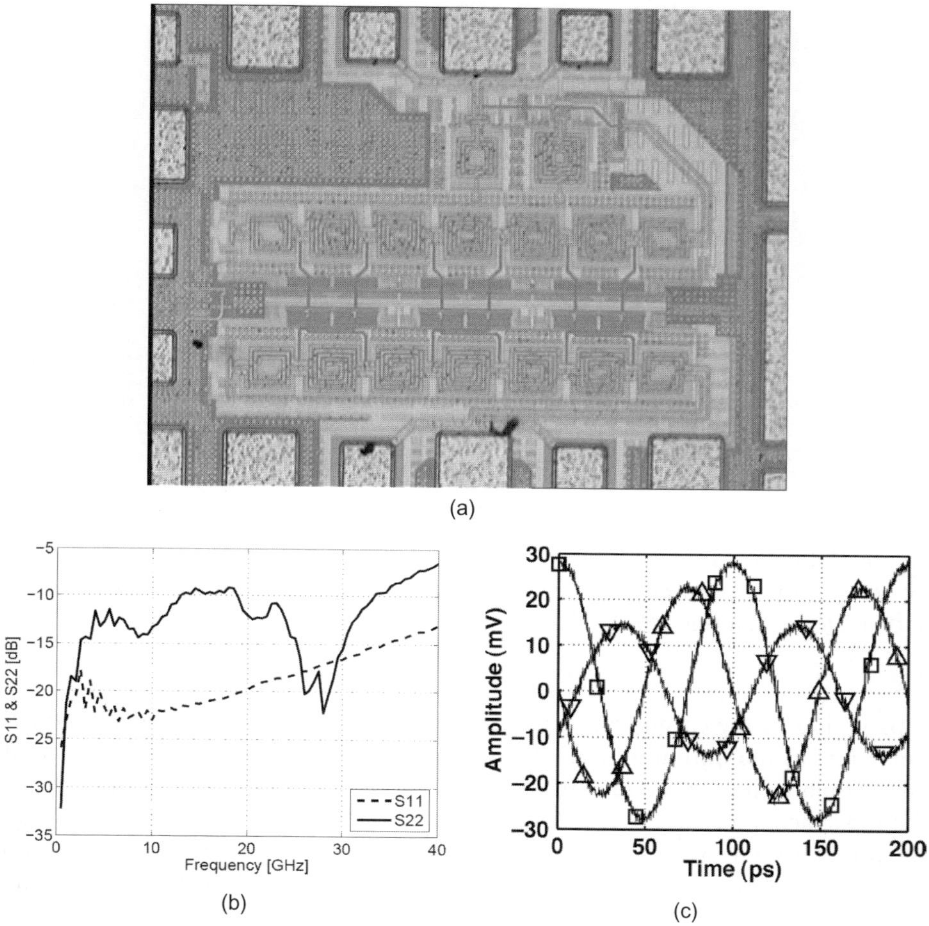

Figure 21.14. TWF crossover topology. **(a)** Die photo. **(b)** Measured input and output return loss. **(c)** Measured tap response for a 10-GHz sinusoid. (Copyright IEEE 2006.)

The inductances $L'$are implemented by differential spiral inductors. The capacitances C' at each node are composed of the device capacitances attached to the node supplemented by a variable capacitance created by a digitally controlled varactor. The varactor allows tuning of C' to compensate for model inaccuracies and process variation. The characteristic impedance and delay can both be tuned by this varactor.

The equalizer was implemented in a 90-nm CMOS process. A die photo of the circuit layout is given in Figure 21.14a. The overall dimensions of the equalizer IC are 0.6 mm × 0.5 mm. All circuit measurements were made on wafer and were single-ended. Figure 21.14b shows the measured input and output return losses. Integrated 50-Ω resistors provided better than 16 dB of input return loss up to 30 GHz. The output return loss was better than only 9 dB up to 30 GHz.

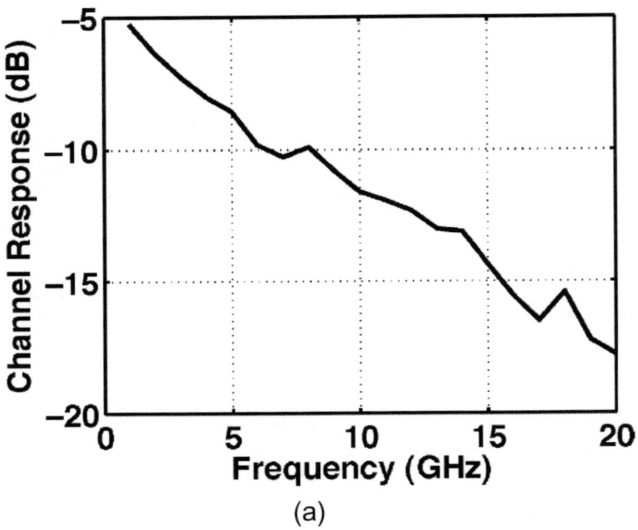

Figure 21.15. Time domain measurements for TWF crossover topology. **(a)** Channel Response. **(b)** Measured input and output eyes at 25 Gbit/s. **(c)** Measured input and output eyes at 30 Gbit/s. (Copyright IEEE 2006.)

This is due to the mismatch in the characteristic impedance of the integrated delay line at the output and the measurement instruments as a consequence of larger-than-estimated node capacitance of the delay lines. Figure 21.14c shows the tap spacing measured at 10 GHz which was obtained by measuring each tap separately on an oscilloscope and superimposing the traces. The measured tap spacing was 37 ps, which was considerably larger than the 25 ps targeted for the filter design. Again this was due to larger-than-estimated node capacitances of the delay lines. For equalization experiments, a lossy channel was constructed with coaxial cables, connectors, and an attenuator. The response of the channel is shown in Figure 21.15a. Input and output eyes to the equalizer are shown in Figures 21.15b and 21.15c for data data rates of 25 Gbit/s and 30 Gbit/s.

The equalizer consumes approximately 25 mW from a 1-V supply. This is significantly lower than both of the 40-Gbit/s FFEs previously reported (820 mW [19] and 750 mW [20]) and nearly all of the 10-Gbit/s FFEs previously reported.

Folded-Cascade TWF Topology. The crossover TWF topology allows the implementation of a symbol-spaced equalizer as a TWF while decreasing the lumpiness of the transmission lines by a factor of two. It is not practical, however, to decrease the lumpiness of the transmission lines by a factor greater than two using the crossover TWF topology. The crossover routing between two taps cannot easily be reproduced for three or more taps without introducing asymmetries and skew between the different paths.

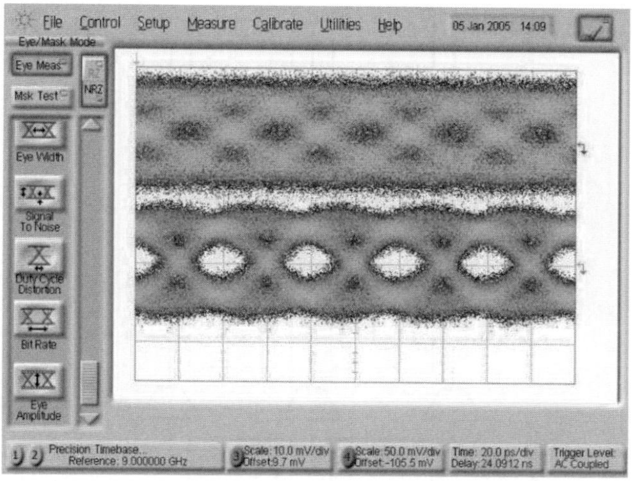

(b)

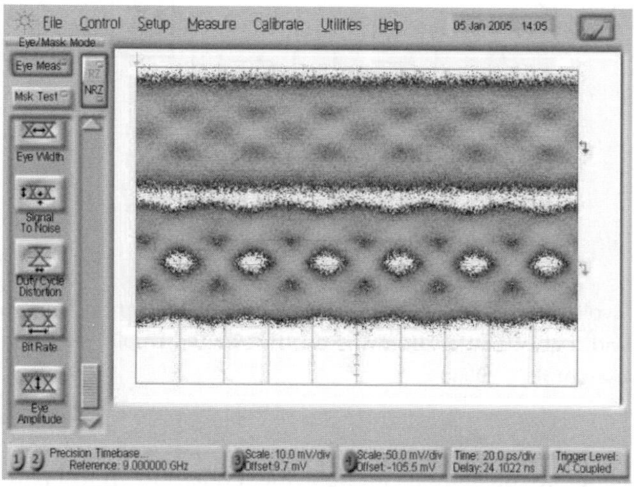

(c)

Figure 21.15. *Continued*

Theoretically, it should be possible to increase the bandwidth of the crossover TWF topology by further subdividing the delay lines and using more than two ampliers per tap. For instance, Figure 21.16a shows a single tap of a crossover TWF using three ampliers per tap. Unfortunately, the crossover routing is no longer symmetric. The resulting mismatch and skew between the paths of different length may introduce unacceptable ripple into the frequency response. By using an intermediate folded transmission line to perform the crossover

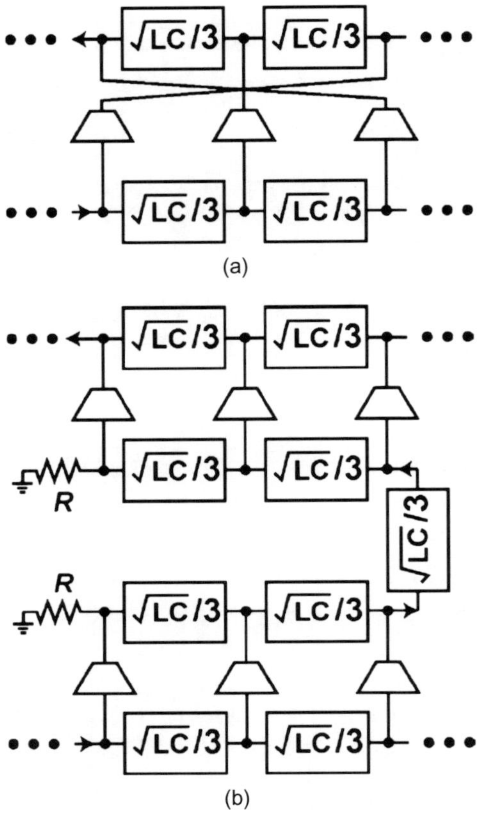

Figure 21.16. Evolution of the folded cascade TWF. **(a)** Using three amplifiers per tap to boost the bandwidth. **(b)** Using a cascade of two distributed amplifiers per tap to boost the bandwidth. (Copyright IEEE 2006.)

routing minimal delay mismatch can be achieved as shown in Figure 21.16b. The tap spacing is the same as in Figure 21.16a. The delay lines are all composed of one-third-sized *L–C* sections, so that the bandwidth is three times greater than a conventional TWF for the same tap spacing. Thus if each tap of a symbol-spaced folded-cascade TWF is composed of two cascaded stages of M gain elements each, the bandwidth of the equalizer is given by

$$f_{3\text{dB}} = \frac{2M}{\pi\tau} \tag{21.17}$$

Using this topology results in the three-tap FIR filter of Figure 21.17a [14]. Note that any two paths through neighboring tap amplifiers differ in length by six *L–C* sections. Hence, the tap spacing is $6\sqrt{(L/3)(C/3)} = 2\sqrt{(L)(C)}$, just as in the conventional and the crossover TWF topology. The values for the inductance of

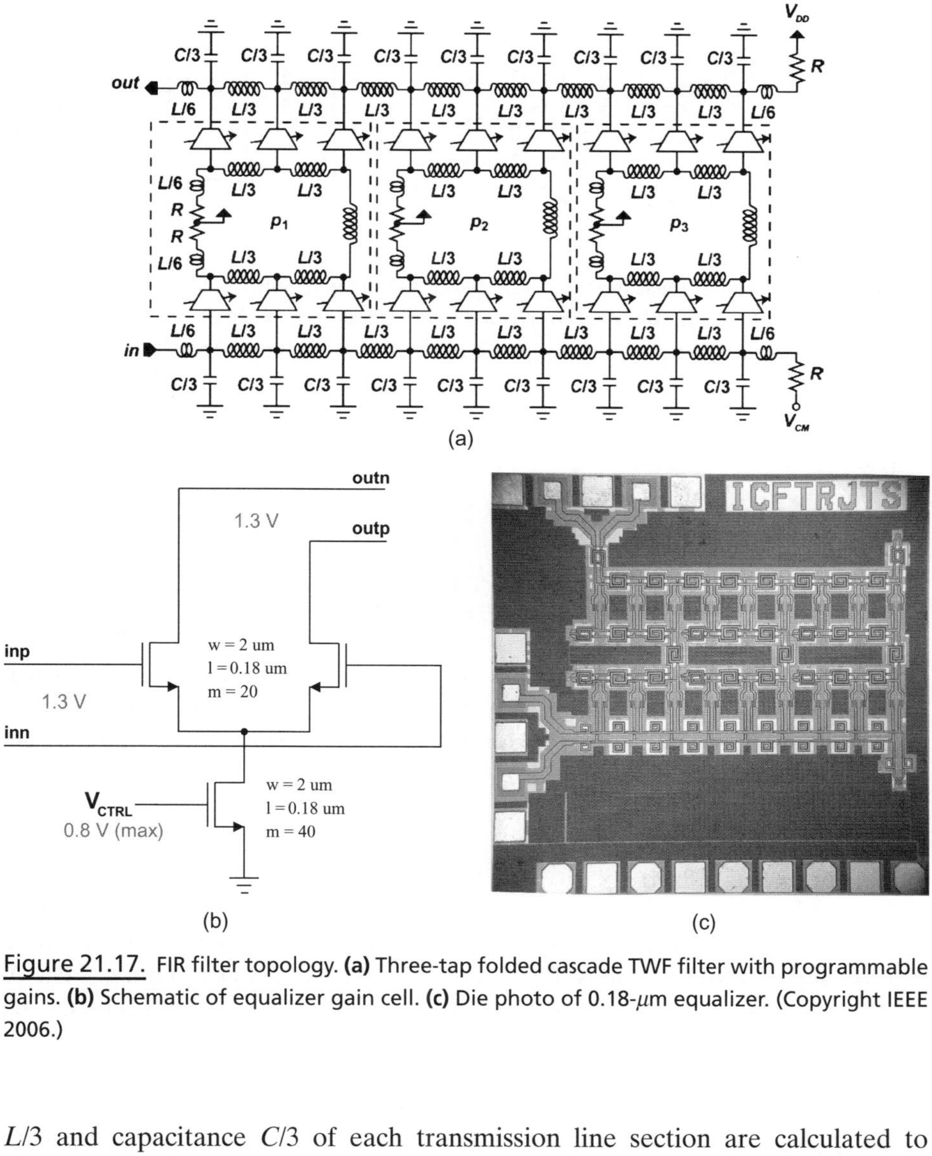

Figure 21.17. FIR filter topology. **(a)** Three-tap folded cascade TWF filter with programmable gains. **(b)** Schematic of equalizer gain cell. **(c)** Die photo of 0.18-μm equalizer. (Copyright IEEE 2006.)

$L/3$ and capacitance $C/3$ of each transmission line section are calculated to provide a characteristic impedance of 50 Ω per side (100 Ω differential) and a tap spacing of 25 ps (one bit period at 40 Gbit/s). Since the tap spacing is determined by six L–C sections, $\sqrt{(L/3)(C/3)}$ must equal to one-sixth of a bit period, or 4.17 ps. The resulting values of $L/3$ and $C/3$ are 209 pH and 83 fF per side, respectively. The node capacitances are made up purely of transistor and inductor parasitics. The inductances are spiral coils. Additional half-sized inductors $L/6 = 105$ pH are required at the ends of each transmission line as shown in Figure 21.17a. The gain cell used in this equalizer is a differential pair, as shown in Figure 21.17b. Six of these differential pairs make up each of the three equalizer taps. Each of

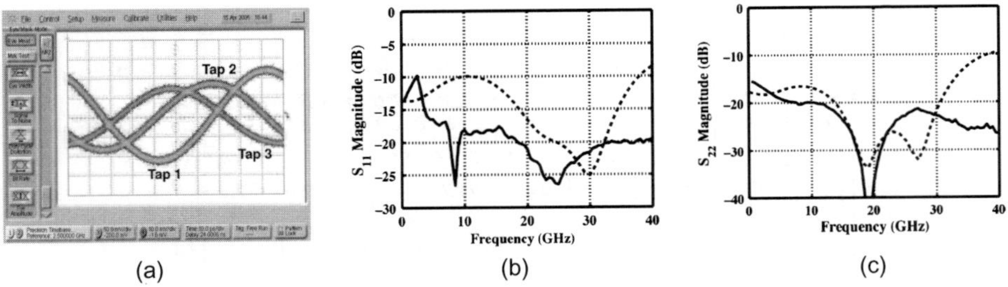

Figure 21.18. FIR filter measurements. **(a)** Measured tap delays with a 200-mV peak-to-peak 10 GHz sinusoidal input. **(b)** Measured (solid) and simulated (dashed) input return loss. **(c)** Measured (solid) and simulated (dashed) output return loss. (Copyright IEEE 2006.)

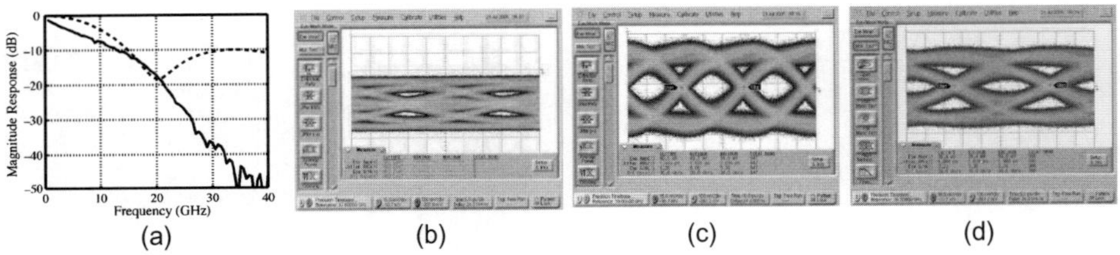

Figure 21.19. Eye diagram measurements. **(a)** Measured cable frequency response (solid) along with the modeled response of PMD channel with $\gamma = 0.4$ and $\Delta\tau = 25$ ps (dashed). **(b)** Channel output (equalizer input) at 40 Gbit/s. **(c)** Equalizer output at 30 Gbit/s. **(d)** Equalizer output at 40 Gbit/s. (Copyright IEEE 2006.)

the three taps has a fixed polarity. The first tap is positive, the second is negative, and the third is positive. These polarities correspond to a high-pass response, which would be the typical response of such an equalizer when used with most chip-to-chip or PMD channels. The filter was implemented in 180-nm CMOS. A die photo of the circuit layout is given in Figure 21.17c. The overall dimensions of the equalizer IC are 1 mm × 1 mm.

All circuit measurements were made on-wafer. Figure 21.18a shows the output of the filter with a 10-GHz sinusoidal input. The measurement was made by turning on each tap individually, with all other taps turned off. The tap spacing is 23 ps, just under one baud interval at 40 Gbit/s. Note that the polarity of tap 2 is reversed compared with taps 1 and 3. Figures 21.18b and 21.18c plot the magnitude of the input and output return losses, respectively, both simulated and measured with a two-port network analyzer. The input return loss is greater than 15 dB from 5 to 40 GHz, while the output return loss is better than 16 dB up to 40 GHz. A 4-m coaxial cable was used for eye diagram measurements. Figure 21.19a plots the frequency response of the cable along with the frequency response of a PMD limited channel model with $\gamma = 0.4$ and $\Delta\tau = 25$ ps and transmitter and receiver bandwidths of 25 GHz. Figure 21.19b shows a 40-Gbit/s single ended unequalized eye pattern at the end of the 4-m cable. The measured output eyes are shown in Figure 21.19c and 21.19d at 30 Gbit/s and 40 Gbit/s respectively.

Research on broadband FIR equalizers is ongoing and active. This section has discussed topologies that distribute parasitic capacitances along *L–C* delay lines [13, 14]. Low-power FIR architectures are shown in Tiruvuru and Pavan [21]. Others have employed small active circuits to mitigate losses in passive delay lines [22]. Research on these topics will no doubt continue; but for longer links and/or higher data-rates, more sophisticated methods for EDC are being researched.

21.5 IIR FILTERS FOR EDC

This section considers the use of traveling-wave filter techniques for the implementation of an infinite impulse response (IIR) filter. Figure 21.20a shows a top-level block diagram of a direct form 2-tap IIR filter that uses a traveling-wave topology with only poles in its transfer function. Due to the absence of a feed-forward path, zeros are eliminated from the transfer function of the filter, thus simplifying the hardware required to implement the filter. The total loop delay from the input to the output through the first tap a_1 is τ, and the total loop delay from the input to the output through the second tap a_2 is 2τ. Ideally the frequency at which peaking is observed in the magnitude response of the all-pole IIR filter is given by

$$f_{\text{peak}} = \frac{1}{2\tau_d} \tag{21.18}$$

where τ_d is the total delay around a feedback loop that includes an active tap amplifier. Thus, ideally tap a_1 produces a peak at $1/(2\tau)$, whereas tap a_2 produces a peak at $1/(4\tau)$. A drawback of the direct-form all-pole IIR topology is the

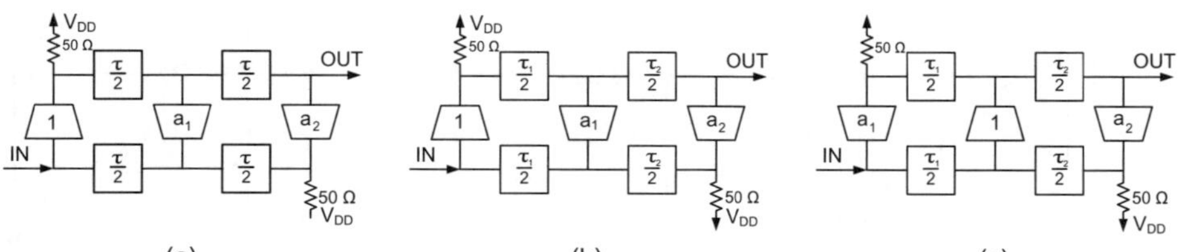

Figure 21.20. All pole traveling wave IIR filter topologies. **(a)** Direct form (uniform delay). **(b)** Multi-delay. **(c)** Double loop multi-delay. (Copyright IEEE 2009.)

double peaking behavior observed when utilizing the low-frequency tap, a_2, with the high-frequency tap, a_1, turned off. This arises from the periodic nature of the filter peaks and the finite bandwidth of the delay line segments. Moreover, with the high-frequency peak at twice as large as the low-frequency peak, it becomes difficult to place both peaks within the frequency band of interest. Practically, it would be desirable to bring the low- and high-frequency peaks close enough to each other to permit them to create a single peak, providing frequency boosting over a broader range. To bring the peak frequencies closer, the direct-form all-pole IIR can be modified into the multi-delay IIR section as shown in Figure 21.20b. Here delay section τ_1 determines the peak location for tap a_1, while delay section $\tau_1 + \tau_2$ determines the peak location for tap a_2. By allowing different segment delays, the multi-delay IIR filter allows two peak frequencies to be placed in close proximity to each other while suppressing any undesired secondary peak with the low-frequency tap on. A major drawback of this topology is that τ_2 cannot be selected independently from τ_1. Consequently, the delay τ_2 would be forced to be small in order to bring the peaking frequencies close together. To alleviate this problem, the multi-delay IIR filter can be modified as shown in Figure 21.20c to form a double loop. By introducing two independent feedback loops, the delays τ_1 and τ_2 can be sized independently according to the desired peak frequencies. Figure 21.21 shows the top-level block diagram of the double-loop multi-delay all-pole IIR filter that uses a traveling-wave architecture

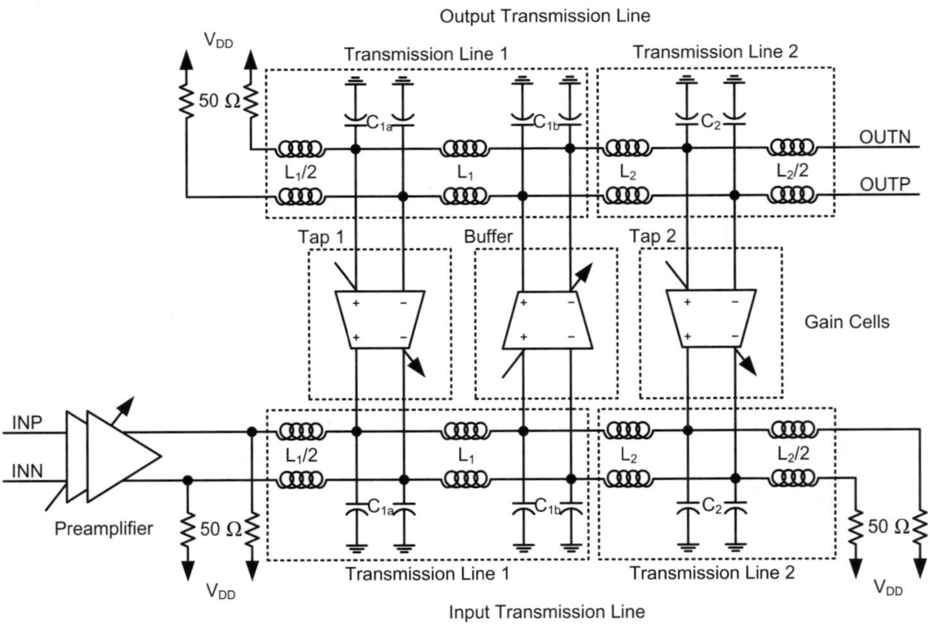

Figure 21.21. Top-level schematic of the double-loop multi-delay IIR TWF equalizer. (Copyright IEEE 2009.)

[23]. The filter delays are realized using lumped L–C transmission sections. The loop containing transmission line 1, tap1, and buffer forms the low-frequency peaking path, while the loop containing transmission line 2, tap2, and buffer forms the high-frequency peaking path. Both transmission line segments were designed to have 50-Ω characteristic impedance. The first segment, transmission line 1, is designed to have 9.375-ps delay. Using Eqs. (21.12) and (21.13), initial values for L_1 and C_{1a} (or C_{1b}) were chosen as 468 pH and 178 fF, respectively. Similarly, transmission line 2 is designed to have 5-ps delay resulting in 250 pH and 100 fF for L_2 and C_2, respectively. The filter utilizes two differently sized gain cells. The center buffer gain cell is larger than the feedback gain cells. The gain cells are sized such that device and parasitic capacitances account for a maximum of 60% of the total node capacitance. The final values of inductance and capacitances were: $L_1 = 370.1$ pH, $L_2 = 220.1$ pH, $C_{1a} = 79.6$ fF, $C_{1b} = 35.4$ fF, and $C_2 = 26$ fF. A lumped two-stage preamplifier that is matched to a 100-Ω differential system impedance precedes the filter. The tail currents of the preamplifier and the gain cells were made adjustable through current mirrors to control their gains.

The IIR filter was implemented in a 90-nm CMOS process and occupied an area of 0.85 mm × 0.625 mm. Figure 21.22 shows the die photo. All circuit measurements were done on wafer. Figure 21.23 plots the measured S-parameters of the filter with both taps on. The filter produces a 12.1-dB peak at 24 GHz. Under these conditions the filter consumes 55.2 mW from a 1.2-V supply. 30-Gbit/s eye diagrams at the filter input and output are shown in Figure 21.24.

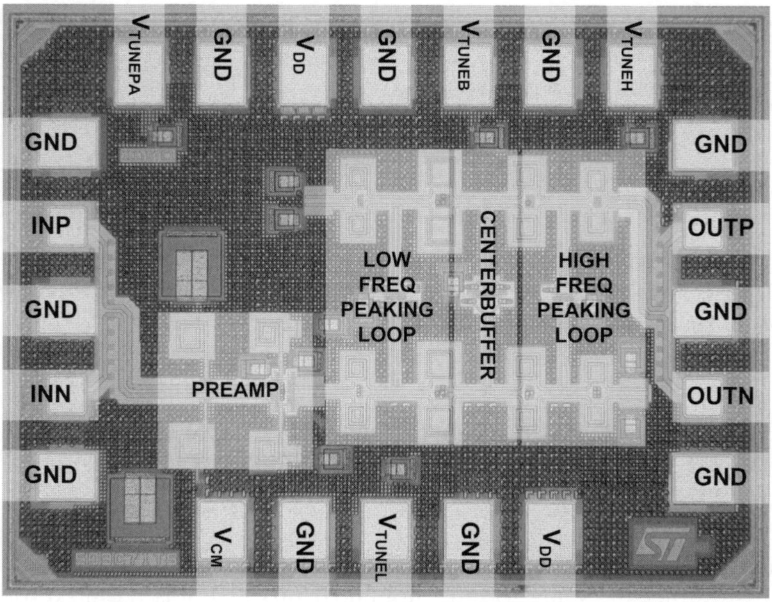

Figure 21.22. Die photo of the 90-nm IIR equalizer. (Copyright IEEE 2009.)

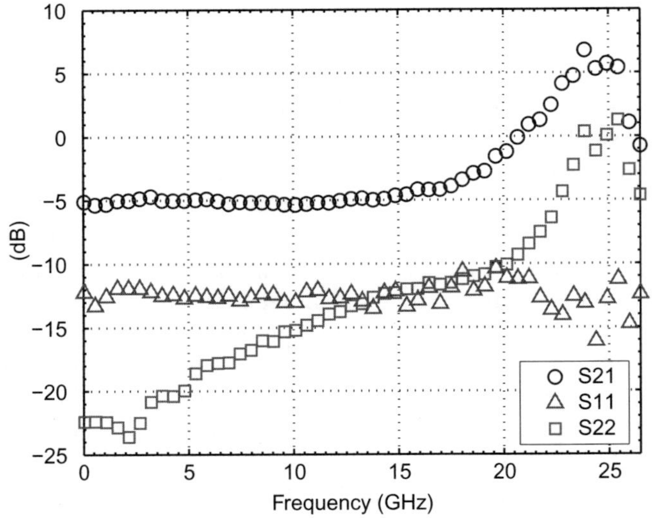

Figure 21.23. Measured examples of mixed peaking. $V_{TUNEL} = 0\,V$, $V_{TUNEH} = 0.4\,V$; S_{21} (circle), S_{11} (triangle), S_{22} (square). (Copyright IEEE 2009.)

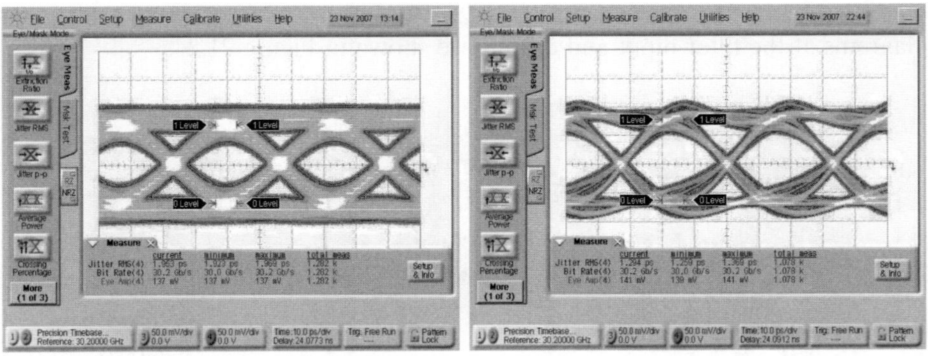

(a) (b)

Figure 21.24. Measured 30.2-Gbit/s channel equalization. **(a)** Unequalized eye input to the filter. **(b)** Equalized eye output of the filter.

21.6 ELECTRONIC DISPERSION COMPENSATION USING NONLINEAR EQUALIZATION: DECISION FEEDBACK EQUALIZATION (DFE)

Ultimately, if the channel has deep nulls in its frequency response, linear equalization is insufficient. Some form of nonlinear processing is required to restore the lost portion of transmitted spectrum. The most common examples are deci-

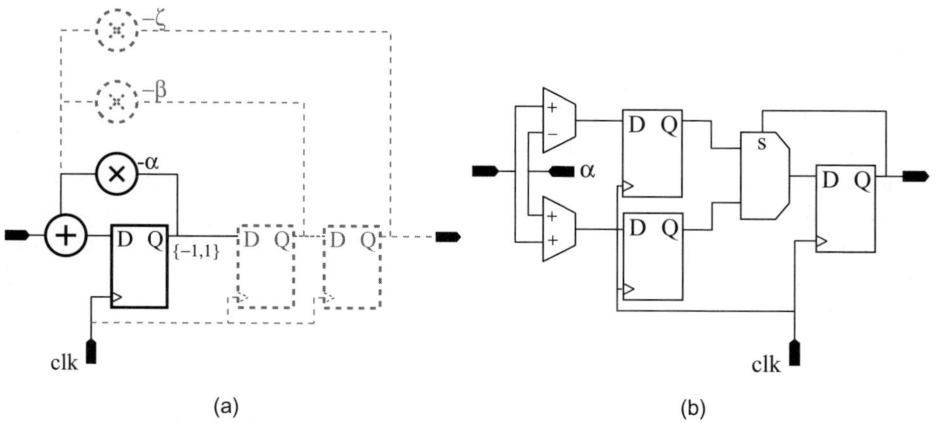

Figure 21.25. DFE architectures. **(a)** Direct feedback architecture. **(b)** Look-ahead architecture reported in references 24 and 25.

sion feedback equalization and maximum likelihood sequence estimation. A decision feedback equalizer (DFE) is a nonlinear receiver that can restore lost portions of transmit spectrum using past decisions. A DFE does not significantly amplify the system noise because it outputs a digital decision instead of an analog voltage. For most practical systems, a DFE is often paired with a feed-forward equalizer (FFE) to correct for a variety of channels.

The direct feedback equalizer is the conventional approach to implementing a DFE as shown in Figure 21.25a. The filter processing is computed in the feedback path and added at the summing node. The implementation allows for multiple taps to be added as seen in Figure 21.25a by the inclusion of the dotted components. From a system viewpoint, direct feedback equalizers still allow for a well-known proven clock and data recovery (CDR) circuit to be used with little added overhead for clock distribution. The major drawback, however, involves the timing bottleneck of the feedback path which must be completed within a bit period. To shorten this critical path and enable higher-speed operation, parallel processing can be employed in a "speculative" or "look-ahead" DFE architecture [24]. By removing the filter stages in the feedback path of the direct form equalizer in Figure 21.25a, the timing constraints are relaxed in the look-ahead architecture shown in Figure 21.25b. The look-ahead architecture equalizes the input signal making multiple decisions. One path makes a tentative decision assuming that the previous bit was a 1, and the other path assumes that the previous bit was a 0. The correct result is then selected, and the other decision is discarded via a decision-selective feedback (DSF) loop. This relaxes the timing constraint within the filter processing as well as removes the summing node for multiple tap designs allowing for high-speed implementation. The addition of the look-ahead architecture, however, complicates the overall system by increasing the complexity of the CDR circuit as well as the clock distribution. This results in an overall increase in the area and power consumption. This technique has been demonstrated in silicon at data rates up to 40 Gbit/s [25].

DFEs can also be realized in the digital domain. This would require an ADC front end to convert the incoming analog signal to a digital signal. The digital signal can then be sent to an FFE in cascade with a DFE. The clock recovery would also be performed digitally. This type of DSP-based architecture has been reported for an MMF link at 10-Gbit/s in 65-nm CMOS [26]. Note that DSP-based transceivers provide robust performance in the presence of severe channel impairments and can be easily scaled with the process, thus resulting in reduced power and area.

21.7 ALTERNATIVE APPROACHES TO DISPERSION COMPENSATION

There are many other alternatives to dispersion compensation. From a research perspective, two of the most noteworthy approaches are discussed here.

21.7.1 Maximum Likelihood Sequence Estimation (MLSE)-Based Receiver

Maximum likelihood sequence estimation (MLSE) recognizes specific patterns of ISI and uses them to inform its decision. This is superior to a DFE-based receiver since a DFE essentially "throws away" signal energy by canceling post-cursor ISI. For example, Figure 21.26 shows a performance improvement of roughly 1.5 dB for an ideal implementation of MLSE compared with a two-tap DFE. However, the ideal implementation of MLSE implies perfect *a priori* knowledge of the channel response at the receiver and signal detection using the energy from the entire pulse response. The complexity of such an implementation would be massive compared to a DFE implementation.

A partial response maximum likelihood (PRML) receiver employs a linear equalizer preceding MLSE to partially equalize the channel response. This results in a shorter channel pulse response that can be processed by an MLSE with practical complexity. However, the performance of a PRML receiver lies somewhere between that of the DFE and ideal MLSE as shown in Figure 21.26, which is due to the noise amplification and correlation of the linear equalizer. Fortunately, noise amplification is minimized and performance close to that of ideal MLSE is obtainable if the partial response target can be chosen to closely match the channel's inherent response.

The Viterbi algorithm is a practical realization of MLSE. Analog implementations of the Viterbi algorithm have been reported for relatively short partial response targets at data rates of hundreds of Mbps [27]. However, most research effort on MLSE receivers at 10 Gbit/s has focused on digital implementations for two main reasons. Firstly, analog Viterbi detectors were not well-suited to the longer and adaptive partial response targets sought in optical communication

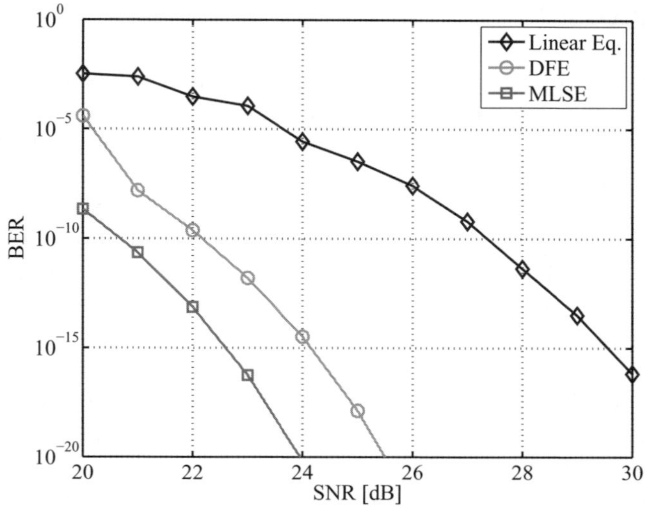

Figure 21.26. Bit error rate versus SNR for the split-pulse channel response in Fig. 21.3c using a six-tap 1/2-UI-spaced linear equalizer, a DFE comprising the same six-tap 1/2-UI-spaced linear equalizer with a two-tap feedback equalizer, and an MLSE receiver. (Copyright IEEE 2008.)

applications. Secondly, advancing CMOS process technologies favor mostly digital architectures. Hence, it would seem that a digital receiver is the ultimate solution for EDC provided that a suitable ADC can be designed. Digital MLSE receivers have already been reported [28]. The current state of the art provides 4- to 6-bit ADCs for 10-Gbit/s systems, but research towards 40 Gbit/s systems is ongoing, with the required ADC having already been demonstrated [29].

21.7.2 Adaptive Optics in Dispersion Compensation

Multiple transmit and receive antennas can be used to combat multipath fading in wireless links. Similarly, multiple lasers and photodetectors can be employed to turn a single optical fiber into a multiple-input multiple-output (MIMO) communication channel with an attendant increase in channel capacity [30]. At the transmitter, the light source is spatially modulated and confined to only one mode of propagation in order to maintain signal integrity [31]. At the receiver, "spatial equalization" is achieved by combining the outputs of an array of photdetectors and canceling interfering propagation modes [32]. Whether at the transmitter or receiver, the optics must be adapted to track time variations in the channel.

Although promising, this avenue of research is still in early stages. Segmented light sources and detectors that are robust and inexpensive are still needed. Advances in signal processing are also required because MIMO methods are currently limited to wireless communication operating at a small fraction of the data rates in optical fiber links.

21.8 CONCLUSION

Dispersion is a major impairment in optical fibers that needs to be compensated in order to achieve high data rates. Integrated circuits can provide cost-effective and adaptive solutions to dispersion compensation by employing improved versions of well-known equalizer topologies. Both linear and nonlinear equalizers were discussed in this chapter. Linear equalizers provide high-frequency boost and are typically implemented using finite impulse response (FIR) or infinite impulse response (IIR) filters. Implementation and measurement results of various FIR filters are presented in Section 21.4. An example of an IIR filter is presented in Section 21.5 with measurement results. Nonlinear architectures such as DFEs are suitable for removing nulls in the frequency response and are presented in Section 21.6. However, these solutions just scratch the surface on EDC, and several difficult challenges are not even touched upon. For instance, adaptation and timing recovery are two key issues that need to be explored. A complete solution to the problem of dispersion is the key to developing reliable broadband and long-reach communication systems.

REFERENCES

1. P. Pepeljugoski, S. E. Golowich, A. J. Ritger, P. Kolesar, and A. Risteski, Modeling and simulation of next-generation multimode fiber links, *J. Lightwave Technol.*, Vol. 21, No. 5, pp. 1242–1255, May 2003.
2. B. Razavi, *Design of Integrated Circuits for Optical Communications*, McGraw-Hill, New York, 2003.
3. C. D. Poole, R. W. Tkach, A. R. Chraplyvy, and D. A. Fishman, Fading in lightwave systems due to polarization-mode dispersion, *IEEE Photon. Technol. Lett.*, Vol. 3, No. 1, pp. 68–70, January 1991.
4. H. Bülow, Operation of digital optical transmission system with minimal degradation due to polarisation mode dispersion, *Electron. Lett.*, Vol. 31, No. 3, pp. 214–215, 1995.
5. G. J. Foschini and C. D. Poole, Statistical-theory of polarization dispersion in single-mode fibers, *J. Lightwave Technol.*, Vol. 9, No. 11, pp. 1439–1456, November 1991.
6. N. Gisin, R. Passy, J. C. Bishoff, and B. Perny, Experimental investigations of the statistical properties of polarization mode dispersion in single mode fibers, *IEEE Photon. Technol. Lett.*, Vol. 5, No. 7, pp. 819–821, July 1993.
7. H. Bülow, W. Baumert, H. Schmuck, F. Mohr, T. Schulz, F. Kuppers, and W. Weiershausen, Measurement of the maximum speed of PMD fluctuation in installed field fiber, in *Optical Fiber Communication Conference, 1999, and the International Conference on Integrated Optics and Optical Fiber Communication. OFC/IOOC '99. Technical Digest*, Vol. 2, 1999, pp. 83–85.
8. R. Ramaswami and K. N. Sivarajan, *Optical Networks—A Practical Perspective, 2nd edition*, Morgan Kaufmann Publishers, San Francisco, 2002.
9. F. Buchali and H. Bülow, Adaptive PMD compensation by electrical and optical techniques, *J. Lightwave Technol.*, Vol. 22, No. 4, pp. 1116–1126, April 2004.

10. H. Bülow, Polarisation mode dispersion (PMD) sensitivity of a 10 Gbit/s transmission system, in *Optical Communication, 1996. ECOC '96. 22nd European Conference on*, Vol. 2, 1996, pp. 211–214.

11. S. Pavan, Continuous-time integrated FIR filters at microwave frequencies, *IEEE Trans. Circuits Syst. II*, Vol. 51, No. 1, pp. 15–20, January 2004.

12. J. Jaussi, G. Balamurugan, D. Johnson, B. Casper, A. Martin, J. Kennedy, N. Shanbhag, and R. Mooney, 8-Gb/s source-synchronous I/O link with adaptive receiver equalization, offset cancellation, and clock de-skew, *IEEE J. Solid-State Circuits*, Vol. 40, No. 1, pp. 80–88, January 2005.

13. J. Sewter and A. Chan Carusone, A CMOS finite impulse response filter with a crossover traveling wave topology for equalization up to 30 Gb/s, *IEEE J. Solid-State Circuits*, Vol. 41, No. 4, pp. 909–917, April 2006.

14. J. Sewter and A. Chan Carusone, A 3-tap FIR filter with cascaded distributed tap amplifiers for equalization up to 40 Gb/s in 0.18-μm CMOS, *IEEE J. Solid-State Circuits*, Vol. 41, No. 8, pp. 1919–1929, August 2006.

15. G. Ng and A. C. Carusone, A 38-Gb/s 2-tap transversal equalizer in 0.13-μm CMOS using a microstrip delay element, in *IEEE RFIC Conference*, Atlanta, Georgia, June 2008.

16. W. Jutzi, Microwave bandwidth active transversal filter concept with MESFETs, *IEEE Trans. Microwave Theory Tech.*, Vol. MTT-19, No. 9, pp. 760–767, September 1971.

17. H. Wu, J. A. Tierno, P. Pepeljugoski, J. Schaub, S. Gowda, J. A. Kash, and A. Hajimiri, Integrated transversal equalizers in high-speed fiber-optic systems, *IEEE J. Solid-State Circuits*, Vol. 38, No. 12, pp. 2131–2137, December 2003.

18. C. Pelard, E. Gebara, A. J. Kim, M. G. Vrazel, F. Bien, Y. Hur, M. Maeng, S. Chandramouli, C. Chun, S. Bajekal, S. E. Ralph, B. Schmukler, V. M. Hietala, and J. Laskar, Realization of multigigabit channel equalization and crosstalk cancellation integrated circuits, *IEEE J. Solid-State Circuits*, Vol. 39, No. 10, pp. 1659–1670, October 2004.

19. M. Nakamura, H. Nosaka, M. Ida, K. Kurishima, and M. Tokumitsu, Electrical PMD equalizer ICs for a 40-Gbit/s transmission, in *Optical Fiber Communication Conference and Exhibit, 2004, OFC 2004*, Vol. TuG4, 2004.

20. A. Hazneci and S. P. Voinigescu, A 49-Gb/s, 7-tap transversal filter in 0.18 μm SiGe BiCMOS for backplane equalization, in *IEEE Compound Semiconductor Integrated Circuits Symposium*, Monterey, CA, October 2004.

21. R. Tiruvuru and S. Pavan, Transmission line basd FIR structures for high speed adaptive equalization, in *IEEE International Symposium on Circuits and Systems*, Kos, Greece, May 2006, pp. 1051–1054.

22. S. Reynolds, P. Pepeljugoski, J. Schaub, J. Tiemo, and D. Beisser, A 7-tap transverse analog-fir filter in 0.13 μm CMOS for equalization of 10 Gb/s fiber-optic data systems, in *IEEE ISSCC Digest Technical Papers*, February 2005, pp. 330–331.

23. G. Ng, F. A. Musa, and A. C. Carusone, A 2-tap traveling wave infinite impulse response (IIR) filter with 12-dB peaking at 24-GHz, *Electron. Lett.*, Vol. 45, No. 9, pp. 463–464, 2009.

24. S. Kasturia and J. H. Winters, Techniques for high-speed implementation of nonlinear cancellation, *IEEE J. Selected Areas Commun.*, Vol. 9, No. 5, pp. 711–717, June 1991.

25. A. Garg, A. Chan Carusone, and S. P. Voinigescu, A 1-tap 40-Gbps look-ahead decision feedback equalizer in 0.18-μm SiGe BiCMOS technology, *IEEE J. Solid-State Circuits*, Vol. 41, No. 10, pp. 2224–2232, October 2006.
26. J. Cao, B. Zhang, U. Singh, D. Cui, A. Vasani, A. Garg, W. Zhang, N. Kocaman, D. Pi, B. Raghavan, H. Pan, I. Fujimori, and A. Momtaz, A 500 mw digitally calibrated AFE in 65 nm CMOS for 10 Gb/s serial links over backplane and multimode fiber, in *Solid-State Circuits Conference, 2009, Digest of Technical Papers, ISSCC. 2009 IEEE International*, 2009, pp. 370–371.
27. B. Zand and D. Johns, High-speed CMOS analog Viterbi detector for 4-PAM partial-response signaling, *IEEE J. Solid-State Circuits*, Vol. 37, No. 7, pp. 895–903, July 2002.
28. H. min Bae, J. Ashbrook, J. Park, N. Shanbhag, A. Singer, and S. Chopra, An MLSE receiver for electronic-dispersion compensation of OC-192 fiber links, in *International Solid-State Circuits Conference, Digest of Technical Papers*, San Francisco, CA, February 2006, pp. 874–883.
29. S. Shahramian, S. P. Voinigescu, A. Chan Carusone, "A 35-GS/s, 4-bit flash ADC with Active Data and Clock Distribution Trees," *IEEE Journal of Solid-State Circuits*, Vol. 44, Issue. 6, pp. 1709–1720, June 2009.
30. H. Stuart, Dispersive multiplexing in multimode optical fiber, *Science*, Vol. 289, No. 5477, pp. 281–283, July 2000.
31. J. Kahn, Compensating multimode fiber dispersion using adaptive optics, in *Optical Fiber Conference*, Anaheim, California, March 2007, p. OTuL1.
32. K. Patel, A. Polley, K. Balemarthy, and S. Ralph, Spatially resolved detection and equalization of modal dispersion limited multimode fiber links, *J. Lightwave Technol.*, Vol. 24, No. 7, pp. 2629–2636, July 2006.

22

HIGH-END SILICON PHOTODIODE INTEGRATED CIRCUITS

Bernhard Goll, Robert Swoboda, and Horst Zimmermann

Optoelectronic integrated circuits (OEICs) [1–3] are used for very compact designs and low-cost application—for example, for optical storage systems such as CD-ROM, DVD, digital video recording, enhanced video disc (EVD), and Blu-ray Disc (BD) or for fiber receivers such as low-cost data comm receivers for plastic optical fibers (POFs). They are important because the inherent compactness of monolithic optical integrated circuits gives the following advantages: good immunity against electromagnetic interference (EMI) due to very short interconnects between photodetectors and amplifiers, reduced chip area because of elimination of bondpads, improved reliability due to the elimination of bondpads and bond wires, cheaper mass production compared to discrete-, wire-bonded-, or some hybrid integrated circuits, and a larger –3-dB bandwidth compared to discrete- or wire-bonded circuits, because of the avoidance of parasitic bondpad capacitances.

Inexpensive optical receivers with top performance are in great demand. Especially the low-cost requirement permanently drives and pushes silicon photodiode integrated circuts (PDICs) instead of expensive III/V semiconductor receivers. For high data rates of more than 10 Gbit/s, III/V receivers have a better performance and it is useful to accept their high costs; but for lower data rates

Convergence of Mobile and Stationary Next-Generation Networks, edited by Krzysztof Iniewski

up to about 10 Gbit/s, silicon PDICs with their lower production costs combined with the advanced circuit design of silicon chips are more attractive for mass market and are true competitors.

This chapter concerns silicon photodiode integrated circuits (PDICs), where the focus is on silica fiber receivers and on low-cost data comm receivers for plastic optical fiber (POF) applications. After a short explanation of some basics and a typical integrated pin-photodiode, circuit concepts of fiber receivers with data rates up to 11 Gbit/s as well as a parallel optical receiver for data rates of 100 + Gbit/s and a multilevel receiver are described.

22.1 OPTICAL ABSORPTION OF IMPORTANT SEMICONDUCTOR MATERIALS

Light can be described by a wave formalism and, due to the work of Max Planck and Albert Einstein, also by a quantum-mechanical light particle (photon) formalism. A photon does not possess a quiescent mass and can be characterized by its frequency f and by its wavelength λ as well as by its energy E [see Eqs. (22.1) and (22.2)], where c is the speed of light in a medium, c_0 is the speed of light in vacuum, n is the index of refraction of the medium, f the frequency, λ is the wavelength in the medium, λ_0 is the vacuum wavelength, and h is Planck's constant.

$$\lambda = \frac{c}{f} = \frac{c_0}{nf} \tag{22.1}$$

$$E = hf = \frac{hc}{\lambda} = \frac{hc_0}{\lambda_0} \tag{22.2}$$

In semiconductors the bandgap energy E_g, which is the distance between the conduction band and the valence band, determines the boundary wavelength of light, which can be absorbed by the material. Silicon has a bandgap energy of E_g = 1.1 eV and a boundary wavelength λ_c according to Eq. (22.3) can be stated.

$$\lambda_c = \frac{hc_0}{E_g} \tag{22.3}$$

When the photon energy E is larger than the bandgap energy E_g, this energy can be transferred to an electron in the valence band of a semiconductor. These photons are absorbed, electron–hole pairs are generated, and thus a photocurrent is able to flow. Only light with shorter wavelengths than λ_c can be detected, because then absorption increases rapidly according to the so-called fundamental absorption. For wavelengths longer than λ_c the semiconductor is transparent. For silicon, λ_c = 1110 nm.

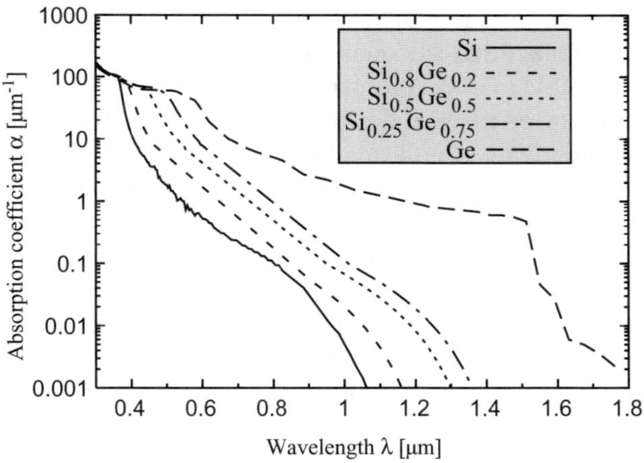

Figure 22.1. Absorption coefficient α of silicon, germanium, and several SiGe alloys versus light wavelength λ.

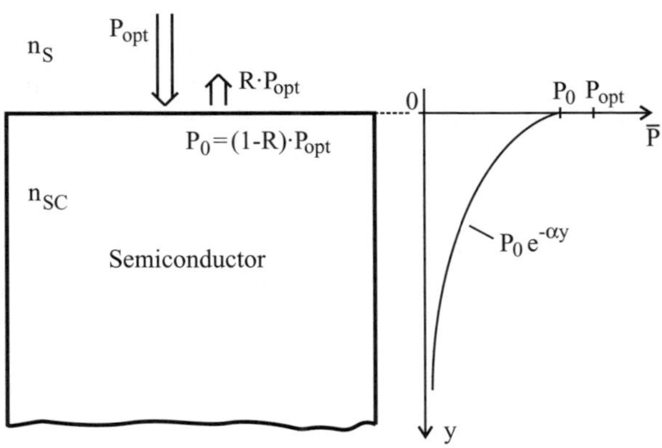

Figure 22.2. Decay of the optical power in the semiconductor ($P_0 = (1 - R)P_{opt}$) including reflection at the surface.

The most important optical constant for photodetectors is the optical absorption coefficient α (see Figure 22.1) or the reciprocal value $1/\alpha$, which is the $1/e$-penetration depth according to Lambert–Beer's law in Eq. (22.4), where I_0 is the light intensity at $y = 0+$ and $I(y)$ is the light intensity in a depth y below the surface of the semiconductor (see Figure 22.2). The absorption coefficient strongly depends on the wavelength of the light. As mentioned before, the absorption coefficient gets large for wavelengths shorter than λ_c. At λ_c, the steepness of the onset of absorption is large for direct band–band transition like in GaAs ($E_g^{dir} = 1.42$ eV at 300 K), in InP ($E_g^{dir} = 1.35$ eV at 300 K), or in Ge ($E_g^{dir} = 0.81$ eV at 300 K) and small for indirect band–band transition as in Si ($E_g^{ind} = 1.12$ eV at 300 K) or in Ge ($E_g^{ind} = 0.67$ eV at 300 K). In Figure 22.1 the optical absorption coefficients α for Si, Ge, and several SiGe alloys versus the wavelength are depicted. Due to the fact that silicon is an indirect semiconductor, the absorption coefficient in the visible (≈ 400–780 nm) and near-infrared spectrum (780–850 nm) is one or two orders of magnitude lower than that of the semiconductor Ge, where its direct bandgap is only a little bit larger than its indirect bandgap. Although it is worthwhile to investigate silicon photodiode integrated circuits because it is the economically most important semiconductor. Equation (22.4) is multiplied at both sides with the cross-section area A of light incidence to get the optical power [Eq. (22.5)]. In Figure 22.2 the decay of the optical power in the semiconductor with considering reflection of the incoming optical power $P_{\text{opt}} = hf\Phi A$ (hf is the energy of a photon and Φ is the number of photons incident per time interval on an area A) according to Eq. (22.5) is shown.

$$I(y) = I_0 e^{-\alpha y} \tag{22.4}$$

$$P(y) = P_0 e^{-\alpha y}, \qquad P_0 = P_{\text{opt}}(1 - R) \tag{22.5}$$

The reflection of the surface is caused by the different refraction indices between air ($n_s = 1$ and the semiconductor ($n_{\text{SC}} \approx 3.5$ for SI). The reflectivity R depends on [see Eq. (22.6)] the index of refraction n_{SC} and on the extinction coefficient κ of an absorbing medium, for which the dielectric function $\varepsilon = (n_{\text{SC}} + j\kappa)^2$ is valid, where ε is the complex permittivity of the medium and j the imaginary unit.

$$R = \frac{(1 - n_{\text{SC}})^2 + \kappa^2}{(1 + n_{\text{SC}})^2 + \kappa^2} \tag{22.6}$$

For calculating the absorption coefficient α, the extinction coefficient κ is sufficient [see Eq. (22.7)].

$$\alpha = \frac{4\pi\kappa}{\lambda_0} \tag{22.7}$$

To avoid reflection, the reflectivity R has to vanish. This can be done for distinct frequencies by adding an anti-reflection coating (ARC) layer between air and semiconductor with refractive index $n_{\text{ARC}} = \sqrt{n_S n_{\text{SC}}}$ and with thickness $d_{\text{ARC}} = \lambda_0/(4n_{\text{ARc}})$.

22.2 *P–I–N* PHOTODIODE

A principal drawing of a *p–i–n* photodiode is shown in Figure 22.3. It consists of a *p–n* junction with an intrinsic (undoped) layer in between. To achieve a strong electric field E in the intrinsic layer, the photodiode is reverse biased with about 3–10 V [3]. The incoming photons (each with energy hf) create electron–hole pairs

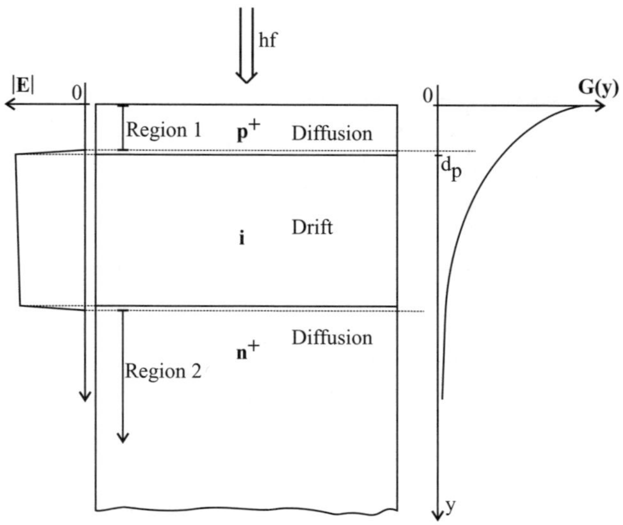

Figure 22.3. Schmatic drawing of a *p–i–n* photodiode.

with a generation rate per volume $G(y)$ due to the internal photoeffect with the condition that $hf > E_g$. With the optical power $P(y)$ [see Eq. (22.5)] $G(y)$ is equal to $-\frac{dP(y)}{dy}\frac{1}{Ahf}$. The result of this differential equation is shown in Eq. (22.8) and in Figure 22.3.

$$G(y) = \frac{\alpha P_0}{Ahf} e^{-\alpha y} \tag{22.8}$$

The generated electron–hole pairs are seperated in the intrinsic layer because of the high electic field E and an electrical current begins to flow. To characterize photodetectors with respect to their speed, the most important processes are carrier drift and minority carrier diffusion. In a typical semiconductor the carrier drift is a much faster process than the minority carrier diffusion. The influence of carrier drift and minority carrier diffusion in the time and frequency domain of the photocurrent is sketched in Figure 22.4. Besides junction capacitances, additional parasitic capacitances and series resistances are also important for calculation of a frequency response.

22.2.1 Carrier Drift

A photogenerated electron–hole pair in the intrinsic region is separated by the electric field. Due to this electric field in the reversed biased diode, the electron drifts toward region 2 (n^+) while the hole drifts toward region 1 (p^+) as shown in Figure 22.3. The drift velocity of electrons υ_n is proportional to the electric field

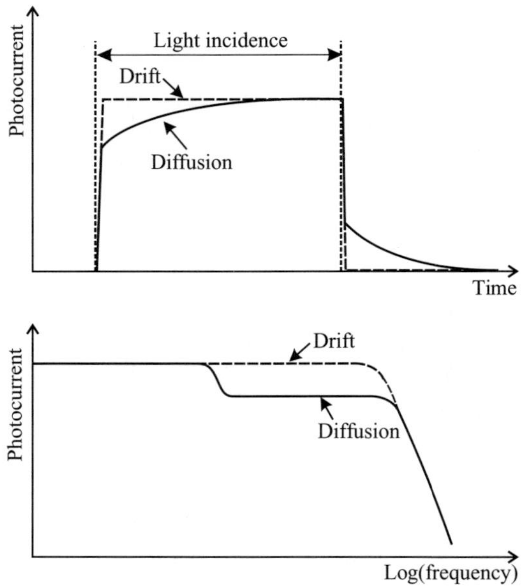

Figure 22.4. Schematic influence of carrier drift and minority carrier diffusion in the time and frequency domain of the photocurrent.

E, $v_n = \mu_n E$, where μ_n is the mobility for electrons. The drift velocity of holes v_p is proportional to E, $v_p = \mu_p E$, where the proportionality factor is the mobility of holes μ_p. The mobilities of electrons and holes depend, for example, on the temperature (higher temperature causes lower mobilities), on the doping concentration, and on the strength of the electric field. In silicon with low doping concentration ($<10^{-16}\,\text{cm}^{-3}$), and electric field strengths of $|E| < 2000\,\text{V/cm}$, we have typical values of mobilities of $\mu_n \approx 1200\,\text{cm}^2/\,\text{Vs} \approx \mu_{n,0}$ and $\mu_p \approx 450\,\text{cm}^2/\text{Vs} \approx \mu_{p,0}$ at a temperature of $T = 300\,\text{K}$, If the temperature is increased, the mobilities get smaller due to larger thermal vibration of lattice atoms and impurity atoms. When the doping concentration is higher the mobilities get smaller due to more scattering centers. The intrinsic region of a *p–i–n* photodiode is designed to have a low doping concentration and $\mu_n = \mu_{n,0}$ as well as $\mu_p = \mu_{p,0}$. If the field strength E is raised, the mobilities also degrade, as can be seen in Eqs. (22.9) and (22.10) [4].

$$\mu_n = \frac{\mu_{n,0}}{1 + \frac{\mu_{n,0} E}{v_n^{sat}}} \tag{22.9}$$

$$\mu_p = \frac{\mu_{p,0}}{1 + \frac{\mu_{p,0} E}{v_p^{sat}}} \tag{22.10}$$

For values larger than approximately 10,000 V/cm the electron drift velocity v_n saturates to v_n^{sat}. On the contrary 25 the hole drift velocity v_p saturates not until a value of about 100,000 V/cm. The values for the saturation velocities v_n^{sat} and v_p^{sat} in cm/s are nearly equal and can be computed from Eq. (22.11) [5], where T is the temperature in K.

$$v_{sat} = v_n^{sat} = v_p^{sat} = \frac{2.4 \times 10^7}{1 + 0.8e^{\frac{T}{100}}} \tag{22.11}$$

Ideally, to a *p–i–n* photodiode a sufficient reverse voltage is applied to get such a large electric field in the intrinsic zone that the drift velocities of electrons and holes saturate to v_{sat}. To estimate rise and fall time (t_r and t_f, respectively) of the photocurrent, it is assumed that one of the charged particles of an electron–hole pair, which is generated at one of the borders of the intrinsic zone to the heavily doped areas, has to drift the whole length d_I of the intrinsic zone and lasts the drift time t_I. This marks the end of the rise time or the fall time, which therefore can be calculated as stated in Eq. (22.12).

$$t_I = t_r = t_f = \frac{d_I}{v_s} \tag{22.12}$$

Rise time and fall time limit the data rate, which can be received by the photodiode. Usually, drift times are much shorter than diffusion times, which are explained next.

22.2.2 Carrier Diffusion

As can be seen in Figure 22.4, minority carrier diffusion may degrade the bandwidth of a photodiode. Electron–hole pairs are generated not only in the intrinsic region of a pin photodiode, but also in the heavily doped p^+ region 1 and in n^+ region 2 (see Figure 22.3), where no electric field is present. In the p^+ region the hole will be absorbed by the voltage source and the electron will first of all diffuse as minority carrier toward the intrinsic zone and then drift through this region and finally be absorbed by the voltage source, when it reaches region 2 and becomes a majority carrier. Diffusion of minority carriers in the p^+ region is in most cases not critical due to the small thickness. On the other hand, a generated hole from the deep n^+ region 2, which diffuses toward the intrinsic zone, has to overcome more distance and therefore needs more time. So critical for speed is usually minority carrier diffusion in the deep region 2. A characteristic value of what average distance a carrier is able to diffuse without recombination are the diffusion lengths L_n or L_p for electrons or holes, respectively, which can be seen in Eqs. (22.13) and (22.14) [2].

$$L_n = \sqrt{D_n \tau_n}, \qquad D_n = \mu_n \frac{k_B T}{q} = \mu_n U_T \tag{22.13}$$

$$L_p = \sqrt{D_p \tau_p}, \qquad D_p = \mu_p \frac{k_B T}{q} = \mu_p U_T \tag{22.14}$$

D_n and D_p are the diffusion coefficients for electrons and holes, respectively, and can be calculated via the Einstein relation [see Eqs. (22.13) and (22.14)], where k_B is Boltzmann's constant, q is the elementary charge, and U_T is the thermal voltage. In silicon, $D_n \approx 35\,\text{cm}^2/\text{s}$ and $D_p \approx 12.5\,\text{cm}^2/\text{s}$ for low doping concentrations. These values reduce for high doping. The physical lifetime of a carrier (τ_n or τ_p for an electron or a hole, respectively) is due to dopants and unwanted impurities, which act as additional recombination centers in the semiconductor. These unwanted impurities are unavoidable because they are introduced in quite small but noticeable amounts in electronic devices during the fabrication process. The sizes of minority carrier lifetimes can reach several milliseconds in unprocessed grown silicon crystal. The fabrication process of integrated circuits or photodetectors can reduce the carrier lifetimes to the order of microseconds. Usually, *p–i–n* photodiodes are designed such that carrier drift is dominant over carrier diffusion, so usually diffusion times do not deteriorate the dynamical behavior. As mentioned above, the main influence of minority carrier diffusion to the photocurrent (see Figure 22.4) has its origin in the deep n^+ region of the pin photodiode due to the larger thickness compared to the thin p^+ region at the surface. For estimation of the time t_{diff} for which a hole diffuses in region 2 via a length d_n (e.g., the thickness of region 2), an equation for a time-dependent sinusoidal electron density due to photogeneration in a p^+ layer of a *p–n* photodiode was calculated in references 6 and 7, which is here adapted for an n^+ region [see Eq. (22.15)].

$$t_{\text{diff}} = \frac{d_n^2}{2D_p} = \frac{d_n^2 q}{2\mu_p k_B T} \tag{22.15}$$

In reference 7 the equation for t_{diff} has been derived for a time-dependent sinusoidal electron density due to photogeneration in the p^+ layer from the electron diffusion equation. There the time response of a low-pass filter with time constant $\tau_r = d_p^2/(2D_n)$ was obtained. This time constant was interpreted as the time that electrons need to diffuse to the space-charge region of a p^+n junction.

22.2.3 Quantum Efficiency and Responsivity

The total, external, or overall quantum efficiency η is defined as the number of photogenerated electron–hole pairs, which contribute to the photocurrent, divided by the number of incident photons [3]. This overall quantum efficiency can be determined when the photocurrent of a photodetector is measured for a known incident optical power. Because a fraction of the incident optical power P_{opt} (see Figure 22.2) is reflected at the surface of a semiconductor, the external quantum efficiency η is the product of the optical quantum efficiency $\eta_0 = 1 - R$ and of the internal quantum efficiency η_i, which is a measure of how many electron–hole pairs contribute to the photocurrent [see Eq. (22.16)]. For example, generated electron–hole pairs may be lost due to recombination; or if there is a fast switching of optical power at high data rates, the slower minority carrier diffusion tail (see Figure 22.4) do not contribute to the photocurrent and the amplitude gets smaller.

$$\eta = \eta_0 \eta_i \tag{22.16}$$

The responsivity R_λ [see Eq. (22.17)] of a photodetector is a useful quantity for the development of photoreceiver circuits and especially of transimpedance amplifiers, because it is interesting to know how large the photocurrent is for a specified power of the incident light with a certain wavelength λ_0. The responsivity is defined as the photocurrent i_{ph} divided by the incident optical power P_{opt}.

$$R_\lambda = \frac{i_{ph}}{P_{opt}} = \frac{q\lambda_0}{hc}\eta \tag{22.17}$$

In Eq. (22.17), q is the elementary charge, h is Planck's constant, and c is the speed of light in a medium. R_λ depends on the wavelength; therefore the wavelength has to be mentioned if a responsivity value is given.

Figure 22.5 shows the reponsivity of real photodetectors versus the wavelength. The dashed line represents the maximum responsivity of an ideal photodetector with $\eta = 1$ (100%). The responsivity of real detectors is always lower due to partial reflection of the light at the semiconductor surface and due to partial recombination of photogenerated carriers in the semiconductor or at its surface.

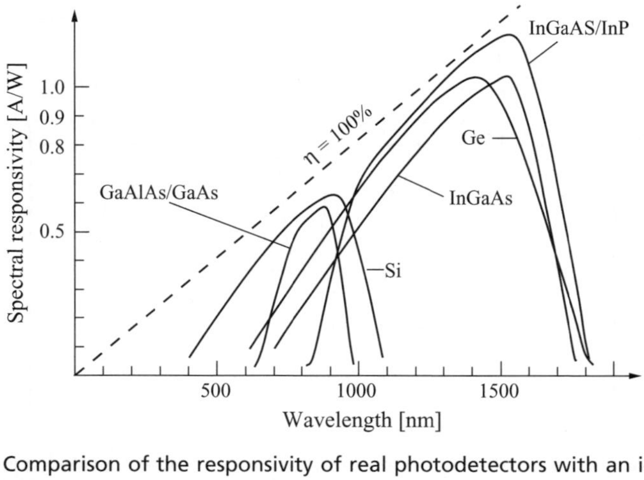

Figure 22.5. Comparison of the responsivity of real photodetectors with an ideal photodetector with a quantum efficiency $\eta = 1$ (100%).

22.2.4 Implementation of an Integrated *p–i–n* Pliotodiode

All integrated photodiode receivers, which are described here, are implemented in a 0.6-μm BiCMOS process with a nominal supply voltage of 5V. The integrated *p–i–n* photodiode that was used for these receivers is similar to the design of the diode presented in reference 8, where the cross section is shown in Figure 22.6. Deep n^+ plugs after low-doped *n*-epitaxial steps form the connection from the Si surface to the n^+ buried-layer cathode. The electrical isolation between the *n–p–n* transistors and the *n* wells is formed by a *p*-isolation layer, which separates them. In reference 8 the reverse bias voltage was 3V, and at 780-nm light a bandwidth of 300MHz was achieved. To improve the speed of the photodiode with speeding up the carrier drift and as a consequence the bandwidth, the reverse bias voltage can be raised to values larger than the chip supply voltage of typically 5V. This is because the *n* cathode is realized in a *p*-type substrate.

22.2.5 Equivalent Circuit of a Photoreceiver

A small-signal equivalent circuit of the input of a photoreceiver considering parasitics due to mounting is shown in Figure 22.7. The current source i_{ph} represents the photocurrent, C_D is the capacitance of the space-charge region with intrinsic region of a *p–i–n* photodiode. The parallel resistor R_D models the reverse, leakage, or dark current. It is usually very large and can be neglected. The series resistor R_S may not be neglected when the photocurrent flows through low-doped regions. But for a *p–i–n* photodiode with highly doped *p* and *n* regions it can be neglected when carrier drift dominates and the intrinsic zone is fully depleted. L_B represents the inductance of the bondwire and C_B is the capacitance of the bond wire. The stray capacitance of the detector package is considered in C_P.

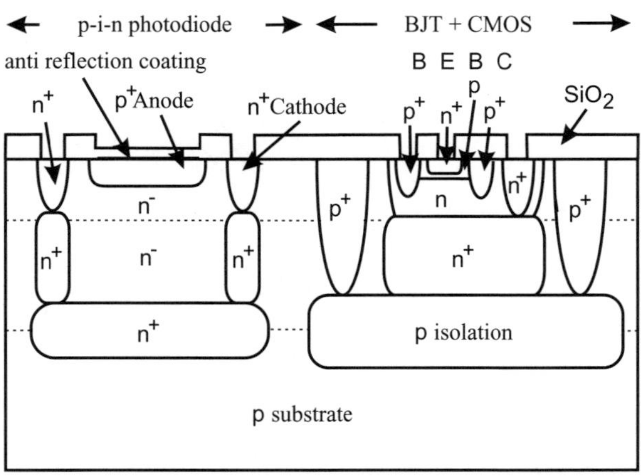

Figure 22.6. Integration of a *p–i–n* photodiode in a standard buried collector BiCMOS (bipolar + CMOS) technology [8] (BJT, bipolar-junction transistor).

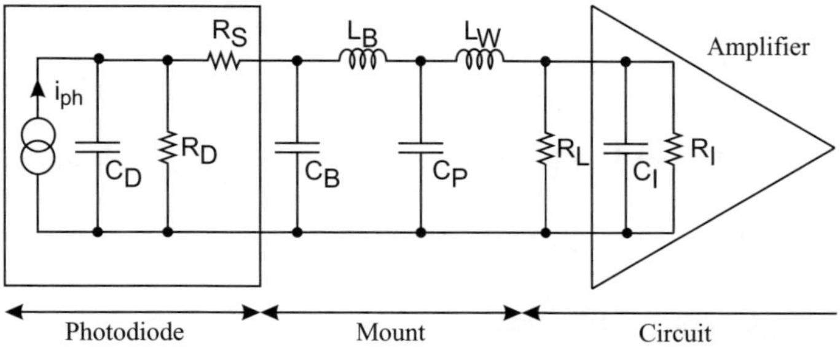

Figure 22.7. Small-signal equivalent input circuit of a photoreceiver.

The lead wires and the wiring on electronic boards introduce the additional inductance L_W. For high frequencies, the inductances L_B and L_W become important. R_L represents a load resistor; and C_I and R_I are the input capacitance and the input resistance, respectively of an amplifier. R_I is important for amplifiers with a bipolar transistor input but can be neglected in most cases for JFET or MOS input transistors. It is clear that in a fully integrated photoreceiver the parasitics caused by mounting (C_B, L_B, C_P, and L_W) between the photodiode and the amplifier are saved, which enhances the overall bandwidth. This can be seen in Eq. (22.18), where the –3-dB bandwidth f_{3dB} of the equivalent circuit of Figure 22.7 with the assumption that L_B and L_W can be neglected is depicted.

$$f_{3\,dB} = \frac{1}{2\pi(R_L \| R_I)(C_D + C_B + C_P + C_I)} \tag{22.18}$$

22.2.6 Bit Error Ratio and Sensitivity

In the design of photodiode receivers, many types of noise have to be considered—for example, either (a) electronic noise caused by the photodiode and the transimpedance amplifier or (b) received optical noise, which is overlaid to optical data signal. For more details on the different noise sources here, see references 2 and 3. Here the overall noise will be represented by an input-referred noise current with variance $\langle \delta i_{\text{ges}}^2 \rangle$ at the transimpedance amplifier input, which can be obtained by integration of the spectral noise density $\langle \delta i^2(f) \rangle$ over a bandwidth B [see Eq. (22.19)], which defines the effective noise bandwidth. Typically this bandwidth is between 0.5× to 0.75× of the bit rate [9].

$$\langle \delta i_{\text{ges}}^2 \rangle = \int_0^B \langle \delta i^2(f) \rangle \, df \tag{22.19}$$

It is assumed that the instant current values have a Gaussian distribution around the expected values $\langle i_0 \rangle$ and $\langle i_1 \rangle$, as can be seen in Figure 22.8. The resulting probability density function for a logical zero ($p_0(i_{\text{ph}})$) and a logical one ($p_1(i_{\text{ph}})$) are depicted in Eqs. (22.20) and (22.21).

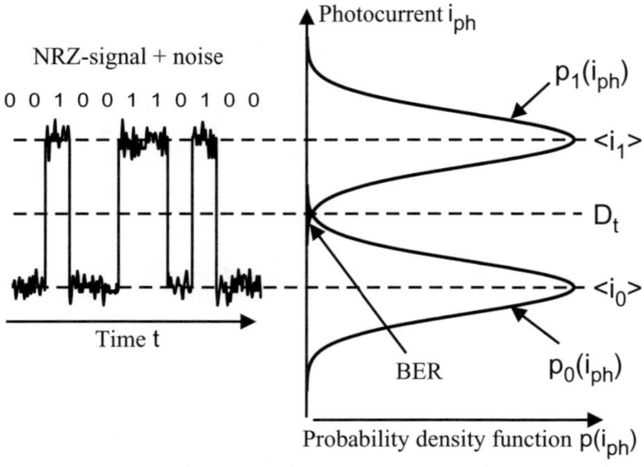

Figure 22.8. Noisy non-return-to-zero (NRZ) input data signal (photocurrent i_{ph}) with noise statistics (probability density functions $p_0(i_{ph})$ and $p_1(i_{ph})$) with variance $\langle \delta i_{ges}^2 \rangle$ of the noise). Here the decision threshold D_t is the mean value of $\langle \delta i_0 \rangle$ and $\langle \delta i_1 \rangle$ and the logical zeros and ones are assumed to be equally distributed.

$$p_0(i_{ph}) = \frac{1}{\sqrt{2\pi \langle \delta i_{ges}^2 \rangle}} e^{-\frac{(i_{ph} - \langle i_0 \rangle)^2}{2\langle \delta i_{ges}^2 \rangle}} \tag{22.20}$$

$$p_1(i_{ph}) = \frac{1}{\sqrt{2\pi \langle \delta i_{ges}^2 \rangle}} e^{-\frac{(i_{ph} - \langle i_1 \rangle)^2}{2\langle \delta i_{ges}^2 \rangle}} \tag{22.21}$$

Furthermore, it is assumed that the logical zeros and the logical ones are equally distributed. Then the decision threshold for minimum decision failures is $D_t = (\langle i_0 \rangle + \langle i_1 \rangle)/2$, where for a current $i_{ph} < D_t$ the detected signal is considered to be logical zero and for a current $i_{ph} > D_t$ it is considered to be logical one. With these assumptions, the bit error ratio (BER) or the probability of a wrong decision is calculated by adding the probabilities that a zero is recognized by the receiver in the case that a one has been sent and the probability that a one is recognized in the case a zero has been sent. This probability is the black area in Figure 22.8 in the diagram with the probability density function. The probability of a false decision of the receiver is equal to the BER and can be calculated as depicted in Eq. (22.22).

$$\text{BER} = \int_{-\infty}^{D_t} p_1(i_{ph})\, di_{ph} + \int_{D_t}^{\infty} p_0(i_{ph})\, di_{ph} = \frac{1}{\sqrt{2\pi}} \int_{d_n}^{\infty} e^{\frac{-u^2}{2}}\, du \tag{22.22}$$

The noise distance d_n is depicted in Eq. (22.23).

$$d_n = \frac{\langle i_1 \rangle - \langle i_0 \rangle}{2\sqrt{\langle \delta i_{\mathrm{ges}}^2 \rangle}} \tag{22.23}$$

For instance, BER = 10^{-9} holds at $d_n = 6$ and BER = 1.3×10^{-12} for $d_n = 7$. The ratio $r = \langle i_0 \rangle / \langle i_1 \rangle$ of the two photocurrent levels can be used to determine the mean optical power $\langle P \rangle$ of a binary signal with equally distributed logical zeros and ones, which is necessary to achieve a certain required BER [see Eq. (22.24)]. The BER is measured with the help of a fast oscilloscope which is able to catch eye-diagrams and amplitude statistics. Another possibility is to use a BER analyzer station, which compares the sent data stream with the received one and counts the errors. For the data stream for measurements on an oscilloscope or with an analyzer station a pseudo-random bit sequence with a certain length (e.g., $2^{31} - 1$) is used, which simulates random mean-free bit sequence, where the logical states are equally distributed.

$$\langle P \rangle = \frac{1+r}{1-r} \frac{hf}{\eta q} d_n \sqrt{\langle \delta i_{\mathrm{ges}}^2 \rangle} \tag{22.24}$$

Equation (22.24) was achieved by combining $D_t = (\langle i_0 \rangle) + \langle i_1 \rangle)/2$ with Eq. (22.23) by using $D_t = \langle P \rangle q\eta / hf$. The quantity $\langle P \rangle$ in Eq. (22.24) is called sensitivity of the optical receiver and is usually expressed in dBm ($10 \times \log(\langle P \rangle / 1\,\mathrm{mW})$). It becomes lowest for $r = 0$—that is, $\langle i_0 \rangle = 0$ or a vanishing optical power for a logical zero. As can be seen in Eq. (22.24), the sensitivity depends on the frequency f and therefore on the wavelength λ.

22.3 INTEGRATED PHOTODIODE RECEIVERS

22.3.1 An Integrated Silicon Photodiode Receiver for 5 Gbit/s with Voltage Up Converter

This integrated photodiode receiver uses a similar *p–i–n* photodiode as shown in Figure 22.6 integrated in a 0.6-μm silicon BiCMOS process with a nominal supply voltage of 5 V [10]. To improve speed and therefore the bandwidth of the photodiode the reverse bias voltage of the diode was set to $V_{\mathrm{PD}} = 11\,\mathrm{V}$, a much larger voltage than the nominal supply voltage. To generate such a voltage an additional voltage-up-converter was integrated. In Figure 22.9 the simplified schematic of the optical receiver part is depicted. The *p–i–n* photodiode which converts the incident light into an appropriate photocurrent i_{ph} in the integrated optical receiver part has a diameter of 50 μm. The TIA (transimpedance amplifier) converts this current into a voltage V_{data} with high bandwidth. A dummy TIA together with a proportional-integral controller (OPA with C_r and R_r) creates an averaged voltage of V_{data}, which is the reference voltage V_{ref} for the following differential amplifier (post-amplifier). This controller defines also the lower cutoff frequency, which is

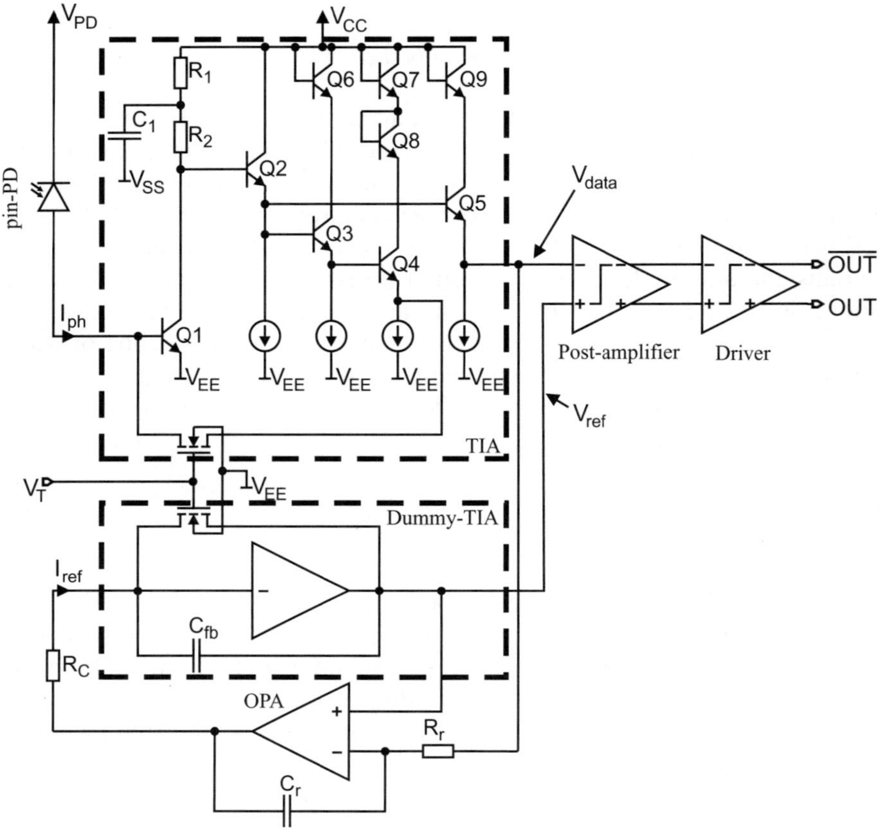

Figure 22.9. Simplified schematic of the optical receiver part.

on the order of 50 kHz. The difference $V_{data} - V_{ref}$ is then amplified via the post-amplifier and the driver, which is able to drive an off-chip 50 Ω measurement system. The post-amplifier and the driver have also a symmetrical limiting behavior to clip the logical voltage levels symmetrically in case of overdrive to avoid large jitter. The MOSFETs used as active feedback resistors were biased by an external bias voltage V_T to have the possibilitiy to adapt the receiver optimally to different data rates and furthermore to enhance the dynamic range; for example, a high optical power at the input requires a low feedback resistance. The value of the feedback resistance can be adjusted with V_T. The TIA itself consists of an input common-emitter amplifier (transistor *Q*1), which provides the voltage gain. The collector network was added to suppress high-frequency supply noise and high-frequency feedback via supply lines, which improves stability. On the other hand, at lower frequencies the gain is higher, which introduces as a drawback a slight intersymbol interference and reduces the sensitivity. Several emitter followers (*Q*2–*Q*5) are used for level shifting and impedance transformation.

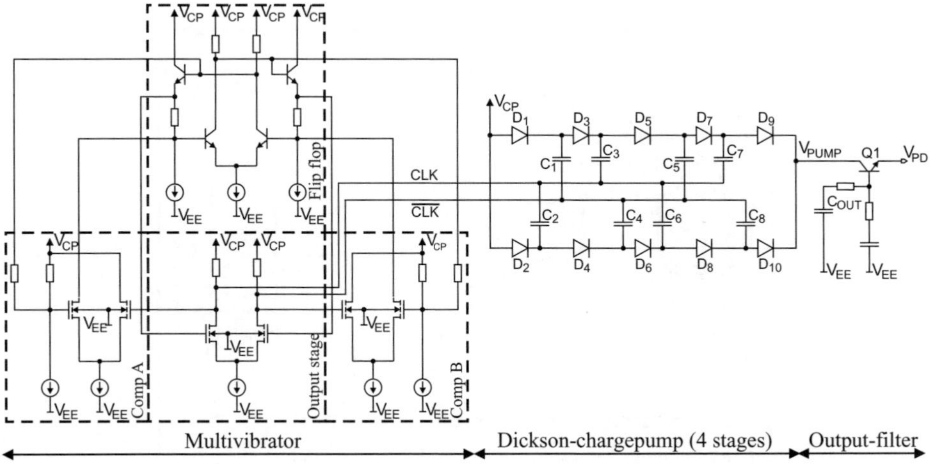

Figure 22.10. Schematic of the voltage up-converter (VUC).

To prevent breakdown of transistors (*Q*3–*Q*5), several diodes are added at the collectors to reduce emitter–collector voltages.

The *p–i–n* photodiode voltage is generated by an additional on-chip voltage up converter (VUC), which is shown in Figure 22.10. It consists of an oscillator, a four-stage Dickson multiplier, and a low-pass filter to reduce the output voltage ripple when biasing the photodiode. The oscillator operates with a symmetrically built multivibrator, whereas the timing capacitance is formed by the pumping network. The Dickson charge pump is also built symmetrically. The multivibrator itself consists of two comparators (A and B), a flip flop, and an output stage to drive the clock lines CLK and $\overline{CLK}$. Depending on the state of the flip flop, it switches when either comparator A detects signals CLK failing below a threshold level or comparator B detects the same for signal $\overline{CLK}$. The threshold levels for both comparators change with the state of the flip-flop, therefore forming a Schmitt trigger behavior to prevent the circuit from unwanted oscillations. When only a low photocurrent is flowing, the operating frequency is about 60 MHz. At higher currents the stronger discharging for every cycle of the pumping capacitors leads to longer time to fully recharge the pump capacitors and the frequency decreases. This is because the charge pump capacitors are the load capacitances of the output stage of the multivibrator, which determines the switching frequency. Therefore a stronger discharging of the pumping capacitors occurs for every clock cycle with increasing load current, which leads to a longer time to fully recharge the pump capacitors in turn. The four-stage Dickson multiplier generates a voltage level V_{PUMP} = 11.9 V, which is filtered with transistor *Q*1, capacitor C_{OUT}, and two resistors to determine voltage V_{PD}. The low-pass filter consists of an emitter–follower, whereby the base voltage is filtered by a simple lead-lag filter with the corner frequencies of 1 MHz and 10 MHz. The microphotograph of the integrated photodiode receiver is shown in Figure 22.11. The chip

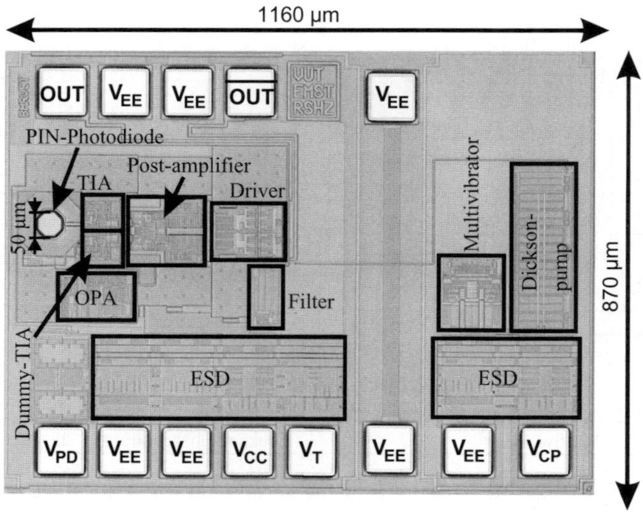

Figure 22.11. Microphotograph of the integrated photodiode receiver.

was fabricated in a modified 0.6-μm BiCMOS technology with a supply voltage of 5V and with a transit frequency $f_t = 25$ GHz of available *n–p–n* transistors. The total chip area was $1160\,\mu\text{m} \times 870\,\mu\text{m}$.

The complete integrated photoreceiver consumes a current of 37 mA + 20 mA of the output 50-Ω driver, which results in a total power consumption of 285 mW (including 50-Ω driver). The receiver itself consumes 260 mW and the VUC 25 mW. The TIA itself consumes 9 mW [10]. The overall transimpedance of the receiver ranges from 248 kΩ to 58 kΩ. The measurement of the sensitivity and an eye diagram is shown in Figure 22.12. At a wavelength of 660 nm at data rates of 2.5, 3, 4, and 5 Gbit/s the sensitivities (BER = 10^{-9}) were measured to be –24.9, –24.3, –22.9, and –20.5 dBm, respectively. At a wavelength of 850 nm the sensitivity for a BER = 10^{-9} was –22.8 dBm at a data rate of 2.5 Gbit/s. This is because the responsivity of the *p–i–n* photodiode was 0.36 A/W at a wavelength of 660 nm and 0.26 A/W at 850 m.

22.3.2 An Integrated Silicon Optical Receiver for 850-nm Wavelength with a Data Rate Up to 11 Gbit/s

A high-speed all-integrated silicon optical receiver, which reaches data rates up to 11 Gbit/s, was designed in reference 11 in a 0.6-μm BiCMOS process with a nominal supply voltage of 5V. The integrated *p–i–n* photodiode had a structure similar to the one shown in Figure 22.6. In a manner similar to that described in Section 22.3.1, the photodiode was biased with a voltage larger than the supply voltage. V_{PD} = 17V was chosen in reference 11. This enhances speed, and the –3 dB cutoff frequency was raised to 2.2 GHz at a wavelength of 850 nm.

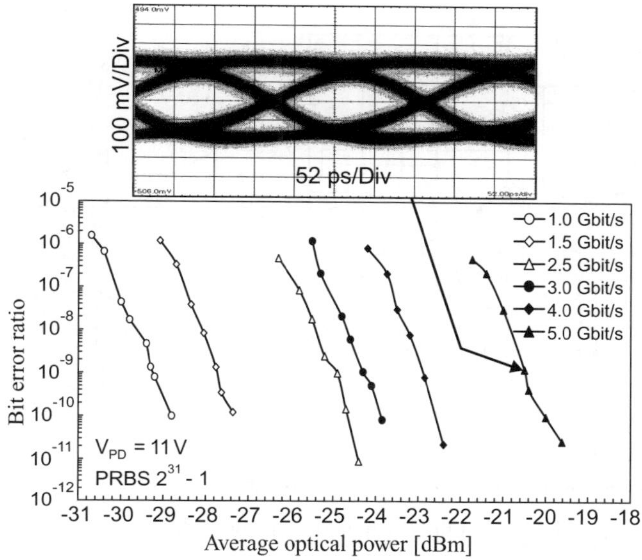

Figure 22.12. Bit error ratio versus the average optical input power at $\lambda = 660$ nm for different data rates. The eye diagram is for the case 5 Gbit/s, $V_T = 4$ V and an average input power of –20.5 dBm.

To make data reception of more than about 4 GHz possible, the behavior of the diode was corrected with the help of an analog equalizer. Equalization of the photocurrent was successfully demonstrated with an *n*-well/*p*-substrate diode in reference 12, where the data rate of a standard optical receiver in a CMOS process was raised to 3 Gbit/s. The simplified schematic of the 11 Gbit/s photoreceiver with the detailed front end is shown in Figure 22.13.

The photocurrent I_{ph} is converted into a voltage with the help of the transimpedance amplifier (TIA) Q1, Q3, R_I, R3, R5, C_I, C3, and R_{fb1}. The TIA has a transimpedance of 500 Ω. This voltage minus a reference voltage, which is generated by a controlling circuit with differential integrator, is amplified with a frequency-dependent characteristic proportional to the inverse characteristic of the photodiode response. To achieve such an inverse characteristic, a cascoded difference amplifier (Q5, Q6, Q7, Q8, R_7, and R_8) that has frequency-dependent coupling elements is used. The drift time of the photodiode is compensated with R_d and C_d, and the small diffusion part is compensated with C_l, R_{l1}, and R_{l2}. After this equalizing amplifier the overall transfer function has a nearly flat frequency response and linear phase. After the TIA with the equalizer a three-stage limiting amplifier follows to improve the gain, limit the signal amplitude, and drive a differential 100-Ω load. The controlling circuit generates together with the differential integrator and a dummy TIA (Q2, Q4, R_2, R_4, R_6, C_2, and R_{jb2}) the reference voltage for the equalizing amplifier by injecting the average photocurrent I_{avg} into the dummy TIA via R_c. Thus, symmetrical clipping of the

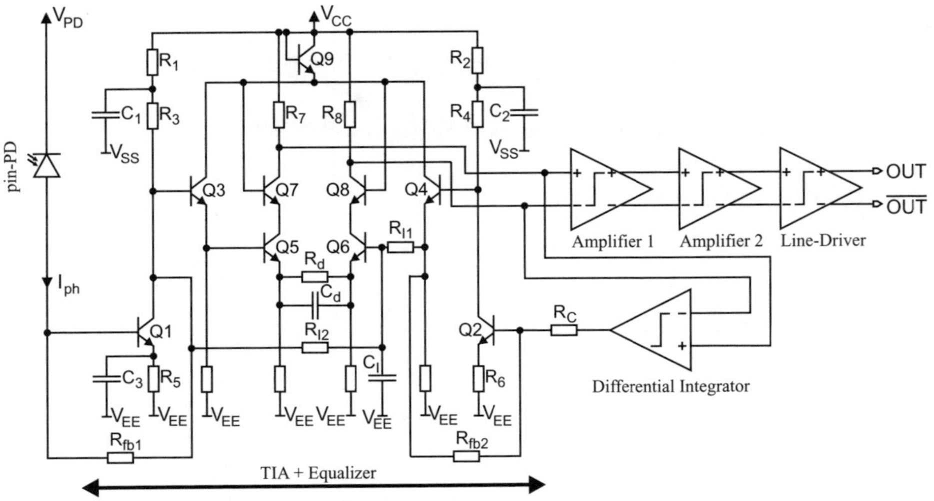

Figure 22.13. Simplified schematic of the 11-Gbit/s photoreceiver.

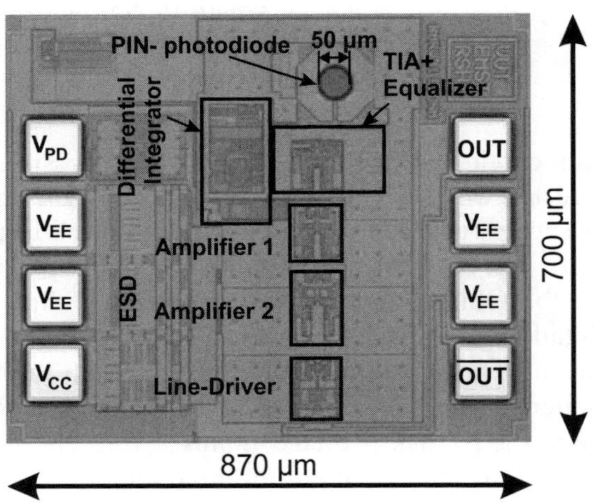

Figure 22.14. Chip micrograph of the complete integrated silicon optical receiver.

logical voltage levels in case of optical overdrive is ensured. The controlling circuit introduces a lower cutoff frequency of 30 kHz, eliminating the influence of background light. The chip micrograph of the 11-Gbit/s integrated silicon optical receiver is depicted in Figure 22.14. The *p–i–n* photodiode has a diameter of 50 μm and matches to a 50-μm multimode fiber. The whole chip needs an area of 870 μm × 700 μm, where bond pads and supply blocking capacitors are included. The results of a BER measurement to obtain the sensitivity of the optical receiver and an eye diagram at 11 Gbit/s, where an optical input power of –8.9 dBm is

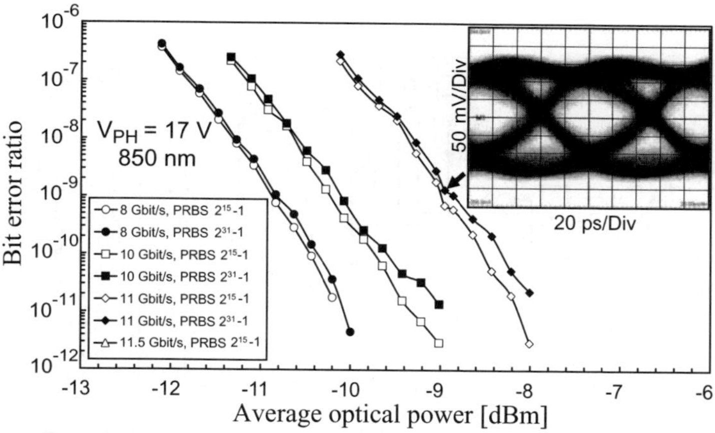

Figure 22.15. Bit error ratio versus average optical power at the input of the receiver. The eye diagram is for the case of 11-Gbit/s and –8.9-dBm optical input power.

needed to achieve a BER of 10^{-9}, can be seen in Figure 22.15. The analog upper –3-dB cutoff frequency of the flat transfer function of the overall optical receiver chip is 7.7 GHz at a wavelength of 850 nm together with an overall transimpedance of 3 kΩ. The power consumption of the whole chip including the 100-Ω driver is 310 mW at a supply voltage of 5 V [11]. It should be noted that the rather moderate transitor transist frequency of about 25 GHz limited the performance at 11 Gbit/s.

22.3.3 Integrated Optical Silicon Receiver Front End for Multilevel Signaling

Reducing the analog bandwidth of a data stream with a distinct bit rate can be done by increasing the number of signal levels, where therefore the duration of a bit can be made longer. A common time-discrete signal, where the bit information is modulated into amplitude levels, is called an M-level pulse-amplitude-modulated signal (M-PAM). For an M amplitude-level scheme, the bandwidth is reduced by a factor of N under the assumption that the bit rate is held constant, where N is the number of bits per symbol: $N = \log_2(M)$ [14]. On the other hand, breaking down an available voltage swing into a larger number of voltage levels reduces the voltage margin of the received symbol levels by a factor of $1/(M-1)$. Therefore it is a higher challenge to design a multilevel receiver, because a better sensitivity and more receiver complexity compared to a usual two-level receiver for binary signals are needed. Due to the larger amount of voltage levels, also an automatic gain controller (AGC) has to be implemented in order to keep the different symbol levels equally spaced. A constant transimpedance or typical limiting post-amplifiers in a binary receiver will not be adequate in a multilevel receiver.

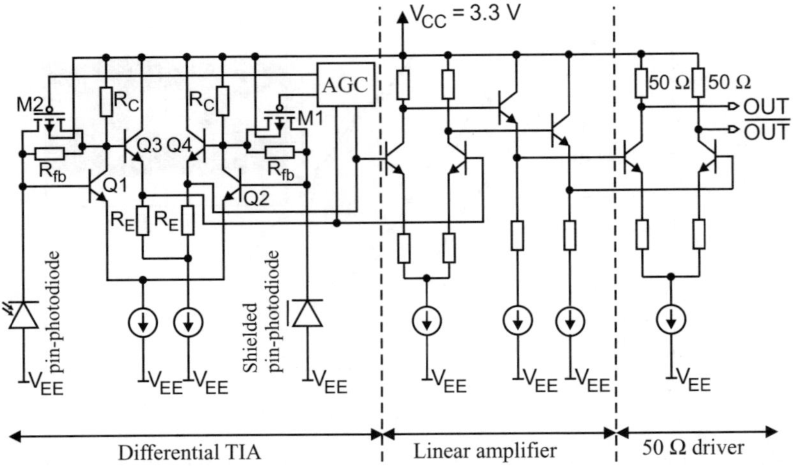

Figure 22.16. Schematic of the front end of an integrated optical multilevel receiver (4-PAM).

The schematic of the front end of an integrated optical multilevel receiver is shown in Figure 22.16 [13]. The test chip was fabricated in a 0.6-μm BiCMOS process and occupies a chip area of 1 mm^2. The chip consumed 100 mW and was designed for a single supply voltage of 3.3 V. The front end consists of an integrated *p–i–n* photodiode, a transimpedance amplifier (TIA), and a 50-Ω driver. A second identical *p–i–n* photodiode, which is shielded, is added to balance the differential TIA. The *p–i–n* photodiode for optical reception has a large diameter of 300 μm so that a plastic optical fiber (POF) directly can be attached. The responsivity of the photodiode is 0.48 A/W for λ = 660 nm. As shown in Figure 22.16 the photocurrent, which is coming from the photodiode, is converted into a voltage by a differential TIA. The AGC circuit decreases the transimpedance of the TIA when the input optical power is increased to ensure that the output voltage of the TIA is constant. The AGC measures the peak voltage at the output of the differential TIA and keeps it constant for the highest level in the multilevel signal by adjusting the transimpedance via transistors Ml and M2. When the optical input power gets higher, the gate voltage at these transistors is reduced, thus reducing the transimpedance. The lower cutoff frequency of the AGC circuit is 7.9 kHz [13].

The measurement results of the chip are depicted in Figure 22.17. For a highest level of the input optical power of larger than 25 μW the output voltage is constant and equal to 750 mV. For M = 4 the logical voltage levels are equally spaced with 250 mV in between the output voltage range of 750 mV. Due to the low cutoff frequency of the AGC circuitry, a correct detection of the multilevel signal at constant threshold voltages is enabled. This optical receiver reaches sensitivities of –14.8 and –11.9 dBm for a data rate of 1 and 1.4 Gbit/s,

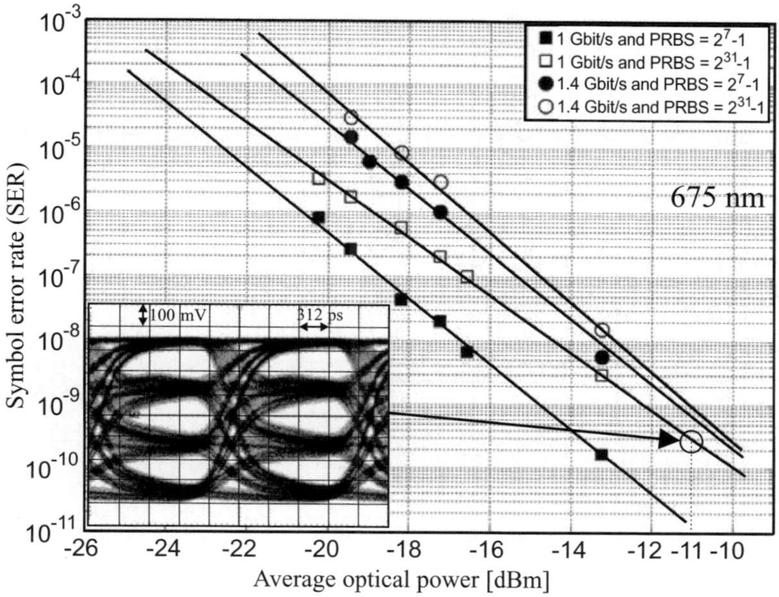

Figure 22.17. Measured 4-PAM symbol error rate (SER) versus average optical power at the input and eye diagram for 1.4-Gbit/s and −11-dBm optical input power.

respectively, at a wavelength of the received light at 675 nm with a SER of 10^{-9} for a pseudorandom bit sequence (PRBS) with a length of $2^7 - 1$. With a PRBS of $2^3 - 1$ and under the same conditions, sensitivities of −12.1 and −11.2 dBm for data rates of 1 and 1.4 Gbit/s were measured [13].

22.3.4 36-Channel Integrated Silicon Optical Receiver with 3 Gbit/s per Channel

The growing use of the Internet and the demand for multi-Gbit/s I/O performance creates a need of 100+ Gbit/s data transmissions and therefore appropriate short-reach high-speed infrastructure for within high-end Telco and Datacom routers, switches, servers, and other chassis-to-chassis links. Solutions for low-cost high-speed optical receivers are also needed in consumer-electronic applications like HDTV interconnection standards (HDMI). Here a 36-channel integrated silicon optical receiver is described with a data rate of 3 Gbit/s per channel which results in an overall data rate of 108 Gbit/s [15]. The receiver chip was designed for a 0.6-μm BiCMOS technology with a supply voltage of 3.3 V and with integrated *p–i–n* photodiode with a responsivity of 0.33 A/W at 850 nm because the antireflection coating on the diode was optimized for 650 nm. However, if the ARC layer is optimized for 850 nm, the responsivity will get better. The simplified schematic of one channel of the receiver with a detailed schematic of

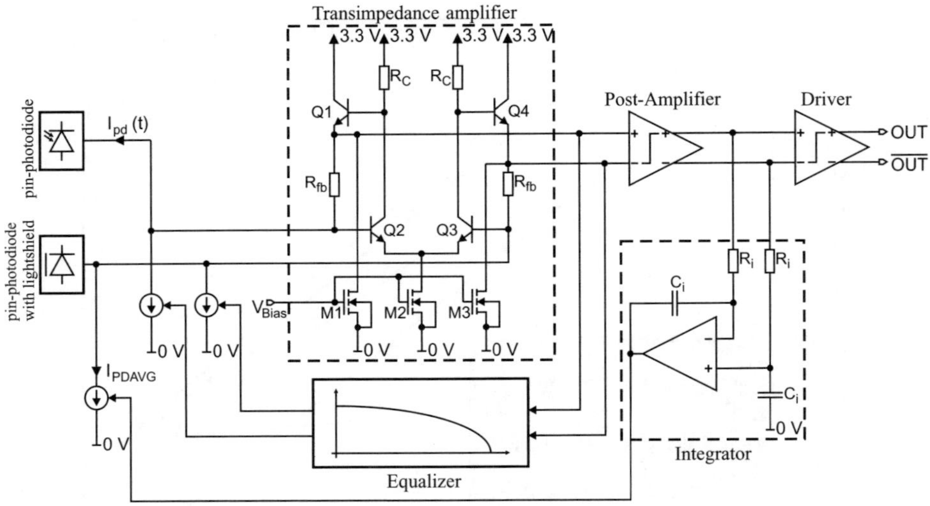

Figure 22.18. Simplified schematic of one channel, which includes a detailed schematic of the differential TIA.

the differential TIA included is shown in Figure 22.18. The whole circuit is fully differential to achieve a good power supply rejection. This begins with a light-shielded *p–i–n* photodiode as reference, which has the same structure as the *p–i–n* photodiode for receiving optical data signals. The transimpedance amplifier (TIA) converts the photocurrent I_{pd}(t) to a signal voltage, which is further amplified by a post-amplifier and a final 50-Ω driver. An equalizer is implemented to correct the frequency-dependent response of the *p–i–n* photodiode by using steered current sources at the inputs of the TIA. An integrator with a lower cutoff frequency of 300 kHz is implemented to set the decision level via the reference input of the differential TIA. Additionally implemented but not shown in Figure 22.18 was a loss-of-power detector with adjustable threshold to detect whether one of the 36 channels receives not enough optical power.

A chip photo of the 36-channel 108-Gbit/s integrated silicon optical receiver and a layout drawing of one channel (3 Gbit/s) are shown in Figure 22.19. The chip was fabricated in a 0.6-μm BiCMOS process with a supply voltage of 3.3V. The diameter of a *p–i–n* photodiode is 90 μm and the photodiodes are arranged in a 250-μm raster. The whole chip consumes a power of 809 mW, which also includes an on-chip laser driver for a return channel and a loss-of-power detector. The crosstalk between two channels was measured to be below −40 dB. In Figure 22.20 the eye diagrams for one channel at 3 Gbit/s for an average optical input power of (a) −14 dBm and (b) −17.8 dBm for BER = 10^{-9} at a wavelength of 850 nm and with a laser extinction ratio of 6 using a PRBS $2^{15} - 1$. If the extinction ratio is assumed to be infinity (no optical power for the zero logical level) the sensitivity is calculated to −19.3 dBm for one channel at a data rate of 3 Gbit/s.

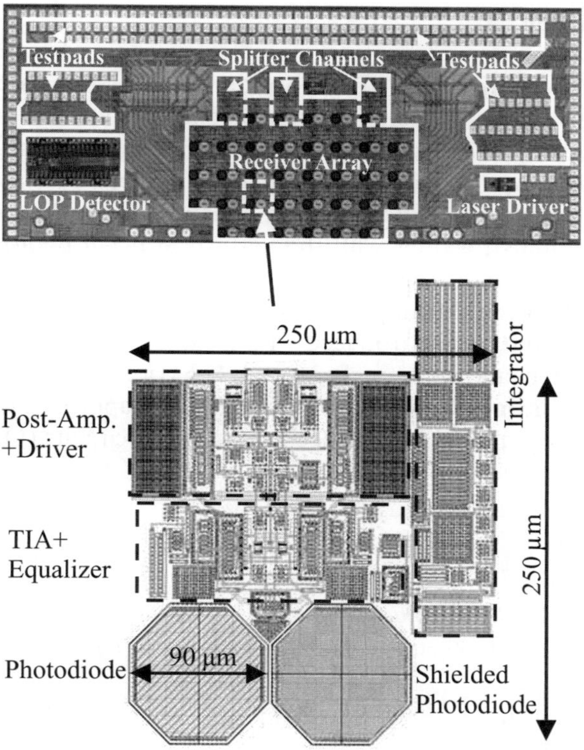

Figure 22.19. Photo of the 36-channel integrated silicon optical receiver and layout drawing of one channel.

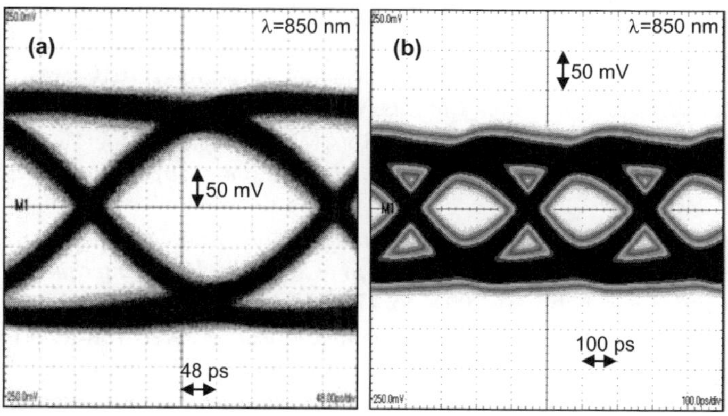

Figure 22.20. Eye diagrams at 3 Gbit/s (PRBS $2^{15}-1$): **(a)** −14-dBm optical input power **(b)** −17.8-dBm optical input power.

22.4 CONCLUSION

The mighty capability of silicon PDICs was demonstrated in mature and reliable industrial silicon chip manufacturing technologies. The vertical integrated *p–i–n* photodiode realizes high bandwidth and high responsivity. Integrated voltage-up converters were shown to increase the data rate of silicon optical receivers up to more than 5 Gbit/s. Implementation of an equalizer in the TIA allowed even a data rate of 11 Gbit/s together with the vertical Si *p–i–n* photodiode. Another new approach—multimode data transmission—was successfully verified with a 4-PAM silicon PDIC. And finally a highly parallel high-speed PDIC up to a total data rate of 108 Gbit/s was introduced with a rather high sensitivity, rather low power consumption, and very good crosstalk behavior.

Altogether, silicon BiCMOS PDICs are very well appropriate for highspeed, high-sensitivity optical application-specific integrated circuits (OPTO-ASICs).

REFERENCES

1. H. Zimmermann, *Integrated Silicon Optoelectronics*, Springer, Berlin, 2000.
2. H. Zimmermann, *Silicon Optoelectronic Integrated Circuits*, Springer, Berlin, 2004.
3. K. Schneider and H. Zimmermann, *Highly Sensitive Optical Receivers*, Springer, Berlin, 2004.
4. D. M. Caughey and R. E. Thomas, Carrier mobilities in silicon empirically related to doping and field, *Proc. IEEE*, Vol. 55, pp. 2192–2193, 1967.
5. S. M. Sze, *Physics of Semiconductor Devices*, John Wiley & Sons, New York, 1981.
6. J. M. Senior, *Optical Fiber Communications*, Prentice-Hall, New York, 1992, p. 437.
7. G. Winstel and C. Weyrich, *Optoelektronik II*, Springer, Berlin, 1986, p. 76.
8. M. Yamamoto, M. Kubo, and K. Nakao, Si-OEIC with a built-in PIN-photodiode, *IEEE Trans. Electron Devices*, Vol. 42, No. 1, pp. 58–63, 1995.
9. K. J. Ebling, *Integrated Optoelectronics*, Springer, Berlin, 1993, p. 76.
10. R. Swoboda, J. Knorr, and H. Zimmermann, A 5-Gb/s OEIC with Voltage-Up-Converter, *IEEE J. Solid-State Circuits*, Vol. 40, No. 7, pp. 1521–1526, 2005.
11. R. Swoboda and H. Zimmermann, 11 Gb/s Monolithically integrated silicon optical receiver for 850 nm, in *IEEE International Solid State Circuit Conference*, February 2006, pp. 240–241.
12. S. Radovanovic, A.-J. Annema, and B. Nauta, 3 Gb/s Monolithically integrated photodiode and pre-amplifier in standard 0.18 μm CMOS, in *IEEE International Solid State Circuit Conference*, February 2004, pp. 472–473.
13. M. Atef, R. Swoboda, and H. Zimmermann, Optical receiver front-end for multilevel signalling, *IET Electron. Lett.* Vol. 45, No. 2, pp. 121–122, 2009.
14. S. Hranilovic and D. A. Johns, A multilevel modulation scheme for high-speed wireless infrared communications, in *IEEE International Symposium on Circuits and Systems*, Vol. 6, 1999, pp. 338–341.
15. R. Swoboda, M. Förtsch, and H. Zimmermann, 3 Gbps-per-channel highly-parallel silicon receiver OEIC, in *33rd European Conference on Optical Communication, September 2007*, pp. 255–256.

23

MIMO WIRELESS TRANSCEIVER DESIGN INCORPORATING HYBRID ARQ

Dimitris Toumpakaris, Jungwon Lee, Edward W. Jang, Hui-Ling Lou, and John M. Cioffi

23.1 INTRODUCTION

Hybrid automatic repeat request (HARQ) combines ARQ and forward error correction (FEC) in order to benefit from the advantages offered by both schemes [1]. ARQ aims at improving reliability by retransmitting packets that arrive at the receiver with errors. FEC schemes are based on a different approach: The objective is to avoid retransmission by including sufficient parity data into the packets so that not only errors be detected (as in ARQ), but also packets be corrected at the receiver. ARQ is preferable in good channels because few retransmissions occur and there is no need to include additional error-correcting parity in each transmitted packet. On the other hand, FEC works better over channels where errors are frequent and many retransmissions would result in low throughput and long delays. Moreover, in some systems, the use of feedback, which is required by ARQ, may not be desirable, especially for channels with long delays. For HARQ, both error correction and retransmissions are employed. When errors at the receiver cannot be corrected by the FEC part, a retransmission is requested. Moreover, contrary to ARQ, previous packets are kept at the receiver and are used together with the retransmitted ones to decode the data. By appropriately

Convergence of Mobile and Stationary Next-Generation Networks, edited by Krzysztof Iniewski

choosing the FEC scheme, HARQ can achieve better throughput performance compared to ARQ for a given error rate. The price for this improvement is the additional system complexity and the memory that is required compared to ARQ, mainly at the receiver. However, recent advances in circuit speeds and size have made it possible to incorporate HARQ in latest wireless systems, such as those based on the 802.16e standard [2], which includes support for HARQ.

This chapter focuses on the use of HARQ in Multiple-Input Multiple-Output (MIMO) systems, not only because Single-Input Single-Output (SISO) can be viewed as a special case of MIMO, but also because MIMO transmission is nowadays used in most commercial wireless systems, including IEEE 802.11 [3] and 802.16 [2]. Furthermore, the use of HARQ in MIMO systems presents additional challenges compared to their SISO counterparts. At the transmitter side, HARQ can be used together with spatial multiplexing or can be combined with space-time coding, as will be explained in the following. At the receiver, joint processing of the signals received by all antennas is performed. The optimal receiver employs a maximum- likelihood (ML) detector that takes into account the transmit scheme (including HARQ) and the channel estimates. In order to simplify the receiver and reduce the memory requirements, suboptimal implementations can be used. One option is to use (spatial) equalization [4]. Linear equalizers may be employed, based on the Zero-Forcing (ZF) or the Minimum-Mean Squared Error (MMSE) criterion [5–7]. In the context of V-BLAST [8], receivers with linear equalizers are often referred to as decorrelators (for ZF) or MMSE receivers. Alternatively, Decision Feedback Equalization (DFE) can be used [9–11], that performs successive interference cancellation. The receiver can be simplified not only along the spatial direction (i.e., the antennas), but also along the time direction, depending on how the received signals corresponding to different retransmissions are used to produce estimates of the transmitted data. Again, the optimal receiver needs to store all received signals and use them to compute the ML estimate of the transmitted data. In some cases, as will be explained in more detail in this chapter, it is possible to reduce the complexity and the memory of the receiver without performance penalty. However, there are scenarios where it is not possible to simplify the receiver without affecting its performance. Nevertheless, in some cases the performance loss may not be large, and suboptimal architectures may be attractive because of the savings in complexity and memory, which results in smaller fabrication costs and lower power consumption.

This chapter is organized as follows. In Section 23.2 a physical layer model for wireless transmission using HARQ is presented and HARQ is discussed in more detail. In Section 23.3 the design of the transmitter is addressed, whereas in Section 23.4 the focus moves to the receiver. In Section 23.5 the performance of different HARQ schemes is compared in terms of the implementation of the receiver, as well as the wireless channel through which transmission takes place. Finally, Section 23.6 contains concluding remarks.

Throughout the chapter it is assumed that the receiver can obtain perfect channel estimates (this is often referred to as Receiver CSI). HARQ is usually employed when the transmitter has no knowledge of the channel.[1] The only feedback information that is passed from the receiver to the transmitter are the positive acknowledgments (ACKs) or negative acknowledgments (NACKs) indicating correct or incorrect reception of the data, respectively. Therefore, no transmit optimization can be performed [12]. Moreover, the narrowband case is considered and, hence, flat fading is assumed. In the latest wireless standards that are based on Orthogonal Frequency Division Multiplexing (OFDM) or Orthogonal Frequency Division Multiple Access (OFDMA), each symbol comprises narrowband subcarriers. Therefore, the OFDMA/OFDMA symbol can be seen as a set of parallel narrowband flat-fading channels. Hence, the transmitter and receiver designs that are presented in this chapter can be applied to each subcarrier of an OFDM/OFDMA system.

In the following, $(\cdot)^T$ denotes transpose, whereas $(\cdot)^*$ denotes conjugate transpose. $\mathbf{x}$ denotes a column vector. $\mathbf{A}$ denotes a matrix. Depending on the context, $\mathbf{0}$ denotes either a vector or a matrix with all elements equal to zero.

23.2 HARQ AND ITS APPLICATION TO MIMO WIRELESS SYSTEMS

23.2.1 The HARQ Protocol. CC- and IR-HARQ

As was mentioned previously, HARQ combines ARQ with a FEC subsystem. By enabling retransmissions, the additional error-control parity can be reduced, therefore increasing throughput and reducing the complexity of the decoder at the receiver. Moreover, the presence of FEC can help avoid the delay and the throughput reduction caused by frequent retransmissions. HARQ schemes are typically divided into Chase-Combining (CC-HARQ) and Incremental Redundancy (IR-HARQ).[2] When CC-HARQ is employed, the same packet is sent every time a retransmission is requested [13]. On the other hand, a different packet is sent through the channel in the case of IR-HARQ. The packet may result from a different encoding of the original user data, or it may contain a different subset of the encoded bits. An example of an IR-HARQ scheme is given later in this section. For either type of HARQ, block or convolutional codes can be used. Moreover, the retransmission schemes employed by ARQ—namely, stop-and-wait, go-back-N and selective-repeat—can be used directly in HARQ [1].

Clearly, CC-HARQ can be thought of as a special case of IR-HARQ. Therefore, it is expected that IR-HARQ will perform better than CC-HARQ. This is, indeed the case, except for some specific scenarios. However, this does not necessarily mean that CC-HARQ should not be considered as an option when designing a system. First, transmitter design is easier when CC-HARQ is employed.

[1]HARQ can also be used when the channel is known partially at the transmitter. Transceiver design with partial channel state information is a very broad and currently active research topic that merits a chapter on its own. For this reason, it is not dealt with here.

[2]An alternative classification is Type I, II, and III HARQ, Type I corresponding to CC-HARQ, and Types II/III being variants of IR-HARQ.

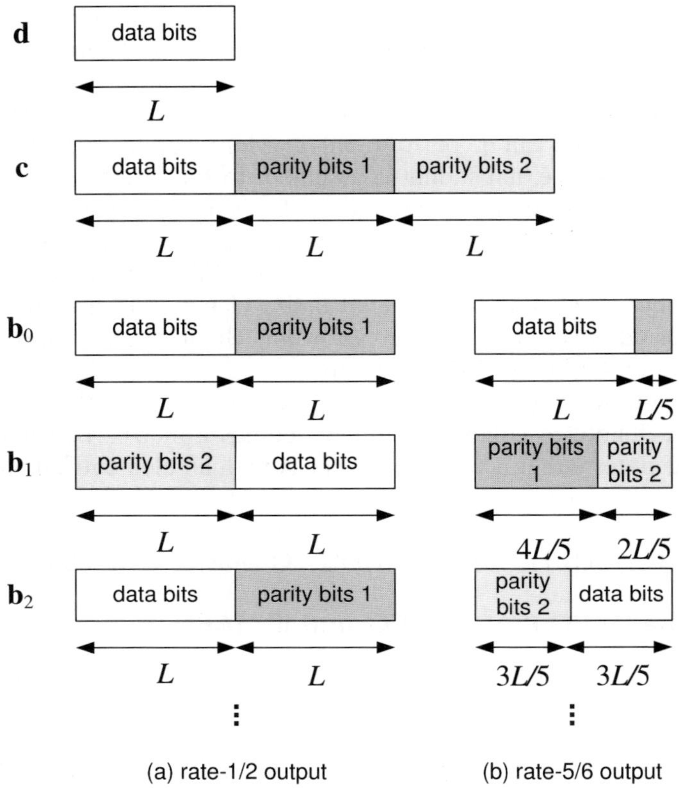

Figure 23.1. Selection of bits during each retransmission in IEEE 802.16 systems.

More importantly, the receiver complexity can be reduced. As will be explained in more detail in Section 23.4, this happens not only because decoding using more parity requires more operations, but also because, when CC-HARQ is used, *symbol combining* can simplify the structure of the receiver without loss in performance. On the other hand, although it may be possible to use symbol combining in some IR-HARQ scenarios, this is not true, in general. This gain in complexity is more pronounced in MIMO systems where, in some cases, the coding gain of IR-HARQ may be attained only by complex maximum-likelihood decoders. This performance-complexity tradeoff will be examined more thoroughly in Section 23.4 where different receiver architectures will be considered.

Assume that a user data packet comprises a sequence, $\mathbf{d} = [d_0, d_1, \ldots, d_{L-1}]^T$, of L bits. During each transmission, i, the sequence $\mathbf{d}$ is mapped to a HARQ packet $\mathbf{s}_i$. As explained above, the mapping depends on the HARQ scheme. A common approach that is used in wireless systems consists of first encoding $\mathbf{d}$ using a rate-r mother code to produce an encoded bit sequence $\mathbf{c} = [c_0, c_1, \ldots, c_{L/r-1}]^T$. This is shown in Figure 23.1. For example, the IEEE 802.16e standard specifies a rate-1/3. Convolutional Turbo Code (CTC) as an option [2] that produces $3L$ bits at the output. The first L (systematic) bits of $\mathbf{c}$ are the original input bits of $\mathbf{d}$. However, if interleaving is used before encoding, the order of the first L bits of $\mathbf{c}$ may not be the same as in $\mathbf{d}$. At each retransmission, i, a subset of the mother code bits is used to form a HARQ packet. When CC-HARQ is used, the bit selection module always outputs the same sequence, $\mathbf{b}$. For IR-HARQ, the bits that form an HARQ packet depend on the transmission index. An example of IR-HARQ bit selection for IEEE 802.16 systems is given in Figure 23.1. Although the figure focuses on the CTC case, the bit selection is similar in other scenarios, such as Convolutional Coding, Block Turbo Coding, or LDPC Coding.

After the bits sequence $\mathbf{b}_i$ is formed, it is mapped to a symbols sequence, $\mathbf{s}_i = [s_{i,0}, s_{i,1}, \ldots, s_{i,M}]^T$, using a bits-to-symbols mapper.[3] In practical systems, the symbols belong to a discrete constellation. In wireless systems, BPSK, QPSK, 16-QAM and 64-QAM are typically used. The length, M, of the symbol sequence depends on the modulation scheme and is equal to the length of $\mathbf{b}_i$ divided by the number of bits mapped to each symbol $s_{i,m}$. The symbol sequence $\mathbf{s}_i$ is then mapped to the transmit antennas before it is sent to the channel. The mapping to the antennas depends not only on the index, i, of the transmission, but also on whether a space-time coding scheme is used. The mapping $\mathbf{b}_i$ to the transmit antennas is covered in more detail in Section 23.3.

23.2.2 MIMO System Model

A MIMO system is considered with N_T transmit and N_R receive antennas. Transmission occurs through a flat block fading channel, that is the channel is constant for the duration of the transmission of one symbol, but may have changed when a subsequent transmission takes place. In general, the MIMO HARQ packet $\mathbf{X}_i$ during each transmission, i, can be thought of as a sequence of K symbol vectors $\mathbf{x}_{i,1}, \mathbf{x}_{i,2}, \ldots, \mathbf{x}_{i,K}$, each of size $N_T \times 1$. The lth entry of each element $\mathbf{x}_{i,k}$ corresponds to the output of the l th transmit antenna during transmission i and slot k. The value of K depends on the channel matrix and the transmit scheme that is used. When the matrix is full rank and full spatial multiplexing is used, $K = \lceil M/N \rceil$, where $N = \max\{N_T, N_R\}$. $\mathbf{X}_i$ is produced either directly from $\mathbf{b}_i$ or from $\mathbf{s}_i$, as will be elaborated in Section 23.3. Because in wireless systems transmission occurs around a central frequency, f_c, the baseband equivalent representation is often used to analyze and design systems. Thus, in this chapter, all (baseband equivalent) quantities will be complex. The values that each element of $\mathbf{x}_{i,k}$ can take depend on the modulation and the space-time coding scheme that is employed. For example, if no coding is used at the transmitter and independent streams are sent from each antenna, each element of $\mathbf{x}_{i,k}$ may be a value of a QAM constellation. If a space-time code is used, it may be a linear combination of QAM symbols. The $N_T \times 1$ vectors $\mathbf{x}_{i,k}$ that form $\mathbf{x}_i$ may be transmitted during

[3]As will be shown in Section 23.3, in some cases $\mathbf{b}_i$ is mapped directly to the transmit antennas instead of first forming $\mathbf{s}_i$.

different time periods or in different frequencies. If each vector is transmitted during a different time instant, it will experience a different channel, in general, because of the time variation of the channel. Similarly, in a wideband system the channel will be different in each subcarrier because of the frequency selectivity of fading resulting from multipath. Because of the block fading model, the transmission of each vector $\mathbf{x}_{i,k}$ can be considered separately.

As explained previously, in the case of IR-HARQ a different HARQ packet $\mathbf{X}_i$ is sent during each retransmission. Therefore, assuming, without loss of generality, that the length of all transmitted packets $\mathbf{X}_i$ corresponding to a given data packet, $\mathbf{d}$, is the same, the received signal in the kth slot of transmission i is equal to

$$\mathbf{y}_{i,k} = \mathbf{H}_{i,k}\mathbf{x}_{i,k} + \mathbf{z}_{i,k} \tag{23.1}$$

where $\mathbf{y}_{i,k} \in \mathcal{C}^{N_R \times 1}$ is the received signal vector resulting from the transmission of $\mathbf{x}_{i,k}$, $\mathbf{H}_{i,k}$ is the $N_R \times N_T$ channel matrix in the timeslot or the subcarrier k used for the transmission of $\mathbf{x}_{i,k}$, and $\mathbf{z}_{i,k}$ denotes the $N_R \times 1$ complex noise vector. If space-time coding is not used at the transmitter, the components of $\mathbf{x}_{i,k}$ can be chosen to be uncorrelated, that is, $\mathbf{E}[\mathbf{x}_{i,k}\mathbf{x}_{i,k}^*] = \mathbf{\Lambda}$, where $\mathbf{\Lambda}$ is a diagonal matrix. As will be explained in Section 23.3, this condition can also be satisfied when linear space-time codes are used if an equivalent matrix $\mathbf{H}_{eq,i,k}$ is constructed that depends on the space-time code. The noise $\mathbf{z}_{i,k}$ is assumed to be independent identically distributed zero-mean circularly symmetric complex Gaussian with covariance matrix $\sigma_z^2\mathbf{I}_{N_R}$.

23.3 TRANSMITTER IMPLEMENTATION

23.3.1 Symbol Selection

Figure 23.2a depicts the transmitter of a MIMO system employing Bit Interleaved Coded Modulation (BICM) and HARQ. As was mentioned in the previous section, the bit sequence $\mathbf{d}$ is encoded to the bit sequence $\mathbf{c}$. At each transmission, a subset, $\mathbf{b}_i$, of the bits of $\mathbf{c}$ are chosen that are sent to a bits-to-symbols mapper to produce the symbols sequence $\mathbf{s}_i$. Finally, a symbols-to-antennas mapper creates the sequence $\mathbf{X}_i$ of K vectors that are sent to the channel using N_T antennas and K slots, either in time or in frequency.

In general, the bit selection, the bits-to-symbols mapping, and the symbols-to-antennas mapping depend on the transmission index, i. More specifically, a different subset of bits of $\mathbf{c}$ may be chosen at each i; the bits of $\mathbf{c}$ may be interleaved and/or aligned differently at each transmission, i; and a given symbol, $s_{i,m}$ of $\mathbf{s}_i$, may be mapped to a different antenna and/or a different slot during each transmission, as when a space-time code is used. This will be discussed in more detail in Section 23.2.. In fact, as was mentioned in Section 23.2.1, in general, when IR-HARQ is employed, the original bit sequence $\mathbf{d}$ may be re-encoded at each transmission i resulting in a sequence $\mathbf{c}_i$, as, for example, in reference 14. By appropriate design of the coded sequence $\mathbf{c}_i$ and the bit-selection process, the

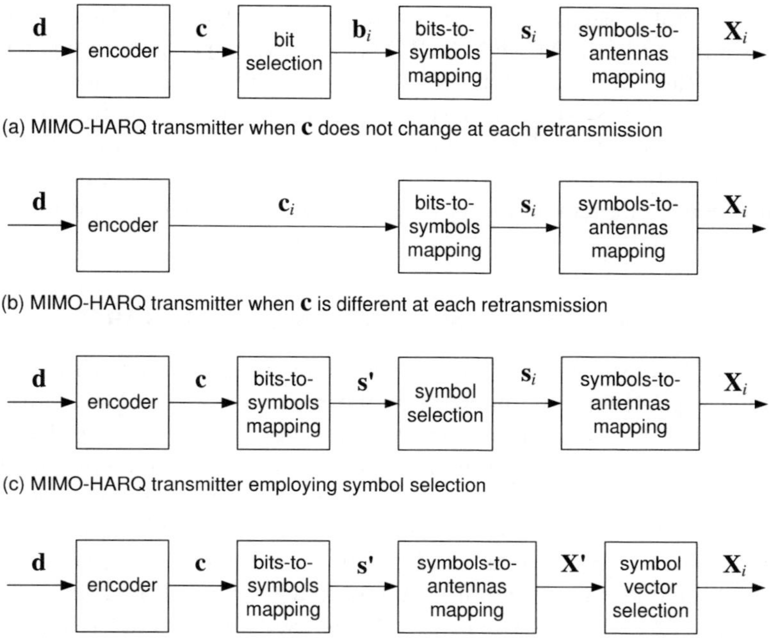

(a) MIMO-HARQ transmitter when **c** does not change at each retransmission

(b) MIMO-HARQ transmitter when **c** is different at each retransmission

(c) MIMO-HARQ transmitter employing symbol selection

(d) MIMO-HARQ transmitter employing symbol vector selection

Figure 23.2. Possible transmitter implementations for systems employing MIMO-HARQ.

coding gain of IR-HARQ can be improved. Moreover, the fact that a MIMO system is used can be taken into account to improve the performance of HARQ [15]. The design of the transmitter for this case is shown in Figure 23.2b. In this chapter, it is investigated how the transceiver should be designed to accommodate CC-HARQ or IR-HARQ variants without getting into the details of HARQ code design, which is a very interesting research topic per se.

In some cases, the design of the transmitter can be simplified with respect to the general case. For example, given a coded bits sequence $\mathbf{c}$, instead of performing *bit selection* at each retransmission i, it is possible to generate all possible symbols of the sequences $\mathbf{s}_i$ in advance, as shown in Figure 23.2c. In the figure, the sequence $\mathbf{s}'$ contains all symbols that may be included in the sequences $\mathbf{s}_i$. Then, the transmitter performs *symbol selection* to form $\mathbf{s}_i$ instead of bit selection. Clearly, whether this approach is preferable to bit selection, as in Figure 23.2, depends on the HARQ scheme and the number of allowed retransmissions. If the bit-to-symbols alignment does not change during each retransmission, if most of the data bits are included in each HARQ packet, and if several retransmissions are needed, on the average, then it may be preferable to store $\mathbf{s}'$ at the transmitter to avoid regenerating symbols at each transmission. On the other hand, if, on average, a small percentage of the symbols of $\mathbf{s}'$ is used for a given transmission, it may be better to generate $\mathbf{s}_i$ or $\mathbf{X}_i$ directly from $\mathbf{c}$ or $\mathbf{b}_i$ using bit selection. Symbol selection can be extended to symbol vector selection. If the mapping of symbols to antennas does not vary significantly, the MIMO HARQ packets, $\mathbf{X}_i$, can be generated in advance and stored at the transmitter. One obvious scenario where the structure of Figure 23.2d can be used is CC-HARQ where $\mathbf{b}_i = \mathbf{b}$, $\mathbf{s}_i = \mathbf{s}'$ and $\mathbf{X}_i = \mathbf{X}'$ for all i.

Preserving the bits-to-symbol vectors alignment between consecutive transmissions not only simplifies the transmitter in some cases, but can also lead to efficient receiver implementations without any performance penalty, as will be shown in Section 23.4.

23.3.2 Combining Space-Time Block Codes with HARQ

When HARQ is used in MIMO systems it can be combined with space-time codes. One such scheme is included in the most recent version of the 802.16 standard [2] and is based on the concept presented in reference 16. The scheme exploits the fact that multiple antennas can be used either to increase the number of signals that are transmitted simultaneously or to improve robustness of the transmission. When the channel is good, it may be preferable to use more than one spatial dimensions, whereas, when it is likely that some of the streams be not transmitted reliably, it may be more prudent to use a space-time code and reduce the code rate. In scenarios where HARQ is used the channel is typically not known a priori. Therefore, a decision has to be made regarding whether to use many spatial dimensions or increase diversity. However, an intermediate approach can be followed. More specifically, during the first transmission, an effort is made to use more than one spatial dimensions. In 802.16 terminology, this means using Matrix B (space-time code rate 2) for a 2×2 system and Matrix C (code rate 4) for a 4×4 system. If no error is found at the receiver, 2 or 4 spatial dimensions, respectively, have been used successfully. Then, the transmitter can proceed with the transmission of new data. On the other hand, if errors are found, the beamforming during the second transmission is chosen so that a space-time code be implemented. More specifically, when two transmit antennas are employed, the matrix that is chosen during the second transmission are the second column of the matrix of Alamouti's code [17]. No new data are sent and a total of two spatial dimensions is used over two slots. For the case of four transmit antennas, two Alamouti codes are used by considering antenna pairs. If there are errors even after the second transmission, the matrices of the first transmission are employed, and the original symbols are retransmitted, and so on. Hence, two beamforming matrices are defined, one for odd- and one for even-numbered transmissions.

Combining HARQ with space-time coding enables adaptation of the space-time coding rate. No decision needs to be made a priori regarding whether to employ a space-time code for each symbol. If symbols get received without error, all the spatial dimensions can be used. On the other hand, when errors occur, the rate can be dropped without affecting the spectral efficiency of previous transmissions. Furthermore, when space-time codes such as the one included in the current IEEE 802.16 standard that operate on the symbol level are used, the design of the receiver can be simplified. As will be seen in Section 23.4 both the number of operations and the memory that are required can be reduced without any performance loss. The reason for this is because use of symbol-level combining is possible.

When the space-time code that is used is linear and operates on symbols rather than bits, the same model as in Section 23.2.2 can be employed for analysis. More specifically, assume that, during each transmission i, a subset of the symbol sequence $\mathbf{s}'$ is mapped to the transmit antennas. Then, $\mathbf{x}_{i,k} = \mathbf{W}_{i,k}\mathbf{s}'$, where $\mathbf{W}_{i,k}$ is an $N_T \times L_s$ matrix that selects N_T out of the L_s symbols of $\mathbf{s}'$ and depends on the transmission index, i, and the slot, k. Then, for every slot, $\mathbf{y}_{i,k} = \mathbf{H}_{i,k}\mathbf{x}_{i,k} = \mathbf{H}_{i,k}\mathbf{W}_{i,k}\mathbf{s}' = \mathbf{H}_{eq,i,k}\mathbf{s}'$. Hence, the equivalent matrix, $\mathbf{H}_{eq,i,k}$, can be used that also incorporates the space-time encoder, and the scheme can be thought of as CC-HARQ.

The use of the equivalent matrix can also be extended to the case of the Alamouti code with a slight modification. For example, when HARQ is used together with the 2×2 Alamouti code in a slot k, the received signal after the first transmission (where full spatial multiplexing is used) is equal to

$$\mathbf{y}_{1,k} = \begin{bmatrix} y(1)_{1,k} \\ y(2)_{1,k} \end{bmatrix} = \begin{bmatrix} h(1,1)_{1,k} & h(1,2)_{1,k} \\ h(2,1)_{1,k} & h(2,2)_{1,k} \end{bmatrix} \begin{bmatrix} x(1) \\ x(2) \end{bmatrix} = \mathbf{H}_{1,k}\mathbf{x}$$

After the second transmission,

$$\mathbf{y}_{2,k} = \begin{bmatrix} y(1)_{1,k} \\ y(2)_{1,k} \end{bmatrix} = \begin{bmatrix} h(1,1)_{2,k} & h(1,2)_{2,k} \\ h(2,1)_{2,k} & h(2,2)_{2,k} \end{bmatrix} \begin{bmatrix} -x^*(2) \\ x^*(1) \end{bmatrix} \Rightarrow \mathbf{y}^*_{2,k} = \mathbf{H}'_{2,k}\mathbf{x}$$

where $\mathbf{H}'_{2,k} = \begin{bmatrix} h^*(1,2)_{2,k} & -h^*(1,1)_{2,k} \\ h^*(2,2)_{2,k} & -h^*(2,1)_{2,k} \end{bmatrix}$

For subsequent transmissions, $\mathbf{H}_{eq,i,k} = \mathbf{H}_{i,k}$ for odd i, and $\mathbf{H}_{eq,i,k} = \mathbf{H}'_{2,k}$ for even i.

23.4 RECEIVER ARCHITECTURES

23.4.1 Optimal Receiver Implementation and Distance-Level Combining

When IR-HARQ is used, different vector sequences $\mathbf{X}_i$ are sent at each retransmission. Using the notation of Section 23.2.2, after N transmissions, N vector sequences $\mathbf{Y}_i$ will be available at the receiver, where $\mathbf{Y}_i = [\mathbf{y}_{i,1}, \ldots, \mathbf{y}_{i,k}, \ldots, \mathbf{y}_{i,K}]$. For a given bit, λ, of the encoded sequence $\mathbf{c}$ (or $\mathbf{c}_i$, in the general case), let X be the set of symbol vectors that include at least one symbol where bit λ is contained. For example, if bit λ is sent during the first and third transmission and in slots 3 and 5, respectively, $X = \{\mathbf{x}_{1,3}, \mathbf{x}_{3,5}\}$. It is reasonable to assume that, during one transmission, each bit is mapped to only one symbol, so, to each i corresponds a single slot, k. Let also $\mathcal{X}^{(0)}_{\lambda}$ and $\mathcal{X}^{(1)}_{\lambda}$ denote all the possible values of the pair $\{\mathbf{x}_{1,3}, \mathbf{x}_{3,5}\}$ when $\lambda = 0$ or 1, respectively (in general, all the possible values of the cartesian product of the elements of X). Finally, let the set Y contain the $\mathbf{y}_{i,k}$ that are received when the $\mathbf{x}_{i,k}$ that belong to X are transmitted, and the set H of the channel matrices $\mathbf{H}_{i,k}$ corresponding to the transmission of the $\mathbf{x}_{i,k} \in X$.

After each transmission, the optimal, Maximum-Likelihood decoder calculates the LLR of bit λ using

$$\begin{aligned}
LLR_{\text{opt}}(\lambda) &= L(b_\lambda \,|\, \mathbf{Y}_1, \mathbf{Y}_2, \ldots, \mathbf{Y}_N, \mathbf{H}_1, \mathbf{H}_2, \ldots, \mathbf{H}_N) \\
&= L(b_\lambda \,|\, \mathbf{y}_{i,k} \in Y, \mathbf{H}_{i,k} \in H) \\
&= \ln \frac{\Pr\{b_\lambda = 1 | \mathbf{y}_{i,k} \in Y, \mathbf{H}_{i,k} \in H\}}{\Pr\{b_\lambda = 0 | \mathbf{y}_{i,k} \in Y, \mathbf{H}_{i,k} \in H\}} \\
&= \ln \frac{\sum_{X^{(1)} \in \mathcal{X}_\lambda^{(1)}} \Pr\{\mathbf{y}_{i,k} \in Y | X^{(1)}, \mathbf{H}_{i,k} \in H\}}{\sum_{X^{(0)} \in \mathcal{X}_\lambda^{(0)}} \Pr\{\mathbf{y}_{i,k} \in Y | X^{(0)}, \mathbf{H}_{i,k} \in H\}} \\
&= \ln \frac{\sum_{X^{(1)} \in \mathcal{X}_\lambda^{(1)}} \prod_{i,k:\mathbf{y}_{i,k} \in Y} \Pr\{\mathbf{y}_{i,k} | \mathbf{x}_{i,k}^{(1)}, \mathbf{H}_{i,k}\}}{\sum_{X^{(0)} \in \mathcal{X}_\lambda^{(0)}} \prod_{i,k:\mathbf{y}_{i,k} \in Y} \Pr\{\mathbf{y}_{i,k} | \mathbf{x}_{i,k}^{(0)}, \mathbf{H}_{i,k}\}} \\
&= \ln \frac{\sum_{X^{(1)} \in \mathcal{X}_\lambda^{(1)}} \prod_{i,k:\mathbf{y}_{i,k} \in Y} \frac{1}{\pi^{N_r}} \exp\left\{-\|\mathbf{y}_{i,k} - \mathbf{H}_{i,k}\mathbf{x}_{i,k}^{(1)}\|^2\right\}}{\sum_{X^{(0)} \in \mathcal{X}_\lambda^{(0)}} \prod_{i,k:\mathbf{y}_{i,k} \in Y} \frac{1}{\pi^{N_r}} \exp\left\{-\|\mathbf{y}_{i,k} - \mathbf{H}_{i,k}\mathbf{x}_{i,k}^{(0)}\|^2\right\}} \\
&= \ln \frac{\sum_{X^{(1)} \in \mathcal{X}_\lambda^{(1)}} \prod_{i,k:\mathbf{y}_{i,k} \in Y} \exp\left\{-\|\mathbf{y}_{i,k} - \mathbf{H}_{i,k}\mathbf{x}_{i,k}^{(1)}\|^2\right\}}{\sum_{X^{(0)} \in \mathcal{X}_\lambda^{(0)}} \prod_{i,k:\mathbf{y}_{i,k} \in Y} \exp\left\{-\|\mathbf{y}_{i,k} - \mathbf{H}_{i,k}\mathbf{x}_{i,k}^{(0)}\|^2\right\}} \\
&= \ln \frac{\sum_{X^{(1)} \in \mathcal{X}_\lambda^{(1)}} \exp\left\{-\sum_{i,k:\mathbf{y}_{i,k} \in Y} \|\mathbf{y}_{i,k} - \mathbf{H}_{i,k}\mathbf{x}_{i,k}^{(1)}\|^2\right\}}{\sum_{X^{(0)} \in \mathcal{X}_\lambda^{(0)}} \exp\left\{-\sum_{i,k:\mathbf{y}_{i,k} \in Y} \|\mathbf{y}_{i,k} - \mathbf{H}_{i,k}\mathbf{x}_{i,k}^{(0)}\|^2\right\}}
\end{aligned} \tag{23.2}$$

In the equation, N is the number of transmissions. Note that the calculation of the LLR using the above formula has to be performed for each bit λ. Clearly, in the general case, the ML decoder can be very complex. A block diagram of an optimal ML receiver is given in Figure 23.3.

The ML receiver can be simplified if the approximation $\log \Sigma \exp(a_i) \approx \max a_i$ is used [18]. Then

$$\text{LLR}_{\text{DLC}}(\lambda) = \min_{X^{(1)} \in \mathcal{X}_\lambda^{(1)}} \left\{ \sum_{i,k:\mathbf{y}_{i,k} \in Y} \|\mathbf{y}_{i,k} - \mathbf{H}_{i,k}\mathbf{x}_{i,k}^{(1)}\|^2 \right\} - \min_{X^{(0)} \in \mathcal{X}_\lambda^{(0)}} \left\{ \sum_{i,k:\mathbf{y}_{i,k} \in Y} \|\mathbf{y}_{i,k} - \mathbf{H}_{i,k}\mathbf{x}_{i,k}^{(0)}\|^2 \right\} \tag{23.3}$$

This simplified scheme performs *distance-level* combining. After each transmission, the Euclidean distances between the received symbol $\mathbf{y}_{i,k}$ and $\mathbf{H}_{i,k}\mathbf{x}_{i,k}$ are added for each possible sequence of $\mathbf{x}_{i,k}^{(0)}$ and $\mathbf{x}_{i,k}^{(1)}$ that contains bit λ. Then the aggregate distance of the most likely transmitted sequence is found for $\lambda = 0$ and

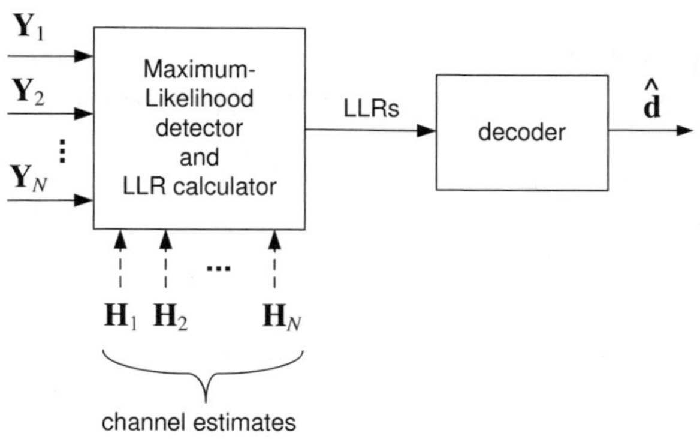

Figure 23.3. Generic block diagram for the optimal receiver based on ML detection.

1. The difference of the two distance sums is used to determine the LLR. The performance of the distance-level combining receiver is very close to that of its optimal counterpart, except for low SNRs [18]. The same receiver architecture can be used as for the optimal receiver in Figure 23.3. What is different is the way the LLRs are calculated, using, Eq. (23.3) instead of Eq. (23.2).

23.4.2 Bit-Level and Symbol-Level Combining

The receiver design can be simplified further by calculating the LLRs from each $\mathbf{Y}_i$ at the end of each transmission. The resulting values are stored, combined with the LLRs of each subsequent transmission and sent to the decoder. In this chapter, this will be referred to as *bit-level* combining in contrast to the joint ML decoding of Eq. (23.2) that uses all transmissions. Bit-level combining can also be used for CC-HARQ systems (since, in essence, they can be thought of as a special case of IR-HARQ). However, as should become clearer later in this section, in general, when CC-HARQ is used, symbol combining is preferable. When bit-level combining is employed, the LLRs corresponding to different transmissions are simply added:

$$\begin{aligned}
\mathrm{LLR}_{\mathrm{BLC}}(\lambda) &= \sum_{i:\exists k:\mathbf{x}_{i,k}\in X} \mathrm{LLR}_i \\
&= \sum_{i:\exists k:\mathbf{x}_{i,k}\in X} L(b_\lambda \,|\mathbf{y}_{i,k}, \mathbf{H}_{i,k}) \\
&= \sum_{i:\exists k:\mathbf{x}_{i,k}\in X} \ln \frac{\Pr\{b_\lambda = 1|\mathbf{y}_{i,k}, \mathbf{H}_{i,k}\}}{\Pr\{b_\lambda = 0|\mathbf{y}_{i,k}, \mathbf{H}_{i,k}\}} \\
&= \sum_{i:\exists k:\mathbf{x}_{i,k}\in X} \ln \frac{\sum_{\mathbf{x}_{i,k}^{(1)}\in\mathcal{X}_\lambda^{(1)}} \Pr\{\mathbf{y}_{i,k}\,|\mathbf{x}_{i,k}^{(1)}, \mathbf{H}_{i,k}\}}{\sum_{\mathbf{x}_{i,k}^{(0)}\in\mathcal{X}_\lambda^{(0)}} \Pr\{\mathbf{y}_{i,k}\,|\mathbf{x}_{i,k}^{(0)}, \mathbf{H}_{i,k}\}} \\
&= \sum_{i:\exists k:\mathbf{x}_{i,k}\in X} \ln \frac{\sum_{\mathbf{x}_{i,k}^{(1)}\in\mathcal{X}_\lambda^{(1)}} \exp\left\{-\left\|\mathbf{y}_{i,k} - \mathbf{H}_{i,k}\mathbf{x}_{i,k}^{(1)}\right\|^2\right\}}{\sum_{\mathbf{x}_{i,k}^{(0)}\in\mathcal{X}_\lambda^{(0)}} \exp\left\{-\left\|\mathbf{y}_{i,k} - \mathbf{H}_{i,k}\mathbf{x}_{i,k}^{(0)}\right\|^2\right\}}
\end{aligned} \tag{23.4}$$

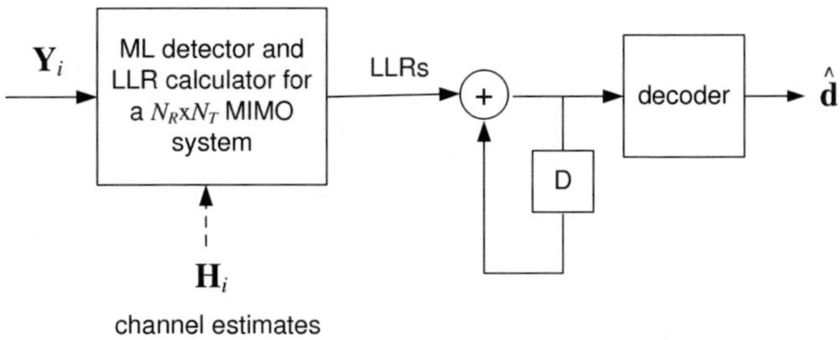

Figure 23.4. Generic block diagram for the bit-level combining receiver.

Use of bit-level combining simplifies the receiver, because it is not necessary to store the received sequences, $\mathbf{Y}_i$, and the channel estimates, $\mathbf{H}_{i,k}$, corresponding to previous transmissions. Moreover, in Eq. (23.4), the sum includes fewer terms compared to Eq. (23.2) where all the terms of the Cartesian product of N terms need to be considered. A block diagram of a bit-level combining receiver is given in Figure 23.4.

Similar to the optimal ML receiver, distance-level combining can be used together with bit-level combining. In this case,

$$\begin{aligned}\mathrm{LLR}_{\mathrm{BLC,\,DLC}}(\lambda) = \sum\nolimits_{i:\exists k:\mathbf{x}_{i,k}\in X} \Big\{ &\min\nolimits_{\mathbf{x}_{i,k}^{(1)}\in\mathcal{X}_\lambda^{(1)}} \exp\Big\{-\big\|\mathbf{y}_{i,k}-\mathbf{H}_{i,k}\mathbf{x}_{i,k}^{(1)}\big\|^2\Big\} \\ &-\min\nolimits_{\mathbf{x}_{i,k}^{(0)}\in\mathcal{X}_\lambda^{(0)}} \exp\Big\{-\big\|\mathbf{y}_{i,k}-\mathbf{H}_{i,k}\mathbf{x}_{i,k}^{(0)}\big\|^2\Big\}\Big\}.\end{aligned} \tag{23.5}$$

Naturally, bit-level combining results in performance loss compared to the optimal ML decoder. In SISO systems this difference is usually small. Nevertheless, this is not necessarily the case for MIMO channels where interference between streams may result in per-stream SNR loss even when the overall channel is good [19].

In contrast to bit-level combining, in some cases it is possible to combine symbols from different HARQ transmissions without a penalty in performance. To better illustrate how this can be done, CC-HARQ is considered first. Then the description is extended to IR-HARQ scenarios where symbol-level combining can be used.

Consider the first transmission. Then from Eq. (23.1), for slot k,

$$\mathbf{y}_{1,k} = \mathbf{H}_{1,k}\mathbf{x}_k + \mathbf{z}_{1,k} \tag{23.6}$$

It is well known that the ML estimate of $\mathbf{x}_{1,k}$ can either be obtained directly from the above equation, or after passing $\mathbf{y}_{1,k}$ through the matched filter, $\mathbf{H}_{1,k}^*$. In the latter case, $\mathbf{x}_{1,k}$ is estimated using

$$\tilde{\mathbf{y}}_{1,k} = \mathbf{H}_{1,k}^*\mathbf{H}_{1,k}\mathbf{x}_k + \mathbf{H}_{1,k}^*\mathbf{z}_{1,k} = \tilde{\mathbf{H}}_{1,k}\mathbf{x}_k + \tilde{\mathbf{z}}_{1,k} \tag{23.7}$$

Note that the noise $\tilde{\mathbf{z}}_{1,k}$ is no longer white: $E[\tilde{\mathbf{z}}_{1,k}, \tilde{\mathbf{z}}_{1,k}^*] = \mathbf{H}_{1,k}^*\mathbf{H}_{1,k} = \tilde{\mathbf{H}}_{1,k}$. One way to proceed is to whiten the noise using $\tilde{\mathbf{H}}_{1,k}^{-1/2}$. In the case of one transmission, this operation will reconstruct $\mathbf{y}_{1,k}$. However, this approach is useful when retransmissions take place as is shown in the following.

Omitting the slot index, k, to simplify the notation, and concatenating the received symbols after N transmissions in a single vector $\mathbf{y}_{c,N}$, the following input–output relation can be written for each slot.

$$\mathbf{y}_{c,N} = \mathbf{H}_{c,N}\mathbf{x} + \mathbf{z}_{c,N} \tag{23.8}$$

where

$$\mathbf{y}_{c,N} = \left[\mathbf{y}_1^T \cdots \mathbf{y}_N^T\right]^T \in \mathcal{C}^{N \cdot N_R \times 1} \tag{23.9}$$

$$\mathbf{H}_{c,N} = \left[\mathbf{H}_1^T \cdots \mathbf{H}_N^T\right]^T \in \mathcal{C}^{N \cdot N_R \times N_T} \tag{23.10}$$

$$\mathbf{z}_{c,N} = \left[\mathbf{z}_1^T \cdots \mathbf{z}_N^T\right]^T \in \mathcal{C}^{N \cdot N_R \times 1} \tag{23.11}$$

From a mathematical point of view, this $N \cdot N_R \times N_T$ MIMO model is the same as Eq. (23.1), except for the dimensions of the variables. The model includes all information available at the receiver and, therefore, can be used to estimate $\mathbf{x}$. Similar to the case of one transmission, the model can either be used directly, or after applying the matched filter. It turns out that the latter approach leads to an efficient receiver implementation.

For the model of Eq. (23.8), the matched filter is the matrix $\mathbf{H}_{c,N}^*$. Therefore, the output of the matched filter is equal to

$$\begin{aligned}\tilde{\mathbf{y}}_{c,N} &= \mathbf{H}_{c,N}^*\mathbf{y}_{c,N} + \mathbf{H}_{c,N}^*\mathbf{z}_{c,N} = \mathbf{H}_{c,N}^*\mathbf{H}_{c,N}\mathbf{x} + \tilde{\mathbf{z}}_{c,N} \\ &= \left(\sum_{n=1}^{N}\mathbf{H}_n^*\mathbf{H}_n\right)\mathbf{x} + \sum_{n=1}^{N}\mathbf{H}_n^*\mathbf{z}_n = \tilde{\mathbf{H}}_N\mathbf{x} + \sum_{n=1}^{N}\mathbf{H}_n^*\mathbf{z}_n\end{aligned} \tag{23.12}$$

Two goals have been achieved by the use of the matched filter, $\mathbf{H}_{c,N}^*$. First, the size of the model has been reduced. The output of the matched filter is now a vector of size $N_T \times 1$ rather than $N \cdot N_R \times 1$, and $\tilde{\mathbf{H}}_N$ is an $N_T \times N_T$ rather than an $N \cdot N_R \times N_T$ matrix. Therefore, estimation can be performed with smaller complexity when Eq. (23.12) is used in place of Eq. (23.8). Furthermore, $\tilde{\mathbf{y}}_{c,N}$ and $\tilde{\mathbf{H}}_N$ can be calculated iteratively after the end of each retransmission.

$$\tilde{\mathbf{y}}_{c,N} = \tilde{\mathbf{y}}_{c,N-1} + \mathbf{H}_N^*\mathbf{y}_N \tag{23.13}$$

$$\tilde{\mathbf{H}}_N = \tilde{\mathbf{H}}_{N-1} + \mathbf{H}_N^*\mathbf{H}_N \tag{23.14}$$

Therefore, the required memory is also reduced compared to the generic ML receiver: N_T complex values are required to store $\tilde{\mathbf{y}}_{c,N}$ and $N_T(N_T - 1)/2$ complex

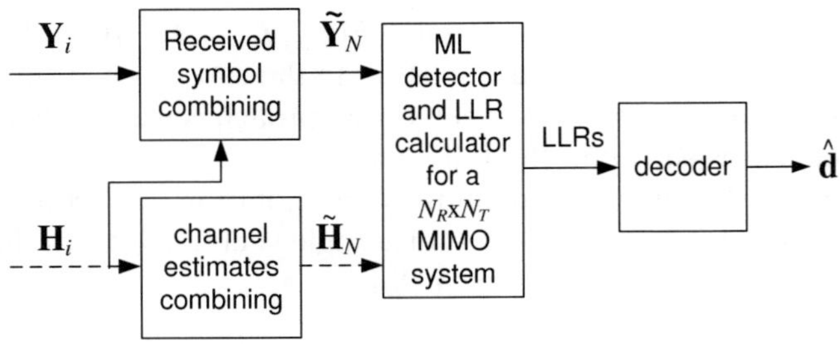

Figure 23.5. Generic block diagram for the symbol-level combining receiver.

values are required for the Hermitian matrix $\tilde{\mathbf{H}}_N$. Note that $\tilde{\mathbf{y}}_{c,N}$ is formed using *symbol-level combining*. As has been mentioned previously, because the use of the matched filter retains all information relevant to the estimation of $\mathbf{x}$, the use of symbol-level combining as detailed above does not affect the performance of the receiver.

As in the case of one transmission, the noise $\tilde{\mathbf{z}}_{c,N}$ is not white: $E[\tilde{\mathbf{z}}_{c,N}\tilde{\mathbf{z}}_{c,N}^*] = \sum_{n=1}^{N} \mathbf{H}_n^*\mathbf{H}_n = \tilde{\mathbf{H}}_N$. Hence, whitening with $\tilde{\mathbf{H}}_N^{-1/2}$ should be applied before using a standard ML receiver or an equalizer. $\tilde{\mathbf{H}}_N$ can be updated efficiently using Eq. (23.14).

Symbol-level combining can also be used in some IR-HARQ scenarios. More specifically, the condition that needs to be satisfied is that bit-to-symbol vector alignment be preserved, even if the symbols are mapped to different antennas at each retransmission. When symbols are mapped to different antennas, a permutation matrix can be employed that is multiplied with $\mathbf{H}_{i,k}$, resulting in an equivalent matrix, $\mathbf{H}_{eq,i,k}$, that is used to update $\tilde{\mathbf{y}}_{c,N}$ and $\tilde{\mathbf{H}}_N$. Symbol-level combining can even be used when bit-to-symbol vector alignment is not preserved, as long as some or all symbol vectors are eventually re-sent through the channel.

Clearly, if the symbol vectors never repeat, or if they repeat after a large number of retransmissions, it may not be possible to use symbol-level combining. In this case, a choice needs to be made between implementing a complex yet optimal receiver or using distance-level and/or bit-level combining. As mentioned previously, distance- and bit-level combining can also be used with CC-HARQ. However, they result in performance loss, and symbol-level combining is preferable unless its use is not feasible because of other system constraints.

The implementation of a receiver employing symbol-level combining is depicted in Figure 23.5.

23.4.3 Receiver Implementation when Space–Time Coding Is Used at the Transmitter

Having described symbol-level combining for CC-HARQ, it is now easy to see how the receiver can be simplified when linear, symbol-level space–time codes are used at the transmitter. Using the equivalent matrix $\mathbf{H}_{eq,1,k}$,

$$\mathbf{y}_{1,k} = \mathbf{H}_{eq,1,k}\mathbf{x}_k + \mathbf{z}_{1,k} \tag{23.15}$$

and symbol-level combining can be performed using $\mathbf{H}_{eq,1,k} = \mathbf{H}_{1,k}\ \mathbf{W}_{1,k}$ in place of $\mathbf{H}_{1,k}$.

As an example, for the 2 × 2 Alamouti code considered in Section 23.3.2, symbol-level combining at the receiver is done using $\mathbf{H}_{i,k}$ for odd i and by conjugating $\mathbf{y}_i$ and using $\mathbf{H}'_{i,k}$ for even i.

23.4.4 Using Equalization to Simplify the Receiver

Given the MIMO model

$$\mathbf{y} = \mathbf{H}\mathbf{x} + \mathbf{z}$$

and assuming equiprobable $\mathbf{x}$, the estimator of $\mathbf{x}$ given $\mathbf{y}$ and $\mathbf{H}$ that minimizes the probability of symbol error is the Maximum-Likelihood detector. When $\mathbf{z}$ is a white Gaussian random vector, as is well known, the ML estimate, $\hat{\mathbf{x}}$, is the symbol among the allowed values that minimizes the Euclidean distance

$$\|\mathbf{y} - \mathbf{H}\hat{\mathbf{x}}\|$$

When $\mathbf{z}$ is not white, a common approach is to apply a whitening filter $\mathbf{W}$ to $\mathbf{y}$ and then find the symbol $\hat{\mathbf{x}}$ that minimizes the distance $\|\mathbf{W}\mathbf{y} - \mathbf{W}\mathbf{H}\hat{\mathbf{x}}\|$.

Although the principle of the ML decoder is simple, in practice, it may be complex to implement. If a brute-force implementation is used, all candidates $\mathbf{x}$ have to be tried in order to calculate the minimizing $\hat{\mathbf{x}}$. One way to simplify the receiver is by using sphere-decoding techniques that attempt to find $\hat{\mathbf{x}}$ by searching more efficiently and/or in a smaller space [20, 21]. Another very common approach is the use of equalization along the spatial direction. Similar to time-domain equalization, linear or nonlinear equalizers may be employed. Moreover, equalizers can be Zero Forcing (ZF) or they may be minimizing the Minimum Mean-Square Error (MMSE). In this section, a brief overview of MIMO equalization is given in order to facilitate the discussion in the remainder of the chapter. For a more detailed treatment of the topic the reader is referred to MIMO texts such as reference 4. Again, a single transmission will be considered. The combination of equalization and HARQ will be described in greater detail in the following section.

MIMO equalizers attempt to estimate the elements of $\mathbf{x}$ separately rather than jointly. One way that this can be done is using the matched-filter corresponding to each dimension of the vector. More specifically, if $\mathbf{y} = \mathbf{H}\mathbf{x} + \mathbf{z} = \sum_{l=1}^{N_R} \mathbf{h}_l x_n + \mathbf{z}$, then, if the dimension of interest is r, $y = \mathbf{h}_r x_r + \Sigma_{l \neq r} \mathbf{h}_l x_l + \mathbf{z}$ and the matched filter $\mathbf{h}_r^*$ can be used. Hence, each of the elements x_r of $\mathbf{x}$ an be recovered by taking the inner product $\mathbf{h}_r, \mathbf{y}$ and using a single-input single-output slicer or LLR calculator. However, the matched filter approach is suboptimal, because, as can be easily shown, the interference $\Sigma_{l \neq r} \mathbf{h}_l x_l + \mathbf{z}$ is not white.

Another frequently used approach, is to employ a linear filter to eliminate all unwanted dimensions. This is often referred to as Zero Forcing (ZF) or interference nulling. Using, again, $\mathbf{y} = \mathbf{h}_r x_r + \Sigma_{l \neq r} \mathbf{h}_l x_l + \mathbf{z}$, if $\mathbf{v}_r$ belongs to a subspace that is orthogonal to the subspace spanned by $\mathbf{h}_l$, $l \neq r$, then we contain $\mathbf{v}_r\mathbf{y} = \mathbf{v}_r\mathbf{h}_r x_r + \mathbf{v}_r\mathbf{z}$. It can be shown that the rows of the pseudoinverse of $\mathbf{H}$, $\mathbf{H}^\dagger = (\mathbf{H}^*\mathbf{H})^{-1}\,\mathbf{H}^*$ can be used as $\mathbf{v}_r$. Although ZF is simple, forcing the interference to zero is not optimal, in general. When $\mathbf{h}_r$ is parallel to some other $\mathbf{h}_l$ or almost parallel, its projection along the direction of $\mathbf{v}_r$ will be small. Therefore, the resulting signal-to-noise ratio may be small, even if the magnitude of $\mathbf{h}_r x_r$ is large.

An improvement to the performance of the equalizer can be attained if the design focuses on maximizing the signal-to-noise ratio. This criterion is satisfied by the Minimum Mean-Square Error (MMSE) equalizer.[4] For the MMSE, the rows of $(\alpha\mathbf{I} + \mathbf{H}^*\mathbf{H})^{-1}\,\mathbf{H}^*$ can be used, where it is assumed that the elements of $\mathbf{x}$ are uncorrelated and have equal power, and $1/\alpha$ is the signal-to-noise ratio (SNR) per dimension. Unlike the ZF equalizer, the MMSE allows part of the interference to "leak" into the equalized signal. However, the resulting signal-to-noise-and-interference-ratio is smaller than in the case where a ZF equalizer is used.

The performance of equalization can be improved further by employing *successive interference cancellation* (SIC). After each element, x_r, of $\mathbf{x}$ is estimated, its effect is removed from the received signal before proceeding with the next element. For example, if x_1 is estimated first, for the second step of the equalization $\mathbf{y}' = \mathbf{y} - \mathbf{h}_1\,\hat{\mathbf{x}}_1$ is used, and so on. SIC can be combined with both ZF and MMSE equalizers. Because a decision needs to be taken regarding the value of x_1, equalizers using SIC are not linear. Although they perform better than their linear counterparts, they are subject to *error propagation*, which results from erroneous decisions. Moreover, the quality of the equalized symbols depends on the *ordering* of SIC. The optimal approach is to start from the dimension with the best signal-to-noise ratio (SNR) and proceed to the direction of decreasing SNR. Determining the optimal ordering adds complexity to the receiver [7]. Equalizers employing SIC are also called Decision Feedback Equalizers (DFEs), especially when equalization takes place along the time direction.

It has been shown that use of the MMSE-SIC equalizer can attain capacity in Gaussian MIMO channels [4], and, therefore, the MMSE-SIC can be used to implement an optimal receiver. However, capacity is attained for Gaussian input signals and for encoded sequences. In practical systems employing signals belonging to finite constellations, the MMSE-SIC is suboptimal.

23.4.5 Pre-equalization and Post-equalization Combining Approaches

Having described MIMO equalization for the single-transmission case, it is now easy to extend to multiple transmissions using the observations of Section 23.4.2.

[4]Strictly speaking, the MMSE equalizer presented here minimizes the Minimum Mean-Square Error when the vector $\mathbf{x}$ is Gaussian. However, it is commonly used in practice, where the assumption usually does not hold.

Returning to Eq. (23.8) and assuming that bits-to-symbol vectors alignment is preserved, we obta in

$$\mathbf{y}_{c,N} = \mathbf{H}_{c,N}\mathbf{x} + \mathbf{z}_{c,N}$$

where $\mathbf{H}_{c,N}$ may also contain the effect of a space-time code. Then all equalizers can be derived using the standard expressions presented in the previous section.

The Zero-Forcing Linear Equalizer (ZF-LE) (not using SIC) that corresponds to Eq. (23.8) is the linear filter $(\mathbf{H}_{c,N}^*\mathbf{H}_{c,N})^{-1}\mathbf{H}_{c,i}^*$ Using the fact that $\mathbf{H}_{c,N}^*\mathbf{y}_{c,N} = \sum_{i=1}^{N}\mathbf{H}_i^*\mathbf{y}_i$ and that $\mathbf{H}_{c,N}^*\mathbf{H}_{c,N} = \tilde{\mathbf{H}}_N = \sum_{i=1}^{N}\mathbf{H}_i^*\mathbf{H}_i$, an incremental implementation for the ZF-LE can be derived: After each transmission, $\tilde{\mathbf{H}}_N$ is formed and its inverse is found. As explained in Section 23.4.2, $\tilde{\mathbf{H}}_N$ can be updated using $\tilde{\mathbf{H}}_N = \tilde{\mathbf{H}}_{N-1} + \mathbf{H}_N^*\mathbf{H}_N$. The matched filter output, $\tilde{\mathbf{y}}_{c,N}$, is also updated using the output of the matched filter corresponding to each transmission: $\tilde{\mathbf{y}}_{c,N} = \tilde{\mathbf{y}}_{c,N-1} + \mathbf{H}_N^*\mathbf{y}_N$. Then the soft estimate $\mathbf{u}_N$ of $\hat{\mathbf{x}}$ is calculated using

$$\mathbf{u} = \tilde{\mathbf{H}}_N^{-1}\tilde{\mathbf{y}}_{c,N} \tag{23.16}$$

Therefore, the receiver *pre-combines* the $\mathbf{y}_i$ to form $\tilde{\mathbf{y}}_N$, and the $\mathbf{H}_i$ to form $\tilde{\mathbf{H}}_N$, before applying the ZF-LE. Clearly, use of pre-equalization combining does not cause any loss in performance compared to when the equalizing filter is performed directly on Eq. (23.8).

In the remainder of the chapter, $\mathbf{u}$ is used to distinguish from the hard decision, $\hat{\mathbf{x}}$ In practical systems, $\mathbf{u}$ is either sliced to yield $\hat{\mathbf{x}}$, or used for LLR calculation. In the above equation, the subscript N denotes that $\mathbf{u}_N$ is the soft estimate after N transmissions.

The pre-equalization combining structure was proposed in reference 22 and is shown in Figure 23.6. Pre-equalization combining helps reduce the memory of the receiver of MIMO systems using HARQ and employing ZF-LEs. Pre-equalization combining is also optimal when MMSE-LEs are used [22].

For the DFE, Eq. (23.8) is used, again, as a starting point. As shown below, this leads naturally to pre-equalization combining. In the following, the Cholesky

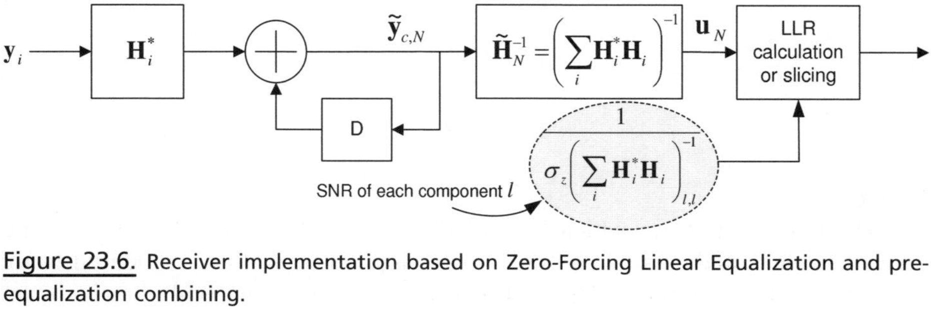

Figure 23.6. Receiver implementation based on Zero-Forcing Linear Equalization and pre-equalization combining.

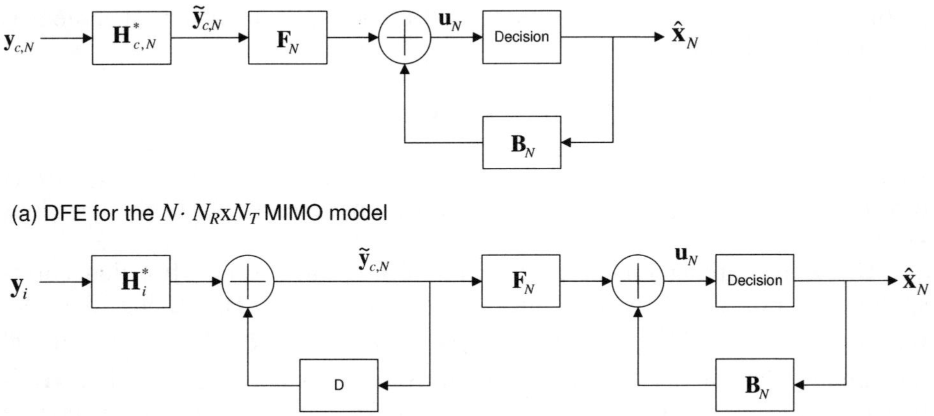

(a) DFE for the $N \cdot N_R \mathrm{x} N_T$ MIMO model

(b) DFE implementation using pre-equalization combining

Figure 23.7. Receiver using Decision Feedback Equalization.

factorization approach is used to derive the DFE. Similar to references [23 and 24], for the model in Eq. (23.8) the ZF-DFE has the form of Figure 23.7a.

As can be seen in the figure, the first operation that is performed on the received signal is matched filtering:

$$\tilde{\mathbf{y}}_{c,N} = \mathbf{H}^{*}_{c,N}\mathbf{y}_{c,N} \tag{23.17}$$

Then the matched-filtered signal $\tilde{\mathbf{y}}_N$ is equalized by a feedforward filter $\mathbf{F}_N$ and a feedback filter $\mathbf{B}_N$. The two filters can be calculated using the Cholesky factorization $\tilde{\mathbf{H}}_N$. Let the Cholesky factorization produce

$$\tilde{\mathbf{H}}_N = G^{*}_{N}\boldsymbol{\Gamma}_N\mathbf{G}_N \tag{23.18}$$

where Γ_N is diagonal with positive elements and $\mathbf{G}_N$ is upper triangular and monic. Then the feedforward filter and the feedback filter are equal to

$$\mathbf{F}_N = \boldsymbol{\Gamma}^{-1}_{N}\mathbf{G}^{-*}_{N} \tag{23.19}$$

and

$$\mathbf{B}_N = \mathbf{I} - \mathbf{G}_N \tag{23.20}$$

respectively.

The matrix $\tilde{\mathbf{H}}_N$ from which $\mathbf{F}_N$ and $\mathbf{B}_N$ are calculated is updated as in the case of linear equalizers. Then, the DFE is applied to $\tilde{\mathbf{y}}_N$ that is updated using pre-equalization combining. The DFE using pre-equalization combining is shown in Figure. 23.7b. Since this implementation was derived by the optimal ZF-DFE design applied to the case of MIMO HARQ without any approximations, pre-equalization combining is, indeed, optimal, not only for linear but also for decision-feedback equalizers.

The MMSE-DFE can be derived in an analogous fashion by performing Cholesky factorization on $\tilde{\mathbf{H}}_{c,n} + \sigma_z^2 \mathbf{\Lambda}^{-1}$ [7].

The memory requirements for receivers employing pre-equalization combining are the same as for the generic symbol-level combining receiver. After each transmission, i, N_T values of the vector $\mathbf{y}_N = \sum_{i=1}^{N} \mathbf{H}_i^* \mathbf{y}_i$ and $\frac{N_T(N_T+1)}{2}$ values of the Hermitian matrix $\tilde{\mathbf{H}}_N = \sum_{i=1}^{N} \mathbf{H}_i^* \mathbf{H}_i$ need to be stored.

After equalization, the N_T estimates of the components of $\mathbf{x}$ are output. In order to reduce the required memory at the receiver even more, it is interesting to investigate whether equalization can precede combining of the estimated symbols. Moreover, in systems where bit-to-symbol vector alignment is not preserved, it is not possible to use symbol-level combining, which means that pre-equalization combining is not an option either. Therefore, it is of interest to explore the performance-memory tradeoff when post-equalization combining is employed.

In reference 22, a post-equalization combining scheme was proposed that combines estimates, $\mathbf{u}_i$, from each transmission using

$$\mathbf{u}_{N,\text{post-IGC}} = \frac{1}{N} \sum_{i=1}^{N} \mathbf{u}_i \tag{23.21}$$

where $\mathbf{u}_i$ is the output of the equalizer designed for $\mathbf{H}_i$, and whose input is $\mathbf{y}_i$. As an example, for the MMSE-LE [25], we have

$$\mathbf{u}_{N,\text{post-IGC,MMSE-LE}} = \frac{1}{N} \sum_{i=1}^{N} (\mathbf{H}_i^* \mathbf{H}_i + \sigma^2 \mathbf{\Lambda}^{-1})^{-1} \mathbf{H}_i^* \mathbf{y}_i \tag{23.22}$$

As can be seen from Eq. (23.22), the matched-filter outputs, $\tilde{\mathbf{y}}_i = \mathbf{H}_i^* \mathbf{y}_i$, are combined using weights that depend on the matrix $\mathbf{H}_i$. In the ZF case, the inverse gain $(\mathbf{H}_i^* \mathbf{H}_i)^{-1}$ is used, thus the name inverse-gain combining (IGC).

The IGC scheme reduces the storage requirements at the receiver. As can be seen from Eq. (23.21), only a $N_T \times 1$ vector needs to be stored after each reception. This results in considerable savings when many slots are used, such as in the case of OFDM systems. The memory savings are the same when a DFE is used, since each received vector is equalized immediately, and, therefore, only the soft equalizer output is stored at the receiver.

In reference 22 it was proven that, when Linear Equalization is used at the receiver, the performance of the post-equalization IGC scheme (23.21) cannot exceed that of pre-equalization combining. It can be shown that this also holds for the DFE. Nevertheless, it is possible to improve performance of post-equalization combining. In the following, the optimal post-equalization combining method is derived for Linear and Decision Feedback Equalizers. It is shown that the method outperforms IGC, but requires more storage. The required storage is still less than the optimal, pre-equalization combining approach, but the latter has better performance. As was explained previously, even in systems where memory reduction is not critical, it may not be possible to use pre-equalization symbol-level combining. For such systems, the optimal post-equalization symbol combining method that is derived in the following yields the best receiver performance.

Returning to the system model (23.1) and using the ZF-LE as an example, we obtain

$$\mathbf{u}_i = (\mathbf{H}_i^*\mathbf{H}_i)^{-1}\mathbf{H}_i^*\mathbf{y}_i = \mathbf{x} + \tilde{\mathbf{z}}_i \tag{23.23}$$

with $\mathbf{E}[\tilde{\mathbf{z}}_i\tilde{\mathbf{z}}_i^*] = \sigma_z^2(\mathbf{H}_i^*\mathbf{H}_i)^{-1}$. Then for the components of $\mathbf{u}_i$, we have

$$u_{i,l} = x_l + \tilde{z}_{i,l} \tag{23.24}$$

and $\tilde{\sigma}_{i,l}^2 = \mathbf{E}[\tilde{z}_{i,l}\tilde{z}_{i,l}^*] = \sigma_z^2[(\mathbf{H}_i^*\mathbf{H}_i)^{-1}]_{l,l}$. As can be seen from Eq. (23.24), the SNR depends on the symbol index, l. Assuming, without loss of generality, that all the x_l have equal power, we obtain

$$\text{SNR}_{i,l} = \frac{1}{\tilde{\sigma}_{i,l}^2} \tag{23.25}$$

A fundamental result in signal processing states that the optimal way to combine the $u_{i,l}$ is by weighting them according to the square root of their SNR (Maximal Ratio Combining—MRC). Thus, MRC post-combining will lead to better performance compared to IGC that simply sums the $\mathbf{u}_i$ (and uses the "inverse" gains $(\mathbf{H}_i^*\mathbf{H}_i)^{-1}$ instead of the MRC gains $1/\tilde{\sigma}_{i,l}^2$).

To obtain an expression for the MRC combiner, the power of the noise is first normalized to 1 for each component l

$$u'_{i,l} = \frac{1}{\tilde{\sigma}_{i,l}}u_{i,l} = \frac{1}{\tilde{\sigma}_{i,l}}x_l + \tilde{z}'_{i,l} = w_{i,l}x_l + \tilde{z}'_{i,l}, \tag{23.26}$$

with $w_{i,l} = \dfrac{1}{\tilde{\sigma}_{i,l}}$ and $\mathbf{E}[\tilde{z}'_{i,l}\tilde{z}'^*_{i,l}] = 1$. Therefore, MRC combining can be achieved using

$$u_{l,\text{optimal post,ZF-LE}} = \frac{1}{\sum_{i=1}^{N}|w_{i,l}|^2}\sum_{i=1}^{N} w_{i,l}u'_{i,l} = \frac{1}{\sum_{i=1}^{N}|w_{i,l}|^2}\sum_{i=1}^{N}|w_{i,l}|^2 u_{i,l} = \frac{1}{\sum_{i=1}^{N}\left(1/\tilde{\sigma}_{i,l}^2\right)}\sum_{1}^{N}\frac{u_{i,l}}{\tilde{\sigma}_{i,l}^2} \tag{23.27}$$

Furthermore,

$$\mathrm{SNR}_{l,\text{optimal post,ZF-LE}} = \sum_{i=1}^{N} \frac{1}{\tilde{\sigma}_{i,l}^2} \tag{23.28}$$

For the Inverse Gain Combining (IGC) ZF-LE of [22],

$$\mathrm{SNR}_{l,\text{IGC-post,ZF-LE}} = \frac{1}{\frac{1}{N^2}\sum_{i=1}^{N} \tilde{\sigma}_{i,l}^2} \tag{23.29}$$

From the convexity of $f(x) = 1/x$ for $x > 0$ and Jensen's inequality, it can be easily shown that $\frac{1}{N}\sum_{i=1}^{N} \frac{1}{\tilde{\sigma}_{i,l}^2} \geq \frac{1}{\frac{1}{N}\sum_{i=1}^{N} \tilde{\sigma}_{i,l}^2}$, and, therefore, MRC post-equalization combining always performs at least as well as IGC. The performance of the two combining schemes is the same only when the SNRs corresponding to each retransmission are equal.

Compared to Inverse Gain Combining, MRC requires N_T more variables, namely the normalization constants $\sum_{i=1}^{N} |w_{i,l}|^2$ for all components l.

The optimal post-combining scheme when DFEs are used can be derived in the same way. The derivation is described in the following mainly because there are two variants for the design of the post-equalization combining DFE that merit some discussion. First, Cholesky factorization of $\mathbf{H}_i^*\mathbf{H}_i$ is performed to determine $\mathbf{F}_i$ and $\mathbf{B}_i$.

$$\mathbf{H}_i^*\mathbf{H}_i = G_i^*\mathbf{\Gamma}_i\mathbf{G}_i \tag{23.30}$$

Assuming that the estimates of all components of $\mathbf{x}$ are correct, the equalized signal $\mathbf{u}_i$ can be written as

$$\mathbf{u}_i = \mathbf{x} + \mathbf{z}_i' \tag{23.31}$$

where $\mathbf{z}_i' = \mathbf{\Gamma}_i^{-1}\mathbf{G}_i^{-*}\mathbf{H}_i^*\mathbf{z}_i$ and $\mathbf{E}[\mathbf{z}_i'\mathbf{z}_i'^*] = \sigma_z^2\mathbf{\Gamma}_i^{-1}$. Assuming, again, that the energy of each component x_l of $\mathbf{x}_i$ is equal to 1

$$\mathrm{SNR}_{i,l} = \frac{\gamma_{i,l,l}}{\sigma_z^2} \tag{23.32}$$

where $\gamma_{i,l,l}$ is the (l, l)th element of the diagonal matrix Γ_i. Maximal ratio combining can now be performed using the equalized values from each transmission.

$$\frac{1}{\sum_{i=1}^{N} \gamma_{i,l,l}} \sum_{i=1}^{N} \gamma_{i,l,l} u_{i,l}. \tag{23.33}$$

Therefore, after MRC, for the estimate of the l th component of $\mathbf{x}$, we have

$$\mathrm{SNR}_{\mathrm{MRC},N,l} = \frac{\sum_{i=1}^{N} \gamma_{i,l,l}}{\sigma_z^2} \tag{23.34}$$

Similar to the pre-equalization combining DFE, and as can be seen from Eq. (23.33), the post-equalization combining scheme can also be implemented incrementally as follows:

$$\tilde{\mathbf{u}}_N = \tilde{\mathbf{u}}_{N-1} + \mathbf{\Gamma}_N \mathbf{u}_N \tag{23.35}$$

$$\tilde{\mathbf{\Gamma}}_N = \tilde{\mathbf{\Gamma}}_{N-1} + \mathbf{\Gamma}_N, \tag{23.36}$$

with $\tilde{\mathbf{u}}_0 = \mathbf{0}$ and $\tilde{\mathbf{\Gamma}}_0 = \mathbf{0}$. Before deciding on the components of $\mathbf{x}$, $\tilde{\mathbf{u}}_N$ is normalized: $\mathbf{u}_{\mathrm{optimal,post}} = \tilde{\mathbf{\Gamma}}_N^{-1} \tilde{\mathbf{u}}_N$. Similar to when linear equalization is used, the post-equalization combining DFE needs to store a $N_T \times 1$ vector $\tilde{\mathbf{u}}_N$ and a diagonal matrix $\tilde{\mathbf{\Gamma}}_N$ after each iteration.

The post-equalization combining method described above forms each equalized vector $\mathbf{u}_i$ by only considering transmission i. Then, all $\mathbf{u}_i$ are combined using MRC before being sent to the slicer that produces the final estimate $\tilde{\mathrm{x}}_{\mathrm{optimal,post}}$. The performance of the post-equalization combining DFE can be improved by maximal ratio combining of information from all transmissions 1 to N when forming $\mathbf{u}_N$. Then the decision is based directly on $\mathbf{u}_N$ rather than the a posteriori maximal ratio combination of the $\mathbf{u}_i$'s. This "layered" method is described below using a simple example.

Let $N = 2$ transmissions and $N_T = 2$. For this example, $a_i(l)$ denotes the lth element of the vector $\mathbf{a}_i$. For the first transmission, for both post-equalization combining schemes, $u_1\,(2) = t_1\,(2)$, and $\hat{\mathbf{x}}_1\,(2)$ results from slicing $t_1\,(2)$ in Figure 23.8. Then $u_1\,(1) = t_1\,(1) + b_1\,\hat{\mathbf{x}}_1\,(2)$, where b_1 is element $(1, 2)$ of the 2×2 matrix $\mathbf{B}_1$. $\hat{\mathbf{x}}_1$ (1) is obtained by slicing $u_1(1)$. For the second transmission, the first, a posteriori (non layered) post-equalization combining method obtains $u_2\,(2)$ and $u_2\,(1)$ by exactly the same way, that is, $u_2\,(2) = t_2\,(2)$, and $u_2\,(1) = t_2\,(1) + b_2 \lfloor u_2\,(2) \rfloor$ (where $\lfloor \cdot \rfloor$ denotes the slicing operation). The soft estimates $\mathbf{u}_1$ and $\mathbf{u}_2$ are then combined using MRC to produce $\hat{\mathbf{x}}_2$. Hence,

$$\hat{x}_2(1) = \left\lfloor \frac{\gamma_{1,1,1} u_1(1) + \gamma_{2,1,1} u_2(1)}{\gamma_{1,1,1} + \gamma_{2,1,1}} \right\rfloor = \left\lfloor \frac{\gamma_{1,1,1}(t_1(1) + b_1 \lfloor t_1(2) \rfloor) + \gamma_{2,1,1}(t_2(1) + b_2 \lfloor t_2(2) \rfloor)}{\gamma_{1,1,1} + \gamma_{2,1,1}} \right\rfloor \tag{23.37}$$

and

$$\hat{x}_2(2) = \left\lfloor \frac{\gamma_{1,2,2} u_1(2) + \gamma_{2,2,2} u_2(2)}{\gamma_{1,2,2} + \gamma_{2,2,2}} \right\rfloor = \left\lfloor \frac{\gamma_{1,2,2} t_1(2) + \gamma_{2,2,2} t_2(2)}{\gamma_{1,2,2} + \gamma_{2,2,2}} \right\rfloor \tag{23.38}$$

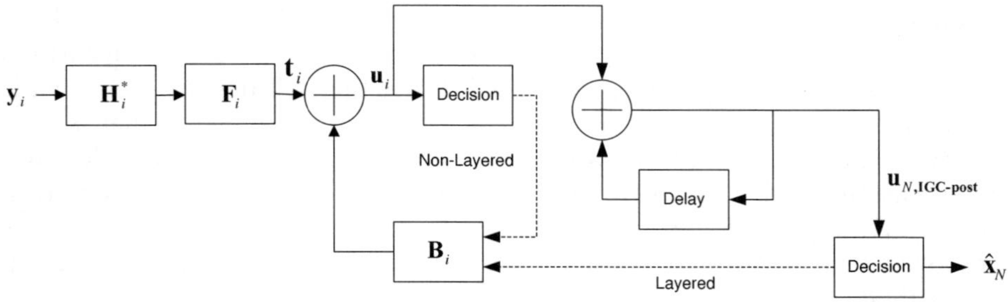

(a) IGC post-equalization symbol-level combining ZF-DFE

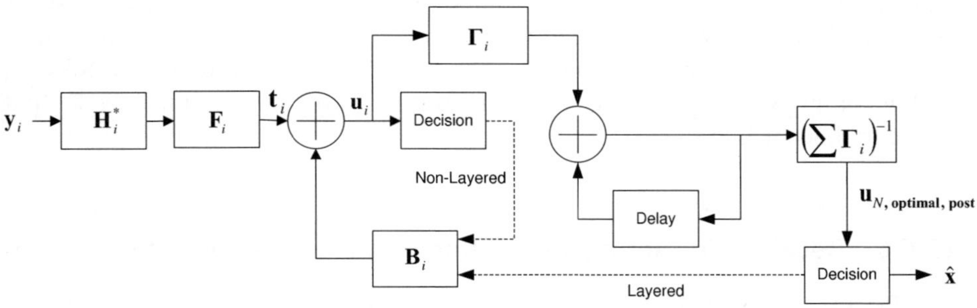

(b) Optimal post-equalization symbol-level combining ZF-DFE

Figure 23.8. Receiver implementation based on Decision Feedback Equalization and post-equalization symbol-level combining.

For the improved "layered" scheme, the soft estimate $\mathbf{u}_2'$ corresponding to the second transmission is formed by maximal ratio combining of all soft estimates. Therefore,

$$u_2'(2) = \frac{\gamma_{1,2,2}u_1'(2) + \gamma_{2,2,2}t_2(2)}{\gamma_{1,2,2} + \gamma_{2,2,2}} = \frac{\gamma_{1,2,2}t_1(2) + \gamma_{2,2,2}t_2(2)}{\gamma_{1,2,2} + \gamma_{2,2,2}}$$
$$\hat{x}_2'(2) = \left\lfloor \frac{\gamma_{1,2,2}t_1(2) + \gamma_{2,2,2}t_2(2)}{\gamma_{1,2,2} + \gamma_{2,2,2}} \right\rfloor \tag{23.39}$$

Similarly,

$$\hat{x}_2'(1) = \lfloor u_2'(1) \rfloor = \left\lfloor \frac{\gamma_{1,1,1}u_1'(1) + \gamma_{2,1,1}(t_2(1) + b_2\hat{x}_2'(2))}{\gamma_{1,1,1} + \gamma_{2,1,1}} \right\rfloor$$
$$= \left\lfloor \frac{\gamma_{1,1,1}(t_1(1) + b_1\lfloor t_1(2) \rfloor) + \gamma_{2,1,1}(t_2(1) + b_2\hat{x}_2'(2))}{\gamma_{1,1,1} + \gamma_{2,1,1}} \right\rfloor \tag{23.40}$$

Note that, although the expression for $\hat{x}_2(1)$ is similar for both schemes, the value by which $\gamma_{2,1,1}$ is multiplied is different. For the first scheme it is just the sliced value of $t_2(2)$, whereas in the second one, it is the sliced value of the maximal ratio combination of the soft estimates of $x(2)$ of each transmission and is given by (39). $\hat{\mathbf{x}}_2(2)$ is equal to $\hat{x}_2'(2)$, but this will only be true for the N_Tth element in the general case. Since the SNR of the quantity that is used for the calculation of $\hat{x}_2'(2)$ cannot be smaller than the SNR of $t_2(2)$, $\hat{x}_2'(2)$ will be at least as reliable as $\lfloor t_2(2) \rfloor$. Thus, it is expected that, in general, the performance of the second, "layered" post-equalization combining scheme will be improved compared to the a posteriori, "nonlayered" post-equalization combining method.

Although the "layered" post-equalization combining scheme may be slightly more complex to implement, the memory requirements do not increase, since, after weighting by the $\gamma_{i,k,k}$ and updating the sum, the $\gamma_{i,k,k}$ can be discarded. Therefore, similar to the previous case, what needs to be stored after processing each incoming vector $\mathbf{y}_i$ are the soft estimates $\mathbf{u}_i$ and the diagonal matrix $\tilde{\boldsymbol{\Gamma}}_i$ that is updated according to Eq. (23.36).

23.4.6 Receiver Implementation Based on QR Factorization

The review of the receiver design concludes with an overview of a practical approach for the implementation of the *symbol-level combining* receiver that relies on QR decomposition [26].

Consider the received signal $\mathbf{y}_1$ after the first transmission. QR decomposition is performed on the channel matrix $\mathbf{H}_1$ and a signal $\tilde{\mathbf{w}}_1$ is calculated using projection on $\mathbf{Q}_1$:

$$\mathbf{H}_1 = \mathbf{Q}_1\mathbf{R}_1 \tag{23.41}$$

and

$$\tilde{\mathbf{w}}_1 = \mathbf{Q}_i^*\mathbf{y}_1 = \mathbf{R}_1\mathbf{x} + \tilde{\mathbf{z}}_1 \tag{23.42}$$

For the processing of the second transmission, the receiver stores only the sufficient $N_T \times 1$ statistic $\tilde{\mathbf{w}}_1$ and the equivalent channel $\mathbf{R}_1(N_T \times (N_T+1)/2$ values). After the Nth transmission, the following relation is satisfied:

$$\begin{bmatrix} \tilde{\mathbf{w}}_{N-1} \\ \mathbf{y}_N \end{bmatrix} = \tilde{\mathbf{H}}_N\mathbf{x} + \begin{bmatrix} \tilde{\mathbf{z}}_{N-1} \\ \mathbf{z}_N \end{bmatrix} \tag{23.43}$$

where

$$\tilde{\mathbf{H}}_N = \begin{bmatrix} \tilde{\mathbf{R}}_{N-1} \\ \mathbf{H}_N \end{bmatrix} \tag{23.44}$$

$$E[\tilde{\mathbf{z}}_{N-1}\tilde{\mathbf{z}}_{N-1}^*] = \sigma_z^2\mathbf{I}_{N_T} \tag{23.45}$$

and the QR decomposition of $\tilde{\mathbf{H}}_N$ is given by

$$\tilde{\mathbf{H}}_N = \tilde{\mathbf{Q}}_N \tilde{\mathbf{R}}_N \tag{23.46}$$

Then the sufficient statistic $\tilde{\mathbf{w}}_N$ is formed using

$$\tilde{\mathbf{w}}_N = \tilde{\mathbf{Q}}_N^* \begin{bmatrix} \tilde{\mathbf{w}}_{N-1} \\ \mathbf{y}_N \end{bmatrix} = \tilde{\mathbf{R}}_N \mathbf{x} + \tilde{\mathbf{z}}_N \tag{23.47}$$

It can be shown that use of this incremental, QR-based symbol-level combining approach does not cause a loss in performance [27].

When ML decoding is used, the value of $\mathbf{x}$ is found that minimizes $\|\tilde{\mathbf{w}}_N - \tilde{\mathbf{R}}_N\mathbf{x}\|$. When linear equalizers are employed, they can be designed based on $\tilde{\mathbf{R}}_N$ and using $\tilde{\mathbf{w}}_N$ as the input instead of $\tilde{\mathbf{y}}_N$. For example, the ZF-LE filters $\tilde{\mathbf{w}}_N$ using $(\tilde{\mathbf{R}}_N^*\tilde{\mathbf{R}}_N)^{-1}\tilde{\mathbf{R}}_N^*$ to produce the soft equalized vector $\mathbf{u}_N$.

When the receiver is based on a ZF-DFE, the feedforward filter equals

$$\mathbf{K}_N = \left[\operatorname{diag}\left(\tilde{\mathbf{R}}_N\right)\right]^{-1} \tag{23.48}$$

and the feedback filter is given by

$$\mathbf{B}_N = \mathbf{I} - \mathbf{K}_N \tilde{\mathbf{R}}_N \tag{23.49}$$

When an MMSE-DFE, is used, the matrices $\tilde{\mathbf{Q}}_N$ and $\tilde{\mathbf{R}}_N$ are the factors of the QR decomposition of the augmented matrix $\begin{bmatrix} \tilde{\mathbf{H}}_N \\ \sqrt{\alpha}\mathbf{I} \end{bmatrix}$, where $1/\alpha$ is the SNR (assumed equal for all elements of $\mathbf{x}$).

The implementation of the receiver using QR decomposition does not offer any memory savings compared to straightforward symbol-level combining. Again, what needs to be stored after each iteration are the N_T $(N_T + 1)/2$ elements of the triangular matrix $\tilde{\mathbf{R}}_N$ and the $N_T \times 1$ vector $\tilde{\mathbf{w}}_N$. However, use of QR decomposition may be attractive because it can be calculated quite efficiently using Givens rotations because of the zero elements in $\tilde{\mathbf{R}}_N$.

The QR-based approach can also be used when post-equalization symbol combining is employed at the receiver [27].

23.5 SIMULATED PERFORMANCE OF HARQ IN MIMO WIRELESS SYSTEMS

Having presented different possible implementations for the receiver, their performance is compared using simulations. A system based on IEEE 802.16e [2] is considered.

23.6.1 Performance of Receivers Based on ML Demodulation and Linear Equalization

The simulation was performed with the following settings. The number of transmit antennas, N_T, and the number of receive antennas, N_R, are both set to 2. The number of subcarriers of the OFDM symbols is 1024, and the full-usage of subcarriers (FUSC) mode is chosen. One packet consists of multiple coded blocks, each spanning 3 orthogonal frequency-division multiplexing (OFDM) symbols, and the duration of one OFDM symbol is equal to approximately 102.85 μs. Each OFDM symbol consists of 16 subchannels, and each subchannel comprises 48 subcarriers. The carrier frequency, the sampling rate, and the subcarrier spacing are equal to 2.5 GHz, 11.2 MHz, and 10.9375 kHz, respectively. The channel model used for the simulations is the 3GPP UMTS case III with mobile speed 3 km/h [28].

Figures 23.9 and 23.10, respectively, show the bit-error rate (BER) versus the SNR when the modulation scheme is 4-QAM with coding rate $r = ½$, and when the modulation scheme is 16-QAM with coding rate $r = ¾$. In the figures, "ML" and "ZF" denote the case where only one transmission occurs. In all other cases, the number of transmissions is equal to 2. The received signal vectors are combined as described in the previous sections. "DLC" denotes Distance-Level Combining, whereas "BLC" denotes Bit-Level Combining.

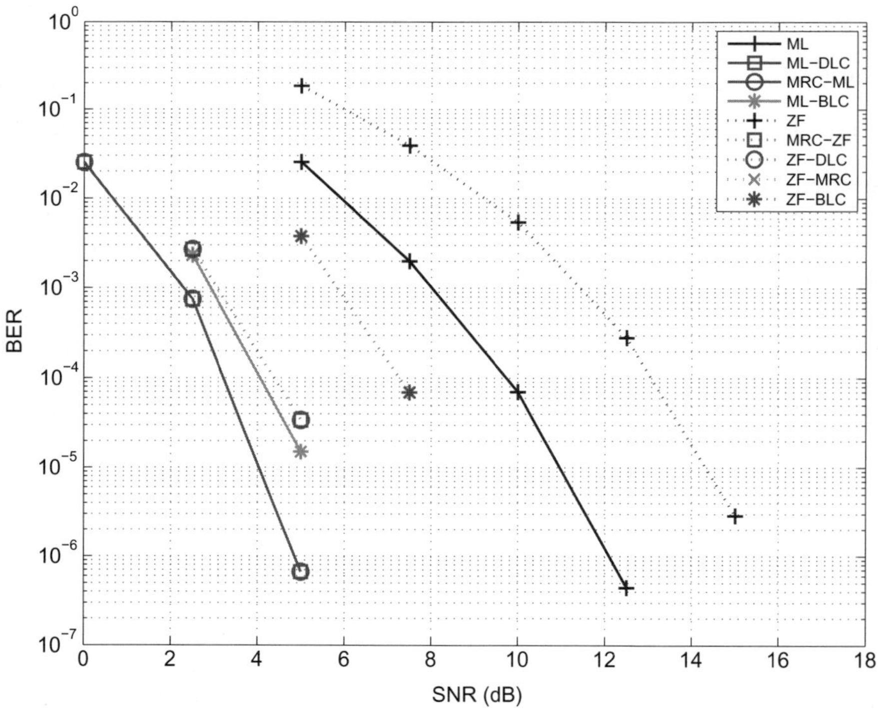

Figure 23.9. Bit eror rates for 4-QAM with code rate 1/2 for 3GPP UMTS channel cases III.

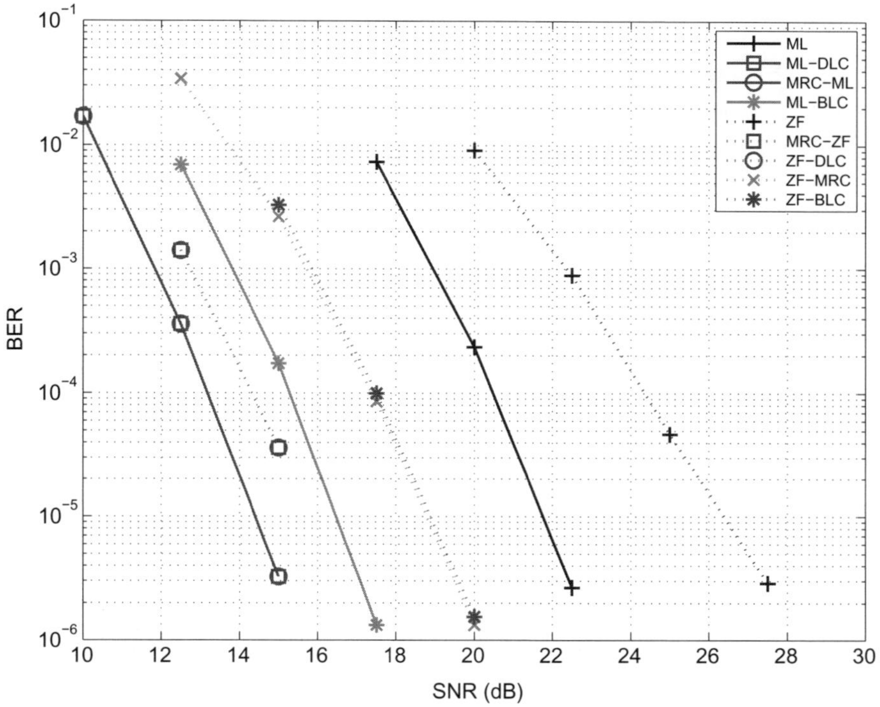

Figure 23.10. Bit error rates for 16-QAM with code rate 3/4 for 3GPP UMTS channel cases III.

The figures show that the BER decreases significantly when two transmissions are used. Moreover, the ML-DLC scheme and the MRC-ML scheme achieve the best performance, whereas the ML-BLC scheme exhibits performance loss. The results agree well with the mathematical analysis in the previous sections.

The figures also show that the performance gap between the MRC-ML and the ML-BLC scheme is small for the 4-QAM case, but becomes larger as the modulation size increases. As can also be seen, the performance gap is much wider for the ZF case, i.e., when comparing the MRC-ZF and the ZF-BLC scheme. This suggests that, using ZF, which ignores the noise correlation, for combining schemes results in significant loss of information. This observation is also supported by the fact that the performance gap between the ZF-BLC scheme and the ZF-MRC scheme is negligible.

23.6.2 Performance of DFE-Based Receivers

In this subsection, the difference in performance of DFE-based receivers employing different combining methods is verified and quantified using simulation. Because there are many different possibilities of feedback schemes in the case

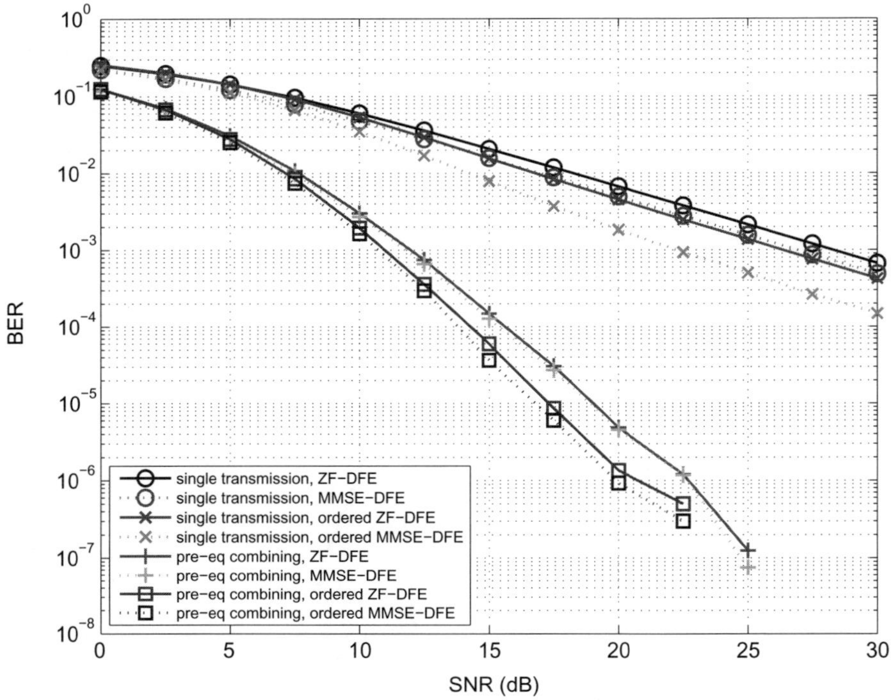

Figure 23.11. Simulated performance of the optimal, pre-equalization combining DFE.

of a coded system, unlike the previous subsection, an uncoded system is tested in order to focus on the performance difference of the combining schemes. A system with $N_T = 2$ transmit and $N_R = 2$ receive antennas is considered. The channel experiences Rayleigh fading and each of the (complex) elements of the matrix $\mathbf{H}_{i,k}$ are independent of each other. Moreover, the values of a given element $h_{i,k,(j,r)}$ are independent for different transmissions. The transmitter uses 4-QAM modulation with $E[\mathbf{xx}^*] = \mathbf{I}_{N_T}$. The noise is identically distributed circular symmetric Gaussian $\mathbf{z} \sim CN\ (0, \sigma_z \mathbf{I}_2)$. Hence, the average SNRs of the two transmitted streams are equal. When retransmissions occur, the receiver processes all received symbols before making a decision on the 4-QAM symbols of each transmit stream.

Figure 23.11 depicts the improvement in performance when two transmissions are used and an optimal, pre-equalization combining DFE is employed at the receiver. Both the Zero Forcing and the MMSE cases are considered, and the effect of ordering is also examined. As can be seen, retransmission results in a significant performance improvement because time diversity is exploited. Use of ordering can also provide some SNR gain. As was shown in Section 23.4, the choice of a particular implementation (QR-based or Cholesky-based) for the pre-equalization combining DFE does not have any impact on performance. This is also verified by simulation. The resulting BER curves are shown in Figure 23.12.

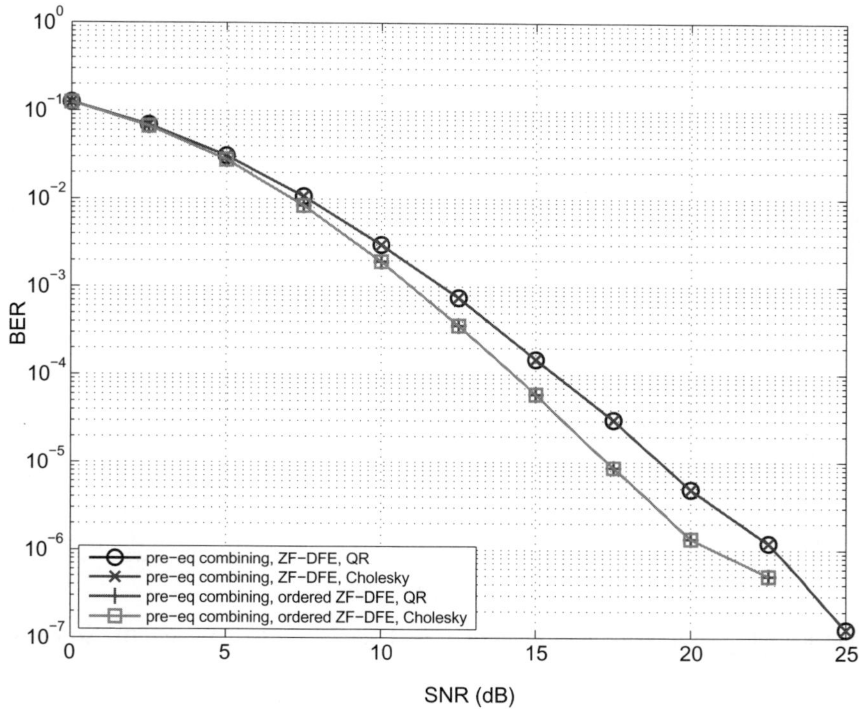

Figure 23.12. Performance of QR- and Cholesky-based pre-equalization combining DFE schemes.

The performance of the optimal, pre-equalization combining DFE is compared to that of the post-equalization combining schemes mentioned in this chapter in Figure 23.13. Again, two transmissions are used for the simulation. As expected, the improved layered MRC post-equalization combining DFE that forms soft estimates based on all received symbols performs better than the a posteriori, non-layered MRC post-equalization combining DFE that combines soft estimates each of which is based on one received symbol. The same is true for the post-equalization combining DFE that uses inverse gains instead of the MRC weights, albeit with a much less pronounced difference in performance. As predicted, the performance of the inverse gain post-equalization combining DFE is inferior compared to its MRC counterpart. In general, as can be deduced from the curves of the figure, the post-equalization combining schemes fail to fully exploit time diversity, and this is the price paid for the reduced storage requirements.

Finally, the performance of the pre-equalization combining DFE is compared to that of Linear Equalization. As can be seen from Figure 23.14, Decision Feedback Equalization provides a gain over Linear Equalization. However, for the single-transmission case, even when ordering is used, there is a limit on

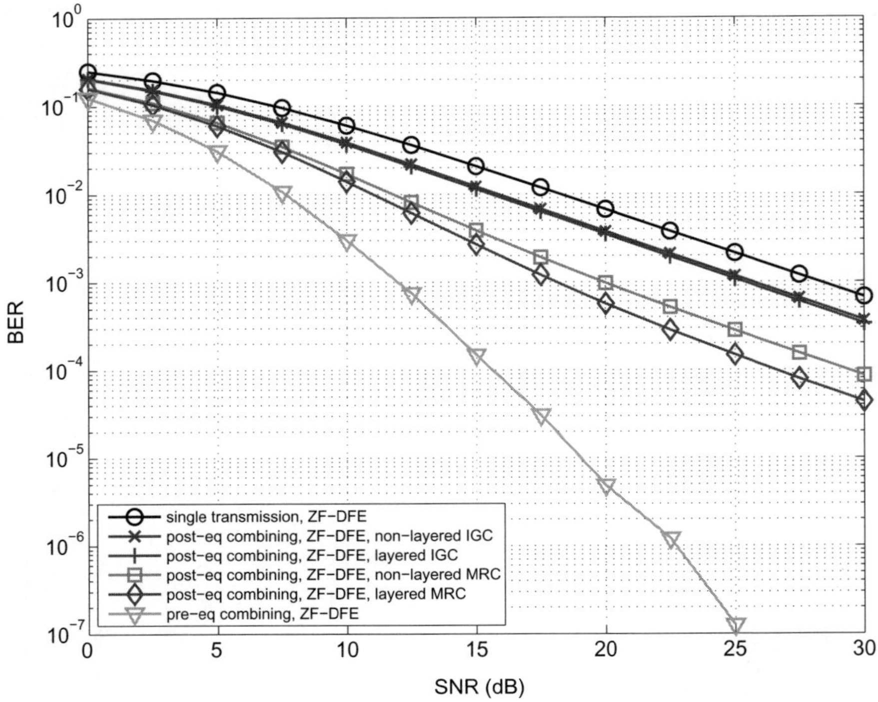

Figure 23.13. Simulated performance of post-equalization combining DFE schemes.

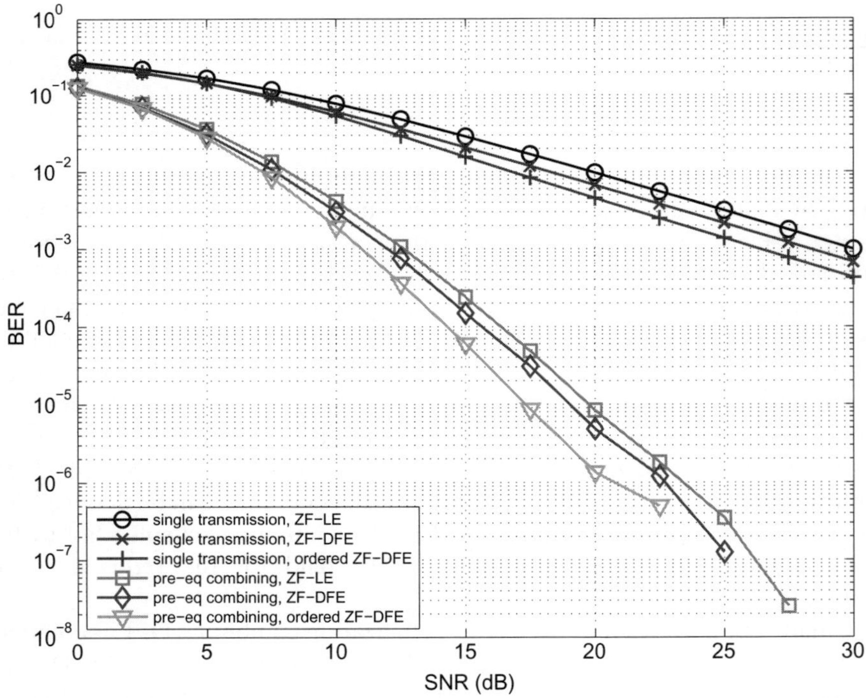

Figure 23.14. Improvement in system performance when retransmission is used.

the achievable performance due to Rayleigh fading. Further performance improvement can only be achieved by taking advantage of diversity, in this case by retransmitting the signals through the channel.

23.5.3 Receiver Performance when Space–Time Block Coding Is Combined with HARQ

In this subsection, the performance of the combination of HARQ and Alamouti's 2×2 space–time code is compared with a scheme employing no precoding and a basis hopping technique [22] that applies a random precoding matrix for each transmission. As in previous simulations, an uncoded MIMO system with two transmit antennas and two receiver antennas is considered. The space–time block code used along with the HARQ is the 2×2 Alamouti code that was presented in Section 23.3. The modulation used is 4-QAM. For the receiver, a Zero-Forcing Linear Equalizer (ZF-LE) and a Minimum Mean-Square error Linear Equalizer (MMSE-LE) are considered to combat inter-stream interference. The combining scheme employed at the receiver is pre-equalization symbol-level combining. For the MIMO channel, it is assumed that there is no correlation between each element of the matrix and that each element is complex Gaussian. Two cases are considered for the channel generation. In the first case, it is assumed that the channel does not change at all over all HARQ retransmissions. In the second case, it is assumed that the channel is constant for the duration of a transmission, but that it varies independently between two transmissions. For the basis hopping technique, the unitary matrix is chosen as in reference 22.

Figures 23.15 and 23.16 plot the bit error rate (BER) for the second transmission and for a time-invariant channel. In Figure 23.15, a ZF-LE is used, whereas in Figure 23.16, the receiver is implemented using an MMSE-LE. As can be seen in the figures, the basis hopping technique achieves significant gains compared to when no precoding is used in this time-invariant channel case. Furthermore, the technique combining HARQ with space–time coding achieves even better performance than the basis hopping technique. Comparing Figures 23.15 and 23.16, it can be seen that the BER for ZF-LE and the BER for MMSE-LE are the same for the space–time coding scheme after the second transmission. This is because the space–time code does not introduce any inter-stream interference at the second transmission and both equalizer matrices for the ZF-LE and the MMSE-LE are (scaled) identity matrices. On the other hand, the performance of the basis hopping technique and the technique that does not use precoding improves slightly when the MMSE-LE is used in place of the ZF-LE. However, the performance of the scheme using space–time coding when the MMSE-LE is used is still much better than the other two schemes.

Figure 23.17 shows the BER after the second transmission for various HARQ transmission techniques and a block-fading channel. The channel during the first transmission is independent from the channel during the second transmission. In this case, there is no difference in the performance of all three techniques.

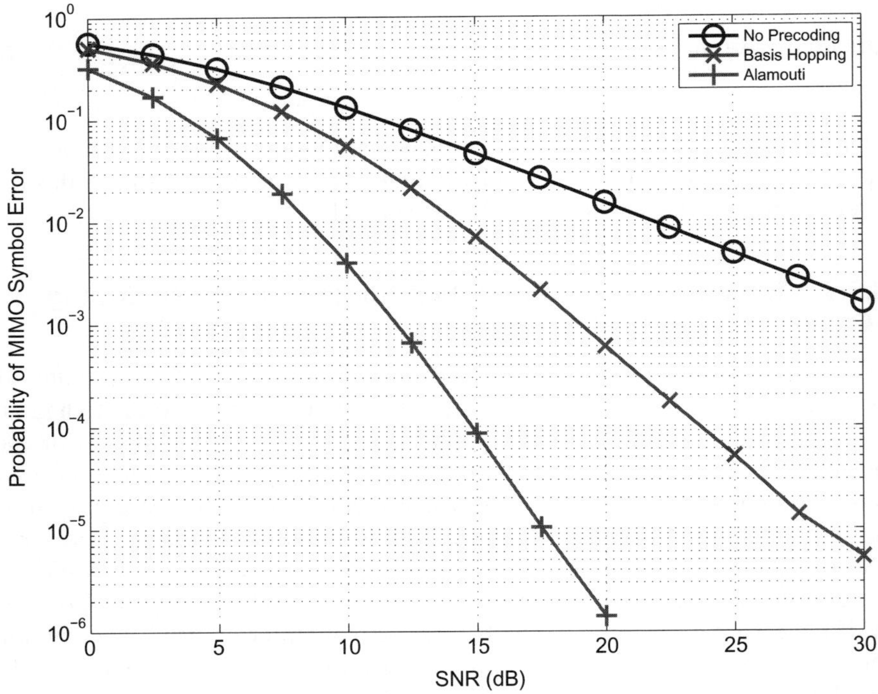

Figure 23.15. Bit error rate at the 2nd transmission for various HARQ transmission techniques in time-invariant channel. Zero-Forcing (ZF) Linear Equalizer (LE) has been used as a MIMO equalizer.

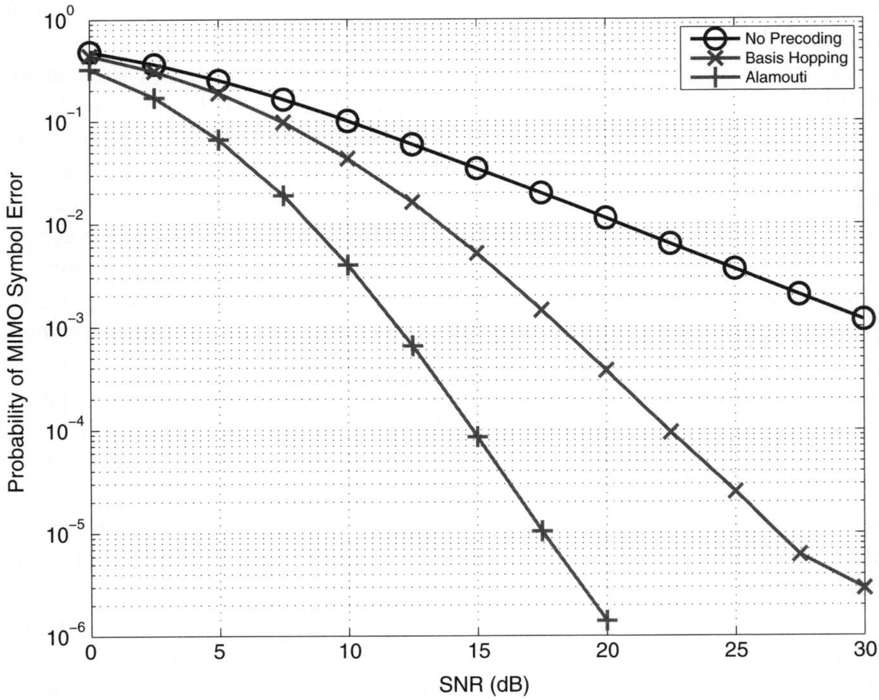

Figure 23.16. Bit error rate at the 2nd transmission for various HARQ transmission techniques in time-invariant channel. Minimum Mean Square Error (MMSE) Linear Equalizer (LE) has been used as a MIMO equalizer.

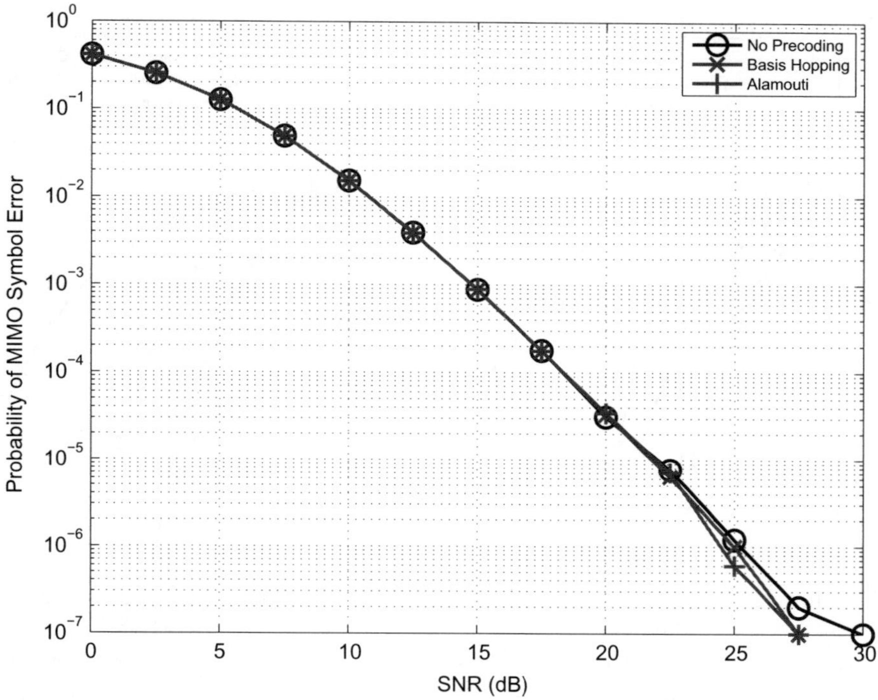

Figure 23.17. Bit error rate at the 2nd transmission for various HARQ transmission techniques in block fading channel, where the channel for the first transmission and the channel for the second transmission are independent. Zero-Forcing (ZF) Linear Equalizer (LE) has been used as a MIMO Equalizer.

23.5.4 CC- Versus IR-HARQ

In Section 23.4, it was argued that the receiver implementation depends on the transmission scheme (CC- or IR-HARQ), on whether the alignment between bits and symbol vectors is maintained, and on the constraints in memory and complexity. Simplifying the receiver may come at a price.

As an example of the performance degradation caused by suboptimal receiver implementations, an IEEE 802.16e-compliant system using Partial Usage of Subchannels (PUSC) and Spatial Multiplexing (Matrix B) is considered [2]. Two transmit and two receive antennas are employed, communicating through a pedestrian Type-A channel with a high degree of spatial correlation and Doppler speed equal to 120 km/h. The data are encoded using the mother rate-1/3 Convolutional Turbo Code (CTC). Bits are punctured sequentially to produce sequences of equal length as in Figure 23.1.

In Figure 23.18a, a bit-level combining receiver is employed using Zero-Forcing Linear Equalization (ZF-BLC). 64-QAM and a rate-1/2 code are considered. IR-HARQ has a coding gain of more than 1 dB compared to CC-HARQ

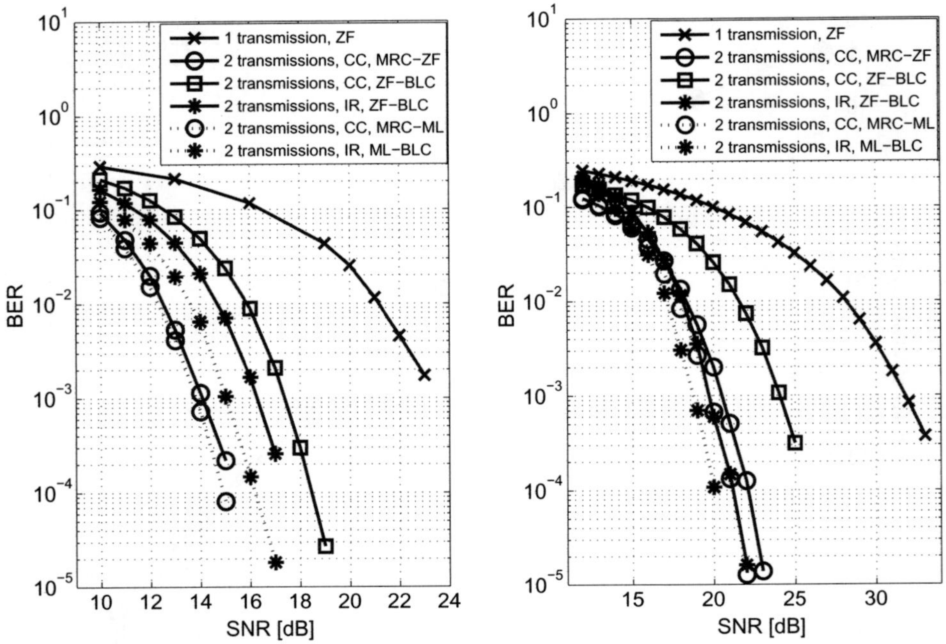

(a) 64−QAM, code rate = 1/2, packet size = 54 bytes

(b) 64−QAM, code rate = 5/6, packet size = 60 bytes

Figure 23.18. 2 × 2 MIMO system, Type-A vehicular channel, 120 km/h, high inter-stream correlation, PUSC, spatial multiplexing.

because of the additional parity bits that are transmitted. However, when the optimal, pre-equalization symbol-combining receiver of Figure 23.8 is used with CC-HARQ (MRC-ZF), the system exhibits a gain of almost 2 dB compared to IR-HARQ. CC-HARQ also outperforms IR-HARQ when ML detection is used instead of equalization, as seen from the curves MRC-ML and ML-BLC. The performance advantage of IR-HARQ can be recaptured using the receiver of Figure 23.3 at the cost increased complexity and memory requirements.

When a rate-5/6 code is used, the coding gain of IR-HARQ over CC-HARQ is much larger compared to the rate-1/2 code (of the order of 4 dB, as shown in Figure 23.18b. Therefore, although symbol-level combining improves the performance of CC-HARQ. IR-HARQ still achieves a gain of approximately 1 dB. The gain is attained for both equalizer-based and ML-based implementations.

23.6 CONCLUSION

In this chapter, an overview of the HARQ scheme was given, as well as a description of how it can be used in MIMO wireless systems. It was seen that, in the general case, the optimal receiver may become very complex. Therefore, although it is possible to design HARQ schemes with good coding gains, the associated receiver memory and complexity requirements should also be considered. In some cases, namely when bit-to-symbols vector alignment does not change, it is possible to simplify the receiver using symbol-level combining schemes without any performance penalty. Further simplification is possible if equalization along the spatial dimensions is employed instead of a Maximum-Likelihood decoder. However, this simplification is attained at the cost of performance degradation. Even simpler receivers can be used by implementing post-equalization combining, which causes an additional performance loss.

When bits-to-symbol vectors alignment is not maintained, symbol-level combining cannot be used, in general. In this case, if receiver simplification is required, a loss in performance needs to be tolerated. The loss depends on the rate of the code, on the channel, and on whether maximum-likelihood decoding or equalization is employed at the receiver.

Therefore, depending on the system requirements, the designer can determine the best design tradeoff that attains a good compromise between performance and complexity and cost constraints.

REFERENCES

1. S. Lin, D. J. Costello, Jr, and M. J. Miller, Automatic-repeat request error-control schemes, *IEEE Commun. Mag.*, Vol. 22, No. 12, pp. 5–17, December 1984.
2. IEEE Std 802.16-2009, IEEE Standard for Local and Metropolitan Area Networks, Part 16: Air Interface for Broadband Wireless Access Systems, May 2009.
3. IEEE P802.11—Task Group N, http://grouper.ieee.org/groups/802/11/Reports/tgn_update.htm.
4. D. Tse and P. Viswanath, *Fundamentals of Wireless Communications*, Cambridge University Press, New York, 2004.
5. D. Wübben, R. Böhnke, V. Kühn, and K.-D. Kammeyer, MMSE extension of V-BLAST based on sorted QR decomposition, in *Proceedings, IEEE Vehicular Technology Conference (VTC) 2003-Fall*, Orlando, Florida, October 2003, pp. 508–512.
6. R. Böhnke, D. Wübben, V. Kühn, and K.-D. Kammeyer, Reduced complexity MMSE detection for BLAST architectures, in *Proceedings, IEEE Global Telecommunications Conference (GLOBECOM)*, Vol. 4, San Francisco, CA, December 2003, pp. 2258–2262.
7. B. Hassibi, An efficient square-root algorithm for BLAST, in *Proceedings, IEEE International Conference Acoustics, Speech, Signal Processing*, Istanbul, Turkey, June 5–9, 2000, pp. 737–740.
8. P. W. Wolniansky, G. J. Foschini, G. D. Golden, and R. A. Valenzuela, V-BLAST: An architecture for realizing very high data rates over the rich-scattering wireless channel, in *Proceeding IEEE ISSSE '98*, Pisa, Italy, September 1998.
9. A. Duel-Hallen, Equalizers for multiple input/multiple output channels and PAM systems with cyclostationary input sequences, *IEEE J. Selected Areas Commun.*, Vol. 10, No. 3, pp. 630–639, April 1992.

10. J. Yang and S. Roy, Joint transmitter-receiver optimization for multi-input multi-output systems with decision feedback, in *IEEE Trans. Inf. Theory*, Vol. 40, No. 5, pp. 1334–1347, September 1994.
11. M. K. Varanasi, Decision feedback multiuser detection: A systematic approach, *IEEE Trans. Inf. Theory*, Vol. 45, No. 1, pp. 219–240, January 1999.
12. H. Sun and Z. Ding, Iterative transceiver design for MIMO ARQ retransmissions with decision feedback detection, *IEEE Trans. Signal Proc.*, Vol. 55, No. 7, July 2007.
13. D. Chase, Code combining—A maximum-likelihood decoding approach for combining an arbitrary number of noisy packets, *OEEE Trans. Commun.*, Vol. COM-33, No. 5, May 1985.
14. K. R. Narayanan and G. Stüber, A novel ARQ technique using the turbo coding principle, *IEEE Commun. Lett.*, Vol. 1, No. 2, pp. 49–51, March 1997.
15. Z. Ding and M. Rice, Hybrid-ARQ code combining for MIMO using multidimensional space–time Trellis codes, in *Proc. IEEE ISIT '07*, Glasgow, Scotland, UK, June 2007.
16. K. Acolatse and Y. Bar-Ness, An Alamouti-based hybrid-ARQ scheme for MIMO systems, *in Proceedings 14th IST Mobile & Wireless Communications Summit 2005*.
17. S. M. Alamouti, A simple diversity technique for wireless communications, *IEEE J. Selected Areas Commun.*, Vol. 16, No. 8, pp. 1451–1458, October 1998.
18. E. W. Jang, J. Lee, H.-L. Lou, and J. M. Cioffi, Optimal combining schemes for MIMO systems with hybrid ARQ, *in Proceedings, International Symposium Inf. Theory.*, Nice, France, June 24–29, 2007, pp. 2286–2290.
19. D. Toumpakaris, J. Lee, A. Matache, and H.-L. Lou, Performance of MIMO HARQ under receiver complexity constraints, *in Proceedings, IEEE GLOBECOM 2008*, New Orleans, LA, November 30–December 4, 2008.
20. H. Vikalo, B. Hassibi, and T. Kailath, Iterative decoding for MIMO channels via modified sphere decoder, *IEEE Trans. Wireless Commun.*, Vol. 3 No. 6, pp. 2299–2311, November 2004.
21. B. Hassibi and H. Vikalo, On sphere decoding algorithm. I. Expected complexity, *IEEE Trans. Signal Processing*, Vol. 53, No. 8, pp. 2806–2818, August 2005.
22. E. N. Onggosanusi, A. G. Dabak, Y. Hui, and G. Jeong, Hybrid ARQ transmission and combining for MIMO systems, in *Proceedings, IEEE International Conference on Communications 2003*, Vol. 5, pp. 3205–3209, May 2003.
23. J. M. Cioffi and G. D. Forney, Generalized decision-feedback equalization for packet transmission with ISI and Gaussian noise, Chapter 4 in *Communications, Computation, Control and Signal Processing*, A. Paulraj, V. Roychowdhury, and C. Schaper, editors, Kluwer, Boston, 1997, pp. 79–127.
24. G. Ginis and J. M. Cioffi, On the relation between V-BLAST and the GDFE, *IEEE Commun. Lett.*, Vol. 5, No. 9, pp. 364–366, September 2001.
25. D. Toumpakaris, J. Lee, E. W. Jang, and H.-L. Lou, Storage-performance tradeoff for receivers of MIMO systems using hybrid ARQ, in *Proceedings, SPAWC 2008*, Recife, Brazil, July 6–9, 2008, pp. 381–385.
26. E. W. Jang, J. Lee, L. Song, and J. M. Cioffi, Concatenation-assisted symbol-level combining scheme for MIMO systems with Hybrid ARQ, *in Proceedings, GLOBECOM 2007*, Washington, D.C., 26–30 November 2007.

27. J. Lee, D. Toumpakaris, E. W. Jang, and H.-L. Lou, DFE-based receiver implementation for MIMO systems employing Hybrid ARQ, *in Proceedings, GLOBECOM 2008*, New Orleans, LA, November 30–December 4, 2008.
28. ETSI TR 125 996 V6.1.0, Universal Mobile Telecommunications system (UMTS); Spatial Channel Model for Multiple Input Multiple Output (MIMO) Simulations (3GPP TR 25.996 version 6.1.0 Release 6), September 2003.

24

RADIO-FREQUENCY TRANSCEIVERS

Alireza Zolfaghari, Hooman Darabi, and Henrik Jensen

The increasing demand for portable communication systems has motivated extensive research and development on wireless transceivers. The necessity for low-cost and low-power solutions along with the limited spectrum availability have introduced numerous challenges in the design of radio-frequency (RF) transceivers for wireless applications. Benefiting from the momentum of the digital market, complementary metal-oxide semiconductor (CMOS) technology has become attractive to provide affordable solutions for RF integrated circuits (IC). In this chapter, we first provide an overview of transceiver architectures, with more emphasis on the transmitters. Then, in the second part, we present examples of transceiver implementations.

24.1 TRANSMITTER ARCHITECTURES

24.1.1 Modulation Overview

In order to have better understanding of transmitter architectures, first we briefly study the modulation types by categorizing them into two groups: constant-envelope and variable-envelope modulations.

Convergence of Mobile and Stationary Next-Generation Networks, edited by Krzysztof Iniewski

24.1.1.1 Constant-Envelope Modulation. If the transmitter signal is specified as $A(t)\cos(\omega_c t + \Phi(t))$, for a constant envelope modulation, as the name implies, $A(t)$ is fixed. This gives significant flexibility in the implementation of the transmitter because the transmitter signal is not affected by any nonlinearity in the transmitter. This especially provides considerable advantage for the power amplifier (PA) using a nonlinear class for higher efficiency. One good example of this type of modulation is the Gaussian Minimum Shift Keying (GMSK) used in the "global system for mobile communication" (GSM) and "general packet radio service" (GPRS). Since that nonlinearity in this case does not change the zero crossing of the signal, it only distorts the amplitude of the carrier. This distortion does not produce any "spectral regrowth" and only generates harmonics of the carrier that can be substantially attenuated by the amplifier low-Q filters (LC tank).

24.1.1.2 Variable-Envelope Modulation. Although constant-envelope modulation is very attractive by relaxing the linearity requirements and improving the power efficiency, it does not utilize the spectrum bandwidth efficiently. On the contrary, variable-envelope modulations make better use of the spectrum. An example of this type of modulation is the 8-PSK (phase shift keying) modulation used in "enhanced data rate for GSM evolution" (EDGE). Having similar bandwidth as GMSK, this modulation increases the data rate by a factor of three. Although the constellation points may remain on a constant-envelope circle in this type of modulation, the amplitude can be variable in transition from one symbol to another because of the limited bandwidth of signal spectrum. The variation in the envelope is more evident as the transition is closer to the origin of the constellation. With the spectrum bandwidth being a valuable asset, variable envelope modulations are becoming more appealing in new wireless communication systems. The major challenge for this type of modulation is the linearity requirements in the transmitter and PA.

24.1.2 Cartesian Transmitters

Also known as an I-Q (in-phase and quadrature-phase) upconverter, a Cartesian transmitter mixes the quadrature components of the baseband signal with a quadrature local oscillator (LO) signal. The upconversion can be done in one or more steps, and accordingly the transmitter can be categorized into different types.

24.1.2.1 Direct-Conversion Transmitters. If the LO frequency for the upconversion is the same as the carrier frequency, the architecture is called *direct conversion*. Shown in Figure 24.1 is the transmitter architecture. As illustrated, the baseband quadrature signals (I-Q) that are usually generated in the digital domain are transferred to the analog domain by digital-to-analog converters (DAC). Then low-pass filters (LPF) provide the analog baseband signals for the upconversion mixers.

The LO signal for the upconversion needs to be very accurate and have low phase noise. This is usually achieved using a phased-lock loop (PLL). A PLL is a feedback system that compares the phase of the output signal with a reference and makes the alignment. Shown in Figure 24.2 is a basic PLL consisting of a phase detector, an LPF, and a voltage-controlled oscillator (VCO). The phase detector measures the phase difference between the output and the reference signal and applies the correction to the LPF to correct the phase of the VCO. The PLL is locked when the phase difference becomes constant, making the input and output frequencies identical.

A variant of the PLL used to generate the LO signal is an RF frequency synthesizer as shown in Figure 24.3. Very similar to the basic one, this PLL uses a phase/frequency detector (PFD) to speed up the acquisition and locking time. Then the measured phase difference is applied to a charge-pump (CP) circuit followed by a loop filter (LF) operating as an LPF to control the VCO. The feedback signal here is then divided by an integer factor in the feedback path. This results in output signal with frequency N times as large as the reference signal.

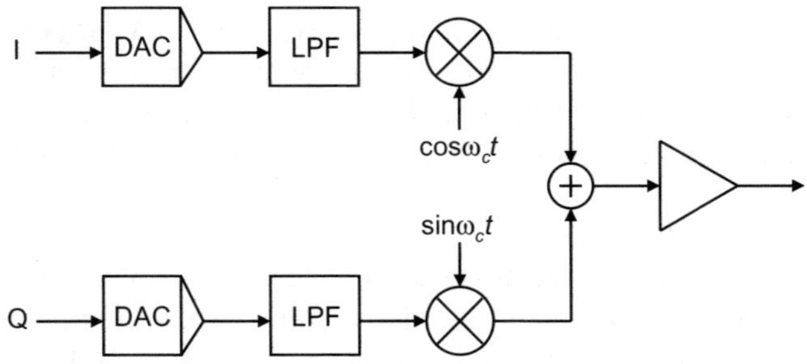

Figure 24.1. Direct-conversion architecture.

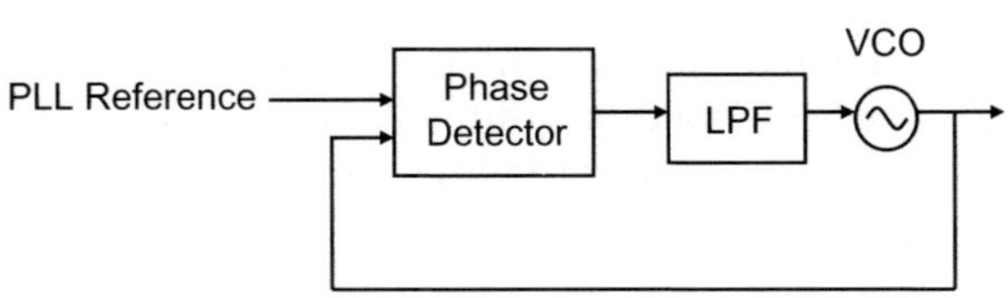

Figure 24.2. Basic PLL.

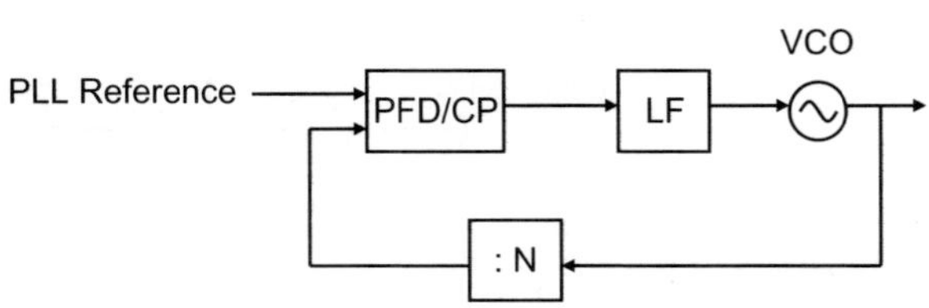

Figure 24.3. Frequency synthesizer.

The quadrature LO signals can be realized using different methods. One common method is using a 90° phase shifter after a voltage-controlled oscillator (VCO) implemented in a phased-locked loop (PLL). Alternatively, a quadarture VCO can be employed to generate quadrature phases [2].

Despite simplicity, direct-conversion transmitters suffer from a major problem if the VCO runs at the carrier frequency. Known as "injection pulling," "injection locking," or "LO pulling," the issue arises when the PA output which is a high-power-modulated signal "pulls" the VCO running at the same frequency, corrupting the LO signal. To mitigate this effect, the VCO can be implemented at twice the LO frequency followed by a divide-by-two circuit to generate the LO frequency. Since the divide-by-two circuit generates two signals 90° apart in phase, this will also produce the I-Q LO signals. While this method addresses the injection pulling problem to the first order, still the VCO can be pulled by the second harmonic of the PA output. In addition to running the VCO at twice the LO frequency, there are other techniques to operate the VCO at different frequencies. An example will be shown in the case studies.

In practice, direct-conversion transmitters face other challenges as well. Any deviation of the phase and amplitude of the I-Q path as a result of mismatch corrupts the transmit signal. The mismatch can be in the I-Q path of the baseband signals, LO signals, or upconversion mixers. Moreover, any DC offset in the baseband signal can result in "LO leakage" to the output, thereby degrading the transmit signal quality. These nonidealities can be partially corrected by calibration circuits.

24.1.2.2 Two-Step Transmitters. In order to resolve the problems of a direct conversion transmitter, an alternative approach is to upconvert the baseband signal in two steps. Figure 24.4 illustrates a two-step I-Q upconverter. As shown, the baseband I-Q signals are first mixed with quadrature LO components (ω_{C1}) and then upconverted again (ω_{C2}). This architecture does not suffer from the injection pulling problem and alleviates the I-Q mismatch but requires more stages in the transmit path. In addition, the second upconversion creates an image signal (ω_{C2}–ω_{C1}), requiring a band-pass filter (BPF).

Normally, a two-step transmitter requires two oscillators to generate the LO signals (ω_{C1} and ω_{C2}). This may make this architecture less attractive, but there

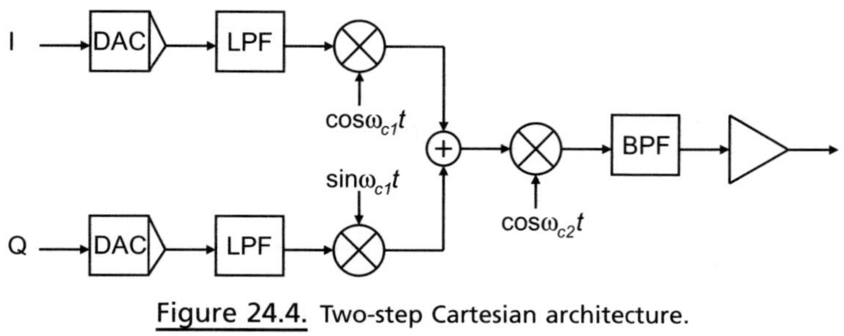

Figure 24.4. Two-step Cartesian architecture.

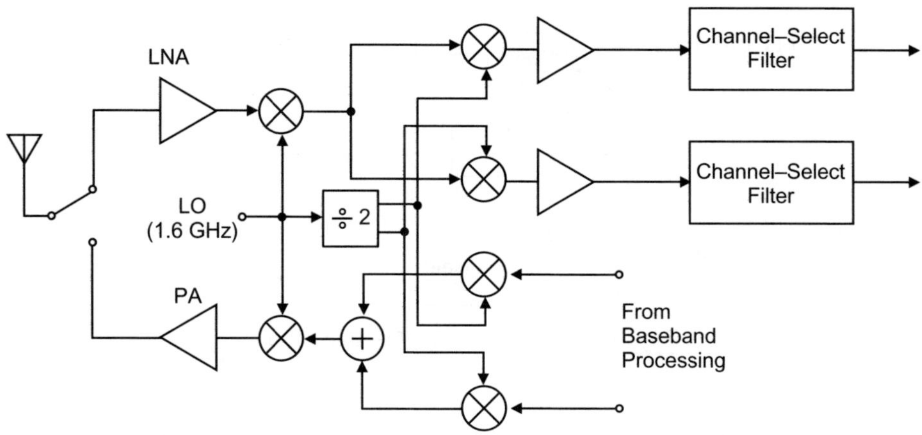

Figure 24.5. Two-step Cartesian architecture using one oscillator [4]. (Copyright IEEE 2003).

are techniques to employ only one oscillator to generate the LO signals. One example is the transceiver architecture for the 2.4-GHz receiver/transmitter in reference 4. As shown in Figure 24.5, the VCO running at 1.6 GHz (2/3 RF) is divided by two to generate the quadrature LO signals to upconvert the baseband signals to 800 MHz (1/3 RF). Then, the VCO output is employed for the second upconversion stage to generate the transmitter output.

24.1.3 Constant-Envelope Transmitters Using Phase-Modulated Loops

As discussed earlier, some communication systems use constant-envelope modulation, providing more flexibility to implement the transmitter. In fact, one easy approach is applying the phase-modulated signal to the control input of a VCO. Since the VCO frequency is a function of its control voltage, varying the control voltage will modulate the VCO phase. Although this method is simple, there are practical issues because the VCO characteristic can vary considerably as temperature or process changes. The nonlinearity of the VCO characteristic further deteriorates the performance.

24.1.3.1 Direct Modulation Using Fractional-N PLL Transmitters. In order to resolve the issues associated with the direct modulation of the VCO in the open loop, a feedback system can be employed. Figure 24.6 shows a phase-modulated PLL to control the VCO in a closed loop. As shown, the PLL uses a fractional-N divider modulated with the phase information of the transmitter signal. This approach resolves most of the issues the I-Q modulator suffers from. Since there is no I-Q LO signals, there is no concern for the I-Q mismatch and LO feedthrough. Furthermore, the VCO and the PA signals are the same; as a

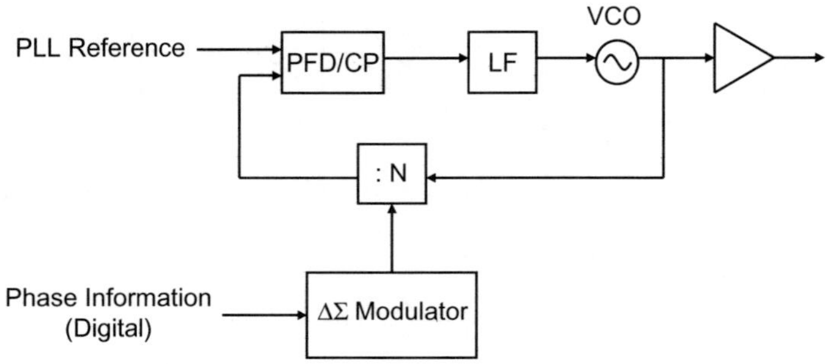

Figure 24.6. Fractional-N PLL transmitter.

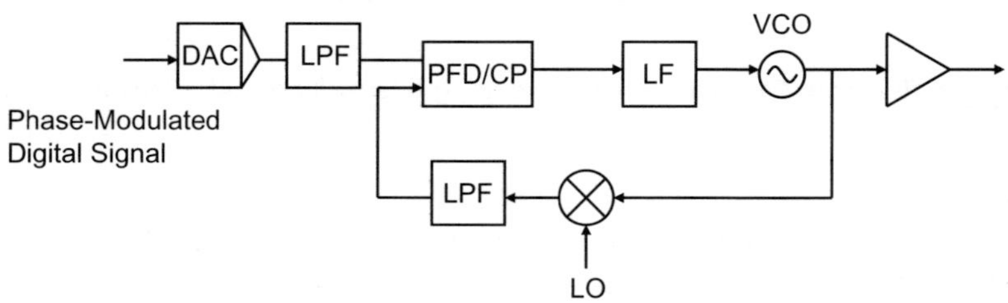

Figure 24.7. Translational-loop transmitter.

result, the injection pulling does not occur. This topology also achieves lower transmitter noise in the receiver band, eliminating the need for an RF transmitter filter. However, recall that this architecture is only applicable to constant-envelope modulations.

While resolving the issues of the I-Q upconverter, a fractional-N PLL in practice has its own challenges. When the reference frequency of the PLL is much lower than the carrier, (e.g., a 26-MHz reference for a carrier of about 1.8–1.9 GHz), the PLL gain will be considerable. This can potentially increase the in-band noise of the PLL coming from the input reference, the delta-sigma modulator, the phase-frequency detector, and the charge pump. We may reduce the PLL bandwidth to improve the in-band noise performance, but the loop needs to be wide enough to accommodate the bandwidth of the transmitter phase signal. Although a digital equalizer can be employed to increase the effective bandwidth, analog variations and sensitivity issues limit the minimum required bandwidth of the PLL. This is a major challenge for this architecture.

24.1.3.2 Translational Loop Transmitters. In order to relax the noise requirements of the fractional-N PLL, a translational loop, also known as "offset PLL," can be employed. As shown in Figure 24.7, instead of dividing the frequency of the VCO, this topology uses a mixer followed by a LPF to downconvert the VCO output to an intermediate frequency. The phase of the downconverted signal is then aligned with an IF-modulated signal in the feedback loop. The modulated

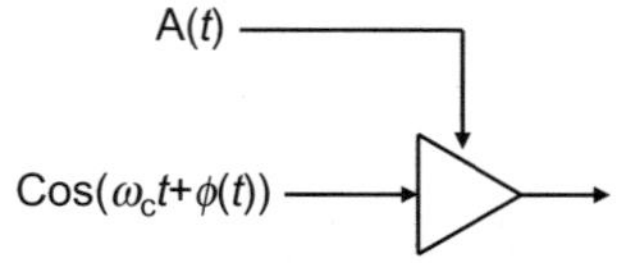

Figure 24.8. Basic polar transmitter.

reference can be generated in the digital domain and then applied to a DAC followed by a LPF. Then, the modulated reference in the analog domain is applied to the PFD input.

A translational loop has two main advantages compared to a phase-modulated fractional-N PLL. First, it employs a downconversion mixer rather than a frequency divider, thereby reducing the PLL gain that lowers the in-band noise. Second, the delta–sigma noise is not directly added to the loop in the feedback path. However, the PLL needs a modulated reference. This requires more analog circuits (i.e., DAC, LPF, etc.), adding to the die area and power consumption in order to keep the quantization noise of the modulated reference low enough.

Translational loops can be implemented differently as well by performing the modulation inside the loop. Although this type does not require any phase-modulated reference, baseband I-Q signals need to be generated very similar to the reference input.

24.1.4 Polar Transmitters

As described before, in a Cartesian topology, the transmitter baseband signals are generated in the quadrature form and then upconverted with quadrature LOs. Alternatively, the transmit signal can be produced using a polar transmitter. As the name implies, in this type of transmitter the signal is decomposed into its polar components: phase and amplitude. This is similar to a complex number that can be represented in a polar form rather than real and imaginary parts. As before, if the modulated signal is specified as $A(t)\cos(\omega_c t + \Phi(t))$, the phase modulation (PM) can be first applied using a phase-modulated PLL (described before) to produce a constant-envelope signal, $\cos(\omega_c t + \Phi(t))$. Then, as shown in Figure 24.8, the amplitude modulation (AM) can be added by modulating the envelope of the RF signal using an amplifier whose output amplitude is a function of $A(t)$.

In order to generate the phase signal, $\cos(\omega_c t + \Phi(t))$, we can use any transmitter topology to generate a constant-envelope signal. An example is shown in Figure 24.9, where a fractinal-N PLL is used. Alternatively, a translational loop can be used to generate the phase signal.

A polar transmitter does not suffer from many of the issues in a Cartesian topology, and more importantly, it also enhances the power efficiency. For a nonconstant envelope modulation, the amplitude is a variable signal; as a result, the envelope peak value is higher than its average or root mean square (RMS).

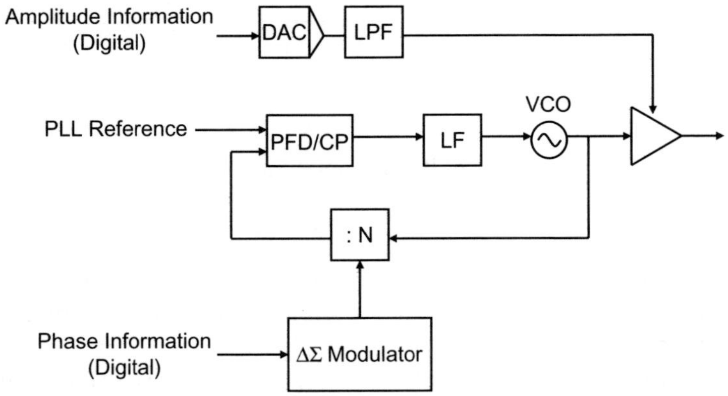

Figure 24.9. Polar transmitter using a fractional-N PLL for the phase modulation.

In order to measure this variation, we can define a quantity called "peak-to-average ratio" (PAPR) to specify the instantaneous peak power with respect to the average power. For instance, for the 8-PSK modulation of the EDGE signal, this ratio can be as high as 3 dB. If this signal is applied to a linear amplifier in the I-Q upconverter, the amplifier needs to be linear enough not only for average power of the carrier, but also to accommodate the instantaneous peak value. Therefore, the amplifier needs to be biased at a high operation point to work well at the peak power, reducing the power efficiency of the amplifier. In contrast, in a polar transmitter, the amplifier does not need to be backed off by PAPR. The phase component that is a constant-envelope signal can be applied to a nonlinear amplifier. Thus the amplitude component can be used to dynamically modulate the output envelope. Thus, at any given amplitude value, the amplifier is operating close to the saturation power, thereby improving the power efficiency. Note that the overall linearity is not degraded, provided that the amplitude signal linearly modulates the output envelope.

Since a polar transmitter does not deal with the quadrature signals, it does not suffer from the I-Q issues of the Cartesian architecture, but instead it faces its own challenges as follows.

24.1.4.1 Delay Mismatch Between Phase and Amplitude. To combine the modulated signal correctly, the amplitude signal, $A(t)$, and the phase signal, $\Phi(t)$, need to be aligned accurately. This in practice can cause a problem when the delay in the phase and amplitude paths are not exactly the same, causing misalignment between the phase and amplitude signals $(A(t - t_{da})\cos(\omega_c t + \Phi(t - t_{dp}))$. This effect may seem similar to the I-Q mismatch, but the delay mismatch between the phase and amplitude can be more serious. The reason is that in an I-Q upconverter, the I-Q nonidealities are mainly caused by the process mismatch between the I-Q analog paths while in the ideal case they are identical. In contrast, the phase and amplitude paths in a polar transmitter

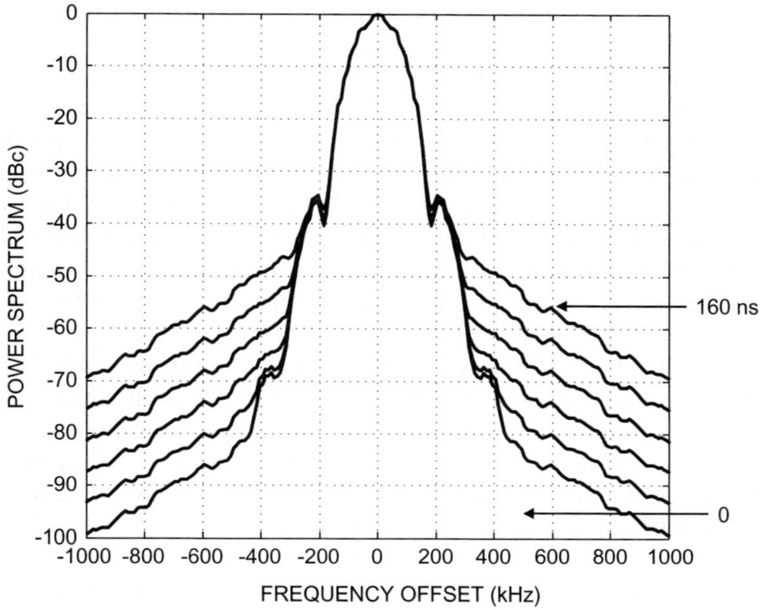

Figure 24.10. Spectral regrowth due to delay mismatch between the phase and amplitude components for 0, 5, 10, 20, 40, 80, and 160 ns (EDGE signal spectrum). (Copyright © IEEE 2008.)

are generated in different paths: The phase uses a phase-modulated loop, while the amplitude is usually implemented using a variable-envelope amplifier driven by the amplitude information.

The delay mismatch between AM and PM can significantly degrade the quality of the transmitter signal. An example is shown in Figure 24.10 where this effect is illustrated on the EDGE transmitter signal. The plots show how the delay mismatch produces spectral regrowth. For instance, for about 80-ns delay mismatch, the spectrum mask at 400-kHz offset becomes about −54 dBc, the minimum level required by the standard. In practice, there are other factors impacting the spectral mask of the transmitter signal. So, the delay mismatch needs to be adjusted to provide margin for other impairment as well.

Different techniques have been developed to adjust the delay. One approach is to provide a loop-back signal to measure the signal quality and then adjust the delay mismatch between the two paths. The loop-back signal can be continuously provided to a feedback circuit for delay adjustment [6], increasing the complexity and power consumption. An alternative method is using the loop-back signal only momentarily to calibrate the delays. This technique consumes less power. Finally, a third option is calibrating the bandwidth of the phase and amplitude signals with enough accuracy so that the delay mismatch does not exceed a certain level. This method has the advantage of eliminating the loop-back transmit signal but requires high-precision calibration schemes.

24.1.4.2 Wide Phase and Amplitude Bandwidth. To relax the noise requirements, it is usually preferred to have smaller bandwidth in the phase and amplitude paths of the analog circuits. The polar components of the signal, however, can have wider bandwidth compared to their Cartesian counterparts (i.e., I-Q components). This is principally due to the nonlinear operations to derive the polar components from the I-Q elements that can be specified as $\Phi(t) = \tan^{-1}(Q/I)$, $A(t) = (I^2 + Q^2)^{1/2}$. Figure 24.11 is an example to illustrate how the phase and amplitude spectrum compare with the composite signal in EDGE. As the plots show, the phase component in particular has much broader bandwidth and even the amplitude spectrum does not roll off as fast as the composite signal.

Widening the bandwidth of the phase and amplitude in the analog paths can have close-in and far-out noise penalty. One solution is using digital equalizers to increase the effective bandwidth for the two paths. In order for the digital equalizers to compensate the bandwidth appropriately, the analog bandwidth should be well-controlled. Nevertheless, the analog bandwidth cannot be arbitrarily reduced because of sensitivity and nonideality issues.

24.1.4.3 AM-to-PM Conversion. This effect represents the variations of the phase signal as a function of the amplitude level. Since in a polar transmitter the phase signal is applied independently, this variation distorts the phase com-

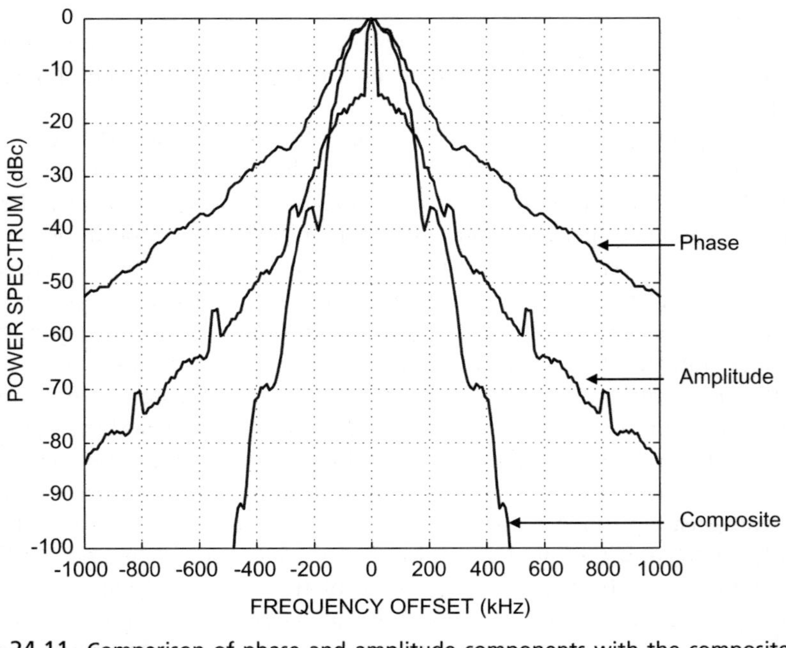

Figure 24.11. Comparison of phase and amplitude components with the composite signal (EDGE signal spectrum). (Copyright © IEEE 2008.)

ponents. Note the AM–PM effect is a nonideality in any type of amplifier, especially for PAs; and even in a Cartesian topology, it can corrupt the signal quality.

24.2 RECEIVER ARCHITECTURE

In this section we briefly study the common receiver architectures. Unlike the transmitters that can be categorized into different groups based on the quadrature or polar representations, receivers mostly employ downconversion mixers in one or more stages.

24.2.1 Single-Step Receivers

24.2.1.1 Zero-IF Receivers. Also known as "direct conversion" or "homodyne" architecture, in this type of receiver the LO frequency is equal to the RF carrier; as a result, the RF signal is *directly* downconverted to *zero IF*. Since the modulated signal spectrum can be asymmetric around the carrier, quadrature LO signals are needed to generate quadrature IF signals. Figure 24.12 shows the simplified architecture of such a receiver. The RF signal is first amplified by a low-noise amplifier (LNA) and then downconverted to IF using I-Q mixers. At IF, LPFs are employed to suppress the unwanted interferers. Then the IF signals can be demodulated in the analog domain; or as shown in Figure 24.12, they are converted to digital signals using analog-to-digital converters (ADC) and then demodulated and processed in the digital domain. The latter is more common in most modern solutions.

While the direct conversion receiver is very simple, in practice it can be very challenging for implementation. Similar to a direct conversion transmitter, any mismatch in the I-Q LOs, mixers, and IF circuits can degrade the quality of the received signal. In addition, DC offsets (voltage or current) and the low-frequency noise, known as flicker noise, fall in the middle of the desired signal at IF and can substantially corrupt the signal, depending on the bandwidth of the signal. Even-order nonlinearity can also generate low-frequency components at IF for in-band and out-of-band interferes. These issues can be partially mitigated using analog circuits and calibration techniques.

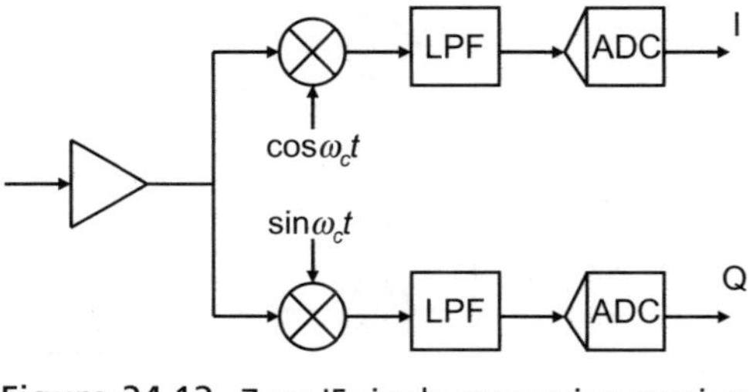

Figure 24.12. Zero-IF single-conversion receiver.

24.2.1.2 Low-IF Receivers. Similar to a direct-conversion receiver, the RF signal is downconverted in one step, but in contrast the LO frequency is not equal to the carrier frequency, resulting in a nonzero IF. Compared to a zero-IF architecture, this receiver resolves the DC offset issue and can relax the requirements on the low-frequency noise and the even-order linearity. However, it introduces a new issue known as *image signal*. Since each mixer does not distinguish between the signal at $LO + IF$ and $LO - IF$, the image of the signal with respect to the LO frequency is also downconverted to the same IF. Having the I-Q IF signals provides enough information to reject the unwanted image signal, but the rejection in practice is limited by nonidealities adding mismatch between the quadrature signals. The image signal can be rejected in either the digital or analog domain. Figure 24.13 shows an example of a low-IF architecture using a polyphase BPF to provide complex filtering and reject the image signal.

24.2.2 Two-Step Receivers

Figure 24.14 shows a two-step receiver. Since the second downconversion is performed with a lower LO frequency, the matching requirements on the quadrature LO signals are more relaxed compared to a direct conversion architecture. Similar to a singe-step receiver, the second IF can be zero or at low frequencies. Adding another downconversion stage at RF causes the image problem for the first downconversion. In this architecture, the image signal for the first mixer is with respect to the first LO, which can be quite far from the signal depending on the LO frequencies planning. To circumvent this problem, a filter in the RF path (before

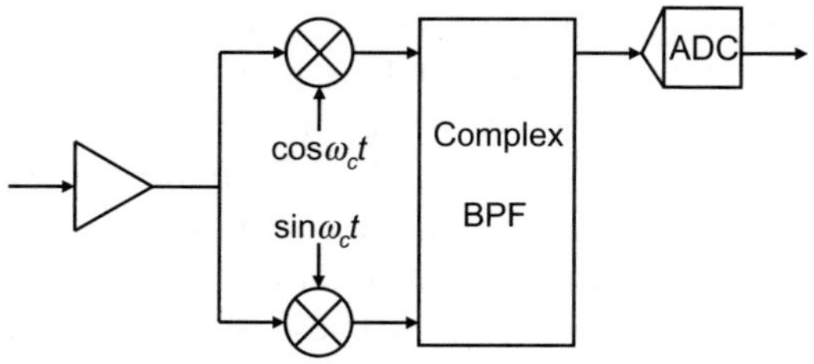

Figure 24.13. A low-IF receiver using a complex BPF.

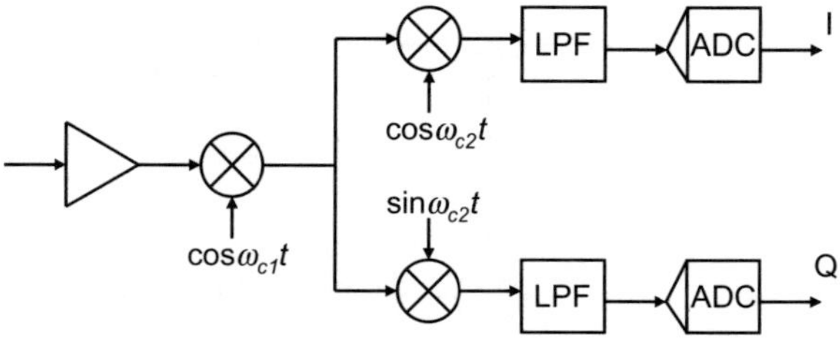

Figure 24.14. Two-step receiver.

or after the LNA) can be employed. There are also other topologies to reject the image signal in the first mixer, such as Hartley and Weaver architectures [1].

24.3 CASE STUDIES

In order to study different transmitter architectures in actual applications, we describe two examples of transceivers implemented for wireless systems focusing on monolithic integration.

24.3.1 A 2.4-GHz CMOS Transceiver for Bluetooth

Bluetooth is a wireless protocol to transfer data over short range for wireless personal area networks (WPAN) and uses the 2.4-GHz Industrial, Scientific, and Medical (ISM) band. The modulation scheme is Gaussian frequency shift keying (GFSK) with frequency deviations of 160 kHz around the carrier to support 1 Mbit/s. A newer version of Bluetooth (2.0) released in 2004, adds an "enhanced data rate" (EDR) mode to increase the data rate up to 3 Mbit/s using 8-PSK modulation. A Bluetooth device must satisfy some certain requirements. It should be low cost and low power to integrate with other portable devices efficiently, and yet it must have robust performance to function properly along with interferers. Such interferers exist in a noisy RF environment in which several powerful radio signals are present in the proximity of the Bluetooth radio, such as GSM or CDMA signals. This clearly spells great design challenges to realize such a high-performance radio.

Shown in Figure 24.15 is the transceiver architecture in [3]. The transceiver uses a time-division duplexing (TDD) system with a direct-conversion architecture. Using this architecture allows low power consumption, and high level of integration. An on-chip modulator produces GFSK signal with 160-kHz

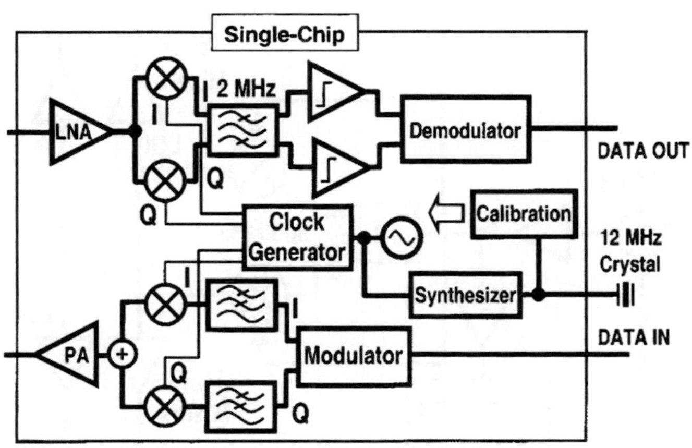

Figure 24.15. Transceiver architecture in reference 3. (Copyright © IEEE 2001.)

frequency deviation at baseband, followed by quadrature Gaussian LPFs to filter the spectrum. Single sideband mixers upconvert the GFSK spectrum from baseband to 2.4 GHz. This transceiver only supports the GFSK modulation. Therefore, a class-AB power amplifier (PA) is employed to deliver a typical output power of 4 dBm to the antenna. The PA linearity is not important, because the upconverted spectrum has no information in the amplitude. Also, an *RC* calibration circuit adjusts the frequency response of the transmitter filters.

Since the transmitter signal is upconverted to the ISM band in one step from the zero-IF baseband, image rejection requirement is relaxed. The in-band spurs, however, can be produced due to the mixer nonlinearity on the IF signal. Therefore, the mixers need to be designed linear enough to avoid spectral regrowth. With a spurious-free LO signal driving the transmitter mixers, out-of-band spurs are mainly generated by the harmonics of the upconverted signal. These spurs are quite far from the carrier and can be suppressed by on-chip tuned filters in the transmitter path.

As discussed before, a direct conversion transmitter can potentially suffer from "LO pulling." The LO in a direct-conversion transmitter coincides in frequency with the large modulated signal at the PA output (on the order of 500 mV in this case). This problem is more severe in this system where the VCO and the PA are integrated on the same chip. To address the issue, an LO generator scheme is used.

Figure 24.16 illustrates the LO generator consisting of a divide-by-two to produce 800-MHz quadrature signals from the VCO, followed by two mixers to produce the I-Q LO signals at 2.4 GHz. Buffers between the stages are added for the signal isolation, amplification, and filtering. The VCO frequency is 800 MHz away from the PA output frequency, thereby making the transceiver insensitive to direct or harmonic pulling issues.

Measurements show that the VCO frequency remains robust even with an output power of up to 20 dBm, using an external power amplifier.

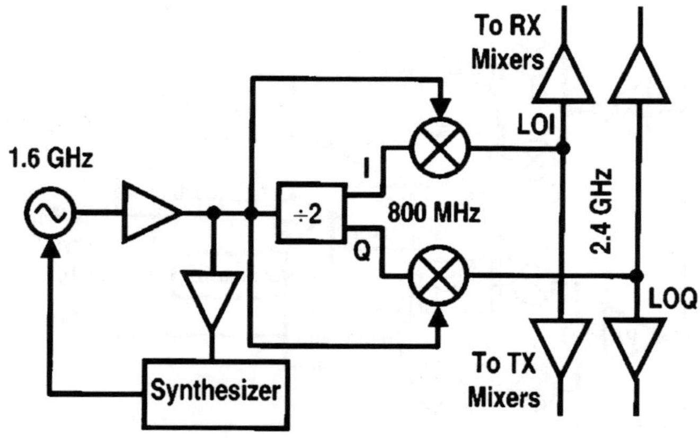

Figure 24.16. LO generator in [3]. (Copyright © IEEE 2001.)

Due to the hard-switching action in the mixer, spurious signals accompany the desired 2.4-GHz LO. This is mainly caused by mixing the VCO signal with the divider outputs. The closest spurs are the lower sideband at 800 MHz and the VCO third harmonic mixed with the divider output, producing a spur at 4 GHz. The spurs, however, are attenuated by the on-chip *LC*-tuned circuits at the output of the LO generator mixer and the buffer. With on-chip inductors having a quality factor of 5.5 at 2.4 GHz, each *LC* circuit attenuates the spur at 1.6 GHz away by about 20 dB. These spurs will be suppressed further by the on-chip *LC* filters in the receiver and transmitter paths. Measurements indicate that all the out-of-band spurs are within the requirements set by the Bluetooth standard, without any off-chip preselect filter. Compared to a conventional two-step transceiver, the advantage of this architecture is that one step of frequency translation is performed inside the LO generator, where noise and linearity are not an issue.

Unlike the transmitter, which uses a direct conversion, the receiver is a low-IF single-step architecture. This is mainly due to the spectrum of the Bluetooth signal having considerable energy around the carrier. As mentioned before, DC offset and flicker noise can significantly degrade the performance. Therefore, the receiver downconverts the RF signal to a 2-MHz IF. Following the mixer, a complex BPF selects the desired the signal and rejects the image. Since the GFSK modulation has no information in the amplitude, limiters are used to amplify the desired signal to a well-defined power and then the signal is demodulated.

24.3.2 A Fully Integrated Quad-Band GPRS/EDGE Radio in 0.13-μm CMOS

In this section we study a transceiver architecture in a fully integrated quad-band GSM/GPRS/EDGE radio [5]. Implemented in a low-cost digital 0.13-µm CMOS technology, this radio achieves high performance and integrates all the receiver and transmitter functions required to support a quad-band GSM/GPRS/EDGE application into a single chip. The radio incorporates several real-time and autonomous calibration schemes to overcome the CMOS impairments. Integrating the digital blocks for signal processing and calibrations in the same die introduces major challenges for the high-performance analog radio. The digital signals can potentially couple to the analog circuits, adding noise and switching spurs. The coupling can be through the die substrate, package, or supply and ground lines, therefore, careful isolation and decoupling are need to avoid the issue.

Figure 24.17 shows the overall radio architecture. The transmitter in the constant-envelope mode (GMSK) employs a fractional-N PLL described before. In order to relax the problem of high in-band noise, a reference signal of around 250 MHz is used as input to the PLL. Note that the reference clock is generally produced by a crystal oscillator at much lower frequencies (e.g., 26 MHz). In order to produce the reference signal for this PLL, the main frequency

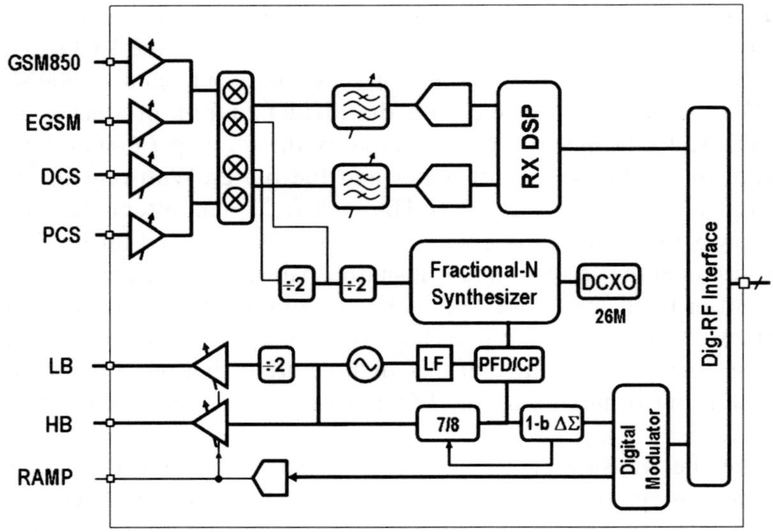

Figure 24.17. Transceiver architecture in reference 5. (Copyright © IEEE 2008.)

synthesizer is utilized. The primary task of the main synthesizer is generating LO signals for the receiver, but the GSM/EDGE standard uses "time-division duplexing" (TDD). Therefore, the main synthesizer can be employed during the transmit time without any conflict with the receiver operation. Besides, the transmitter PLL is a fractional-N PLL, capable of producing the output carrier frequency with a wide range of the input frequency. Having a high-frequency reference also provides more flexibility in the frequency planning and relaxes spurious requirements.

In order to support the variable envelope modulation in EDGE (8PSK), the digital modulator creates a separate AM path, providing the amplitude modulation for the PA driver. While meeting the stringent noise requirements, this polar architecture significantly reduces the die area and power consumption. The four outputs of the transmitter required for the quad-band operation are paired into two lines: high band (HB) and low band (LB). LB output covers GSM-850 from 824 to 849 MHz and "extended GSM" (E-GSM) band ranging from 880 to 915 MHz and HB output includes the other two bands: "digital cellular system" (DCS) from 1710 to 1785 MHz and "personal communication services" (PCS) from 1850 to 1910 MHz.

To reduce the phase noise and power consumption, a low-gain and narrowband PLL is advantageous, but the narrow loop bandwidth requires an equalizer block in the digital domain. The equalizer characteristic needs to match the analog PLL bandwidth. Also, in order to align the delay mismatch between the AM and PM paths in the EDGE mode, the PLL bandwidth must be tightly controlled. This requires a very precise calibration scheme for the PLL bandwidth. The calibration is performed in two steps before the actual transmit data. First, the VCO frequency, which is divided into 256 sub-bands, is tuned using a digital counter to perform the coarse frequency adjustment within a worst case of ±2.5 MHz from the desired channel frequency. This also covers the wide tuning range needed for the quad-band transmitter. Next, the calibration circuit locks the VCO to a known set of frequencies of $\pm\Delta f$ around the center, applied through the $\Delta\Sigma$ modulator. The variation in the control voltage is monitored by a 10-bit analog-to-digital converter, and the VCO gain is estimated accordingly. The PLL bandwidth variation is then compensated by changing the charge-pump current proportionally. This results in measured precision of better than ±2.5% in the loop bandwidth, minimizing the degradation in the spectrum mask and EVM in spite of the narrow PLL bandwidth. This is a key factor to enhance the performance in a polar transmitter to establish a successful modulated transmitter signal without any power or area overhead.

The receiver uses a low-IF architecture with four inputs for the cellular bands. The RF input is amplified by one of the LNAs, depending on the band of operation, and is then downconverted by the I-Q mixers. To relax the low-frequency and even-order nonlinearity requirements, the IF is set at 200 kHz. This imposes stringent requirements on the image rejection. Therefore, two ADCs are used to perform the image rejection in the digital domain, and calibration schemes are employed to correct the I-Q imbalance. A fractional-N frequency synthesizer creates the LO and reference signal for both receiver and transmitter. The on-chip VCO runs at around 4 GHz. Similar to the transmitter VCO, it is divided into 256 sub-bands to reduce the VCO gain, and its coarse frequency is adjusted by a digital calibration circuitry. The main synthesizer uses a reference frequency of 26 MHz produced by an on-chip digitally controlled crystal oscillator (DCXO).

REFERENCES

1. B. Razavi, *RF Microelectronics*, Prentice-Hall, Upper Saddle River, NJ, 1998.
2. A. Rofougaran, J. Rael, M. Rofougaran, and A. Abidi, A 900 MHz CMOS LC oscillator with quadrature outputs, in *International Solid-State Circuits Conference, Digital Technical Papers*, February 1996, pp. 392–393.
3. H. Darabi, S. Khorram, E. Chien, et al., A 2.4 GHz CMOS transceiver for Bluetooth, in *International Solid-State Circuits Conference, Digital Technical Papers*, February 2001, pp. 200–201.
4. A. Zolfaghari and B. Razavi, A low-power 2.4 GHz CMOS transmitter/receiver CMOS IC, *IEEE J. Solid-State Circuits*, Vol. 38, pp. 176–183, February 2003.
5. H. Darabi, A. Zolfaghari, H. Jensen, et al., A fully integrated quad-band GPRS/EDGE radio in 0.13-μm CMOS, in *International Solid-State Circuits Conference, Digital Technical Papers*, February 2008, pp. 206–207.

6. T. Sowlati, D. Rozenblit, E. MacCarthy, et al. Quad-band GSM/GPRS/EDGE polar loop transmitter, *International Solid-State Circuits Conference, Digital Technical Papers*, February 2004, pp. 186–187.
7. M. Elliott, T. Montalvo, F. Murden, et al., A polar modulator transmitter for EDGE, *International Solid-State Circuits Conference, Digital Technical Papers*, February 2004, pp. 190–191.

INDEX